The Mollusca

VOLUME 6

Ecology

The Mollusca

Editor-in-Chief

KARL M. WILBUR

Department of Zoology
Duke University
Durham, North Carolina

The Mollusca

VOLUME 6

Ecology

Edited by

W. D. RUSSELL-HUNTER

Department of Biology
Syracuse University
Syracuse, New York

1983

ACADEMIC PRESS, INC.
(Harcourt Brace Jovanovich, Publishers)
Orlando San Diego San Francisco New York London
Toronto Montreal Sydney Tokyo São Paulo

ACADEMIC PRESS, INC.
Orlando, Florida 32887

United Kingdom Edition published by
ACADEMIC PRESS, INC. (LONDON) LTD.
24/28 Oval Road, London NW1 7DX

Library of Congress Cataloging in Publication Data

Main entry under title:

The Mollusca.

Includes index.
Contents: v. 1. Metabolic biochemistry and molecular biomechanics / edited by Peter Hochachka -- v. 2. Environmental biochemistry and physiology / edited by Peter W. Hochachka -- [etc.] -- v. 6. Ecology of mollusca.
1. Mollusca--Collected works. I. Wilbur, Karl M.
QL402.M6 1983 594 82-24442
ISBN 0-12-751406-6 (V6)

PRINTED IN THE UNITED STATES OF AMERICA

84 85 86 87 9 8 7 6 5 4 3 2 1

Contents

3. Mangrove Bivalves

BRIAN MORTON

4. Coral-Associated Bivalves of the Indo-Pacific

BRIAN MORTON

5. Reproductive and Trophic Ecology of Nudibranch Molluscs

CHRISTOPHER D. TODD

6. Physiological and Trophic Ecology of Cephalopods

WILLIAM C. SUMMERS

7. Physiological Ecology of Freshwater Bivalves

ALBERT J. BURKY

8. Physiological Ecology of Freshwater Prosobranchs

D. W. ALDRIDGE

11. Actuarial Bioenergetics of Nonmarine Molluscan Productivity

W. D. RUSSELL-HUNTER AND DANIEL E. BUCKLEY

12. Ecology of an Invasive Pest Bivalve, *Corbicula*

ROBERT F. MCMAHON

13. Population Genetics of Marine Gastropods and Bivalves

EDWARD M. BERGER

14. Ecology and Ecogenetics of Terrestrial Molluscan Populations

A. J. CAIN

15. Life-Cycle Patterns and Evolution

PETER CALOW

Contributors

Numbers in parentheses indicate the pages on which the authors' contributions begin.

D. W. Aldridge (329), Department of Biology, North Carolina Agricultural and Technical State University, Greensboro, North Carolina 27411

J. A. Allen (29), University Marine Biological Station, Millport, Isle of Cumbrae, Scotland

Edward M. Berger (563), Biology Department, Dartmouth College, Hanover, New Hampshire 03755

Daniel E. Buckley (463), Department of Biology, Syracuse University, Syracuse, New York 13210

Albert J. Burky (281), Department of Biology, University of Dayton, Dayton, Ohio 45469

A. J. Cain (597), Department of Zoology, University of Liverpool, Liverpool L69 3BX, United Kingdom

Peter Calow (649), Department of Zoology, University of Glasgow, Glasgow G12 8QQ, Scotland

Robert F. McMahon (359, 505), Section of Comparative Physiology, Department of Biology, University of Texas, Arlington, Texas 76019

Brian Morton (77, 139), Department of Zoology, University of Hong Kong, Hong Kong

Wayne A. Riddle (431), Department of Biological Sciences, Illinois State University, Normal, Illinois 61761

W. D. Russell-Hunter (1, 463), Department of Biology, Syracuse University, Syracuse, New York 13210

William C. Summers (261), Western Washington University, Bellingham, Washington 98225

Christopher D. Todd (225), Department of Marine Biology, Gatty Marine Laboratory, The University of St. Andrews, St. Andrews, Fife KY16 8LB, Scotland

General Preface

This multivolume treatise, *The Mollusca,* had its origins in the mid 1960s with the publication of *Physiology of Mollusca,* a two-volume work edited by Wilbur and Yonge. In those volumes, 27 authors collaborated to summarize the status of the conventional topics of physiology as well as the related areas of biochemistry, reproduction and development, and ecology. Within the past two decades, there has been a remarkable expansion of molluscan research and a burgeoning of fields of investigation. During the same period several excellent books on molluscs have been published. However, those volumes do not individually or collectively provide an adequate perspective of our current knowledge of the phylum in all its phases. Clearly, there is need for a comprehensive treatise broader in concept and scope than had been previously produced, one that gives full treatment to all major fields of recent research. *The Mollusca* fulfills this objective.

The major fields covered are biochemistry, physiology, neurobiology, reproduction and development, evolution, ecology, medical aspects, and structure. In addition to these long-established subject areas, others that have emerged recently and expanded rapidly within the past decade are included.

The Mollusca is intended to serve a range of disciplines: biological, biochemical, paleontological, and medical. As a source of information on the current status of molluscan research, it should prove useful to researchers of the Mollusca and other phyla, as well as to teachers and qualified graduate students.

Karl M. Wilbur

Preface

Widespread interest in molluscan biology, and hence the justification for the present series of volumes, stems in large part from two unlikely aspects of the ecology of molluscs. First, the phylum Mollusca ranks second only to the phylum Arthropoda, and well ahead of the vertebrates, both in global biomass and in species numbers. Second, the success of certain molluscs in our planet's bioeconomy has been achieved despite a pattern of animal construction that employs ciliary and mucous mechanisms within a soft permeable body moved hydraulically from place to place. These distributional and physiological unlikelihoods require scientific investigation.

The present volume surveys the current status of molluscan studies in 15 chapters, arranged into three levels of ecological perspective: (a) distributional studies, (b) physiological ecology and bioenergetics, and (c) population genetics and dynamics. After a preliminary overview in the first chapter, Chapters 2–6 are concerned with the life styles and distribution of molluscs on the deep-sea bottom, in mangroves, and on coral reefs, and with the trophic and reproductive ecology of those intrinsically fascinating molluscan groups, the nudibranchs and cephalopods. Chapters 7–11 present physiological ecology in land snails and in freshwater bivalves, prosobranchs, and pulmonates, with a survey of the techniques of actuarial bioenergetics as applied to nonmarine molluscs. Chapters 12–15 cover population dynamics and biology in an introduced pest species, population genetics of marine molluscs, ecogenetics of land snails, and life-cycle patterns throughout the major molluscan taxa.

These 15 chapters cannot claim to present a final comprehensive account of molluscan ecology, because there are enormous gaps in our field of living molluscs, and because so many questions of ecological theory remain open ended. In all chapters, therefore, many potential new and profitable research areas in molluscan ecology are indicated. In the central group of chapters, definitions of physiological ecology vary from author to author; as with biophysics, different scientists claim distinct missions. This reflects a more general diversity of viewpoints. No attempt has been made to force conceptual uniformity on the contributors to this volume. Contrasting opinions can be found herein on the evolution of reproductive strategies, on the maintenance of genetic diversity, and on a working delimitation of physiological ecology. Despite this, we are all attempting to increase understanding of the interactions (internal and external, biotic and abiotic) that determine the patterns of distribution and abundance of living molluscs. If questions left open or in controversy provoke a number of our readers to begin rewarding investigations of molluscan ecology, then we shall have succeeded.

For valued counsel from the start of the enterprise, I am most grateful to Dr. Karl M. Wilbur of Duke University. Once again I must thank Myra Russell-Hunter and Peregrine D. Russell-Hunter for both stimulus and succor during the assembly of this volume.

W. D. Russell-Hunter

Contents of Other Volumes

Volume 1: Metabolic Biochemistry and Molecular Biomechanics

Volume 2: Environmental Biochemistry and Physiology

Volume 3: Development

Volume 4: Physiology, Part 1

Volume 5: Physiology, Part 2

Volume 7: Reproduction

1

Overview: Planetary Distribution of and Ecological Constraints upon the Mollusca

W. D. RUSSELL-HUNTER

Department of Biology
Syracuse University
Syracuse, New York

I. Introduction

One of the most successful patterns of animal construction is the molluscan plan. There are more than twice as many species of molluscs as there are of vertebrates, and only the arthropods are clearly a more numerous and more successful group (Russell-Hunter, 1979, 1982).

Thus, part of the justification for publishing a series of volumes on the Mollusca comes from extant ecological data on the phylum. It is not merely that the molluscan plan of animal construction has been highly successful in terms of numbers of individuals and numbers of species (perhaps 110,000 living molluscan species), but that the biomass of these species dominates the lower trophic levels of many marine (and a few nonmarine) ecosystems. By any ecological measure, the molluscs constitute a major phylum. Despite this, no authoritative or totally comprehensive volume can now be prepared as a pandect of molluscan ecology. The reasons for this inability lie not only in the enormous gaps which

The Mollusca, Vol. 6
Ecology

ISBN 0-12-751406-6

exist in our field knowledge of the interactions that determine patterns of molluscan distribution and abundance (although we know somewhat more of the patterns *per se*) but also in the recent history of conceptual and methodological changes in the science of ecology. Few current lines of investigation will receive any final summary in this volume; more potential new ones will be indicated. Sections III, IV, and V of this introductory chapter provide synoptic outlines for all the other chapters in this volume. Apart from this preliminary overview, the volume is somewhat arbitrarily arranged to move from distributional studies, by way of physiological ecology and bioenergetics, to population genetics and dynamics. This could be construed as reflecting historical shifts of ecological perspective (if not of paradigm), although all three subdisciplines should be synergic in any complete ecology.

As modern animal ecology developed, classic distributional studies became complemented by work both on physiological ecology and on population dynamics. The design of mathematical models for ecological interactions has become important, with earlier deterministic representations being gradually replaced by stochastic ones. Almost independently, modern population genetics has emerged to become as significant to the rethinking of autecology and population ecology as to that of evolutionary studies. In common with all other sciences, progress in ecology has come, and will continue to come, from the design and execution of experimental tests appropriate to each ecological hypothesis. A grave disadvantage to this progress results from the greater ease of designing and interpreting experiments on laboratory cultures compared to natural populations in the field. Therefore, an earlier stage of scientific inquiry remains important in molluscan ecology, as in the ecology of other invertebrate groups. G. E. Hutchinson has characterized this stage as the search for a pattern. This research, largely field-based, involves searching for geographical patterns of distributional categories from which ecological processes can be deduced.

Ecological perspectives and methods are closely related. The initial investigation of any ecological problem can occur in the field, in the laboratory, or at the computer. Despite certain pundits, there is no ''best'' starting method. All three methods must subsequently revert to field conditions for verification, to the dynamics of naturally age-structured populations in their natural environments. This necessity follows from the primary aim of studies on molluscan ecology, which is to increase our understanding of the interactions (internal and external, biotic and abiotic) that determine the patterns of distribution and abundance of molluscs in the manifold habitats of Earth. Ultimately, this understanding must be of dynamic relationships rather than static faunal structure, that is, it must be concerned with functional processes through time rather than with detailed descriptions at any single point in time. That it will always require field testing, if molluscan ecology is to become a predictive science, is obvious.

There are enormous gaps in our field knowledge of molluscan ecology, and no volume can be comprehensive at this time. The present volume is even less comprehensive than originally planned, due to several accidents of scheduling and authorship. For example, it was hoped that Drs. Bruce A. Menge and Jane Lubchenco would contribute not only an account of the distributional ecology of molluscs in the littoral zones of temperate seas but also a chapter on the role in marine community ecology of molluscan grazers and predators. Unfortunately, they had to withdraw from the present volume, but their monograph on community ecology in the rocky littoral remains available (Lubchenco and Menge, 1978; see also Lubchenco, 1978, 1980; Menge, 1976). Similarly, because Dr. Alec C. Brown was unable to describe the molluscan faunas of sandy shores, that topic must be omitted. Population bioenergetics of nonmarine molluscs is covered in this volume, although a similar treatment of marine molluscs has been placed in two chapters in one of the physiological volumes of this series (see Bayne and Newell, Volume 4). In addition, the general editor, Dr. Karl M. Wilbur, earlier placed certain marginally ecological topics—including host–parasite relationships and molluscs as disease vectors—in other volumes of this series. An up-to-date survey of functional morphology in the Mollusca is given by Seed (Chapter 1 in Volume 1 of this series).

II. The Molluscs of Planet Earth

Molluscs are largely marine, but this does not affect their status as one of the four most abundant and successful patterns of animal morphology. The view we now have of our planet from near space emphasizes the fact that about 71% of its surface area is covered by the sea and suggests that its name should be not Terra (Earth) but Oceanus. The most diverse molluscs are marine gastropods; the most abundant molluscs are marine bivalves. No marine habitat lacks molluscs, and in many marine communities the dominant organisms of the second trophic level (primary consumers of plant productivity) are molluscs. In the global economy of the oceans, certain bivalves are second only to planktonic copepods such as *Calanus* in annual turnover of calorific value for animal tissues in food chains.

However, there are also nonmarine molluscs. A few bivalve genera are very important in the faunas of brackish and fresh waters, as are the much more diverse groups of freshwater gastropods. The only land molluscs belong to two major subdivisions of the Gastropoda, but they are surprisingly successful as land animals. Their soft, hydraulically moved bodies and relatively permeable skins, typical of all molluscs, limit most land snails to more humid terrestrial habitats and cause their numerical and ecological success to seem unlikely. Despite these physiological limitations, the gastropod subclass Pulmonata brings the phylum Mollusca into third place (behind Chordata and well behind Arthropoda) in bioenergetic terms of animal turnover in terrestrial ecosystems (Russell-

Hunter, 1979, 1982), and it may be claimed that molluscs come close to the phylum Annelida (including the earthworms) in significance to terrestrial ecology. Later in this volume, Cain (Chapter 14) points out that, in terms of the number of species occurring in terrestrial faunas, land molluscs outnumber terrestrial vertebrates and rank second only to the Arthropoda.

Although most measures of ecological success involve considerable human subjectivity, it is clear that in global terms some kinds of molluscs are more successful than others. Detailed classification of the phylum Mollusca is given elsewhere in this series (Volume 1), but a brief survey of the three major groups and a number of smaller ones which together comprise the molluscs presently living on Earth will serve as an introduction to their ecological interactions.

The phylum Mollusca is usually divided into eight clearly distinguishable classes, seven of which include extant species and three of which can be regarded as major (Gastropoda, Bivalvia, and Cephalopoda). These seven classes encompass a bewildering variety of external body forms, all based on a remarkably uniform basic plan of structure and function. Oysters, clams, chitons, snails, slugs, sea butterflies, squid, and octopuses are all molluscs, and all share a basically tripartite anatomical plan. The soft body, usually contained within a hard, calcareous shell, is the near-diagnostic feature upon which the phylum name is based. Three anatomically distinct regions show different modes of growth and function (modifications of their relative importance in each form of mollusc reflect ecological capabilities). These regions are (1) the *head-foot* with some nerve concentrations, most of the sense organs, and all of the locomotory organs; (2) the *visceral mass* (or hump) containing the organs of digestion, reproduction, and excretion; and (3) the *mantle* (*pallium*) hanging from the visceral mass, enfolding it and secreting the shell. In its development and growth, the head-foot shows a bilateral symmetry with an anteroposterior axis of growth. Over and around the visceral mass, however, the mantle-shell shows a biradial symmetry and always grows by marginal increment around a dorsoventral axis. It is of considerable functional importance that a space is left between the mantle-shell and the visceral mass, forming a semiinternal cavity; this is the mantle-cavity or pallial chamber within which the *ctenidia*, the typical gills of the mollusc, develop. This mantle-cavity is essentially diagnostic of the phylum; it is primarily a respiratory chamber housing the ctenidia, but with alimentary, excretory, and genital systems all discharging into it. Although its basic functional layout is always recognizable, it has undergone some remarkable modifications of structure and function in different groups of molluscs, and these reflect paramount ecological constraints. In addition to its respiratory function (aquatic or aerial), the mantle-cavity provides the feeding chamber of bivalves and some gastropods, a "marsupial" brood pouch in some forms, and an organ of locomotion in a few bivalves and in the most highly organized cephalopods. Several authors, including Seed (Chapter 1 in Volume 1 of this series) and Russell-Hunter (1979, 1982), have presented detailed reconstructions of the mantle-cavity and associated organs in an archetypic mollusc.

The head-foot functions largely by muscles which show many of the usual fast reflexes of other bilaterally symmetrical animals, whereas the visceral mass and the mantle-cavity function slowly and continuously, using mucus and cilia. The organs responsible for locomotion and retraction into the shell are contained in the head-foot; those responsible for respiration, excretion, feeding, and all other visceral activities are those found in the other two regions. These different functions, and the corresponding and clearly distinct patterns and axes of growth, assume different levels of relative importance in different kinds of molluscs. The ultimate adult form of a mollusc may result from dominance of the mantle-shell over the head-foot in growth, as in clams, or vice versa, as in slugs. Obviously, the ecological potentialities of these two forms of mollusc are markedly different.

Each of the eight classes of the Mollusca has a characteristic body form and shell shape. It is important to realize that whatever the shape of the shell, it is always underlain by the mantle, a fleshy fold of tissues which has secreted it. The genes that control the patterns of cell division along the growing mantle edge determine the shell shape that is generated. Of the seven extant classes, two are enormous (Gastropoda and Bivalvia), one is of moderate extent (Cephalopoda), and the other four (Polyplacophora, Monoplacophora, Scaphopoda, and Aplacophora) are minor by comparison. The class Gastropoda constitutes a diverse group (probably including more than 74,000 species), with the shell usually of one piece. This shell may be coiled as in typical snails, that is, helicoid or turbinate, or it may form a flattened spiral, or a short cone as in the limpets, or be secondarily absent as in marine nudibranchs and land slugs.

Most gastropod species are marine, but many are found in fresh waters and on land; in fact, they are the only successful nonmarine molluscs. The marine gastropods are also the most diverse: living between tide marks on the seashore, on the bottom of the ocean at all depths, and even (with reduced shells) drifting in the plankton near the sea surface. Ecologically, gastropods may be microherbivores, macroherbivores, or carnivores, and they exhibit a wide variety of feeding mechanisms. In contrast, the Bivalvia are a more uniform group (about 31,000 species), with the shell in the form of two calcareous valves united by an elastic hinge ligament. Extension of the mantle and its shell has occurred, followed by its lateral compression, so that all parts of the body (including the visceral mass and foot) lie within the mantle-cavity and the head is lost.

Mussels, clams, and oysters are familiar bivalves. The group is mainly marine, with a few highly successful genera in estuaries and fresh waters. There can be no land bivalves, because their basic functional organization is as filter feeders.

The third major group of the Mollusca, the class Cephalopoda, includes the most active and most specialized molluscs: the octopods, cuttlefish, and squids. There is a chambered, coiled shell in *Nautilus* and in many fossil forms; this becomes an internal structure in cuttlefish and squids and is usually entirely absent in octopods. All cephalopods are marine, and almost all are predaceous carnivores. Although there are probably only about 550 living species of cephalopods, they are of disproportionate importance in the ecology of the sea and are dominant organisms of the fourth or fifth trophic level in many marine food chains. Of course, there are at least 7000 known fossil species of cephalopods, some of which were among the most common marine animals of their times.

In addition to these three major classes of living molluscs, extant types occur in four minor classes, all marine with relatively few species that represent only a tiny fraction of bioenergetic turnover in marine communities. The largest is the class Polyplacophora, which consists of about 600 species placed in 43 genera of rather stereotyped chitons, or "coat-of-mail shells," with eight plates. The chitons are clearly adapted for life on hard and uneven surfaces in the low intertidal, where the suctorial foot and the eight articulated shell plates allow them to resist the strongest surf action. All chitons are slow, continuously grazing microherbivores. The class Monoplacophora encompasses several fossil families and the living deep-water genus, *Neopilina,* all with limpet-like shells. The class Scaphopoda consists of about 250 species, all of which live at moderate depths in the sublittoral (or epibenthic) zone and are known as "elephant's-tusk shells." Scaphopods clearly have a relatively unsuccessful pattern of molluscan construction with no obvious adaptive radiation. All have a strong radula, which is used along with prehensile cephalic tentacles to feed on foraminiferans. Finally, the class Aplacophora consists of about 20 genera living in moderately deep water, all with worm-like bodies whose mantles secrete discrete calcareous spicules but never a shell.

Aplacophorans are classified into two distinct orders, both with somewhat circumscribed ecology. Those of the order Chaetodermatoidea live in burrows in certain bottom oozes and are probably detritus feeders, whereas the more numerous species of the order Neomenioidea live among certain deep-sea hydroids and other colonial cnidarians on which they graze.

In our systematic survey, a little more detail on the two largest classes is required to help elucidate ecological constraints. The usual taxonomic arrangement of the class Gastropoda involves three somewhat unequal subclasses. It must be noted that all gastropods at some time in their phylogeny and at some stage in their development have undergone torsion, a process whereby the visceral mass and the mantle-shell that covers it have become twisted through 180° in relation to the head-foot, which results in an anterior mantle-cavity (for further discussion, see Volumes 3 and 8 in this series).

The first subclass of gastropods, the Prosobranchia, is the largest and most diverse, with marine, freshwater, and land representatives all of which retain internal evidence of torsion. Prosobranchs are divided into four orders on the basis of increasing asymmetry of their mantle-cavity complex, corresponding to greater hydrodynamic efficiency of their respiratory and sanitary arrangements. Note that in all molluscs, not only in gastropods, the internal structures of the cardiac and renogenital complex are closely linked to the complex of organs in the mantle-cavity. In general, if there is a symmetrical pair of ctenidia (gills), there will be a symmetrical pair of auricles on either side of the muscular ventricle of the heart: if there is a single ctenidium, then there is one auricle; if there are four ctenidia, there are four auricles. With asymmetry of ctenidia and heart structures in the evolution of higher gastropods, various degrees of asymmetry (and hence of functional separation) have developed in the renogenital structures. Such functional separation can be of great ecological significance (in potentially allowing large eggs or viviparity). Relatively few living gastropods are zygobranch, with two ctenidia and a symmetrical pallial and cardiac complex; they are placed in the order Archaeogastropoda, along with certain more numerous snails which have a single aspidobranch (two-sided, feather-like) ctenidium.

Of the four orders of the subclass Prosobranchia, the Archaeogastropoda and the Neritacea have aspidobranch gills and retain certain other archaic features. In these forms, there is some danger of particulate material clogging the dorsal part of the mantle-cavity above the two-sided gill or gills and reducing the respiratory water current. All such species are ecologically limited to relatively clean water over hard substrata and are unable to invade areas of the sea bottom or seashore covered with mud or silt. In contrast, snails in the order Mesogastropoda, the largest group of prosobranchs, have a single pectinibranch (one-sided, comb-like) ctenidium whose axis is fused to the mantle-cavity wall. This structure provides greater hydrodynamic efficiency with less risk of clogging and frees these snails from the ecological limitations of the two more primitive orders. The pectinibranch gill is accompanied by simplified asymmetric pallial and cardiac structures involving complete separation of genital and renal ducts. Mesogastropod species live everywhere in the sea and show a wide variety of feeding mechanisms. On the coasts of temperate and colder seas, one mesogastropod family (Littorinidae) has successful species at different levels of the seashore. In species of higher intertidal levels, various terrestrial adaptations including air breathing and viviparity have evolved anacoluthically (see Section VI). Other mesogastropod stocks are successful in estuaries, in fresh waters, and on land (especially in certain tropical areas). The fourth order, the Neogastropoda (or Stenoglossa), is again asymmetric and pectinibranch, and includes the most highly specialized prosobranchs, marine carnivores with a long extensible proboscis and a flesh-tearing radula.

Each of the other two gastropod subclasses (Opisthobranchia and Pulmonata) are considerably more uniform than the subclass Prosobranchia, and in both the effects of torsion are reduced or obscured by secondary processes of development and growth. The marine subclass Opisthobranchia consists largely of the beautiful sea slugs (or nudibranchs), in which the shell and mantle-cavity are reduced or lost; the adult is bilaterally symmetrical with a variety of secondary (neomorphic) gill conditions. Ecologically, some nudibranch species are highly specialized in diet and microhabitat (see Chapter 5 in this volume). The final subclass, Pulmonata, consists of gastropods with the mantle-cavity modified into an air-breathing lung with no ctenidia. There are a few littoral marine forms, but the order Basommatophora is mainly composed of the freshwater lung-snails (see Chapter 9, this volume), and the order Stylommatophora consists of the successful land snails such as *Helix* and *Cepaea* (see Chapters 10 and 14) along with a few shell-less families of land slugs.

As already noted, species in the class Bivalvia are generally more uniform in structure and function than are gastropod species, but there are three distinct subclasses, unequal in extent and in ecological significance. The great majority of living bivalves are lamellibranchs; that is, they have enormously enlarged gills used as organs of water propulsion and of food collection in filter-feeding as microphages on suspended material. Each lamellar gill has many elongated filaments, and although it is homologous both functionally and morphologically with the ctenidium in gastropods and other molluscs in terms of its blood vessels, arrangement of cilia, and so on, it is far more extensive than is required for the respiratory needs of the animal. Apart from the major subclass Lamellibranchia, there are two groups of extant bivalves usually ranked by neontologists as subclasses, neither of which are filter feeders. The smaller of these, the subclass Septibranchia, is composed of specialized bivalves with the ctenidia replaced by a muscular septum; pumping by this septum allows them to feed as macrophages on such things as copepod crustaceans of moderate size. Septibranchs are relatively rare in moderately deep water throughout the world's oceans. The other group, the subclass Protobranchia, is of greater evolutionary interest, because its pallial arrangements are in many ways functionally intermediate between those of true lamellibranchs and more generalized molluscan stocks. Their ctenidia are relatively small and, in most protobranchs, the important feeding structures are mobile ciliated extensions around the mouth forming palp proboscides and labial palps. Such a bivalve is essentially a deposit feeder, using these extensible palps to bring in organic detritus from bottom deposits outside the body. Although protobranchs usually comprise somewhat less than 10% of the bivalves in the shallower inshore seas, they can be more than 70% of the bivalves in deep-sea samples (see Chapter 2, this volume). This may, in part, be a true relict distribution.

In terms of numbers of individuals and ecological bioenergetics, the subclass Lamellibranchia, comprising the majority of bivalves, is the most successful of all molluscan groups in the sea. In addition, all nonmarine bivalves are lamellibranchs. All the most primitive marine bivalves are infaunal, burrowing into soft substrates, but several superfamilies are epifaunal, mostly attached to rocks.

Success in marine habitats results from incorporation into the lamellibranch body form of the most efficient mechanism ever evolved in any benthic animal group to collect the largest plant crop in the world—the marine phytoplankton. Although the number of species of marine gastropods is larger, the biomass of marine bivalves is much greater. Dense beds of epifaunal mussels and oysters are obvious features of the lower intertidal and the shallower seas in many parts of the world, and mussel beds are clearly seen in many aerial and even in some satellite photographs. Less visible infaunal clams can be almost as abundant. One estimate gives densities of mactrids (small ''surf clams'') reaching 8000/m^2 and covering about 2500 km^2 of the Dogger Bank in the North Sea. Lamellibranch bivalves are clearly important in the marine ecology of this planet.

The global distribution may be summarized as follows: all classes of living molluscs occur in the sea, and all habitats have molluscs. Protobranch bivalves are found at ocean depths of more than 9000 m, whereas species of the mesogastropod genus *Littorina* and the basommatophoran genus *Melampus* live high on the seashore in zones merely wetted by sea spray. The major class Cephalopoda is limited to the oceans (see Section VI, and Chapter 6, this volume), as are the four minor extant classes. Fresh waters support certain genera of bivalves, several stocks of prosobranch gastropods, and the abundant species of freshwater pulmonates (see Chapters 7, 8, and 9, this volume). Land molluscs are all gastropods, with the subclass Pulmonata dominating most habitats, and with certain stylommatophoran pulmonates occurring on suitable mountains at altitudes of nearly 6000 m. Several stocks of prosobranch snails are also represented on land, particularly in certain tropical rain forests, and include mesogastropods and neritaceans. There is considerable evolutionary interest in tracing the origins and intermediate adaptations of nonmarine stocks (see Section VI).

III. Distributional and Trophic Ecology

Somewhat arbitrarily, the studies presented in Chapters 2–6 of this volume may be grouped under this heading. They are principally concerned with descriptive ecology and comparative analysis of habitats for certain groups of molluscs, and with their trophic resources and feeding mechanisms.

Among the least known and most difficult to study of all environments on Earth are the extensive abyssal plains and hadal depths of the ocean floor. Together they cover 54.5% of our planet's surface. The survey of deep-sea molluscs (see Allen, Chapter 2, this volume) begins with an account of this environment, both abiotic and trophic, passes to a consideration of faunal assemblages, and thence to a class-by-class examination of the functional morphology and ecology of abyssal molluscs. Inevitably, this is largely concerned with bivalves and, within the bivalves, with protobranchiate forms. Allen describes some remarkable carnivores as well as the detritus feeders and discusses the fact that not only the successful stocks of deposit-feeding protobranchs but also septibranchs and two eulamellibranch families (Thyasiridae and Xylophagidae) show extreme evolutionary radiation in the deep sea. He postulates that this is related to the absence of competition from other lamellibranchs, and also to the general stability of the deep-sea environment as earlier characterized by Sanders (1968, 1977, 1979; see also Stanley, 1976).

Deep-sea gastropods belong to a much wider array of families and genera (with no groups predominating), but although represented by large numbers of species, they are ecologically much less important than bivalves in energetic and biomass terms. There is still a lack of physiological studies on the direct effects of pressure but, as Allen points out, individual catabolism is certainly slowed in many deep-sea molluscs and population recolonization takes at least 10 times longer (Grassle, 1977) than it does on comparable sediments in shallow water. Generalizations on life-cycle evolution are difficult, and Allen's survey contrasts the slow growth of some protobranch bivalves of the genus *Tindaria,* which are believed to live for at least a century and mature at 50–60 years of age, to others such as *Nucula cancellata,* which can mature in 2 years. Allen notes that even greater contrasts in deep-sea ecology are provided by those molluscs which live in two kinds of "islands" of high productivity in the generally rather sterile conditions of the deep sea: thermal vents (Boss and Turner, 1980) and submerged wood (Knudsen, 1961, 1970; Turner, 1973).

In general accounts of marine biology, it is often noted that exceptions are provided to the generally low bioenergetic turnover (albeit with high species diversity) in tropical seas by cold-water upwellings and by coral reef and mangrove communities. Mangal ecosystems, found in intertidal and shallow estuarine waters throughout the tropics, are known to support a relatively rich fauna. The first review of the distributional ecology of mangrove bivalves (see Morton, Chapter 3, this volume) draws attention to the existence of precise microhabitats within the rich organic turnover of the mangal, and discusses the genera and species (relatively few) that show such remarkable adaptations for life in this highly productive but difficult environment. Earlier papers on mangrove molluscs (Coomans, 1969; Bouchet, 1977) were concerned mainly with gastropods and did not attempt, as Morton does here, to analyze either the physiological and behavioral specializations of mangrove bivalves or certain broader aspects of the faunal economy of mangroves which clarify ecological interactions. These biotic interactions, in turn, determine the microhabitat distributions both of the few endemic species and of the larger number of more widely distributed bivalve species which are associated with the seaward fringes of mangroves.

It should be emphasized that mangroves are formed in the upper half of the intertidal zone (supralittoral fringe down to mean tidal level) and that mangal plants both prefer and accumulate soft mud deposits. In general, filter-feeding lamellibranch molluscs are best adapted for lower tidal levels and for sandy or firmer deposits. As Morton points out, adaptation of such bivalves for life in mangroves must involve problems of gill clogging and silt removal. In other aspects of their physiological ecology, they resemble bivalve species specialized for life in the high salt marshes of temperate coasts where the growth patterns of *Spartina, Distichlis,* and other grasses produce similar accumulations of soft, highly organic deposits. Peculiarly adapted forms have evolved in mangroves, including *Enigmonia aenigmatica,* an anomiid "saddle oyster" which has become motile like a limpet (Yonge, 1968). In addition, Morton points out that a case can be made for the mangal environment as the original habitat of wood-boring bivalves, including the highly modified shipworms (Teredinidae, plus a few genera of Pholadidae). In general, the true mangrove bivalves (totaling perhaps 36 species in the Caribbean and 86 species in the Indo-Pacific), along with the allochthonous bivalves of the seaward fringes which often occur in enormous densities, are together very important in the bioenergetic recycling which maintains the high local productivity of the mangal community.

Much of the earlier literature on the bivalves associated with coral reefs is anecdotal, and there have been few investigations of bioenergetic relationships and even less experimental work on known ecological associations. The review (Morton, Chapter 4, this volume) of the coral habitats of Indo-Pacific bivalves may help stimulate needed further studies in quantitative and experimental ecology. Morton reminds us that boring bivalves are of central importance in the biodestruction of reef material (and hence in the necessary recycling that perpetuates a dynamic reef system). However, there is little quantitative information on either their clastic activities or their demography and life cycles, despite recognition of their significance to reef ecosystems by investigators ranging from Otter (1937) to Bromley (1978). Besides considering the bivalve borers of living and dead coral, Morton provides detailed analyses of the life style and distribution of three other categories of bivalves: epizooic forms on living and dead coral; crevice and coral cave dwellers; and other coral-associated bivalves (including the giant clams and forms living in coral sands and small rubble). Many cases of evolutionary convergence can be noted within each of these categories of life style, reminding this editor of the "guilds" that are so useful a concept (Root, 1967) in the ecology of certain birds and insects. The functional morphology and commensal association in *Tridacna* species is certainly unique. The expanded siphonal tissues of these giant clams serve as culture "farms" in which zooxanthellae are exposed to light (Yonge, 1953); Morton provides an up-to-date survey of the physiological and ecological investigations of this peculiar relationship. The evolution of coral-associated bivalves has resulted in low familial and generic but high species diversity, which suggests that wide adaptive radiation has occurred *within* the coral ecosystem.

The beauty of nudibranchs has attracted scientific naturalists for more than a century (while presumably inhibiting potential predators in many cases), but ecologists have ignored this opisthobranch gastropod group until relatively recently. The review of the reproductive and trophic ecology (Todd, Chapter 5, this volume) of the true nudibranchs supplements an excellent account of opisthobranch natural history (Thompson, 1976) and complements a parallel review of the specific diets and prey associations of British nudibranchs (Todd, 1981). Nudibranchs are, almost without exception, exclusively predators of sessile marine invertebrates. Furthermore, the majority of nudibranch species reproduce by pelagic larvae, and these plankotrophic veliger larvae must show selective settlement on or near the prey species before beginning their benthic life as juvenile and then adult nudibranchs. Todd points out that this results in a synergism between life-cycle tactics and adult trophic ecology in these species.

A set of models of life-cycle strategies for nudibranchs which correspond to four patterns of prey availability are outlined by Todd, who develops theories of selection conditions favoring subannual, annual, and biennial life cycles. His review provides a detailed survey of the variations in egg sizes, duration of larval life, larval types, and patterns of metamorphosis; many of these are related to the constraints of particular predator–prey associations. The somewhat arbitrary arrangement of this volume is again revealed by the fact that Todd's discussion of life-cycle patterns could have been placed equally appropriately next to either the general evolutionary discussion of life-cycle in molluscs (Calow, Chapter 15) or the account of actuarial bioenergetics in nonmarine molluscs with a variety of life cycles (Russell-Hunter and Buckley, Chapter 11). The evolution of life-cycle patterns remains an important subject for ecologists, and future experimental workers could turn profitably to some of the systems described by Todd for different types of nudibranch.

Squid, cuttlefish, and octopuses, as representative living cephalopods, are relatively large, fast-moving, "brainy" animals, notably nonmolluscan (not sluggish) in their behavior. Most swim by jet propulsion and are predaceous carnivores. All are marine and dioecious. Their ecology obviously differs markedly from that of other molluscan groups, but scientific investigation of their life styles (particularly in the case of pelagic squid) is beset with practical difficulties. The review of the physiological and trophic ecology of cephalopods (Summers, Chapter 6, this volume) begins with a survey of types and a frank account of these difficulties. Only relatively few species have been studied alive, and fewer still have been maintained long enough to carry out the simpler physiological measurements of respiration, digestion, excretion, and so on. Demography and population dynamics can be deduced for a few species on the bases of diverse (but mostly biased) sampling techniques. These summaries of methods of laboratory maintenance and difficulties in field sampling by Summers are backed by his pathfinding experience in both fields. Consideration of the physiology of locomotion and buoyancy leads to a general review of behavior in relation to habitat. The feeding methods of different forms of cephalopods (pelagic, demersal, and cryptic-benthic) are all based on a strike, or attack, sequence upon the prey organism which involves complex innate patterns along with considerable capacity to learn (documented mostly in inshore forms such as *Sepia* and *Octopus*). These higher molluscs have an indisputable claim to be the most highly organized of all invertebrate animals, and behavioral aspects of their ecology are based on individual neural capacities exceeded only by those of some birds and mammals. The cephalopod habit of mincing food on ingestion makes gut analyses difficult. On the other hand, our knowledge of the predators upon cephalopods comes from specific identification and counts of cephalopod beaks (jaws) in the gut contents of sperm and other toothed whales. As observational methods improve (particularly by using small submersibles and automatic cameras), many problems of cephalopod ecology could become less difficult.

IV. Bioenergetic and Physiological Ecology

In the current state of ecological studies, it is relatively easy to define bioenergetic ecology as being concerned with trophic and reproductive transfers and efficiencies in populations or communities of organisms. It is much more difficult to delimit physiological ecology as a subdiscipline. Definitions (whether explicit or not) used in Chapters 6 through 12 of this volume vary from author to author. In the broadest sense, physiological ecology is that part of the study of the interrelations between organism and environment which is especially concerned with physiological mechanisms or is largely based on physiological measurements. This exegesis is obviously rather vague. As in the similar case of biophysics, for which different groups of scientists claim distinct missions, it may not be necessary or desirable to limit inquiry by strict definition of the field. It is important, however, to try to ensure that the experimental controls and restricted factor analyses of sound physiological work are paralleled in field investigations. It should not be merely physiological work "done under the worst of conditions" (Fogg, 1965). Use of organisms in laboratory culture can determine mean values for the rates of any physiological process over a range of values for some physical factor, but only studies based on natural populations can establish the range of individual variation in these rates. It is upon this variation that the ecological (and evolutionary) significance of each process depends. The physiological and bioenergetic ecology of various nonmarine molluscs is discussed in Chapters 7–11 of this volume.

Freshwater bivalve molluscs belong to a limited number of (largely cosmopolitan) genera classified in three lamellibranch superfamilies. Two of these, the Unionacea (the large freshwater mussels) and the Corbiculacea (the small fingernail and pea clams) are more important. In the review of physiological ecology of these bivalves, Burky (Chapter 7, this volume) notes that unionacean mussels are generally associated with major, relatively permanent drainage systems, whereas pisidiid (formerly known as sphaeriid) clams of the Corbiculacea live in small, transient bodies of fresh water. Differences in the physiological ecology of these two groups result in part from the fact that most species of pisidiids probably evolved in transient habitats, as is also claimed for many pulmonate gastropods living in fresh water (Russell-Hunter, 1978). After a brief introduction to the distributional ecology of these clams, Burky uses physiological measures of metabolic responses (both short-term and long-term) in a discussion of the results of acclimation and environmental stress, particularly in respiration and feeding. Once again, the seasonal fit of shifts in physiology to each style of life cycle is important.

A survey of reproductive tactics and life-cycle strategies is complemented by extensive and important data on bioenergetics and population dynamics in these freshwater clams. Burky (Chapter 7) then relates this body of comparative data on life cycles to current ecological theories of evolutionary strategies. He develops a paradoxical concept that habitats such as temporary ponds or places with regularly recurring dry seasons represent sets of predictable climatic conditions which correspond to stable environments in the evolutionary sense. Some data on genetic polymorphism and heterozygosity are becoming available to test this concept and others. Simple interpretation of much of this information in terms of *r* and *K* selection theory is further complicated by populations whose genetic monomorphism suggests sustained tactics of self-fertilization. Another problem with the ecogenetics and population dynamics of freshwater bivalves remains the relative significance of passive dispersal by aerial means and, in the unionids, the use of a parasitic larval stage carried by fish. Freshwater bivalves also provide systems appropriate for many kinds of experimental ecology.

In Chapter 8 on freshwater prosobranchs (in this volume), Aldridge begins with a statement on the value of the physiological ecology viewpoint. It is worth emphasizing that brackish and fresh waters have been colonized from the sea many times by different stocks of mesogastropod and neritacean prosobranchs. Accordingly, there is considerable interest in comparative studies of feeding, excretion, and ionic regulation in these prosobranch stocks. Aldridge, in discussing demographic variables in growth and feeding, is also concerned with sex differences in metabolism and ecology, because nonmarine prosobranch species are mostly dioecious. Of particular interest are assessments of differential catabolism (from concurrent assessments of oxygen uptake and nitrogenous excretion) which can be related to reproductive output in metabolic balance sheets. Short- and long-term processes of acclimation are important in many aspects of the physiology of freshwater prosobranchs and can involve periods of diapause. Differences in types and rates of nitrogenous excretion may be linked to these capacities.

Aldridge questions the simplistic application of current theories on reproductive effort to some semelparous and iteroparous prosobranch species. In particular, the association of high effort with semelparity (Williams, 1966; Browne and Russell-Hunter, 1978; Calow, 1978) cannot apply in certain longer-lived prosobranchs such as *Leptoxis carinata* (Aldridge, 1982). In general terms, precisely optimal strategies in model life histories will be greatly modified by the physiological and ecological constraints appropriate to each species (see also Chapters 11 and 15, this volume).

In addition, Aldridge provides a detailed survey of the preferred and principal diets of nonmarine prosobranchs. In the past, questions of distinct trophic resources have often been answered with the general statement that the main food supply was *Aufwuchs*. To merely define certain questions on *K* versus *r* selection patterns in modeling the life-cycle dynamics of such prosobranchs, we need more specific knowledge of actual items and quantities of food intake and the ability to assess seasonal and other shifts in food availability. Aldridge adds some suggestions for future ecological research on nonmarine prosobranchs.

In certain ancient lakes, including Tanganyika, Baikal, and Ochrida, peculiar freshwater prosobranchs are found with more elaborate secondary adaptations and, in some cases, bizarre ''thalassoid'' shells. Their ecology is not reviewed in this volume, but discussions of their morphology and systematics (Boss, 1978; Russell-Hunter, 1978) have concluded that this diversity results from processes of adaptive radiation of endemic stocks which have occurred in environments much less transitory than the majority of our planet's freshwater habitats.

The dominant molluscs of fresh waters are, of course, pulmonate snails of the order Basommatophora. McMahon begins his review (Chapter 9, this volume) of freshwater pulmonates with an examination of working definitions of physiological ecology and a survey of the systematics and evolution of the Basommatophora. Unlike the vast majority of higher organisms living in fresh waters, limnic pulmonates have had an evolutionary history marked by increasingly greater adaptation to an aquatic schesis from a more terrestrial one (see also Russell-Hunter, 1978, 1979). This fact of evolutionary ecology has left an imprint on all aspects of physiology in the group. McMahon surveys respiratory adaptations from diving air breathers (and the utilization of gas bubbles as nonstructural physical gills) to freshwater limpets and planorbids which have again evolved anatomically neomorphic gills. He then reviews nitrogenous excretion, ionic regulation, calcium metabolism, and thermal tolerance in freshwater pulmonates.

The advantages of freshwater pulmonates as research subjects in physiological ecology stem both from their field accessibility (their biomass is greatest in shallower marginal waters) and from their general tractability in both field and laboratory experimentation. Perhaps the most significant and extensive part of McMahon's review concerns work on the short-term and long-term physiological shifts which occur in response to environmental changes in temperature or oxygen concentration. In particular, a full discussion is provided of both posthypoxic recovery and hypotheses regarding the anomalous reverse acclimation to temperature that characterizes several of these forms (hypotheses adumbrated in part elsewhere; see Burky, 1971; McMahon, 1973; McMahon and Russell-Hunter, 1978, 1981, and references therein). McMahon hopes that the more speculative parts of this survey of unusual adaptive physiology will stimulate further investigations. It is already clear that our understanding of metabolic-compensating processes in ectotherms is being enhanced by studies involving these freshwater pulmonates.

Unlike all aquatic habitats, the terrestrial environment is characterized by large daily and seasonal variations in temperature and by limitations on the availability of free water. As noted above (Section II), however, pulmonate gastropods are remarkably successful as land animals despite their retention of such typically molluscan features as relatively permeable skins, ciliary–mucous mechanisms, and soft, hydraulically moved bodies. In his review of the physiological ecology of land snails and slugs (Chapter 10, this volume), Riddle concentrates on the physiological and behavioral adaptations that are required by the abiotic constraints and stresses of terrestrial habitats. Water balance and thermal relations are central to the ecology of all land snails. Once again the capacity for acclimation is important, and once again periods of diapause may be necessary for survival. Riddle points out that even the tissue biochemistry involved in nitrogenous excretion must be related to the temporal shifts between periods of desiccation and rehydration which characterize the life style of land snails and slugs. Many behavioral shifts (and corresponding detectable neural activity) depend upon body fluid osmolality, and the capacity (Riddle, 1981) to retain fluid in and transfer fluid from the mantle-cavity lung can be of considerable ecological significance.

Locomotor activity in land snails and slugs is often restricted to favorable daily periods, and Riddle reviews the evidence for both endogenous (circadian) controls and direct responses to various environmental factors such as light, relative humidity, and temperature. Photoperiodism appears to influence all aspects of somatic physiology in these pulmonates, in addition to having obvious effects on reproductive maturation and activity. As Riddle emphasizes, regulation of catabolic metabolism (including locomotor activity) over daily and seasonal time periods ensures that the essential activities of land snails and slugs occur under the most favorable environmental conditions. This review concludes with several areas of physiological ecology that deserve more intensive investigation in the future.

Nonmarine molluscs provide suitable natural populations for bioenergetic studies. Quantitative balance sheets can be prepared relating the energy gained from the environment to the necessary catabolic energy expenditures for maintenance, along with the bioenergetic surplus that is partitioned between individual growth and reproduction. In their review of these in conjunction with nonmarine molluscan productivity (Chapter 11, this volume), Russell-Hunter and Buckley assert that sound assessment of the evolutionary implications of many ecological processes (including differential reproductive effort) must be based on an *actuarial* presentation quantifying all inputs and outputs. Recent bioenergetic studies have moved closer to the current commercial usage of actuarial (as relating to the computation of premiums which involve interaction of interest rates with certain statistical probabilities) and have considered differences in bioenergetic rates corresponding to the demography of natural populations. We cannot procure the best data for actuarial bioenergetics, which would be those of energy-flux levels obtained instantaneously and nondestructively. Instead, much of the work surveyed by Russell-Hunter and Buckley is based on determinations of total organic carbon in sequential samples from large natural populations. A discussion of methods and rate limits is followed by reviews of the available data on standing crop biomass and on rate productivities for nonmarine molluscs in three operational categories. Presentation of these data leads to consideration of the value of ratios and dimensionless index numbers in making ecological comparisons, for example, of trade-offs between growth and reproduction.

Russell-Hunter and Buckley continue with an examination in these terms of available data on the internal components of balance sheets for the population or modal individual, and subsequently of those data allowing productivity comparisons between populations and between species, including the development of measures of differential reproductive effort. Turnover times and productivity rate constants can also be employed in comparisons, and detailed data on turnover times for 41 species populations show that (1) productivity rates are obviously related to length of life and (2) levels for turnover times of organic carbon can be defined that correspond to triennial, biennial, annual (univoltine), and bivoltine life-cycle patterns. The need for a dynamic measure of reproductive effort in terms of rate fraction (Russell-Hunter and Romano, 1981) is discussed and related to physiological constraints (see also Chapter 15, this volume). Several field experiments involving transfer of stocks or segregation of year classes allow assessment of demographic effects in bioenergetic terms. A brief consideration of causality and generalizations in evolutionary ecology emphasizes the need for substantive data in appropriate rate terms to be collected from natural populations. Bodies of fresh water that have experienced varying degrees of spatial and temporal limitation provide suitable experimental populations for comparative studies (Russell-Hunter, 1978). Different populations of a single freshwater species can vary greatly in panmictic size and in their temporal history as steady state populations.

V. Population Dynamics and Genetics

The final group of chapters in this volume deal with the twin bases of molluscan autecology—studies on genetic variation and population dynamics. The fundamental purpose of such studies, to increase our understanding of the evolutionary process in quantitative terms, and their value to all other kinds of ecological studies, is evident.

Both the geographic distribution of genetic variation within a species and the rate at which the species can colonize (or recolonize) new habitats depend upon the relative time that a dispersal stage exists in the life span of the species. In theory at least, a sedentary molluscan species which is viviparous and bears young which immediately take up an adult life style should have many localized races exhibiting considerable genetic differences and a low colonizing ability. In contrast, a molluscan species which spends a long time as a planktonic veliger should show a relatively uniform genome over wide geographic areas and considerable ability to colonize new habitats. Because both kinds of species exist, intuitively we deduce that the adaptive advantages of different levels of dispersal capacity can be balanced by other properties that increase fitness in different life-cycle patterns. Examples of the inverse relation between egg numbers and egg sizes in groups of molluscan species are presented in Chapters 7, 8, 9 and 11 of this volume, and the seeming tautology of the above predictions and deduction is defended in Chapter 11. A new contribution on marine larval dispersal is missing from this volume, but an earlier review (Scheltema, 1971) provides an analysis of larval dispersal between geographically separated populations of shallow-water benthic marine gastropods and allows levels of genetic exchange between distant populations to be deduced. Scheltema found that long-lived larvae (*teleplanic*) were literally crossing oceans in surface water movements such as the equatorial currents and the North Atlantic drift to the east-northeast. Potential life as a "refractory" larva could extend to several months (as many as 10). Scheltema deduced that this long-distance dispersal had permitted certain tropical mollusc species restricted as adults to shallow water near coasts, to achieve amphi-Atlantic distribution (that is, to live on both African and American coasts).

Dispersal in nonmarine molluscs is usually less effective between continents. However, several cases of new colonization of fresh waters by invasive species have been detected historically, and we have some documentation for at least four of these. For example, from details of the spread of the small prosobranch snail *Potamopyrgus jenkinsi* throughout Britain (Russell Hunter and Warwick, 1957), it was possible to distinguish between rapid and continuous spread through "canal basins" where bodies of fresh water had been linked by humans and less regular colonization along major migration routes of water birds. A most extensive literature (starting in 1944 with records begun in 1938) reports the spread of the freshwater clam, *Corbicula fluminea,* throughout habitats in the United States. This is reviewed and discussed by McMahon (Chapter 12, this volume), who deduces considerable human involvement in several extensions of the species range. It is now established on both coasts and throughout the Mississippi and Ohio drainage systems, and apparently it is a permanent member of the freshwater fauna from southern Canada to northern Mexico. McMahon provides the first complete account of its physiological ecology and extends the account of its life-cycle bioenergetics published earlier (Aldridge and McMahon, 1978), in addition to reviewing its colonization. Much of the interest in (and support for research on) this invasive bivalve of Asian origin derives from its status as a pest that fouled water intake pipes and jammed condensers in power stations and other human facilities utilizing raw water. Although *Corbicula* fouling is most hazardous in nuclear reactor cooling systems, enormous costs and inconvenience have resulted from blockage of cooling lines for fossil-fueled power stations and bearing cooling systems in factories and of raw water pipe lines for fire protection systems and irrigation networks. McMahon points out that no single control method, such as screening intakes, back-flushing, periodic heat shock, periodic chlorination, or manual shell removal, is applicable to all problems. In a few cases, the only solution may be the expensive development of closed-loop cooling systems, which do not draw in raw water and can be continuously chlorinated at levels which would otherwise be unacceptably high.

In addition to these applied problems, McMahon points out many features of the adaptive physiology and local ecology of *C. fluminea* which make it interesting in its own right, including its high capacity to resist desiccation and its lack of resistance to hypoxia (in contrast to the many freshwater molluscs that exhibit considerable capacity to regulate oxygen uptake rates after exposure to low oxygen conditions). McMahon concludes that *C. fluminea* has only recently evolved as a freshwater species from a stock living in estuarine conditions. However, it displays life-cycle and reproductive characteristics that diverge greatly from those of its closest estuarine relatives, including hermaphroditism, earlier maturity, larval incubation, reduction of life span, and provision for passive dispersal of well-developed juveniles ready for benthic life. Essentially, it is these life-cycle tactics that make this particular freshwater bivalve both a successful invasive species and an exasperating pest of manmade water systems.

Very little classical genetic analysis has been carried out on molluscs other than a few nonmarine species. In his review of population genetics (Chapter 13, this volume) of marine gastropods, Berger draws principally on data surveying polymorphisms of shell structure and, most notably, of enzyme systems, along with smaller data bases from cytogenetics (of the prosobranch *Thais*) and growth rates and differential salinity tolerance in commercial bivalves such as oysters and mussels. Berger begins with a succinct outline of four lines of inquiry which provide a conceptual background for contemporary studies in population genetics, and then presents the data base for marine molluscs. The most extensive data are on allozyme variation in about 40 marine molluscs, with major surveys of gene– enzyme variation having been completed on 10 species. For the benefit of readers, Berger has tabulated the extensive literature referring to this data on polymorphism (including 23 papers on *Mytilus edulis* alone), and has extracted values for average heterozygosity and the proportions of polymorphic loci from the 10 major surveys.

From an ecological viewpoint, data on the geographical distribution of genetic variation in natural populations is most interesting, and Berger summarizes the work on littorinid snails and on *Mytilus*. Of particular interest are the major genetic differences between European and North American stocks of *Littorina littorea* and his own observations that the level of heterozygosity in North American *L. littorea* is very low (Berger, 1977). This suggests a *founder effect* and relatively recent immigration from Europe. In *Mytilus edulis,* marked allele frequency clines have been detected in one locus, and a series of biochemical and physiological experiments have demonstrated that in this case the genetic variation has adaptive significance in relation to environmental differences in water temperature and salinity. Berger also discusses the use of electrophoretic studies to supplement morphological work in the systematics of littorinids and details special examples of genetic strategies.

Among the few convincing demonstrations of selection in action on natural animal populations are studies over the last 3 decades on the land pulmonate *Cepaea*. The review of ecology and ecogenetics in terrestrial molluscan populations provided here (Cain, Chapter 14, this volume) is written by a principal founder (and sustainer) of such studies. In introducing his review, Cain points out that land snails have many of the advantages of plants for ecogenetic research in that they occur in highly localized populations which can readily be marked and sampled. However, panmictic units of snail populations may be relatively small, and moderately long life spans with overlapping generations also create difficulties. The first part of the review is concerned with the physiological ecology and population interactions of these land snails and provides a badly needed basis for the interpretation of their genetic variation. As Cain notes, some workers on population genetics have tended to treat land snails merely as packets of genetic data sitting around in spatial patterns and to ignore their ecology. In the provocative middle section of his chapter, Cain considers conceptual questions of selection and on maintenance of polymorphism, in terms of the simplest hypothesis and justifiable scientific skepticism. The remainder of the review is concerned with recent developments in ecogenetics of land pulmonates. Some cytogenetics has been investigated and some classical breeding work completed, and considerable data on allozymes have accumulated. Cain notes that further studies on wild populations of *Cepaea* have shown massive linkage disequilibria, additional evidence for selection of some sort. Medium-term (historical) and long-term (based on subfossil samples) changes in morph frequencies in *Cepaea* are discussed, as is the enormous body of geographical survey data on color and banding patterns. Cain discusses the abiotic and biotic factors involved, and notes that genetic drift is often invoked without evidence to explain peculiar distributions of morph frequencies, usually when little is known about the ecology of the populations concerned. He concludes that the ecology and behavior of land snails are sufficiently complex that many of the difficulties in interpretation of genetic variation arise from local differences in population biology, and he again emphasizes the role of natural selection in wild populations.

Much recent discussion in theoretical ecology has involved evolutionary models of life-cycle strategies (Stearns, 1976, 1977, 1980), but in many invertebrate groups including the Mollusca there is a lack of substantive data both on the demography of natural populations and on physiological constraints on reproduction. The review (Calow, Chapter 15, this volume) of life-cycle patterns throughout the major molluscan taxa is largely concerned with the interaction of endogenous (physiological) constraints with those extrinsic factors which vary under different environmental circumstances. Endogenous factors involve the trade-offs between reproductive investment and parental survivorship, and Calow discusses various ways of quantifying reproductive effort (drawing largely on his own earlier research publications: Calow, 1978, 1979, 1981; see also Chapter 11, this volume). Trade-offs involving growth to larger adult sizes are always important in animals with indeterminate growth patterns, and those involving egg size and numerical fecundity are peculiarly significant in many molluscs. Calow analyzes the effects of these trade-offs in terms of fitness merits. The form of survivorship curves remains important, as does the extent of density-dependent and density-independent controls on the processes of growth and reproduction. As Calow points out (see also Calow, 1978), the alternatives of producing a few relatively hardy offspring per parent (as in *Ancylus fluviatilis*) or of producing a larger number of more mortality-prone offspring (as in *Lymnaea peregra*) can balance out to produce equi-fit strategies even in coexisting species. In general, certain aspects of the plasticity of molluscan reproductive pattern defeat simple r and K analyses and emphasize the significance of differential survival in molluscs as a determinant of life-cycle tactics (see also Hart and Begon, 1982).

Calow points out some systematic biases explicable in terms of the major taxonomic groups of the Mollusca but notes that no absolute limitations are obvious. Ecological trends related to particular series of microenvironments are more readily detected. In their life-cycle patterns, as in their morphology and physiology, the Mollusca display an impressive diversity based on a common ground plan. As already noted, comparative data on life-table statistics for closely related species of molluscs with different patterns of life cycle are rarely available. Further elucidation of the ways in which natural selection has shaped life-cycle patterns in molluscs to fit the trophic resources of different environments could require such detailed demographic studies. Like most questions of molluscan ecology, these on life-cycle pattern remain open-ended.

VI. Ecological Constraints on Molluscan Evolution

The present overview of molluscan ecology began with a paean to the success of some molluscs in our planet's bioeconomy (Sections I and II). Emphasis was placed on the unlikelihood of this ecological success in a pattern of animal construction employing ciliary and mucous mechanisms for most life-sustaining functions, these mechanisms being enclosed within a soft, permeable body moved hydraulically from place to place. Despite this unlikelihood, certain molluscs dominate marine ecosystems and others follow arthropods and vertebrates as successful land animals. To end this introductory chapter, a brief consideration of some ecological and physiological limitations which have constrained molluscan evolution may be instructive. In certain instances, these limitations can be explained as historical consequences of earlier adaptive trends in stocks of molluscs. A general account of adaptive radiation is provided by Seed (Chapter 1 in Volume 1 of this series).

Several constraints of this sort are obvious. Among the very successful lamellibranch bivalves, there are no terrestrial representatives, and none could evolve retaining that body form. The enormous biomass of marine lamellibranchs results from filter feeding, using the hypertrophied gills as organs of water propulsion and food collection. Anatomical changes during their evolution, principally involving these lamellar ctenidia, have allowed these bivalves to outcompete all rival types of filter feeders in the richest waters of shallower inshore seas, including the lampshells of the phylum Brachiopoda and their own protobranch relatives (but see Chapter 2). The same anatomical features effectively remove all potential for the evolution of a land stock of lamellibranchs.

The cephalopods provide a number of pertinent examples of constraints. The modern subclass Coleoidea (encompassing squid, cuttlefish, and octopods) includes the largest, fastest, and brainiest invertebrates. The outstanding efficiency obtained by the modified mantle-cavity as an organ of jet propulsion in all the pelagic cephalopods occurred when that prime mover competed with the myotome muscle blocks of those chordates which contemporaneously swam in the sea: various fishes, predaceous Mesozoic reptiles, and, finally, the dolphins and toothed whales. Size and speed ratios seem to have kept pace by selection pressures of alternate predation and competition in the evolution of the larger pelagic cephalopods along with the swimming vertebrates (Packard, 1972). Without the pressure to keep up with the competitive vertebrates, the molluscan body plan would probably not have evolved the speed, size, and brains of the coleoid cephalopods.

The fastest of the users of these water-jet motors, the squid, are limited to the open waters of the oceans, and the largest to middle oceanic depths. Ecologically, all cephalopods are limited to the sea. The functional efficiencies of jet propulsion and of relatively massive brains have not been paralleled in other physiological systems, such as the excretory organs and the respiratory functions of the blood. These could not be readily adapted for life in estuaries or fresh waters. Evolution of terrestrial forms is also prevented by the lack of a rigid skeleton and the delicacy of cephalopod skin layers. An even more significant constraint is that the reproduction of most cephalopods involves large numbers of only moderate-sized eggs (a female *Octopus* can produce as many as 150,000 eggs; a female squid of the genus *Loligo,* 70,000). These are not truly cleidoic eggs, because the cephalopod embryos normally take up considerable quantities of inorganic salts from the sea in the course of their prehatching development (Russell-Hunter and Avolizi, 1967). Such numerical fecundity (and "open" eggs) not only precludes nonmarine life but also prevents the development of any parental care (Russell-Hunter, 1979). Thus, despite their big brains and ability to learn, there can be no nongenetic informational communication between generations in cephalopods, and some ways of developing individual intelligence open to man and even to dolphins cannot be employed by these brainy molluscs.

In stocks of nonmarine molluscs, and in their estuarine and littoral predecessors, several physiological constraints displayed by their marine relatives have had to be circumvented. General textbook accounts of the evolution of land animals often indicate the intertidal zone of the sea as providing an important series of increasingly terrestrial habitats in which animal stocks could have acquired various physiological "preadaptations" for land life. Although this deduction about appropriate adaptive selection is true in part, physiological evolution in extant littoral molluscs is clearly an irregular process. For example, five species of periwinkles belonging to the mesogastropod family Littorinidae are found at different levels of the intertidal zone on both sides of the temperate North Atlantic. The lowest in the series, *Lacuna vincta,* also lives sublittorally and ecologically as a truly marine form; the highest, *Littorina neritoides,* is practically a land snail; the three other species of *Littorina* living at different intermediate levels are intermediate in most aspects of form and function. Primary physiological adaptations for life on land concern water control with modification of the excretory system for conservation of water, loss of gills with conversion of the mantle-cavity to a lung for air breathing, temperature regulation, and modification of the reproductive system for the production of larger eggs (with suppression of planktonic larval stages) or for viviparity.

Recent reinvestigation of the respiratory schesis in these littorinids (McMahon and Russell-Hunter, 1977, 1978) reveals that adaptations for aerial respiration (along with a capacity to regulate oxygen consumption in relation to thermal and hypoxic stresses) are not matched exactly with appropriate reproductive adaptations or precisely with vertical distribution within the intertidal zone. It seems that the evolution of increasingly terrestrial adaptations in these five periwinkle species has proceeded anacoluthically, or in discontinuous series not necessarily matched with each other. This is not surprising, given modern views of processes of natural selection (blind variation and recombination, followed by systematic elimination and selective retention). At any higher level in the littoral, evolution of the ''best'' land eggs and land lungs should not be expected to have occurred contemporaneously in any single species. Processes of preadaptation do occur, but are displayed only in matched sets in fully terrestrial forms that have moved elsewhere on land.

Considering the ecology of nonmarine stocks in the broadest terms, there are several kinds of physiological limitations that involve historical consequences of earlier adaptive trends. It is worth reemphasizing that in their evolution the pulmonate snails ran counter to the higher vertebrates and paralleled the freshwater insects in deriving their freshwater stocks from an air-breathing land stock. The higher limnic pulmonates show progressive readaptation of an air-breathing stock to aquatic life (Chapter 9, this volume; see also Russell-Hunter, 1978). In contrast, most stocks of freshwater prosobranchs (along with the bivalve genera in fresh waters) are directly derived from, and in some cases closely related to, estuarine forms (and, through those, derived from marine stocks). Land prosobranchs (the so-called operculate land snails) are less successful than pulmonate land snails and represent many independent invasions of terrestrial habitats. Several such stocks are limited to certain warm, damp forest areas of the Greater Antilles. Many cases of convergent evolution have occurred and would repay modern investigation of aspects of their comparative physiology and ecology. On a smaller scale, much good ecological work has been done on convergent, or recently separated, species pairs. Cain (Chapter 14) presents some of this work on species of the land pulmonate *Cepaea,* particularly in regard to microclimatic separation; and Kohn (1971, and other papers) has studied species of the neogastropod genus *Conus,* particularly on partitioning of their animal prey. Species pairs utilizing the same trophic resources should show spatial separation of their niches; species pairs with different food resources can be more completely sympatric. Much still remains to be learned about the ecology of both species pairs resulting from evolutionary convergence and the more often studied congeneric species pairs.

A final example of large-scale ecological limitation concerns land pulmonates. A major physiological problem of land snails is control of water loss; ecologically, therefore, habitats with high relative humidities are desirable. However, as demonstrated long ago by Hogben and Kirk (1944), slugs and snails can have a considerable capacity for temperature regulation by continued evaporation from the moist skin. Thus, at high (even lethal) temperatures, these pulmonates can survive, but only if the relative humidity is lower (e.g., below 55%). Pulmonate slugs are certainly polyphyletic, but the slug body form can be regarded as an adaptation to cryptic life in narrow crevices and soil cavities. Those diverse pulmonate stocks which have evolved into slugs have each accepted a physiological limitation with marked behavioral and ecological consequences. In relation to survival under different environmental conditions, a shelled snail has a choice which a slug has not. Except in saturated air, a snail withdrawn into its shell loses less water by evaporation but lessens its capacity to keep cool at high temperatures (Hogben and Kirk, 1944; but see also Chapter 10, this volume). The slug is compelled to lose water by evaporation when external temperatures are low, that is, in circumstances which confer no advantage to offset depletion of its water reserve.

Clearly, ecological constraints have been operative in the evolution of certain major molluscan groups. Elucidation of small-scale limitations with a similar history could be highly rewarding to evolutionary ecologists. Apparent exceptions to optimal adaptation in molluscs may be particularly interesting. After all, in the broader aspects of traditional biology, the continued existence in man of caudal vertebrae or of the many nonfunctional muscles of the ear pinna (no longer concerned with direction finding or tactile detection of air currents) provides considerable evolutionary interest. Many similar cases in molluscan physiological ecology, in which past histories of specialization create present constraints detectable as suboptimalities in structures or processes, must await investigation by sufficiently curious ecologists.

In conclusion, I hope that all of the chapters in this volume suggest areas of molluscan life that could repay more intensive investigation by ecologists. I am sincerely grateful to my fellow students of the Mollusca, the other 12 authors of these chapters, for their diverse contributions. For help with this chapter, I owe more than usual thanks to my different drummers, Myra Russell-Hunter and Peregrine D. Russell-Hunter.

References

Aldridge, D. W. (1982). Reproductive tactics in relation to life-cycle bioenergetics in three natural populations of the freshwater snail, *Leptoxis carinata*. *Ecology* **63,** 196–208.

Aldridge, D. W., and McMahon, R. F. (1978). Growth, fecundity, and bioenergetics in a natural population of the asiatic freshwater clam, *Corbicula manilensis* Philippi, from north central Texas. *J. Moll. Stud.* **44,** 49–70.

Berger, E. M. (1977). Gene–enzyme variation in three sympatric species of *Littorina*. II. The Roscoff population, with a note on the origin of North American *L. littorea*. *Biol. Bull. (Woods Hole, Mass.)* **153,** 255–264.

Boss, K. J. (1978). On the evolution of gastropods in ancient lakes. *In* "Pulmonates" (V. Fretter and J. F. Peake, eds.), Vol. 2, pp. 385–428. Academic Press, New York.

Boss, K. J., and Turner, R. D. (1980). The giant white clam from the Galapagos Rift, *Calyptogena magnifica* species novum. *Malacologia* **20,** 161–194.

Bouchet, P. (1977). Distribution des Mollusques dans les mangroves du Senegal. *Malacologia* **16,** 67–74.

Bromley, R. G. (1978). Bioerosion of Bermuda reefs. *Palaeogeogr., Palaeoclimatol., Palaeoecol.* **23,** 169–197.

Browne, R. A., and Russell-Hunter, W. D. (1978). Reproductive effort in molluscs. *Oecologia* **37,** 23–27.

Burky, A. J. (1971). Biomass turnover, respiration, and interpopulation variation in the stream limpet, *Ferrissia rivularis* (Say). *Ecol. Monogr.* **41,** 235–251.

Calow, P. (1978). The evolution of life-cycle strategies in freshwater gastropods. *Malacologia* **17,** 351–364.

Calow, P. (1979). The cost of reproduction—a physiological approach. *Biol. Rev. Cambridge Philos. Soc.* **52,** 23–40.

Calow, P. (1981). Adaptational aspects of growth and reproduction in *Lymnaea peregra* (Gastropoda: Pulmonata) from exposed and sheltered aquatic habitats. *Malacologia* **21,** 5–13.

Coomans, H. E. (1969). Biological aspects of mangrove molluscs in the West Indies. *Malacologia* **9,** 79–84.

Fogg, F. E. (1965). "Algal Cultures and Phytoplankton Ecology." Athlone Press, London.

Grassle, J. F. (1977). Slow recolonization of deep-sea sediment. *Nature (London)* **265,** 618–619.

Hart, A., and Begon, M. (1982). The status of general reproductive strategy theories, illustrated in winkles. *Oecologia* **52,** 37–42.

Hogben, L., and Kirk, R. L. (1944). Studies on temperature regulation. I. The Pulmonata and Oligochaeta. *Proc. R. Soc. London, Ser. B* **132,** 239–252.

Knudsen, J. (1961). The bathyal and abyssal *Xylophaga* (Pholadidae Bivalvia). *Galathea Rep.* **5,** 163–208.

Knudsen, J. (1970). The systematics and biology of abyssal and hadal Bivalvia. *Galathea Rep.* **11,** 7–241.

Kohn, A. J. (1971). Diversity, utilization of resources, and adaptive radiation in shallow-water marine invertebrates of tropical oceanic islands. *Limnol. Oceanogr.* **16,** 332–348.

Lubchenco, J. (1978). Plant species diversity in a marine intertidal community: importance of herbivore food preference and algal competitive abilities. *Am. Nat.* **112,** 23–39.

Lubchenco, J. (1980). Algal zonation in the New England rocky intertidal community: an experimental analysis. *Ecology* **61,** 333–344.

Lubchenco, J., and Menge, B. A. (1978). Community development and persistence in a low rocky intertidal zone. *Ecol. Monogr.* **48,** 67–94.

McMahon, R. F. (1973). Respiratory variation and acclimation in the freshwater limpet, *Laevapex fuscus*. *Biol. Bull. (Woods Hole, Mass.)* **145,** 492–508.

McMahon, R. F., and Russell-Hunter, W. D. (1977). Temperature relations of aerial and aquatic respiration in six littoral snails in relation to their vertical zonation. *Biol. Bull. (Woods Hole, Mass.)* **152,** 182–198.

McMahon, R. F., and Russell-Hunter, W. D. (1978). Respiratory responses to low oxygen stress in marine littoral and sublittoral snails. *Physiol. Zool.* **51,** 408–424.

McMahon, R. F., and Russell-Hunter, W. D. (1981). The effects of physical variables and acclimation on survival and oxygen consumption in the high littoral salt-marsh snail, *Melampus bidentatus* Say. *Biol. Bull. (Woods Hole, Mass.)* **161,** 246–269.

Menge, B. (1976). Organization of the New England rocky intertidal community: role of predation, competition, and environmental heterogeneity. *Ecol. Monogr.* **46,** 355–393.

Otter, G. W. (1937). Rock destroying organisms in relation to coral reefs. *Sci. Rep. Great Barrier Reef Exped.* **1,** 323–352.

Packard, A. (1972). Cephalopods and fish: The limits of convergence. *Biol. Rev. Cambridge Philos. Soc.* **47,** 241–307.

Riddle, W. A. (1981). Hemolymph osmoregulation and urea retention in the woodland snail, *Anguispira alternata* (Say) (Endodontidae). *Comp. Biochem. Physiol.* **69A,** 493–498.

Root, R. B. (1967). The niche exploitation pattern of the blue-grey gnatcatcher. *Ecol. Monogr.* **37,** 317–350.

Russell-Hunter, W. D. (1978). Ecology of freshwater pulmonates. *In* "Pulmonates" (V. Fretter and J. F. Peake, eds.), Vol. 2, pp. 337–383. Academic Press, New York.

Russell-Hunter, W. D. (1979). "A Life of Invertebrates." Macmillan, New York.

Russell-Hunter, W. D. (1982). Mollusca. *In* "Encyclopedia of Science and Technology," 5th ed. Vol. 8, pp. 696–701. McGraw-Hill, New York.

Russell-Hunter, W. D., and Avolizi, R. J. (1967). Organic content in developing squid eggs assessed from carbon, nitrogen, and ash, and its evolutionary significance. *Biol. Bull. (Woods Hole, Mass.)* **133,** 470–471.

Russell-Hunter, W. D., and Romano, F. A. (1981). Reproductive effort of molluscs in bioenergetic terms: some computational methods. *Biol. Bull. (Woods Hole, Mass.)* **161,** 316.

Russell Hunter, W., and Warwick, T. (1957). Records of "*Potamopyrgus jenkinsi*" (Smith) in Scottish fresh waters over fifty years (1906–1956). *Proc. R. Soc. Edinburgh, Sect. B: Biol.* **66,** 360–373.

Sanders, H. L. (1968). Marine benthic diversity: a comparative study. *Am. Nat.* **102,** 243–282.

Sanders, H. L. (1977). Evolutionary ecology and deep sea benthos. (The changing scenes in natural sciences, 1776–1976.) *Philadelphia Acad. Sci., Spec. Publ.* **12,** 223–243.

Sanders, H. L. (1979). Evolutionary ecology and life history patterns in the deep sea. *Sarsia* **64,** 1–7.

Scheltema, R. S. (1971). Larval dispersal as a means of genetic exchange between geographically separated populations of shallow-water benthic marine gastropods. *Biol. Bull. (Woods Hole, Mass.)* **140,** 284–322.

Stanley, S. M. (1976). Stability of species in geologic time. *Science* **192,** 267–268.

Stearns, S. C. (1976). Life-history tactics: a review of the ideas. *Q. Rev. Biol.* **51,** 3–47.

Stearns, S. C. (1977). The evolution of life history traits: a critique of the theory and a review of the data. *Annu. Rev. Evol. Syst.* **8,** 145–171.

Stearns, S. C. (1980). A new view of life-history evolution. *Oikos* **35,** 266–281.

Thompson, T. E. (1976). "Biology of Opisthobranch Molluscs." Ray Society, London.

Todd, C. D. (1981). The ecology of nudibranch molluscs. *Oceanogr. Mar. Biol.* **19,** 141–234.

Turner, R. D. (1973). Wood-boring bivalves, opportunistic species in the deep sea. *Science* **180,** 1377–1379.

Williams, G. C. (1966). "Adaptation and Natural Selection." Princeton Univ. Press, Princeton, New Jersey.

Yonge, C. M. (1953). Mantle chambers and water circulation in the Tridacnidae (Mollusca). *Proc. Zool. Soc. London* **123,** 551–561.

Yonge, C. M. (1968). *Enigmonia aenigmatica* Sowerby, a motile anomiid (saddle oyster). *Nature (London)* **180,** 765–766.

2

The Ecology of Deep-Sea Molluscs

J. A. ALLEN

University Marine Biological Station
Millport, Isle of Cumbrae, Scotland

I. Introduction

Edward Forbes (1815–1854) may have given the first considered opinion as to whether or not living organisms occurred in the deepest parts of the sea, but undoubtedly other nineteenth-century naturalists wondered to what depth sea animals could penetrate, because a great surge of interest in marine biology took place in Europe and America in early Victorian times. Indeed, I write from one of the first centers whose establishment resulted from this activity, for it was from Millport in the 1840s that Robertson, the Cumbrae naturalist, and Sir John Murray, in his steam yacht "Medusa," began their studies by investigating the

The Mollusca, Vol. 6
Ecology

ISBN 0-12-751406-6

deeper waters of the Clyde (200 m) a short distance from the Marine Station. Although Forbes did not sample in water deeper than 421 m, his sampling, significantly, was carried out in the Aegean. Here, almost more than anywhere else in the world, the number of macrofaunal animals decreases dramatically with depth, so much so that Forbes (1844) extrapolated that no animals could exist in the deepest water.

His opinion and the ensuing debate were instrumental in initiating the great Deep Sea Expeditions and, in particular, that of *HMS Challenger*. At his deepest stations in the Aegean, Forbes (1844) collected *Dentalium quinquiangulare, Kellia abyssicola, Pecten hoskynsi,* and *Arca imbricata,* all highly significant in that the samples showed that bivalves dominated the molluscs; the only other member of the Mollusca present was a scaphopod. Furthermore, although these bivalve genera are not the most common in the deep sea, they are typical of the larger suspension-feeding bivalves of deep water, which reflects the coarseness of the mesh of the Naturalist's Dredge that Forbes used. This sampling instrument was introduced by Müller in 1799 and is used to this day. Use of this dredge and the style of the *Challenger* Expedition, which involved a ship sampling with no experimental design other than stations along the line of a transoceanic cruise from port to port, persisted until the 1950s (Menzies et al., 1973). The last great expedition of this type was that of the Danish research ship *Galathea* in 1950, which focused on the Pacific Trenches. These expeditions confirmed a logarithmic diminution of numbers of animals with increasing depth and distance from land and the occurrence of organisms in the deepest waters. In the early 1960s, the United States and the Soviet Union initiated programs that were more scientifically probing. Their objectives went far beyond the questions about the kinds of animals found in the deep oceans which had been asked for more than 100 years; they were quantitative, recording faunal changes with depth and distance, and asking why the animals they found could exploit great depths, how they lived and reproduced, and how they had evolved. To do this, it was necessary to design more efficient methods of sampling (Sanders et al., 1965; Hessler and Jumars, 1974; Rice and Collins, 1980).

Many of the earlier workers who studied the collections of deep sea animals were malacologists—Jeffreys, Dall, Michael Sars, and Smith, to name only a few—partly because shelled animals survived the long journey to the surface much better than did Peracarida and Annelida. In fact, these three groups (Mollusca, Peracarida, and Annelida) account for 80% of the deep-sea macrofauna, although all other marine invertebrate groups are represented. The Mollusca, because they are relatively common and robust, soon became relatively well identified; possibly 40% of deep-sea molluscs were described before 1950 and most discovered within the last 20 years have now also been described. But, until the last decade or so, it was the shells and not the animals that were known.

II. The Environment (Physical and Chemical)

The sea covers about 70% of the world's surface to an average depth of 3000 m. The area of sea bed from the low-water mark to the edge of the continental shelf at approximately 200–300 m accounts for only 15% of the total. The shelf–slope break marks a change in sea bed slope from about 1 : 1000 on the shelf to 1 : 40 on the continental slope of the escarpment of the continents. Thereafter, from depths of about 1000 m, the gradient becomes progressively more gentle as the foot of the continental slope (the abyssal rise) is approached. At 2000 m the gradient decreases from about 1 : 100 to 1 : 700, and farther still across the abyssal plain to 1 : 1000 to 1 : 10000 to a depth of about 5000 m. Along fringing areas of the western Pacific, and to a lesser extent off the Antilles, South Sandwich Islands, Sumatra, and New Zealand, delineated on one side by the lower continental slope and on the other by the edge of the abyssal plain, extend deep, narrow, elongate trenches which mark the collision lines of tectonic plates. These are active, perturbed areas of the ocean bed that penetrate to hadal depths below 5000 m to as deep as 10,863 m at the southern limit of the Marianas Trench (Carruthers and Lawford, 1952). Abyssal plains are separated from each other by ridges and seamounts, so that oceans may be said to be divided into some 47 basins (Clarke, 1961).

Although the bathyal Mollusca of the continental slope will be mentioned, this chapter is principally concerned with the abyssal and hadal Mollusca below 1000 m. At this depth seasonal factors largely disappear, and those that persist may be so attenuated as to be imperceptible to the fauna. Thus, the abyssal fauna is below the deepest thermocline, with temperatures throughout the year between 1° and 2.5°C, varying by as little as 0.1°C. Solar light is absent. The sediments of the abyssal plain are oozes that are uniform in character over many thousands of square kilometers. Possibly the input of organic matter is largely uniform, seasonal planktonic cycles being dissipated by trophic activities and environmental factors before the 1000-m isobath is reached, although a seasonal rapidly deposited planktonic pulse may be widespread (Billett et al., 1983). It has been reported that faint transmission of diurnal tidal fluctuations can be detected by sophisticated pressure instruments at 4400 m (Filloux, 1970), but whether these can be detected by the fauna is debatable. Similarly, long-term current meter measurements taken at 50 m above the bottom at depths as great as 4100 m indicate increased current speeds where there is a raised featre such as a slope, bottom unconformity, or abyssal hill (Gould et al., 1981), which may relate to wind speeds at the surface. Thus, such currents may vary from year to year, usually peaking in winter during February–March and at the time of the equinoctial gales during October–November, there being about a 1-month delay between surface signal and depth. These have not been recorded on the abyssal plain

TABLE I

Estimated Particle Composition of the Material Taken from Sediment Traps[a]

Depth (m)	Clay	$CaCO_3$[b]		Opal	Organic matter	
398	12.1	53.5	(62.7)	11.1	23.3	(19.7)
998	21.0	51.1	(55.1)	10.0	17.9	(17.5)
3755	37.5	48.5	(56.3)	6.5	7.5	(10.3)
5086	45.9	41.2	(49.2)	4.4	8.5	(10.4)
1000	9.4	65.6	(43.7)	6.5	18.2	(17.6)
4000	25.9	62.6	(69.0)	0	11.5	(10.0)

[a] From Brewer et al., 1980.
[b] Numbers in parentheses are the values measured by Honjo.

(Dickson et al., 1982). Current speeds about 2–5 cm/sec are common but may be enhanced to as much as 44 cm/sec (Schwartzlose and Isaacs, 1969). Hills have a doming effect as much as three times their height and may be of profound influence on mixing processes in deep water. Some of these enhanced flows may be related to turbidity currents (Heezen et al., 1955; Webster, 1969; Schmitz and Hogg, 1978).

Sedimentation is a continuous feature that is enhanced by downslope turbidity flows which are generated by sediment slides, particularly along the courses of canyons that transverse the continental shelf. Coarser particles quickly sediment in the canyons and trenches (Ericson et al., 1951), and only the finest particles are carried to the abyssal plain in density currents that behave somewhat as does mercury underwater. The clay fraction of the abyssal oozes, particles less than 1 μm in diameter, increases with depth (Saski et al., 1962). Such particles settle at an average speed of about 21 m/day and total about 1800 $\mu g/(cm^2 \cdot yr)$ below the clear surface waters of the Sargasso Sea. Much higher rates have been suggested for "marine snow" (Silver and Bruland, 1981). The increase in particles close to the sea bed, caused partly by turbidity flows and partly by bioturbation within the nephaloid layer (Ewing and Thorndike, 1965), is an environmental feature of profound significance to the benthic fauna. At abyssal depths, particulate $CaCO_3$ is a significant part of the sedimentary flux (see Table I) and may vary from equal to or double the clay fraction. Together these fractions form 80% or more of the sedimentary particles. The remainder are of organic origin, dominated by fecal pellets (Honjo, 1978, 1980). In fact, sediment traps above the bottom show a much higher percentage of organics than do the bottom deposits, which is to be expected if the material in suspension is the primary source of food for both deposit and suspension feeders.

High pressures (1-atm increase for every 10 m in depth) affect the dissolution of the sedimentary $CaCO_3$. The rate of dissolution increases with the pressure (aragonite being more soluble than calcite), and the pH of the seawater at abyssal depth is enhanced to 8.1–8.2 at the usual abyssal salinity of 34.9 ‰. Except for very limited areas of the sea bed, some of which will be referred to later (pp. 39 and 61), dissolved oxygen (5 ml/liter) is not depleted either immediately above or within the sediments. The abyss is undoubtedly one of the most physically stable of the Earth's environments and occupies the greatest area.

III. The Environment (Organic)

Clearly, the quantity of living matter that can exist in the deep sea is dependent on the organic input into the system; that this input is the key to sustained life in the deep sea was foreseen by Agassiz in 1888. Because no photosynthesis can occur, food for organisms must be secondarily derived. Deep-sea marine sediments usually contain less than 1% organic matter (Kuenen, 1950). The main organic inputs consist of particulate matter from planktonic material, carcasses of large nekton, marine macrophyte debris, terrigenous matter, and chemotrophic production (Fig. 1). Rowe and Starensinic (1979) estimated that the organic carbon input in g C/($m^2 \cdot yr$) to the sediments below the Sargasso Sea was (1)

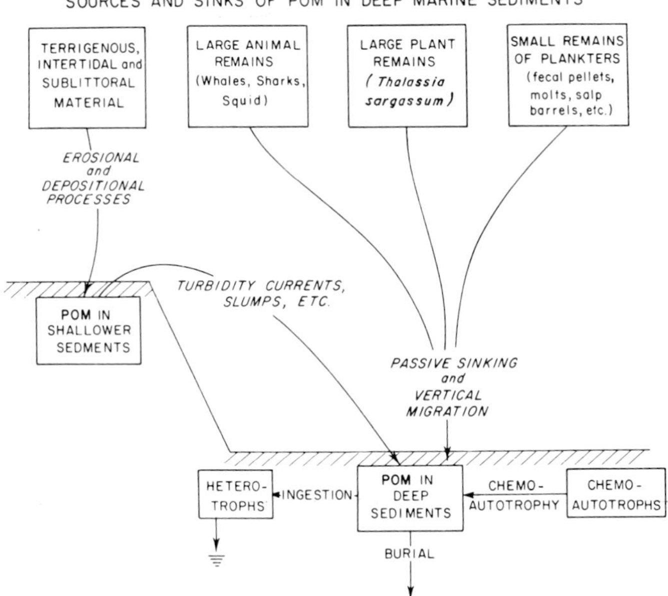

Fig. 1. Conceptual model of the potential sources and transport mechanisms of particulate organic matter (POM) to the deep sea. (From Rowe and Staresinic, 1979, with kind permission of the Royal Swedish Academy of Sciences.)

small planktonic remains (4.0), (2) sargassum weed (0.4), (3) nekton (0.05), and (4) chemotrophic production (0.0001). Another source of organic material for the fauna is dissolved amino acids. Park et al. (1962) identified 17 amino acids from bottom water at 3500 m in the Gulf of Mexico. Of these, glutamic acid, lysine, glycine, and aspartic acid were reported in concentrations >1 mg/m^3. Turbidity currents undoubtedly commonly carry large quantities of terrigenous and shallow-water plant remains down onto the abyssal rise and into the trenches adjoining the shelf (George and Menzies, 1972; Heezen et al., 1955; Wolff, 1976). Farther out from land, terrigenous plant material is not particularly common. A descending chain of living vertical migrations also has been proposed as a source of living plankton close to the bottom (Vinogradov, 1979; Zenkevitch and Birstein, 1956), but at 2000–4000 m the living planktonic biomass probably is less than 2% of that at the surface. Various authors have reported intact phytoplankton in very deep water, up to 3750 cells/liter, have been recorded. These include coccolithophores (Bernard, 1948), diatoms (Wiseman and Hendey, 1953; Wood, 1956), and flagellates (Billett et al., 1983). Some, such as *Nitzschia seriata, Navicula* spp., and several unnamed flagellates, have been cultured after capture (Kimball et al., 1963). Because these species are not found in shallow water, it has been suggested that they are autochthonous (Wood, 1966). Wood (1964) also reports dividing cells in upwelling Antarctic water off Australia. Morita (1979) reports bacterial counts of 10^6 cells/ml at the sediment–water interface at 2810 m, and 10^5 cells/ml at 30 cm above the bottom from the same station off the Oregon coast.

Of the various sources listed above, it seems clear that the major input is the organic rain of planktonic (Eppley and Peterson, 1979) and larger debris consisting of the carcasses of fish, whales, and squids (McCave, 1975) and byproducts of human activities on the surface; these fall to the bottom and are quickly consumed by epinatant Crustacea such as *Eurythenes gryllus* (Amphipoda) which then spread the remains further via their feces. In fact, fecal material, both planktonic and demersal, is probably one of the most important organic inputs to the sea bed (Wiebe et al., 1976; Honjo and Roman, 1978); planktonic fecal material may carry intact algal cells into deep water. Degradation of surface plankton is rapid, and 90% is recycled in the photic zone (Riley, 1951; Broeker, 1974). Nevertheless, Rowe and Gardner (1979), using a column of sedimentation traps down to 3650 m (the deepest was 518 m off the bottom), collected 14–30 mg/(m^2·day) organic carbon which was 38–60% of the total material collected (20% was pelletized particles >63 μm diameter). The pellets were larger and more numerous in the nephaloid layer, suggesting that resuspension was taking place. Silver and Aldredge (1981), in their investigation of the ''marine snow'' present throughout the water column that was first seen by Beebe (1935) from his submersible, found that it consists of mucous sheets from larvacean housings and organic exoskeletal flakes of unknown origin. Associated with this material are vast numbers of bacteria (as much as 69% of the biomass), cells,

TABLE IIA

Contents of Fecal Pellets of Salps (*Salpa fusiformis* and *Pegea socia*) and Pteropods (*Corolla spectabilis*): Average Biomass of Intact Cells per mm³ Fecal Pellet Volume[a]

Cell type	Salps: Total cell volume ($\times 10^4$ μm³)	Salps: Total cell carbon (μg C)	Pteropoda: Total cell volume ($\times 10^4$ μm³)	Pteropoda: Total cell carbon (μg C)
Bacteria	1.2	0.04	1.0	0.03
Blue-green algae	0.1	0.01	0.8	0.02
Eukaryotic algae[b]	34.0	1.19	14.4	0.38
Total intact cells	35.3	1.24	16.2	0.43
Olive-green cells	13.8	0.48	18.1	0.26

[a] From Silver and Bruland, 1981.

[b] Mainly dinoflagellates (*Procentrum minimum*), mixed diatom species, and various nannoplankton-sized green algae.

and biogenic debris, in particular fecal pellets about 50 μm long (Table II). At least three-quarters of the cells are healthy and are mostly composed of diatoms, cyanobacteria, and green algae. A smaller fraction, possibly associated with the remains of pseudothecostomatous pteropods (Silver and Bruland, 1981), are the "olive green cells" of Fournier (1972) which have been frequently observed at abyssal depths. This marine snow comprises approximately 8–16 μg C/ml and has a sinking rate of 50–300 m/day. Such a flux is more than adequate to fuel the benthos of the abyssal plain, although not that of the trenches and the abyssal rise and slope. It must be remembered that the average residence time of the deep

TABLE IIB

Carbon and Numbers of Picoplankton Sized (0.2–2.0 μm) Cells on Flocculent Marine Snow[a]

	Average diameter (in TEM) section	Cell volume × 10³ μm³/ml snow: East Cortez 1000 m	Cell volume × 10³ μm³/ml snow: San Clemente 1650 m	μg C/ml: East Cortez 1000 m	μg C/ml: San Clemente 1650 m	Numbers/ml: East Cortez 1000 m	Numbers/ml: San Clemente 1650 m
Heterotrophic bacteria	0.8 μm	33.7	206.8	4.4	26.9	0.9×10^5	1.5×10^5
Cyanobacteria	0.9 μm	9.1	14.4	1.2	1.9	1.6×10^4	2.6×10^4
Coccoidal green eukaryote	1.6 μm	16.6	38.6	4.3	10.0	0.8×10^4	1.9×10^4

[a] Silver and Aldredge, 1981.

water is ~1200 years, and also that microbial degradation within bottom sediments is extremely slow (Jannasch and Wirsen, 1973, 1977; Jannasch et al., 1971; Wirsen and Jannasch, 1975).

IV. The Fauna

The last 20 years of enhanced postexploratory deep-sea research with more quantitatively defined objectives have shown that the benthos, whether quantified by density, species composition, or biomass, is not spatially uniform but resembles the world map of primary productivity (Koblentz-Miske et al., 1970; Belyaev et al., 1973; Hinga et al., 1979). The slope and abyssal fauna is much much more abundant than the earliest expeditions had shown, largely because of modern improved sampling methods such as the epibenthic sledges of various designs (Sanders et al., 1965; Rice et al., 1982), and the box corer of Hessler and others (Hessler and Jumars, 1974). With the exception of the trenches, canyons, and volcanic rifts, the slope–abyssal fauna is one of the most diverse in the world (Hessler and Sanders, 1967). It is composed, for the most part, of small animals less than 5 mm in length. The larger animals which caught the attention of the earlier naturalists are sparse and largely epifaunal. Despite the fact that vast numbers have been collected in individual hauls in recent years, the logarithmic diminution in the density of the fauna distributed downslope and across the abyss remains evident (Allen, 1978).

The Mollusca are one of three dominant phyla (Annelida, Crustacea) in the deep sea, and all Recent classes are present. They represent about 10% of the fauna of every basin that has been investigated and are present at the greatest depths. Their diversity is astonishing; there are more molluscan species recorded from the Woods Hole–Bermuda traverse across the Northwest Atlantic Basin than from the entire East European intertidal and shelf seas. In comparison with shelf species, the proportionate familial composition is strikingly different (Table III); the deep-sea molluscs are dominated by representatives of ancient groups, many well established in the early Ordovician. Deposit feeders dominate; carnivores and ectoparasites are less common by an order of magnitude. Suspension feeders, which are poorly represented, are mostly epifaunal, with an occasional endobyssate species living in the top few millimeters of the surface sediment.

Before the present surge of interesting and detailed analyses of the fauna of the deep sea, Clarke (1962b) compiled a list of more than 1000 species of deep-sea molluscs; although now superseded in many ways, it provides general information that remains largely true. Clarke recorded that (1) all abyssal chitons belong to the primitive genus *Lepidopleurus;* (2) in contrast to shallow water fauna, the Archaeogastropoda outnumber the Mesogastropoda; (3) the Scaphopoda are relatively abundant in deep water; (4) the Protobranchia are the dominant bivalves,

TABLE III

Distribution of the Major Groups of the Bivalvia in Shallow and Deep Seas of the World[a]

Group	Shelf	Slope	Abyss
Protobranchia			
Solemyoida	*	*	*
Nuculoida	**	**	**
Nuculanoida	*	*	****
Lamellibranchia			
Arcacea	**	*	*
Limopsidacea	*	*	***
Mytilacea	****	*	* (1 g)
Pinnacea	*	—	—
Pectinacea	**	*	**
Pteracea	**	—	—
Limacea	**	*	*
Ostreacea	*	*	—
Astartacea	**	*	—
Glossacea	*	*	* (2 g)
Lucinacea	**	*** (1 f)	* (1 f)
Carditacea	*	—	—
Cardiacea	***	*	—
Galeommatacea	*	*	—
Solenacea	*	—	—
Veneracea	****	*	—
Tellinacea	****	*	* (1 g)
Mactracea	***	* (1 s)	—
Myacea	***	—	—
Pholadacea	***	*	* (1 g)
Anomalodesmacea	***	*	—
Poromyacea	*	*	***

[a] The number of asterisks gives some indication of the species and numerical dominance of the group (s, g, f = species, genus, family).

(5) the Septibranchia outnumber the Eulamellibranchia; and (6) the families Littorinidae, Veneridae, Cardiidae, and Tellinidae, which dominate shallow seas, are absent from the deep sea. The percentage composition of feeding types given by Clarke (1962a) (detritus feeders, 24%; suspension feeders, 15%; scavengers and carnivores, 54%; and commensals and parasites, 4%) does not now hold; his list consists mainly of relatively large species, and there has been a spectacular increase in the number of species of small deposit-feeding bivalves recorded from the deep sea.

Within individual oceans (Pacific, Atlantic, Arctic, and Antarctic), many molluscs have a wide distribution. Clarke (1962b) records that 35% of all Atlantic species at depths >2000 m have an amphioceanic distribution. The abyssal molluscs of the oceans are unequally distributed, and the least dense populations occur beneath the impoverished gyres where productivity is low (Okutani, 1967, 1968; Hessler, 1974; Sokalova, 1965; Vinogradova, 1979). As might be expected, the depth ranges of deep-water molluscs vary considerably. Most species appear to have a preferred depth which extends to either extreme varying in terms of depth and attenuation of population density. Vinogradova (1979) reports that 60% of the gastropods and bivalves but only 1% of the scaphopods are stenobaths. Some species are much more common than others; e.g., *Ledella ultima* is present in large numbers throughout the Atlantic (Allen and Hannah, in preparation), but a species of *Spinula,* at the other extreme, is represented by only three specimens taken from the frequently sampled Gay Head–Bermuda transect (Allen and Sanders, 1982).

A. Infauna

Since 1963, biological oceanographers of many countries have made a major effort to quantify the benthic communities of the abyss (Filotova, 1960; Sanders et al., 1965; Sanders, 1977; Rowe, 1972; Gage, 1977, 1979; Laubier and Sibuet, 1979). From these studies, it is possible to extract information particularly pertinent to the Mollusca. It is clear that there are marked differences in the density, diversity, and composition of the Mollusca (as well as other groups) between shelf and abyss. The eulamellibranch-dominated molluscan fauna of the shelf gives way to the protobranch-dominated molluscan fauna of the abyss, and the density of the macrobenthos, which is 500–2000 specimens/m^2 on the shelf, drops to 1–10 specimens/m^2 in the abyss and even less in areas below central oceanic gyres (Jumars and Hessler, 1976). This is reflected in total figures for the biomass ($\sim$170 g/m^2 for the shelf and $<$2.0 g/m^2 for the abyss) (Gage, 1977); the mollusc fraction follows the whole. A further feature is the decrease in size with depth of the infauna; few abyssal molluscs have any dimension greater than 5 mm. Low density and small size result from the quantity and quality of the food resources; they are also related to the fact that small animals, although tending to have higher metabolic rates than larger animals, can exploit small amounts of energy more efficiently. As a result, the proportion of meiofaunal species in the benthos increases with depth (Thiel, 1972, 1979b), and the biomass of macrobenthic Mollusca decreases more rapidly than the abundance (Rowe and Menzel, 1971; Rowe et al., 1974). Despite the wide occurrence of many species in each of the four major oceans, few species are truly cosmopolitan; Vinogradova (1979) gives the figure as 4%. This appears to be particularly true of the Arctic, which is isolated by relatively shallow sills and has a particularly impoverished molluscan fauna (Clarke, 1963). The latter is true even for the deep-water Gastropoda, which, although known to have planktotrophic larvae, have only 9 of 27

Arctic species occurring south of the Wyville Thomson Ridge (Bouchet and Warén, 1979b).

In other areas of exceptional conditions, such as the Red Sea where the temperature is 21.5°C at 2000 m, the Mollusca account for less than 1% of an infauna dominated by nematodes (81%) (Thiel, 1979a).

At depths greater than 6000 m, there is a marked change in the molluscan infauna. This distinct, largely endemic hadal/ultraabyssal community is affected not only by high pressures but also by food and sediment conditions (Wolff, 1970). Trench faunas, unlike the abyss fauna, have low species diversity (Hessler and Jumars, 1974). In some trenches there is a relatively high standing crop, a result of rapid sedimentation caused by slumping (Jumars and Hessler, 1976) which brings down shallow-water plants and other organic material. Under such conditions, infaunal species must be able to unbury themselves; this partly explains the extraordinarily high number of aplacophorans and the marked dominance of bivalves compared with gastropods (Table IV).

In the Pacific Trenches, stenobathic *Spinula* species characterize the infaunal communities (Filotova, 1976; Levenstein and Pasternak, 1976; Vinogradova, 1979). Other protobranch genera reported are *Pristigloma* and *Ledella,* as well as characteristic lamellibranchs such as *Cuspidaria myasiri* and species of *Kelliella* (Filotova, 1960). Horikoshi (1970) suggested that *Spinula* exhibits allopatric speciation, but this conclusion was based on inadequate sampling. In fact, some trench species also occur at abyssal depths, including some of the few cosmopolitan species such as *Malletia cuneata, Bentharca pernula,* and *Spinula calcar* (Okutani, 1974). Vinogradova (1979) supports Wolff (1970) and suggests that as many as 75% of hadal species may be endemic.

The gastropods that are present in the trenches are mostly associated with large plant debris, which may sometimes be present in sufficient quantity to cause eutrophic conditions. For instance, George and Higgins (1979) report eutrophic conditions in parts of the Puerto Rico Trench, where there is a reverse in *within*

TABLE IV

Percentage Composition of the Infaunal Molluscan Classes of the Total Abyssal Invertebrate Benthos of Different Parts of the World's Oceans

	Pacific trenches		Central Pacific	Sargasso	Rockall
	Hessler (1974)	Wolff (1970)	Hessler and Jumars (1974)		Gage (1971)
Aplacophora	10.5	?	0.4	0.3	?
Gastropoda	0.7	2.5–8.5	0.4	0.6	0.9–1.9
Bivalvia	11.0	8.0–19.5	7.1	4.2	8.4–9.3
Scaphopoda	?	?	2.5	0.2	2.9–3.7

class proportions, with bivalves constituting $<1\%$ of the fauna whereas gastropods are 4.2%. All are wood-dwelling species, and as in the Red Sea, nematodes and harpacticoids dominate. Nevertheless, apart from these exceptional conditions, the trench fauna is similar in general composition to that of the abyss, except that the percentages of bivalves, amphipods, polychaetes, and holothurians are somewhat higher than in the abyss and those of echinoderms other than holothurians are less frequent. Wolff (1970) reports that 65.4% of the bivalves (26 species) and 87.5% of the gastropods (16 species) in the Pacific trenches at depths >6000 m are endemic. As would be expected, bysally attached bivalves are common, particularly mytilids, as are also species of *Xylophaga* that bore into pieces of wood and sea grass rhizomes (see pp. 45–46); the mytilids frequently inhabit the old tubes made by *Xylophaga.* Epifaunal browsing gastropods that associate with plant debris can be abundant and, like the boring bivalves, are probably opportunistic. Cocculinid gastropods predominate, and many new species of the two new genera *Caymanabyssia* and *Fedikovella* have been described (Moskalev, 1976). In addition, species belonging to the families Turridae and Trochidae, as well as *Tonna maculosa,* have been reported by Wolff (1979), who also records gastropod limpets and polyplacophorans including two new species of *Lepidopleurus* and two species of *Pectinodonta* (Acmaeidae).

B. Epifauna

In deep water, seamounts and ridges with clean rock surfaces, isolated rocks of various sizes rafted and dropped by icebergs or resulting from volcanic activity, manganese nodules and pavement, clinkers, and coal are abundant (Heezen and Hollister, 1971; Kidd and Huggett, 1981), as, on occasion, are large plant remains. All provide surfaces to which molluscs can attach.

As would be expected, it is the byssally attached species of bivalves that are predominantly epifaunal. The families Mytilidae, Arcidae, Limopsidae, and Pectinidae dominate; however, within these families there are particular genera that appear to be restricted for the most part to abyssal depths where they are common (e.g., *Dacrydium, Bentharca, Limopsis,* and *Pseudoamussium*). Unlike most infaunal bivalves, these are suspension-feeders. The epifaunal bivalves are well able to migrate away from conditions that would smother them, because the attached species have a well-developed foot.

The pectinids have especially light shells, and it is possible that they may have a habit of more or less continuously swimming; they are one of the most neglected of the deep-sea groups. They occur in large numbers, as both the Challenger Expedition (Smith, 1885) and recent sampling show (J. A. Allen, personal observation), and the number of species is great. They are relatively small in size, extremely thin-shelled, and unlike the deep-sea arcoids (Oliver and Allen,

1980a,b) do not support an epifauna. They represent the culmination of a shallow- to deep-water trend of increasingly light shells which was observed earlier in the shelf pectinids (Allen, 1954), and as yet nothing is known of their biology and habits.

V. Functional Morphology

A. Monoplacophora

The first living monoplacophoran was taken in 3570 m in the Pacific Ocean off Central America (Lemche, 1957) and has been the subject of a major monograph (Lemche and Wingstrand, 1959). However, other species of *Neopilina* have been recorded in both the Pacific and the Atlantic (Menzies, 1968). Suffice to say that *Neopilina* has an untorted limpet form, although it occurs on soft sediment and is a deposit feeder. One of the features of *Neopilina* is the long coiled gut, reminiscent of the deposit-feeding bivalves discussed below (Fig. 2). It can be only assumed that, like these bivalves, it can digest highly refractive scleroproteins and that the elongate but compact gut is an adaptation to prolong the digestive processes. The sexes are separate, and *Neopilina* probably has a short-lived

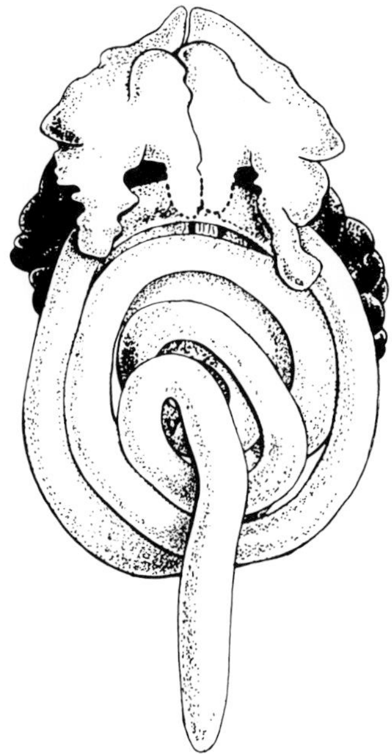

Fig. 2. Dorsal view of the gut of *Neopilina*. (From Lemche and Wingstrand, 1959.)

lecithotrophic larval stage. The anatomical features are unique but can be individually linked with those of all the molluscan classes, which supports the conclusion that these classes evolved from a common stem, probably in the early Cambrian.

B. Bivalvia

The Bivalvia numerically dominate the Mollusca of the deep sea, and dominating the Bivalvia is the class Protobranchia. Representatives of all the protobranch divisions are present, and many are exclusive to the deep sea. They have been the subject of detailed analyses by Allen (1971, 1978), Allen and Sanders (1969, 1973, 1982), Filotova (1976), and Sanders and Allen (1973, 1977). In fact, the bivalves of the deep sea are the best studied of all the molluscan classes, and the anatomical descriptions and distributions of many species have been completed. Approximately 400 abyssal and hadal species have been described to date (Knudsen, 1967, 1970, 1979; Allen and Morgan, 1981; Oliver and Allen, 1980a,b; Payne and Allen, in preparation; Allen and Hannah, in preparation). Species of the following groups represent more than 90% of the bivalve infauna in number and in species: Protobranchia (deposit feeders), Septibranchia (carnivores), and Thyasiridae (suspension-feeding in the nephaloid layer?). The bivalve epifauna is largely restricted to the following groups: Mytilidae, Arcoida, Pectinacea, and Kelliellidae (suspension feeders); borers are restricted to the subfamily Xylophaginae. Of the common and dominant shallow-water infaunal groups, species of the Cardiidae, Veneridae, and Tellinacea are for the most part absent; only one species of *Abra* occurs at abyssal depths (Allen and Sanders, 1966). This striking change in composition (Table III) takes place at the shelf/slope break and may be related to the relatively abrupt change in temperature fluctuation from about 10.5°C at 100 m to 1.5°C at 500 m (Sanders and Hessler, 1969).

Adaptations to life at great depths are the result of direct and indirect pressures. For example, direct pressures include temperature, atmospheric pressure, food, and sediment type, and indirect pressures include the age of the ecosystem and its stability, competition, and niche availability. The effects of direct pressures have been documented, but those of indirect pressures are poorly understood. In uniform environments of large extent (such as the abyssal soft oozes) with a rich diversity of species some of which are closely related but live in close proximity, the concept of the *niche* becomes difficult to sustain. However, by the nature of their morphological adaptations, variations in locomotion, and depth of burial, ecological separations may result although the species appear to feed in the same way and on the same food, as do many protobranch species.

Morphological adaptations more often emphasize existing structures than evolution of new ones. It is evident that all infaunal bivalves, the protobranchs in particular, have a well-developed foot and are capable of large-scale movements through the sediment. This ability must be important not only in food gathering but also in escaping from the covering sediment that follows a slump. The scale of such slumps is impressive. The Grand Banks current alone probably carries some 220 billion m^3 of shallow-water sediment over an area of 200,000 km^2 of the abyss. In contrast, there is evidence that burial of hard objects lying on abyssal muds occurs more rapidly than is explained by sedimentation rates even at the overly generous figure of 10 cm/10^3 yr (Kidd and Huggett, 1981). Bioturbation has been suggested as the reason for this, in which case protobranch molluscs can be expected to play a significant role. The importance of nuculanid protobranchs in bioturbation has long since been proven in shallow water [e.g., the work of Rhoads (1973)] on the effect of reworking of surface sediments by *Yoldia limatula*].

The deep-sea protobranchs themselves are dominated both in species and in numbers by the Nuculanacea, a consequence of adopting a posterior position for the inhalant current into the mantle, instead of the anteroventral ingress of the Nuculacea. Such a change of ingress, which also occurred in the Lamellibranchia, allows for greatly varied exploitation of the available environmental niches. But the overwhelming success of the Protobranchia as compared with the Lamellibranchia is a direct result of differences in their digestive physiology and their ability to exploit the organic content of the oozes. Even so, the problem of having to digest such refractory organic matter quickly leads to modifications in form, the most obvious of which is the increase in length (and sometimes also in diameter) of the hindgut to allow for a longer retention time of the material to be digested (Fig. 3). It is also probable that the rate of movement of the material through the gut may be slower than it is in species in shallow water. This is probably a consequence of the physiological effects of high pressure (Section VI), although this cannot be confirmed until living specimens are brought to the surface under pressure (Allen, 1978).

In the case of the Lamellibranchia, the sole deep-sea tellinacean *Abra* has a lengthened gut; for the rest, the gut is of typical length, reflecting the ability to feed on the living organisms and organic matter that are present. Thus, their anatomy differs little from that of their shallow-water relatives, except that many species of the deep-sea eulamellibranch bivalves have been reported to have enlarged kidneys. Apart from the physiological stress of high pressure on basic metabolism and the resultant bodily response, there is no clear reason for the enlargement (Oliver and Allen, 1980b; Allen and Turner, 1974).

One of the more striking features of the bivalve fauna of the deep sea is the extent to which the Septibranchia have exploited this environment. They not only persist but are present in large numbers of species, although, with the exception of the Atlantic species *Cuspidaria parva* and *C. (Myonera) atlantica,* not with any appreciable number of specimens. Of the three septibranch families, the

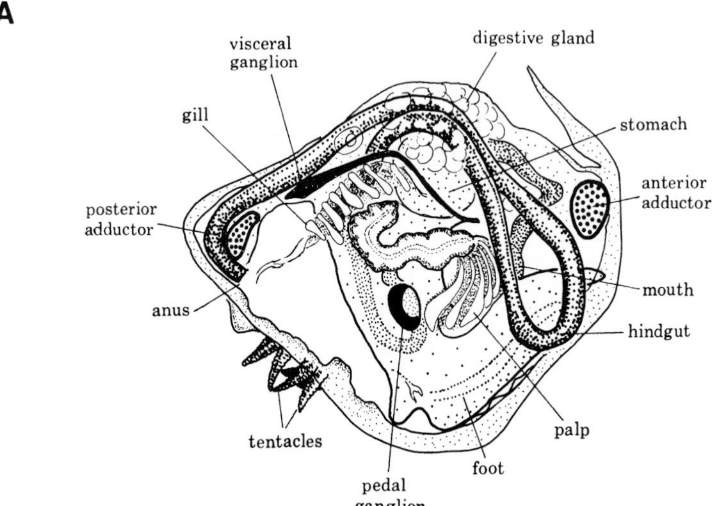

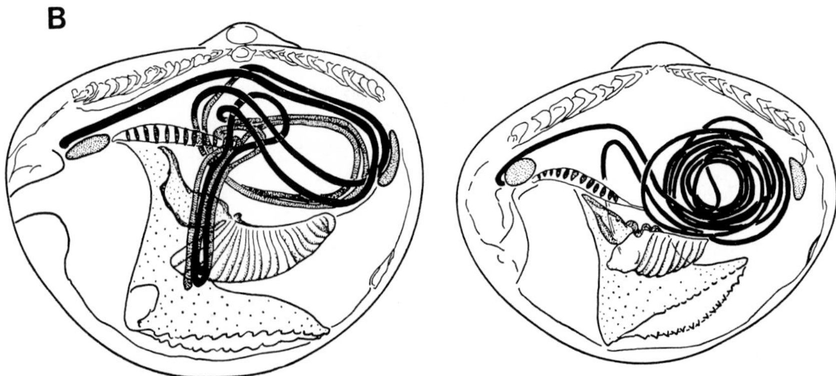

Fig. 3. (**A**) Lateral view of the right side of *Tindaria callistiformis* with the shell removed. Shell length, 4 mm. (From Allen, 1978, with kind permission of the Royal Society.) (**B**) Comparison of the hindgut configuration of *Yoldiella* species K (right) and *Yoldiella* species L (left). (From Allen, 1978, with kind permission of the Royal Society.)

Verticordiidae and the Poromyidae occur only in deep water. The Cuspidaridae are present on the shelf, but are not particularly common (Allen and Turner, 1974; Allen and Morgan, 1981). In the deep Atlantic alone there are more than 100 septibranch species, many described for the first time within the last 20 years (Soot-Ryen, 1958, 1966; Allen and Turner, 1974; Allen and Morgan, 1981). These studies have shown ctenidial morphologies intermediate between the extreme septibranch condition and the eulamellibranch condition as exhibited by the lyonsiid anomalodesmacean bivalves. This confirms that the septum is largely derived from the eulamellibranch gill and is not a structure formed *de novo*. Ecological interest lies in the success of these bivalve carnivores in the deep sea. They appear to feed largely on copepods and ostracods, each family having its own particular method for capturing prey. The cuspidariids rely on sensing and locating their prey by means of specialized cilia on the edge of the inhalant aperture and explosively sucking them into their mantle cavity (Allen and Morgan, 1981; Reid and Reid, 1974). In contrast, the verticordiids extend the sticky tentacles that surround the inhalant aperture over the surface of the sediment and "lick" off any organisms that adhere to them by means of a valve across the inhalant aperture (Allen and Turner, 1974). In the case of the poromyids, the tentacles are not adhesive; a hood-like structure is extended outside the inhalant aperture which "scoops" the prey into the mantle cavity (Morton, 1981) (Fig. 4). All three families have an extremely large, dilatable mouth with flap-like palps and a voluminous, muscular, chitin-lined stomach to accommodate and crush their prey.

One of the more remarkable discoveries has been the large concentrations of species of the subfamily Xylophaginae in submerged wood at abyssal depths (Knudsen, 1961; Turner, 1965, 1966, 1973, 1977). Anatomically, these are

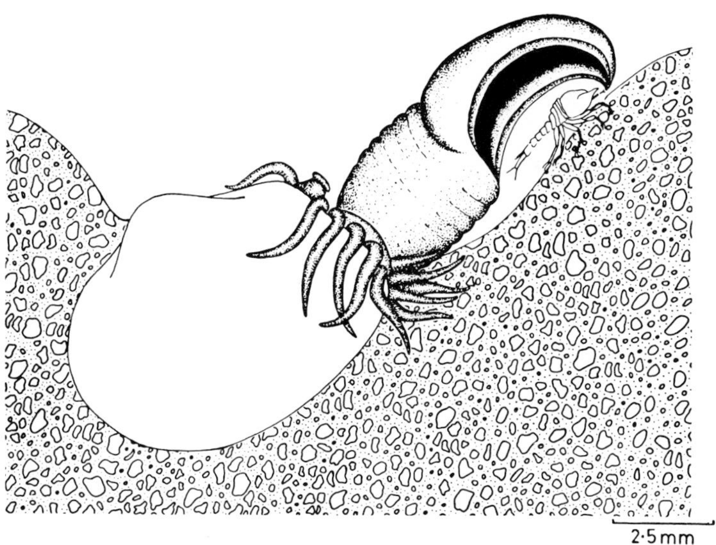

Fig. 4. *Poromya granulata*. The animal is in a natural position in the sand, with the inhalant siphon fully extended. It is illustrated here capturing a bottom-dwelling cumacean. This was not observed in life, but is the postulated feeding mechanism. (From Morton, 1981, with kind permission of *Sarsia*.)

similar to shallow-water species, but their life histories are in sharp contrast to those of other deep-sea bivalves. Turner (1973) demonstrated that wood panels placed on the sea floor off the Bahamas at 1830 m for 104 days became completely riddled by *Xylophaga* spp. and *Xyloredo ingolfia*. The indications were that two settlements had occurred in each panel during the test period (Culliney and Turner, 1976), and that after that time the burrows were 3 cm in length and the animals were mature, containing as many as 35,000 extremely small eggs (45 μm). It is rare for the eggs of bivalves to be less than 65 μm in diameter. The heaviest settlement occurs in the first 20 cm above the sea bed. In similar studies off the California coast, three other species of *Xylophaga* in densities up to $125/in^2$ were found in test panels of wood and manila rope on the bottom; boards 3 ft from the bottom were only slightly attacked, and those between 25 and 50 ft from the bottom were free of *Xylophaga* (Turner, 1965). Thus, boring molluscs that can utilize wood for food are opportunists (Cole and Turner, 1977); they grow quickly and produce enormous numbers of small larvae within weeks of settlement. It is clear that these larvae must disperse close to the sea bed and that they are able to delay metamorphosis until a suitable substratum appears. Nevertheless, in keeping with what little is known of the rates of processes of deep-sea bivalves, Tipper (1968) showed that in the case of *X. washingtonia*, penetration of wood panels was only 31.5% as fast at 1000 as at 200 m. Turner (1973) suggests that the breeding of *Xylophaga* may be related to seasons when freshwater runoff from the land carries plant material into the deep sea, supporting this view by observations, albeit limited, that larvae occur in spring but not in midsummer.

In addition to the Septibranchia and Xylophaginae, there is a limited number of other lamellibranch genera that are predominantly or exclusively restricted to deep water. These include members of the Thyasiridae, Vesicomyidae, and Kelliellidae; the last may include neotenous venerids (Boss, 1969), two species of *Abra* (Tellinidae) (Allen and Sanders, 1966; Wikander, 1980), at least 10 species of *Dacrydium* (Mytilidae) (Allen, 1979), and two or three other mytilid species. Finally, a few members of the family Arcidae, all known species of the family Limopsidae, and numerous species of the family Pectinidae (Oliver and Allen, 1980a,b; Oliver, 1979) are included. Many of these are byssally attached and occupy specialized niches.

The species of the deep-sea genus *Bathyarca* show a progressive change in habit from epibyssate nestling to endobyssate partial burial in soft sediments. With this change is a corresponding reduction in the strength and size of the byssus and an increase in posterior heteromyarian enlargement of the mantle-shell (Oliver and Allen, 1980a). A similar progression is seen in the abyssal family Limopsidae (Oliver and Allen, 1980b). Despite their colonization of soft sediments, limopsids remain suspension feeders. They are able to extend the gill and mantle considerably beyond the posterior margin of the shell. Unlike many other deep-sea bivalves, the arcoid gill of limopsids is not greatly reduced, although some slight reduction has occurred in the deepest living species. This must be a reflection of the part played by the gill in collecting sparse food in suspension. However, the total body volume relative to the volume of the shell is reduced with the viscera confined to the most dorsal part of the shell cavity (Allen and Oliver, 1980b) (Fig. 5). Although body size is clearly related to the amount of energy available, it may be that the relatively large shell deposited by the small biomass of an attenuated mantle requires only a very small proportion of the energy input, and confers protection in a community dominated by very small animals where predators are "attuned" to prey of that size.

At least 50 species of the family Thyasiridae are present in deep water in the Atlantic; they are predominantly found on the lower slope (C. Payne and J. A. Allen, unpublished). They are highly specialized bivalves; all are small and infaunal, feeding via an inhalant tube that is formed by a highly modified vermiform foot (Allen, 1958). As they are suspension feeders, their gut is simple, and there is evidence that they are relatively nonselective feeders and accept larger particles than do most other bivalves. Shallow-water species tend to live in stressed habitats (Allen, 1958) and are present in the finer sediments of the shelf. Their pattern of distribution suggests that they are at a disadvantage in the abyssal basins, and it seems likely that unlike other eulamellibranchs they are able to take advantage of the stress-producing turbidity currents of the slope and

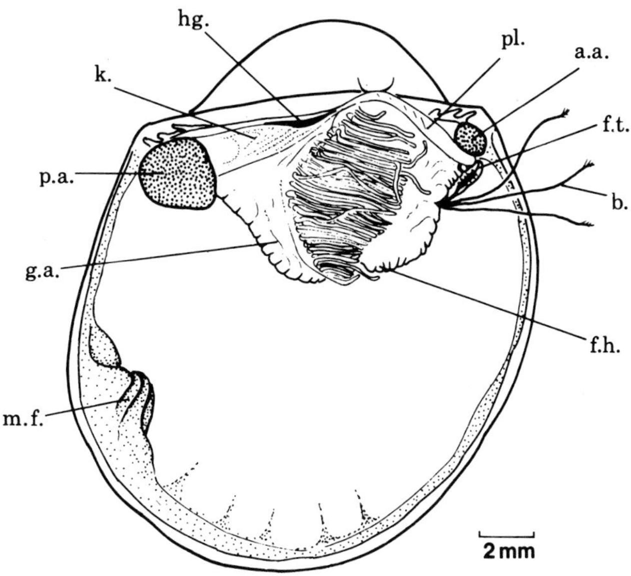

Fig. 5. *Bathyarca corpulenta*. Anatomy as seen from the right side. a.a, anterior adductor; b., byssus; f.h., heel of foot; f.t., toe of foot; g.a., gill axis; hg., hindgut; k., kidney; m.f., mantle fold; p.a., posterior adductor. (From Oliver and Allen, 1980b, with kind permission of the Royal Society.)

extract adequate food from them. There is still much to discover about the feeding biology of the thyasirids, but like the other major deep-sea bivalve groups they have been able to radiate in bathyal depths in the absence of competition from other lamellibranchs. Whether the possession of an unusual morphology (most of the digestive diverticula and gonads are contained in pouches connected by a narrow neck to the rest of the body) has adaptive advantage for life in the deep sea is far less certain. It may possibly have significance in that the exploitation of mantle space in this manner enables a small animal to maximize digestive and reproductive effort.

Two deposit-feeding species of *Abra* (*A. profundorum* and *A. longicalis*, from slope and abyssal depths, respectively) are able to survive in the deep sea because of their specialized methods of feeding and digestion. It is probable that the output of pseudofeces by *A. longicalis* at the surface of the sea bed and of feces below the surface gives rise to bacterial "gardens" which are then cropped (Wikander, 1980) (Fig. 6). *A. profundorum* (and possibly *A. longicalis*), by enlarging the hindgut and pelletizing the fecal material passing through it, provide a culture surface for internal bacterial decomposition of skeletal proteins (Allen and Sanders, 1966) (Fig. 7). These tellinids, in common with other lamellibranchs and the deposit-feeding protobranchs, reduce the size of their gills

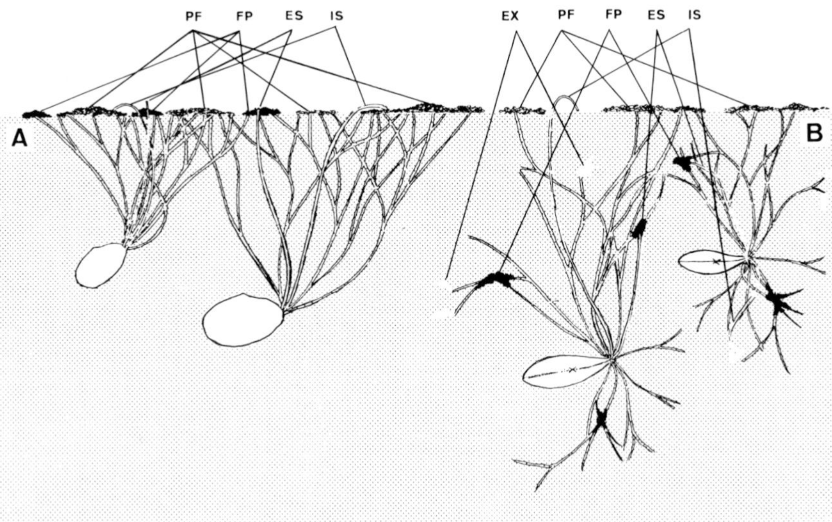

Fig. 6. Elements of niche separation between *Abra nitida* and *A. longicallus,* showing their position in the substrate and siphonal activity zones indicated by networks of abandoned siphonal channels. Note the differences in deposition of fecal pellets and inhalant siphonal activity. (**A**) *A. nitida*. (**B**) *A. longicallus*. PF = pseudofeces; FP = fecal pellets; ES = exhalant siphon, IS = inhalant siphon; EX = subsurface excavations made by inhalant siphon. (From Wikander, 1980, with kind permission of *Sarsia*.)

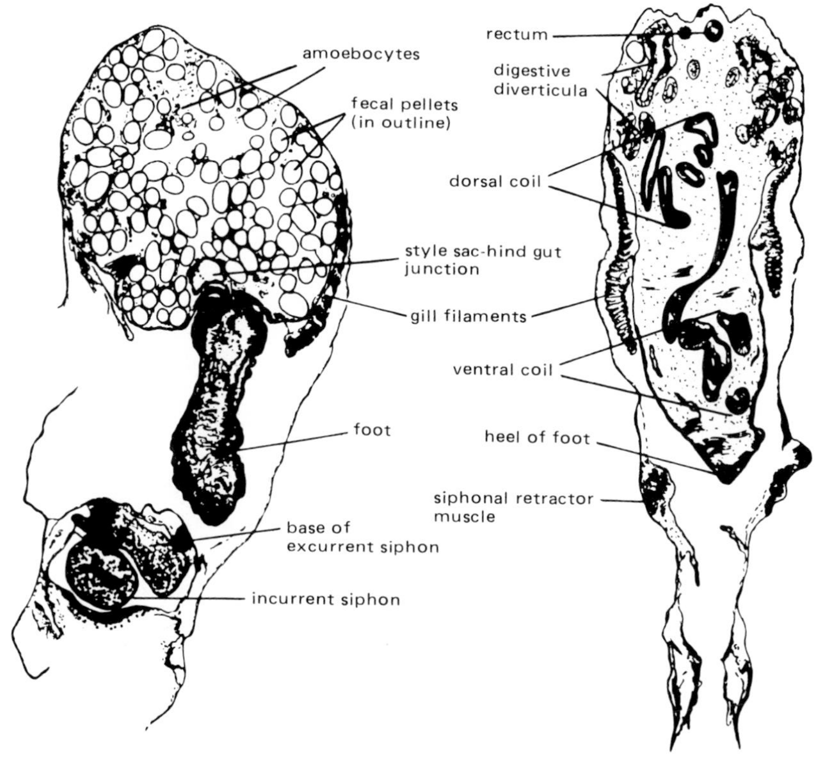

Fig. 7. Transverse sections through the hindgut of *Abra profundorum* (abyss) and *A. ovata*, extreme high water of spring tides (EHWST). (From Allen and Sanders, 1966, with kind permission of Pergamon Press.)

and enlarge their palps progressively with the progression of species to deeper dwelling. It is thought that this is related to metabolic physiology at high pressures in which diminished oxygen requirements are coupled with the need to maintain a level of mantle ciliation. As an aside, it could be that this reduction, correlated with adoption of large-particle feeding, may have been instrumental in initiating the evolution of the septibranch condition. The septum might be considered as the extreme end point of gill reduction. Because gills are no longer able to filter food efficiently, evolutionary pressure would initiate alternative methods. Clearly, food collection in the septibranchs has evolved in at least three ways. In the protobranchs, although the reduction of the size of the gills can be extreme (as shown for *Pristigloma alba* with its two gill plates) (Sanders and Allen, 1973), a septum has not been evolved; it is unnecessary because food collection by the palps is mostly separated from respiratory function.

Although the protobranchs remain deposit feeders [and variation in man-

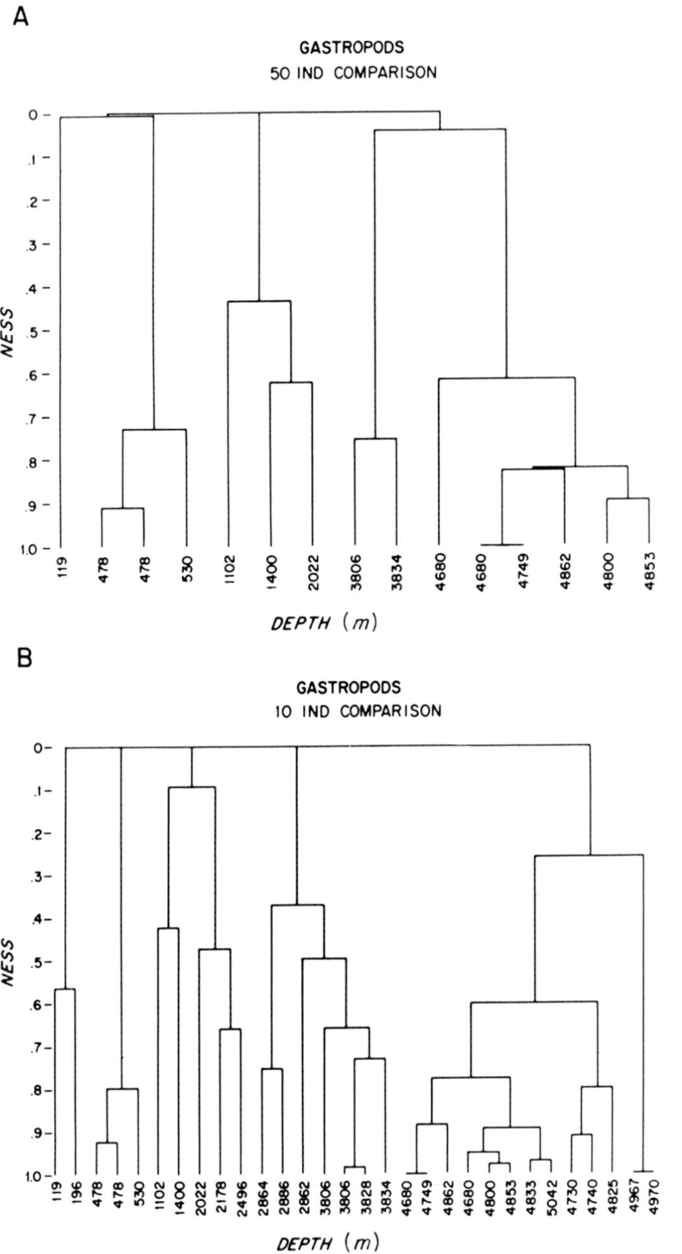

Fig. 8. (**A**) NESS (normalized expected species shared) clustering at $n = 50$ for epibenthic sled samples of gastropods taken at various depths on the Gay Head–Bermuda transect. (**B**) NESS clustering at $n = 10$ for epibenthic sled samples taken at various depths on the Gay Head–Bermuda transect. (From Grassle et al., 1969, with kind permission of the Swedish Academy of Sciences.)

tleshell morphology indicates that they have evolved a variety of different modes of life (Allen, 1978)], they have not achieved the wide exploitation of available habitats attained by the lamellibranchs. They have remained dedicated to deposit feeding in soft sediments, leaving exploitation of hard substrates, and of suspension feeding and the carnivorous habit, to the lamellibranchs. Hence, their success in the vast areas of soft sediments in the deep sea is related to their efficiency as converters of scleroproteins to metabolic energy.

Before leaving morphological considerations, one final feature of the Bivalvia (whether lamellibranch or protobranch) that must be stressed is the subtle variation in shell shape of the various populations of pan-oceanic deep-sea species. Clearly, this is related to slow gene flow with the consequent emergence of varieties and subspecies (Oliver and Allen, 1980b; Allen and Sanders, 1973, 1977; Sanders and Allen, 1973).

C. Gastropoda

The overwhelming difference between the bivalve and gastropod species of the abyss is the greater variety of form and habit of the latter. No one group predominates. Members of the following families of Gastropoda have been reported from the deep sea: Trochidae (*Seguenzia*), Lepetidae, Rissoidae, Naticidae, Triphoridae, Buccinidae, Muricidae, Columbellidae, Cancellariidae, Volutidae, Turidae, Actaeonidae, Diaphanidae, Philinidae, Scaphandridae, Retusidae, Eulimidae, Pyramidellidae, Architectonidae, Pediculariidae, and Epitonidae. Many are carnivores or ectoparasites (Bouchet and Warén, 1979b, 1980; Knudsen, 1964, 1973; Clarke, 1962b). In addition, a few epifaunal browsers belonging to the little-known families Cocculinidae and Planilabiata (Moskalev, 1976) are restricted to plant debris carried downslope onto the abyssal rise and into canyons and trenches.

Rex (1976) reports 93 prosobranch species and 30 opisthobranch species with a wide range of feeding types from the Woods Hole–Gay Head Traverse. Deposit feeders include 20 archaeogastropods, 15 primitive mesogastropods, and 5 primitive opisthobranchs; predators of bivalves and annelids include 1 mesogastropod, 46 neogastropods, and 20 opisthobranchs (including 17 cephalaspids and 3 nudibranchs); ectoparasites of echinoderms and anthozoans include 11 eulimids, 8 epitonids, and 5 pyramidellids.

As in the case of the bivalves, cluster analysis shows that the greatest change in the composition of the gastropod fauna occurs at the shelf edge and at mud slope depths (Rex, 1977) (Fig. 8). However, the gastropods may be more stenobathic than the bivalves (Fig. 9). In fact the gastropods, although not numerically dominant, are one of the most, if not the most, diverse of the major groups in the deep sea. The molluscan fauna for the Scottish Marine Biological Station's permanent station at 2900 m in the Rockall Trough and the Gay Head–Bermuda

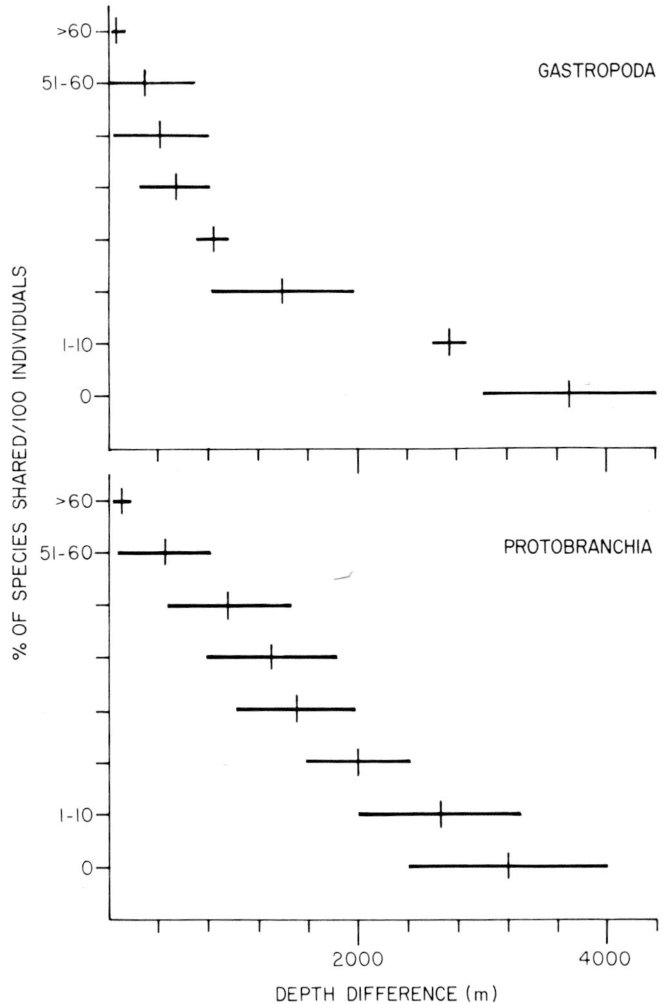

Fig. 9. The patterns of zonation for two major deep-sea benthic molluscan taxa from the Gay Head–Bermuda transect. Mean depth differences and standard deviations are compared for station pair groupings on the basis of the percentage of species shared. (From Sanders, 1979, with kind permission of the Philadelphia Academy of Natural Sciences.)

Traverse at abyssal depths (Gage et al., 1980) give some indication of this (Table V).

Despite their narrow zonation, some species are widespread. *Epitonium nitidium, Cithria tenella, Adeorbis umbilicatus, Mangelia bandella,* and *Lacuna cossmanni* all have pan-Atlantic distributions (Rex and Boss, 1973). *E. nitidium* is unusual in that it is an alloiostrophic (open-coiled) abyssal species, a condition that occurs in numerous gastropod lineages. Although its adaptive significance is not clear, in this case it may assist in the maintenance of the animal's position in the soft sediment.

Despite their differences in diversity, their narrow depth ranges, and, for the most part, their markedly different feeding habits, gastropods show basic adaptations to life in the deep sea similar to those of bivalves. Rex (1979), in his study of five populations of the shelf and slope species *Alvania pelagica,* points out that populations at 800 m devote more energy to growth and less to reproduction, and show greater longevity, than do those at 200 m. Similarly, Rex et al. (1979), in their analysis of *Benthonella tenella,* the most common abyssal deposit-feeding gastropod in the North West Atlantic (3806–5047 m) which they had collected at different times of the year, showed that (1) in individual males and females, developing sperm and ova were at differing states of maturity throughout the year, and (2) size frequency histograms were dominated by larger individuals, indicating that recruitment is infrequent and variable. But, unlike most bivalves, the eggs are small (82.5 μm) and the larval shell is large (790 μm), indicating a long planktotrophic existence before metamorphosis. This latter has been confirmed by Bouchet (1977) (Section VII).

As can be seen from the list of families, ectoparasitic species are present in the deep sea. Many of these are associated with echinoderms. For example, *Ophieulima minima* is parasitic on the ophiuroid *Ophiactis abyssicola* (Warén and Sibuet, 1981). It adheres to the host by a buried "snout" that is a modification of the foot, inserting the suctorial proboscis into the host's tissues. Sexes are separate, and in this case the female attaches eggs to the shell of the male (Fig. 10). A similar account is given by Warén and Carney (1981) of another eulimid, *Ophiolamia armigeri,* ectoparasitic on *Ophiomusium armagerum.* In this species the egg capsules were found on the host close to the adults. The capsules contain some 75 embryos 200 μm in diameter; the associated adults contained immature eggs, indicating that they reproduce more than once during their lifetime. The frequency of parasitization is not great; only 23 cases were recorded out of 3000 ophiuroids examined.

The relative abundance of predatory gastropods is positively and significantly

TABLE V

The Number of Species and of Individuals of the Major Molluscan Classes Present at Abyssal Depths at Locations in the North East and North West Atlantic

	Bivalvia	Scaphopoda	Gastropoda
Rockall Permanent Station			
Number of species	21	6	19
Number of individuals	14,908	777	110
Gay Head Traverse			
Number of species	131	?	132
Number of individuals	22,520	?	6,347

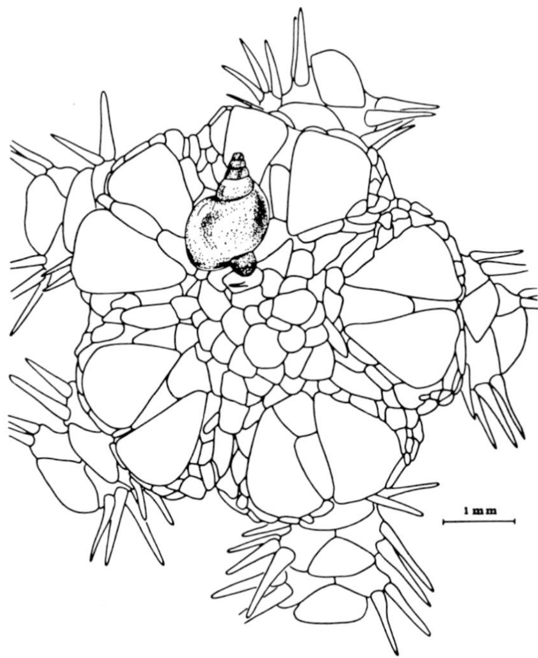

Fig. 10. *Ophieulima minima in situ* on its host, *Ophiactis abyssicola*. Two egg capsules have been removed to show the shape of the shell more clearly; a third is attached at the base of the shell. (From Warén and Sibuet, 1981, with kind permission of *Sarsia*.)

correlated to their diversity. It is lowest on the abyssal plain, highest on the lower slope and abyssal rise, and low on the upper slope. The variation in gastropod diversity is similar to that of their bivalve and polychaete prey and is also related to faunal density, which decreases exponentially with depth. The relationship between diversity, predation, and production is consistent with the theory that predators exert a diversifying influence on communities and that the degree to which they do so depends on the rate and stability of the production (Rex, 1976).

Where croppers and predators of a group dominate, as do the epibenthic gastropod macrofauna of the slope and abyssal rise, they tend to zone with depth more rapidly than do the infaunal deposit feeders. In fact, the zonation of gastropods is more rapid than that of any other deep-sea group including the bivalves (Fig. 9) (Rex, 1976). The zonation is enhanced by the greater compression of vertical ranges on the slope, which in part results from interspecific competition between members at higher trophic levels. Predators may alleviate competition between the infaunal lower trophic groups and allow overlapping distributions, diminishing the rate of faunal change with depth (Rex, 1973, 1976).

One of the features of the recent studies on deep-sea Mollusca has been the discovery of unusual morphologies, many of which have filled evolutionary gaps (pp. 41 and 45) (Lemche, 1957; Allen and Sanders, 1969; Allen and Morgan,

1981). *Neomphalus fretterae* one of several externally similar limpets present in the Galapagos Rift community (Section VIII), is a significant discovery (McLean, 1981; Fretter et al., 1981). As Yonge (1952) pointed out, the limpet form has occurred in many different groups, mostly as a response to epifaunal life on hard sediments; in the case of the deep-sea genus *Neopilina,* for example, this is not always so. However, the morphological significance of *Neomphalus* lies not so much in its limpet shape as in the fact that it presents a mixture of archaeogastropod and mesogastropod features, emphasizing the fact that the mesogastropods are not a single taxonomic group. Thus, *Neomphalus* has an aspidobranch gill of considerable length, a rhipidoglossate radula, epipodial tentacles, and a nervous system with primitive features; at the same time, it has a reduced number of muscles in the buccal mass, the cerebral ganglia have a labial lobe, the right kidney and right auricle have almost disappeared, there is an enlarged left kidney and the remnants of the right are incorporated into the reproductive tract, and the gut does not enter the pericardial cavity. Torsion is 270° rather than the usual 180°. Fretter et al. (1981) and McLean (1981) conclude that it is probably the ultimate abyssal survivor of an archaic group, possibly a derived euomphalacean, in this case persisting in a very special habitat.

D. Scaphopoda, Aplacophora, and Polyplacophora

Although members of these three groups are known from the deep sea, few studies of them have been carried out. Among the deep-sea Mollusca, the Scaphopoda are common numerically (as much as 4%) and occur at all depths (Knudsen, 1964), which is not surprising. The class feeds on small testate organisms, particularly Foraminifera, which are common in abyssal muds.

Similar observations are true for the chaetodermatoidean aplacophorans, which probably have much the same feeding habits as the scaphopods. These too are present in number in abyssal muds. Neomeniids, a few of which have also been reported from abyssal depths (Clarke, 1959; Wolff, 1960), presumably are associated with particular epifaunal species on which they feed.

The polyplacophorans are in contrast. Few chitons have been recorded in deep waters, and those few have are restricted to manganese nodules and plant materials. As far as is known, all those that have been described belong to the genus *Lepidopleurus,* which has the longest evolutionary history of any in the class (Wolff, 1960; Clarke, 1962a; Paul, 1976).

E. Cephalopoda

Relatively little is known of the cephalopods that occur at abyssal depths. Insofar as they are large, active, swimming carnivores, they are in marked contrast to the other epifaunal or infaunal Mollusca. Analysis of the literature shows that remarkably few species have been recorded from depths exceeding 3000 m, although Haedrich et al. (1980) believe that, with the pectinids, they may form a significant part of the megafauna in both numbers and biomass at depths exceeding 2100 m. Although this may be true of pectinids, the great majority of the oceanic Cephalopoda occur in water shallower than 1000 m (Voss, 1967). When they penetrate deeper, it is because they are occasional wanderers, [e.g., *Galiteuthis phyllura* (Roper and Young, 1975)]; or, more usually, because their downward vertical migration has carried them into these depths (Roper, 1969; Lu and Roper, 1979). The latter is almost certainly true of the three histioteuthid species, *Histioteuthis altanium, H. macrolista,* and *H. altana,* that have been reported at abyssal depths (Voss, 1969). Even though present in deep water, they occur many hundreds of meters above the bottom (Robson, 1933; Voss, 1967). Possible notable exceptions recorded in the earlier literature are *Vampyroteuthis infernalis* (Pickford, 1946, 1949) and *Bathyteuthis abyssicola* (Roper, 1969), which may be restricted to water of very low temperature. Even these are known to occur at ~600 m on occasion. Only the deep-sea cirrate Octopoda, recently recorded photographically, truly seem to be permanently abyssal (Roper and Brundage, 1972); they feed on bottom-living animals.

It is not difficult to advance a reason why fast-swimming cephalopds are not present in the deepest part of the sea. They require a frequent diet of protein in large quantity, mostly in the form of fish, and the necessary densities of prey occur in the upper part of the sea. Cephalopods, because of their sophisticated nerve–muscle physiology and high degree of sensory function, may well be restricted in their penetration of the deep sea by the effects of high pressure. Thus, those species in the deep sea are restricted to a less active life-style with atypical food. In addition to *Bathyteuthis* and *Vampyroteuthis,* other octopods present in abyssal depths belong to the genera *Japetella, Valbyteuthis, Taningia, Opisthiteuthis, Grimpoteuthis, Cirrothauma,* and *Cirroteuthis.* In fact, the cirrate octopods are not known in shallow water. Thus, the Cephalopoda, like the other molluscan classes, possess representatives that are entirely restricted to deep water.

The cephalopods are among the least known deep-sea animals; even their systematics are fragmentary. This is largely because of the poor condition of the few captured specimens, which further deteriorates on subsequent fixation. Apart from the extremely fragile structure and semigelatinous state, a noteworthy characteristic is their relatively long arms, forming two-thirds of the total body length and joined together by an extensive web (Fig. 11). In some species the body may be depressed. Eyes may be well developed and in some dorsally situated; they are used to detect the light of photophores common to the deep-sea natant fauna. They themselves have a complex system of photophores. It appears that most are neutrally buoyant. They move by three main modes: by using their webbed arms as a pulsating bell (*Vampyroteuthis*), by the ejection of water through the mantle funnel, or by undulation of the fins.

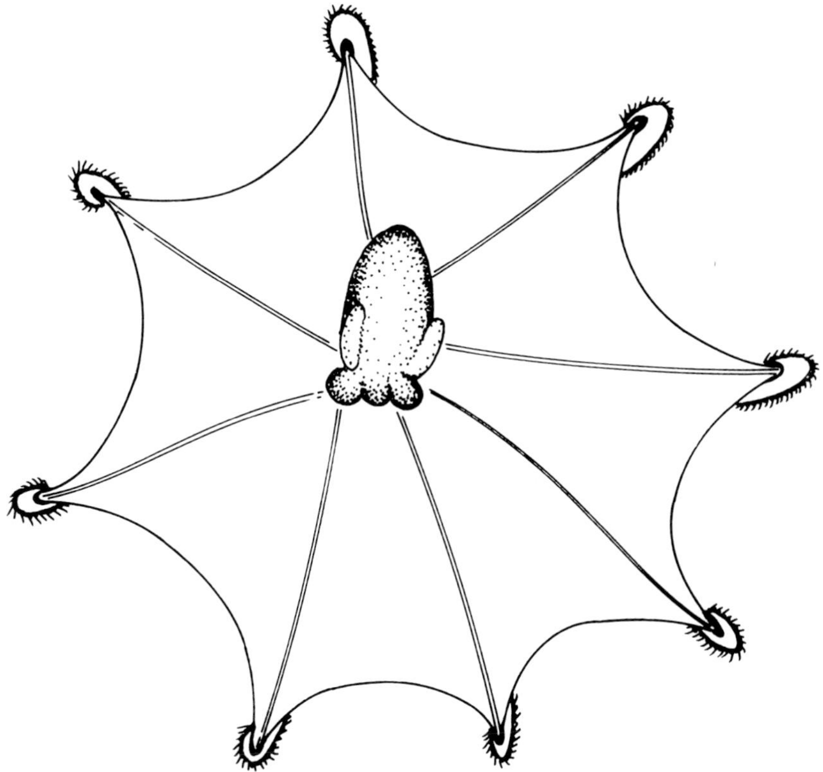

Fig. 11. Dorsal view of a cirrate octopod.

In species other than the cirrate octopods, natant crustaceans are common food organisms. It appears that the cirrate octopods are demersal, with the bell and mouth directed downward. Benthic polychaetes, isopods, and other crustaceans have been recorded from their stomachs. They may use the sensory cirri to seek out their prey. As might be expected, the deep-water octopods are not particularly large oceanic cephalopods, rarely being more than 30 cm in the greatest dimension.

Of all the molluscs, these few cephalopods might be regarded as "croppers" in the true sense of Dayton and Hessler (1972), who postulated that biological disturbance helps to maintain the diversity of the deep sea. One needs to look to other phyla to supply major predatory impact and to reduce the probability of competitive exclusion that this theory demands.

VI. Physiology

Pressure changes of significant magnitude can affect virtually every aspect of biochemical reaction and biological structure (Zimmerman and Zimmerman,

1972). In particular, changes to 200 atm involve excitable tissues; beyond that pressure, contractile tissue and microtubular structures are affected. At trench pressures (>600 atm) there are increasingly significant effects on protein structure, electrolytes, and diffusion systems, and even on the structure of water itself (Hamann, 1957; MacDonald, 1972; Hochachka, 1975). These steps may correspond to the changes in faunal composition that occur at about 2000–3000 m (abyss) and 6000–7000 m (trench) (Menzies and George, 1972).

High pressure may be responsible for the reduced metabolic rates of "lethargic" deep-sea animals (George, 1979). It enhances the effect of low temperature and may be a significant factor in the exploitation by animals of the deep sea; the response to pressure is temperature dependent (Menzies and George, 1972). Deep-sea molluscs must have a pressure-stable molecular basis for their life processes; although no experiments at pressure have been carried out on any deep-sea mollusc, in some Crustacea (such as *Gigantocypris*), it is known that locomotor activity ceases at 1000 atm and 30°C, and heart rate is reduced by 75%. If this is generally true, it is an added factor in the success of molluscs living in an environment with little organic input.

From direct experiments with pressure on molluscs, it is known that the shallow-water gastropod *Littorina irrorata* can survive pressures as great as 1309 atm (Menzies and George, 1972), and the ciliary activity of the excised gill of *Mytilus edulis* is greatly modified at high pressures (Pease and Kitching, 1939). Other physiological evidence comes from a few observations that indicate rate processes in the deep sea are slow. Thus, metabolism as measured in *in situ* respiration experiments indicates that oxygen uptake is two orders of magnitude less at 1850 m than on the shelf [0.39–0.55 ml/(m^2·h) and 46.8 ml/(m^2·h), respectively (Smith and Teal, 1973)]. Sediment boxes devoid of life placed on the deep-sea bed for 25 months have one order of magnitude less in number of animals colonizing, and these are smaller in size than those in similar experiments in shallow water (Grassle and Sanders, 1973). Turekian et al. (1975) show that specimens of the protobranch *Tindaria callistiformis* 8.4 mm in length are at least 100 years old, and gonads are first evident at 4 mm at an age of 50–60 years.

VII. Reproduction

What is known of biological processes in the deep sea suggests that low reproductive potential, and long maturation and life span, are to be expected. Ockelmann (1965) demonstrated a positive correlation between increasing depth (up to 400 m) and the percentage of bivalve species with lecitrotrophic development. Thorson (1950) predicted that most deep-sea molluscs would have a nonpelagic development. Knudsen (1970), Scheltema (1972), Grassle and Sanders (1973), and Sanders and Allen (1973, 1977) have shown that lecithotrophy with

TABLE VI

Comparison of Reproductive Potential between Shoal- and Deep-Water Lecithotrophic Species of the Genus *Nucula* Offshore the Coast of the Northeastern United States[a]

Species	Vertical distribution	Shell length (mm)	Gonad volume (mm^3)	Gonad volume/ shell length	Total number of eggs	Eggs/ shell length
Nucula proxima	Shelf	6.6	2.06	0.31	4120	624
Nucula annulata	Shelf	3.3	0.80	0.24	1233	374
Nucula granulosa	Slope	2.2	0.11	0.05	217	99
Nucula cancellata	Slope	3.3	0.24	0.07	194	59
Nucula verrilli	Abyss	4.3	0.24	0.06	260	60

[a] After Scheltema, 1972.

a brief nonfeeding planktonic stage, usual in deep-sea bivalves, is a response to the long-term stability of the deep-sea habitat. Thus, Knudsen (1967, 1970) shows that of the molluscs taken by the John Murray Expedition, 58% have lecithotrophic, 24% direct, and 8% planktotrophic development; those taken by the Galathea Expedition have 78% lecithotrophic, 13% direct, and 9% planktotrophic. Their broad geographic distribution is achieved by a slow step-wise dispersal. Where the environment is unpredictable in restricted areas such as thermal vents and wood ''islands'', molluscs do not follow the usual deep-sea pattern of development but have instead large numbers of eggs and a long planktonic larval stage. Nevertheless, even in these situations, eggs may be yolky; for example, those of *Calyptogena* are reported to be 300 μm in diameter (Boss and Turner, 1980).

Most (if not all) deep-sea molluscs have reduced fecundity compared with close relatives in shallow water. Scheltema (1971, 1972) clearly shows this in his comparative investigation of the reproductive state of shelf, slope, and abyssal species of *Nucula* (Table VI).

Under stable environmental conditions, to maintain a population that has reduced fecundity, there must be (1) better survival and/or (2) a longer period of reproduction during life; that is, deep-sea species must live longer and/or reproduce more often. If this is true, there must be a high initial survivorship in comparison with shelf species, which has been confirmed for many species of bivalves (Fig. 12) (Oliver and Allen, 1980b; Sanders and Allen, 1977). Periodicity may be advantageous in that successful fertilization in a scattered population probably will be enhanced. There is a little evidence that this occurs in the mollusc. Lightfoot et al. (1979) have produced some evidence to the contrary in the case of a *Ledella* and a *Yoldiella* species, although *Nucula cancellata* is in a reproductive state throughout the year.

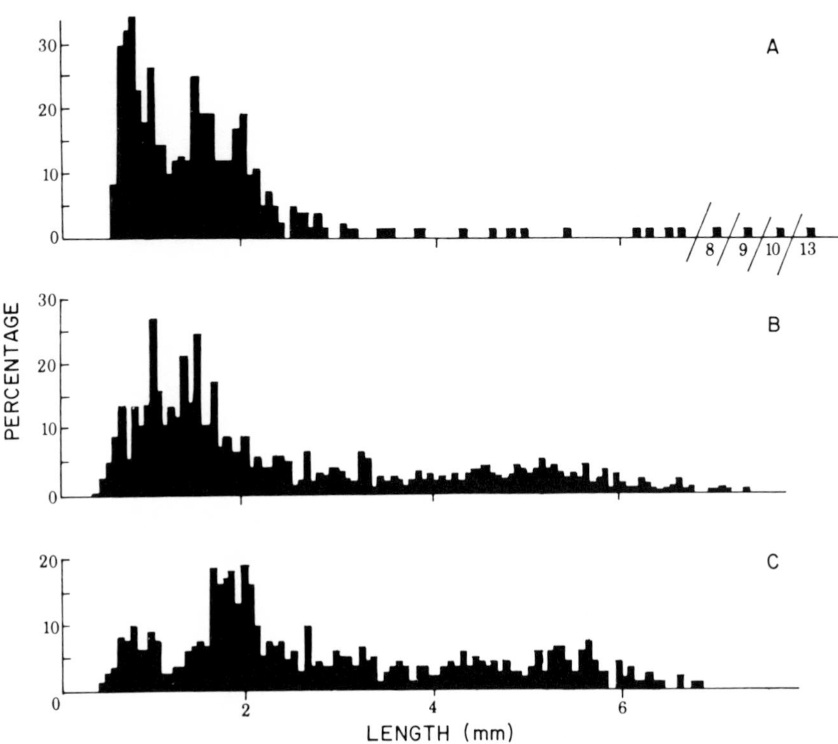

Fig. 12. Size frequency histograms of three species of *Limopsis* from widely different depths: (**A**) *L. surinamensis,* 500 m; (**B**) *L. cristata,* 1500 m; (**C**) *L. galathea,* 3500 m. (From Oliver and Allen, 1980a, with kind permission of the Royal Society.)

There is some evidence to suggest that the larvae of a few deep-sea bivalves are present in surface-water oceanic plankton (Lutz et al., 1980). Observations indicate that few species of bivalve larvae are present in surface zooplankton of the Atlantic Ocean (J. A. Allen, personal observation); on the other hand, there is little evidence that bivalve larvae occur at great depths, although Mileikovsky (1968) recorded lamellibranch larvae at 1800–2000 m near the Kurile–Kamtchatka Trench. From Turner's work (1973), there clearly must be high densities close to the bottom at times. For the most part, work on the recolonization of deep-sea sterile muds appears to indicate that the numbers of molluscan larvae settling must be low or perhaps patchy (Grassle, 1977). Experiments at 1800, 2000, and 3600 m with animal-free sediments retrieved 3 months to 3 years later all show low rates of colonization in comparison with similar shallow-water experiments. Most colonizers are juveniles even after 2 years; an exception is *N. cancellata,* which has been shown to grow to maturity in 2 years.

In contrast to the bivalves, there is considerable evidence that veligers of some deep-sea gastropods reach the surface. Bouchet (1977), Bouchet and Warén

(1979a, 1980), and Bouchet and Fontes (1981) have shown that a protoconch typical of a planktonic larva is present in a number of abyssal prosobranchs, and that analysis of the $^{18}O/^{13}O$ ratio in the calcium carbonate of the shells indicates that the larval shell was laid down in warmer water than that in which the postlarvae live. The protoconchs of the adults show calcite corrosion following the dissolution effects of high pressure and high pH. Despite this, it is evident that deep-sea molluscs are able to lay down thick shells. Both *Benthonella tenella* and *Benthomangelia macro* larvae (Richter and Thorson, 1974) have been recorded in surface plankton, but other deep-sea gastropod larvae may well feed demersally on debris and bacteria or, if bathypelagic or mesopelagic, they must be able to obtain sufficient suspended food to carry forward their development to metamorphosis. Nothing is known of the larval development of deep-sea representatives of other molluscan classes.

One of the effects of small body size is the consequential inability to produce large numbers of eggs because of lack of space. Thus, a number of protobranch bivalves, such as *Microgloma* spp., adopt the meiobenthic strategy of continuous reproduction; they produce one or two eggs at a time (Sanders and Allen, 1973). Some go further, with miniaturization of the body cells (except for the egg). In fact, in most deep-sea protobranch species of small size, the maturing adults contain eggs of varying degrees of development, and although discrete peaks may be recorded in size frequency histograms (Fig. 12), for the most part these probably relate to sporadic larval settlement rather than to year classes. The eggs of eulamellibranchs of the soft sediments and of hard surfaces (other than plant remains and thermal vents), similar to those of the protobranchs, are few in number (<100 per adult) and of moderate size (~ 120 μm). The prodissoconch II is relatively small, which indicates that the larval phase is short and lecithotrophic.

VIII. Thermal Vent Mollusca

One of the more remarkable recent discoveries has been the fauna associated with thermal vents in deep water. Those of the Galapagos Rift have been the subject of intensive research in the last 4 years (Spiess et al., 1980). Plate tectonics predicted heat input into the oceans in areas of sea-floor spreading, and this led to the discovery of hydrothermal vents. Associated with these vents is a bizarre fauna, individual species of which are both large and numerous. They live at a temperature of $\sim 20°C$ (Enright et al., 1981). More than 100 species have been collected; among the most conspicuous of these animals are huge (>300 cm in length), gutless vestimentiferan worms and huge vesicomyid and mytilid bivalves (>26 cm in length) (Fig. 13). Other molluscs include the limpet *Neomphalus* (Section V,C).

The food that sustains these exceptional animals differs from that of the usual

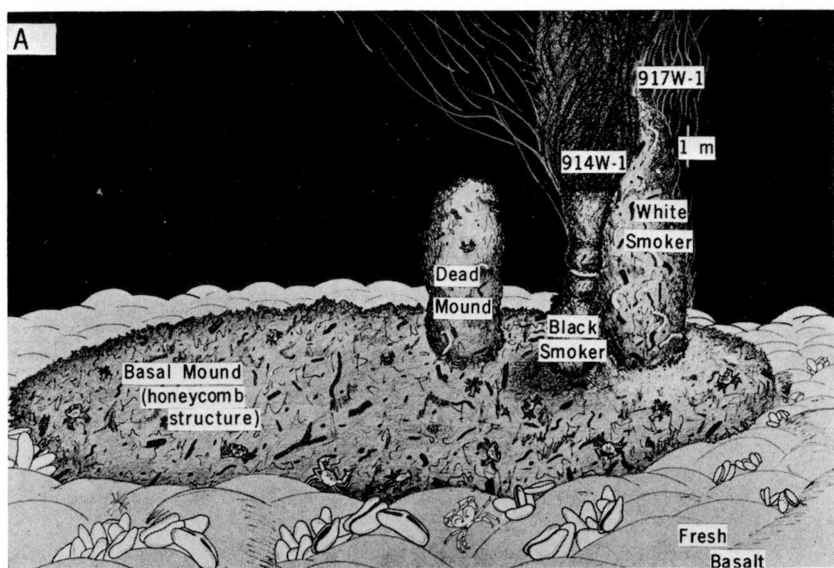

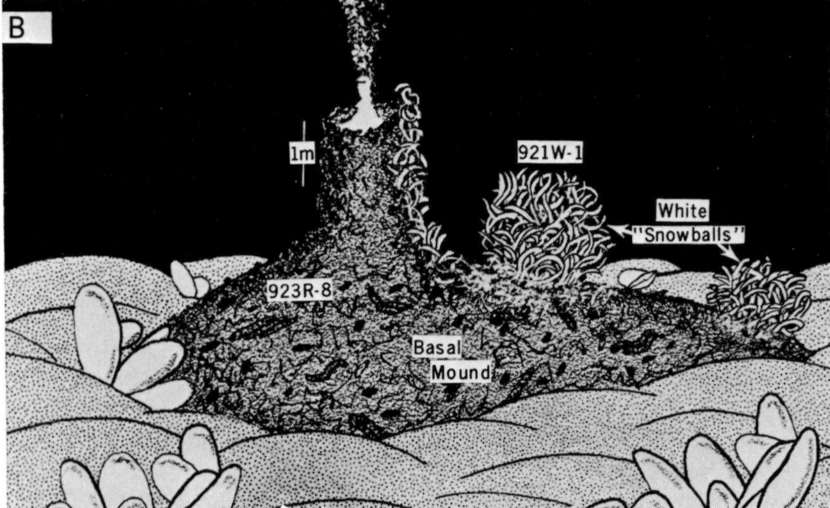

Fig. 13. Sketches of hot-water vents by (**A**) T. Juteau and (**B**) C. Rangin during observational dives in which 380 ± 30°C water temperatures were measured. *Calypogena* colonies surround the vent area. (From Spiess et al., 1980. Copyright American Association for the Advancement of Science.)

abyssal benthic fauna. Vent water is sulfide rich, and bacteria capable of chemosynthesis in a thiosulfate medium probably provide the food for the filter-feeding bivalves. Large concentrations (10^9 cells/ml) of bacteria are reported in pure vent water. Such concentrations would seem to dwarf any possible concentration of food due to advection currents associated with the vents (Lonsdale, 1977; Ballard and Grassle, 1979; Corliss, 1979; Karl et al., 1980). Nevertheless, there are problems in accepting the hypothesis that these bivalves depend on the bacteria of vent water for food. For instance, they are also found at normal deep-sea temperatures many meters from the nearest vent, as well as close to vents with outflows as hot as 350°C. Recent observations suggest that the concentration of bacteria may not be quite as large as was first thought; the bacteria are three orders of magnitude less concentrated. An advective source of food from the water–substratum interface where organic matter accumulates may be important. Calculations suggest that for every cubic meter of vent discharge 350 mg of particulate organic carbon would be advected upward to rain down again close to the vents. This process must resuspend particles every 2–3 days. That the food source is exceptional is supported by the fact that tissue $^{13}C/^{12}C$ ratios differ significantly from those of other deep-sea animals. A possible third nutrient source is via symbiotic microorganisms present in the gills of the vesicomyid *Calyptogena* (Boss and Turner, 1980). The gills are typically fleshy and dark in color and may contain sulfide oxidation enzymes (Felbeck et al., 1981). Their physiology probably resembles that of *Solemya panamaensis,* although unlike that species they have a typical digestive tract through which particulate material passes. A close resemblance, perhaps, would be to the Lucinidae that occur in the sulfide-rich layer under sea grass beds and also have dark fleshy gills (Allen, 1958).

As would be expected, under these conditions growth is rapid (Rhoads et al., 1981). At 2500 m, *Calyptogena magnifica* grows at a rate of about 4 cm/yr, and the as-yet-unnamed mytilid grows approximately 1 cm/yr, as estimated by shell abrasion experiments in which the mussels were held and marked using the mechanical arm of the submersible *Alvin.* Like the vents with which they are associated, Galapagos Rift bivalves communities are relatively transient. The bivalves live approximately 10–20 years, and dead shells are abundant at "dead" vent areas (Corliss, 1979; Turekian et al., 1979).

Thus, a life cycle is required differing from that of the protobranchs, which dominate the molluscan fauna of the abyssal plain. The life style of the mytilid is similar in some ways to that of another deep-sea mytilid occupying an ephemeral environment, *Idasola argentea,* which occurs on wood attacked by *Xylophaga.* Like *Idosola* and shallow-water opportunistic species, the vent mytilid has a planktotrophic larva. It is not clear whether these larvae occur in the surface plankton, or are deep-water pelagic and planktotrophic, or are demersal, bottom-feeding on bacteria and other debris. The prodissoconch of the thermal vent mytilid indicates not only a planktotrophic larval stage but also one with long-range dispersal characteristics, possibly on the order of hundreds of kilometers (Lutz et al., 1980). In this case, the unusual environmental conditions required by adults would probably act as stimuli for larval settlement and, judging by the disposition of the adults, would be enhanced by a gregarious settlement behavior. The prodissoconch I measures about 95 μm, and the prodissoconch II is >400 μm. Because metabolic rates are depressed in deep-sea animals, it is possible that the larvae have an exceptionally long planktonic life. As deep-sea currents of more than 30 cm/sec have been recorded, these larvae could be carried very great distances; this is an unusual mode of development in deep sea (Section VII).

In contrast, *Calyptogena* has large yolky eggs ~300 μm in diameter (Turner, 1981). It was Keen (1977a,b) who recognized that this very large species of bivalve associated with the thermal vents belonged to the family Vesicomyidae and was a new species closely related to *Calyptogena modioliforma* (Boss) and *C. elongata* Dall. The family is typically infaunal but *C. magnifica* is epifaunal, nestling without byssal attachment in crevices and among mussels. Anatomically, it resembles the shallow-water species *C. pacifica* and *C. kilmeii* (Bernard, 1974). The fused inner lobe of the mantle forms an extrusable velum, and there is a sensory knob below an incurrent siphon. The sensory papillae at the mantle edge may be involved in mantle respiration of the type suggested for the Lucinidae (Allen, 1958), but are more likely to be involved with cleansing of the mantle. The labial palps are reduced in size, but this is also typical of the shallow-water species and may not necessarily be an adaptation to a new feeding habit. However, the general feeding habits of the genus may favor life in the vent environment. The gut is not greatly extended, and digestive diverticula are large. All these adaptations suggest that external particulate matter is taken in as food, at least in part, although the gills do not have a marginal food grove (Turner, 1981).

IX. Evolution

It has been pointed out that the great majority of molluscs that occur in the abyss are representatives of groups with extremely long geological histories, and many of the genera (if not species) can be recognized from as long ago as the Ordovician (Allen, 1978). Furthermore, the assemblage of species in the Atlantic is largely distinct from that of the Pacific and the Arctic. Although an ocean existed in Cambrian times, the present floor of the oceans is relatively young (Bullard, 1969). In terms of antiquity, the Atlantic is relatively recent (Berggren and Hollister, 1974). The initial opening of the Atlantic was in the Mesozoic some 180–190 million years ago, and separation of Africa and South America may have taken place about 95 million years ago, thus allowing water and faunal exchange between oceans. Only in the last 50 million years has the present deep-water circulation become established. There is great similarity in North-South Atlantic molluscan faunas, with the fauna of the Argentine Basin perhaps an exception. It should also be remembered that Pleistocene glaciation must have produced a disastrous decrease in temperature from 10 to 2°C in the abyss (Emiliani and Edwards, 1953).

Clearly, within the last 50 million years there has been a great radiation of these ancient groups, and this is particularly true of the protobranch molluscs. They are incredibly diverse and dominate the deep-sea molluscan fauna. In a sense they have been opportunists on a grand scale and have not as yet been swamped by modern groups as was the ancient, diverse, protobranch-dominated fauna. Presumably, the low quantity of food in suspension and the unique ability of the protobranch to digest the scleroproteins of skeletal remains have prevented this.

The rapid diversification that occurred in the Atlantic protobranchs is not unique. Thus, in the early Ordovician there was a 5-million-year period when silty benthic environments experienced a sharp rise in invertebrate diversity. This was an invasion of a depauperate environment by established taxa which then produced new species; this finally resulted in a steady state community (Rosenzweig and Taylor, 1980). A similar picture of evolution following the establishment of the Atlantic Ocean, which gave rise to a very similar protobranch-dominated fauna, is suggested. Presumably, the invasion followed downslope from existing shelf species. This concept is supported by the endemism of the Arctic Ocean and the Argentinian and Indo-Pacific basins and might suggest that the Atlantic was not invaded from the deep-water fauna of more ancient oceans. This concept is not new; as early as 1899, Perrier had suggested that the ancient groups were relicts of shelf faunas. In this context, it is worth remembering a feature of molluscan distribution that has been known for some time, the phenomenon of tropical submergence; populations of cosmopolitan and pan-oceanic species live at greater depths in equatorial waters than they do in polar seas (Filotova, 1976). This is predominantly a temperature effect, although there may be a relict component from the Pliocene, when there was a freer connection between the North Atlantic and the Arctic Ocean and when some species (e.g., *Yoldia* spp.) extended much farther south (Ockelmann, 1954).

Molluscs of high latitudes are characterized by a slow growth rate, low basal metabolism, and reduced reproductive effort. Cold adaptation signifies a reduction in the individual total annual energy uptake (Clarke, 1980) [not an increase, as Dunbar (1968) predicted], and thus high-latitude invertebrates have a response to their environmental conditions similar to that of deep-sea molluscs. However, in high latitudes there is ample primary production and, consequently, a high standing crop of suspension feeders. A theory involving the origin of the deep-water fauna from shallow seas in high latitudes must therefore be tempered with caution. Although the molluscan fauna in these regions is impoverished (Ellis, 1959; Clarke, 1960, 1963; Ockelmann, 1954; Soot-Ryen, 1958), relatives of the dominant deep-sea groups commonly occur. Many molluscan groups are found only in deep water (Verticordiidae, Monoplacophora), but even they are thought to be derived from shallow-water forms (Lemche, 1957; Clarke and Menzies, 1959; Bruun and Wolff, 1961; Allen and Turner, 1974).

That molluscan diversification on a grand scale could be established in the deep sea no doubt results from the physical stability of the environment (Sanders, 1968, 1969, 1977, 1979; Grassle and Sanders, 1973; Slobodkin and Sanders, 1969). It also reflects the genetic variability of the molluscs (Grassle, 1972), which in turn relates to the reproductive isolation of an established low-density population from a single reproductive event, particularly when growth is slow and the age of maturity great. As Stanley (1976) points out, a rapid source of evolutionary divergence is the appearance of polymorphism within species under conditions of relaxed selection pressure. This frequently occurs after the invasion of "ecologic islands" lacking predators and competitors. The abyssal Atlantic was a very large "island." In fact, it would seem that because cropping pressure is proportional to prey abundance, it is improbable that predation has played a part in the proliferation of individual mollusc species in the deep sea as great as Dayton and Hessler (1972) suggested. Temporal stability, spatial configuration, the amount of energy available for growth and metabolism, and time span available for evolution are far more important.

Another important factor allowing for evolution and diversification of the fauna is that the food supply should be not only continuous, as is the rain of fecal and skeletal proteins, but also varied. Pressure is no barrier to more recent specialized molluscs if their food source is present. Thus, three genera and about 40 species of deep-sea obligate wood-boring bivalves of the subfamily Xylophaginae have evolved since the Upper Cretaceous. Similarly, when large quantities of bacteria and nutrients in solution exist, other specialized species evolve, (the epifaunal vent species of the genus *Calyptogena* and the as-yet-unnamed mytilid). Nevertheless, these wood islands and thermal vents are transient in relation to the sea floor as a whole so that few mollusc species exploit them; however, as in the estuarine faunas, those that do occur in large numbers grow rapidly and produce large numbers of eggs.

The large number of molluscan carnivores, bivalve as well as gastropod, may seem one of the more unusual features of the deep-sea molluscan fauna. On further consideration this is perhaps understandable, for they are particularly well adapted to exploiting those prey species that live in the abyss; megabenthic echinoderms by suctorial prosobranchs and opisthobranchs, natant surface meiofauna by septibranchs, and infaunal bivalves by naticid and muricid gastropod drills.

In fact, one of the exciting evolutionary aspects of the studies on deep-sea molluscs has been the elucidation of evolutionary pathways at all levels. Thus, the discovery of the living Monoplacophora reopened questions of the segmental origin of the Mollusca; new genera of septibranchs have clearly shown the course of evolution of the septum from the eulamellibranch gill; the study of *Nucinella* shows a possible derivation for the solemyid protobranchs from the Ordovician actinodonts and, not least, the Galapagos Rift limpet *Neomphalus* gives rise to new speculation on the complexity of gastropod evolution.

References

Agassiz, A. (1888). Three cruises of the United States Coast and Geodetic Survey Steamer BLAKE. *Bull. Mus. Comp. Zool.* **14,** 1–314.

Allen, J. A. (1954). Observations on the epifauna of the deep-water muds of the Clyde Sea Area, with special reference to *Chlamys septemradiata* (Müller). *J. Anim. Ecol.* **22,** 240–260.

Allen, J. A. (1958). On the basic form and adaptations to habitat in the Lucinacea (Eulamellibranchia). *Philos. Trans. R. Soc. London* **241,** 421–484.

Allen, J. A. (1971). Evolution and functional morphology of the deep water protobranch bivalves of the Atlantic. *Proc. Jt. Oceanogr. Assembly, 1970,* pp. 251–253.

Allen, J. A. (1978). Evolution of the deep sea protobranch bivalves. *Philos. Trans. R. Soc. London, Ser. B* **284,** 387–401.

Allen, J. A. (1979). The adaptations and radiation of deep-sea bivalves. *Sarsia* **64,** 19–27.

Allen, J. A., and Morgan, R. E. (1981). The functional morphology of Atlantic deep water species of the families Cuspidariidae and Poromyidae (Bivalvia): an analysis of the evolution of the septibranch condition. *Philos. Trans. R. Soc. London, Ser. B* **294,** 413–546.

Allen, J. A., and Sanders, H. L. (1966). Adaptations to abyssal life as shown by the bivalve *Abra profundorum* (Smith). *Deep-Sea Res.* **13,** 1175–1184.

Allen, J. A., and Sanders, H. L. (1969). *Nucinella serrei* Lamy (Bivalvia: Protobranchia), a monomyarian solemyid and possible living actinodont. *Malacologia* **7,** 381–396.

Allen, J. A., and Sanders, H. L. (1973). Studies on deep-sea Protobranchia (Bivalvia); the families Siliculidae and Lametilidae. *Bull. Mus. Comp. Zool.* **145,** 263–310.

Allen, J. A., and Sanders, H. L. (1982). Studies on the deep-sea Protobranchia; the subfamily Spinulinae (family Nuculanidae). *Bull. Mus. Comp. Zool.* **150,** 1–30.

Allen, J. A., and Turner, J. F. (1974). On the functional morphology of the family Verticordiidae (Bivalvia) with descriptions of new species from the abyssal Atlantic. *Philos. Trans. R. Soc. London, Ser. B* **268,** 401–536.

Ballard, R. D., and Grassle, J. F. (1979). Return to oases of the deep. *Natl. Geogr. Mag.* **156,** 689–705.

Beebe, W. (1935). "Half Mile Down." John Lane, London.

Belyaev, G. M., Vinogradova, N. G., Levenstein, R. Y., Pasternak, P. A., Sokolova, M. N., and Filatova, Z. A. (1973). Distribution laws of the deep sea bottom fauna in the light of the idea of the biological structure of the ocean. *Okeanologiya (Moscow)* **3,** 149–157.

Berggren, W. A., and Hollister, C. D. (1974). Paleography, paleobiogeography and the history of circulation in the Atlantic Ocean. *Stud. Paleo-Oceanogr., Econ. Paleontol. Miner., Spec. Publ.* **20,** 126–186.

Bernard, F. (1948). Recherches sur le cycle du *Coccolithus fragilis* Lohm., flagellé dominant des mers chaudes. *J. Cons., Cons. Int. Explor. Mer* **15,** 177–188.

Bernard, F. R. (1974). The genus *Calyptogena* in British Columbia with a description of a new species (Bivalvia, Vesicomyidae). *Venus* **33,** 11–22.

Billett, D. S. M., Lampitt, R. S., Rice, A. L. and Mantura, R. F. C. (1983). Seasonal sedimentation of phytoplankton to the deep-water benthos. *Nature (London)* **302,** 520–522.

Boss, K. J. (1969). Systematics of the Vesicomyidae (Mollusca: Bivalvia). *Malacologia* **9,** 254–255.

Boss, K. J., and Turner, R. D. (1980). The giant white clam from the Galapagos Rift. *Calyptogena magnifica* species novum. *Malacologica* **20,** 161–194.

Bouchet, P. (1977). Mise en évidence de stades larvaires planctoniques chez des gastéropodes prosobranches des étages bathyal et abyssal. *Bull. Mus. Natl. Hist. Nat., Zool.* **400,** 947–972.

Bouchet, P., and Fontes, J. C. (1981). Migrations verticales des larves de gastropodes abyssaux arguments nouveaux dûs à l'analyse isotopique de la coquille larvaire et postlarvaire. *C.R. Hebd. Seances Acad. Sci.* **192,** 1005–1008.

Bouchet, P., and Warén, A. (1979a). Planktotrophic larval development in deep water gastropods. *Sarsia* **64,** 37–40.

Bouchet, P., and Warén, A. (1979b). The abyssal molluscan fauna of the Norwegian Sea and its relation to other faunas. *Sarsia* **64,** 211–243.

Bouchet, P., and Warén, A. (1980). Revision of the Northeast Atlantic bathyal and abyssal Turridae (Mollusca: Gastropoda). *J. Moll. Stud., Suppl.* **8,** 1–116.

Brewer, P. G., Nozaki, Y., Spencer, D. W., and Fleer, A. P. (1980). Sediment trap experiments in the deep N. Atlantic: isotopic and elemental fluxes. *J. Mar. Res.* **38,** 703–728.

Broeker, W. S. (1974). "Chemical Oceanography." Harcourt, Brace, Jovanovich, New York.

Bruun, A. F., and Wolff, T. (1961). Abyssal benthic organisms, origin, distribution and influence on sedimentation. *In* "Oceanography," Publ. No. 67, pp. 391–397. Am. Assoc. Adv. Sci., Washington, D.C.

Bullard, E. (1969). The origin of the oceans. *Sci. Am.* **221,** 66–75.

Carruthers, J. N., and Lawford, A. L. (1952). The deepest oceanic sounding. *Nature (London)* **169,** 601.

Clarke, A. (1980). A reappraisal of the concept of metabolic cold adaptation in polar marine invertebrates. *Biol. J. Linn. Soc.* **14,** 77–92.

Clarke, A. H. (1959). Preliminary report on abyssal Atlantic mollusks from the Theta and Vema expeditions. *Bull. Am. Malacol. Union* pp. 3–4.

Clarke, A. H. (1960). Arctic, arctic benthal and abyssal mollusks from drifting Station Alpha. *Breviora* **119,** 1–19.

Clarke, A. H. (1961). Structure, zoogeography and evolution of the abyssal mollusk fauna. *Bull. Am. Malacol. Union* pp. 20–21.

Clarke, A. H. (1962a). On the composition, zoogeography, origin and age of the deep-sea mollusk fauna. *Deep-Sea Res.* **9,** 291–306.

Clarke, A. H. (1962b). Annotated list and bibliography of the abyssal marine molluscs of the world. *Bull.—Natl. Mus. Can.* **181,** 1–114.

Clarke, A. H. (1963). On the origin and relationships of the arctic ocean abyssal mollusca fauna. *Proc. Int. Congr. Zool., 16th, 1963* Vol. 1, p. 202.

Clarke, A. H., Jr., and Menzies, R. J. (1959). *Neopilina (Vema) ewingi,* a second living species of the Paleozoic class Monoplacophora. *Science* **129,** 1026–1027.

Cole, T., and Turner, R. D. (1977). Genetic relations of deep sea wood borers. *Bull. Am. Malacol. Union* pp. 19–25.

Corliss, J. B. (1979). Submarine thermal springs on the Galapagos Rift. Science, **203,** 1073–1083.

Culliney, J. L., and Turner, R. D. (1976). Larval development of the deep-water wood boring bivalve, *Xylophaga atlantica* Richards (Mollusca, Bivalvia, Pholadidae). *Ophelia* **15,** 149–161.

Dayton, P. K., and Hessler, R. R. (1972). Role of biological disturbance in maintaining diversity in the deep sea. *Deep-Sea Res.* **19,** 199–208.

Dickson, R. R., Gould, W. J., Gurbutt, P. A., and Killworth, P. D. (1982). A seasonal signal in ocean currents to abyssal depths. *Nature (London)* **195,** 193–196.

Dunbar, M. J. (1968). "Ecological Development in Polar Regions: A Study in Evolution." Prentice-Hall, Englewood Cliffs, New Jersey.

Ellis, D. V. (1959). The benthos of soft sea-bottom in Arctic North America. *Nature (London)* **184,** B.A.79–B.A.80.

Emeliani, C., and Edwards, C. (1953). Tertiary ocean bottom temperatures. *Nature (London)* **171,** 887–888.

Enright, J. T., Newman, W. A., Hessler, R. R., and McGowan, J. A. (1981). Deep-ocean hydrothermal vent communities. *Nature (London)* **289,** 219–220.

Eppley, R. W., and Peterson, B. J. (1979). Particulate organic matter flux and planktonic new production in the deep ocean. *Nature (London)* **282,** 677–680.

Ericson, D. B., Ewing, M., and Heezen, B. C. (1951). Deep-sea sands and submarine canyons. *Geol. Soc. Am. Bull.* **62,** 961–965.

Ewing, M., and Thorndike, E. M. (1965). Suspended matter in deep ocean water. *Science* **147,** 1291–1294.

Felbeck, H., Childress, J. J., and Somero, G. N. (1981). Calvin-Benson cycle and sulphide oxidation enzymes in animals from sulphide rich habitats. *Nature (London)* **293,** 291–293.

Filloux, J. (1970). Deep sea tides 1250 kilometers off Baja California. *Science* **169,** 862–864.

Filotova, Z. A. (1960). Quantitative distribution of bivalves in the far Eastern seas of the USSR and in the western Pacific. *Tr. Inst. Okeanol. im. P. P. Shirshova, Akad. Nauk SSSR* **41,** 132–145.

Filotova, Z. A. (1976). Composition of the genus *Spinula* (Dall, 1908) family Malletidae and its distribution in the ocean Deep Sea bottom fauna of the Pacific Ocean. *Trans. P.P. Shirshov. Inst. Oceanogr., Acad. Sci. USSR* **99,** 219–240 (in Russian).

Forbes, E. (1844). Report on the Mollusca and Radiata of the Aegean Sea, and their distribution, considered as bearing in Geology. *Rep. Br. Assoc., 1843* pp. 30–193.

Fournier, R. O. (1972). The transport of organic carbon to organisms living in the deep oceans. *Proc. R. Soc. Edinburgh, Sect. B: Biol.* **73,** 203–211.

Fretter, V., Graham, A., and McLean, J. H. (1981). The anatomy of the Galapagos Rift limpet *Neomphalus fretterae. Malacologia* **21,** 337–361.

Gage, J. D. (1977). Structure of the abyssal macrobenthic community in the Rockall Trough. *In* "Biology of Benthic Organisms" B. F. Keegan, P. O. O'Ceidigh, and P. J. S. Boaden, eds.), pp. 247–260. Pergamon, Oxford.

Gage, J. D. (1979). Macrobenthic community structure in the Rockall Trough. *Ambio Spec. Rep.* **6,** 43–46.

Gage, J. D., Lightfoot, R. H., Pearson, M., and Tyler, P. A. (1980). An introduction to a sample time-series of abyssal macrobenthos: methods and principal sources of variability. *Oceanol. Acta* **3,** 169–176.

George, R. Y. (1979). What adaptive strategies promote immigration and speciation in deep-sea environment. *Sarsia* **64,** 61–65.

George, R. Y., and Higgins, R. P. (1979). Eutrophic hadal benthic community in the Puerto Rico Trench. *Ambio Spec. Rep.* **6,** 51–58.

George, R. Y., and Menzies, R. J. (1972). Deep sea faunal zonation of benthos along Beaufort-Bermuda Transect in the north-western Atlantic. *Proc. R. Soc. Edinburgh, Sect. B: Biol.* **73,** 183–194.

Gould, W. J., Hendry, R., and Huppert, H. E. (1981). An abyssal topographic experiment. *Deep-Sea Res.* **28A,** 409–440.

Grassle, J. F. (1972). Species diversity, genetic variability and environmental uncertainty. *Proc. Eur. Mar. Biol. Symp., 5th, 1972* pp. 19–26.

Grassle, J. F. (1977). Slow recolonization of deep-sea sediment. *Nature (London)* **265,** 618–619.

Grassle, J. F., and Sanders, H. L. (1973). Life histories and the role of disturbance. *Deep-Sea Res.* **20,** 643–659.

Grassle, J. F., Sanders, H. L., and Smith, W. K. (1969). Faunal changes with depth in the deep-sea benthos. *Ambio Spec. Rep.* **6,** 47–50.

Haedrich, R. L., Rowe, G. T., and Polloni, P. T. (1980). The megabenthic fauna in the deep sea south of New England, USA. *Mar. Biol. (Berlin)* **57,** 165–179.

Hamann, S. D. (1957). "Physico-Chemical Effects of Pressure." Butterworth, London.

Heezen, B. C., and Hollister, C. D. (1971). "The Face of the Deep." Oxford Univ. Press, London and New York.

Heezen, B. C., Ewing, M., and Menzies, R. J. (1955). The influence of submarine turbidity currents on abyssal productivity. *Oikos* **6,** 170–182.

Hessler, R. R. (1974). The structure of deep benthic communities from central oceanic waters. *In* "The Biology of the Oceanic Pacific" (C. B. Mitler, ed.), pp. 79–93, Oregon State Univ. Press, Corvallis.

Hessler, R. R., and Jumars, P. A. (1974). Abyssal community analysis from replicate box corers in the central North Pacific. *Deep-Sea Res.* **21,** 185–209.

Hessler, R. R., and Sanders, H. L. (1967). Faunal diversity in the deep-sea. *Deep-Sea Res.* **14,** 65–78.

Hinga, K. R., Sieberth, J.McN., and Heath, G. R. (1979). The supply and use of organic material at the deep sea floor. *J. Mar. Res.* **37,** 557–579.

Hochachka, P. W. (1975). Why study proteins of abyssal organisms? *Comp. Biochem. Physiol. B* **52B,** 1–2.

Honjo, S. (1978). Sedimentation of materials in the Sargasso Sea at 5367 m in deep station. *J. Mar. Res.* **36,** 469–492.

Honjo, S. (1980). Material fluxes and modes of sedimentation in the mesopelagic and bathypelagic zones. *J. Mar. Res.* **38,** 53–97.

Honjo, S., and Roman, M. R. (1978). Marine copepod fecal pellets; production, preservation and sedimentation. *J. Mar. Res.* **36,** 45–57.

Horikoshi, M. (1970). Deep-sea benthos, its zonation and physical environment. *Mar. Sci.* **2,** 131–136.

Jannasch, H. W., and Wirsen, C. O. (1973). Deep-sea microorganisms; *in situ* response to nutrient enrichment. *Science* **180,** 641–643.

Jannasch, H. W., and Wirsen, C. O. (1977). Microbial life in the deep sea. *Sci. Am.* **236,** 42–52.

Jannasch, H. W., Eimhjellen, K., and Wirsen, C. O. (1971). Microbial degradation of organic matter in the deep sea. *Science* **171,** 672–675.

Jumars, P. A., and Hessler, R. R. (1976). Hadal community structure: Implications from the Aleutian Trench. *J. Mar. Res.* **34,** 547–560.

Karl, D. M., Wirsen, C. O. and Jannasch, H. W. (1980). Deep-sea primary production at the Galapagos thermal vents. *Science* **207,** 1345–1347.

Keen, A. M. (1977a). A deep water paradox. *Annu. Rep. West Soc. Malacol.* **10,** 10.

Keen, A. M. (1977b). New sea-floor oasis. *Veliger* **20,** 179–180.

Kidd, R. B., and Huggett, Q. J. (1981). Rock debris on abyssal plains in the North east Atlantic: a comparison of epibenthic sledge hands and photographic surveys. *Oceanol. Acta* **4,** 99–104.

Kimball, J. F., Jr., Corcoran, E. F., and Wood, E. J. F. (1963). Chlorophyll containing microorganisms in the aphotic zone of the oceans. *Bull. Mar. Sci. Gulf Caribb.* **13,** 574–577.

Knudsen, J. (1961). The bathyal and abyssal *Xylophaga* (Pholadidae Bivalvia). *Galathea Rep.* **5,** 163–208.

Knudsen, J. (1964). Scaphopoda and Gastropoda from depth exceeding 6000 meters. *Galathea Rep.* **7,** 125–136.

Knudsen, J. (1967). The deep sea Bivalvia. *Sci. Rep. John Murray Exped.* **11,** 235–343.

Knudsen, J. (1970). The systematics and biology of abyssal and hadal Bivalvia. *Galathea Rep.* **11,** 7–241.

Knudsen, J. (1973). *Guivillea alabastrina* (Watson 1882) an abyssal volutid (Gastropoda: Mollusca). *Galathea Rep.* **12,** 127–131.

Knudsen, J. (1979). Deep sea bivalves. *In* "Pathways in Malacology" (S. van der Spoel, A. C. van Buggen, and J. Lever, eds.), pp. 195–224. Bohn, Schellema & Holkema, Utrecht.

Koblentz-Miske, O. J., Volkovinsky, V. V., and Kabanova, J. G. (1970). Plankton primary production of the world ocean. *In* "Scientific Exploration of the South Pacific" (W. Wooster, ed.), pp. 183–193. Natl. Acad. Sci., Washington, D.C.

Kuenen, P. (1950). "Marine Geology." Wiley, New York.

Laubier, L., and Sibuet, M. (1979). Ecology of the benthic communities of the deep North East Atlantic. *Ambio. Spec. Rep.* **6,** 37–42.

Lemche, H. (1957). A new living deep sea mollusc of the Cambro-Devonian Class Monoplacophora. *Nature (London)* **179,** 413–416.

Lemche, H., and Wingstrand, K. G. (1959). The anatomy of *Neopilina galatheae* Lemche, 1957 (Mollusca Tryblidiacea). *Galathea Rep.* **3,** 9–72.

Levenstein, R. Y., and Pasternak, F. A. (1976). Some peculiarities of the distribution of the bottom fauna of the lower and middle ultra-abyssal zone of the Philippine Trench. *Tr. Inst. Okeanol. im. P.P. Shirshova, Akad. Nauk SSSR* **99,** 211–217.

Lightfoot, R. H., Tyler, P. A., and Gage, J. D. (1979). Seasonal reproduction in deep-sea bivalves and brittlestars. *Deep-Sea Res.* **26A,** 967–973.

Lonsdale, P. (1977). Clustering of suspension feeding macrobenthos near abyssal hydrothermal vents at oceanic spreading centers. *Deep-Sea Res.* **24,** 857–863.

Lu, C. C., and Roper, C. F. E. (1979). Cephalopods from deep water Dumpsite 106 (Western Atlantic): vertical distribution and seasonal abundance. *Smithson. Contrib. Zool.* **288,** 1–36.

Lutz, R. A., Jablonski, D., Rhoads, D. C., and Turner, R. D. (1980). Larval dispersal of a deep sea hydrothermal vent bivalve from the Galapagos rift. *Mar. Biol. (Berlin)* **57,** 127–133.

McCave, I. N. (1975). Vertical flux of particles in the ocean. *Deep-Sea Res.* **22,** 491–502.

MacDonald, A. G. (1972). The role of high hydrostatic pressure on the physiology of marine animals. *Symp. Exp. Biol. Symp.* **26,** 209–231.

McLean, J. H. (1981). The Galapagos Rift limpet *Neomphalus. Malacologia* **21,** 291–336.

Menzies, R. J. (1968). New species of *Neopilina* of the Cambro-Devonian class. Monoplacophora from the Milne–Edwards Deep of the Peru Chile Trench, R/V Anton Bruun. *Proc. Symp. Mollusca (India)* **1,** 1–9.

Menzies, R. J., and George, R. Y. (1972). Hydrostatic pressure temperature effects on deep sea colonization. *Proc. R. Soc. Edinburgh Sect. B: Biol.* **73,** 195–202.

Menzies, R. J., George, R. Y., and Rowe, G. T. (1973). "Abyssal Environment and Ecology of the World Oceans." Wiley, New York.

Mileikovsky, S. A. (1968). Distribution of pelagic larvae of bottom invertebrates of the Norwegian and Barents Seas. *Mar. Biol. (Berlin)* **1,** 161–167.

Morita, R. Y. (1979). Deep sea microbial energetics. *Sarsia* **64,** 9–12.

Morton, B. (1981). Prey capture in the carnivorous septibranch *Poromya granulata* (Bivalvia: Anomalodesmata: Poromyacea). *Sarsia* **66,** 241–256.

Moskalev, L. (1976). On the generic classification in Cocculinidae (Gastropoda Prosobranchia). *Tr. Inst. Okeanol. im. P.P. Shirshova, Akad. Nauk SSSR* **99,** 59–70.

Ockelmann, K. W. (1954). On the interrelationship and zoography of northern species of *Yoldia*

Möller s. str. (Mollusca, Fam. Ledidae) with a new subspecies. *Medd. Groenl.* **107,** No. 7, 1–33.

Ockelmann, K. W. (1965). Developmental types in marine bivalves and their distribution along the Atlantic coast of Europe. *Proc. Eur. Malacol. Congr., 1st, 1962* pp. 25–35.

Okutani, T. (1967). Characteristics and origin of archibenthal molluscan fauna on the Pacific coast of Honshu, Japan. *Venus* **25,** 136–146.

Okutani, T. (1968). Systematics, ecological distribution and paleoecological implication of archibenthal and abyssal mollusca from Sogami Bay and adjacent areas. *J. Fac. Sci., Univ. Tokyo, Sect. 2* **17,** 1–98.

Okutani, T. (1974). Review and new records of abyssal hadal molluscan fauna in Japanese and adjacent waters. *Venus* **33,** 23–39.

Oliver, P. G. (1979). Adaptations of some deep sea suspension-feeding bivalves (*Limposis* and *Bathyarca*). *Sarsia* **64,** 33–36.

Oliver, P. G., and Allen, J. A. (1980a). The functional and adaptive morphology of the deep sea species of the Arcacea (Mollusca: Bivalvia) from the Atlantic. *Philos. Trans. R. Soc. London, Ser. B* **291,** 45–76.

Oliver, P. G., and Allen, J. A. (1980b). The functional and adaptive morphology of the deep sea species of the family Limopsidae (Bivalvia: Arcoida) from the Atlantic. *Philos. Trans. R. Soc. London, Ser. B* **291,** 77–125.

Park, K., Williams, W. T., Prescott, J. M., and Wood, D. W. (1962). Amino acids in deep water. *Science* **138,** 531–532.

Paul, A. Z. (1976). Deep sea bottom photographs show that benthic organisms remove sediment cover from manganese nodules. *Nature (London)* **263,** 50–51.

Pease, D. C., and Kitching, J. A. (1939). The influence of hydrostatic pressure on ciliary frequency. *J. Cell. Comp. Physiol.* **14,** 135–142.

Perrier, E. (1899). "Les explorations sous-marines." Hachette, Paris.

Pickford, G. E. (1946). *Vampyroteuthis infernalis* Chun, an archaic dibranchiate cephalopod. I. Natural history and distribution. *Dana Rep.* **29,** 1–40.

Pickford, G. E. (1949). *Vampyroteuthis infernalis* Chun, an archaic dibranchiate cephalopod. II. External anatomy. *Dana Rep.* **32,** 1–132.

Reid, R. G. B., and Reid, A. M. (1974). The carnivorous habit of members of the septibranch genus *Cuspidaria* (Mollusca: Bivalvia). *Sarsia* **56,** 47–56.

Rex, M. A. (1973). Deep sea species diversity: decreased gastropod diversity at abyssal depths. *Science* **181,** 1051–1053.

Rex, M. A. (1976). Biological accommodation in the deep sea benthos: comparative evidence on the importance of predation and productivity. *Deep-Sea Res.* **23,** 975–987.

Rex, M. A. (1977). Zonation in deep sea gastropods: The importance of biological interactions to rates of zonation. *In* "Biology of Benthic Organisms" (B. F. Keegan, P. O. Ceidigh, and P. J. S. Boaden, eds.), pp. 521–530. Pergamon, Oxford.

Rex, M. A. (1979). r- and K-selection in a deep sea gastropod. *Sarsia* **64,** 29–32.

Rex, M. A., and Boss, K. J. (1973). Systematics and distribution of the deep sea gastropod *Epitonium (Eccliseogyra) nitidum. Nautilus* **87,** 93–98.

Rex, M. A., van Ummersen, C. A., and Turner, R. D. (1979). Reproductive pattern in the abyssal mud snail *Benthonella tenella* (Jeffreys). In "Reproductive Ecology of Marine Invertebrates" (S. E. Stancyk, ed.), pp. 173–188. Univ. of South Carolina Press, Columbia.

Rhoads, D. C. (1973). Rates of sediment reworking by *Yoldia limatula* in Buzzards Bay, Massachusetts and Long Island Sound. *J. Sediment. Petrol.* **33,** 723–727.

Rhoads, D. C., Lutz, R. A., Revelas, E. C., and Cerrato, R. M. (1981). Growth of bivalves at deep sea hydrothermal vents along the Galapagos Rift. *Science* **214,** 911–913.

Rice, A. L., and Collins, E. P. (1980). The face of the deep sea. *Br. J. Photogr.* 9 May, 449–451.

Rice, A. N., Aldred, R. G., Darlington, E., and Wild, R. A. (1982). The quantitative estimation of the deep sea megabenthos: a new approach to an old problem. *Oceanol. Acta* **5,** 63–72.

Richter, G., and Thorson, G. (1974). Pelagische Prosobranchier-Larven des Golfs von Neapel. *Ophelia* **13,** 109–195.

Riley, G. A. (1951). Oxygen, phosphate, and nitrate in the Atlantic Ocean. *Bull. Bingham Oceanogr. Collect.* **13,** 1–126.

Roper, C. F. E. (1969). Systematics and Zoogeography of the worldwide bathypelagic squid *Bathyteuthis* (Cephalopoda: Oegopsida). *Bull.—U.S. Natl. Mus.* **291,** 1–210.

Roper, C. F. E., and Brundage, W. L. (1972). Cirrate octopods with associated deep sea organisms: new biological data based on deep benthic photographs (Cephalopoda). *Smithson. Contrib. Zool.* **121,** 1–46.

Roper, C. F. E., and Young, R. E. (1975). Vertical distribution of pelagic cephalopods. *Smithson. Contrib. Zool.* **209,** 1–51.

Rowe, G. T. (1972). The exploration of submarine canyons and their benthic faunal assemblages. *Proc. R. Soc. Edinburgh, Sect. B: Biol.* **73,** 159–169.

Rowe, G. T., and Gardener, W. D. (1979). Sedimentation rates in the slope water of the northwest Atlantic Ocean measured directly with sediment traps. *J. Mar. Res.* **37,** 581–600.

Rowe, G. T., and Menzel, D. W. (1971). Quantitative benthic samples from the deep Gulf of Mexico with some comments on the measurement of deep-sea biomass. *Bull. Mar. Sci.* **21,** 556–566.

Rowe, G. T., and Staresinic, N. (1979). Sources of organic matter to the deep sea benthos. *Ambio Spec. Rep.* **6,** 19–23.

Rowe, G. T., Polloni, P. T., and Horner, S. G. (1974). Benthic biomass estimates from the northwestern Atlantic Ocean and the northern Gulf of Mexico. *Deep-Sea Res.* **21,** 641–650.

Rozenzweig, M. L., and Taylor, J. A. (1980). Speciation and diversity in Ordovician invertebrates: filling niches quickly and carefully. *Oikos* **35,** 236–243.

Sanders, H. L. (1968). Marine benthic diversity: a comparative study. *Am. Nat.* **102,** 243–282.

Sanders, H. L. (1969). Benthic marine diversity and the stability-time hypothesis. *Brookhaven Symp. Biol.* **22,** 71–81.

Sanders, H. L. (1977). Evolutionary ecology and deep sea benthos. *Acad. Nat. Sci. Philadelphia, Spec. Publ.* **12,** 223–234.

Sanders, H. L. (1979). Evolutionary ecology and life history patterns in the deep sea. *Sarsia* **64,** 1–7.

Sanders, H. L., and Allen, J. A. (1973). Studies on the deep sea Protobranchia (Bivalvia); prologue and the Pristiglomidae. *Bull. Mus. Comp. Zool.* **145,** 237–262.

Sanders, H. L., and Allen, J. A. (1977). Studies on the deep sea Protobranchia (Bivalvia); the family Tindariidae and the genus Pseudotindaria. *Bull. Mus. Comp. Zool.* **148,** 23–59.

Sanders, H. L., and Hessler, R. R. (1969). Ecology of deep sea benthos. *Science* **163,** 1419–1424.

Sanders, H. L., Hessler, R. R., and Hampson, G. R. (1965). An introduction to the study of deep sea faunal assemblages along the Gay Head–Bermuda transect. *Deep-Sea Res.* **12,** 845–867.

Saski, T., Okami, N., Oshiba, G., and Watanabe, S. (1962). Studies on suspended particles in deep water. *Sci. Pap. Inst. Phys. Chem. Res. (Jpn.)* **56,** 77–83.

Scheltema, R. S. (1971). Reproduction and population dynamics of some protobranch bivalves from the continental shelf, slope and abyss of the Northeastern United States. *Thalassia Jugosl.* **7,** 361–362.

Scheltema, R. S. (1972). Reproduction and dispersal of bottom dwelling deep sea invertebrates: a speculative summary. *In* "Barobiology and the Experimental Biology of the Deep Sea" (R. W. Brauer, ed.), pp. 58–66. Univ. of North Carolina Press, Chapel Hill.

Schmitz, W. J., and Hogg, N. G. (1978). Observations of energetic low frequency current fluctuations in the Charlie-Gibbs Fracture Zone. *J. Mar. Res.* **36,** 725–734.

Schwartzlose, R. A., and Isaacs, J. D. (1969). Transient circulation event near the deep sea floor. *Science* **165,** 889–891.

Silver, M. W., and Aldredge, A. L. (1981). Bathypelagic marine snow: deep sea algal and detrital community. *J. Mar. Res.* **39,** 501–530.

Silver, M. W., and Bruland, K. W. (1981). Differential feeding and fecal pellet composition of salps and pteropods, and the possible origin of the deep water flora and olive geen "cells." *Mar. Biol. (Berlin)* **62,** 263–272.

Slobodkin, L. B., and Sanders, H. L. (1969). On the contribution of environmental predictability to species diversity. *Brookhaven Symp. Biol.* **22,** 82–93.

Smith, E. A. (1885). Report on the Lamellibranchiata collected by *HMS Challenger* during the years 1873–76. *Rep. Sci. Res. 'Challenger'* **13,** 1–341.

Smith, K. L., and Teal, J. M. (1973). Deep-sea benthic community respiration, an *in situ* study at 1850 m. *Science* **179,** 282–283.

Sokolova, M. N. (1965). On the irregularity in the distribution of nutrition groupings of deep sea benthos in connection with irregularity of sedimentation. *Okeanologiya (Moscow)* **3,** 498–506.

Soot-Ryen, T. (1958). Pelecypods from East Greenland. *Skr., Nor. Polarinst.* **113,** 1–32.

Soot-Ryen, T. (1966). Revision of the pelecypods from the Michael Sars North Atlantic Deep Sea Expedition 1910, with notes on the family Verticordiidae and other interesting species. *Sarsia* **24,** 1–31.

Spiess, F. N., MacDonald, K. C., Atwater, T., Ballard, R., Carranza, A., Cordoba, D., Cox, C., Diaz Garcia, V. M., Francheteau, J., Guerro, J., Hawkins, J., Haymon, R., Hessler, R., Juteau, T., Kastner, M., Larson, R., Luyendyk, B., Macdougall, J. D., Miller, S., Normak, W., Orcutt, J., and Rangin, C. (1980). East Pacific Rise: Hot springs and geophysical experiments. *Science* **207,** 1421–1433.

Stanley, S. M. (1976). Stability of species in geologic time. *Science* **192,** 267–268.

Thiel, H. (1972). Die Bedeutung der Meiofauna in küstenfernen benthischen Lebensgemeinschaften verschiedener geographischer Regionen. *Verh. Dtsch. Zool. Ges.* **65,** 37–42.

Thiel, H. (1979a). First quantitative data on Red Sea deep benthos. *Mar. Ecol.: Prog. Ser.* **1,** 347–350.

Thiel, H. (1979b). Structural aspects of the deep sea benthos. *Ambio Spec. Rep.* No. 6, 25–31.

Thorson, G. (1950). Reproductive and larval ecology of marine bottom invertebrates. *Biol. Rev. Cambridge Philos. Soc.* **25,** 1–45.

Tipper, R. (1968). Ecological aspects of two wood-boring molluscs from the continental terrace off Oregon. Dept. of Oceanography, School of Science, Oregon State University. (Unpublished doctoral dissertation.)

Turekian, K. K., Cochran, J. K., Kharkar, D. P., Cerreto, R. M., Vaisnys, J. R., Sanders, H. L., Grassle, J. F., and Allen, J. A. (1975). Slow growth rate of a deep sea clam determined by 228 Ra chronology. *Proc. Natl. Acad. Sci. U.S.A.* **72,** 2829–2939.

Turekian, K. K., Cochran, J. K., and Nozaki, Y. (1979). Growth rate of clam from Galapagos Rise hot spring field using natural radionuclide ratios. *Nature (London)* **280,** 385–387.

Turner, R. D. (1965). Some results of deep water testing. *Bull. Am. Malacol. Union* pp. 9–11.

Turner, R. D. (1966). Implication of recent research in the Teredinidae. *In* "Holz und Organismen" (G. Becker and W. Liese, eds.), pp. 437–446. Duncker & Humbolt, Berlin.

Turner, R. D. (1973). Wood-boring bivalves, opportunistic species in the deep sea. *Science* **180,** 1377–1379.

Turner, R. D. (1977). Wood, mollusks and deep sea food chains. *Bull. Am. Malacol. Union* pp. 13–19.

Turner, R. D. (1981). "Wood islands" and "thermal vents" as centers of diverse communities in the deep sea. *Biologiya, Morya* **1,** 3–10.

Vinogradov, M. E. (1968). "Vertical distribution of oceanic zooplankton." Nauka, Moscow.

Vinogradova, N. G. (1979). The geographical distribution of the abyssal and hadal (ultra-abyssal) fauna in relation to the vertical zonation in the ocean. *Sarsia* **64,** 41–50.

Voss, G. L. (1967). The biology and bathymetric distribution of deep sea cephalopods. *Stud. Trop. Oceanogr.* **5,** 511–535.

Voss, N. (1969). A monograph of the Cephalopoda of the North Atlantic. The family Histioteuthlidae. *Bull. Mar. Sci.* **19,** 705–867.

Warén, A., and Carney, R. S. (1981). *Ophiolamia armigeri* gen. et sp. n. (Mollusca, Prosobranchia) parasitic on the abyssal ophiuroid *Ophiomusium armigerum. Sarsia* **66,** 183–193.

Warén, A., and Sibuet, M. (1981). *Ophieulima* (Mollusca, Prosobranchia), a new genus of ophiuroid parasites. *Sarsia* **66,** 103–107.

Webster, F. (1969). Vertical profiles of horizontal ocean currents. *Deep-Sea Res.* **16,** 85–98.

Wiebe, P. H., Boyd, S. H., and Winget, C. (1976). Particulate matter sinking to the deep sea floor at 2000 m in the Tongue of the Ocean, Bahamas, with a description of a new sedimentation trap. *J. Mar. Res.* **34,** 341–354.

Wikander, P. B. (1980). Biometry and behaviour in *Abra nitida* (Müller) and *A. longicallus* (Scacchi) (Bivalvia: Tellinacea). *Sarsia* **65,** 255–268.

Wirsen, C. O., and Jannasch, H. W. (1975). Activity of marine psychrophilic bacteria at elevated hydrostatic pressure and low temperatures. *Mar. Biol. (Berlin)* **31,** 201–208.

Wiseman, J. D. H., and Hendey, N. I. (1953). The significance and diatom content of a deep sea floor sample from the neighbourhood of the greatest oceanic depth. *Deep-Sea Res.* **1,** 47–59.

Wolff, T. (1960). The hadal community, an introduction. *Deep-Sea Res.* **6,** 95–124.

Wolff, T. (1970). The concept of the hadal or ultra abyssal fauna. *Deep-Sea Res.* **17,** 983–1003.

Wolff, T. (1976). Utilization of seagrass in the deep sea. *Aquat. Bot.* **2,** 161–174.

Wolff, T. (1979). Macrofaunal utilization of plant remains in the deep sea. *Sarsia* **64,** 117–136.

Wood, E. J. F. (1956). Diatoms in the ocean deeps. *Pac. Sci.* **10,** 377–381.

Wood, E. J. F. (1964). Studies in microbial ecology of the Australisian region. *Nova Hedwigia* **8,** 5–54, 453–569.

Wood, E. J. F. (1966). Plants of the deep oceans. *Z. Allg. Mikrobiol.* **6,** 177–179.

Yonge, C. M. (1952). The mantle cavity in *Siphonaria alternata* Say. *Proc. Malacol. Soc. London* **29,** 190–199.

Zenkevitch, L. A., and Birstein, Ya.A. (1956). Studies of the deep water fauna and related problems. *Deep-Sea Res.* **4,** 54–64.

Zimmerman, A. M., and Zimmerman, S. B. (1972). Commentary on high pressure effects on cellular systems. *In* "Barobiology and the Experimental Biology of the Deep Sea" (R. W. Brauer, ed.), pp. 179–210. Univ. of North Carolina Press, Chapel Hill.

3

Mangrove Bivalves

BRIAN MORTON

Department of Zoology
University of Hong Kong
Hong Kong

I. Introduction

Mangroves are found in shallow, brackish-water estuaries throughout the tropics, their distribution rarely extending north or south of latitudes 25°N and S.

The Mollusca, Vol. 6
Ecology

ISBN 0-12-751406-6

Generally, two major regions are recognized: the Indo-Pacific, extending from East Africa eastward, and the Occidental, encompassing the tropical Caribbean and West Africa. As with the hermatypic corals, greater species diversity is found in the Indo-Pacific. Abel (1926) recognized four mangrove plants from the Caribbean belonging to the genera *Avicennia, Rhizophora, Laguncularia,* and *Conocarpus.* The same author recognized 23 species from the Indo-Pacific, although Van Steenis (1962) increased this to 43 (with 10 species from the Caribbean).

Mangroves are high zoned, typically occupying the upper half of the eulittoral and dominating the supralittoral fringe. They grow best in soft mud, and these two aspects alone partially explain the lack of data on mangrove bivalves; the Bivalvia are in general best adapted to lower tidal levels and to firmer deposits. *Lasaea rubra,* for example, is one of the few bivalves capable of colonizing the high intertidal almost worldwide (Morton, 1960), although, as will be seen, the *Spartina* marsh associate *Geukensia demissa* has similar adaptations to a high-zoned life (Lent, 1969), as do species of *Polymesoda* in the Indo-Pacific (Morton, 1976b). Colonization of soft deposits by a filter feeder enhances the very real problems of gill clogging and sediment removal.

Thus, little is known of mangrove bivalves, especially those few species which appear to be endemic components of the mangrove forest. As will be seen, large numbers of bivalves have been recorded from the seaward fringe of the mangroves, and their status as true mangrove associates is debatable. Apart from the obvious difficulties of working in a mangrove forest, numerous authors (Warner, 1969; Sasekumar, 1974; Murty and Balaparameswara Rao, 1977) ignored the bivalves in favor of the more active and therefore more conspicuous mangrove associates (the Gastropoda) (Robertson, 1960; Brown, 1971; Vermeij, 1974). Much of the literature on mangrove bivalves is thus scattered throughout obscure journals or occurs as isolated records in broader papers dealing in general terms with the mangrove community (Rodriguez, 1959, 1963). Coomans (1969) has drawn attention to the inherent interest in mangrove molluscs and Bouchet (1977) has provided data on West African mangrove molluscs, drawing on the data by Binder (1968) on Ivory Coast mangroves and by Coomans (1969) on the Caribbean fauna to compare the molluscan faunas of various mangrove regions. However, even these authors emphasize the mangrove gastropods, although there are mangrove bivalves of some interest and occasionally, such as the mangrove oysters (especially *Crassostrea rhizophorae* in the Caribbean), of much wider economic potential.

This chapter has consolidated some of the mangrove bivalve literature to encourage greater attention to this group in broad surveys of the mangrove system, and also to promote special interest in a few animals which have remarkable adaptations to a difficult environment.

II. Is There a Specific Mangrove Fauna?

The above question, posed by Macnae (1963), is central to any discussion of a faunistic component of the mangrove ecosystem. Macnae (1963) points out that major faunal elements of the mangroves, e.g., *Saccostrea cucullata, Littorina scabra,* and *Balanus amphitrite,* can and do occur outside mangrove areas; for example, on rocks at an equivalent tidal level. This applies to many other important mangrove associates including fiddler crabs, sesarmid crabs, shrimps, and a wide variety of snails. Conversely, Macnae gives the example of *Holothuria parva* in South Africa, which is found only in the mud among the roots of *Rhizophora.*

Among the bivalves, especially representatives of the more common and noticeable groups (e.g., in the Indo-Pacific, *S. cucullata, Isognomon, Pteria* and *Brachidontes*), most can and do occur elsewhere. Thus, Taylor (1976), describing the rocky shore fauna of the Aldabra atoll, shows that *S. cucullata* and its nestling associate *B. variabilis* are exceptionally common and the specific prey of a range of nonmangrove thaid gastropods.

In southwest Trinidad, *Mytella falcata* occurs on the surface of mudflats, and in the Gulf of Paria it attaches to the intertidal zone of buoys and pilings. *M. guyanensis* either attaches to stones or lies nearly buried in muddy sand (Soot-Ryen, 1955). In the west coast mangroves of Trinidad, two adjacent but totally different habitats are occupied by these species; *M. guyanensis* is buried in humic mud below mangrove trees on the banks of the estuary channels and lagoons, whereas *M. falcata* is attached to rhizophores of *Rhizophora mangle* (Bacon, 1975).

Gerlach (1958) found *M. guyanensis* in the swamps of Cananeia, Brazil, but Rios (1970) records this species and *M. falcata* from rocky bottoms. *M. falcata* occurs in the splash zone on rock or sandstone, but not mangrove, at Maracaibo, Venezuela (Rodriguez, 1963). Altena (1969) recorded *M. falcata* in Surinam on mangrove roots, driftwood, and stone walls of drainage ditches when these were free of silt.

Conversely, as will be seen later, a list of mangrove associates does include species that probably would not be associated with mangroves under typical conditions. *Septifer bilocularis,* which attaches to the roots, is the best example; in the Indo-Pacific this is the common bivalve of surf-exposed rocky shores (Loi, 1967; Taylor, 1971). It may occur in mangroves at some localities, e.g., Guam (Stojkovich, 1977), because of more constant marine conditions. Also, a young *S. bilocularis* is extremely difficult to distinguish, without opening the shell to reveal the myophore plate or septum, from species of *Brachidontes.* Similarly, *Asaphis deflorata* and *Caecella horsfieldi* were recorded from Southeast Asian

TABLE I

Important Families of Bivalves and Their Representative Species Recorded from Mangroves of Various Regions[a]

Family	Region					
	West coast of the Americas	Southern coast of North America	Caribbean and northern coast of South America	West Africa	East Africa	Indo-Pacific
Arcidae	*Anadara tuberculosa*	—	*Anadara notabilis*	*Anadara senilis*	*Arca gibba* *Arca obliquata*	*Anadara granosa*
Ostreidae	*Ostrea columbiensis* *Ostrea corteziensis*	*Crassostrea virginica*	*Crassostrea rhizophorae*	*Crassostrea gasar*	*Saccostrea cucullata*	*Saccostrea cucullata*
Mytilidae	*Mytella falcata* *Modiolus* spp.	*Brachidontes exustus* *Guekensia demissa*	*Brachidontes exustus* *Mytella falcata* *Modiolus* spp. *Mytella guyanensis*	*Brachidontes niger* *Modiolus* spp.	*Brachidontes variabilis* *Modiolus philippinarum*	*Brachidontes variabilis* *Modiolus metcalfei*
Isognomonidae	—	*Isognomon alatus*	*Isognomon alatus*	—	—	*Isognomon isognomon* *Isognomon ephippium*

Anomiidae	—	*Pododesmus rudis*	—	—	—	*Enigmonia aenigmatica*
Corbiculidae	*Polymesoda mexicana* *Polymesoda inflata*	*Polymesoda caroliniana*	*Polymesoda inflata* *Polymesoda floridiana*	—	—	*Geloina erosa* *Geloina proxima* *Geloina bengalensis*
Veneridae	—	*Anomalocardia cunimeris*	*Anomalocardia brasiliana*	*Callista floridella* *Dosinia isocardia*	*Dosinia hepatica*	*Gafrarium* spp.
Dreissenidae	*Mytilopsis adamsi*	*Mytilopsis leucophaeta*	*Mytilopsis sallei*	—	—	—
Mactridae	*Rangia mendica*	*Rangia cuneata* *Mulinia lateralis*	—	*Mactra diolensis* *Mactra glabrata*	—	—
Lucinidae	—	—	*Phacoides pectinatus*	*Loripes aberrans*	*Loripes clausus* *Lucina edentula*	—
Psammobiidae	—	—	*Asaphis deflorata*	*Tagelus angulata*	*Psammotellina capensis*	*Gari togata* *Elizia orbiculata*
Teredinidae	—	—	*Neoteredo reynei*	*Neoteredo reynei* *Teredo senegalensis* *Teredo adami* *Bankia bipennata*	*Teredo manni*	*Teredo manni*
Pholadidae	—	—	*Martesia striata*	—	—	*Martesia striata*

[a] Data taken from references listed in Tables II–VII.

mangroves by Gomez (1980b). This is not the typical habitat of these bivalves (Narchi, 1980); elsewhere they occur high zoned on cobble-gravel beaches. Given an understandable enthusiasm to locate as a wide a range of organisms as possible from the mangrove to bolster well-deserved efforts to protect this fast-diminishing habitat (Gomez, 1980b), it is clear that many bivalves can at least survive in a mangrove system, although their optimal niche may be elsewhere. Similarly, species that are often regarded as unmistakable mangrove associates can and do survive in estuarine environments where the trees have been cut or for some reason do not occur. Thus, at least relative to the bivalves, the mangrove fauna is complex, deriving its many components from other areas and possessing species that are not wholly dependent upon it for survival.

The question posed by Macnae thus has not yet been answered, but in the Indo-Pacific a number of genera are clear mangrove associates occurring, as far as is known, nowhere else. Macnae (1968) only briefly mentions four bivalves from the Indo-Pacific mangrove: *Enigmonia, Geloina, Laternula,* and *Glauconome*. Species of *Polymesoda* (*Geloina*) (Morton, 1976b,d) occur, infaunally buried, at the back of mangroves, typically inhabiting the banks of the small streams that drain them. *Enigmonia* (Yonge, 1957; Berry, 1975; Morton, 1976c)

TABLE II

Bivalves Recorded from the Mangroves of the Southern Shores of the United States

Family and species	References
Ostreidae	
Crassostrea virginica	Harry (1942); Odum and Heald (1972); Siung (1980)
Mytilidae	
Brachidontes exustus	Odum and Heald (1972, 1975)
Geukensia demissa	Harry (1942)
Isognomonidae	
Isognomon alatus	Siung (1980)
Corbiculidae	
Polymesoda caroliniana	Harry (1942); Tabb and Moore (1971); Duobinis and Hackney (1979)
Dreissenidae	
Mytilopsis leucophaeta	Odum and Heald (1972, 1975)
Veneridae	
Anomalocardia cunimeris	Breuer (1957); Odum and Heald (1972, 1975)
Mactridae	
Rangia flexuosa	Harry (1942)
Rangia cuneata	Harry (1942)
Mulinia lateralis	Harry (1942); Breuer (1957)

occurs byssally attached to the tree trunks or prop roots, as is typical of pioneer species. (As will be shown later, probably only a single species is present which, through colonization of slightly different microhabitats, occurs in a number of morphological forms.) *Laternula* and *Glauconome* are deep burrowers of the seaward fringe muds (Morton, 1973b; Owen, 1959). Much less information is available on the specificity of other bivalves to the mangroves of other regions, but a comparison of faunal lists shows that a number of groups are consistently recorded from most areas (Table I). This list excludes families with representatives in one system only, e.g., *Glauconome* [Glaucomyidae (Owen, 1959)] and *Laternula* [Laternulidae (Morton, 1973b)] in the Indo-Pacific. However, it shows that certain groups colonize essentially similar microhabitats within the mangrove system everywhere.

The mangrove bivalve fauna thus seems to include a number of elements:

1. Those species generally associated with tropical estuarine areas and thus with mangroves
2. Species capable of surviving in mangroves as a habitat at the limit of their ecological range
3. Species that have intimately evolved within a specific mangrove system

III. Mangrove Bivalve Communities

A. Southern Coast of North America

It is perhaps convenient to begin a discussion of mangrove bivalve communities by examining them at the limits of their range, where the plants themselves are diminutive (both in numbers and in size) and interact with temperate marsh grasses and sedges (Table II). The southern coast of the United States is such a zone of interaction; the dominant high-zoned plant, *Spartina,* forms mixed stands with *Avicennia* in south Texas. In Florida, *Juncus* interacts with *Rhizophora* and *Laguncularia* (Tabb and Moore, 1971). A number of bivalves occur in this transition zone, typically associating with the marsh grasses but also, in some areas, living within what can be broadly defined as a mangal.

The marshes are characterized by the high-zoned mytilid *Geukensia demissa.* This bivalve is highly eurytropic, tolerating temperatures of −22 to 40°C (Kanwisher, 1955; Lent, 1969) and salinities of 5 to 75‰ (Wells, 1961; Lent, 1969). This allows *G. demissa* to occupy a very high intertidal habitat, where it is exposed to terrestrial conditions for as much as 83% of the time (Kuenzler, 1961; Bertness, 1980). As do so many marsh and mangrove bivalves, *Geukensia* can live outside the preferred habitat. When it occurs on sea walls and pilings the height : length ratio of the shell is altered (Seed, 1980). Lutz and Castagna (1980) have suggested that *Geukensia* can live for up to 23 years. Stiven and Kuenzler (1979) have shown that crowding increases the mortality of *G. demissa* but that intraspecific competition for food is an additional limiting process.

In deeper waters beyond the marsh occurs the brackish-water mactrid *Rangia cuneata* (Olsen, 1976), which seems (Peddicord, 1976) to be reoccupying its old (Pleistocene) range (i.e., from New Jersey to Mexico). This has been described by Hopkins and Andrews (1970).

Fairbanks (1963) correlated growth and population densities in *R. cuneata* with various physical and biological factors and attributed differences in abundance, size, and growth to variations in sediment property. Growth was faster in sandy areas, clams attaining a length of 15 to 20 mm in the first year and adding 5–10 mm in the second and 4–5 mm in the third year. In muddy areas, the clams were smaller but more numerous. These conclusions were supported by Tenore et al. (1968) but not by Godwin (1968), who believed that salinity was the most important factor limiting the occurrence of *R. cuneata* although bottom type influenced distribution and density. According to Wolfe and Petteway (1968), *R. cuneata* can live for 10 years.

The leaves of marsh grasses cannot support oysters, and *Crassostrea virginica,* typically, is low zoned, attached to stones and empty shells at the marsh edge, often forming "reefs" (Manzi et al., 1977). Nestling among the oysters are the epibyssate mytilid *Brachidontes exustus* (Seed, 1980) and the laterally flattened *Isognomon alatus.* At the mudline, the roots and stems of *Spartina* may be colonized by the byssally attached dreissenid *Mytilopsis leucophaeta.* The marine offshore muds are characterized by three low salinity-adapted bivalves, *Mulinia lateralis, Anomalocardia cunimeris,* and *Polymesoda caroliniana* (Harry, 1942; Olsen, 1976).

B. The Caribbean and South America

The Caribbean mangrove includes only four species (Abel, 1926), the typical pioneer being *Rhizophora apiculata.* The prop roots and stems of this plant are colonized by the oyster *Crassostrea rhizophorae* (Table III). In Florida, this species co-occurs with *C. virginica. C. rhizophorae* is being extensively cultivated throughout the Caribbean and the shores of South America (Nikolic and Melendez, 1968; Bacon, 1970; Angell, 1973). With the oyster a number of nestling species can co-occur including *Barbatia cancellaria* (Perez, 1974) and *Lasaea bermudensis* (Robertson, 1960), a number of species of *Pinctada* and *Pteria,* and a number of species of *Brachidontes* including *B. exustus* which was noted earlier from North American shores. *Musculus lateralis* occurs in the tests of the ascidian *Bothryllus,* densely attached to mangrove roots (Coomans, 1969). *Isognomon radiata* and *I. alatus* (also recorded earlier from Florida) (Siung,

TABLE III

Bivalves Recorded from the Mangroves of the Caribbean and Northern Shores of South America

Family and species	References
Arcidae	
Anadara notabilis	Coomans (1969); Bouchet (1977)
Arca umbonata	Perez (1974)
Barbatia cancellaria	Perez (1974)
Anadara sp.	Castaing et al. (1980)
Isognomonidae	
Isognomon alatus	Robertson (1960); Coomans (1969); Kolehmainen et al. (1973); Bouchet (1977); Perez (1974); Siung (1980); Sutherland (1980)
Isognomon radiata	Coomans (1969)
Ostreidae	
Crassostrea rhizophorae	Robertson (1963); Coomans (1969); Bacon (1970); Kolehmainen et al. (1973); Bouchet (1977); Sutherland (1980); Wedler (1980)
Ostrea equestris	Kolehmainen et al. (1973)
Ostrea frons	Coomans (1969)
Anomiidae	
Pododesmus rudis	Coomans (1969); Kolehmainen et al. (1973); Perez (1974); Bouchet (1977)
Pteriidae	
Pteria colymbus	Coomans (1969)
Pinctada radiata	Coomans (1969)
Pinctada imbricata	Perez (1974)
Mytilidae	
Brachidontes exustus	Coomans (1969); Kolehmainen et al. (1973); Bouchet (1977)
Brachidontes recurvus	Coomans (1969); Perez (1974)
Brachidontes spp.	Rodriguez (1959); Perez (1974)
Brachidontes citrinus	Perez (1974)
Modiolus modiolus	Rodriguez (1959)
Modiolus americanus	Rodriguez (1963); Coomans (1969); Bacon (1970); Bouchet (1977)
Musculus lateralis in tests of *Bothryllus*	Coomans (1969); Kolehmainen et al. (1973)
Mytella guyanensis	Gerlach (1958); Altena (1969); Bacon (1975)
Mytella falcata	Bacon (1975)
Chamidae	
Chama macerophylla	Rodriguez (1959); Coomans (1969); Perez (1974)

(Continued)

TABLE III *Continued*

Family and species	References
Chama congregata	Coomans (1969); Perez (1974)
Pseudochama radians	Perez (1974)
Limidae	
Lima scabra	Perez (1974)
Veneridae	
Anomalocardia brasiliana	Coomans (1969); Bouchet (1977)
Chione cancellata	Coomans (1969); Bouchet (1977)
Corbiculidae	
Polymesoda inflata	Castaing et al. (1980)
Dreissenidae	
Mytilopsis sallei	Robertson (1963); Escarbassiere and Almeida (1976)
Lucinidae	
Phacoides pectinatus	Bacon (1970)
Erycinidae	
Lasaea bermudensis	Robertson (1960)
Psammobiidae	
Asaphis deflorata	Coomans (1969); Bouchet (1977)
Cardiidae	
Trachycardium muricatum	Coomans (1969)
Pholadidae	
Martesia striata	Bacon (1970)
Teredinidae	
Teredo spp.	Coomans (1969); Bouchet (1977)
Neoteredo reynei	Turner (1966); Turner and Johnson (1971)

1980) occur attached in clusters in the forks of prop roots close to the mud line. The solitary anomiid *Pododesmus rudis* is byssally attached to the tree trunks, as are several species of Chamidae (Coomans, 1969). *Lima scabra* has been reported from mangroves by Perez (1974), but limids are more typical of clean, deeper waters and have not been recorded anywhere else from mangrove areas, although they do occur in crevices, usually building a loosely woven byssal nest. The nest-building habit is also seen in some species of the genus *Mytella*. *M. guyanensis* (Bacon, 1975) occupys a cocoon-like nest in mud at the base of the mangrove roots and *M. falcata* attaches to the prop roots above the mudline. Robertson (1963) and Escarbassiere and Almeida (1976) record *Mytilopsis sallei* from the lagoons of Venezuela. In Mexico the same species occurs attached to stones and within algal mats, beyond the limits of the mangroves (Marelli and Berrend, 1978).

A very wide array of mud-dwelling bivalves has been reported from the Caribbean mangrove; most important are members of the Arcidae (*Anadara notabilis*), Corbiculidae (*Polymesoda inflata*), Veneridae (*Anomalocardia brasiliana* and *Chione cancellata*), Lucinidae (*Phacoides pectinatus*), Psammobiidae (*Asaphis deflorata*), and Cardiidae (*Trachycardium muricatum*) (Coomans, 1969). Narchi (1972, 1976) has studied the morphology and sexual cycle of *A. brasiliana;* two periods of gametogenesis occur, in spring and in autumn.

The tree roots are bored by the hardwood pholad *Martesia striata* and the softwood teredinids *Teredo* spp. (Coomans, 1969; Bacon, 1970), possibly *Neoteredo reynei* (Turner and Johnson, 1971).

C. The Gulf of California

In the Gulf of California and the other regions of the western coast of the Americas where stands of mangroves occur, the bivalves have some degree of commonality with those of the Caribbean (Table IV). Parker (1963, 1964) has investigated the mangrove fauna of the Gulf of California and has shown that of the 19 species (of all invertebrates) recorded from this area, 11 have their counterparts in the lagoons of the Gulf of Mexico, although *Anadara tuberculosa* and the large corbiculids are typical of the Gulf of California alone. Groups of essentially similar species do occur, however, most notably the oysters (*Ostrea columbiensis* and *O. corteziensis*). Among the oysters occur small nestling modiolines and *Mytilopsis adamsii*. There is apparently a notable lack, however, of the thin, plate-like isognomonids and pteriids seen in the Caribbean.

The infauna seems to be dominated by the Corbiculidae, with possibly as many as nine species of *Polymesoda* (Parker, 1963, 1964; Castaing et al., 1980), *Cyrenoida panamensis* (Cyrenoididae), *Rangia mendica* (Mactridae), *Corbula inflata* (Corbulidae), and the burrowing mytilid *Mytella falcata,* also found in the Caribbean (Bacon, 1975). Mangrove cockles (*Anadara*) are common along the Pacific coast of the Americas from Baja California to Peru (the Panamic Zoogeographic Province). On the coast of Colombia the common species is *Anadara tuberculosa,* although two other species, *A. similis* and *A. multicostata,* are also mangrove associates and were recorded from estuaries of Costa Rica and Nicaragua by Ellis (1968). From the intertidal muds beyond the estuaries has been recorded *A. grandis* (Squires et al., 1975). Most recently, Contreras and Cantera (1978) have described the molluscan fauna of Colombian mangroves and conclude that *Ostrea corteziensis, O. colombiensis, Anadara tuberculosa, A. similis, Protothaca asperrima, Polymesoda inflata* and *Mytella guyanensis* are true mangrove associates although other species occur in the muds beyond.

A. tuberculosa occurs among mangrove roots in shaded areas, mostly buried in the soft mud but sometimes attached to the roots by a byssus (Squires et al., 1975). This species seems to be slow-growing, averaging about 1 mm/month up to a maximum length of approximately 110 mm. The species matures at a length

TABLE IV

Bivalves Recorded from the Mangroves of the Gulf of California and Other Sites on the Tropical Western Coast of the Americas

Family and species	References
Arcidae	
Anadara tuberculosa	Parker (1963, 1964); Squires et al. (1975); Baquiero (1980)
Anadara sp.	Castaing et al. (1980)
Anadara similis	Squires et al. (1975)
Anadara multicostata	Squires et al. (1975)
Anadara grandis	Squires et al. (1975)
Mytilidae	
Mytella falcata	Parker (1963, 1964)
Mytella guyanensis	Contreras and Cantera (1978)
Modiolus sp.	Castaing et al. (1980)
Ostreidae	
Ostrea columbiensis	Parker (1963, 1964); Contreras and Cantera (1978)
Ostrea corteziensis	Parker (1963, 1964); Contreras and Cantera (1978)
Corbiculidae	
Polymesoda mexicana	Parker (1963, 1964)
Polymesoda inflata	Contreras and Cantera (1978); Castaing et al. (1980)
Polymesoda (7 spp.)	Parker (1963, 1964)
Cyrenoididae	
Cyrenoida panamensis	Parker (1963, 1964)
Dreissenidae	
Mytilopsis adamsi	Parker (1963, 1964)
Mactridae	
Rangia mendica	Parker (1963, 1964)
Corbulidae	
Corbula inflata	Parker (1963, 1964)
Veneridae	
Protothaca asperrima	Contreras and Cantera (1978)

of 32 and 36 mm for females and males, respectively. The average density was one cockle/m^2.

D. West Africa

Little research has been undertaken on the mangroves of West Africa, but Binder (1968) and Bouchet (1977) have researched the molluscs of the Ivory Coast and Senegal mangroves, respectively (Table V). As elsewhere, the trees

TABLE V

Bivalves Recorded from the Mangroves of West Africa[a]

Family and species	References
Nuculanidae	
Leda bicuspidata	Binder (1968)
Arcidae	
Anadara senilis	Yoloye (1974)
Ostreiidae	
Crassostrea gasar	Sandison (1967); Hunter (1969); Ajana (1979, 1980)
Mytilidae	
Brachyodontes (sic) *niger*	Binder (1968)
Modiolus nigeriensis	Binder (1968)
Modiolus elegans	Binder (1968)
Veneridae	
Callista floridella	Binder (1968)
Dosinia isocardia	Binder (1968)
Mactridae	
Mactra diolensis	Binder (1968)
Mactra glabrata	Binder (1968)
Psammobiidae	
Tagelus angulata	Binder (1968)
Lucinidae	
Loripes aberrans	Binder (1968)
Tellinidae	
Tellina nymphalis	Binder (1968)
Tellina distorta	Binder (1968)
Semelidae	
Abra pilsbryi	Binder (1968)
Ungulinidae	
Diplodonta globosa	Binder (1968)
Diplodonta diaphana	Binder (1968)
Condylocardiidae	
Cuna gambiensis	Binder (1968)
Soleniidae	
Solen guineensis	Binder (1968)
Teredinidae	
Neoteredo reynei	Turner (1966)
Teredo petitii (a variety of *T. senegalensis*)	Turner (1966)
Teredo adami	Turner (1966)
Bankia bagidaensis (probably = young *B. bipennata*)	Turner (1966)

[a] After Bouchet (1977), with additions as indicated.

are densely colonized by an oyster; in this area it is *Crassostrea gasar* [also reported by Hunter (1969) from Sierra Leone]. Nestling species are apparently few, with *Brachyodontes* (=*Brachidontes*) *niger* and two species of *Modiolus*. As on the West Coast of North America, no species of Isognomonidae or Pteriidae occur. The paucity of epifaunal nestlers in the West African mangroves (as in western America) is countered by the great diversity of burrowing species. As in other mangrove areas, the Arcidae are important (*Anadara senilis*). This is the only mangrove from which a protobranch, the nuculanid *Leda bicuspidata*, has been recorded (Binder, 1968). Nine families of burrowing lamellibranchs have been recorded: the Veneridae, Mactridae, Psammobiidae, Lucinidae, Tellinidae, Semelidae, Ungulinidae, Condylocardiidae, and Solenidae. None of the representative species of these families is common to the Caribbean despite the floral similarity.

The West African mangrove shipworms are better known, and Bouchet (1977) recorded three species: *Teredo petitii, T. adami,* and *Bankia bagidaensis*. The previous work of Turner (1966) adds *Neoteredo reynei* to this list and suggests that the first and last of the previous three were misidentified and should be recorded as *T. senegalensis* and *B. bipennata*.

E. East and South Africa

The mangroves of the eastern and southern coasts of Africa are better known (Macnae, 1963; Macnae and Kalk, 1962), but without a correspondingly greater knowledge of the associated fauna, including the molluscs (Table VI). Macnae (1968), following Ekman (1953), acknowledges that the Indo-West Pacific region, including East and South Africa, constitutes a unit. Macnae (1968), however, divides this unit into eight subzones, only one of which (East and South Africa) is treated separately here. This is because throughout the Indo-Pacific, such divisions are generally recognizable by their endemic fauna. Nevertheless, there is a high degree of floral and faunal commonality, especially with regard to the mangroves. Many Indo-Pacific mangrove bivalves do not occur on East African shores (Tables VI, VII), and for this reason alone this area is treated separately within the framework of the Indo-Pacific. Four species of oysters have been recorded (Branch and Grindley, 1979) from East and South Africa, but *Saccostrea cucullata* is of most interest and assumes much more importance in the Indo-Pacific mangroves. The nestling species include an array of mytilids including *Brachidontes variabilis, Perna perna,* and *Septifer bilocularis* (Day, 1974; Branch and Grindley, 1979). The same family has a number of infaunal representatives including *Musculus virgiliae, Arcuatula (Lamya) capensis,* and *Modiolus philippinarum* (which also occurs in the Indo-Pacific). Two species of the Arcidae occur, *Arca gibba* and *A. obliquata*. Other burrowers include species of the Psammobiidae, Tellinidae, Veneridae, Solenidae, and Lucinidae.

TABLE VI

Bivalves Recorded from the Mangroves of South and East Africa

Family and species	References
Arcidae	
Arca gibba	Branch and Grindley (1979)
Arca obliquata	Branch and Grindley (1979)
Ostreiidae	
Ostrea algoensis	Branch and Grindley (1979)
Ostrea cf. *ritrefacta*	Day (1974)
Saccostrea cucullata	Macnae and Kalk (1962); Macnae (1963); Day (1974); Branch and Grindley (1979)
Saccostrea margaritacea	Branch and Grindley (1979)
Mytilidae	
Brachidontes variabilis	Branch and Grindley (1979)
Musculus virgiliae	Branch and Grindley (1979)
Perna perna	Day (1974); Branch and Grindley (1979)
Septifer bilocularis	Day (1974); Branch and Grindley (1979)
Arcuatula (Lamya) capensis	Day (1974); Branch and Grindley (1979)
Modiolus phillipinarum	Day (1974)
Psammobiidae	
Psammotellina capensis	Macnae (1963); Branch and Grindley (1979)
Tellinidae	
Tellina gilchristi	Branch and Grindley (1979)
Macoma litoralis	Macnae (1963); Day (1974); Branch and Grindley (1979)
Macoma retroversa	Day (1974)
Macoma sp.	Macnae (1963)
Veneridae	
Dosinia hepatica	Macnae (1963); Day (1974); Branch and Grindley (1979)
Eumarcia paupercula	Day (1974); Branch and Grindley (1979)
Solenidae	
Solen corneus	Day (1974); Branch and Grindley (1979)
Lucinidae	
Loripes clausus	Macnae (1963); Branch and Grindley (1979)
Lucina edentula	Macnae (1963); Branch and Grindley (1979)
Thyasiridae	
Cryptodon eutornus	Macnae (1963)
Teredinidae	
Teredo (= *Dicyathifer*) cf. *manni*	Day (1974)

TABLE VII

Bivalves Recorded from the Mangroves of the Indo-Pacific

Family and species	References
Arcidae	
Anadara granosa	Purchon (1956); Pathansali and Soong (1958); Lim (1963); Coomans (1969); Berry (1972); Bouchet (1977); Anonymous (1980); Nateewathna and Tantichodok (1980); Piyakarnchana (1980); Sasekumar (1980); Gomez (1980b); Wells and Slack-Smith (1981)
Anadara nodifera	Frith et al. (1976); Gomez (1980b)
Anadara antiquata	Taylor (1971); Gomez (1980a,b); Soegiarto (1980)
Anadara auriculata	Anonymous (1980); Gomez (1980b)
Anadara cuneata	Anonymous (1980); Gomez (1980b)
Striarca lacerata	Anonymous (1980); Gomez (1980b)
Striarca sculptilis	Anonymous (1980); Gomez (1980b)
Striarca tenebrosa	Anonymous (1980); Gomez (1980b)
Barbatia fusca	Berry (1963); Anonymous (1980); Gomez (1980b)
Barbatia sp.	Gomez (1980a,b)
Striarca pectunculiformes	Anonymous (1980); Gomez (1980b); Nateewathna and Tantichodok (1980)
Ostreiidae	
Saccostrea cucullata	Taylor (1971); Sasekumar (1974; 1980); Berry (1975); Frith et al. (1976); Stojkovich (1977); Gomez (1980a,b); Nateewathna and Tantichodok (1980); Pinto and Wignarajah (1980); Piyakarnchana (1980); Soegiarto (1980)
Crassostrea parasitica	Coomans (1969); Bouchet (1977)
Crassostrea tuberculata	Gomez (1980a)
Crassostrea lugubrius	Gomez (1980a)
Crassostrea sp.	Anonymous (1980)
12 additional oyster species	Gomez (1980b)
Isognomonidae	
Isognomon isognomon	Lim (1963); Coomans (1969); Bouchet (1977); Anonymous (1980); Gomez (1980b)
Isognomon ephippium	Berry (1963, 1975); Frith et al. (1976); Anonymous (1980); Gomez (1980a,b); Nateewathna and Tantichodok (1980); Piyakarnchana (1980)

TABLE VII *Continued*

Family and species	References
Isognomon cf. *vitrea*	Wells and Slack-Smith (1981)
Isognomon dentifer	Taylor (1971)
Anomiidae	
Enigmonia aenigmatica	Iredale (1939); Frith et al. (1976); Morton (1976b); Anonymous (1980); Gomez (1980b); Sasekumar (1980); Soegiarto (1980); Wells and Slack-Smith (1981)
Enigmonia rosea	Berry (1963, 1975); Lim (1963); Coomans (1969); Bouchet (1977); Anonymous (1980); Gomez (1980b)
Mytilidae	
Brachidontes variabilis	Taylor (1971); Berry (1975); Anonymous (1980); Gomez (1980b); Pinto and Wignarajah (1980); Sasekumar (1980)
Brachidontes rostratus	Frith et al. (1976); Piyakarnchana (1980)
Brachidontes spp.	Sasekumar (1974); Gomez (1980b)
Modiolus philippinarum	Anonymous (1980); Gomez (1980b)
Modiolus spp.	Lim (1963); Coomans (1969); Bouchet (1977)
Septifer spp.	Anonymous (1980); Gomez (1980b)
Septifer bilocularis	Stojkovich (1977); Gomez (1980a,b)
Modiolus albicostatus	Anonymous (1980); Gomez (1980b)
Modiolus aratus	Anonymous (1980); Gomez (1980b)
Modiolus auriculatus	Anonymous (1980); Gomez (1980b)
Modiolus metcalfei	Morton (1977a); Anonymous (1980); Gomez (1980b)
Modiolus senhausia	Anonymous (1980); Gomez (1980b)
Perna viridis	Anonymous (1980); Gomez (1980b)
Trapeziidae	
Trapezium sublaevigatum	Berry (1963); Anonymous (1980); Gomez (1980b); Piyakarnchana (1980)
Trapezium angulatum	Frith et al. (1976); Gomez (1980b)
Corbiculidae	
Geloina ceylonica (= *G. erosa*)	Lim (1963); Piyakarnchana (1980); Morton (in manuscript)
Geloina coaxans (= *G. erosa*)	Gomez (1980b); Soegiarto (1980); Morton (in manuscript)
Geloina expansa	Gomez (1980b); Soegiarto (1980)
Geloina erosa	Frith et al. (1976); Morton (1976d); Anonymous (1980); Gomez (1980b)
Geloina proxima (= *G. expansa*)	Morton (1976b); Morton (in manuscript)
Geloina bengalensis	Morton (in manuscript)

(*Continued*)

TABLE VII *Continued*

Family and species	References
Tellinidae	
Tellina palatum	Taylor (1971)
Tellina capsoides	Frith et al. (1976); Gomez (1980b)
Tellina opalina	Frith et al. (1976); Gomez (1980b)
Tellina sp.	Piyakarnchana (1980)
Macoma sp.	Wells and Slack-Smith (1981)
Asaphidae	
Sanguinolaria sp.	Anonymous (1980)
Psammobiidae	
Gari togata	Lim (1963); Coomans (1969); Bouchet (1977)
Elizia orbiculata	Lim (1963)
Glauconomidae	
Glauconome rugosa	Owen (1959); Lim (1963); Anonymous (1980); Gomez (1980b)
Glauconome virens	Frith et al. (1976); Anonymous (1980); Gomez (1980b); Piyakarnchana (1980)
Glauconome cf. *cumingi*	Wells and Slack-Smith (1981)
Glauconome straminea	Anonymous (1980); Gomez (1980b)
Veneridae	
Gafrarium divaricatum	Anonymous (1980); Gomez (1980b); Piyakarnchana (1980)
Gafrarium tumidum	Taylor (1971); Frith et al. (1976); Stojkovich (1977); Gomez (1980a,b)
Gafrarium pectinatum	Taylor (1971)
Katelysia striata	Frith et al. (1976); Gomez (1980b); Piyakarnchana (1980)
Paphia luzonica	Lim (1963); Coomans (1969); Bouchet (1977); Anonymous (1980); Gomez (1980b)
Paphia hiantinus	Anonymous (1980); Gomez (1980b)
Meretrix meretrix	Lim (1963); Coomans (1969); Bouchet (1977); Anonymous (1980); Gomez (1980b)
Dosinia rustica	Lim (1963); Anonymous (1980); Gomez (1980b)
Dosinia sp.	Gomez (1980a,b)
Caecella horsfieldi	Frith et al. (1976); Anonymous (1980); Gomez (1980b)
Caecella sp.	Piyakarnchana (1980)
Marcia sp.	Gomez (1980a)
Anomalocardia squamosa	Anonymous (1980); Gomez (1980b)
Solenidae	
Solen delesserti	Lim (1963); Frith et al. (1976); Gomez (1980b); Piyakarnchana (1980)
Sinonovacula constricta	Frith et al. (1976); Gomez (1980b); Piyakarnchana (1980)

TABLE VII *Continued*

Family and species	References
Cultellidae	
Pharella acuminata	Owen (1959)
Laternulidae	
Laternula truncata	Lim (1963); Morton (1973b); Frith et al. (1976); Anonymous (1980); Gomez (1980b)
Lucinidae	
Ctena divergens	Taylor (1971)
Ungulinidae	
Diplodonta globosa	Frith et al. (1976); Gomez (1980b)
Diplodonta cumingii	Anonymous (1980); Piyakarnchana (1980)
Pholadidae	
Xylophaga spp.	Frith et al. (1976); Gomez (1980b); Piyakarnchana (1980)
Martesia striata	Ganapati and Lakshmana Rao (1959); Berry (1963); Anonymous (1980); Gomez (1980b); Pinto and Wignarajah (1980)
Barnea birmanica	Balasubramanian et al. (1979)
Teredinidae	
Teredo (=*Dicyathifer*) *manni*	Lim (1963); Coomans (1969); Rajagopal (1970); Bouchet (1977); Anonymous (1980); Gomez (1980b)
Teredo spp.	Berry (1963); Frith et al. (1976); Gomez (1980b); Pinto and Wignarajah (1980); Piyakarnchana (1980)
Teredo edax	Anonymous (1980); Gomez (1980b)
Teredo thoracites	Roonwal (1954a,b); Rajagopal (1970); Anonymous (1980); Gomez (1980b)
Bankia rochi	Anonymous (1980); Gomez (1980b)
Bankia campenellata	Ganapati and Lakshmana Rao (1959); Rajagopal (1970)
Bankia roonwali (= *B. rochi*)	Rajagopal (1970); Turner (1966)
Nausitoria lanceolata (= *N. dunlopei*)	Rajagopal (1970); Rajagopal and Daniel, (1973); Turner (1966)
Nausitoria sajnakhaliensis (= *Bankia nordi*)	Rajagopal (1970); Turner (1966)
Teredo (*Teredo*) *furcillatus* (= *T. furcifera*)	Ganapati and Lakshmana Rao (1959); Turner 1966
Teredo (*Dactyloteredo*) *juttingae* (=*Nototeredo edax*)	Ganapati and Lakshmana Rao (1959); Turner 1966
Bankia (*Bankiella*) *edmondsoni* (=*Bankia carinata*)	Ganapati and Lakshmana Rao (1959); Turner (1966)

The only wood borer recorded from South Africa is *Teredo* cf. *manni* (Day, 1974).

F. The Indo-Pacific

The most extensive and most comprehensively studied mangroves are those of the Indo-Pacific. The center of this faunistic region is located on the southern tip of the Malaysian Peninsula. The flora and fauna become steadily impoverished as one moves away from this center, improved only slightly by the addition of endemic species. The checklist of bivalves recorded from the mangroves of this vast region is correspondingly long (Table VII), although, as will be discussed later, the confused taxonomy of some groups has almost certainly led to the erection of too many species names, and some authors have clearly placed inhabitants of other habitats among the mangroves (Gomez, 1980b).

The most important mangrove bivalve is *Saccostrea cucullata,* earlier recorded from West Africa. Gomez (1980a,b) has attempted to show that 12 other species of oysters occur in the mangroves. This is unlikely, with the possible exception of *Crassostrea parasitica* (Coomans, 1969; Bouchet, 1977), which, like its counterpart, *C. frons,* in the Caribbean (Coomans, 1969), is apparently specialized for attaching to the thin twigs of gorgonians and occurs fortuitously in the mangroves.

Both *Isognomon isognomon* and *I. ephippium* typically occur between the clefts of prop roots; the saddle oyster, *Enigmonia aenigmatica* (also *E. rosea*), is a typical mangrove associate, often red in color and looking like a "branch scar" on the stems particularly of *Avicennia* and *Rhizophora.* Nestling among the oysters is an array of heteromyarian, byssally attached bivalves, including species of *Barbatia* and *Striarca* (Arcidae), *Brachidontes, Septifer, Modiolus,* and *Perna* (Mytilidae), and *Trapezium* (Trapeziidae).

The mud-dwelling members of the community, especially members of the Corbiculidae [*Polymesoda* (*Geloina*)], are of particular interest because of the latter's high-zoned position in the mangal and its unusual physiological adaptations to this environment. This bivalve seems to be the ecological equivalent of *Geukensia* in the marshes of the southern coast of North America. On the seaward fringe of the mangrove occur representatives of a wide range of families including the Arcidae (notably *Anadara granosa*), Tellinidae, Psammobiidae, Glauconomidae, Veneridae, Solenidae, Cultellidae, Solecurtidae, Laternulidae (notably *Laternula truncata*), and Ungulinidae.

The mangrove trees are bored by representatives of two families, the Pholadidae (broadly including *Xylophaga* spp., *Martesia striata,* and *Barnea birmanica*) and Teredinidae (notably *T. manni,* also recorded from East and South Africa), including four species of *Bankia* and *Nausitoria,* recorded only from the Sundarbans of the Ganges Delta by Roonwal (1954a,b) and Rajagopal (1970).

Again, Turner (1966) has suggested that many of these names are either incorrect or invalid, although it is clear that the greatest array of mangrove borers has been recorded from this area.

IV. Vertical Zonation on Mangrove Trees

The vertical zonation of fauna on mangroves has been described for Thailand (Frith et al., 1976), Singapore (Morton, 1976c), Malaysia (Berry, 1963), and Sri Lanka (Pinto and Wignarajah, 1980), and Morton (1976c) has pointed out the essential similarities between the fauna and flora of rocky shores throughout the world and those of mangroves. The zonation of plants and animals on a rocky shore is created by the interplay of the tides and other environmental factors such as the degree of exposure to wave action, the influence of fresh water, and so on. Stephenson and Stephenson (1949) divided shores vertically into three biological zones, which although they vary in extent are always characterized by particular groups of organisms:

1. The lichen and littorine zone
2. The barnacle and limpet zone
3. The algal (temperate climates) or coral (tropical climates) zone

The mangrove stems, however, also constitute a hard shore island in an otherwise soft mud sea and are colonized by more typical representatives of hard shores (Fig. 1). It is thus most convenient to look at mangrove tree organisms from the point of view of Stephenson and Stephenson.

The mangrove leaves are held out of water above extreme high water spring (tide) (EHWS). The bark at these levels is often covered with a profusion of lichens, and there are littorine snails characteristics of the mangrove (*Littorina angulifera* in the Caribbean (Coomans, 1969) and West Africa (Bouchet, 1977), *L. melanostoma* in Singapore (Morton, 1976c), *L. carinifera* and *L. scabra* in Thailand (Frith et al., 1976), and the latter two species and *L. undulata* in Malaysia (Lim, 1963). Only *L. scabra* occurs in South African mangroves (Macnae, 1963). At about mean high water spring (tide) (MHWS), the tree trunks become progressively covered by encrusting organisms. Higher up, this is a chthamalid barnacle [*Chthamalus rhizophorae* in Trinidad (Bacon, 1975) and *C. withersi* in the Indo-Pacific (Berry, 1963, 1975; Frith et al., 1976)]. Below this occurs another barnacle, typically one of the numerous subspecies of *Balanus amphitrite,* although others may also occur, e.g., *B. eburneus* in Trinidad (Bacon, 1975). Sharing the trunk with the balanoid barnacles are oysters [as we have seen, *C. rhizophorae* in the Caribbean (Coomans, 1969), *C. gasar* in West Africa (Bouchet, 1977), and *S. cucullata* in the Indo-Pacific (Frith et al., 1976;

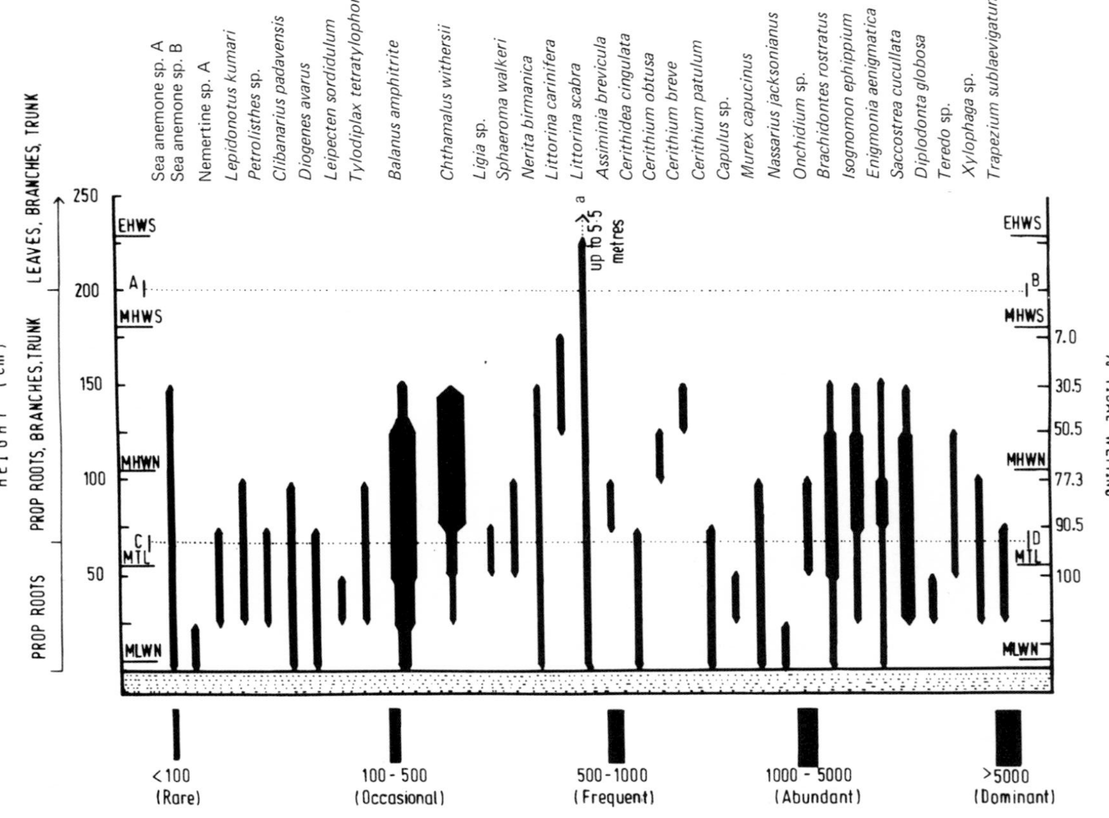

Fig. 1. Vertical zonation and abundance of animals on *Rhizophora apiculata* trees within the mangrove forest at Ao Nam-Bor, Phuket Island, Thailand. A . . . . B = mean level of lower leaf line; C . . . . D = mean level of bank surface; ▩ = mean level of channel water at low tide. (Redrawn after Frith et al., 1976.)

Berry, 1975; Morton, 1976c; Pinto and Wignarajah, 1980)]. True limpets (Patellacea) are rarely capable of adaptation to estuarine environments; in these situations the oysters are usually far more numerous than the typically marine barnacles which form a scattered band above and overlapping with them. In the mangal, the limpets are replaced by a range of other gastropods [e.g., *Nerita birmanica* and species of *Cerithium* in Thailand (Frith et al., 1976), *Neritina virginea* and *C. litteratum* in the Caribbean (Coomans, 1969), and *Nerita glabrata* and *N. adansoniana* in West Africa (Bouchet, 1977)]. Morton (1976c) has pointed out the basic similarity in form between a limpet and the exclusively mangrove bivalve, *Enigmonia aenigmatica,* in the Indo-Pacific and *Pododesmus rudis* in the Caribbean (Coomans, 1969).

Because mangrove trees are high zoned, the rich oyster community typically extends down to the mudline. Beyond it, the barnacles and oysters are replaced by infaunal filter feeding bivalves, e.g., species of *Arca* and *Anadara* in the Arcidae and a very wide array of other bivalves (Tables II–VII). The third zone of Stephenson and Stephenson (1949), i.e., the algal or coral zone, is replaced in these situations by a sublittoral fringe of either algae (e.g., *Caulerpa*) or submerged grass (e.g., *Halophila*) (Berry, 1975; Morton, 1976c).

The zonation of mangrove organisms that are equivalent in status to the zoning organisms of rocky shores can be defined as follows:

EHWS		1. Lichens and littorines
EULITTORAL ZONE		2a. Barnacles, oysters, anomiids; neritiid and cerithiid gastropods
	Mudline	
		2b. Venerid, mytilid, and arciid bivalves; cerithiid gastropods
ELWS		3. Algae or halophytic grasses

This diagram of zonation in a mangrove forest appears to be as universal as rocky shore zonation, and adequately defines, in broad terms, the major faunistic groups characterizing mangroves everywhere. In this system, it is clear that the Mollusca are important zoning organisms (as they are on rocky shores) and account for much of the biomass of the eulittoral zone especially. The oysters in particular, replacing barnacles at low salinities, create a well-defined zone; although replaced by less obvious, buried filter feeders in the lower half of the eulittoral, they constitute an important niche for a wide array of other nestling organisms, among which the bivalves are also notable. Almost universally, these are species of *Brachidontes* and *Isognomon,* although the latter do not seem to occur on the west coasts of the Americas and Africa. The reasons for this apparent absence are unknown.

The interface between the roots and the mud is a further niche that deserves mention, especially because of its importance to the bivalves. The mytilid genera *Modiolus, Perna, Mytella,* and *Musculus* are all byssally attached, and thus require a site for firm anchorage. Yet they also occur partially buried in mud; a few are embedded in ascidian tests, e.g., *Musculus lateralis* (Coomans, 1969); and sometimes they build byssal nests, e.g., *Mytella guyanensis* in the Caribbean (Bacon, 1975) and *Musculista senhausia* in the Indo-Pacific (Morton, 1973a). However, all are typically covered by loose mud. It has been suggested by Stanley (1970) and Morton (1981) that these genera, being essentially modioliform, represent an important transition in the subsequent evolution of the heteromyarian form and the colonization of exposed, rocky shores; there, as in the mangroves, mytilids are also important zoning organisms.

V. Patterns of Species Distribution through a Mangrove

Transects through a mangrove from shore to seaward fringe are most valuable in revealing the distribution not only of the trees but also of the associated fauna. Profiles of a mangrove stand reveal basic distribution patterns. Unfortunately, because mangroves are often inpenetrable, there are few such studies, and the bivalves are very often ignored (Sasekumar, 1974) in favor of the more obvious snails and crustaceans. However, Day (1974) has given a very good picture of distribution patterns in the Morrumbeene estuary in Mozambique.

Studies of this kind, especially on the bivalves, have been undertaken mostly in Southeast Asia. Frith et al. (1976) have made a thorough study of zonation in a Phuket (Thailand) mangrove showing that the distribution of the bivalves can be related to rather specific biotopes. Thus, *Geloina erosa* was the only species to occur at the highest levels of the shore and nowhere else. The distribution of the other bivalve mangrove associates commenced approximately halfway along the mangrove profile. Within the seaward half occurred those species either byssally attached or cemented to the roots and stems, e.g., *Brachidontes rostratus, Isognomon ephippium, Enigmonia aenigmatica, Saccostrea cucullata,* and *Trapezium angulatum.* The mud in this region of the mangrove also supported a number of burrowers, e.g., *Glauconome virens* and *Laternula truncata; Diplodonta globosa* and *Solen delesserti* occurred within the mangrove but also extended beyond onto the sand. A number of species, e.g., *Gafrarium tumidum, Katelysia striata,* and *Caecella horsfieldi* occurred in the sand but not within the mangrove proper; *Anadara nodifera, Tellina capsoides, T. opalina,* and *Sinonovacula constricta* occurred in mud, again separated from the mangrove forest. Shipworms were restricted to the trees. Nateewathna and Tantichodok (1980) have basically confirmed the observations of Frith et al. (1976) by showing a very clear pattern of distribution in a mangrove at Ko Yao Yi, southern Thailand. Only *Anadara granosa, Scapharca inequivalvis, S. pectunculiformes, Isognomon ephippium, Saccostrea cucullata, Tellina opalina, Sinonovacula constricta, Solen delesserti,* and *Laternula truncata* occurred within the mangroves (some of them also extending onto the mud beyond), but a much greater number of species could be found on either sand, mud, or grass biotopes lower down the shore.

Berry (1963) has studied faunal zonation of a Malaysian mangrove swamp in broader terms and divided the ecosystem into five zones characterized by certain biological types of organisms (other than the trees themselves) (Fig. 2). A first High Tree Zone, characterized by the presence of *Littorina* spp., may be conveniently termed the Littorine Zone, corresponding to an equivalent zone on hard shores (Stephenson and Stephenson, 1949); the Lower Tree Zone (second) is characterized by other gastropods, notably species of *Nerita*. Both these zones extend from the seaward fringe of the mangrove well back to the landward edge. A Marginal Zone (third) close to the sea edge is characterized by the presence of bivalves which typically do not occur within the forest [an exception to this, not noted by Berry (1963), are species of *Polymesoda* (*Geloina*)]. The Ground Zone

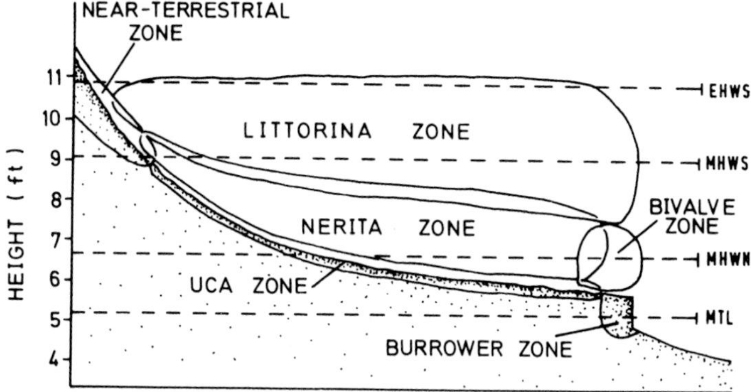

Fig. 2. Diagrammatic profile of a mangrove swamp showing the essential features of animal zonation. Horizontal dimensions are greatly shortened compared with vertical ones. The height in feet refers to chart datum, and the tidal levels are derived from the Admiralty Tide Tables for Singapore Inner Harbor. (Redrawn from Berry, 1963.)

(fourth) is rich in crabs, especially *Uca,* and the fifth or Burrower Zone comprises the coastal bank.

The bivalves therefore occupy the marginal fringe (with the exception of *Geloina,* noted above), colonizing the trees and the sediments of this region and extending onto the sand or mud beyond. Berry's classification recognizes the importance of the bivalves as a major faunal component of the mangrove community accounting for much of the biomass of the coastal fringe, an observation that has escaped most other researchers.

VI. Horizontal Zonation down an Estuary

Mangroves are plants adapted for estuarine conditions: when they occur elsewhere, their importance and stature are reduced. Down an estuary, species composition varies according to a changing substrate composition and differences in water quality, notably with regard to salinity (Branch and Grindley, 1979). In the *Spartina* marsh community of Uruguay, the euryhaline mytilids *Brachidontes darwinianus* and *Mytella charruana* are replaced under more marine conditions by *B. rodriguez* and *Mytilus platensis* (Scarabino et al., 1975).

Arca gibba, A. obliquata, Brachidontes variabilis, Ostrea olgoensis, Perna perna, Saccostrea margaritacea, and *Septifer bilocularis* were attached to rocks at the mouth of the Mngazana estuary in Transkei. *Saccostrea cucullata* was distributed throughout the estuary and *Musculus virgiliae* dominated the rocks at its head (Branch and Grindley, 1979). *Loripes clausus, Donax serra,* and *Tellina gilchristi* were common in the sand at the mouth of the same estuary; *Arcuatula (Lamya) capensis* colonized rocks in mud at the middle reaches. *Tapes corrugatus* occurred in sand with rocks in the middle reaches, and *Eumarcia paupercula* occurred in mud at the mouth of the estuary. Unfortunately Branch and Grindley apparently did not collect these bivalves in the mangrove itself, but rather from the sediments beyond. Nevertheless, their study does illustrate estuarine faunal changes in response to a changing environment.

As noted by Branch and Grindley, rocky substrates in the lower reaches of an estuary support a far higher percentage of epibyssate, stenohaline species, which explains the rapid decrease in rock fauna upstream. The smaller proportion of stenohaline soft substrate species similarly declines just as suddenly, but this decrease is more than offset by the large number of euryhaline species in this muddier substrate. Such observations are in accord with the view that rocks are more common in exposed marine conditions (and therefore are colonized by marine species), whereas muds are typical of brackish waters and subject to fewer marine influences (e.g., waves) and are thus colonized by euryhaline species. The bivalves, typically burrowers or immobile, byssal or cemented colonizers of mangrove roots or stones, must be considered good indicators of environmental change both natural and artificial within an estuary (and of a particular mangrove stand). The latter has been demonstrated by Kolehmainen et al. (1973).

Bertness (1980) has shown that the high-zoned marsh bivalve *Geukensia demissa* showed progressively decreasing growth rates and maximum size but increasing longevity from the open coast to the more severe estuarine environment. The inverse relationship between size and longevity between those *G. demissa* populations that are probably genetically continuous suggests that the principle of allocation is operating at the phenotypic level. Mussels that are able to achieve fast growth rates by virtue of their habitat may channel more of their resources into reproduction at the expense of maintenance, whereas mussels in habitats that do not allow high growth rates may invest more in general maintenance to ensure reproductive success over a longer life span (Bertness, 1980). It seems at least possible that this sort of phenotypic plasticity in life-history characteristics is typical of many mangrove bivalves. Baquiero (1980) has shown that significant differences exist between populations of *Anadara tuberculosa* in eight mangrove swamps in Baja California Sur, Mexico. He uses height–width differences to postulate a counterclockwise circulation system within Almejas Bay which could also explain patterns of larval dispersion.

VII. Bivalve Predators

Although many mangrove bivalves are abundant in their chosen microhabitat and must thus constitute a valuable potential food resource, little is known of their predators. The high-zoned *Spartina* associate *Geukensia demissa* apparently has no predators, except possibly the racoon *Procyon lotor*. The absence of predators, however, leads to intraspecific competition for space and thus eventually to mortality through crowding (Bertness, 1980).

In Florida, *Mytilopsis leucophaeta* and *Brachidontes exustus* are important components of the diet of a number of fishes, especially the sheepshead *Archosargus probatocephalus* (Odum and Heald, 1972). In the marshes of Texas, *Mulinia lateralis* and *Anomalocardia cunimeris* are similarly important in the diet of the black drum *Pogonias cromis* (Breuer, 1957).

In the Caribbean mangal, the mud-dwelling habit of *Mytella guyanensis* exposes the animals to a variety of predators, although the extent of predation is unknown (Bacon, 1975). Predators include the whelks *Melongena melongena* and *Pugilina morio,* which feed without removing the mussels from their burrows, and the crabs *Goniopsis cruentata* and *Eurytium limosum,* which pull the mussels out. Several species of swamp birds, e.g., herons, sandpipers, and rails, have been observed feeding on the mussels at low tide. The main predators of *Mytella falcata* are the puffer fish, *Sphaeroides testudineus,* and the crab, *Pan-*

opeus herbstii, although both species take only small individuals. *Melongena melongena* and *Pugilina morio* may feed on *M. falcata* but they also feed on oysters (*Crassostrea rhizophorae*) (Bacon, 1975). In Colombia, *C. rhizophorae* is the prey of *Callinectes sapidus, C. danae, Eurypanopeus depressus, Panopeus occidentalis* and *P. herbstii,* and of the snails *Thais haemostoma, T. trinitatensis,* and *M. melongena* (Wedler, 1980). Palacio (1977) has shown that adolescent oysters (up to 4 months old) are extremely vulnerable to predation. The normal mortality rate for oysters in the Cienega Grande is 40–50% (in periods of extremely low salinity, 80–90%). In culture baskets, however, only 10–15% of the oysters were victims of their predators (Wedler, 1980). The puffer fish *Sphaeroides testudineus* also feeds on *C. rhizophorae* (Bacon, 1970). Coomans (1969) reports that *Murex brevifrons* is a common predator on *C. rhizophorae* in the Caribbean, as are *M. melongena, Pugilina morio,* and (in Florida) *M. corona.* In Tobago, *Murex brevifrons* drills into *Isognomon alatus* (Bacon, 1970), but in Jamaica *I. alatus* apparently has few predators (Siung, 1980).

In Sierra Leone, Hunter (1969) reports that *T. haemostoma* and *T. coronata* are the main predators of the oyster *Crassostrea gasar.* These thaids drill through the oyster shells and devour the meat. The gastropods occurred in the ratio of some 1 : 500–2000 oysters and thus were thought to have a negligible impact upon the population. In Senegal, Bouchet (1977) reports that *Thais forbesi* and *T. callifera* are the predators of *C. gasar* and the mytilid *Brachidontes niger.*

In Malaysia, Berry (1975) reports that *Murex capucinus* feeds primarily on *Saccostrea cucullata* and rarely on *Enigmonia rosea,* but Frith et al. (1976) report that in Thailand the same snail feeds upon *Balanus amphitrite.*

Vermeij (1980) has shown that the Indo-Pacific *Anadara granosa* is drilled and eaten by the muricid *Bedeva blosvillei.* The mortality due to this predator is extraordinarily high (88%), but the incidence of incompleted drill holes increases with increasing size of the bivalve. The right valve is attacked about twice as often as the left; the reason for this is unknown.

Bertness (1980) has arrived at an interesting conclusion with regard to predation and the high-zoned *Spartina* marsh bivalve *Geukensia demissa.* The ability of this mussel to withstand wide environmental variability (Wells, 1961; Lent, 1969) and to evolve specific adaptations for an extremely harsh high intertidal life (Lent, 1969; Pierce, 1970) permits avoidance of marine predators and interspecific competitors unable to survive at this height and allows the attainment of high population densities in a physically stressful environment. However, having mastered life in the high intertidal zone to the point of attaining high population densities, *G. demissa* populations are located too high for predators, competitors, or both to alleviate the effect of intraspecific competition. *G. demissa* is a gregarious species, but it would be interesting to see what factors are influencing population density and mortality in the similarly high-zoned but solitary species of *Polymesoda* in the Indo-Pacific.

VIII. Mangrove Bivalves

A. *Enigmonia aenigmatica*

Perhaps the most characteristic bivalve of the Indo-Pacific mangrove community is the monomyarian anomiid *Enigmonia aenigmatica* (Yonge, 1957; Berry, 1975; Morton, 1976c). In the Caribbean, *Pododesmus rudis* occupies a similar niche but is not restricted to the mangal. The closely related *P. cepio* has been studied by Yonge (1977).

Yonge (1957, 1977) also reported that *E. aenigmatica* was a white, highly mobile saddle oyster, actively crawling over the leaves of *Avicennia*. Its more usual habit, however, is byssal attachment to the stems and prop roots of a variety of mangrove plants, notably the pioneering *Avicennia* and *Rhizophora*. Berry (1975) has made a study of *E. rosea* (a morph of *E. aenigmatica*, which will be discussed later) and found (Fig. 3) that it was most abundant near the forest edge, occurring in very low numbers 70–165 m within the forest and absent beyond. On the forest edge, *E. rosea* occurred on 23–35% of trees with a trunk diameter greater than 6 cm at average densities of 0.45–0.36/tree. Just

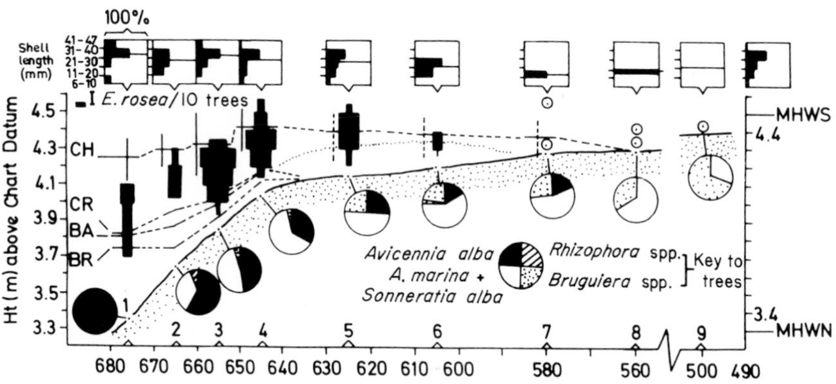

Fig. 3. A profile showing the distribution of attached animals near a mangrove forest edge at Selangor, Malaysia. CH = *Chthamalus withersii*, with mean upper limits (horizontal lines joined by dashed lines) and vertical ranges of upper limits (vertical lines, dashed where animals were few). Mean upper limits of *Crassostrea* (CR) (*Saccostrea cucullata*), *Balanus* (BA), and *Brachidontes* (BR) are shown. Position and depth of the conspicuous algal mat are indicated by a dotted line. Black "kites" show numbers of *Enigmonia*; ○ indicates individuals. Upper histograms show percentage of *Enigmonia* by shell length together with mean shell length (horizontal line) at nine stations; the overall distribution by shell length is at extreme top right. "Rose" diagrams indicate the relative abundance of mangrove tree species. (Redrawn after Berry, 1975.)

within the forest, the species occurred on 50–55% of the trees at average densities of 1.23–0.95/tree. The densest concentrations were found approximately 4.2 m above chart datum. The distribution of *E. rosea* on the trees differed significantly from the calculated random distribution, suggesting a preference for settlement on trees that already bore one or more *Enigmonia* individual. Large individuals occur throughout the vertical range of the species, with smaller animals occurring lower down. This distribution may reflect greater opportunities for settlement during longer submergence times rather than conditions especially favorable for growth.

Morton (1976c) has also investigated aspects of the ecology of *Enigmonia* and shown (Fig. 4) that on the trees it is oriented with the umbones pointing downward. Moreover, it seems to occur on the sides of the trees facing toward the flowing estuary. *Enigmonia* is clearly highly adapted to life in the Indo-Pacific mangrove. These ecological adaptations are, however, matched by other adaptations to both the shell and the internal anatomy (Morton, 1976c).

Berry (1975) reported upon *E. rosea;* earlier (Berry, 1972) he had discussed both *E. aenigmatica* and *E. rosea*. Morton (1976c) has, however, shown that there is only a single polymorphic species. Gray (1849) recognized four conchological varieties (Fig. 5). Variety 2, described as *Anomia naviformis* Jonas 1846, is purple-brown and assumes its elongate shape after settling upon narrow stems, sticks, and the prop roots of *Rhizophora* (Yonge, 1977). Variety 3 (also Yonge, 1957) is small, cream-colored, often with a dash of purple at the upper left umbo, and is usually found attached to the undersurface of the leaves of *Avicennia* shoots and *Scyphiphora hydrophyllacea* (Morton, 1976c). Variety 1, described as *Tellina aenigmatica* Chemnitz 1795, occurs farther back in the mangrove and is colored deep purple. Variety 4, the *Enigmonia rosea* of Berry (1972, 1975), is large, rounded, purple in color, and occurs on the seaward fringe of the mangroves (Berry, 1975). *Enigmonia aenigmatica* is thus capable of subtle morphological adaptation to some of the different microhabitats of the mangrove shore system.

Enigmonia has an unusual body plan. The animal lies on the very fragile right shell valve, which possesses a deep byssal indentation, so that the byssus seems to emerge from its center. The single adductor muscle is small in comparison to the solitary, enlarged, left posterior byssal retractor muscle, which serves to pull the upper, slightly thicker left valve down over the right. The upper valve is, however, transparent, so that pallial eyes unusually developed under this valve are capable of photoreception (Bourne, 1907), possibly for defensive purposes but also to assist orientation at least in relation to light. Well-developed anterior and posterior pedal retractors allow mobility especially of the white, unattached form and of the juveniles before they choose a habitat. Water flows into and out of the mantle cavity in an anterior-posterior direction, which may explain the orientation of the animal on the tree. Because *Enigmonia* lives high up on the

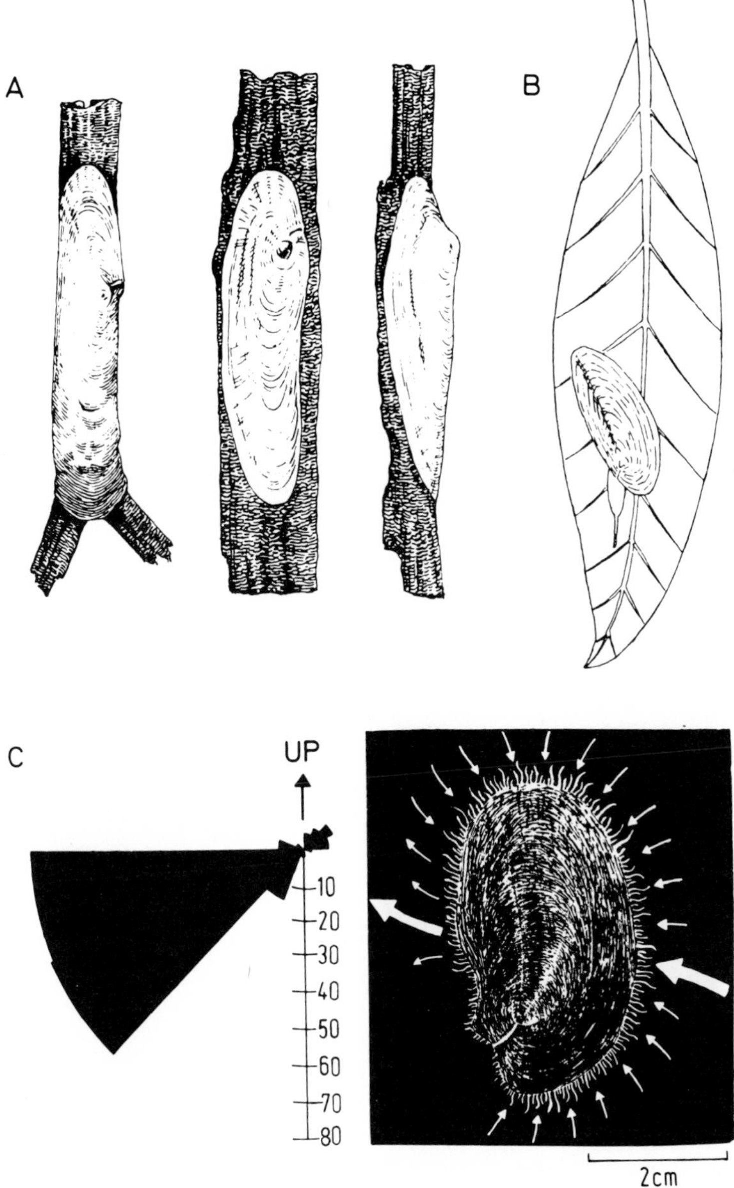

Fig. 4. *Enigmonia aenigmatica*. (**A**) Drawings of specimens of "*E. rosea*" permanently attached to mangrove branch. (**B**) A motile individual crawling on the surface of an *Avicennia* leaf. (**C**) An animal in its normal orientation on mangrove trees, viewed from above, with the inhalant and exhalant currents represented by arrows. The orientation of the population on the trees in the Singapore mangroves is shown as a rose diagram to indicate the direction in which the umbones are typically pointed. [(A) and (B) after Yonge, 1977; (C) after Morton, 1976c.]

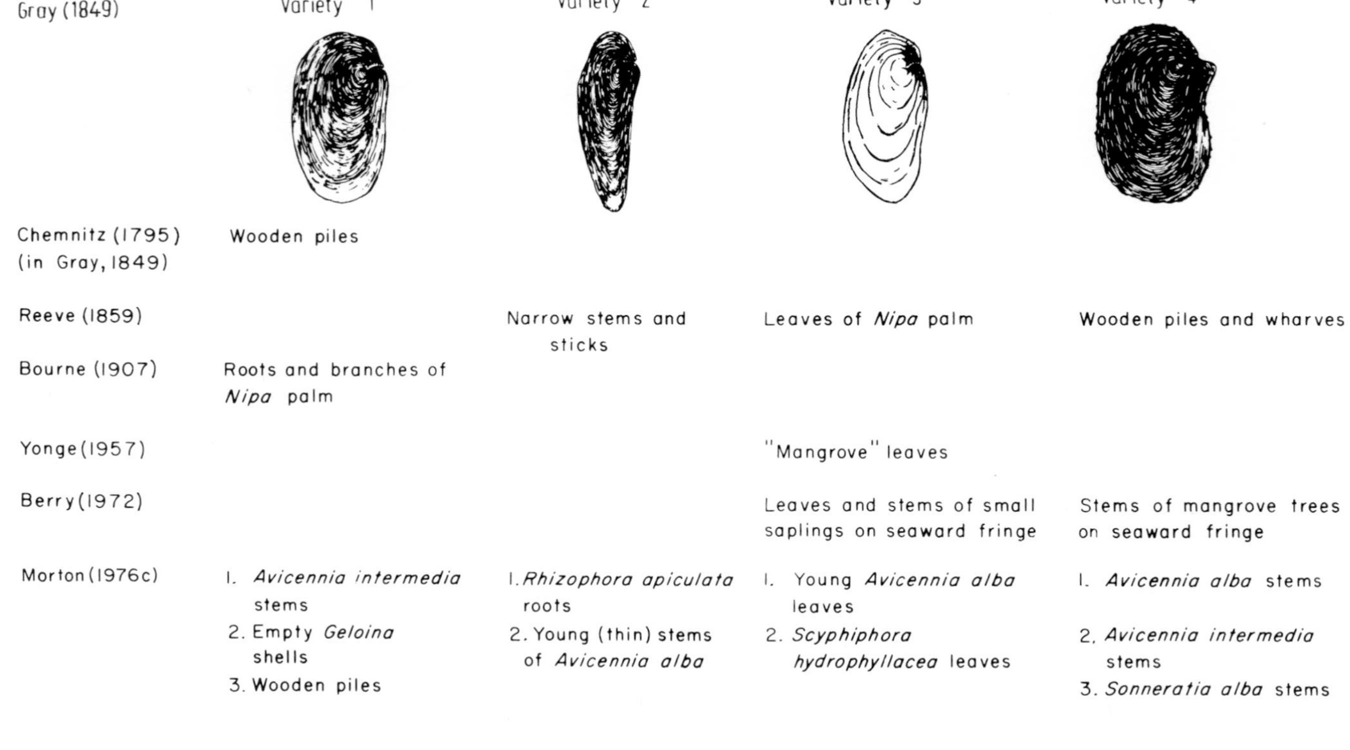

Gray (1849)
Variety 1
Variety 2
Variety 3
Variety 4
Chemnitz (1795) (in Gray, 1849)
Wooden piles
Reeve (1859)
Narrow stems and sticks
Leaves of *Nipa* palm
Wooden piles and wharves
Bourne (1907)
Roots and branches of *Nipa* palm
Yonge (1957)
"Mangrove" leaves
Berry (1972)
Leaves and stems of small saplings on seaward fringe
Stems of mangrove trees on seaward fringe
Morton (1976c)
1. *Avicennia intermedia* stems
2. Empty *Geloina* shells
3. Wooden piles
1. *Rhizophora apiculata* roots
2. Young (thin) stems of *Avicennia alba*
1. Young *Avicennia alba* leaves
2. *Scyphiphora hydrophyllacea* leaves
1. *Avicennia alba* stems
2. *Avicennia intermedia* stems
3. *Sonneratia alba* stems
Tellina aenigmatica Chemnitz 1795
Anomia naviformis Jonas 1846
Anomia aenigmatica Sowerby 1825
Aenigma roseum (Gray 1846)

Fig. 5. *Enigmonia aenigmatica*. The growth varieties (Gray, 1849) correlated with their habitats. (After Morton, 1976c.)

trees, it can filter the water, which is of course very rich in nutrients, for short periods only. Particle selection is by the gills; the labial palps, unusually, accept all material passed to them. The short bursts of feeding activity result in a digestive system that possesses a distinct rhythm in which the dissolution of the style in the stomach and the cytological changes in the digestive tubules are related to the feeding cycle and thus to the tidal flows within the mangrove.

B. *Polymesoda*

Members of the veneroid family Corbiculidae are characteristic of lentic and lotic systems throughout much of the world. In particular, the genus *Polymesoda* is characteristic of the two major mangrove systems, the Caribbean and the Indo-Pacific. In the Caribbean (and the Gulf of California), the genus is represented by two subgenera, *Pseudocyrena* in the Caribbean, e.g., *Polymesoda* (*Pseudocyrena*) *floridana,* and *Polymesoda,* e.g., *Polymesoda* (*Polymesoda*) *caroliniana,* in North America. In the Indo-Pacific, the subgenus is *Geloina* (Morton, in manuscript). This system of classification follows Keen and Casey (1969), but it is this author's opinion that, at the very least, the subgenus *Geloina* will eventually have to be separated from *Polymesoda.* Morton (in manuscript) has revealed, for example, an inherent difference between *Geloina* and its American relatives. The former is very high zoned in mangroves; the latter are typically low zoned, even into the sublittoral, well beyond the mangrove fringe. This suggestion has the support of Rehder (in Tabb and Moore, 1971).

From the Gulf of California, Parker (1963, 1964) records *Polymesoda mexicana* (plus seven other unidentified species), and Castaing et al. (1980) record *P. inflata* from the Pacific coast of Costa Rica. Parker also records the closely related *Cyrenoida panamensis* from the Gulf of California.

Polymesoda caroliniana was thought by Van der Schalie (1933) to be a disjunct species occurring well north of Florida, e.g., Virginia (Andrews and Cook, 1951) and along the northern shores of the Gulf of Mexico, although not in southern Florida. Tabb and Moore (1971) later suggested, however, that this interrupted distribution is probably artificial, resulting from the drainage of marshes for development of cities; they point out that the species now occurs in the Everglades, where hitherto it did not; *P. caroliniana* thus seems to be reoccupying its former continuous range. On the northern shores of the Gulf of Mexico, *P. caroliniana* is characteristic of *Spartina* marshes, occurring low down the shore, typically beyond the lower levels of the plants (Harry, 1942). Along the Texas coasts, where *Spartina* forms mixed stands with *Avicennia, P. caroliniana* is also present, occupying a position essentially similar. In the North River of the Cape Sable region of South Florida, the dominant emergent vegetation is a mixture of *Juncus roemerianus* and stunted red and white mangroves (*Rhizophora mangle* and *Laguncularia racemosa*), and here *P. caroliniana* occurs in the soft mangrove peat deposits along the edges of the brackish-water creeks (Tabb and Moore, 1971).

In a Mississippi tidal marsh, Duobinis and Hackney (1979) have shown that the largest populations of *P. caroliniana* were found in a regularly flooded area (293 individuals/m^2) and in the upper reaches of a tidal creek enervating the marsh (380 individuals/m^2). Few were found in areas that were flooded for only 0.2% of the year. Juvenile recruitment occurred in February and July, with a high mortality following each peak in abundance.

Tabb and Moore (1971) have investigated *P. caroliniana* in the Florida Everglades. The optimum habitat is a narrow belt of shallow marshland where the mangrove forests intergrade with the freshwater flora of the interior. The normal water depth here ranges between 0.25 and 1.0 m; the salinity range is 0–10 ‰ but may rise to 20 ‰. Adult *P. caroliniana* can tolerate salinities as high as 26.3 ‰ for short periods but will be adversely affected when it rises above 18–20 ‰. South Florida populations apparently spawn during the spring and early summer, coincident with the onset of the rainy season, so that young individuals probably live in almost fresh water (Van der Schalie, 1933; Tabb and Moore, 1971).

Little research has been undertaken on this genus in the Caribbean, but from the Pacific coast of Costa Rica, Castaing et al. (1980) record *Geloina* [almost certainly *Polymesoda* (*Pseudocyrena*) *inflata*], occurring at the bases of the prop roots of pioneer mangroves (*Rhizophora mangle*). This species is thus low zoned as is *P. caroliniana*. The same authors found that the numbers of *P. inflata* increased during the months of May and June, indicating an early summer spawning season. More is, however, known about *Polymesoda* in the Indo-Pacific, where it is represented by the subgenus *Geloina*.

Table VII indicates that an array of species have been recorded, mostly from Malaysia, Thailand, Indonesia, and Singapore. The status of the species within the Indo-Pacific mangrove has, however, been investigated by Morton (1976b,d, 1983a, in manuscript). Indo-Pacific *Geloina* have been reviewed by Prashad (1932), who described three species: *Geloina erosa* (Solander 1786), *Geloina bengalensis* (Lamarck 1818), and *Geloina expansa* (Mousson 1849). Bentham-Jutting (1954) similarly records these species only from Java.

The most recent revision of the group by Brandt (1974) concludes that there are four species: *Geloina coaxans* (Gmelin 1791), *Geloina bengalensis* (Lamarck 1818), *Geloina proxima* (Prime 1864), and *Geloina galatheae* (Mörch 1850). Brandt, however, acknowledged that the last species may have been only an elongate specimen of *G. bengalensis*. Thus, ignoring *G. galatheae*, Brandt essentially agrees with Prashad and Bentham-Jutting, i.e., there are three species.

Morton (in manuscript) has investigated *Geloina* throughout the Indo-Pacific and shown that Prashad was probably correct in his nomenclatorial assessment of the genus. From the literature and from the collections of these species housed in the British Museum (Natural History), Morton has further shown that the three species are indeed recorded only from the mangroves of the Indo-Pacific (Fig. 6).

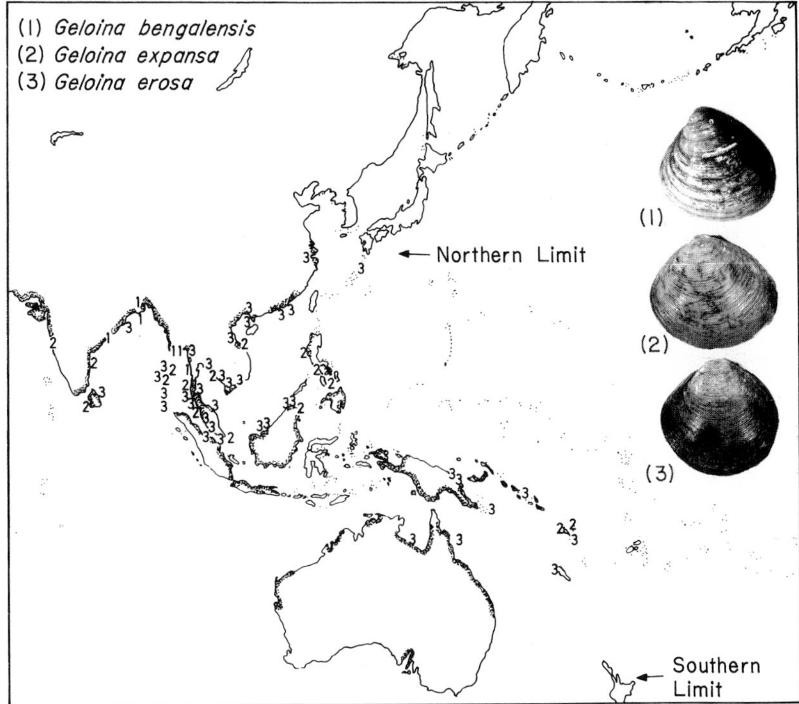

Fig. 6. The distribution of (1) *Geloina bengalensis,* (2) *Geloina erosa,* and (3) *Geloina expansa* in the Indo-Pacific. Also shown are photographs of the three species and the distribution of the mangroves (after Tranter, 1974). (Redrawn from Morton, in manuscript)

G. bengalensis is of particular interest because there are few authenticated reports of its occurrence outside the Bay of Bengal, and it appears to be particularly associated with the mangroves of the Ganges and Irrawaddy deltas (the Sundarbans) and the smaller estuaries of the western coast of Thailand. *Geloina erosa* and *G. expansa,* however, appear to be sympatric species with a widely overlapping distribution, although the former ranges much farther north and south than the latter. Morton (in manuscript) records both species in the same ponds in the Singapore mangrove. In Hong Kong, only *G. erosa* occurs; in Japan, the species is referred to as *P. luchea,* although authenticated specimens are indistinguishable from *G. erosa.*

Little research has been undertaken on *Geloina.* Morton (1976d) described the general ecology of *G. erosa* in Singapore, showing that individuals occur on the landward fringe (wholly unlike *Pseudocyrena* in the Caribbean, which occurs on the seaward fringe) with such plants as *Brownlowia tersa, Heritiera littoralis, Excoecaria agallocha, Nipa fruticans,* and *Cocos nucifera.* Between the roots of these plants, small streams cut channels and form fetid-looking pools. Here, *G.*

erosa and *G. expansa* occur together. The pH of the mangrove mud from these regions ranged from 5.35 to 6.28. During spring tides the salinity of the pools was 20.72–21.56 ‰ and at neap tides 11.67–19.87 ‰, although after a cloudburst it fell to 13.17 ‰. The acid soils cause considerable erosion of the shell, especially at the umbones (the oldest part of the shell), explaining much of the difficulty involved in separating species and also making aging of the shells (by the conventional means of counting growth rings) virtually impossible. Tabb and Moore (1971) have similarly shown that the shells of *P. caroliniana* dissolve rapidly in the acid mangal soil after the death of the animal. Dissolution is appreciable after 3–4 months.

Because it is so high zoned, *Geloina* is rarely covered by the tides and can withstand considerable periods of emersion. Specimens kept out of water in Hong Kong can survive for more than 90 days, rapidly commencing to filter-feed and immediately gaining weight when returned to water (B. Morton, unpublished data). A further adaptation to emersion is the ability of *G. erosa* to exchange mantle fluids with subterranean water via the pedal gape (Morton, 1976b,d). During periods of emersion, *Geloina* also gapes and slightly exposes the mantle margins posteriorly, which permits, it is thought, some degree of gaseous exchange. This occurs at night (Morton, 1976b), constituting a recordable diurnal rhythm. When inundated by water, by either the tide or simulated rainfall, the animal immediately commences filter feeding. These adaptations to life in the landward margin of mangroves enable *Geloina* to exploit a habitat not normally occupied by other filter feeders.

In most morphological respects, *Geloina* is unspecialized (Morton, 1976d). Members of the Corbiculacea typically are specialized reproductively. The riverine *Corbicula fluminalis* is a protogynous consecutive hermaphrodite, whereas the stream-dwelling *C. fluminea* has variable reproductive strategies, but incubates larvae within the ctenidia until a crawling juvenile emerges (Morton, 1977b, 1982). *G. erosa* is dioecious with no larval incubation (Morton, 1983a) in Hong Kong; it breeds during a single extended phase in the summer. There must, however, be behavioral specializations coordinating the release of gametes, especially because *Geloina* is emersed for considerable periods of time. Could fertilization take place, for example, in the maze of subterranean burrows that connect individuals with each other and not, as in the typical venerid, in the sea? Much more research on *Geloina* is needed. In the Indo-Pacific it is widely utilized as food (Morton, 1976d), and its ability to withstand long periods of desiccation is a major asset in the transfer from shore to market, when bivalve mortality is typically high.

C. *Laternula truncata*

Lim (1963) highlighted three interesting mangrove bivalves: *Enigmonia*, shipworms, and *Laternula truncata*. He noted with regard to the latter that "The valves are unusual in that they are always cracked in one or more places.

. . . perhaps the fragile shell "gives way" when the animal grows or perhaps it is the result of violent muscular contraction."

The Laternulidae have an Indo-Pacific distribution (Morton, 1976a); *L. truncata* is a true mangrove associate in the region (Purchon, 1958; Morton, 1973b, 1976a). Four other species are recorded from muddy shores in the region, although nothing more is known of them. *Laternula truncata,* occurring on the seaward fringe of the mangrove (Frith et al., 1976), has two valves that are cracked at right angles to the hinge line across the umbones. This feature is typical of the family and in effect creates four valves that flex anteriorly and posteriorly about the crack when the adductors contract. This novel method of exchanging the fluids of the mantle cavity enables the animal to lie virtually immobile within a deep burrow. Defense is not provided by the fragile, paper-thin shell or by siphonal retraction; instead, the siphonal crown possesses exceptionally well-developed eyes (Adal and Morton, 1973) and long tentacles. A "shadow reflex" causes the tentacles to flick sand particles over the siphonal crown, thereby possibly camouflaging the animal from wading birds that feed by probing burrow apertures. *L. truncata* is a simultaneous hemaphrodite, but nothing more is known about it.

D. The Pholadidae and Teredinidae

The shipworms are a group of highly modified bivalves that have evolved to tunnel into wood. Teredinids have a nutritional dependency upon the wood they erode from the tunnel heading and a digestive system that breaks down the wood fragments with an endogenous cellulase, although cellulolytic streptomycetes have been isolated from many (Morton, 1978a). Before humans used wood in the sea, the only naturally occurring wood would have been dead trees floating down rivers, and mangroves. *Psiloteredo healdi* is the only known freshwater species occurring in lakes and streams along the northern coast of South America (Turner, 1966). Thus, the coastal fringe mangrove in the tropics is probably not only the normal habitat of many shipworms but also constitutes the evolutionary base for the family as a whole.

This latter view has not been adequately explored but, interestingly enough, Kohlmeyer and Kohlmeyer (1965) make the same point with regard to marine fungi. These authors consider that the most recent (primary) marine fungi have been derived from terrestrial ancestors via the intermediary of (secondary) marine fungi that migrated into the marine habitat by attacking terrestrial plants of the shoreline, i.e., mangroves.

Turner and Johnson (1971) divide the Teredinidae into four main groups based on their geographical distribution, a classification which also reflects their reproductive strategies.

1. Marine larviparous species with a worldwide distribution within the limits of their salinity and temperature tolerance
2. Oviparous marine species which are or may become circumtropical

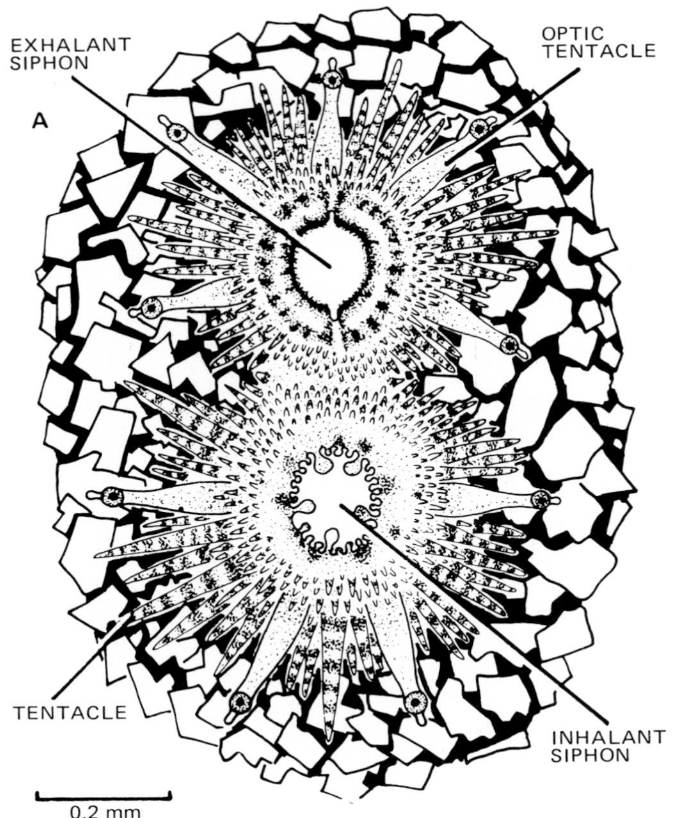
A
EXHALANT SIPHON
OPTIC TENTACLE
TENTACLE
INHALANT SIPHON
0.2 mm

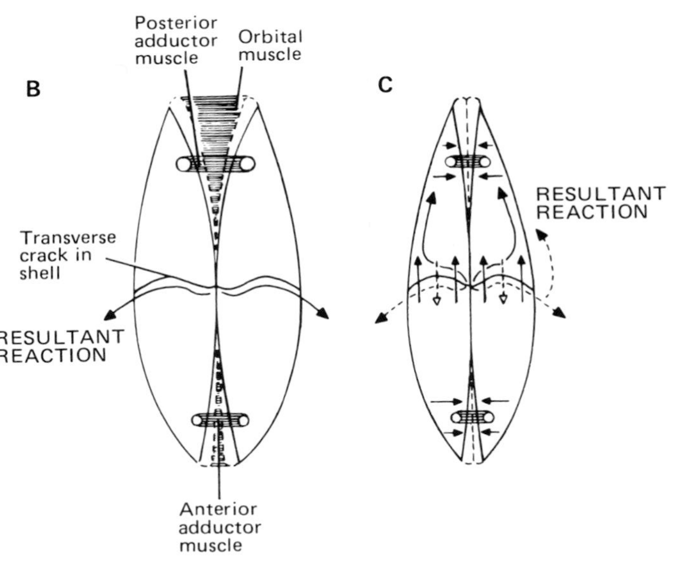
B
Posterior adductor muscle
Orbital muscle
Transverse crack in shell
RESULTANT REACTION
Anterior adductor muscle
C
RESULTANT REACTION

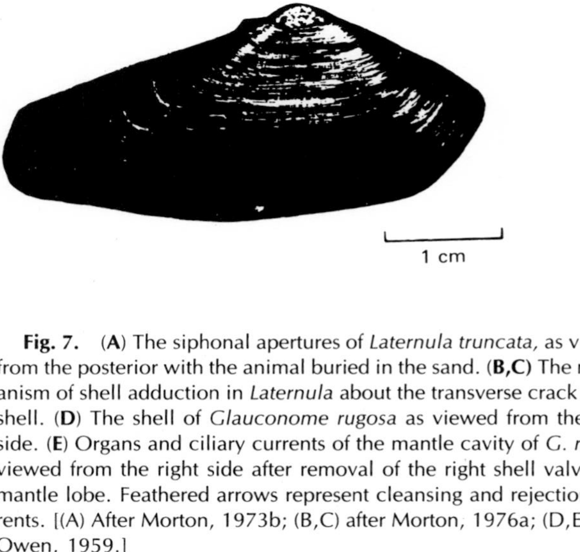

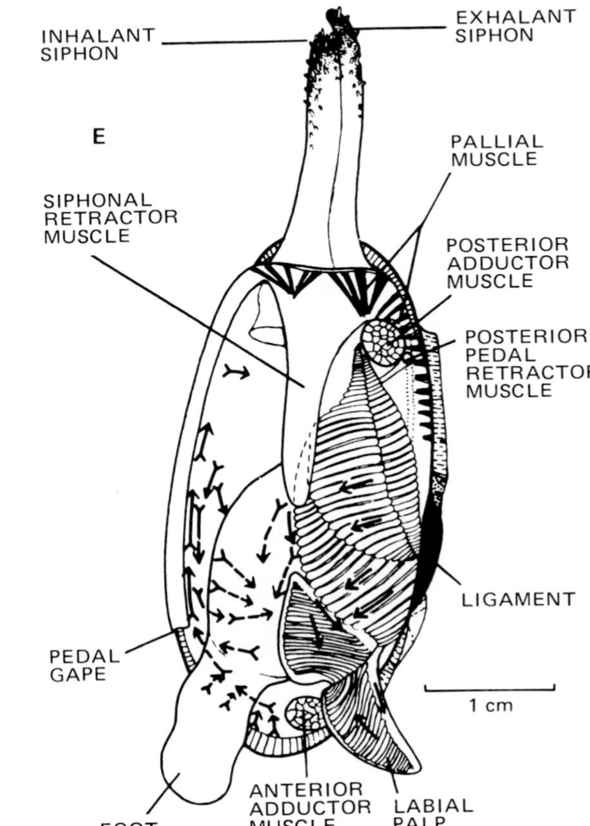

Fig. 7. (**A**) The siphonal apertures of *Laternula truncata,* as viewed from the posterior with the animal buried in the sand. (**B,C**) The mechanism of shell adduction in *Laternula* about the transverse crack in the shell. (**D**) The shell of *Glauconome rugosa* as viewed from the right side. (**E**) Organs and ciliary currents of the mantle cavity of *G. rugosa* viewed from the right side after removal of the right shell valve and mantle lobe. Feathered arrows represent cleansing and rejection currents. [(A) After Morton, 1973b; (B,C) after Morton, 1976a; (D,E) after Owen, 1959.]

3. Oviparous temperate species which are usually restricted to large ocean basins
4. Brackish-water species, both larviparous and oviparous, which generally have a restricted range and a discontinuous distribution

The mangrove shipworms should fall into the last category.

The Pholadidae, a group closely related to the Teredinidae, has representatives that bore mangrove woods. These bivalves cannot digest wood; the most common species recorded from the Caribbean and the Indo-Pacific mangroves is *Martesia striata* (Ganapati and Lakshmana Rao, 1959; Berry, 1963; Bacon, 1970; Pinto and Wignarajah, 1980). This species is widespread throughout the warmer waters of the world (Turner, 1955), typically boring very hard woods and even stone (Annandale, 1924) and plastics (Morton, 1971). From Thailand mangroves a species of *Xylophaga* has been recorded, although this is unlikely because the genus is largely pelagic, occurring mainly in floating and waterlogged wood (Turner, 1973). Balasubramanian et al. (1979) have recorded *Barnea birmanica* from the mangroves of Porto Novo, India; they also isolated three cellulolytic streptomycetes from the gut of this borer, although pholads are not generally believed to ingest the wood they erode from their tunnels.

The Teredinidae are, however, the more common borers of mangroves, and from the Caribbean an unidentified species of *Teredo* has been recorded (Coomans, 1969). From West Africa, Bouchet (1977) has recorded two species of *Teredo* and one species of *Bankia* (specific names were corrected by Turner, 1966). *Teredo manni* has been recorded from South Africa by Day (1974); the same species has been commonly recorded from the Indo-Pacific (Lim, 1963; Coomans, 1969; Bouchet, 1977). An array of other species of *Teredo, Bankia,* and *Nausitoria* from this region has been recorded (Roonwal, 1954a,b; Ganapati and Lakshmana Rao, 1959). Rajagopal (1970), adding further data to earlier work (Rajagopalaiengar, 1961a,b; Rajagopal, 1964), recorded 6 species from the Sundarbans of the Ganges delta alone, although Turner (1966) has synonymized some of these (Table VII). Rajagopal and Daniel (1973) recorded *N. dunlopei* from mangroves in the Nicobar Islands. Tan (1970) records 12 species of *Teredo* and *Bankia* from Singapore waters, nearly all occurring in healthy-looking fallen mangrove tree trunks and wooden piles. *T. thoracites* was collected from the root of a living *Bruguiera* mangrove.

Although little is known of the specific identities and ecologies of the various teredinids recorded from mangroves, Turner (1966) gives an excellent review of the shipworms, including some information on the various groups. Possibly the most primitive teredinid genus is *Kuphus,* a mangrove mud borer restricted to the Indo-Pacific and about which little is known. Turner (1966) records the genera *Bactronophorus* and *Dicyathifer* from the mangroves of the Indo-Pacific; *Nausitoria* from the Indo-Pacific, East Pacific, and West Africa; and *Neoteredo* from the Caribbean and Africa.

As noted earlier, brackish-water teredinids can be either larviparous or oviparous (Turner and Johnson, 1971). *Nausitoria dunlopei* is most active and breeds when the salinity is less than 10 ‰ (Watson, 1936; Smith, 1963). Therefore, the larvae probably are intolerant of high salinities, and those carried into the open sea would perish. Consequently, the various populations are isolated, and colonization of a new area is probably accomplished by adults carried in floating wood which must reach another brackish-water habitat within the life of the adult.

Characteristically, the greatest shipworm attack in mangroves is restricted to the mudline on the seaward fringe of the stand (Turner, 1966), possibly because the larvae are positively geotropic, negatively phototropic, or both. Thus, the larval stages of riverine species of *Nausitoria* and *Teredo* (Greenfield, 1952) are most concentrated near the bottom.

Rajagopal (1970) has suggested a preference for particular trees by different teredinids in the Sundarbans of India. Thus *T. manni* seems to prefer *Avicennia* and *Excoecaria; Bankia roonwali* (= *B. rochi*) (Turner, 1966), and *Nausitoria lanceolata* (= *N. dunlopei*) (Turner, 1966) seem to prefer *Ceriops, Xylocarpus,* and *Heritiera,* whereas all specimens (only three) of *N. sajnakhaliensis* (= *B. nordi*) (Turner, 1966) were collected from *Ceriops. Bankia campanellata* occurred commonly in *Excoecaria,* less so in *Ceriops, Xylocarpus,* and *Heritiera,* and rarely in *Avicennia.* Whether such observation are indicative of true wood selection by these bivalves, or of differential zonation or distribution within the mangrove, is unknown.

Iredale et al. (1932) report the destruction of mangrove timbers in Australian waters within 9 months. Kraemer (1944), however, ascribes the lasting qualities of *Rhizophora mucronata* to its resistance to marine borers. Edmondson (1955) has shown that the bark of *R. mangle* offers marked resistance to *Teredo* attack. Test woods with bark intact were without infestation for periods exceeding 15 months; removal of the bark resulted in a heavy *Teredo* attack within 140 days.

E. Mangrove Oysters

Probably all oysters that colonize mangrove roots can and do occur elsewhere, attached either to stones or shells embedded in the mud or to artificial wharves in estuaries where the mangroves have been cut. They are nevertheless a feature of every mangrove shore. From the Gulf of California, Parker (1963, 1964) records *Ostrea columbiensis* and *O. corteziensis.* Along the eastern and southern coasts of North America, *Crassostrea virginica* occurs as offshore reefs (Manzi et al., 1977), and this habitat has been artificially copied for culture in ponds (Lunz, 1955; Shaw, 1965; Badger, 1968; Anderson, 1976). (The extensive literature on the largely subtidal *C. virginica* is beyond the scope of this review.) Manzi et al. (1977) have confirmed, for example, that growth in this species is greater in ponds than in tidal creeks. Growth means ranged from 1.0 to 2.25 mm/month in creeks to as much as 3.11 mm/month in ponds. Growth was correlated with primary production and phytoplankton biomass, but there was an inverse correlation between growth and nutrient concentration.

In the Caribbean mangroves, *C. rhizophorae* is widespread (Saenz, 1965; Nikolic and Melendez, 1968; Angell, 1973), although Menzel (1973) has tentatively concluded that this animal is a subspecies of *C. virginica.* From West Africa, Bouchet (1977) records *C. gasar,* and throughout the Indo-Pacific and East Africa, *Saccostrea cucullata* is everywhere dominant (Taylor, 1971; Berry, 1975; Stojkovich, 1977; Pinto and Wignarajah, 1980). The taxonomy of the oysters has yet to be properly elucidated, but following Stenzel (1971), the genus *Saccostrea,* with type species *S. cucullata,* can be distinguished from *Crassostrea.* Species of *Saccostrea* are nonincubatory and have chomata on the margins of their shells; there is also a distinct pallial impression of separate small muscles constituting a disjunct pallial line. This description may be extended to include the distinctive features of the prodissoconch, such as orthogyrate umbones and symmetrical teeth. *Crassostrea,* on the other hand, includes oysters lacking chomata and a distinct pallial impression on the shell. In addition, the prodissoconchs show inequilateral growth, with modified posterior teeth and umbones that tend to be opisthogyrate (Stenzel, 1971; Dinamani, 1976). Gomez (1980b) records a wide range of other oysters from rocks and mangroves in the Indo-Pacific; probably most of these are invalid. *Saccostrea glomerata, S. denticulata, S. echinata, S. mordax,* and *S. commercialis,* for example, are either morphs or geographic varieties of *S. cucullata,* the latter probably better regarded as a *superspecies* (Dinamani, 1976). The range of this superspecies is enormous; from East Africa (Day, 1974; Branch and Grindley, 1979) to the Pacific islands, e.g., Guam [as *S. mordax* (Stojkovich, 1977)].

Most research has been undertaken on *C. rhizophorae,* the mariculture potential of this species in the Caribbean having been recently realized. General studies of *C. rhizophorae,* including its ecology and techniques involved in its culture, have been undertaken by Saenz (1965), Nikolic and Melendez (1968), Nikolic and Alfonso (1971), and Nikolic et al. (1976) in Cuba. The species is also cultivated in Jamaica and Puerto Rico (Mattox, 1949; Nikolic and Melendez, 1968), Trinidad (Bacon, 1970), Colombia (Wedler, 1980), Brazil (Nascimento and Pereira, 1980), and Venezuela (Rojas and Ruiz, 1972).

In Colombia, experimental artificial spat collectors have included mangrove roots or branches, automobile tires, oyster shells, and "eternite" or plastic trays. The plastic trays showed by far the best results, with 10,000–12,000 oysters/m^2. The greatest increase in meat concentration coupled with the least mortality was attained by growing the oysters during the first phase (up to the fourth month) in closed baskets, later changing to open suspended baskets. After 8 months, the oysters reached a length of 7 cm with only a 2% mortality (Wedler, 1980).

In Brazil, Nascimento et al. (1980b) have shown that a primary bisexual gonad is formed in *C. rhizophorae* at a size of 7 mm, 45 days after setting. The gonad contains the antecedents of both sexes but 90% of individuals are protandric. Sexual maturity is attained at a size of 20 mm, 120 days after setting; sex reversal is completed at 40 mm. In Trinidad, where *C. rhizophorae* occurs with a mean density of 341/root and a mean size of 25.5 mm (Bacon, 1970), the species seems to have a single period of reproduction extending from late summer (August–October) to March. No larvae were found in the plankton from April to July. Bacon (1970) could not correlate larval release with major seasonal changes or with any other physical factor, except for the fact that it occurred during periods of rising salinity; this was also noted by Wedler et al. (1978). A similar pattern of settlement is exhibited by *C. rhizophorae* in Venezuela (Rojas, 1972), with a peak in September–November and reduced settlement from January to March.

Successive spatfalls result in crowding and stunted growth. Under experimental conditions, oysters reached 40 mm after 4 months of growth during the wet season and 32 mm after 5 months of growth during the dry season (when temperatures were lower and salinities higher); 40–50 mm is considered a marketable size. From three mangrove stakes, Bacon (1970) obtained 135 oysters with a mean size of 35 mm and a total weight of 245 g after 3 months immersion. It is necessary to remove the bark, which exudes a slime, from the stakes, [also noted by Hunter (1969) in Sierra Leone] because this reduces the spatfall. Bark removal, however, makes the stakes more susceptible to borer (Pholadidae and Teredinidae) attack [Edmondson (1955); *Martesia striata,* in Trinidad] and rotting (Wedler, 1980), and the stakes must be discarded after one use. With longer periods of immersion (9 months) there is increased mortality (55%), and from 16 stakes, 809 oysters of mean length of 25 mm and total meat weight of 331.3 g were obtained.

Intense competition for space by *C. rhizophorae* results in a misshapen growth form, premature release from the substrate, and a shorter life span (5–7 months) (Wedler, 1980). Pora et al. (1969) have investigated the resistance of *C. rhizophorae* to starvation and asphyxia and concluded that in its natural environment it has a low resistance to starvation. After 10 to 15 days, a noticeable mortality occurs and the energy reserves of the body are considerably depleted. Long-lasting anoxia hastens mortality, but this can be reduced by allowing access to water from time to time, by enriching the water with calcium, and by lowering the temperature. *C. rhizophorae* thus does not possess the metabolic ability to withstand extended periods of emersion, as do its high-zoned relatives, e.g., *Geukensia* in *Spartina* marsh and *Geloina* in the Indo-Pacific landward margin of the mangrove.

Angell (1973) studied October and December spatfalls of *C. rhizophorae.* At 10 months and a length of 7 cm, growth becomes asymptotic. During the first 4 months, both groups grew equally; thereafter, the October spatfall grew slightly faster, Nascimento et al. (1980a) have similarly shown that at a size greater than 7 cm, growth rate and meat production decreased in *C. rhizophorae* from Brazil, although the size class between 4.1 and 6.0 cm gave the best commercial yields (Nascimento and Pereira, 1980).

Hunter (1969) has investigated the feasibility of culturing *Crassostrea gasar* in Sierra Leone. This author has shown that, as with *C. rhizophorae,* larval release and spatfall occur during the winter dry season beginning in November when the salinity reaches about 21%, and lasts until May when salinity falls to about 31%. The highest spatfall was recorded in January, when some 3000 spat/m^2 were counted on corrugated asbestos collectors left in the water for 2 weeks. Lefevre (quoted in Hunter, 1969) demonstrated that spat in Lagos Lagoon showed a strong moon photopositive reaction, and highest spat settlements of 1108–1342 spat/200 cm^2 occurred in the 7 days around the February and March full moons.

Ajana (1979, 1980) has reviewed the literature and investigated the feasibility of culturing *C. gasar* in Lagos Lagoon, Nigeria. Heavy mortality was encountered during the summer rainy season when salinities fell to zero over an extended period. It was concluded that only a small-scale fishery is possible because growth is limited to the high-salinity season from November to May (Sandison, 1967). Ajana (1979) has shown that settlement of *C. gasar* spat was optimal on collectors placed between 70 and 120 cm below the water surface, and preferentially on shaded hardwood collectors.

F. The Mytilacea

Representatives of the byssally attached superfamily Mytilacea are a consistent feature of every mangrove stand. Generally speaking, four microhabitats are occupied by this group. The first is the cracks and crevices among the mass of mangrove oysters growing on the prop roots. The second is the root–mud interface, in which the bivalves are attached either to roots or to stems (or stones or shells) but are buried in the mud. A third is the habit of a few bivalves of constructing byssal nests in the mud within which they are protectively cocooned. The fourth is lodgment within the tests of mangrove-associated ascidians.

The nestlers are almost universally represented by the genus *Brachidontes. B. exustus* (with *B. recurvus* and *B. citrinus*) occurs throughout the Caribbean and the Gulf of Mexico (Coomans, 1969; Perez, 1974); *B. niger* in West Africa (Bouchet, 1977); and *B. variabilis* (with *B. rostratus*) throughout the Indo-Pacific (Taylor, 1971; Berry, 1975; Sasekumar, 1980; Pinto and Wignarajah, 1980), including East Africa (Branch and Grindley, 1979).

Other nestlers recorded from among the oysters include *Septifer bilocularis* and *Perna perna* (East Africa) (Day, 1974) and *P. viridis* (the Indo-Pacific)

(Gomez, 1980b). The former is either a wrong identification, because this species is a typical inhabitant of exposed rocky beaches (Taylor, 1971), or the mangroves themselves were in a somewhat exposed position. Species of *Perna* are, however, estuarine mytilids that are being extensively cultivated in some areas of India and Asia (Piyakarnchana, 1980). They typically occur, however, lower down the shore from the mangroves, being grown artificially on stakes placed offshore.

Just as ubiquitous as *Brachidontes* but occupying a second microhabitat is the genus *Modiolus: M. americanus* (Bacon, 1970) in the Caribbean, *M. nigeriensis* [with *M. elegans* (Binder, 1968)] in West Africa, *M. philippinarum* (Day, 1974) in East Africa, and *M. phillipinarum* and *M. metcalfei* (Morton, 1977a) [with a range of other species recorded by Gomez (1980b)] in the Indo-Pacific. The genus *Modiolus* is ancient, possibly representative of the stock that ultimately produced the ventrally flattened colonizers of the rocky intertidal (a good example is *Septifer*) and also produced the byssally cocooned mytilids of tropical mudflats. *M. metcalfei* is relatively unspecialized; largely buried in mud, although attached to the mangrove roots, it has powerful cleansing currents in the mantle cavity (Morton, 1977a).

In the Caroni Swamp, Trinidad, Bacon (1975) records that *Mytella falcata* was found only on red mangrove rhizophores. It was most common in a vertical zone about 60–70 cm high between mean low and mean high water levels. The mytilid grew among oysters and barnacles, although it occurred frequently in compact clusters of 200–300 individuals, occupying the rhizophore to the almost total exclusion of other organisms.

The third group of mytilids are buried in the mud, sometimes in dense, gregarious masses [e.g., *Musculista senhausia* in the Indo-Pacific (Morton, 1973a)]. In the Caribbean, this habit has been adopted by *Mytella guyanensis* (Bacon, 1975). This bivalve affords an interesting comparison with the habit of the closely related *M. falcata*. In Trinidad, *M. guyanensis* lives in holes in the mud in the zone between high- and low-water neap tide levels. At low tide the posterior end of the shell is about 1 cm below the mud surface, but when covered by water, the mussel raises itself so that the extended mantle edge is level with the top of the elongated opening in the mud. Byssal attachment is to subterranean rootlets of both *Rhizophora* and *Avicennia* or to buried vegetable detritus. The byssal threads extend all around the mussel, attaching to various points on the burrow, so that the animal becomes enclosed in a ''nest'' of byssal threads. The mass of threads is pressed to the burrow walls and compacted by the regular opening and closing of the shell valves, giving the nest a smooth inside surface and consolidating the walls of the burrow. The byssal threads are very fine (0.03 mm in diameter) in comparison with those of other epibyssate bivalves (0.15 mm in diameter in *Perna perna*).

In East Africa, *Arcuatula* (*Lamya*) *capensis* (Day, 1974: Branch and Grindley,

1979) is apparently a mangrove associate, although in Hong Kong, *A. elegans* never occurs intertidally (Morton, 1980). In the Indo-Pacific, Gomez (1980b) records *Musculista senhausia* as being widely distributed on mangrove shores. This species has been studied by Morton (1973a), who showed it to be an essentially opportunistic bivalve, periodically occurring as dense mats of conjoined byssal cocoons (each containing a bivalve) that have a profoundly detrimental effect upon the normal fauna of a mudflat. The species also occurs on test panels for sublittoral foulers and in sublittoral muds (in Hong Kong); it is characterized by very rapid growth and the ability, because of its high fecundity, to quickly dominate a community. The nest-building habit is presumed to be a defensive measure because the shells of *Arcuatula* and *Musculista* are exceptionally thin, although Merrill and Turner (1963) report that *Musculus discors* protects its eggs within the cocoon. Bacon (1975) has shown, however, that *M. guyanensis* is predated upon by a number of gastropods, crabs, and birds. The byssal nest, the construction of which has been described by Merrill and Turner (1963), also allows these bivalves to colonize soft deposits and serves to keep mud away from the immediate vicinity of the bivalve, so that the mantle cavity, largely open, is not clogged by sediment.

A final (fourth) niche occupied by mytilids has been reported by Coomans (1969). *Musculus lateralis* lives in the tests of the ascidian *Bothryllus,* which shares the mangrove root habitat with the oysters.

G. The Isognomonidae

Isognomon isognomon and *I. ephippium* are commonly found in the Indo-Pacific mangrove (Berry, 1975), generally occurring close to the mud line and as bunched masses in the clefts of mangrove roots, typically at the seaward margin. *Isognomon alatus* (and to a lesser extent *Brachidontes exustus*) is extremely tolerant of high temperatures, as was demonstrated by Kolehmainen et al. (1973) in a thermally altered mangrove area in Puerto Rico. The effects of temperature and emersion upon *I. alatus* have been investigated by Trueman and Lowe (1970), and Yonge (1968) investigated the form and habits of *I. isognomon.* In the Gulf of Mexico and the Caribbean, *I. alatus* is widely distributed (Siung, 1980). The isognomonids are by no means restricted to mangroves, often occurring on wharf pilings and attached to stones intertidally and subtidally. In Port Royal, Jamaica, for example, only 15% of the total population of *I. alatus* occur intertidally, whereas 75% occur subtidally (Siung, 1980). Siung (1980) made a study of *I. alatus* in Jamaica and has shown that a high percentage of individuals have ripe gonads throughout the year. Peak spawning periods occur, however, after the onset of the rainy season when salinity decreases. By contrast, the mangrove oysters (Ostreidae) appear to spawn during the dry winter season, when salinity is high (Hunter, 1969; Bacon, 1970). The presence of eggs and sperms in the water induces other individuals of *I. alatus* to spawn. Growth of *I. alatus* is rapid, a height of some 50 mm being attained within 6 months. Survival rates expressed as a percentage of the total number of individuals growing on each panel were 33% intertidally and 17% subtidally. Siung has also pointed out that *I. alatus* has great potential as a cultivatable species; indeed, the bivalve is already cropped in Jamaica.

Sutherland (1980) has shown that in Venezuela, *I. alatus* appears to be invading new mangrove roots as well as increasing its coverage of old roots, the change in mean species composition for this bivalve ranging from 5% in May–June 1976 to 37% in September 1977. The spread of this species does not, however, result in the extinction of others because its shell is as attractive as the roots to other settling organisms.

H. The Arcidae

The mudflats in front of mangroves are typically colonized by members of the Arcidae (the "bloody cockles," so-called because of a hemoglobin blood pigment), which with the venerid *Anomalocardia* are the equivalent of representatives of the Cardiidae on temperate shores (Morton, 1978b).

In the Gulf of California and other mangrove areas of the west coast of the Americas, the Arcidae are represented by *Anadara tuberculosa* (Parker, 1963, 1964; Squires et al., 1975; Baquiero, 1980); *A. notabilis* (with *Arca umbonata*) in the Caribbean (Coomans, 1969); *Anadara senilis* in West Africa (Bouchet, 1977); *Arca gibba* and *A. obliquata* in South Africa (Branch and Grindley, 1979); and *Anadara granosa* plus a range of other species in the Indo-Pacific (Pathansali and Soong, 1958; Lim, 1963; Bouchet, 1977; Sasekumar, 1980; Piyakarnchana, 1980; Wells and Slack-Smith, 1981).

Purchon (1956) investigated the biology and morphology of *A. granosa;* this information was later supplemented by Lim (1966). The species is extensively cultivated in Southeast Asia (Piyakarnchana, 1980). Pathansali and Soong (1958) and Pathansali (1966) suggest that *A. granosa* is unisexual. Gonads do not develop until a length of 17.5 mm is attained, and the first spawning occurs at a length of 24–25 mm. A peak of spawning occurs in October in Malaysia with a second peak in May/June (Broom, 1983). Yoloye (1974) argues that *A. senilis* is a protandric consecutive hermaphrodite, although the most recent research on this species by Yankson (1982) suggests that it too is dioecious. In mangrove areas on the west coast of Mexico, *A. tuberculosa* is intensively exploited (Baqueiro et al., 1978). Previous studies of this species by Flores (1971) and Squires et al. (1975) have recently been reviewed by Baqueiro (1980). *A. tuberculosa* reaches its highest density between the roots of *Rhizophora mangle* in muddy bottoms rich in organic matter. It was never found in sandy or shelly bottoms even if *R. mangle* was present, indicating that the substrate is more important than the mangroves themselves. Variations in density ranged from $47.75/m^2$ to $0.6/m^2$. The populations in Baja California possessed three age classes, with the largest including individuals greater than 80 mm in length (Baqueiro, 1980).

I. Other Byssally Attached Nestlers

The Mytilidae are the typical nestlers of the mangrove environment. In the Gulf of Mexico, the dreissenid *Mytilopsis leucophaeta* is widely distributed (Odum and Heald, 1972, 1975). *M. adamsi* occurs in the Gulf of California (Parker, 1963, 1964) and *M. sallei* in lagoons along the coast of Venezuela (Escarbassiere and Almeida, 1976). *Mytilopsis* typically occurs beyond the mangrove itself, attached to stones or shells in the lower eulittoral muds. *Barbatia cancellaria* (Arcidae) (Perez, 1974) apparently also occurs attached to mangrove roots. Three species of the Pteriidae, i.e., *Pteria colymbus, Pinctada radiata,* and *P. imbricata* (Coomans, 1969; Perez, 1974), also attach to mangrove roots. The normal habit of *Pteria colymbus* is, however, attachment to the gorgonian *Leptogorgia,* and therefore it is incorrectly included in the mangrove community, except perhaps accidentally (Patton, 1972).

In the Indo-Pacific, other nestlers include two species of *Trapezium,* i.e., *T. sublaevigatum* (Berry, 1963; Piyakarnchana, 1980) and *T. angulatum* (Frith et al., 1976). Morton (1979) has, however, studied *T. sublaevigatum* and shown that the species has a very wide vertical distribution, being an opportunistic colonizer of a wide range of habitats at a variety of depths.

J. Other Infaunal Mangrove Associates

The seaward fringe of sand or mud beyond the mangroves is the home of a wide array of burrowing bivalves, all taking advantage, no doubt, of the rich suspension of nutrients in the mangrove waters. Macnae (1963) points out that the burrowing bivalves he recorded from South African mangroves were all, in fact, collected from *Zostera* meadows below the lowest *Avicennia* pneumatophores. Similarly, Taylor (1971) believes the burrowing bivalves he records from the western Indian Ocean mangroves are simply tolerant of any muddy-sand sheltered shore. With few exceptions, however, it is impossible to say whether or not these bivalves are true mangrove associates (as are *Enigmonia* and *Geloina*) or will occur on any tropical intertidal mud flat, whether mangroves are present or not. Families recorded include representatives of the Veneridae, Lucinidae, Psammobiidae, Cardiidae, Tellinidae, Asaphidae, Glauconomidae, Solenidae, Cultellidae, Solecurtidae, Ungulinidae, Mactridae, Semelidae, Condylocardiidae, Corbulidae, and Nuculanidae. These can be divided into two categories, suspension and deposit feeders; the Lucinidae, Ungulinidae, Tellinidae, Nuculanidae, Semelidae, and Mactridae are representative of the latter.

From this array, a few are noteworthy. Some families are more or less consistently reported from mangroves, e.g., the Mactridae, with *Rangia mendica* in the Gulf of California (Parker, 1963, 1964), *Mulinia lateralis* and *R. cuneata* in the Gulf of Mexico (Harry, 1942), and two species of *Mactra* in West Africa (Binder, 1968). In the Indo-Pacific, no mactrid mangrove associate has hitherto been noted. The Psammobiidae are also widespread, e.g., *Asaphis deflorata* in the Caribbean (Coomans, 1969), *Tagelus angulata* in West Africa (Binder, 1968), *Psammotellina capensis* in East Africa (Macnae, 1963), and *Elizia orbiculata* in the Indo-Pacific (Lim, 1963). The last occurs in very soft, sloshy muds in Malaysia, often beyond the mangroves. In Hong Kong, *Asaphis dichotoma* occurs high zoned on cobble beaches (Narchi, 1980) and not with mangroves. The Lucinidae and Ungulinidae also occur almost universally, these ancient bivalves possessing unusual adaptations for life in soft muds (Allen, 1958). The Solenidae do not seem to have representatives on mangrove shores in the Americas, but these deep-burrowing bivalves are well represented in West Africa [*Solen guineensis* (Binder, 1968)], South and East Africa [*S. corneus* (Day, 1974)] and the Indo-Pacific [*S. delesserti* and *Sinonovacula constricta* (Frith et al. 1976)].

Owen (1959) reports upon *Glauconome rugosa* (Glauconomidae) and *Pharella acuminata* (Cultellidae) from the Asian mangrove. Within the Indo-Pacific at least, *Glauconome* seems to be a true mangrove associate. "They occur intertidally on the banks of streams running through mangrove swamps and although there is a good tidal flow the conditions are calm and well protected from wave action. The water is brackish and almost fresh at high water. They are found in stiff mud at a depth of 7 to 10 in. and the maximum density observed was eighteen to twenty-two animals/sq. yd." (Owen, 1959).

Glauconome is one of the very few deep-burrowing members of the Veneracea, being convergently similar to the Solenacea, and has powerful cleansing currents in the mantle cavity that remove sediment. This is not true of the venerid *Anomalocardia squamosa* from the Indo-Pacific. Although often recorded from the mangrove, for example by Gomez (1980b), this species in fact occurs low-zoned on sand flats (Morton, 1978b). Similarly, *Meretrix meretrix* and *Caecella* spp. do not occur in mangroves in Hong Kong; the former is adapted to high wave energy sand flats and the latter (*C. chinensis*) to cobble beaches, although here it is high zoned (Narchi, 1980). Conversely, the North American *Anomalocardia cunimeris* does appear to be associated at least with the much finer sediments of *Spartina* marshes and mangroves (Breuer, 1957; Odum and Heald, 1972, 1975) as is *A. brasiliana* in the Caribbean (Coomans, 1969); Narchi (1972), however, records the latter species as occurring in sandy deposits in Brazil (as does *A. squamosa* in Hong Kong). In the Indo-Pacific, representatives of the venerid genus *Gafrarium* are consistently recorded from the mangrove fringe, e.g., *G. divaricatum* (Gomez, 1980b), *G. tumidum* (Frith et al., 1976), and *G. pectinatum* (Taylor, 1971); other species of this genus also occur in temperate waters (Ansell, 1961).

Clearly, much more research must be undertaken on the precise ecology of this very wide assemblage of little-known bivalves.

IX. Discussion

The various mangrove areas of the tropics possess a relatively rich community of other plants and animals. Of these the bivalves are probably the least well known, as there is little comprehensive information in the literature about them. Macnae (1963) originally posed the question ''is there a specific mangrove fauna''? and this chapter has perhaps partially answered that question if only for the bivalves.

It is probable that most of the tropical brackish-water shipworms (the Teredinidae and a few Pholadidae) evolved in this environment (Hoagland and Turner, 1981); one of the more primitive, *Kuphus,* is recorded only from mangrove muds, with a wide range of other genera and species boring the often living trees (Rajagopal, 1970). Most modern research investigates shipworms colonizing woods placed placed artificially in the sea, either piers or test panels; very few studies have been made of them in their normal environment. At the present time, it is impossible to assess their impact upon the mangroves with regard to their role in the breakdown of dead wood or their effects upon living trees. Virtually nothing is known of their ecology, distribution patterns, reproductive cycles, conditions for growth, or population dynamics. This is partly understandable because they are the least obvious of mangrove bivalves.

In the Indo-Pacific, it is clear that the widely distributed *Enigmonia aenigmatica* is a specific mangrove associate; it is even camouflaged according to its chosen microhabitat, i.e., cream when attached to leaves and red on the stems. Its distribution, orientation, and zonation (Berry, 1975) within the mangrove clearly establish it as well adapted to this environment, a conclusion also borne out by studies of its anatomy (Morton, 1976c). Nothing is known, however, of its reproduction, population structure, or life span.

In the Caribbean and on the northern shores of the Gulf of Mexico, the Gulf of California, and the Indo-Pacific, the genus *Polymesoda* is another mangrove associate. In the Indo-Pacific, at least, where the subgenus *Geloina* is high zoned (the Caribbean *Pseudocyrena* and North American *Polymesoda* are low zoned), the various species have never been found outside the mangal. Unlike *Enigmonia,* with its unusual modifications, *Geloina* has a rather simple body plan, yet possesses equally remarkable physiological and behavioral adaptations to life high above the level of neap tides. The Corbiculacea probably evolved in the sea, although most species today are freshwater and riverine. In many ways, *Geloina* is colonizing a habitat that is intermediate between these two major systems, with the added complication of its high-zoned, almost semiterrestrial, existence. The acid mangrove soils destroy variable amounts of the shell, which makes aging and any study of the population dynamics of this genus very difficult. Three species occur in the Indo-Pacific, the subgenera *Pseudocyrena* and *Polymesoda* possibly being represented by an even greater array of species in the Americas (Parker, 1963, 1964); although a judicious review of the genus might reduce this figure considerably and elevate some of the various subgenera to generic status. In many ways, *Geloina* is the ecological equivalent of the similarly high-zoned mytilid of the temperate marshes of eastern and southern North America, *Geukensia demissa,* which although it can also withstand considerable periods of emersion (Lent, 1969) and lives for a comparatively long time, is in many morphological ways (like *Geloina*) unspecialized (Pierce, 1973).

In the Indo-Pacific mangrove, the rare anomalodesmatan *Laternula truncata* has remarkable anatomical adaptations to an infaunal mode of life close to the seaward fringe, but again, nothing is known about it other than its anatomy (Morton, 1973b). Species of the genus *Glauconome* (Glauconomidae), veneraceans adapted to a deep, burrowing mode of life also seem to be distinct mangrove associates (Owen, 1959) as are species of the genera *Pharella* (Cultellidae) and *Sinonovacula* (Solenidae).

The above bivalves, all recorded from Asian mangroves, clearly are highly specialized associates. It is significant that these are mostly Indo-Pacific species. This region, with the greatest diversity of mangrove plants and associated fauna, might be expected to have a greater number of species which have evolved intimately within this ecosystem.

The remaining bivalves from all regions are less specific in their choice of habitat. The oysters, a consistent feature of all mangroves, forming a thick crust around the prop roots of the pioneer trees, can and do live elsewhere. Their nestling associates, which are typically members of the Mytilidae and Isognomonidae, occur with the oysters in areas where mangroves are not present or have been cut. The same applies to virtually all of the other attached or infaunal bivalves.

Plante (1964) records different faunas from different genera of mangroves, but Taylor (1971) believes that this is a reflection of the different situations occupied by the mangroves rather than of any microsubstrate differences among them. Thus, except for the Teredinidae, which, as suggested by Rajagopal (1970), may have preferences for different trees, it would seem that in the case of the burrowing mangrove bivalves distribution is sediment related. The trees, coincidentally, have similar preferences, although by their very presence, they progressively adjust the composition of the sediment. Conversely, the mangrove root associates are probably influenced only by the water, especially its salinity, and can occur anywhere on anything solid within their salinity tolerance range. Thus, much of a mangrove bivalve community is not specifically adapted to the ecosystem.

Nevertheless, it is also clear that the great variety and number of bivalves that do occur with the mangroves are there because of the rich food supply that the mangal provides, as demonstrated for *Rangia cuneata* and *Polymesoda caroliniana* by Olsen (1976). Accreting waves and tide bring sea-borne nutrients to the mangrove. The rivers and streams draining through the mangrove provide terrigenous nutrients, whereas falling mangrove leaves and their subsequent biodegradation further add to the productivity of the system as a whole. Occurring in sheltered positions with little wave action, the mangal is thus biologically rich and stable enough to permit colonization by a wide array of bivalves with a variety of adaptations to life either in soft mud or attached to the firm substrate of the prop roots.

Binder (1968) divides the mangrove fauna of the Ebrie Lagoon on the Ivory Coast into three categories. Characteristic species are *Anadara senilis, Modiolus nigeriensis, Mactra diolensis, Loripes aberrans,* and *Tagelus angulatus.* Accompanying species are *Callista floridella, Mactra glabrata, Tellina distorta, Cuna gambiensis,* and *Diplodonta diaphana.* A third category of euryhaline species can occur if sediment and salinity requirements are met. Representatives of this category include *Leda bicuspidata, Modiolus elegans, Dosinia isocardia, Tellina nymphalis, Abra pilsbryi, Diplodonta globosa,* and *Solen guineensis.*

Wells and Slack-Smith (1981) have suggested that there seems to be a paucity of burrowing bivalves within the mangrove, possibly because the mangrove they investigated was at the limits of the tropics. They concluded that the sediment might be too fine for molluscan burrowing (although other burrowing phyla manage, and bivalves are plentiful beyond the seaward fringe where the mud may be even sloppier).

Vermeij (1974) pointed out that the mangrove soils are very acidic and showed that a large percentage of mangrove molluscs exhibit calcium carbonate resorption or conservation. *Geloina,* in Indo-Pacific mangroves, has a very thick shell which is often badly eroded; perhaps thinner-shelled species, unable to survive the acid mangal conditions, are limited to the seaward edge. *Laternula* and *Enigmonia,* however, have very thin shells (Morton, 1973b, 1976c). It should also be pointed out that mangroves are typically high zoned (i.e., occur in the upper half of the eulittoral) and that possibly one would not expect to find very many bivalves here, because in no intertidal habitats are high-zoned filter feeders dominant.

Despite all these arguments, however, it is clear (Tables I–VII) that bivalves do colonize mangroves, possibly because the muds retain water, allowing colonization at levels higher than are typical. In some areas, e.g., the Caribbean and the Indo-Pacific, they occur in large numbers both of species and of individuals. Many, especially the oysters, mussels, and ark-shells, are sufficiently numerous to be of great economic importance. Apart from *Geloina,* which inhabits the back of the mangrove, most species occur on the seaward fringe in what Berry (1963) has termed the "bivalve zone," recognizing the importance of this group in the ecosystem.

Vermeij (1974) has shown that the greater diversity of mangrove gastropods in Singapore relative to Madagascar and the Ivory Coast is not achieved by an increase in the species : family ratio, but rather by an increase in the number of families. Table VIII, although indicating only some of the families, genera, and species of bivalves from the mangrove regions herein discussed (because faunal lists are not complete), lends some support to the conclusions of Vermeij (1974). Thus, a comparison of the mangrove bivalves of East Africa with those of the other regions of the Indo-Pacific shows a doubling in the number of families and genera. The number of species, however, almost quadruples, apparently contradicting Vermeij, for an increase in species richness. This may, however, be misleading because taxonomic decisions at the species level are less satisfactory than those at the generic and family levels. Gomez (1980b), for example, records 12 species of oysters from the Indo-Pacific, which is almost certainly an exaggeration. Similarly, many of the infaunal bivalves reported from the Asian mangal probably do not have an intimate relationship with the mangrove but are occurring here at the more extreme limits of their range.

Comparing the bivalve fauna of marsh and mangrove communities along the southern coast of the United States with that of the Caribbean, there is again a doubling of families and genera but a trebling of species. Although these faunas, because of an overall reduced diversity in comparison with the Indo-Pacific, are better known (consequently rendering taxonomic decisions less problematic), synonomy of some species may reduce an apparent richness of species and thus give more support to the contention of Vermeij (1974) that increasing diversity toward the foci of mangrove development comes about through familial (and generic) increases. Here also some bivalves are wrongly placed in the mangrove. Coomans (1969), for example, records *Pteria colymbus* from the Caribbean

TABLE VIII

The Numbers of Families, Genera, and Species of Bivalves Recorded from Mangrove Areas

	West coast of the Americas	Southern coast of North America	Caribbean and northern coast of South America	West Africa	East Africa	Indo-Pacific
Families	9	7	17	14	10	19
Genera	10	9	27	18	19	40
Species	17	10	37	23	24	86

[a] Data extracted from Tables II–VII.

mangal, but the true habit of this bivalve is attachment to the gorgonian *Leptogorgia* (Patton, 1972). Species diversity thus may be artificially exaggerated here also.

Many of these families (and sometimes genera) have considerable conformity (Table I) in the mangrove areas of the world. It would thus appear that many families have evolved representatives within this system (to lesser or greater degrees of dependence), although these may involve only one or two genera and, in turn, only one or two species (Tables I–VIII). Such a picture argues for narrow adaptive radiation from a relatively wide array of stocks that has attained to varying degrees, identity with the tropical, estuarine, mangrove community. This difficult, high-zoned, muddy habitat is thus characterized by evolutionary conservatism in the Bivalvia.

This picture is in contrast to that seen for the bivalves of coral shores, where there is low familial and generic, although high species, diversity. This argues for wide adaptive radiation specifically within this system from narrow stocks (Morton, 1983b) and for evolutionary dynamicism.

References

Abel, O. (1926). Fossil Mangrovesumpfe. *Palaeontol. Z.* **8,** 130–139.

Adal, M. N., and Morton, B. S. (1973). The fine structure of the pallial eyes of *Laternula truncata* (Bivalvia: Anomalodesmata: Pandoracea). *J. Zool.* **171,** 533–556.

Ajana, A. M. (1979). Preliminary investigation into some factors affecting the settlement of the larvae of the mangrove oyster *Crassostrea gasar* (Adanson) in the Lagos lagoon. *Malacologia* **18,** 271–275.

Ajana, A. M. (1980). Fishery of the mangrove oyster, *Crassostrea gasar,* Adanson (1757), in the Lagos area, Nigeria. *Aquaculture* **21,** 129–137.

Allen, J. A. (1958). The basic form and adaptations to habitat in the Lucinacea (Eulamellibranchia). *Philos. Trans. R. Soc. London, Ser. B* **241,** 421–484.

Altena, C. O. van Regteren (1969). The marine Mollusca of Suriname (Dutch Guiana) holocene and recent. Part 1. General introduction. *Zool. Verh.* **101,** 3–49.

Anderson, W. D. (1976). A comparative study of a salt water impoundment and its adjacent tidal creek pertinent to the culture of *Crassostrea virginica* (Gmelin). Thesis, Old Dominian University, Norfolk, Virginia.

Andrews, J. D., and Cook, K. 1951. Range and habitat of the clam *Polymesoda caroliniana* (Bosc) in Virginia (Family Cycladidae). *Ecology* **32,** 758–760.

Angell, C. L. (1973). Crecimento y mortalidad de la Ostra de mangle cultivada (*Crassostrea rhizophorae*). *Mem. Soc. Cienc. Nat. La Salle* **33**(94/95), 152–162.

Annandale, N. (1924). Bivalve molluscs injuring brickwork in the Calcutta docks. *Resp. Comm. Inst. Civ. Eng.* **4,** 60–61.

Anonymous (1980). "The Present State of Mangrove Ecosystems in Southeast Asia and the Impact of Pollution: Singapore," Working Pap. SCS/80/WP/94d. South China Fisheries Development and Coordinating Programme.

Ansell, A. D. (1961). The functional morphology of the British species of Veneracea (Eulamellibranchia). *J. Mar. Biol. Assoc. U.K.* **41,** 489–515.

Bacon, P. R. (1970). Studies on the biology and cultivation of the mangrove oyster in Trinidad with notes on other shellfish resources. *Trop. Sci.* **12,** 265–278.

Bacon, P. R. (1975). Shell form, byssal development and habitat of *Mytella guyanensis* (Lamarck) and *M. falcata* (Orbigny) (Pelecypoda: Mytilidae) in Trinidad, West Indies. *Proc. Malacol. Soc. London* **41,** 511–520.

Badger, A. G. (1968). Oyster research in South Carolina. *In* "Proceedings of the Oyster Culture Workshop," Contrib. Ser. 6, pp. 67–69. Ga. Game Fish. Comm., Athens, Georgia.

Balasubramanian, T., Lakshmanaperumalsamy, P., Chandramohan, D., and Natarajan, R. (1979). Cellulolytic activity of *Streptomycetes* isolated from the digestive tract of a marine borer. *Indian J. Mar. Sci.* **8,** 111–114.

Baquiero, C. E. (1980). Population structure of the mangrove cockle *Anadara tuberculosa* (Sowerby, 1833) from eight mangrove swamps in Magdalena and Almejas Bays, Baja California Sur, Mexico. *Proc. Natl. Shellfish. Assoc.* **70,** 201–206.

Baquiero, C. E., Alvarez, L. R., and Kensler, C. B. (1978). Analisis de la producción nacional de mariscos para el periodo de 1970–1976. *Congr. Nac. Oceanogr., 1978* pp. 1–18.

Bentham-Jutting, W. S. S. van. (1954). Systematic studies on the non-marine Mollusca of the Indo-Australian Archipelago. IV. Critical revision of the freshwater bivalves of Java. *Treubia* **22,** 19–73.

Berry, A. J. (1963). Faunal zonation in mangrove swamps. *Bull. Natl. Mus. Singapore* **32,** 90–98.

Berry, A. J. (1972). The natural history of West Malaysian mangrove faunas. *Malay., Nat. J.* **25,** 135–162.

Berry, A. J. (1975). Molluscs colonizing mangrove trees with observations on *Enigmonia rosea* (Anomiidae). *Proc. Malacol. Soc. London* **41,** 589–600.

Bertness, M. (1980). Growth and mortality in the ribbed mussel *Geukensia demissa* (Bivalvia: Dreissenacea). *Veliger* **23,** 62–69.

Binder, E. (1968). Répartition des Mollusques dans la laguna Ebrié (Côte d'Ivoire). *Cah. ORSTOM, Ser. Hydrobiol.* **2,** 2–34.

Bouchet, P. (1977). Distribution des mollusques dans les mangroves du Senegal. *Malacologia* **16,** 67–74.

Bourne, G. C. (1907). On the structure of *Aenigma aenigmatica* Chemnitz; a contribution to our knowledge of the Anomiacea. *Q. J. Microsc. Sci.* **51,** 258–295.

Branch, G. M., and Grindley, J. R. (1979). Ecology of southern African estuaries. Part XI. Mngazana: a mangrove estuary in Transkei. *S.-Afr. Tydskr. Dierk.* **14,** 149–170.

Brandt, R. A. M. (1974). The non marine aquatic Mollusca of Thailand. *Arch. Molluskenkd.* **105,** 1–423.

Breuer, J. P. (1957). An ecological survey of Baffin and Alazan bays, Texas. *Inst. Mar. Sci.* **4,** 134–155.

Broom, M. J. (1983). Gonad development and spawning in *Anadara granosa* (L.) (Bivalvia: Arcidae). *Aquaculture* **30,** 211–219.

Brown, D. A. (1971). The ecology of Gastropoda in a South African mangrove swamp. *Proc. Malacol. Soc. London* **39,** 263–279.

Castaing, A., Jiménez, A. M., and Villalobos, C. R. (1980). Observaciones sobre la ecologia de manglores de la Costa Pacifica de Costa Rica y su relación con la distribución del molusco *Geloina inflata* (Philippi) (Pelecypoda: Corbiculidae). *Rev. Biol. Trop.* **28,** 323–339.

Contreras, R., and Cantera, J. R. (1978). Notas Sobre la Ecologia de los Moluscos Asociados al Ecosistema Manglar-Estero en la Costa del Pacifico Colombiano. *Memoirs. Seminar on the South American Pacific Ocean. Univ. Valle, Cali (Colombia)* **2,** 709–770.

Coomans, H. E. (1969). Biological aspects of mangrove molluscs in the West Indies. *Malacologia* **9,** 79–84.

Day, J. H. (1974). The ecology of Morrumbeene estuary, Mocambique. *Trans. R. Soc. S. Afr.* **41,** 43–97.

Dinamani, P. (1976). The morphology of the larval shell of *Saccostrea glomerata* (Gould, 1850) and a comparative study of the larval shell in the genus *Crassostrea* Sacco, 1897 (Ostreidae). *J. Moll. Stud.* **42,** 95–107.

Duobinis, E. M., and Hackney, C. T. (1979). Seasonal and spatial distribution of the Carolina marsh clam, *Polymesoda caroliniana* (Pelecypoda: Corbiculidae), in a Mississippi tidal marsh. *ASB Bull.* **25,** 44.

Edmondson, C. H. (1955). Resistance of woods to marine borers in Hawaiian waters. *Bernice P. Bishop Mus. Bull.* **217,** 1–91.

Ekman, S. (1953). ''Zoogeography of the Sea'' Sidgwick & Jackson, London.

Ellis, R. (1968). Muluscos de Nicaragua y Costa Rica. *Programma Reg. Desarrollo Pesq. Cent. Am. Inf.* pp. 1–8.

Escarbassiere, R. M., and Almeida, P. (1976). Biological and ecological aspects of *Mytilopsis sallei* Recluz (Bivalvia: Eulamellibranchia) in adjacent areas of the Unare Lagoon (Estado Anzoategui, Venezuala). *Acta Biol. Venez.* **9,** 165–194.

Fairbanks, L. D. (1963). Biodemographic studies of the clam *Rangia cuneata* Gray. *Tulane Stud. Zool.* **10,** 3–47.

Flores, M. A. (1971). Contribución al conocimiento biológico de la pata de mula, *Anadara (Anadara) tuberculosa* (Sow., 1833). *Tesis Prof. Esc. Nac. Cienc. Biol. Inst. Pol. Nal. Mex.* pp. 1–130.

Frith, D. W., Tantanasiriwong, R., and Bhatia, O. (1976). Zonation and abundance of macrofauna on a mangrove shore, Phuket Island, Southern Thailand. *Phuket Mar. Biol. Cent. Res. Bull.* **10,** 1–34.

Ganapati, P. N., and Lakshmana Rao, M. V. (1959). Incidence of marine borers in the mangroves of the Godavari estuary. *Curr. Sci.* **28,** 232.

Gerlach, S. A. (1958). Die Mangroveregion tropischer Küsten als Lebensraum. *Z. Morphol. Oekol. Tiere* **46,** 636–730.

Godwin, W. F. (1968). The distribution and density of the brackish water clam, *Rangia cuneata,* in the Altamaha River, Georgia. *Ga. Game Fish. Comm., Mar. Fish. Div. Cent., Ser. 5* pp. 1–10.

Gomez, E. D. (1980a). ''The Present State of Mangrove Ecosystems in Southeast Asia and the Impact of Pollution: Philippines,'' Working Pap. SCS/80/WP/94c. South China Sea Fisheries Development and Coordinating Programme.

Gomez, E. D. (1980b). ''The Present State of Mangrove Systems in Southeast Asia and the Impact of Pollution: Regional, Working Pap. SCS/80/WP/94. South China Sea Fisheries Development and Coordinating Programme.

Gray, J. E. (1849). On the species of Anomiidae. *Proc. Zool. Soc. London* **1849,** 113–124.

Greenfield, L. J. (1952). The distribution of marine borers in the Miami area in relation to ecological conditions. *Bull. Mar. Sci. Gulf. Caribb.* **2,** 448–464.

Harry, H. W. (1942). List of Mollusca of Grand Isle, Louisiana, recorded from the Louisiana State University Marine Laboratory 1929–1941. *Occas. Pap. Mar. Lab., La. State Univ.* No. 1, pp. 1–13.

Hoagland, K. E., and Turner, R. D. (1981). Evolution and adaptive radiation of wood-boring bivalves (Pholadacea). *Malacologia* 21, 111–148.

Hopkins, S. H., and Andrews, J. D. (1970). *Rangia cuneata* on the east coast: Thousand mile range extension or resurgence? *Science* **167,** 868–869.

Hunter, J. B. (1969). A survey of the oyster population of the Freetown estuary, Sierra Leone, with notes on the ecology, cultivation and possible utilisation of mangrove oysters. *Trop. Sci.* **11,** 276–285.

Iredale, T. (1939). Mollusca Part 1. *Scient. Rep. Gt. Barrier Reef Exped.* (1928–1929). **5,** 209–425.

Iredale, T., Johnson, R. A., and McNeill, F. A. (1932). Destruction of timber by marine organisms in the Port of Sydney, *Sydney Harbour Trust.*

Kanwisher, J. W. (1955). Freezing in intertidal animals. *Biol. Bull. (Woods Hole, Mass.)* **109,** 56–63.

Keen, M., and Casey, R. (1969). Superfamily Corbiculacea Gray 1847. *In* "Treatise on Invertebrate Paleontology" (R. C. Moore, ed.), Part N, Mollusca 6 (2 of 3), pp. 664–670. Geol. Soc. Am. and Univ. of Kansas Press, Lawrence.

Kohlmeyer, J., and Kohlmeyer, E. (1965). New marine fungi from mangroves and trees along eroding shorelines. *Nova Hedwigia* **9,** 89–104.

Kolehmainen, S., Morgan, T., and Castro, R. (1973). Mangrove root communities in a thermally altered area in Guayanilla Bay, Puerto Rico. *In* "Aquirre Power Project Environmental Studies 1972 Annual Report," USAEC Rep. PRNC-162, pp. 320–360. Puerto Rico Nuclear Center.

Kraemer, J. H. (1944). "Native Woods for Construction Purposes in the Western Pacific Region." Bureau of Yards and Docks, Department of Navy, Washington, D.C.

Kuenzler, E. J. (1961). Structure and energy flow of a mussel population in a Georgia salt marsh. *Limnol. Oceanogr.* **6,** 191–204.

Lent, C. M. (1969). Adaptations of the ribbed mussel, *Modiolus demissa* (Dillwyn), to the intertidal habitat. *Am. Zool.* **9,** 283–292.

Lim, C. F. (1963). A preliminary illustrated account of mangrove molluscs from Singapore and South-West Malaya. *Malay. Nat. J.* **17,** 235–239.

Lim, C. F. (1966). A comparative study on the ciliary feeding mechanisms of *Anadara* species from different habitats. *Biol. Bull. (Woods Hole, Mass.)* **130,** 106–117.

Loi, T. N. (1967). Peuplements animaux et végétaux du substrat dur intertidal de la Baie de Nha Trang (Viet Nam). *Inst. Oceanogr. Nha Trang, Mem.* **11,** 1–236.

Lunz, G. R. (1955). Cultivation of oysters in ponds at Bears Bluff Laboratories. *Proc. Natl. Shellfish. Assoc.* **46,** 83–87.

Lutz, R. A., and Castagna, M. (1980). Age composition and growth rate of a mussel (*Geukensia demissa*) population in a Virginia salt marsh. *J. Moll. Stud.* **46,** 106–115.

Macnae, W. (1963). Mangrove swamps in South Africa. *J. Ecol.* **51,** 1–25.

Macnae, W. (1968). A general account of the fauna and flora of mangrove swamps and forests in the Indo-West-Pacific region. *Adv. Mar. Biol.* **6,** 73–720.

Macnae, W., and Kalk, M. (1962). The ecology of mangrove swamps of Inhaca Island, Mocambique. *J. Anim. Ecol.* **31,** 93–128.

Manzi, J. J., Burrell, V. G., and Carson, W. Z. (1977). A comparison of growth and survival of subtidal *Crassostrea virginica* in South Carolina salt marsh impoundments. *Aquaculture* **12,** 293–310.

Marelli, D. C., and Berrend, R. F. (1978). A new species record for *Mytilopsis sallei* (Recluz) in Central America (Mollusca: Pelecypoda). *Veliger* **21,** 144.

Mattox, N. T. (1949). Studies on the biology of the edible oyster, *Ostrea rhizophorae* Guilding, in Puerto Rico. *Ecol. Monogr.* **19,** 339–356.

Menzel, R. W. (1973). Hybridization in oysters (*Crassostrea*). *Malacol. Rev.* **6,** 179.

Merrill, A. S., and Turner, R. D. (1963). Nest building in the bivalve genera *Musculus* and *Lima. Veliger* **6,** 55–59.

Morton, B. S. (1971). A note on *Martesia striata* (Pholadidae) tunnelling into plastic piping in Hong Kong. *Malacol. Rev.* **4,** 207–208.

Morton, B. S. (1973a). Some aspects of the biology, population dynamics and functional morphology of *Musculista senhausia* Benson (Bivalvia: Mytilacea). *Pac. Sci.* **28,** 19–33.

Morton, B. S. (1973b). The biology and functional morphology of *Laternula truncata* (Lamarck 1818) (Bivalvia: Anomalodesmata: Pandoracea). *Biol. Bull. (Woods Holes, Mass.)* **145,** 509–531.

Morton, B. S. (1976a). The structure, mode of operation and variation in form of the shell of the Laternulidae (Bivalvia: Anomalodesmata: Pandoracea). *J. Moll. Stud.* **42,** 261–278.

Morton, B. S. (1976b). The diurnal rhythm and the feeding response of the South East Asian mangrove bivalve, *Geloina proxima* Prime 1864 (Bivalvia: Corbiculacea). *Forma Funct.* **8,** 405–418.

Morton, B. S. (1976c). The biology, ecology and functional aspects of the organs of feeding and digestion of the S.E. Asian mangrove bivalve, *Enigmonia aenigmatica* (Mollusca: Anomiacea). *J. Zool.* **179,** 437–466.

Morton, B. S. (1976d). The biology and functional morphology of the S.E. Asian mangrove bivalve *Polymesoda* (*Geloina*) *erosa* (Solander 1786) (Bivalvia: Corbiculidae). *Can. J. Zool.* **54,** 482–500.

Morton, B. S. (1977a). The biology and functional morphology of *Modiolus metcalfei* Hanley 1844 (Bivalvia: Mytilacea) from the Singapore mangrove. *Malacologia* **16,** 501–508.

Morton, B. S. (1977b). The population dynamics of *Corbicula fluminea* (Müller 1774) (Bivalvia: Corbiculacea) in Plover Cover Reservoir, Hong Kong. *J. Zool.* **181,** 21–42.

Morton, B. S. (1978a). Feeding and digestion in shipworms. *Oceanogr. Mar. Biol.* **16,** 107–144.

Morton, B. S. (1978b). The population dynamics of *Anomalocardia squamosa* Lamarck (Bivalvia: Veneracea) in Hong Kong. *J. Moll. Stud.* **44,** 135–144.

Morton, B. S. (1979). Some aspects of the biology and functional morphology of *Trapezium* (*Neotrapezium*) *sublaevigatum* (Lamarck) (Bivalvia: Arcticacea). *Pac. Sci.* **33,** 177–194.

Morton, B. S. (1980). The biology and some aspects of the functional morphology of *Arcuatula elegans* (Mytilacea: Crenellinae). *In* "Proceedings of the First International Workshop on the Malacofauna of Hong Kong and Southern China" (B. S. Morton, ed.), pp. 331–345. Hong Kong Univ. Press, Hong Kong.

Morton, B. S. (1981). The mode of life of *Gregariella coralliophaga* (Gmelin 1791) (Bivalvia: Mytilacea) with a discussion on the evolution of the boring Lithophaginae and adaptive radiation in the Mytilidae. *In* "Proceedings of the First International Marine Biological Workshop: The Marine Flora and Fauna of Hong Kong and Southern China" (B. S. Morton and C. K. Tseng, eds.), pp. 875–895. Hong Kong Univ. Press, Hong Kong.

Morton, B. S. (1982). Some aspects of the population structure and sexual strategy of *Corbicula* cf. *fluminalis* (Bivalvia: Corbiculacea) from the Pearl River, People's Republic of China. *J. Moll. Stud.* **48,** 1–23.

Morton, B. S. (1983a). The reproductive strategy of the mangrove bivalve *Polymesoda* (*Geloina*) *erosa* (Bivalvia: Corbiculacea) in Hong Kong. *Malacol. Rev.*

Morton, B. S. (1983b). Coral-associated bivalves of the Indo-Pacific. In "The Mollusca" (K. M. Wilbur, ed.), Vol. 6. Academic Press, New York. (*In press.*)

Morton, B. S. A review of *Polymesoda* (*Geloina*) Gray 1842 (Bivalvia: Corbiculacea) from Indo-Pacific mangroves. (*In manuscript*).

Morton, J. E. (1960). The responses and orientation of the bivalve *Lasaea rubra* Montagu. *J. Mar. Biol. Assoc. U.K.* **39,** 5–26.

Murty, A. S., and Balaparameswara Rao, M. (1977). Studies on the ecology of Mollusca in a south Indian mangrove swamp. *J. Moll. Stud.* **43,** 223–229.

Narchi, W. (1972). Comparative study of the functional morphology of *Anomalocardia brasiliana* (Gmelin, 1791) and *Tivela mactroides* (Born, 1778) (Bivalvia: Veneridae). *Bull. Mar. Sci. Gulf Caribb.* **22,** 643–670.

Narchi, W. (1976). Ciclo anual da gametogenese de *Anomalocardia brasiliana* (Gmelin, 1791) (Mollusca Bivalvia). *Bol. Zool. Univ. São Paulo* **1,** 331–350.

Narchi, W. (1980). A comparative study of the functional morphology of *Caecella chinensis* Deshayes 1855 and *Asaphis dichotoma* (Anton, 1839) from Ma Shi Chau, Hong Kong. *In* "Proceedings of the First International Workshop on the Malacofauna of Hong Kong and Southern China" (B. S. Morton, ed.), pp. 253–276. Hong Kong Univ. Press, Hong Kong.

Nascimento, I. A., and Pereira, S. A. (1980). Changes in the condition index for mangrove oysters (*Crassostrea rhizophorae*) from Todos os Santos Bay, Salvador, Brazil. *Aquaculture* **20,** 9–15.

Nascimento, I. A., Pereira, S. A., and Souza, R. C. E. (1980a). Determination of the optimum commercial size for the mangrove oyster (*Crassostrea rhizophorae*) in Todos os Santos Bay, Brazil. *Aquaculture* **20,** 1–8.

Nascimento, I. A., da Silva, E. M., Ramos, M. J. S., and dos Santos, A. E. (1980b). Development of the primary gonad in the mangrove oyster *Crassostrea rhizophorae,* age and length at first spawning. *Ciencia e Cultura* **32,** 736–742.

Nateewathna, A., and Tantichodok, P. (1980). Species composition, density and biomass of macrofauna of a mangrove forest at Ko Yao Yai, Southern Thailand. *Asian Symp. Mangrove Environ.: Res. Manage., 1980* pp. 1–32.

Nikolic, M., and Alfonso, S. J. (1971). El ostion del Mangle *Crassostrea rhizophorae* Guilding 1828 (explotacion del recurso y posibilidales para el cultivo). *FAO Fish. Rep.* **71**(2), 209–218.

Nikolic, M., and Melendez, S. A. (1968). El ostion del mangle *Crassostrea rhizophorae* Guilding, 1828 (experimentos iniciales en el cultivo). *Inst. Nac. Pesca Cuba, Nota Sobre Invest.* No. 7, pp. 1–31.

Nikolic, M., Bosch, A., and Alfonso, S. (1976). A system for farming the mangrove oyster (*Crassostrea rhizophorae* Guilding, 1828). *Aquaculture* **9,** 1–18.

Odum, W. E., and Heald, E. J. (1972). Trophic analyses of an estuarine mangrove community. *Bull. Mar. Sci. Gulf Caribb.* **22,** 671–738.

Odum, W. E., and Heald, E. J. (1975). The detritus-based food web of an estuarine mangrove community. *Estuarine Res.* **1,** 265–286.

Olsen, L. A. (1976). Ingested material in two species of estuarine bivalves: *Rangia cuneata* (Gray) and *Polymesoda caroliniana* (Bosc.). *Proc. Natl. Shellfish. Assoc.* **66,** 103.

Owen, G. (1959). Observations on the Solenacea with reasons for excluding the family Glaucomyidae. *Philos. Trans. R. Soc. London, Ser. B* **242,** 59–97.

Palacio, J. (1977). Invertebrados del area estuárica de la Ciénaga Grande de Santa Marta con énfasis en la fauna accompanante de la ostre *Crassostrea rhizophorae* Guilding. *Inst. Invest. Marinas Punta de Betin, Santa Marta, Tesis de Grado* pp. 1–151.

Parker, R. H. (1963). Zoogeography and ecology of some macroinvertebrates, particularly mollusks, in the Gulf of California and the continental slope of Mexico. *Vidensk. Medd. Dan. Naturh. Foren.* **126,** 1–178.

Parker, R. H. (1964). Zoogeography and ecology of macro-invertebrates of Gulf of California and continental slope of western Mexico. *In* "Marine Geology of the Gulf of California—A Symposium," Mem. No. 3. pp. 331–376. Am. Assoc. Pet. Geol., Tulsa, Oklahoma.

Pathansali, D. (1966). Notes on the biology of the cockle, *Anadara granosa* L. *Proc. Indo. Pac. Fish. Counc.* **11,** 84–98.

Pathansali, D., and Soong, M. K., (1958). Some aspects of cockle (*Anadara granosa* L.) culture in Malaya. *Proc. Indo-Pac. Fish. Counc.* **8,** 26–31.

Patton, W. K. (1972). Studies on the animal symbionts of the gorgonian coral, *Leptogorgia virgulata* (Lamarck). *Bull. Mar. Sci. Gulf Caribb.* **22,** 419–431.

Peddicord, R. K. (1976). Effects of substratum on growth of the bivalve *Rangia cuneata* Gray, 1831. *Veliger* **18,** 398–404.

Perez, P. A. (1974). Distribucion de los moluscos en la costa Centro-Occidentale (Patanemo-Punta Tucacas) de Venezuala. Comparacion de los habitats litorales. *Mem. Soc. Cienc. Nat. La Salle* **34**(96/97), 24–52.

Pierce, S. K. (1970). Water balance in the genus *Modiolus* (Mollusca: Bivalvia: Mytilidae): osmotic concentrations in changing salinities. *Comp. Biochem. Physiol.* **36,** 535–545.

Pierce, S. K. (1973). The rectum of "*Modiolus*" *demissus* (Dillwyn) (Bivalvia: Mytilidae): a clue to solving a troubled taxonomy. *Malacologia* **12,** 283–293.

Pinto, L., and Wignarajah, S. (1980). Some ecological aspects of the edible oyster *Crassostrea cucullata* (Born) occurring in association with mangroves in Negombo lagoon, Sri Lanka. *Hydrobiologia* **69,** 11–19.

Piyakarnchana, T. (1980). "The Present Status of Mangrove Ecosystems in Southeast Asia and the Impact of Aquatic Pollution: Thailand," Working Pap. SCS/80/WP/94e. South China Sea Fisheries Development and Coordinating Programme.

Plante, R. (1964). Contribution à l'étude des peuplements de hauts niveaux sur substrats solides non récifaux dans la region de Tuléar. *Recl. Trav. Stn. Mar. Endoume, Suppl.* **2,** 205–315.

Pora, E. A., Wittenberger, C., Suárez, G., and Portilla, N. (1969). The resistance of *Crassostrea rhizophorae* to starvation and asphyxia. *Mar. Biol. (Berlin)* **3,** 18–23.

Prashad, B. (1932). "The Lamellibranchia of the Siboga Expedition," Systematic Part II, Siboga Exped. No. 53, pp. 1–353. Brill, Leiden.

Purchon, R. D. (1956). The biology of "Krang" the Malayan edible cockle. *Proc. Sci. Soc. Malaya* **2,** 61–68.

Purchon, R. D. (1958). The stomach in the Eulamellibranchia; Stomach Type IV. *Proc. Zool. Soc. London* **131,** 487–525.

Rajagopal, A. S. (1964). Two new species of marine borers of genus *Nausitora* (Mollusca: Teredinidae) from West Bengal, India. *J. Bombay Nat. Hist. Soc.* **61,** 108–118.

Rajagopal, A. S. (1970). Field ecology of some marine borers (Mollusca: Teredinidae) of mangroves in Sundarbans, India. *Rec. Zool. Surv. India* **62,** 21–27.

Rajagopal, A. S., and Daniel, A. (1973). Boring organisms of the Great Nicobar Island. Mollusca: Teredinidae. *J. Bombay Nat. Hist. Soc.* **69,** 676–678.

Rajagopalaiengar, A. S. (1961a). A new species of the marine borer, *Bankia* (*Neobankia*) *roonwali* (Mollusca: Teredinidae) from India. *Sci. Cult.* **27,** 550.

Rajagopalaiengar, A. S. (1961b). Fuller description of a recently described species of the marine borer *Bankia* (*Neobankia*) *roonwali* Rajagopalaiengar from West Bengal, India. *Rec. Indian Mus.* **59,** 449–454.

Rios, E. C. (1970). "Coastal Brazilian Seashells." Fundacáo Cidade do Rio Grande.

Robertson, R. (1960). The mollusk fauna of Bahamian mangroves. *Bull. Am. Malacol. Union, Inc., Annu. Rep.* **26,** 22–23.

Robertson, R. (1963). The mollusks of British Honduras. *Proc. Philos. Shell Club* **1,** 15–20.

Rodriguez, G. (1959). The marine communities of Margarita Island, Venezuela. *Bull. Mar. Sci. Gulf Caribb.* **9,** 237–280.

Rodriguez, G. (1963). The intertidal estuarine communities of Lake Maracaibo, Venezuala. *Bull. Mar. Sci. Gulf Caribb.* **13,** 197–218.

Rojas, A. V. (1972). Fijacion de la larva de la ostra de los bancos naturales de Bahia de Mochima y Laguna Grande. *Bol. Inst. Oceanogr., Univ. Oriente* **11,** 97–106.

Rojas, A. V., and Ruiz, J. B. (1972). Variacion estacional del engorda del ostion *C. rhizophorae* da Baia de Mochima y Laguna Grande. *Bol. Inst. Oceanogr., Univ. Oriente* **11,** 39–43.

Roonwal, M. L. (1954a). *Bactronophorus thoracites* (Gould) as a pest of living trees in Sundarbans, Bengal (Mollusca: Teredinidae). *Curr. Sci.* **23,** 301.

Roonwal, M. L. (1954b). The marine borer *Bactronophorus thoracites* (Gould) Mollusca, Eulamellibranchiata, Teredinidae, as a pest of living trees in the mangrove forests of Sundarbans, Bengal, India. *Proc. Zool. Soc. Bengal* **7,** 91–105.

Saenz, B. A. (1965). El ostion *C. rhizophorae* Guilding y su cultivo experimental en Cuba. *Nota Invest. Cent. Invest. Pesq. Bauta* **6,** 1–34.

Sandison, E. E. (1967). The effect of salinity fluctuation on the life cycle of *Gryphaea gasar* ((Adanson) Dautzenberg) in Lagos harbour, Nigeria. *J. Anim. Ecol.* **35,** 379–389.

Sasekumar, A. (1974). Distribution of the macrofauna on a Malayan mangrove shore. *J. Anim. Ecol.* **43,** 51–69.

Sasekumar, A. (1980). "The Present State of Mangrove Ecosystems in Southeast Asia and the Impact of Pollution: Malaysia," Working Pap. SCS/80/WP/94b. South China Sea Fisheries Development and Coordinating Programme.

Scarabino, V., Maytia, S., and Caches, M. (1975). Carta bionomica litoral del departmento de Montevideio. I. Niveles superiores del sistema litoral. *Com. Soc. Malacol. Urug.* **4,** 117–126.

Seed, R. (1980). A note on the relationship between shell shape and life habits in *Geukensia demissa* and *Brachidontes exustus* (Mollusca: Bivalvia). *J. Moll. Stud.* **46,** 293–299.

Shaw, W. N. (1965). Pond culture of oysters—past, present and future. *Trans. North Am. Wildl. Nat. Resour. Conf.* **30,** 114–120.

Siung, A. M. (1980). Studies on the biology of *Isognomon alatus* Gmelin (Bivalvia: Isognomonidae) with notes on its potential as a commercial species. *Bull. Mar. Sci. Gulf Caribb.* **30,** 90–101.

Smith, M. L. (1963). The Teredinidae of the Queensland coast from Cairns to Brisbane. M.Sc. Thesis, University of Queensland, Brisbane.

Soegiarto, A. (1980). "The Present Status of Mangrove Ecosystems in Southeast Asia and the Impact of Aquatic Pollution: Indonesia," Working Pap. SCS/80/WP/94a. South China Sea Fisheries Development and Coordinating Programme.

Soot-Ryen, T. (1955). A report on the family Mytilidae (Pelecypoda). *Allan Hancock Pac. Exped.* **20,** 1–174.

Squires, H. J., Estevez, M., Barona, O., and Mora, O. (1973). Mangrove cockles, *Anadara* spp. of the Pacific coast of Colombia. *Veliger* **18,** 57–68.

Stanley, S. M. (1970). *Relation of shell form to Life Habits in the Bivalvia (Mollusca). Mem. Geol. Soc. Am.* **125,** 1–296.

Stenzel, H. B. (1971). Oysters. *In* "Treatise on Invertebrate Paleontology" (R. C. Moore, ed.), Part N, Mollusca 6 (3 of 3). Geol. Soc. Am. and Univ. of Kansas Press.

Stephenson, T. A., and Stephenson, A. (1949). The universal features of zonation between tide marks on rocky coasts. *J. Ecol.* **37,** 289–305.

Stiven, A. E., and Kuenzler, E. J. (1979). The response of two salt marsh molluscs, *Littorina irrorata* and *Geukensia demissa* to field manipulations of density and *Spartina* litter. *Ecol. Monogr.* **4,** 151–171.

Stojkovich, J. O. (1977). Survey and species inventory of representative pristine marine communities on Guam. *Tech. Rep.—Univ. Guam Mar. Lab.* **40,** 1–183.

Sutherland, J. P. (1980). Dynamics of the epibenthic community on roots of the mangrove *Rhizophora mangle,* at Bahia de Buche, Venezuela. *Mar. Biol. (Berlin)* **58,** 75–84.

Tabb, D. C., and Moore, D. R. (1971). Discovery of the Carolina marsh clam, *Polymesoda caroliniana,* a supposed Florida disjunct species, in Everglades National Park, Florida. *Gulf Res. Rep.* **3,** 265–277.

Tan, W. H. (1970). Some Singapore shipworms (Family Teredinidae). *J. Singapore Acad. Sci.* **2,** 13.

Taylor, J. D. (1971). Reef associated molluscan assemblages in the western Indian Ocean. *In* "Regional Variations in Indian Ocean Coral Reefs" (D. R. Stoddard and C. M. Yonge, eds.), pp. 501–534. Academic Press, New York.

Taylor, J. D. (1976). Habitats, abundance and diets of muricacean gastropods at Aldabra Atoll. *J. Linn. Soc. London, Zool.* **59,** 155–193.

Tenore, K. R., Horton, D. B., and Duke, T. W. (1968). Effects of bottom substrate on the brackish water bivalve *Rangia cuneata. Chesapeake Sci.* **9,** 238–248.

Tranter, D. J. (1974). Marine biology. *UNESCO, Resour. Nat. Asie Trop. Humide* pp. 385–426.

Trueman, E. R., and Lowe, G. A. (1970). The effects of temperature and littoral exposure on the heart rate of a bivalve mollusc, *Isognomon alatus,* in tropical conditions. *Comp. Biochem. Physiol. A.* **38,** 555–564.

Turner, R. D. (1955). The family Pholadidae in the Western Atlantic and the Eastern Pacific. Part. 2. Martesiinae, Jouannetiinae and Xylophaginae. *Johnsonia* **3**(34), 65–160.

Turner, R. D. (1966). "A Survey and Illustrated Catalogue of the Teredinidae." Museum of Comparative Zoology, Harvard University, Cambridge, Massachusetts.

Turner, R. D. (1973). Deep water wood boring mollusks. *Proc. Int. Congr. Mar. Corros. Fouling, 3rd, 1912,* pp. 831–841. Gaithersburg, Maryland.

Turner, R. D., and Johnson, A. C. (1971). Biology of marine wood boring molluscs. *In* "Marine Borers, Fungi and Fouling Organisms of Wood" (E. B. G. Jones and S. K. Eltringham, eds.), pp. 17–64. O.E.C.D., Paris.

Van der Schalie, H. (1933). Notes on the brackish water bivalve *Polymesoda caroliniana* (Bosc). *Occas. Pap. Univ. Mich. Misc. Zool.* **258,** 1–8.

Van Steenis, C.G.G.J. (1962). The distribution of mangrove plant genera and its significance for palaeogeography. *Proc. K. Ned. Akad. Wet., Ser. C* **65,** 164–169.

Vermeij, G. J. (1974). Molluscs in mangrove swamps: Physiognomy, diversity, and regional differences. *Syst. Zool.* **22,** 609–624.

Vermeij, G. J. (1980). Drilling predation in a population of the edible bivalve *Anadara granosa* (Arcidae). *Nautilus* **94,** 123–125.

Warner, G. F. (1969). The occurrence and distribution of crabs in a Jamaican mangrove swamp. *J. Anim. Ecol.* **38,** 379–389.

Watson, C. J. J. (1936). Marine borers destroying timber in the port of Brisbane. *Queens. For. Serv. Bull.* **12,** 10–30.

Wedler, E. (1980). Experimental spat collecting and growing of the oyster, *Crassostrea rhizophorae* Guilding, in the Cienaga Grande de Santa Marta, Colombia. *Aquaculture* **21,** 251–259.

Wedler, E., Perez, L., and Palacio, J. (1978). Ostricultura en la Ciénaga Grande de Santa Marta. Primera etapa. *Inf. Proyecto Esp. Colciencias, Santa Marta, Columbia* pp. 1–64.

Wells, F. E., and Slack-Smith, S. M. (1981). Zonation of molluscs in a mangrove swamp in the Kimberley, Western Australia. *In* "Biological Survey of Mitchell Plateau and Admiralty Gulf, Kimberley, Western Australia," Part. 9, pp. 265–274. Western Australian Museum, Perth.

Wells, H. W. (1961). The fauna of oyster beds, with special reference to the salinity factor. *Ecol. Monogr.* **31,** 239–266.

Wolfe, D. A., and Petteway, E. N. (1968). Growth of *Rangia cuneata* Gray. *Chesapeake Sci.* **9,** 99–102.

Yankson, K. (1982). Gonad maturation and sexuality in the West African Bloody Cockle, *Anadara senilis*(L.). *J. Moll. Stud.* **48,** 294–301.

Yoloye, V. L. A. (1974). The sexual phases of the "West African Bloody Cockle" *Anadara senilis* (Linnaeus). *Veliger* **17,** 70–72.

Yonge, C. M. (1957). *Enigmonia aenigmatica* Sowerby, a motile anomiid (Saddle oyster). *Nature (London)* **180,** 765–766.

Yonge, C. M. (1968). Form and habit of species of *Malleus* (including the "Hammer Oysters") with comparative observations on *Isognomon isognomon. Biol. Bull. (Woods Hole, Mass.)* **135,** 378–405.

Yonge, C. M. (1977). Form and evolution in the Anomiacea (Mollusca: Bivalvia) *Pododesmus, Anomia, Patro, Enigmonia* (Anomiidae): *Placunanomia, Placuna* (Placunidae Fam. Nov.). *Philos. Trans. R. Soc. London, Ser. B* **276,** 453–523.

4

Coral-Associated Bivalves of the Indo-Pacific

BRIAN MORTON

Department of Zoology
University of Hong Kong
Hong Kong

The Mollusca, Vol. 6
Ecology

ISBN 0-12-751406-6

I. Introduction

Hitherto there have been few reviews of coral-associated molluscs, although Yonge (1974) has highlighted some aspects of their biology. Patton (1975) reviewed the coral-associated fauna, but the Mollusca were treated somewhat superficially. Hadfield (1976) reviewed the molluscs associated with living corals in more detail and produced an extensive table of defined associations for some 80 species. Perhaps to be expected, because they are often more active and conspicuous on a reef system, the Gastropoda received much more extensive treatment than the Bivalvia. The conclusion of Hadfield was, however, that much more research was needed to assess the real nature of most such associations.

Austin et al. (1980) described 101 species associated with the coral *Pocillopora damicornis* on the Great Barrier Reef; none was a bivalve. Interest in the Gastropoda has been greater, with a variety of studies investigating, for example, growth and longevity (Frank, 1969) and predator–prey relationships in the coral ecosystem (R. Robertson, 1970; Salvini-Plawen, 1972; Taylor, 1976, 1977, 1978).

Comprehensive studies on the coral-associated Bivalvia are lacking. Although the most characteristic associates of Pacific reefs, the giant clams (Tridacnidae) have aroused wider interest of late because of commercial slaughter for their adductor muscles. Pearson (1977), for example, reports that since 1969 more than 156,000 giant clams have been taken from Swain Reefs in the Australian Great Barrier Reef. Between 1972 and 1975, 98% of *Tridacna gigas* and 97% of *T. derasa* populations were harvested from the reef. Giant clam exploitation at Panope, Palau, has already resulted in the recent extinction of *T. gigas* there (Gwyther and Munro, 1981) and prompted research into the spawning and artificial rearing of larvae.

Similarly, the relatively recent realization that some of the lithophagine borers of the reefs attack living corals as well as dead skeletons has aroused interest in aspects of the fidelity of species to particular corals, their mode of boring, and their methods of surviving habitation among living coral polyps. However, there is still very little quantitative information on the destructive role that these and other borers play in the biodeterioration of a reef system. With increasing interest in coral reef systems, it is to be hoped that there will be accompanying interest in the bivalve components of this community. This chapter attempts to coalesce literature on the coral-associated bivalves of the Indo-Pacific to aid future research.

Much of the literature on coral bivalves is anecdotal, contained in broader accounts of the coral community. With few exceptions (Taylor, 1968), little is known of the ecology of these animals or of their special role in, and relationships with, the other coral reef fauna. This discussion may stimulate research in areas where there is an obvious paucity of information.

II. Definitions

The most recent review of Mollusca associated with living coral reefs was by Hadfield (1976). In common with most previous authors, he principally discussed the Gastropoda, but recognized at the same time the very clear relationship that exists between living corals and the bivalve borers in particular, which are by far the most important agents in the bioerosion of corals (Bromley, 1978). Other species are clearly symbiotic, for example, *Fungiacava eilatensis* in *Fungia* (Goreau et al., 1969, 1970), whereas others attach to living corals, for example, *Pedum spondyloideum* and *Hemipecten forbesianus* (Yonge, 1967b, 1981) in a somewhat less well-defined association. Many others attach to dead coral bases, as do species of the byssally attached Arcacea (*Arca ventricosa, A. avellana, Barbatia helblingi, B. decussata*) (Salvat, 1967; Taylor, 1971b,c; Kay and Switzer, 1974).

It is difficult to generalize about coral-associated assemblages, because a reef is a complex habitat and some of the species found living there are refugees from surrounding substrates. Taylor (1971c) suggested that coral assemblages can be divided into two distinct categories; some species use the coral as a convenient hard substrate, and others feed upon the coral or associate closely with it. The first category varies according to the habitat in which the coral is found; the second is largely independent of the surrounding habitat.

Hadfield (1976) divided the coral-mollusc association into four classes: (1) predators and parasities, (2) borers of living corals, (3) epizoic species, and (4) molluscs which serve as substrate for corals. Clearly, with predatory bivalves confined to abyssal and bathyal depths (Knudsen, 1970), the first category is not applicable here. Similarly, although Morton (1983a) reports that in Hong Kong the coral *Oulastrea crispata* is commonly found attached to *Trisidos semitorta,* it is certain (Hadfield, 1976) that any large, dead mollusc can act as a substrate for coral settlement. In this case, one would expect tridacnid shells to be important (Mastaller, 1978). Perhaps the best examples of such an association are the solitary corals *Heteropsammia* and *Heterocyathus,* which settle on empty shells. The sipunculid *Aspidosiphon* lives with the coral, and each has a commensal bivalve (*Jousseaumiella*) (Bouvier, 1894; Bourne, 1906). Despite these examples, however, it is clear that such a subject is not yet ready for review, and this discussion will be confined largely to those bivalves which fall into the two categories of Taylor (1971c) and which, with the removal of classes 1 and 4, find commonality with classes 2 and 3 of Hadfield (1976).

In this chapter, to demonstrate better habitat subdivisions by the various bivalve lineages, the bivalve–coral associations are separated into different classes based on those of Taylor (1971c) and Hadfield (1976) (as redefined above). These are

1. Bivalves which bore (a) living and (b) dead corals

2. Epizoic bivalves of (a) living scleractinian corals, (b) dead coral substrates, and (c) antipatharian and gorgonian corals
3. Crevice- and gallery-dwelling bivalves
4. Coral reef-associated bivalves occurring in habitats (coral sands and rubble) closely dependent upon the corals

As substrate, coral colonies are very complex; their dead undersides and branches offer shelter and protection for large numbers of animals. There are obvious general differences in the type of substrate provided by various growth form categories of the corals; for instance, branching corals are better for byssate forms, and massive corals are more suitable for cemented or boring bivalves. There are, for example, very few borers recorded from branching corals (Scott, 1980; Highsmith, 1980). In some cases, it is uncertain if epizoic bivalves are definitively (rather than fortuitously) associated with either living or dead corals; settlement upon living corals might, for example, kill the underlying polyps. Some borers may colonize dead coral, only to have their habitat overgrown by corals of the same or a different species; others might bore living corals which subsequently die.

Thus, the above definitions, although not strict because of a present lack of accurate data, will at least allow some generalizations to be made which may stimulate further research.

III. Distribution and General Ecology

A. Indian Ocean

Abbott (1950) has investigated the molluscs of the Cocos-Keeling Islands in the Indian Ocean. Of the 26 species of recorded bivalves, *Barbatia decussata, Isognomon perna, Pinctada margaratifera, Pteria penguin, Electroma smaragdina* (attached to *Seriatopora*), *Pinna saccata, Spondylus hystrix, Saccostrea cucullata, Cardita variegata, Trapezium oblongum, T. bicarinatum,* and *Chama aspersa* were either byssally attached or cemented to coral blocks. *Pinna muricata, Atrina vexillum, Codakia punctata, Lucina edentula, Fragum fragum, Pitar prora, Venus puerpera, Paphia literata, Asaphis deflorata, Arcopagia scobinata,* and *Tellina crassiplicata* occurred in coral sands, and *T. crocea* on coral rubble.

Taylor (1968) has recorded the molluscs of coral reef systems on Mahé Seychelles, and records 202 species of gastropods and 115 bivalves. This comprehensive review of the Mahé fauna is a good starting point for any discussion of reef molluscs.

The fore reef community of the Seychelles has been described by Lewis and

Taylor (1966) and Taylor (1968). The bivalves recorded from this coral sand community included *Arca navicularis, Laevicardium biradiatum, Ctenocardium sueziense, Lioconcha picta, Circe scripta, Ervilia bisculpta, Cadella semen, Tellina fabrefacta,* and *Corbula subquadrata.*

The corals of the lagoon shores of Diego Garcia in the Indian Ocean support an array of molluscs including 54 species of bivalves (Taylor, 1971a,b). In crevices and on dead branches the byssate species *Barbatia helblingi, Isognomon legumen, I. perna,* and *Septifer bilocularis* were very common; other species included *Lima lima* and *Gloripallium pallium. Pedum spondyloideum* occurred solely in narrow crevices in massive *Porites* colonies. The coral colonies themselves were bored by *Lithophaga teres, Botula cinnamomea,* and *Gastrochaena cuneiformis. Tridacna maxima* occurred byssally attached to the upper surface of coral colonies and in crevices. The sand patches between coral heads supported *Fragum fragum.* Boulders encrusted with algae on a clean substrate at Eclipse Point possessed cemented bivalves, e.g., *Chama aspersa* and *Ostrea numisma,* the byssate *Barbatia helblingi,* and the small *Pinctada margaratifera.* Boulders scattered in a *Cymodocea* bed possessed a few bivalves such as *Pinctada margaratifera, Barbatia helblingi, Lima fragilis,* and *T. maxima.*

Taylor (1971c), in reviewing the reef-associated molluscan assemblages of the Indian Ocean, has considered most habitats including mangroves and offshore soft deposits. The reader is referred to this paper for an analysis of these molluscan communities; only the true coral associates will be commented upon here. The noncoralline hard substrates are of interest because (as in Port Sudan) *Tridacna maxima* was byssally attached to the reef rubble crust. The bivalve fauna was limited to suspension feeders, both byssally attached and cemented species. *Modiolus auriculatus* may occur as a byssate crust on platforms with *Pinctada margaratifera* (byssate) on open surfaces and crevices. Other species, including *Cardita variegata, Arca avellana, Barbatia helblingi, A. plicata, Ostrea numisma,* and *Chama asperella,* occur cemented or byssally attached beneath boulders and cobbles.

From the Indian Ocean corals themselves, Taylor (1971c) recorded *T. maxima* and *T. squamosa* byssally attached in crevices between living colonies on the dead upper surface. Typically, a wide range of byssate bivalves attach to the dead part of branches or the undersides of massive corals. These include *B. helblingi, B. fusca, Septifer bilocularis,* several species of *Chlamys, Pteria,* and *Pinctada, Isognomon ephippium, Streptopinna saccata, Cardita variegata,* and *Trapezium oblongum.* Other forms cement to the branches or undersides, including *Spondylus aurantius, S. hystrix, Chama* spp., and several oysters, particularly *Ostrea frons* and *O.* (= *Hyotissa*) *hyotis.* Borers were also common, particularly *Lithophaga, Botula,* and *Gastrochaena.*

Nielsen (1976b) has reported upon the coral-associated bivalves of Phuket, Thailand, facing the Andaman Sea. A total of 91 species were recorded; these included borers and a number of species such as *Donax faba, Atactodea glabrata,* and *Caecella transversalis* which are surf-exposed sand beach burrowers. The coral species could be divided into two major categories, those attached to living or dead coral heads and those occurring in coral sands and among shale and coral rubble. Community patterns, however, were not established.

Thomassin and Ganelon (1977) have analyzed the molluscan assemblages on coral reef boulder tracts in the Tulear coral reefs of Madagascar. These authors showed that the community is characterized by a diverse group of vagilous gastropods (trochids, turbinids, rissoinids, muricids, buccinids, and conids) and bivalves, which are all attached in the adult (arcids and mytilids) stage or in the juvenile (venerids and semelids) stage by a byssus; this is an adaptation to the strength of the surf action. Bivalves recorded by these authors included *Fossularca sculptilis, Cardita variegata, Gafrarium pectinatum, Arca avellana, Modiolus auriculatus, Venus toreuma, Anisodonta lutea,* and *Nesobornia* sp. According to Thomassin and Ganelon, the same faunistic assemblages can be expected from all Indo-Pacific coral boulder tracts or beach cobbles, although the densities change with the edaphic conditions and water levels. The boulder bank fauna essentially comprises species living on the underside of hard substrates and crevices and, at its limit, will be found partly beneath scattered coral colonies on sediments of the inner reef sea grass beds.

Mastaller (1978) has reported upon the molluscan assemblages of the Red Sea at Port Sudan. This major paper covered most of the marine habitats. The intertidal zone included a band of rock oysters (*Ostrea*), with *Brachidontes variabilis* byssally attached in exposed areas. The interstices of sublittoral boulders were colonized by *Arca plicata, Chama* sp., and *Brachidontes variabilis,* whereas the undersurface contained *Pinctada vulgaris, Arca plicata,* and *Cardita variegata.* On sea grass flats were found *Atrina vexillum, Pinna muricata,* and *Arca antiquata.* Sandy bottoms sustained a greater array of highly adapted bottom feeders such as *Asaphis violascens, Laevicardium orbita, Tellina virgata, Pharaonella aurea, Lucina dentifera, Bellucina* sp., *Lioconcha picta,* and *Tapes literata.* The most abundant bivalve, *Tridacna maxima,* is certainly one of the pioneer species in this habitat, being responsible for the eventual formation of a hard bottom community. It settles on any large, solid substrate (often empty valves of *Laevicardium* and *Cyclotellina*), thereby creating the preliminary aggregation point for a number of sedentary reef inhabitants including reef corals. Mastaller (1978) also showed that the bivalves are good indicators of community change. Thus, the fore reef areas (as opposed to the clean sand community noted above) possesses a different bivalve fauna including *Glycymeris pectiniformis, Codakia divergens, Laevicardium orbita, Fragum* sp., and *Cyclotellina scobinata.*

On coral heads and dead coral rocks in Port Sudan, Mastaller (1978) recorded the byssally attached or partially cemented *Arca lacerata, Barbatia setigera, Notirus* cf. *macrophyllus, Chama* spp., *Spondylus gaederopus,* and *T. maxima,* with *Hemicardium* (= *Fragum*) *fragum* occurring partially buried between coral heads among debris. In the crevices of the coral rocks, Mastaller recorded *Pinctada vulgaris, Plicatula plicata,* and *Cardium auricula* (the latter presumably not attaining adulthood in this habitat), with *Modiolus auriculatus* occurring in associated algal mats, and *Vulsella vulsella* presumably occurring in sponges (Reid and Porteous, 1980). The massive coral *Porites* was regularly found to be inhabited by *Pedum* sp. Branched corals such as *Acropora* give protection for several fragile bivalves, including *Pinna bicolor, Streptopinna saccata,* some scallops (*Pecten lividus, P. squamatus*), and *Ctenoides annulatus,* which attach loosely to the corals. Dead corals were bored by *Gastrochaena cuneiformis* and *G. deshayesi.* No bivalves were recorded from the reef platform, but they were found on the reef crest and slope. Mastaller (1978) recorded *Pinctada margaratifera, Lopha folium, Spondylus aculeatus,* and *Plicatula plicata.* In higher exposure occurred *Pteria crocea, Lopha cucullata,* and *L. cristagalli,* all three living epizoically on thorny corals such as *Cirripathes* and *Antipathes.*

Taylor (1971c) has shown that there seems to be a positive correlation between the number of coral bivalves and the area of shallow water around an island. Thus, Maes (1967) recorded 87 species from Cocos-Keeling. On Diego Garcia, Taylor (1971b) recorded only 54 species, and on Aldabra 104 species (Taylor, 1971c). Taylor compared these figures with 224 species recorded from Dar es Salaam (Spry, 1964) and 167 from Inhaca, an island off the Mozambique coast (Boshoff, 1965). The granitic Seychelles, which lie on an extensive, shallow water bank, support 176 species (Taylor, 1968), a faunal size more comparable with that of the continental margins than with the low coralline islands. Although island isolation may prevent colonization from mainland stocks, the islands and atolls also have a reduced diversity of habitats available for colonization. Those bivalves present are more properly coral associated; the greater number recorded from mainland shores are from a greater variety of loosely associated habitats.

Taylor (1971c) concluded, however, that the islands of the western Indian Ocean have a very uniform fauna composed of species with larvae capable of surviving pelagic transport. The continental margins have these same species, but more of them are found at a particular locality. In addition, they have a set of species which are confined to the continental margins and some endemic species. Isolated areas on the fringes of the Indian Ocean show a higher degree of endemicity while experiencing a latitudinal decrease in diversity (Fig. 1).

Mastaller (1978) recorded 77 species of bivalves from the Port Sudan area; Nielsen (1976b) recorded 91 species from Phuket, Thailand, possibly disagreeing with Taylor's concept of higher diversity on continental shores. Mastaller, however, argued that the species list was not exhaustive and, further, that the molluscan fauna is locally impoverished here because there are almost no tidal

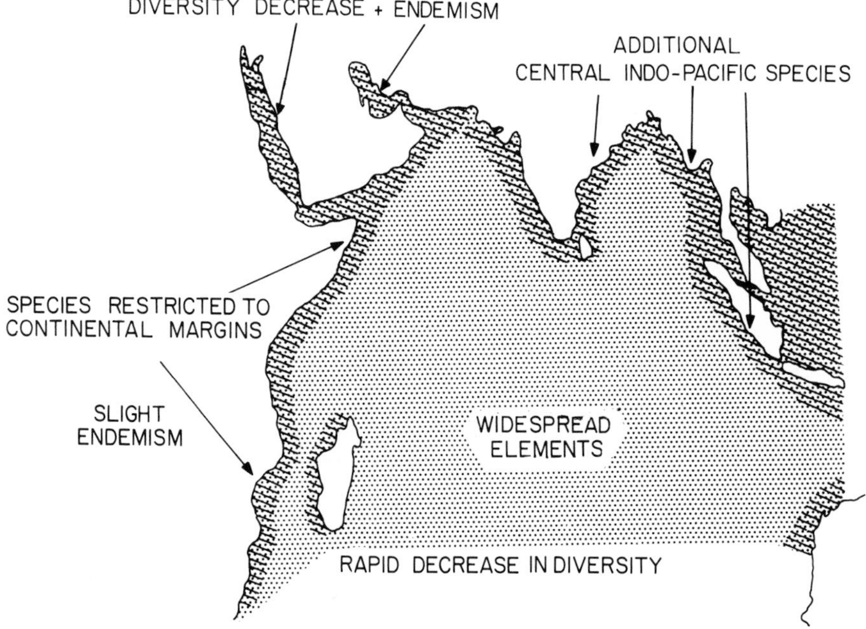

Fig. 1. Generalized diagram suggesting a summary of the major distributive features of the molluscan fauna of the Indian Ocean. There is a dominant widespread element, and a group of species more or less confined to continental margins that is supplemented in the east by central Indo-Pacific species and by endemics around the fringes. Superimposed is a latitudinal diversity gradient. [After Taylor (1971c). With permission of The Zoological Society of London.]

movements in the Central Red Sea and thus no littoral zone. The species are thus restricted to shallow-water assemblages and to coral-dominated habitats (as were those recorded by Nielsen, 1976b). Therefore, it is likely that the 77 species which Mastaller records from Port Sudan are mostly coral associates. This number (and that from Nielsen, 1976b), accords favorably with the numbers of species recorded from coral-dominated island communities (Maes, 1967; Taylor, 1971b,c), adding further weight to the argument presented here that throughout the Indian Ocean most coral habitats (mainland or island) are colonized by approximately the same number of species, and that the apparently greater continental diversity is brought about by an increase in the diversity of other habitats.

B. Pacific Ocean

Ostergaard (1935) described 39 species of bivalves from coral habitats at Tonga in the Pacific. Most of these are clear coral associates (Tables I and II), although many were recorded only as empty valves.

TABLE I

Epibyssate and Cemented Bivalves Recorded from Indo-Pacific Corals

Bivalve (families)	References
Arcidae	
Barbatia plicata	Dudgeon and Morton (1982)
Barbatia bicolorata	Smith (1978)
Barbatia setigera	Mastaller (1978)
Barbatia decussata	Abbott (1950); Taylor (1971b); Kay (1971); Kay and Switzer (1974)
Barbatia tenella	Banner and Randall (1952); Taylor (1968, 1971b)
Barbatia helblingi	Taylor (1968, 1971a,b,c; 1980); Nielsen (1976b); Dudgeon and Morton (1982)
Barbatia fusca	Ostergaard (1935); Taylor (1968, 1971b); Nielsen (1976b)
Arca navicularis	Lewis and Taylor (1966); Taylor (1968)
Arca maculata	Ostergaard (1935)
Arca lacerata	Mastaller (1978)
Arca avellana	Taylor (1968, 1971b,c); Nielsen (1976b); Thomassin and Ganelon (1977); Dudgeon and Morton (1982)
Arca clathrata	Taylor (1968)
Arca ventricosa	Salvat (1967, 1972); Randall et al. (1975); Nielsen (1976b)
Arca plicata	Taylor (1968, 1971b,c); Mastaller (1978)
Fossularca sculptilis	Thomassin and Ganelon (1977)
Striarca tenebrica	Taylor (1971b)
Striarca afra	Nielsen (1976b); Dudgeon and Morton (1982)
Arcopsis symmetrica	Taylor (1968, 1980)
Malleidae	
Malleus regula	Yonge (1968); Nielsen (1976b)
Pinnidae	
Streptopinna saccata	Banner and Randall (1952); Rosewater (1961); Taylor (1968, 1971c); Mastaller (1978)
Pinna bicolor	Nielsen (1976b); Mastaller (1978)
Pinna radiata	Nielsen (1976b)
Mytilidae	
Septifer bilocularis	Taylor (1968, 1971a,b,c, 1980); Sumadhiharga (1971); Randall et al. (1975); Nielsen (1976b); Stojkovich (1977)
Septifer virgatus	Dudgeon and Morton (1982)
Brachidontes variabilis	Mastaller (1978); Taylor (1968, 1971b, 1980)
Modiolus flavidus	Nielsen (1976b)
Modiolus plumescens	Nielsen (1976b)
Modiolus auriculatus	Taylor (1968, 1971a,b); Randall et al. (1975); Thomassin and Ganelon (1977); Mastaller (1978)

(Continued)

TABLE I *Continued*

Bivalve (families)	References
Gregariella coralliophaga	Nielsen (1976b); Scott (1980); Morton (1982a)
Gregariella striatus	Nielsen (1976b); Scott (1980)
Limidae	
Lima lima	Taylor (1971a,b); Nielsen (1976b); Morton (1979a); Dudgeon and Morton (1982)
Limaria fragilis	Ostergaard (1935); Taylor (1968, 1971a,b); Smith (1978); Morton (1979a); Dudgeon and Morton (1982)
Limaria hongkongensis	Morton (1979a); Dudgeon and Morton (1982)
Ctenoides annulatus	Taylor (1968); Nielsen (1976b); Mastaller (1978)
Pteriidae	
Pinctada margaratifera	Abbott (1950); Banner and Randall (1952); Salvat (1969); Taylor (1968, 1971a,b,c); Sumadhiharga (1971); Stojkovich (1977); Mastaller (1978)
Pinctada cumingii	Ostergaard (1935)
Pinctada vulgaris	Mastaller (1978)
Pinctada maxima	Salvat (1972)
Pinctada maculata	Salvat (1967); Smith (1978)
Pinctada chemnitzi	Nielsen (1976b)
Pinctada martensii	Dudgeon and Morton (1982)
Pinctada radiata	Nielsen (1976b)
Electroma smaragdina	Abbott (1950)
Electroma alacorvi	Taylor (1968, 1971b)
Electroma spp.	Kay (1971); Kay and Switzer (1974)
Pteria penguin	Abbott (1950); Sumadhiharga (1971); Dudgeon and Morton (1982)
Pteria castanea	Ostergaard (1935)
Pteria crocea (attached to gorgonians)	Mastaller (1978)
Isognomonidae	
Isognomon perna	Abbott (1950); Taylor (1971a,b); Nielsen (1976b)
Isognomon legumen	Taylor (1971a,b, 1980); Nielsen (1976b); Dudgeon and Morton (1982)
Isognomon ephippium	Taylor (1971c); Nielsen (1976b)
Isognomon isognomon	Nielsen (1976b)
Isognomon dentifer	Taylor (1968)
Pectinidae	
Gloripallium pallium	Ostergaard (1935); Taylor (1968, 1971a,b); Waller (1972)
Chlamys coruscans	Waller (1972)
Chlamys marshallensis	Waller (1972)
Chlamys squamosus	Ostergaard (1935); Smith (1978)
Chlamys albolineata	Nielsen (1976b)
Hemipecten forbesianus	Yonge (1981)

TABLE I *Continued*

Bivalve (families)	References
Pecten squamatus	Mastaller (1978)
Pecten lividus	Mastaller (1978)
Pedum spondyloideum	Taylor (1968, 1971a,b); Waller (1972); Nielsen (1976b); Yonge (1967b, 1981); Smith (1978)
Spondylidae	
Spondylus hystrix	Abbott (1950); Taylor (1968, 1971c); Nielsen (1976b)
Spondylus aurantius	Taylor (1971b,c)
Spondylus gaederopus	Mastaller (1978)
Spondylus aculeatus	Mastaller (1978)
Spondylus ducalis	Dudgeon and Morton (1982)
Plicatulidae	
Plicatula imbricata	Nielsen (1976b)
Plicatula plicata	Mastaller (1978); Taylor (1980); Dudgeon and Morton (1982)
Plicatula chinensis	Taylor (1968, 1971b)
Ostreidae	
Saccostrea cucullata	Abbott (1950); Taylor (1968); Nielsen (1976b)
Ostrea sandvichensis	Kay (1971); Kay and Switzer (1974)
Ostrea numisma	Taylor (1968, 1971a–c)
Ostrea frons (attached to gorgonians)	Coomans (1969); Taylor (1971c)
Ostrea (=*Hyotissa*) *hyotis*	Taylor (1968, 1971c, 1980); Nielsen (1976b)
Lopha folium	Taylor (1968); Mastaller (1978)
Lopha cucullata (often attached to gorgonians)	Mastaller (1978)
Lopha cristagalli (often attached to gorgonians)	Taylor (1968); Nielsen (1976b); Mastaller (1978); Smith (1978)
Alectryonella plicatula	Nielsen (1976b); Taylor (1980); Dudgeon and Morton (1982)
Chamidae	
Chama aspersa	Abbott (1950); Taylor (1968, 1971a,b); Nielsen (1976b)
Chama brassica	Nielsen (1976b)
Chama asperella	Taylor (1971c); Nielsen (1976b)
Chama pacifica	Nielsen (1976b)
Chama imbricata	Salvat (1967, 1972); Taylor (1971b)
Chama multisquamosa	Nielsen (1976b)
Chama reflexa	Nielsen (1976b); Dudgeon and Morton (1982)
Chama iostoma	Ostergaard (1935)
Carditidae	
Cardita variegata	Ostergaard (1935); Abbott (1950); Taylor (1968, 1971a,b,c); Kay (1971); Kay and Switzer (1974); Nielsen (1976b); Thomassin and Ganelon (1977); Mastaller (1978); Dudgeon and Morton (1982)

(Continued)

TABLE I *Continued*

Bivalve (families)	References
Veneridae	
Irus (Notirus) macrophyllus	Mastaller (1978); Dudgeon and Morton (1982)
Irus irus	Nielsen (1976b)
Trapeziidae	
Trapezium angulatum	Nielsen (1976b)
Trapezium oblongum	Ostergaard (1935); Abbott (1950); Taylor (1968, 1971c); Kay (1971)
Trapezium bicarinatum	Abbott (1950)
Coralliophaga coralliophaga	Morton (1980a); Dudgeon and Morton (1982)
Coralliophaga decussata	Taylor (1968)

From Onotoa Atoll in the Gilbert Islands, Banner and Randall (1952) recorded *Isognomon* sp. and *Barbatia tenella* between the inner branches of coral heads. From sand flats, these authors recorded a range of species including *Gafrarium pectinatum, Tellina crassiplicata,* and *Tellina* sp. Other bivalves, recorded from the atoll lagoon, included *Pinna atropurpurea, Streptopinna saccata, Pinctada margaratifera, Hippopus hippopus, Tridacna squamosa, Cardium flavum, Cardium* sp., *Gafrarium tumidum, Venus clathrata, Pitar japonica, Mesodesma striata, Protothaca staminea, Asaphis dichotoma,* and *A. deflorata.*

From Pombo Island in the Moluccas, Sumadhiharga (1971) lists only five bivalves: *Pinctada margaratifera, Pteria penguin, T. squamosa, Septifer bilocularis,* and *Spondylus* sp.

A survey of the littoral marine molluscs of Fanning Island by Kay (1971) reported 305 species, showing a clear distinction between the species composition of the seaward reefs and that of the lagoon. *Modiolus metcalfei* and *T. maxima* characterized the seaward reefs; *Electroma* sp., *Ostrea sandvichensis, Cardita variegata, Barbatia decussata, Trapezium oblongum,* and *Periglypta reticulata* characterized the lagoon. Subsequently, Kay and Switzer (1974) reported upon the molluscs of Fanning Lagoon. From the lagoon reef flat these authors recorded approximately 100 species, of which 33 were large molluscs (16 were alive) and 65 micromolluscs; the total was dominated by the gastropods. Two bivalves, *Fragum fragum* and *Tellina robusta,* were very common, occurring at 25% of the stations; *Fragum* was most frequent (30%) on the northeastern and *Tellina* most frequent (37%) on the southwestern perimeters, respectively. In addition to the dominant micromolluscs of the reef flat, six localized assemblages were recorded, one of which included *Macoma dispar,* with *Gafrarium pectinatum* occurring near brackish-water inlets. On the patch reefs, 32 species of large molluscs and 44 species of micromolluscs were recorded. Of the bivalves, *Electroma* sp. was found most frequently on *Acropora;*

TABLE II

Burrowing Bivalves Recorded from Coral Reef-Associated Habitats

Bivalve (families)	References
Arcidae	
Anadara antiquata	Ostergaard (1935); Taylor (1968); Mastaller (1978)
Anadara uropigymelana	Taylor (1968, 1971b)
Glycymeridae	
Glycymeris pectiniformis	Mastaller (1978)
Malleidae	
Malleus malleus	Iredale (1939); Yonge (1968); Nielsen (1976b)
Malleus albus	Yonge (1968)
Pinnidae	
Atrina vexillum	Abbott (1950); Rosewater (1961)
Pinna muricata	Abbott (1950); Rosewater (1961); Salvat (1967); Taylor (1968, 1971b)
Pinna atropurpurea	Banner and Randall (1952)
Lucinidae	
Codakia tigerina	Ostergaard (1935); Taylor (1971b); Nielsen (1976b)
Codakia ramulosa	Ostergaard (1935)
Codakia punctata	Ostergaard (1935); Abbott (1950)
Codakia divergens	Taylor (1971b); Salvat (1972); Kay and Switzer (1974); Randall et al. (1975); Nielsen (1976b); Mastaller (1978)
Codakia interrupta	Ostergaard (1935)
Lucina edentula	Abbott (1950); Nielsen (1976b)
Lucina dentifera	Mastaller (1978)
Lucina vesicula	Ostergaard (1935)
Bellucina sp.	Mastaller (1978)
Fimbriidae	
Fimbria fimbriata	Ostergaard (1935); Allen and Turner (1970); Morton (1979c)
Mesodesmatidae	
Mesodesma striata	Banner and Randall (1952)
Ervilia bisculpta	Lewis and Taylor (1966); Mastaller (1978)
Cardiidae	
Cardium lyratum	Ostergaard (1935)
Cardium flavum	Banner and Randall (1952); Nielsen (1976b)
Cardium unedo	Ostergaard (1935); Nielsen (1976b)
Cardium auricula	Mastaller (1978)
Laevicardium orbita	Mastaller (1978)
Laevicardium biradiatum	Lewis and Taylor (1966); Taylor (1968)
Ctenocardium sueziense	Lewis and Taylor (1966); Taylor (1968, 1971a,b)
Vasticardium alternatum	Smith (1978)
Fragum fragum	Ostergaard (1935); Abbott (1950); Salvat (1967); Taylor (1971b); Kay and Switzer (1974); Randall et al. (1975); Mastaller (1978)
Fragum carinatum	Nielsen (1976b)
Corculum cardissa	Kawaguti (1950, 1968); Nielsen (1976b)
Veneridae	
Pitar japonica	Banner and Randall (1952)
Pitar prora	Abbott (1950)

(*Continued*)

TABLE II *Continued*

Bivalve (families)	References
Chione lamarckii	Ostergaard (1935)
Venus toreuma	Thomassin and Ganelon (1977)
Venus clathrata	Banner and Randall (1952)
Paphia literata	Abbott (1950); Nielsen (1976b); Mastaller (1978)
Paphia exarata	Smith (1978)
Circe scripta	Lewis and Taylor (1966)
Circe dispar	Ostergaard (1935); Nielsen (1976b)
Circe gibba	Ostergaard (1935)
Circe transversaria	Ostergaard (1935)
Lioconcha fastigiatum	Smith (1978)
Lioconcha ornata	Smith (1978)
Lioconcha picta	Lewis and Taylor (1966); Mastaller (1978)
Protothaca staminea	Banner and Randall (1952)
Periglypta reticulata	Ostergaard (1935); Kay (1971)
Periglypta puerpera	Ostergaard (1935); Abbott (1950); Randall et al. (1975); Nielsen (1976b); Smith (1978)
Gafrarium pectinatum	Banner and Randall (1952); Salvat (1967); Taylor (1968, 1971b); Kay and Switzer (1974); Randall et al. (1975); Nielsen (1976b)
Gafrarium tumidum	Banner and Randall (1952); Taylor (1971b); Nielsen (1976b)
Gafrarium divaricatum	Nielsen (1976b)
Psammobiidae	
Asaphis deflorata	Abbott (1950); Banner and Randall (1952); Taylor (1968)
Asaphis violascens	Taylor (1971b); Mastaller (1978)
Asaphis dichotoma	Banner and Randall (1952); Nielsen (1976b)
Sportellidae	
Anisodonta lutea	Thomassin and Ganelon (1977)
Tellinidae	
Tellina robusta	Kay and Switzer (1974)
Tellina cruciata	Ostergaard (1935)
Tellina crassiplicatula	Abbott (1950); Banner and Randall (1952)
Tellina fabrefacta	Lewis and Taylor (1966); Taylor (1968)
Tellina virgata	Taylor (1968); Mastaller (1978)
Tellina (Cadella) semen	Lewis and Taylor (1966)
Tellina staurella	Taylor (1968); Nielsen (1976b)
Tellina remies	Nielsen (1976b)
Quidnipagus palatum	Taylor (1968, 1971b); Randall et al. (1975); Nielsen (1976b); Smith (1978)
Arcopagia scobinata	Ostergaard (1935); Abbott (1950); Taylor (1971b); Nielsen (1976b); Eldridge et al. (1977); Mastaller (1978)
Macoma dispar	Taylor (1971b); Salvat (1972); Kay and Switzer (1974); Nielsen (1976b)
Corbulidae	
Corbula archaeformis	Nielsen (1976b); Taylor (1968)
Corbula subquadrata	Lewis and Taylor (1966)
Corbula modesta	Taylor (1971b); Nielsen (1976b)
Corbula crassa	Nielsen (1976b)
Mactridae	
Mactra angulifera	Nielsen (1976b)

Ostrea sandvichensis occurred more frequently on *Stylophora* than on *Acropora* although it was not limited to living corals; and *Barbatia decussata* was found only on massive heads of *Porites*. *Tridacna maxima* and *Cardita variegata* were found on *Porites* and *Acropora* but similarly were not limited to living coral. The bivalves *Gafrarium, Pitar,* and *Trapezium* were found in sand, more frequently under rubble than under living coral.

Overall, *Tridacna* and *Barbatia* were limited in their distribution to the massive corals of the clear lagoon, whereas *Gafrarium, Pitar,* and *Trapezium* occurred in greater numbers on the rubble-covered reefs of the turbid southern lagoon. The boring mytilid *Lithophaga nasuta* was more common in corals of the turbid lagoon than in the clear lagoon. *Electroma* and *Ostrea* were distributed throughout the lagoon, but densities of the former appeared to be greater in the turbid regions. On the lagoon floor, *Atrina vexillum* occurred in the clear lagoon and *Codakia divergens* was recorded from the calcareous mud sediments of the turbid lagoon. The clear lagoon sediments were also characterized by a greater proportion of small tellinids (comprising 64% of the bivalves) than occurred in the turbid lagoon (29%).

Salvat (1967), in a broad survey of the important molluscs of Polynesian atolls, showed that the principal bivalves of beach rock, coral, and shell rubble are *Tridacna maxima, Pinctada maculata, Chama imbricata,* and *Arca ventricosa*. The distribution of these bivalves throughout Polynesia is quite variable, apparently related to substrate availability. Throughout this region, *Pinna muricata, Fragum fragum,* and *Gafrarium pectinatum* were the dominant bivalves on lagoon floors.

Salvat (1972) reported upon the lagoon molluscs at Réao, Tuamotu Islands, recording only 28 species. Of the infaunal species, only *Tellina robusta* was also dominant at Fanning Island (Kay and Switzer, 1974); *Macoma dispar* and *Codakia divergens* were also present, but occurred in much higher numbers at Réao. *Electroma,* dominant at Fanning, was replaced at Réao by *Pinctada maxima. Arca ventricosa* and *Chama imbricata,* dominant on the massive corals at Réao, were present at Fanning but were largely replaced by *Barbatia decussata*. From the closed lagoon of Takapoto atoll, Richard et al. (1979) record at least 93 species of molluscs; 34 were bivalves, with *A. ventricosa, C. imbricata,* and *T. maxima* dominant epifaunally and *Fragum fragum* common infaunally. Richard (1976) also recorded 104 species of molluscs from the closed atoll of Taiaro (French Polynesia); 23 were bivalves.

Coral-associated bivalves are often largely ignored in broad surveys of reef systems. For example, from Maug, in the northern Mariana Islands, Eldredge et al. (1977) describe only *Tridacna maxima* and *T. squamosa*. Similarly, from Uruno/Ritidian, Double Reef, Haputo Beach fringing reef, and Anae Island patch reef, on Guam, only *Tridacna maxima* is recorded (Stojkovich, 1977). *T. maxima* and *Scutarcopagia scobinata* occurred on Luminao Barrier reef; *T.*

maxima, Septifer bilocularis, and *Gafrarium* sp. occurred in Cetti Bay; *T. maxima* and *Pinctada margaratifera* occurred at Ajayan Bay. At Fadian Point, *Chama* sp., *Donax* sp., *S. bilocularis, T. maxima,* and *T. squamosa* occurred. However, from Cocos Lagoon on Guam, Randall et al. (1975) recorded a much greater array of bivalves including *Arca ventricosa, Barbatia* sp., *Chione* sp., *Codakia divergens, Fragum fragum, Gafrarium pectinatum, Modiolus auriculatus, Periglypta puerpera, Pinctada* sp., *Quidnipagus palatum, Septifer bilocularis, Spondylus* sp., and *Tellina* sp. From Yap Lagoon, Smith (1978) recorded the following species, either cemented or byssally attached to corals: *Barbatia bicolorata, Barbatia* sp., *Mytilus* sp., *Pinctada maculata, Chlamys squamosa, Pedum spondyloideum, Spondylus* sp., *Dendostrea* sp., *Lopha cristagalli, Tridacna maxima,* and *T. squamosa. Hippopus hippopus* lived freely on sand/coral substrates, and the following species were buried in coral sands; *Vasticardium alternatum, Vasticardium* sp., *Quidnipagus palatum, Lioconcha fastigiata, L. ornata, Paphia exarata,* and *Periglypta puerpera; Lima fragilis* occurred in crevices; and *Pecten* spp. were free swimming.

Most reports of coral-associated molluscs come from island or atoll systems, but Dudgeon and Morton (1982) have investigated the distribution of coral-associated gastropods and bivalves at six stations along a mainland estuary in Hong Kong. They recorded a total of 92 species, of which 38 were bivalves including the coral borers. The borers and the commensal leptonaceans will be discussed later. The remainder were epilithic nestling species, occurring mostly on dead coral heads or the dead bases of the living corals.

Four cemented species, *Chama reflexa, Alectryonella plicatula, Spondylus ducalis,* and *Plicatula plicata,* have been recorded from Tolo Harbor on coral surfaces. The first was the most common, recorded consistently from all stations and in greatest numbers from the central region of the estuary. The oyster, *Alectryonella,* is a sublittoral species recorded only occasionally from most stations. *Plicatula plicata* is largely intertidal, recorded only from central stations; similarly, *Spondylus* was recorded only rarely from central stations. The most common bivalves were the byssally attached members of the Arcacea and Mytilacea, with two species, *Barbatia helblingi* and *Septifer virgatus* (= *S. bilocularis*), widely dominant throughout the inlet. A second mytilid (*Perna viridis*) and two pearl oysters (*Pinctada martensii* and *Pteria penguin*) occurred in the inner harbor area. Other Arcacea (*Striarca afra, Arca avellana,* and *Barbatia plicata*) were common at the mouth of the estuary, where the saddle oyster *Monia umbonata* also occurred. These bivalves are typically large, sometimes gregarious, and generally dominant on dead coral heads.

The coral blocks are, however, suitable for colonization by a wide range of other solitary, byssally attached nestlers. The most common of these are members of the Limidae, especially *Limaria* (*Platilimaria*) *fragilis* and *L.* (*P.*) *hongkongensis*, which build loose byssal nests in coral galleries. *Isognomon legumen* is a small, bean-shaped nestler common throughout the estuary. *Vulsella vulsella* and *Crenatula modiolaris* are exclusively embedded in sponges associated with the coral community (Reid and Porteous, 1980) in its area of greatest diversity at the mouth of the estuary. The former species is associated with the massive, rounded black sponge *Suberites,* and the latter with *Sigmadocia symbiotica* (not *Haliclona,* as reported by Reid and Porteous, 1980), whose green color is derived from the symbiotic alga *Ceratodictyon spongiosum.* Other solitary nestlers included *Gregariella coralliophaga* (and *G. striatus*) (Scott, 1980); each builds a posterior byssal nest and enlarges its chosen burrow anteriorly (Morton, 1982a). *Musculus* sp., *Cardita variegata, Irus macrophyllus,* and *Coralliophaga coralliophaga* are solitary nestlers of coral crevices at the mouth of the estuary, for which a distinct pattern of increasing species diversity from the head to the mouth of the inlet was recorded.

According to Ladd (1960), the Indo-Pacific fauna had the central Pacific as its center of dispersion. Abbott (1960) believed however that this focus had its center in the western Pacific. In a major review of the marine molluscs of Polynesian atolls, Salvat (1967) showed that the region possesses an impoverished Indo-Pacific fauna, with the center of dispersion in the Indonesia/Philippines, New Guinea, New Caledonia region; this supported Abbott (1960). The Polynesian species generally have a vast Indo-Pacific distribution; impoverishment is primarily due to the remoteness of the continental land masses. Currents unfavorable for dispersion of mainland larvae, low nutrient values in the tropical seas, and high temperatures are thought to be the important factors limiting diversity on the Polynesian Islands. To the east, elements of the Indo-Pacific fauna overlap with Panamanian and Peruvian species; this is especially obvious at Clipperton, Galapagos, Cocos-Keeling, and Juan Fernandez islands.

Salvat (1967) studied the distribution of marine molluscs on eight Polynesian atolls in more detail (Fig. 2). The vertical distribution of the species on the exterior of the reefs was essentially the same in all cases. This absence of diversity and regional conformity results from the homogeneous ecological conditions common to all the reefs. Conversely, in the lagoons of the eight atolls, a different situation prevailed. Here there is considerable ecological diversity, created initially by the varying extent to which the lagoons are open to the sea. In the more open atolls, diversity is greater, with a reduced dominance; in the more closed lagoons, reduced diversity with ecological dominance by a few species is apparent, significantly including the bivalves *Tridacna maxima, Pinctada maculata, Chama imbricata, Arca ventricosa,* and *Fragum fragum.* Figure 2 shows the variable distribution of the first four of these species from atoll to atoll in relation to various physical processes. Salvat believes that in some lagoons the bivalves are even more important biologically than are the corals themselves; they are responsible, for example, for sedimentological effects as a result of

	MURUROA	SOUTH MARUTEA	PUKARUA	FANGATAUFA	MATUREI VAVAO	TUREIA	REAO	PUKAPUKA
Exchange between lagoon waters and the open sea	Open atoll with a passage to the sea	No passage, but numerous functional small passages	Closed atoll with functional small passages	Closed atoll with functional small passages	Closed atoll with some small passages functional at high tide	Closed atoll with some small passages functional at high tide	Closed atoll with several small passages occasionally functional at high tide	Completely closed atoll, with lagoon filling up and drying up
Maximum length of the atoll (km)	27	18	15	10	7	13	23	6(?)
Maximum breadth of the atoll (km)	Variable	9	3	5	4	8	5	2(?)
Maximum depth of the lagoon(km)	50	45	35	43	43	65	20	5
Tridacna maxima	•	●	⬤	○	⬤	⬤	⬤	○
Pinctada maculata	•	●	●	⬤	⬤	⬤	⬤	⬤
Chama imbricata	•	●	●	⬤	⬤	●	○	○
Arca ventricosa	•	•	○	●	⬤	○	○	○

○ Absent or very rare • Present ● Abundant ⬤ Characteristic of the lagoon and very dominant

Fig. 2. The relative abundance of the sessile species of bivalves, which in each of eight lagoons constitute more than 90% of the hard substrate molluscan fauna. Several characteristics of each atoll are also given. (After Salvat, 1967.)

empty valve accumulation. The bivalves, especially the borers, also play a role in the destruction, fragmentation, and dissolution of the reefs.

IV. Coral Borers

A. The Lithophaginae

The most important members of the primary cryptobion are the Lithophaginae (a subfamily of the Mytilidae) (Soot-Ryen, 1969), which are responsible for the erosion and eventual destruction of a wide range of calcareous substrates in the sea. In the Mediterranean, calcareous rocks are bored by *Lithophaga lithophaga* (Jaccarini et al., 1968). *L. falcata* bores rocks off the coast of California (Yonge, 1955; Warme and Marshall, 1969). Turner and Boss (1962) have reviewed the Western Atlantic species. Of the six species recorded by these authors, *L. nigra, L. antillarum,* and *L. bisulcata* bore, among other substrates, dead and possibly living corals. Lithophagines also bore the shells of gastropods, bivalves (Amemiya and Ohsima, 1923; Hodgkin, 1962), and chitons (e.g., *L. aristata*) (Bullock and Boss, 1971). They also bore dead coral skeletons (Otter, 1937; Yonge, 1974).

Bromley (1978) has reported upon the bioerosion of Bermuda reefs. A total of seven bivalves were found, six of them common; the most common was *Gastrochaena hians. Spengleria rostrata* was found only in water more than 6 m deep. The mytilid *Lithophaga nigra* was common; *L. bisulcata* also bores dead coral. A single, half-grown individual of *Botula fusca* was found in 11 m of water. The two small boring bivalves *Petricola lapicida* and *Gregariella coralliophaga* produce holes that closely correspond to the shape of their shells. However, Morton (1982a) has shown, for the latter species at least, that *Gregariella* commonly occupies empty *Lithophaga* burrows. Finally, the epilithic *Arca imbricata* was shown to commonly erode the substrate beneath it by physical abrasion with its shell. In this bivalve both boring and embedding processes are involved; epilithic growths are inhibited beneath the shell but accrete on the surrounding substrate.

The following lithophagines were reported by P. B. Robertson (1963) from Florida: *Lithophaga nigra, L. antillarum, L. bisulcata,* and *L. aristata.* None was reported as boring living corals, and it seems clear that in this area these bivalves may be no more important in the bioerosion of coral skeletons than are the Gastrochaenidae or various other bivalve borers. As will be seen, this is in marked contrast to the Indo-Pacific, where a very wide range of species occurs. Although these species can bore dead coral skeletons, they are almost the only ones capable of penetrating living corals, thereby initiating destruction of the living coral heads.

1. The Boring Mechanism

Pojeta and Palmer (1976) have reviewed the methods of boring in the Lithophaginae. Most authors (except Gohar and Soliman, 1963a,c; Soliman, 1969) now agree with the overwhelming evidence (Carazzi, 1892; List, 1902; Pelseneer, 1911; Kühnelt, 1930; Yonge, 1955; Hodgkin, 1962; Jaccarini et al., 1968; Morton and Scott, 1980) that boring in *Lithophaga* is primarily a chemical process. Similarly, it is believed, the calcareous paste with which these bivalves partially line their burrows and deposit on their shells is derived from the products of chemical boring. Barthel (1982), however, believes that a period of boring in *L. obesa* is followed by a period of active calcium secretion from the inflated mantle folds. Such an observation requires further study.

The shell of *Lithophaga lithophaga* is relatively thick, with radial ridges. It bores anteriorly into calcareous rock with the aid of a neutral mucoprotein that has calcium-binding properties; this is secreted by a pallial boring gland (Jaccarini et al., 1968; Morton and Scott, 1980) that extends all the way around each mantle lobe margin (except under the ligament) but is especially enlarged anteriorly and posterior to the ligament. *L. falcata* also has a thick shell, similarly ridged, but with a pronounced dorsal keel; it also bores by means of pallial boring glands, as does *L. lithophaga* (Yonge, 1955), although the animal is attached to the burrow wall by byssal threads aligned anteriorly and posteriorly. In Thailand, Australia, and Hong Kong, *L. teres* similarly is thick shelled, ridged, and a borer of dead coral rock (Nielsen, 1976a; Wilson, 1979; Scott, 1980; Morton and Scott, 1980).

In Hong Kong, *L. malaccana* is also a dead coral borer, but it is capable of withstanding surface recolonization by either new coral polyps or other encrusting organisms. *L. malaccana* also bores rock oysters in the intertidal zone (Scott, 1980), but is the most common lithophagine borer of dead corals in the area (Dudgeon and Morton, 1982). Surface recolonization is countered by secretions from an extra gland located in the posterior regions of the mantle (Morton and Scott, 1980). *L. malaccana* has a smooth, somewhat elongate shell.

Gardiner (1903) thought that entrance into living corals could occur only through a dead portion of the colony. Soliman (1969) believed that the association between *Lithophaga* and living corals resulted from the settlement of the young mussel on the living coral tissue with subsequent penetration. More recent studies by Kleeman (1977, 1980a,b), Wilson (1979), and Morton and Scott (1980) support this view. Morton and Scott have shown that the posterior pallial gland of *L. malaccana* is greatly enlarged in live coral borers (e.g., *L. lima, L. nasuta, L. hanleyana, L. simplex,* and *L. mucronata*); this presumably inhibits coral recalcification around the siphonal aperture. These lithophagines possess additional siphonal glands which may inhibit the coral polyps from attacking, with their nematocysts, the siphons that project among them (Morton and Scott,

1980). The structure of these three glands has been described by Morton and Scott (1980) (Fig. 3); these authors have also demonstrated an increasing glandular sophistication and specialization in borers adapted for life with living corals. The exact chemical nature of the secretions from these glands, however, remains unknown.

2. Dead Coral Borers

Most early authors indicated that lithophagines inhabited only calcareous rocks or dead coral skeleton (Hass, 1943; Yonge, 1955; Turner and Boss, 1962; P. B. Robertson, 1963; Appukuttan, 1973). Others (Bertram, 1936) recognized that they occurred in either dead or living coral skeletons without reference to species. Otter (1937) claimed that *Lithophaga* was unable to bore a living coral surface, but Bertram (1936) found *L. teres* and *L. hanleyana* in both live and dead corals at Ghardaqa on the Red Sea. Nielsen (1976a) found *L. teres, L. nasuta,* and *L. malaccana* all boring dead blocks of *Porites* at Phuket, Thailand; Kleeman (1980a) recorded *L. antillarum* exclusively from dead coral heads. From the Great Barrier Reef, Wilson (1979) reported that *L. teres, L. antillarum, L. nasuta, L. malaccana, L. hanleyana,* and *L. divaricalyx* are borers of dead corals. From Hong Kong, Scott (1980) and Morton and Scott (1980) also recorded that *L. teres* and *L. antillarum* are dead coral borers, but considered *L. malaccana* to be a special case. This species is normally thought to colonize dead areas but can also withstand surface recolonization (Morton and Scott, 1980). Significantly, Iredale (1939) recorded this species (as *L. calcifer*) from living corals. Only comparatively recently, however, has it been fully appreciated that lithophagines also bore living corals.

3. Live Coral Borers

Gohar and Soliman (1963a) first recognized a specific relationship between three species of *Lithophaga* and the coral into which they bored at Ghardaqa. These authors argued that "a specificity of the boring species to the coral attacked obviously exists." They found *L. cumingiana* boring in *Stylophora flabellata; L. hanleyana* in *Cyphastrea microphthalma, C. chalcidicum, Montipora lanuginosa,* and *Goniastrea* sp., and *L. lima* in various species of *Montipora* and *Cyphastrea.*

Highsmith (1980) has provided a list of lithophagines with their known hosts which has been extended by the present author (Table III). It appears that the only branching corals colonized by bivalves are *Stylophora* spp. and *Pocillopora* spp. (Pocilloporidae). These slender, branched species are unsatisfactory for a borer, being easily broken and susceptible to consumption by coral-eating fishes. Much more commonly, lithophagines attack massive or encrusting live corals (Fig. 4). There appear to be two major kinds of live coral borers: those that can live in a wide range of corals (wide-spectrum borers) and those that are specific,

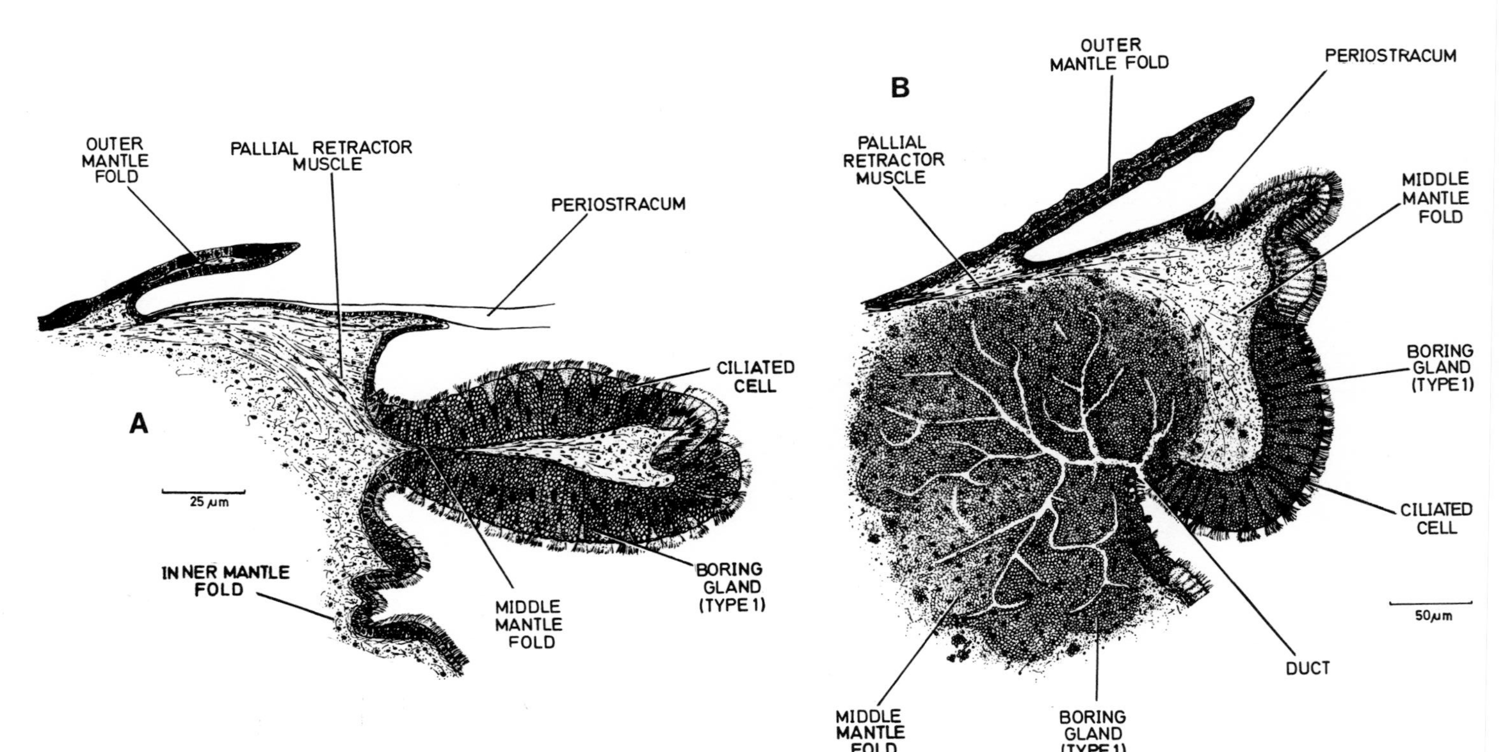
A
OUTER MANTLE FOLD
PALLIAL RETRACTOR MUSCLE
PERIOSTRACUM
CILIATED CELL
25 µm
INNER MANTLE FOLD
MIDDLE MANTLE FOLD
BORING GLAND (TYPE 1)
B
OUTER MANTLE FOLD
PERIOSTRACUM
PALLIAL RETRACTOR MUSCLE
MIDDLE MANTLE FOLD
BORING GLAND (TYPE 1)
CILIATED CELL
50 µm
DUCT
MIDDLE MANTLE FOLD
BORING GLAND (TYPE 1)

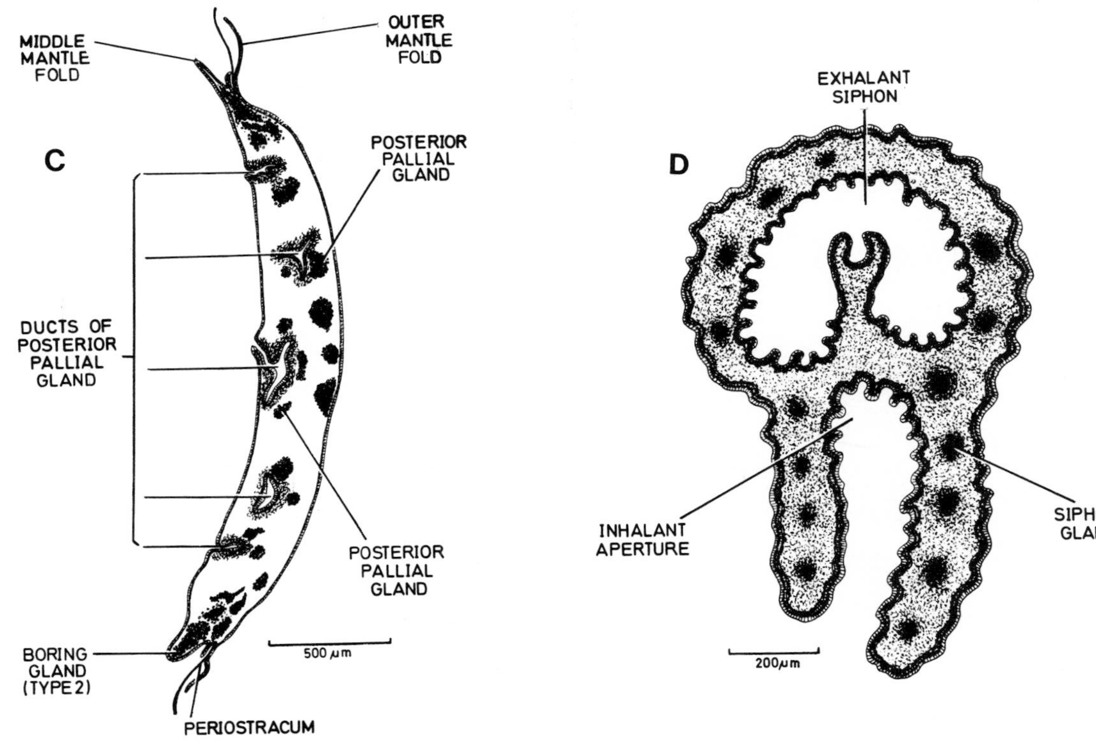

Fig. 3. The various pallial glands of lithophagine borers. (**A**) The dorsal boring gland of the dead coral borer *Lithophaga teres*. (**B**) The posterodorsal boring gland of the live coral borer *L. lima*. (**C**) The posterior pallial gland of the live coral borer *L. simplex*. (**D**) The siphon gland of the live coral borer *L. hanleyana*. [After Morton and Scott (1980). With permission of The Zoological Society of London.]

TABLE III

The Indo-Pacific species of Boring *Lithophaga* and *Fungiacava* Recorded from the Corals with Which They Associate

Bivalve	Coral	Reference
Wide-spectrum borers		
Lithophaga lessepsiana[a]	*Stylophora pistillata*	Vaillant (1865); Fishelson (1973); Kleeman (1980b)
	Stylophora flabellata	Gohar and Soliman (1963a)
	Heteropsammia michelini	Arnaud and Thomassin (1976); Kleeman (1980b)
	Acropora palifera	Kleeman (1977, 1980b); Wilson (1979)
	Pocillopora eydouxi *Porites* sp. *Favia* sp.	Wilson (1979)
	Stylophora mordax *Pocillopora damicornis*	Kleeman (1980b)
Lithophaga simplex[b]	*Porites* spp. *Symphyllia* spp.	Iredale (1939)
	Favia pallida (preferred) *Lobophyllia* spp. *Symphyllia* spp. *Goniastrea retiformis*	Kleeman (1980b)
	Favia spp. (preferred)	Scott (1980); Morton and Scott (1980)
	Favia pallida *Favia favosa* *Favia amicorum* *Favia lizardensis* *Symphyllia nobilis* *Lobophyllia costata*	Kleeman (1980b)
	Astreopora sp. *Lobophyllia corymbosa* *Goniastrea retiformis* *Goniastrea pectinata*	Kleeman (1980b)
Lithophaga hanleyana[c]	*Porites* spp.	Otter (1937)
	Porites spp.	Kleeman (1977)
	Cyphastrea microphthalma *Cyphastrea chalcidicum* *Montipora lanuginosa* *Goniastrea* spp. (few occurrences) *Stylophora* spp. (rare)	Gohar and Soliman (1963a)

TABLE III *Continued*

Bivalve	Coral	Reference
	Alveopora allingi	Scott (1980); Morton and Scott (1980)
	Porites lichen	
	Cyphastrea serailia	
	Favia spp.	
Lithophaga laevigata[d]	*Leptastrea purpurea*	Kleeman (1980b)
	Psammocora profundacella	
	Psammocora contigua	
	Cyphastrea microphthalma	
	Cyphastrea serailia	
	Astreopora myriophthalma	
	Montipora sinensis	
	Porites lichen	
	Porites murrayensis	
	Porites vaughani	
	Goniastrea sp.	
Lithophaga purpurea	*Montipora* cf. *stilosa*	Kleeman (1980b)
	Cyphastrea sp.	
	Cyphastrea microphthalma	
	Echinopora gemmarea	
Lithophaga lima[e]	*Montipora* spp.	Gohar and Solimann (1963a)
	Cyphastrea spp.	
	Porites spp.	Wilson (1979)
	Favites spp.	
	Favia spp.	
	Pleisiastraea spp.	
	Leptoria spp.	
	Favia spp.	Scott (1980); Morton and Scott (1980)
	Platygyra spp.	
	Alveopora allingi	
	Porites lichen	
	Cyphastrea serailia	
	Goniastrea aspera	
	Montipora informis	
	Acanthastrea echinata	Kleeman (1980b)
	Astreopora ehrenbergi	
	Astreopora myriophthalma	
	Cyphastrea serailia	
	Diploastrea heliopora	
	Favia pallida	
	Leptoria phrygia	
	Platygyra daedalea	
	Porites sp.	

(*Continued*)

TABLE III *Continued*

Bivalve	Coral	Reference
Narrow-spectrum borers		
Lithophaga kuehnelti	*Acropora palifera*	Kleeman (1977, 1980b); Evseev (1982)
	Stylophora mordax	Evseev (1982)
Lithophaga curta	*Montipora berryi*	Highsmith (1980)
Lithophaga mucronata[f]	*Stylophora pistillata*	Highsmith (1980)
	Montipora informis	Scott (1980); Morton and Scott (1980)
Lithophaga sp.	*Acropora palifera* *Pocillopora damicormis*	Connell (1973)
Lithophaga obesa	*Cyphastrea* sp.	Wilson (1979)
Fungiacava eilatensis	*Fungia scutaria* *Fungia fungites*	Goreau et al. (1969)
Fungiacava cf. *gardineri*	*Cycloseris fragilis* *Cycloseris sinensis* *Diaseris distorta*	Goreau et al. (1969, 1972)
Dead coral borer capable of withstanding surface recolonization		
Lithophaga calcifer (= *L. malaccana*)[g]	*Favia* spp. *Goniastrea* spp. *Pocillopora* spp.	Iredale (1939); Wilson (1979); Scott (1980); Morton and Scott (1980); all believe this species (= *L. malaccana*) to be a dead coral borer.

[a] Identified as *L. cumingiana* by Gohar and Soliman (1963a), Arnaud and Thomassin (1976), and Kleeman (1980b). Wilson (1979) also believes this to be *L. simplex* (of Iredale, 1939) and *L. hanleyana* [in part (of Otter, 1937)].

[b] Wilson (1979) believes *L. simplex, L. cumingiana* (of Gohar and Soliman, 1963a), and (in part) *L. hanleyana* (of Otter, 1937) to be synonymous with *L. lessepsiana*. Kleeman (1980b) believes *L. simplex* to be a discrete species.

[c] Wilson (1979) believes this species includes *L. laevigata instigans* (of Iredale, 1939) and *L. hanleyana* [in part only (of Otter, 1937)].

[d] Kleeman (1980b) believes this to be *L. hanleyana* [in part (of Otter, 1937)] and the *L. hanleyana* (of Gohar and Soliman, 1963a, and Kleeman, 1977).

[e] Kleeman (1980b) believes the *L. lima* (of Gohar and Soliman, 1963a) is *L. purpurea*.

[f] May be synonymous with *L. lessepsiana* (Vaillant, 1865).

[g] Wilson (1979) believes *L. calcifer* (of Iredale, 1939) to be synonymous with *L. malaccana* (a dead coral borer), *L. laevigata* (of Gohar and Soliman, 1963a), and *L. hanleyana* [in part (of Iredale, 1939)].

Fig. 4. (**A**) The siphons (arrow) of *Lithophaga lima* protruding amid the expanded polyps of the day-feeding coral *Goniopora tenuidens*. (**B**) The siphons (arrows) of *L. hanleyana* opening out onto the polyps of *Favites abdita*. (Scale = 1 cm.) [After Morton and Scott (1980). With permission of The Zoological Society of London.]

often to only one species of coral (narrow-spectrum borers) (Morton and Scott, 1980).

Consider first the former category. Of the nine species of *Lithophaga* recorded by Wilson (1979) from the Great Barrier Reef, only *L. lima* and *L. lessepsiana* occurred in living coral. The former was found in a very wide range of corals, a point also noted for the same species from the same locality by Kleeman (1980b), and from Hong Kong, where it bores representatives of the Faviidae and Acroporidae (Scott, 1980; Morton and Scott, 1980). However, Arnaud and Thomassin (1976) point out that *L. lessepsiana* occurs only in *Stylophora* sp. in the northern sector of the Red Sea, where its normal host (*Heteropsammia michelini*) is absent. Kleeman (1980b) recorded *L. lessepsiana* from *Stylophora mordax* and *Pocillopora damicornis* on the Great Barrier Reef, but Wilson (1979) found this species on a range of other corals in the same locality. Kleeman (1980b) also recorded *L. laevigata* and *L. purpurea* from a range of corals on the Great Barrier Reef. In Phuket, Scott (1980) recorded *L. nasuta* from *Goniastrea* (Faviidae) and *Porites,* although most authors (Nielsen, 1976a; Wilson, 1979) consider this is a dead coral borer. In Hong Kong, *L. hanleyana* occurs in *Goniastrea* and *Porites* as well as *Alveopora,* but Gohar and Soliman (1963a) recorded it from a range of other corals in the Red Sea. Wilson (1979) also recorded this species from dead corals on the Barrier Reef.

In the second category, *L. simplex* and *L. mucronata* (in Hong Kong) seem to be much more specific in their choice of hosts, the former occurring only in *Favia* (Faviidae) and the latter only in *Montipora* (Acroporidae) (Scott, 1980). Highsmith (1980) recorded this species only from *Stylophora pistillata*. Kleeman (1977, 1980b), however, recorded *L. simplex* from a range of other corals but added, as did Scott (1980) and Morton and Scott (1980), that this species seems to "prefer" faviid corals. *L. simplex* may be representative of conditions intermediate between those of the wide- and narrow-spectrum borers. *L. obesa* apparently bores only *Cyphastrea* sp. on the Great Barrier Reef (Wilson, 1979). *L. kuehnelti* is apparently restricted to the massive *Acropora palifera* (Kleeman, 1977) [although Evseev (1982) also records it from *Stylophora mordax*], and *L. curta* appears to associate only with *Montipora berryi* (Highsmith, 1980). Boring in *L. kuehnelti* equals the rate of radial growth of the coral host. After the death of the coral colony, the mortality of the bivalves increases and new settlement does not occur (Evseev, 1982). Arnaud and Thomassin (1976) reported that infestation of *Heteropsammia michelini* by *L. lessepsiana* may be considered as a kind of parasitism because burrowing induces death of the polyp when the mussel outgrows the coral. However, *L. lessepsiana* (Gohar and Soliman, 1963a; Fishelson, 1973), *L. hanleyana* (Otter, 1937; Gohar and Soliman, 1963a; Wilson, 1979), and *L. calcifer* (Iredale, 1939) have also been found in reef substrates other than live coral. Presumably, larvae of these bivalves are normally induced to settle and metamorphose by some stimulus from the preferred coral. In the absence of the appropriate stimulus, metamorphosis can be delayed, but not indefinitely; in time, selectivity may decline forcing the larvae to settle elsewhere, perhaps either with other related corals or in calcareous rock (Highsmith, 1980). Alternatively, corals around the borer may have died, giving the impression of habitation of dead coral, or, as pointed out by Otter (1937), some species such as *L. hanleyana* may settle on dead spots of *Porites* spp. and become incorporated into the live coral surface by subsequent growth of the original coral species or by overgrowth of an invading coral species.

Thus, the fidelity of a particular borer to a particular coral may be influenced by many factors, giving a false impression of the normal mode of life. The footnotes to Table III, however, highlight a much more serious problem in studies of lithophagine borers of corals. Taxonomic decisions at both the generic and species levels in the Lithophaginae are by no means unanimous (Wilson, 1979; Kleeman, 1980b); moreover, such taxonomic problems extend to the corals themselves. Thus, there is a need for very careful recording of well-defined species of borer and "host" coral in order to determine accurately the extent to which particular species of borer associate with different species of coral.

Over the ranges of both borer and coral, preferences may be adjusted, as has been discussed here. Careful records will help elucidate the reasons, but until more studies of the coral-associated Lithophaginae have been undertaken, it is not possible to comment further on the precise nature of the relationship that exists between borer and coral. Clearly, in some cases such as *L. kuehnelti* in *Stylophora* and *Acropora,* it approaches commensalism.

A more clearly defined "host" specificity is demonstrated by *Fungiacava eilatensis* (Goreau et al., 1969, 1970, 1972), which bores the solitary coral *Fungia scutaria* (and also *F. fungites, Cycloseris sinensis,* and *Diaseris distorta*). This highly modified lithophagine (Fig. 5A) is perfectly smooth, with very reduced musculature and byssus, and the middle fold of the mantle is reflected over the shell. Clearly, boring is exclusively chemical (the shell being completely enclosed) and, because the animal fits exactly into the burrow, there must also be selective dissolution and secretion of lime by the mantle tissues. The siphons of *Fungiacava* open not onto the surface of the coral but into the coelenteron of the host, so that in life it is completely hidden (ventral side uppermost). Goreau et al. (1970) postulate that *Fungiacava* is one member of a symbiotic troika (Fig. 5B) in which the bivalve collects plankton drawn into the coelenteron by the feeding currents of the coral and can also utilize excess zooxanthellae released by the coral's mesenteries. The coral benefits by the inorganic salts discharged renally by the bivalve, which are also available to the zooxanthellae. The ctenidia of *Fungiacava* are large but the labial palps are minute, suggesting that selection of food in the mantle cavity is minimal; this may be accomplished for it by the cleansing currents of the coral. *Fungiacava*

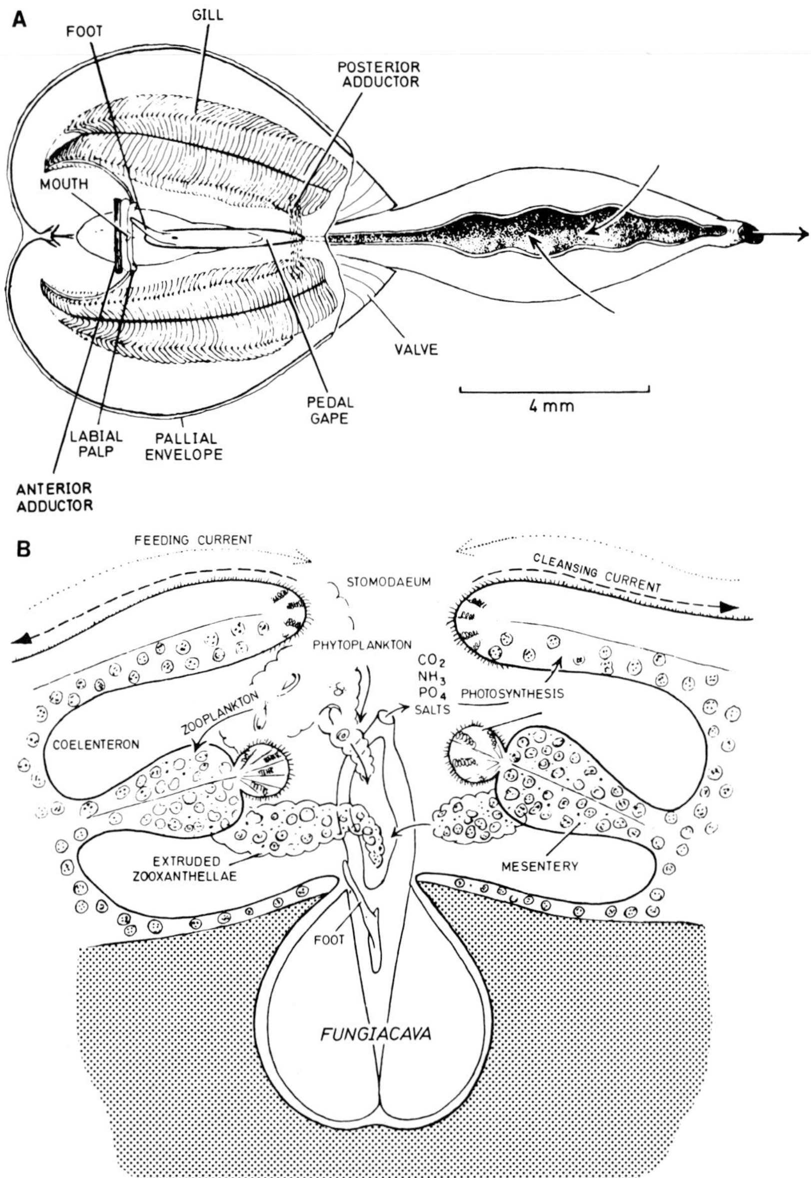

Fig. 5. (**A**) *Fungiacava eilatensis* viewed from above (as it lies in the boring) showing the ventral surface with fully expanded siphons which in life reach into the coelenteron of the fungid coral. Water currents are indicated; inhalant above, exhalant below. Shell valves are almost completely enclosed within pallial envelope; enlarged food-collecting ctenidia with distal oral grooves convey food to mouth flanked by vestigial labial palps. Both adductors are reduced and probably functionless; the foot, with opening of byssal gland behind, projects through pedal gape. (**B**) Schematic diagram showing the trophic relations between *Fungia scutaria* and a contained *Fungiacava eilatensis*. [(A) After Goreau et al. (1969). (B) After Goreau et al. (1970). With permission of The Zoological Society of London.]

seems to be mostly restricted to the Gulf of Eilat, where 60% of *Fungia scutaria* are infected with as many as 16 bivalves to a single coral. It becomes progressively rarer farther south and has never been found off Port Sudan. A few other specimens have been collected from the Maldives, and there is evidence of its occurrence from one locality in the central Pacific (Yonge, 1974). In a discussion on its evolution, Soot-Ryen (in Goreau et al., 1970) suggests that *Fungiacava* might represent a separate line of evolution from the Lithophaginae. This has been disputed by Morton and Scott (1980), who see it as the natural culmination of a recognizable evolutionary trend within this subfamily. Morton has further discussed the evolution of the Lithophaginae (1982a).

4. Corals in Bivalve Burrows and the Theory of Reciprocal Larval Recruitment

Highsmith (1980) discovered five empty *Lithophaga* burrows, each containing a very small colony of *Montipora berryi*. The lithophagine burrow resembled that of *L. curta*, which favors settlement on *M. berryi*. Two of the colonies were only about 3.5 mm wide, indicating recent planular settlement, and the others had expanded to fill the width of the burrow. The advantages to *M. berryi* of such a settlement site are clear. The burrows provide a refuge for juvenile colonies from incidental predation by grazers as well as from specialized coral predators. Similarly, the settling larvae avoid damage, predation, and competition, yet obtain adequate sunlight.

On the basis of this observation, Highsmith (1980) constructed an interesting theory of reciprocal larval recruitment (Fig. 6), in which *M. berryi* and its boring lithophagine, *L. curta*, are dependent on each other for larval recruitment. Adult *L. curta* living in *M. berryi* colonies produce larvae that settle preferentially on *M. berryi*. Some *M. berryi* and the *L. curta* in them die, either at the same time or the coral first. Larvae of *M. berryi* then settle in burrows of dead *L. curta*, eventually producing potential settlement sites for *L. curta*. A possible variation might be that the corals outlive the bivalves in them and seal the burrows. Following the death of the coral, the burrows eventually may be reopened by erosion of the surface by grazing organisms.

5. The Origin of the Lithophaginae

The origin of the Lithophaginae has been discussed by Morton (1982a) (Fig. 7). Pojeta and Palmer (1976) have described a fossil Ordovician lithophagine, *Corallidomus scobina*, which they consider to be a facultative borer, nestling but also boring by ventral dissolution of a slot in stromatoporoid corals (and in crevices of the trepostome bryozoans *Prasopora* and *Hallopora*). Morton (1982a) has researched *Gregariella coralliophaga*, a nestling lithophagine hitherto placed in the Crenellinae (Soot-Ryen, 1969). This bivalve occupies empty *Lithophaga* burrows and weaves a byssal nest around the posterior regions of the

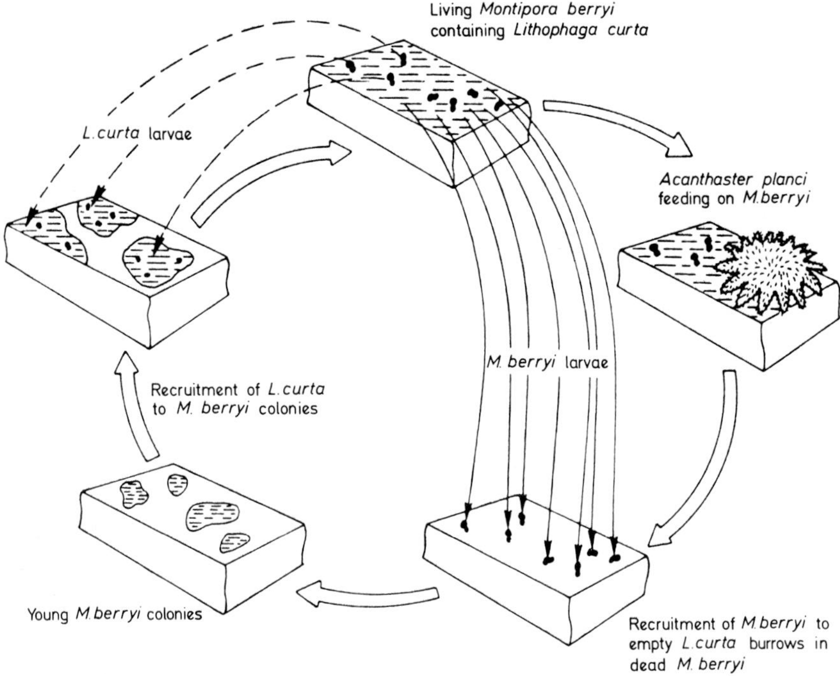

Fig. 6. *Lithophaga curta* and *Montipora berryi*. Diagram of hypothesized reciprocal larval settlement. (After Highsmith, 1980.)

shell to conceal and block the entrance to the burrow (Fig. 14B). The animal can also modify the burrow shape by secretions from pallial glands similar to those of *Lithophaga,* although it does not, as suggested by Bromley (1978), create a closely fitting burrow. The shell of *Gregariella* is thick, strongly ribbed anteriorly and posteriorly, and sharply keeled posterodorsally; it is very similar to the shell of *L. plumula* (Yonge, 1955). *Lithophaga gracilis* (Pelseneer, 1911) possesses a foliose anterior swelling of the mantle that encloses the anterior region of the shell. This tissue of the inner mantle folds is densely ciliated and is probably responsible for sediment removal from the burrow in the mechanical borer *Adula falcata* (Fankboner, 1971a) and in *Gregariella* (Morton, 1982a). Such a swelling also occurs in *L. plumula* (Yonge, 1955). Yonge thought that these tissues were probably the most important adaptation to boring in *Lithophaga,* but it is more likely that they are used for sediment consolidation and removal (especially in *Adula,* where boring is mechanical) (Fankboner, 1971a).

In *Gregariella,* boring is accomplished by secretions from pallial glands developed in the middle mantle fold (as in all species of *Lithophaga*), not in the inner mantle fold, as was reported by Yonge (1955). Thus, the series *Cor-*

allidomus–Gregariella–L. (Diberus) plumula–Lithophaga demonstrates a clear trend toward an increasingly sophisticated boring mode of life. In this trend, the shell becomes progressively thinner, smoother, transversely rounded, and elongate; ridges and the dorsal keel are progressively reduced and eventually disappear.

Morton and Scott (1980) have identified a further trend in the coral-boring Lithophaginae (Fig. 8). Gohar and Soliman (1963a,c) and Soliman (1969) have shown that the direction of boring by *Lithophaga* in living corals is to the outside, the borers keeping pace with the growing coral layer to maintain their burrow aperture. Yonge (1974) has confirmed this for *Fungiacava* and shown that inhabitation of a living coral is fundamentally different from the habits of other lithophagines, which bore inert rock. In the latter, boring is downward in an anterior direction, keeping pace with both shell growth and the progressive

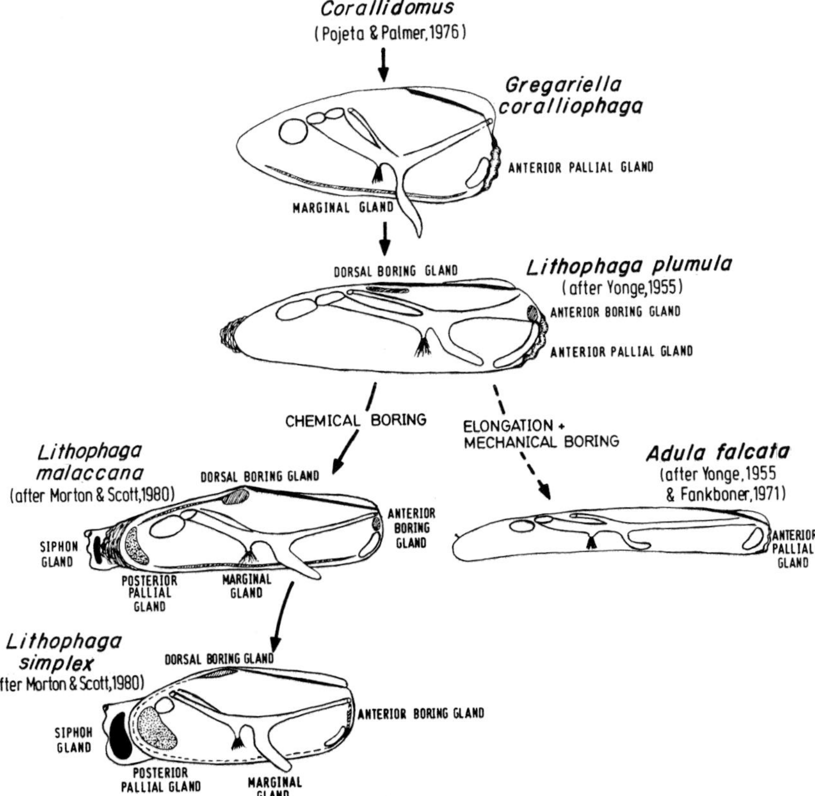

Fig. 7. A proposed system of adaptive radiation in the Lithophaginae, showing the progressive sophistication toward glandular elaboration in *Lithophaga*. (After Morton, 1982a.)

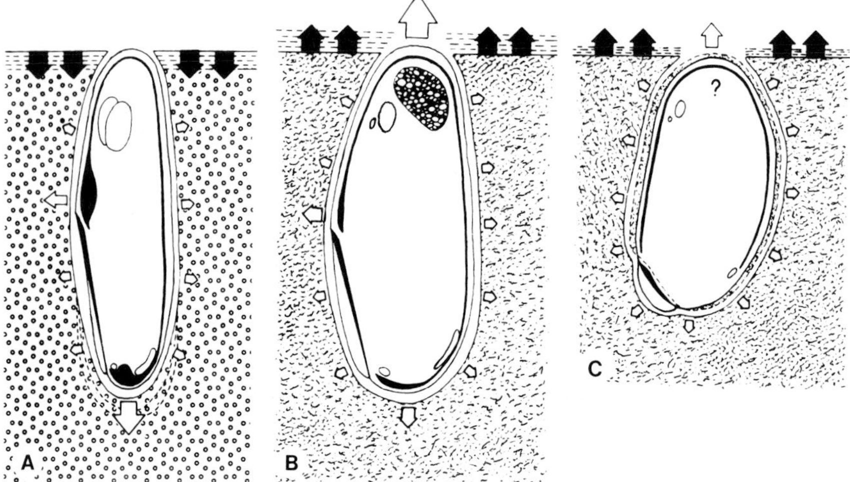

Fig. 8. A comparison of shell form, boring gland distribution, major regions of secretory activity (open arrows), and areas of habitat erosion or calcification (closed arrows) in (**A**) *Lithophaga lithophaga*, (**B**) *L. simplex*, and (**C**) *Fungiacava eilatensis*. [After Morton and Scott (1980). With permission of The Zoological Society of London.]

erosion of the rock surface. In *Fungiacava,* however, calcium deposition by the coral means that boring is posterior as well as anterior. Moreover, because anterior and posterior growth are matched, there must be dissolution of the burrow in all directions to house the growing bivalve. Morton and Scott (1980) suggest, therefore, that there is a trend for the shell to assume a rounded form with an increasing degree of intimacy between borer and coral. The coral borers of Hong Kong, demonstrating differing degrees of intimacy with their hosts, illustrate this; the narrow spectrum borers *L. lima* and *L. simplex* have more rounded shells than their wide-spectrum counterparts. Morton and Scott (1980) believe *Fungiacava* to be the culmination of this concept of evolution, the adoption of a commensalistic relationship with a coral being reflected in a simplification of form and a fragile rounded shell—protection has become the responsibility of the host.

6. *Gregariella and the Lithophaginae in Retrospect*

If by looking forward *Gregariella* can be seen as the link in the evolution of the Lithophaginae, it can also be judged, when looking back, as a link with other possibly less specialized mytilids to explain this particularly successful evolutionary trend (Fig. 9). An important feature of *Gregariella* is its habit of building a posterior byssal nest. This habit has also been adopted by a number of mud-dwelling members of the Mytilinae. Thus, *Musculista senhausia* and *Mytella guyanensis* both build byssal nests (Morton, 1973; Bacon, 1975), which enable

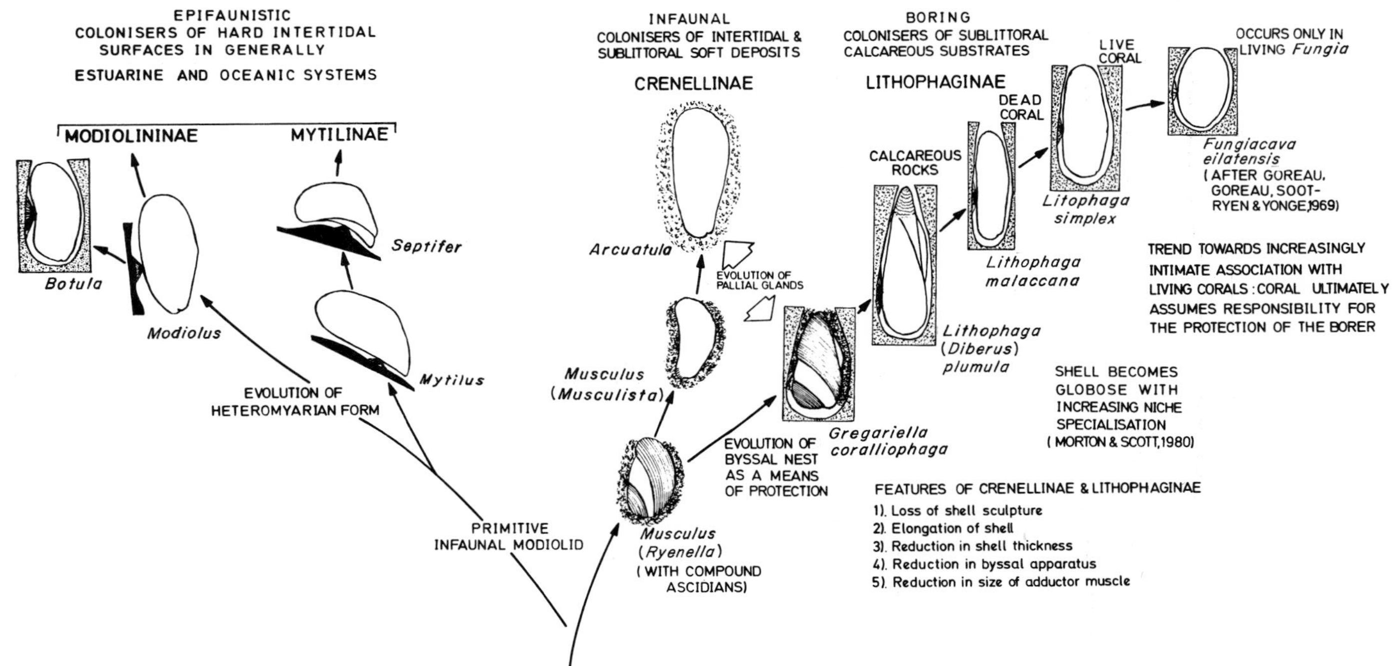

Fig. 9. A proposed pattern of adaptive radiation in the extant Mytilacea, showing the major lineages with emphasis on the Crenellinae and Lithophaginae. *Gregariella* is particularly important to understanding the origin and evolution of these two subfamilies. (After Morton, 1982a.)

them to colonize soft muds. *Arcuatula elegans* also builds a byssal nest (Morton, 1980c), but the threads are exceptionally fine and bound together by a mass of gelatinous mucus secreted from glands in the middle fold of the mantle. Morton (1982a) speculates that *Gregariella* represents a line of evolution from this group of byssal nest-building mytilids that has taken to endolithic nestling in rock and coral crevices. Subsequently, the glands that in *Arcuatula* secrete a mucus cocoon have assumed a quite different function, that of producing a secretion to facilitate chemical boring. In this trend, *Gregariella* is also representative of intermediate conditions.

Thus, the mechanisms prerequisite for the successful adoption of a chemical boring mode of life by the Lithophaginae have been evolved in the Crenellinae, with a long origin in the Modiolinae (Morton, 1982a). An initial adaptation to an endobyssate mode of life is essentially a preadaptation for a change in habit and the exploitation of living corals as an hitherto unoccupied substrate for borers.

7. *Botula*

The mytilid genus *Botula* is represented in the Caribbean by *B. fuscus,* which bores limestone blocks (P. B. Robertson, 1963). In the Indo-Pacific, *B. cinnamomea* is the major species, reported from Australia by Otter (1937), from the Red Sea by Soliman (1969), from Diego Garcia in the Indian Ocean by Taylor (1971a,b), and from Thailand by Nielsen (1976a,b). Warmke and Abbott (1961) believe this species is a junior synonym of *B. fuscus.*

Yonge (1955) has reported upon *B. falcata* from the west coast of North America and shown that boring "is purely mechanical, the rock is not calcareous, acid has no effect on it. Actual abrasion of the rock must almost exclusively be carried out by the dorsal surfaces of the valves, along a broad region running forward from the umbones." Soliman (1969) also believes *B. cinnamomea* is a mechanical borer. Nielsen (1976a), however, believes there is a chemical aspect to boring in *Botula,* possibly by the anterior pallial tissues (inner mantle folds) that encompass the anterior face of the shell. However, as explained earlier, these tissues have the function of sediment removal in the mechanical borer *Adula falcata* (= *B. falcata* of Yonge, 1955; Fankboner, 1971a), and if chemical boring is proven, the responsible glands should be sought in the middle mantle folds, the site of secretion in the Lithophaginae (Morton and Scott, 1980) and the Tridacnidae (Yonge, 1980).

Like *Lithophaga, Botula* attaches byssally to its burrow, which, as it also conforms to the shape of the shell, argues for chemical boring (Nielsen, 1976a). Soliman (1969) also thought that a chemical secretion aided in the process of widening the siphonal part of the boring. Morton (1982a) has argued that *Adula* and *Botula* are on a separate (within the Mytilidae) but convergent line of boring evolution with the Lithophaginae, the former possibly evolving more directly from a modioline ancestor, rather than via the intermediary of a crenelline mytilid as in the Lithophaginae (Fig. 9).

B. The Gastrochaenidae

The Gastrochaenacea Gray 1840 comprise a single family which, according to Keen (1969), can be divided into four genera: *Gastrochaena* Spengler 1783, *Eufistulana* Eames 1951, and the extinct *Gastrochaenopsis* Chavan 1952 and *Kummelia* Stephenson 1937. The genus *Gastrochaena* is divided into two subgenera: *G. (Gastrochaena)* and *G. (Spengleria)* Tryon 1862. Carter (1978) adds to the taxonomic discussion by dividing *Gastrochaena* into three subgenera: *G. (Gastrochaena), G. (Rocellaria)* de Blainville 1828, and *G. (Cucurbitula)* Gould 1861. Keen (1969) earlier had placed the last subgenus in synonymy with *Gastrochaena*.

The Gastrochaenacea are a discrete group of hard substrate borers distributed throughout the tropical, subtropical, and warm temperate waters of the world. The group is taxonomically much less diverse than the Lithophaginae; about 15 species from a number of habitats comprise recent faunas (Boss, 1971).

Carter (1978) has reviewed the ecology and evolution of the group. It is generally agreed that gastrochaenids bore only limestone blocks and dead coral skeletons (Fig. 10) (Purchon, 1954; Yonge, 1955; P. B. Robertson, 1963; Gohar

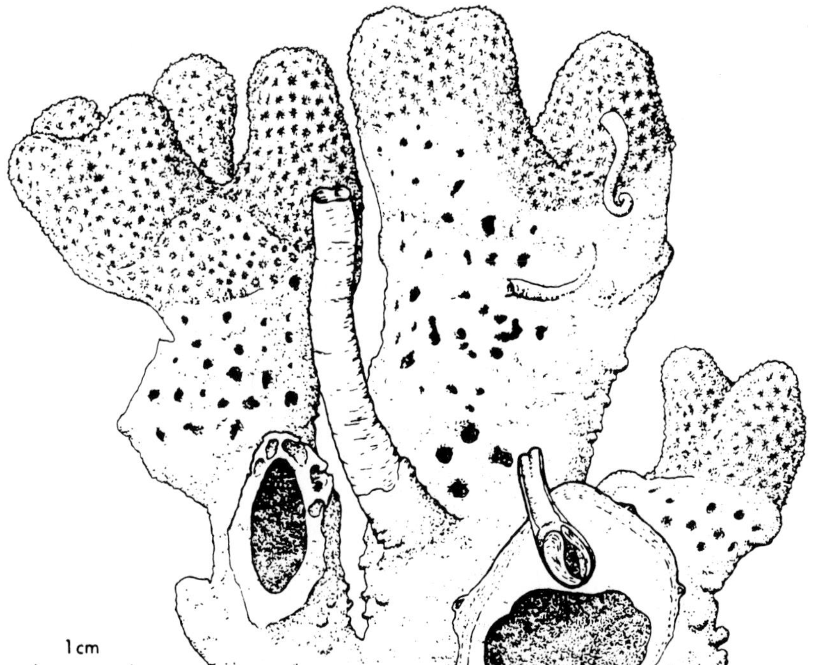

Fig. 10. *Madracis decactis* inhabited by two *Gastrochaena hians*. Extension of the siphonal tube prevents coral overgrowth. [After Bromley (1978). By permission of Elsevier Scientific Publishing Co.]

and Soliman, 1963b; Soliman, 1969, 1973; Ansell and Nair, 1969; Scott, 1980), although Bromley (1978) suggested that *Gastrochaena hians* can bore living corals.

1. The Boring Mechanism

Carter (1978) has reviewed the evidence concerning boring in the Gastrochaenidae. Purchon (1954) thought that boring is mechanical in *Gastrochaena cuneiformis,* effected by the "interaction of the posterior adductor muscle and the anterior retractor muscles of the byssal apparatus," which "in closing the shell valves causes boring into the substratum." Yonge (1963) also thought that boring is mechanical; "(the shell valves) must abrade as a result of water pressure aided by the opening thrust of the long ligament." Carter (1978) noted that the anteroventral regions of gastrochaenid shells are often abraded. Moreover, the gastrochaenid periostracum possesses sharp siliceous spines that possibly also serve to erode the tunnel heading (Carter and Aller, 1975; Carter, 1978).

Carter (1978) and other earlier authors (Otter, 1937; Gohar and Soliman, 1963b; Yonge, 1963) believed, however, that boring may also be accomplished in part by chemical means and that this is the function of the pedal organ located on the anterior face of the visceral mass. Deshayes (1846), Cailliaud (1856), Carazzi (1903), Pelseneer (1911), Kühnelt (1934), and Morton (1982b) reported pallial glands in the Gastrochaenacea similar to those of *Lithophaga* (Morton and Scott, 1980), but Carter (1978) believed these were more involved with pseudofeces consolidation. Carter (1978) concluded that although chemical boring by either the pedal, siphonal, or anterior pallial epithelium seems certain, the exact mechanism and the nature of the boring agent are unknown. Carter (1978) did not understand how glands lining the inner surface of the mantle (Pelseneer, 1911) could be involved in the production of a secretion to effect chemical boring. However, Morton (1982b) has shown that the mantle lobes of *Gastrochaena (Cucurbitula) cymbium* are reflected over the shell, so that inner surfaces become, effectively, outer. In this unusual gastrochaenid, which bores empty bivalve shells and then builds a calcareous "igloo," reflected middle folds with anteriorly associated glands periodically dissolve the capsule to secrete it again by glands situated more posteriorly in the reflected inner folds. The siphons, as noted by Otter (1937), Soliman (1973), and Carter (1978), are also secretory at certain times.

2. Ecology

Carter (1976, 1978) has described the broad ecology of *Gastrochaena* in the western Atlantic. *Spengleria rostrata* and *Gastrochaena hians* are found throughout the Caribbean and in Bermuda. *G. hians* dominates in the Florida Keys, with *S. rostrata* being rarer everywhere. *G. ovata* occurs throughout the western Atlantic except Bermuda, where it is replaced by *G. mowbrayi; G.*

stimpsonii ranges from North to South Carolina, and two other species occur in the Gulf of Mexico and Puerto Rico and the Bahamas, respectively. In the Florida Keys, three species occur: *S. rostrata, G. hians,* and *G. ovata.* The latter two species bore the margins of dead skeletons of *Diploria,* and *S. rostrata* bores the undersurface. Bromley (1978) also reported *G. hians* and *S. rostrata* from Bermuda. Gohar and Soliman (1963b) and Soliman (1973) reported on *G. retzii* and *G. rüppelli,* both dead coral borers, from the Red Sea.

In Hong Kong, the ecology of three other species (*G. cuneiformis, G. interrupta,* and *G. laevigata*) has been more extensively investigated by Scott (1980) and Dudgeon and Morton (1982). Scott (1980) showed that all dead corals, except *Porites,* which has a compact, hard skeleton, were bored by these three gastrochaenids. All species were most common at depths of −5.5 to 6.0 m C.D.

Along the Tolo Harbor estuary in Hong Kong, Dudgeon and Morton (1982) have shown that population numbers of all three species rose progressively toward the head of the estuary, where salinity is lower and most corals are dead (probably because of pollution). Of the three gastrochaenids, *G. interrupta* was the most numerous and was approximately equally distributed throughout the estuary. *G. laevigata* was most common in Tolo Harbor (i.e., the head of the estuary), whereas *G. cuneiformis* was ecologically separated, being more numerous in Tolo Channel (i.e., the mouth). The lithophagine borer *L. malaccana* also bores dead corals in the estuary; it shares its niche with *G. interrupta* throughout its local range, and with the other two species at the extremes of its range.

It has been noted that *L. malaccana* can avoid coral overgrowth, probably by a secretion from the siphons that inhibits calcium deposition around the siphonal apertures (Morton and Scott, 1980). Soliman (1969, 1973), Bromley (1978), and Carter (1978) have shown that gastrochaenids achieve the same result by elongating the siphonal end of their burrow by extra calcium deposition to keep pace with the encroaching coral.

3. Evolution of the Gastrochaenidae

The origins and adaptive radiation of the Gastrochaenacea have been discussed by Carter (1978). The primitive condition seems to be represented by *Spengleria* (Boss, 1967), which has separate siphons opening onto calcareous surfaces from separate orifices. Among modern gastrochaenids, *Spengleria* is also unique (except for juveniles of a few species of *Gastrochaena*) in the possession of periostracal spikes (Carter, 1976). It is likely that such spikes initially functioned to increase friction between the shell and the substratum in shallow-burrowing and semiendolithic nestling ancestors of the Gastrochaenidae. In *Spengleria,* they may assist chemical boring by scraping softened debris from the burrow walls. Carter (1978) believes that *Gastrochaena* has developed mechanical as well as chemical boring.

Subsequent adaptive radiation in the Gastrochaenacea has seen the evolution of burrowers in their juvenile phases only, with the subsequent adult habit of cementation within a protective igloo [e.g., *Cucurbitula* (Morton, 1982b)]. Ultimately, *Eufistulana* adopted a tube-dwelling mode of life deeply embedded in sublittoral sands and muds (Carter, 1978).

Although the exact origins of the Gastrochaenacea are unknown, it seems most probable that evolution proceeded from a nestling ancestor. Closest to *Spengleria* are the Permophoridae and Grammysidae (Carter, 1978), which were most common in the Permian; the Gastrochaenacea evolved its distinctive life-style in the Jurassic.

V. Other Coral Borers

A. The Petricolidae

The temperate-water *Petricola pholadiformis* is a mechanical borer of soft calcareous rocks and mud stones (Purchon, 1955a), although it is not highly adapted for rock boring. *Petricola typica* occurs from North Carolina to Brazil; in the latter locality it bores the (dead) coral blocks of *Mussismilia hispida* and colonies of *Schizoporella unicornis* (Bryozoa); more frequently, it is found inside the sandy reefs built by the polychaete *Phragmatopoma lapidosa*. Initial settlement probably occurs in a preexisting cavity (often of *Lithophaga bisulcata*), which is enlarged to form a hole into which the bivalve fits exactly when its valves are open (Narchi, 1975). A similar habit is recorded for *P. lapicida* by Otter (1937) from the Great Barrier Reef and for *P. carditoides* by Yonge (1958) from California; in Bermuda, *P. lapicida* fits snugly into a burrow that conforms exactly to its shape (Bromley, 1978) (Fig. 14C). Boring in *Petricola* is exclusively mechanical (Ansell and Nair, 1969).

B. The Pholadidae

The Pholadidae are a large group of stone- and wood-boring bivalves closely related to the shipworms (Turner, 1955). Along the tropical west coast of America, *Jouannetia pectinata* bores soft stones in the intertidal. In the Indo-Pacific, *J. globulosa* can be found boring coral blocks (Nielsen, 1976b).

Pholads are mechanical borers (Purchon, 1955b; Ansell and Nair, 1969), but the genus *Jouannetia* is characterized by an almost globular shell, suggesting [if a comparison can be made with the live coral-associated Lithophaginae, in which there is a trend toward a spherical shell (Morton and Scott, 1980)] that *J. globulosa* has a more intimate relationship with coral. Nothing is known of these bivalves, however, and it would be of especial interest to understand how *Jouannetia* bores.

C. The Clavagellidae

The anomalodesmatan family Clavagellidae are generally considered to be tube shells that inhabit soft deposits. Some, it is thought (Soliman, 1973), may attach to some object or become embedded in the holes (probably *Gastrochaena*) of corals and then become overgrown by coral. Soliman (1971), however, has described a species of *Clavagella* living in coral rock overgrown by successive layers of living *Cyphastrea* and *Goniastrea,* which supports the contention of Kühnelt (1933) that these bivalves (e.g., *Clavagella aperta*) bore. Soliman (1971) described this species as occupying empty boreholes of other species of borer. It fixes the left valve to the burrow wall and eventually widens the hole to fit its growing body. Soliman was unsure whether the boring is chemical or mechanical. Evidence for mechanical boring in *Clavagella* includes the rough, abrasive outer surface of the right valve, the calcareous accumulations on it, the eroded periostracum, and the reduction of the hinge. Carazzi (1903) also noted that *Clavagella* can bore siliceous as well as calcareous rocks. *Clavagella* also produces a siphonal tube which, as in *Spengleria* (Gastrochaenacea), prevents coral overgrowth. Growth implies that this tube has to be enlarged regularly to accommodate the siphons, and Soliman (1971) assumed that the siphons of this species are responsible for this enlargement by mechanical friction followed by secretion of a new tube.

D. *Tridacna crocea*

In the byssally attached *Tridacna maxima,* there is a limited degree of downward penetration (Yonge, 1980). This foreshadows the situation in *T. crocea,* whose shell is somewhat longer and also wider than the opening into the boring. The animal fits snugly into its cavity, being massively attached byssally. Yonge (1936, 1953) and Purchon (1955c) originally thought that boring was mechanical, but Yonge (1980) subsequently showed that there is considerable extension of the middle mantle folds around the byssal gape. This presents the obvious possibility of chemical activity. It may be significant that in the Lithophaginae the middle fold houses the glands which produce the secretion for boring (Morton and Scott, 1980). Yonge believed that *T. crocea* used alternate periods of mechanical and chemical activity, the valves grinding against the chelated coral rock.

Hamner and Jones (1976), investigating *T. crocea* on the Great Barrier Reef, showed that densities on unattached coral heads exceed 100 clams/m^2, and a single specimen could produce 200 g sediment/m^2/year. *T. crocea* is sharply zoned on the reef and erodes coral rubble from a discrete area of the reef flat, removing it from the reef surface so that the reef does not become elevated. It is also important in the formation of "micro atolls."

VI. Epizoic Bivalves of Living and Dead Corals

A. Epizoic Bivalves of Living Scleractinian Corals

1. The Pectinidae

The most comprehensive account of coral-associated pectinids (Table I) is that of Waller (1972), who described seven species from Eniwetok Atoll. The four most common were *Chlamys coruscans, C. marshallensis, Gloripallium pallium,* and *Pedum spondyloideum,* all of which are coralliphilous, living byssally attached to the undersides of corals or in their crevices. Three of these common species (excluding *C. marshallensis*) were also the most widely distributed, extending from the east coast of Africa to the Tuamotu Archipelago in the central Pacific. These three widely distributed species (and the others found on Eniwetok) are latitudinally restricted by water temperature. All live between latitudes 35°N and 25°S in waters with surface temperatures generally higher than 25°C (August surface isotherm). Two of the three less common species at Eniwetok, i.e., *Excellichlamys spectabilis* and *Juxtamusium maldivense,* are not primarily coralliphilous. The third, *Comptopallium vexillum,* lives in abundance on terrigenous sediment in association with marine grasses.

Chlamys coruscans has a small, strong shell with the left anterior auricle modified to minimize the byssal gape. *C. marshallensis* has a thin, relatively vulnerable shell, but its low convexity allows it to move deep into coral crevices, where it attaches with a strong byssus. *Gloripallium pallium* (also recorded from the Seychelles by Lewis and Taylor, 1966) possesses unusually thickened interiors on its auricles which greatly reduce the size of the auricular gapes. The shell is relatively thick and strongly plicate, allowing the animal to live partially exposed to predators and high turbulence. Taylor (1968) also records *Chlamys dringi* from branching corals with *G. pallium* in the Seychelles. Mastaller (1978) recorded *Pedum lividus* and *Pecten squamatus* from Red Sea corals; Smith (1978) recorded *Chlamys squamosus* from the corals of Yap lagoon.

Throughout the Indo-Pacific, *Pedum spondyloideum* (Smith, 1978) attaches to living corals at a very early growth stage and subsequently becomes completely enclosed within the coral except for its ventral region, which maintains an elliptical opening (Yonge, 1967b, 1973) (Fig. 11). The growth rates of coral and scallop are precisely balanced, and the form of the scallop is highly modified by its semienclosed existence. *Pedum* usually (perhaps always) occurs on species of *Porites* (Lewis and Taylor, 1966; Nielsen, 1976b). The veliger larva must metamorphose on the living coral surface to make permanent byssal attachment with the free margins of the valves pointed upward. It so influences the growth of the coral that the elongate pectinid comes to live in deep clefts in which the heavily pigmented inner mantle lobes with their glistening eyes are highly conspicuous.

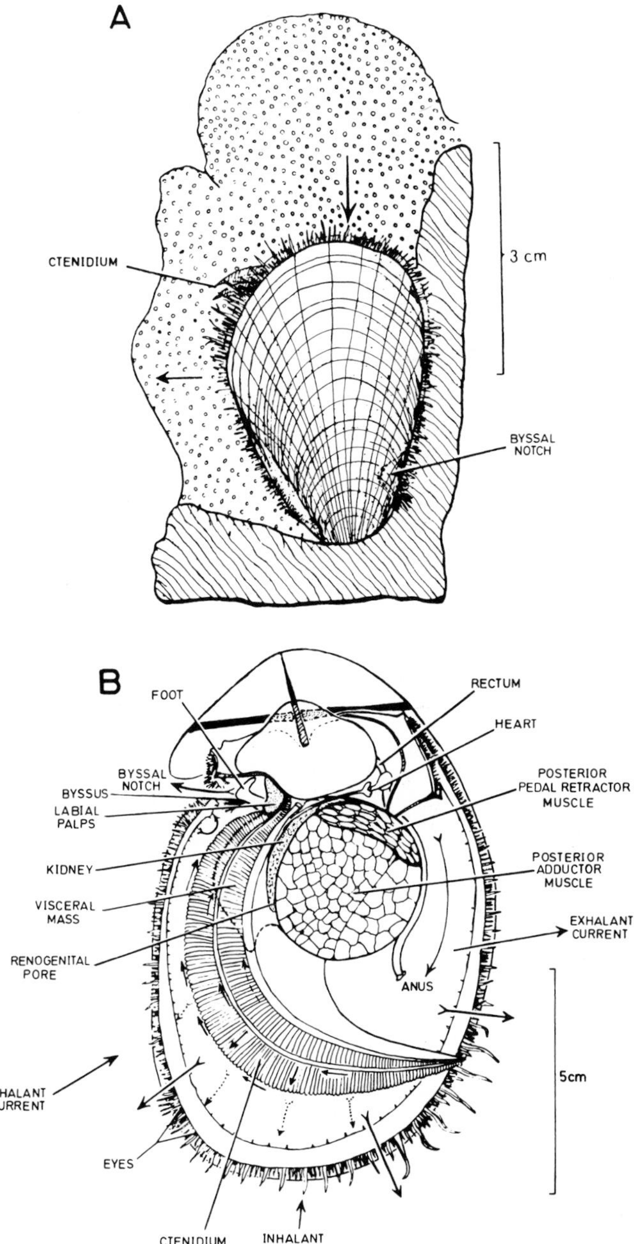

Fig. 11. (**A**) Colony of *Porites* broken open to reveal the left valve of *Pedum spondyloideum* attached within crevice; animal is expanded, showing fringe of tentacles with expanded tips of ctenidia. (**B**) *Pedum spondyloideum;* organs in the mantle-cavity viewed after removal of the left valve and mantle lobe. Plain arrows indicate directions of water and feeding currents; feathered arrows indicate directions of cleansing currents. [After Yonge (1967b); with permission of Royal Society of Edinburgh.]

The left posterior pedal retractor is hypertrophied, and its contraction draws the animal downward into the cleft when the adductor also contracts. To prevent overgrowth by the coral, upward growth results in an extremely elongate shell, and for better food collecting within the enclosing cleft the posterior tips of the ctenidia extend beyond the confines of the shell, being withdrawn by enlarged branchial muscles (Yonge, 1967b).

Yonge (1981) has also described *Hemipecten forbesianus,* found attached to the smooth undersurface of colonies of the scleractinian coral *Turbinaria mesenterina.* These extremely flattened, disc-like pectinids conform perfectly to the coral surface against which they are adpressed. The byssal notch on the lower (right) valve is extremely deep. This mode of byssal attachment resembles that of the Anomiacea (Yonge, 1977), which are rarely recorded from corals [e.g., *Monia umbonata* in Hong Kong (Dudgeon and Morton, 1982)], and clearly marks *Hemipecten* as a specialized associate of living corals.

2. The Pinnidae

The Indo-Pacific members of the Pinnidae have been reviewed by Rosewater (1961). Members of this family are mostly typical of offshore sands and muds, where they lie vertically disposed, endobyssally attached to sand grains (e.g., *Atrina vexillum*). One Indo-Pacific species (*Streptopinna saccata*), however, is reported to attach byssally, protectively confined, within the branches of living corals such as *Acropora* (Abbott, 1950; Banner and Randall, 1952; Rosewater, 1961; Taylor, 1968, 1971b,c; Mastaller, 1978). Mastaller (1978) also recorded *Pinna bicolor* from coral heads in the Red Sea.

B. Epizoic Bivalves of Dead Coral Bases

1. The Arcidae

The dead bases of both living and dead coral heads are often thickly encrusted by a mass of byssally attached or cemented bivalves (Table I). Of the former, members of the Arcacea are the most common, including *Barbatia decussata* (Abbott, 1950; Kay and Switzer, 1974), *B. tenella* (Banner and Randall, 1952), *B. helblingi* (Taylor, 1968, 1971a; Dudgeon and Morton, 1982), *B. setigera* (Mastaller, 1978), *B. fusca* (Taylor, 1968), *B. bicolorata* (Smith, 1978), *B. plicata* (Dudgeon and Morton, 1982), *Arca navicularis* (Lewis and Taylor 1966), *A. avellana* (Taylor, 1971b; Thomassin and Ganelon, 1977; Dudgeon and Morton, 1982), *A. ventricosa* (Salvat, 1972; Randall et al., 1975), *A. antiquata* (Mastaller, 1978), *A. plicata* (Taylor, 1968, 1971b; Mastaller, 1978), *Fossularca sculptilis* (Thomassin and Ganelon, 1977), and *Striarca afra* (Dudgeon and Morton, 1982). The above arcids have all been recorded from the Indo-Pacific, whereas *A. zebra* appears to be one of the very few coral-associated arks in the

Caribbean (Rodriguez, 1959; R. Robertson, 1963). The genus *Barbatia* is typically somewhat heteromyarian in form, broadly resembling a mussel, whereas *Arca* is usually approximately equivalve and much larger, with a wider byssal notch to cater for a massive byssus.

In Takapoto Lagoon, Richard (1982) estimates the total biomass of *Arca ventricosa* to be 240 kg/ha in the zone of highest abundance. Growth rates are slow, however, compared to *T. maxima* and *F. fragum,* reducing the potential of this species for mariculture.

2. The Mytilidae

Mytilids such as species of *Modiolus* (especially *M. auriculatus,* which seems to be very widespread in the Indo-Pacific (Taylor, 1971c; Randall et al., 1975; Thomassin and Ganelon, 1977; Mastaller, 1978) are typical epibyssate colonizers of dead coral bases.

Septifer bilocularis (Lewis and Taylor, 1966; Sumadhiharga, 1971; Randall et al., 1975; Stojkovich, 1977) and *S. virgatus* (Dudgeon and Morton, 1982) are occasionally recorded from different regions, but *S. bilocularis* (at least in the Indo-Pacific) is the common bivalve of exposed rocky shores (Loi, 1967) and probably occurs only with corals in the bottom part of its intertidal range. Other byssally attached bivalves include *Isognomon perna* and *I. legumen* (Lewis and Taylor, 1966; Dudgeon and Morton, 1982). *Malleus regula* attaches to rocky substrates (coral blocks) associated with mud (Yonge, 1968); in the Caribbean, *M. candeanus* attaches to coral bases (R. Robertson, 1963; Boss and Moore, 1967).

3. The Pteriidae

A final, important group of broadly byssally attached coral associates is the pearl oysters (*Pinctada* and *Pteria*). *Pinctada* seems to be much more widely represented in the Indo-Pacific; commonly recorded species are *P. margaratifera* (Lewis and Taylor, 1966; Taylor, 1968, 1971a,b,c; Sumadhiharga, 1971; Stojkovich, 1977), *P. maculata* (Salvat, 1969, 1971; Smith, 1978), *P. maxima* (Salvat, 1972), *P. vulgata* (Mastaller, 1978), and *P. martensii* (Dudgeon and Morton, 1982). The genus *Pteria* is most noticeably represented by *P. penguin* (Abbott, 1950; Sumadhiharga, 1971; Dudgeon and Morton, 1982); other members of this genus are more commonly attached to gorgonians, antipatharians, and *Millepora,* although Taylor (1968) records *P. inguinata* attached to corals in the Seychelles. Species of *Electroma* (e.g., *E. smaragdina*) are also occasionally recorded from coral bases (Abbott, 1950; Kay and Switzer, 1974); *E. alacorvi* attaches to *Acropora* and *Pocillopora* at Cocos-Keeling Island (Maes, 1967).

The above byssate bivalves can occur as large, gregarious clusters. Yonge (1968) has recorded whole beds of *Malleus regula* and *Isognomon isognomon* from a coral rubble and sand-mud habitat. Pearl oysters, of course, are intensively cultivated, especially in Japan, as well as fished over coral reefs. Other smaller and more solitary byssally attached nestlers of coral crevices will be discussed later.

4. *The Ostreidae, Chamidae, and Plicatulidae*

Cemented bivalves of dead coral bases are most commonly represented by members of the Ostreidae, Chamidae, and Plicatulidae (Fig. 12). Cementation in these bivalves has been reviewed by Yonge (1979). The Ostreiidae seem to be more numerous, with the following species having been recorded: *Saccostrea cucullata* (Abbott, 1950), *Ostrea numisma* (Taylor, 1971a,b), *O. frons* (Taylor, 1968, 1971b), *O. (=Hyotissa) hyotis* (Taylor, 1968, 1971b), *Lopha folium* (Mastaller, 1978), *L. cristagalli* (Smith, 1978), *Dendostrea* sp. (Smith, 1978), and *Alectryonella plicatula* (Dudgeon and Morton, 1982). Most species attach to hard substrates in association with coral reefs [e.g., *Hyotissa hyotis* beneath dead coral blocks (Taylor, 1968)], but *Ostrea folium* was only found attached to the spines of the echinoid *Prionocidaris verticillatra* in the Seychelles (Taylor, 1968). Recorded members of the Chamidae (whose structure has been reviewed by Yonge, 1967a) include *Chama aspersa* (Abbott, 1950; Lewis and Taylor, 1966), *C. asperella* (Taylor, 1971b), *C. imbricata* (Salvat, 1972), *C. squamosa* (Smith, 1978), and *C. reflexa* (Dudgeon and Morton, 1982). *Plicatula plicata* has been recorded from coral bases by Mastaller (1978) and Dudgeon and Morton (1982).

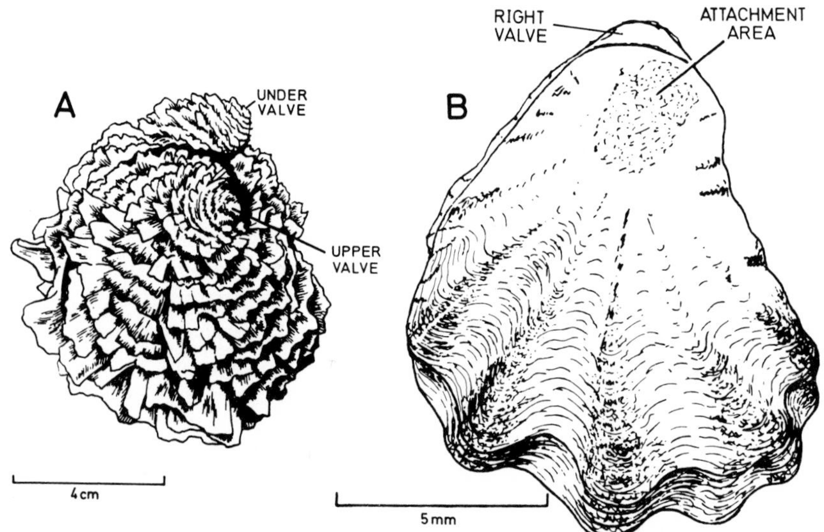

Fig. 12. (**A**) *Chama pellucida* viewed from above, attached by the left valve. (**B**) *Plicatula gibbosa*, viewed from above. [(A) After Yonge, 1967a. (B) After Yonge, 1973.]

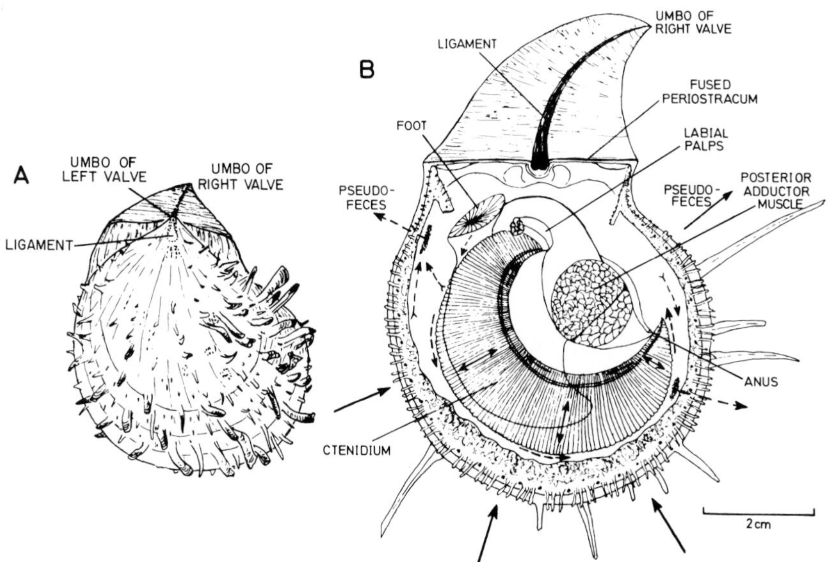

Fig. 13. (**A**) *Spondylus* sp., the shell viewed from above. (**B**) *Spondylus americanus* lying within right valve after removal of left valve and mantle lobe. Arrows indicate directions of major water currents and ciliary currents. Broken arrows, rejection currents; feathered arrows, cleansing currents on mantle. (After Yonge, 1973.)

5. The Spondylidae

The Spondylidae is the second family of the Pectinacea and includes large and conspicuous members of the coral-associated fauna (Fig. 13). In the Caribbean only *Spondylus americanus* occurs (Yonge, 1973), but numerous species such as *S. hystrix* (Abbott, 1950), *S. aurantius* (Taylor, 1971b), *S. gaederopus* (Mastaller, 1978), and *S. ducalis* (Dudgeon and Morton, 1982) occur in the Indo-Pacific. Cementation by the right valve follows prior byssal attachment by the post larva (Yonge, 1979). The upper (left) valve often bears characteristic spines. As in all pectinaceans, there is a single adductor muscle divided into smooth and striated components that, in *Spondylus,* probably assists in maintaining shell closure especially against predators.

C. Epizoic Bivalves of Antipatharian and Gorgonian Corals

Antipatharians and gorgonians are the site of attachment for certain pteriid and ostreid bivalves (Field, 1949; Rees, 1967; Patton, 1972; Mastaller, 1978). The pteriid bivalves attach by a byssus [e.g., *Pteria colymbus* attaches to *Leptogorgia virgulata* (R. Robertson, 1963; Patton, 1972)] in the Caribbean; settlement on the coenenchyme causes damage to the flesh, exposing the axial skeleton. Ostreids also attach to gorgonians, as *Ostrea frons* on *Gorgonaria* (Coomans, 1969). Cementation is by finger-like processes, developed from the under valve, which curl around the coral axial skeleton.

Rodriguez (1959) recorded *Arca zebra* and *Chama macerophylla* as attaching to the false coral *Millepora alcicornis* on the coral community of Margarita Island, Venezuela. From the Red Sea, Mastaller (1978) has recorded *Pteria crocea, Lopha cucullata,* and *L. cristagalli* from *Cirripathes* and *Antipathes;* the pteriid also attaches to *Millepora.*

VII. Crevice and Coral Gallery Bivalves

The holes created by coral-boring bivalves (the primary cryptobion), as well as those bored by other organisms, become the refuge for a number of solitary bivalves (the secondary cryptobion) of limited diversity. These burrows eventually become enlarged, by the further boring activities of other coral eroders, into galleries that riddle the bases of dead coral heads. These galleries, in turn, constitute the microhabitat of a number of highly specialized coral associates (the tertiary cryptobion) of which a number of bivalves are important members, often living in commensal association with other invertebrates.

A. Crevice Dwellers

1. The Trapeziidae

The mytilid *Gregariella coralliophaga,* half nestler with the ability to modify the shape of its burrow, has already been mentioned (Nielsen, 1976a; Morton, 1982a) (Fig. 14B). Another relatively common nestler which, like *G. coralliophaga,* is recorded from both the Caribbean and the Indo-Pacific (Solem, 1954), is *Coralliophaga coralliophaga* (Morton, 1980a) (Fig. 14A), a member of the Trapeziidae (Solem, 1954). Other representatives of this family such as *Trapezium oblongum* and *T. bicarinatum* have been recorded as coral nestlers (Abbott, 1950; Taylor, 1968, 1971b; Nielsen, 1976b). *T. sublaevigatum* nestles in a variety of habitats at a range of depths (Morton, 1979b). *Coralliophaga,* however, seems to be specifically associated with corals. In Bermuda (Bromley, 1978), empty *Lithophaga* boreholes are reoccupied by *Coralliophaga.* Several generations of this species may succeed one another in the hole, each having less space and becoming more stunted than the preceding one, until the boring contains shells of five or six individuals.

The shell of *C. coralliophaga* is very thin, with several raised concentric rings ornamenting each valve (Morton, 1980a). These projections, it is thought, may serve to confuse potential predators by offering them a number of slots resem-

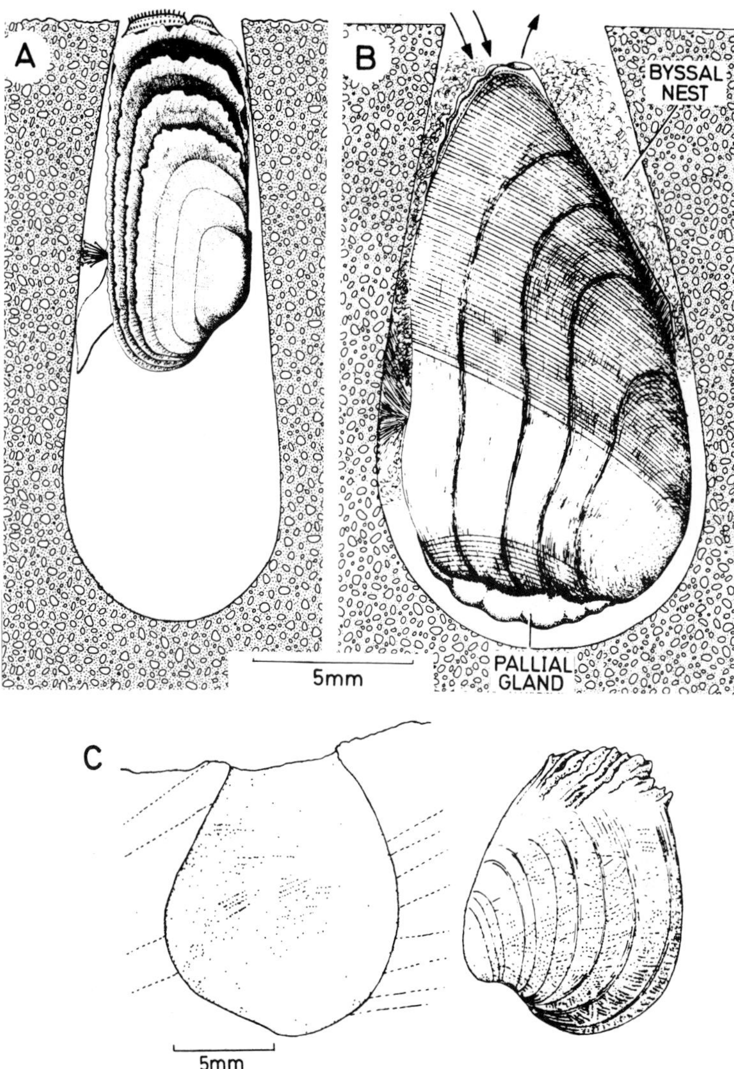

Fig. 14. (**A**) *Coralliophaga coralliophaga*, (**B**) *Gregariella coralliophaga*, and (**C**) *Petricola lapicida* in their burrows. *Coralliophaga* is occupying a *Lithophaga* borehole which it does not modify. *Gregariella* is also occupying a *Lithophaga* borehole, which it can modify; *Petricola* is occupying a borehole of its own construction. (A) After Morton, 1980a. (B) After Morton, 1982a. (C) After Bromley (1978).

bling the margins of the valves, so that unsuccessful penetration might be followed by disinterest. The animal usually occupies empty *Lithophaga* burrows in Hong Kong; as it is byssally attached to the burrow wall which it does not modify, growth results in a deformed shell.

Mastaller (1978) and Dudgeon and Morton (1982) recorded *Irus (Notirus) macrophyllus* from coral heads in the Red Sea and Hong Kong, respectively; this venerid nestler has a shell very similar to that of *C. coralliophaga* although thicker, with the raised concentric rings almost certainly fulfilling a similar function. Nielsen (1976b) records *Irus irus* as a coral crevice dweller from Phuket in Thailand.

2. The Carditidae

The Carditidae (whose structure was reviewed by Yonge, 1969) are also solitary nestlers in a range of habitats. One species in particular, however, has been consistently recorded from Indo-Pacific coral crevices; *Cardita variegata* (Abbott, 1950; Taylor, 1968, 1971a,b,c; Kay, 1971; Kay and Switzer, 1974; Nielsen, 1976b; Thomassin and Ganelon, 1977; Mastaller, 1978). This byssally attached, heteromyarian nestler has a very thick shell. The related *C. semiorbiculata* is laterally flattened and modioliform, often occupying narrow crevices between colonies of solid *Porites* (Yonge, 1969).

B. Gallery Bivalves

1. The Limidae

A characteristic component of the coral gallery community throughout the tropics (but also occurring elsewhere) is the Limidae. One species in particular, *Limaria fragilis,* has been consistently recorded from this microhabitat in the Indo-Pacific (Ostergaard, 1935; Taylor, 1968, 1971a; Smith, 1978) (Fig. 15B). In addition, Mastaller (1978) recorded *Ctenoides annulatus* from the Red Sea and Morton (1979a) recorded *Lima lima* and *Limaria hongkongensis* from coral galleries in Hong Kong; the former species has been recorded from Diego Garcia by Taylor (1971a). For *L. hongkongensis,* an association (of undefined fidelity) with other coral gallery inhabitants seems possible. The Limidae are characterized by a thin shell, a single adductor muscle, and numerous long tentacles that are the primary means of defense by autotomy and release of a presumably noxious secretion (Gilmour, 1967). The thin-shelled species, *L. fragilis* and *L. hongkongensis,* can also swim, although this is uncommon and is probably either a final means of escape from predators when exposed or a rapid means of recessing deeper into the gallery. *L. lima,* in comparison, has a much thicker, ridged shell and can withdraw its tentacles completely between the shell valves. It occurs more often in superficial coral crevices (Morton, 1979a).

2. The Commensal Leptonacea

Boss (1965) has reviewed the occurrence of symbiotic leptonacean bivalves with a wide range of marine invertebrates; very few are recorded from corals. From unnamed corals, Pease (1865) named *Libratula plana* from the Central Pacific; Adams (1868) named *Thyreopsis coralliophila* from Mauritius. Franc (1960) cited these as species of *Galeomma;* Chavan (1969) placed them in the Galeommatidae. Nielsen (1976b) recorded a large number of unnamed species of *Galeomma* from a coral community in Phuket, Thailand. Thomassin and Ganelon (1977) record *Nesobornia* sp. from the corals of Madagascar, but no host was named.

Yamamoto and Habe (1961) and Habe (1973) described *Nipponomontacuta actinariophila,* which lives with the anemone *Halcampella maxima.* Ponder (1971) has made a study of *Montacutona ceriantha,* which is commensal with a species of the burrowing anemone *Cerianthus* from Australia. In Hong Kong, *M. olivacea* is also commensal with the burrowing anemone *Cerianthus* cf. *filiformis* (Morton, 1980b). From Ceylon, Bourne (1906) described two species of *Jousseaumiella* (*J. heteropsammiae* and *J. heterocyathi*) from the burrows of the sipunculid *Aspidosiphon,* which also lives with the corals *Heteropsammia michelini* and *Heterocyathus aequicostatus,* respectively.

From Hong Kong, Morton (1980b) described *Montacutona compacta* as occurring with living corals (Fig. 15A). As will be discussed later, it appears that this species may be unique in actually associating with living hermatypic corals. Arakawa (1960) has shown that *Ephippodonta murakamii* lives attached to the deep-sea coral *Dendrophyllia cribosa.* In Hong Kong, *E. oedipus* occurs in the galleries of dead corals, associating with the ophiuroid *Macrophiothrix* cf. *aspidota* and *Alpheus* spp. (Morton, 1976; Dudgeon and Morton, 1982). This genus is unusual in that the two valves are very widely separated and covered by the reflected middle fold of the mantle so that the animal looks like a limpet. This probably has the adaptive advantage of making the animal less obstructive and thus less easily dislodged by its more active gallery associates. The adult is female, with two dwarf males resident in pockets in the mantle (Fig. 16).

From Tolo Harbor in Hong Kong, Dudgeon and Morton (1982) described three other species of leptonaceans, *Scintilla nitidella, Paraborniola matsumotoi,* and *Kellia porculus,* which (as do *E. oedipus* and *M. compacta*) live in the galleries of dead coral heads associated with some one the numerous gallery inhabitants. These bivalves occurred with the greatest frequency in coral areas with a high incidence of dead coral, being less common in areas of flourishing coral growth.

Other than *Ephippodonta murakamii,* which attaches to the deep-water ahermatypic coral *Dendrophyllia cribosa* (Arakawa, 1960), although nothing more is known, the only true commensal with hermatypic corals is *Montacutona com-*

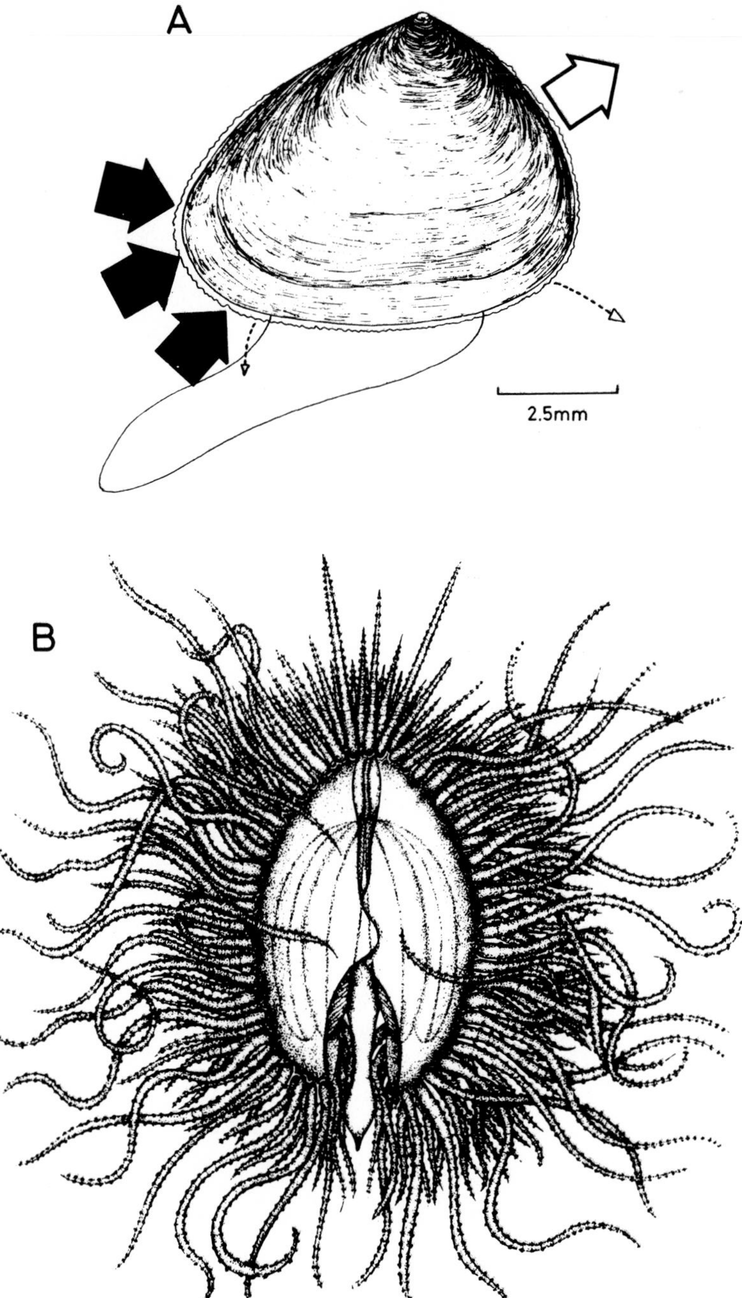

Fig. 15. (**A**) The commensal leptonid *Montacutona compacta* from coral crevices. This species is thought to live in direct association with corals. (**B**) *Limaria fragilis* from coral galleries. [(A) After Morton, 1980b. With permission of The Zoological Society of London. (B) After Morton, 1979a.)]

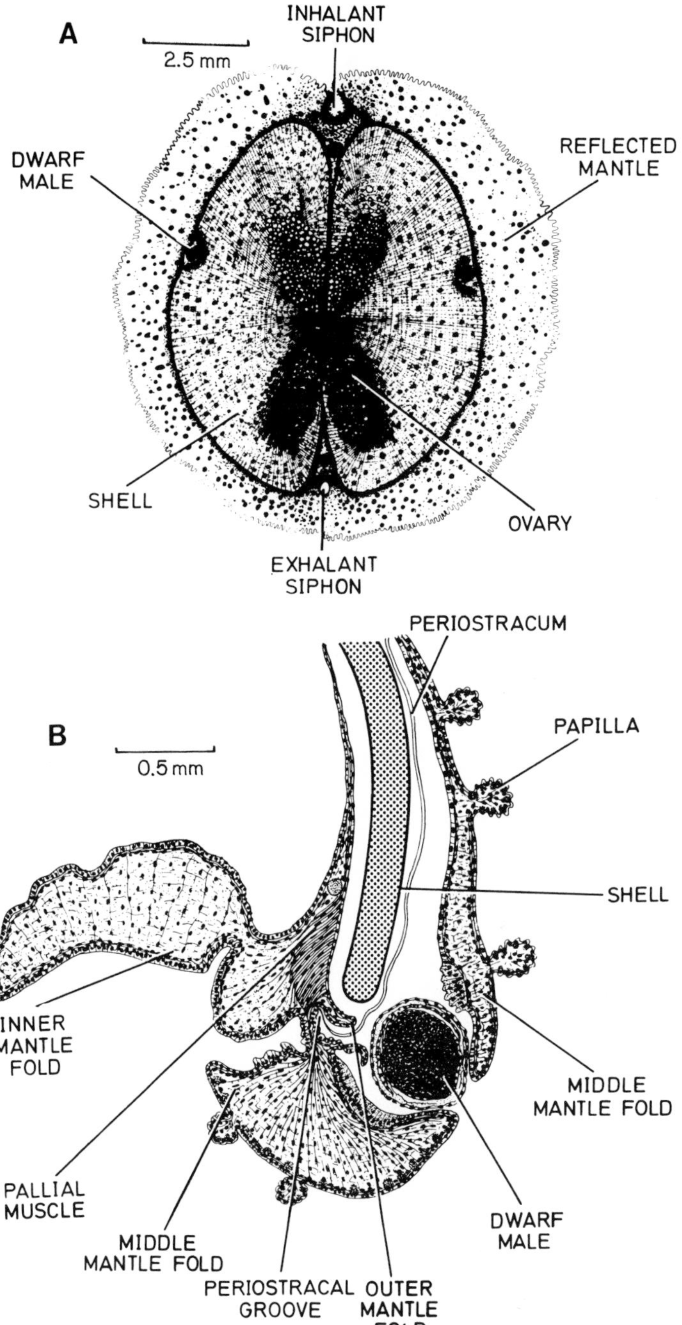

Fig. 16. *Ephippodonta oedipus*. (**A**) The animal viewed from above. (**B**) Section through the mantle margin showing one of the dwarf males. (After Morton, 1976.)

pacta (Morton, 1980b). Related species occur with cerianthid anemones. *M. compacta,* which is very rare, possesses an array of pallial glands that are thought to confer immunity from polyp attack. Dudgeon and Morton (1982) collected only 14, all as solitary individuals, from large samples of living corals with dead bases from six stations in Hong Kong. This species is a self-fertilizing protandric hermaphrodite. The male gonads develop first, and sperm are released to be stored in a ctenidial marsupium until the individual subsequently changes into a female. Release of the eggs results in self-fertilization in the suprabranchial chamber, and it is possible that fertilized eggs are retained and incubated there for release as small adults.

VIII. Bivalves of Coral-Associated Habitats

A. Coral Sand and Rubble Bivalves

The erosion of coral skeletons by borers, by predators, and as a result of wave action in more exposed situations, creates a clean biogenic gravel or rubble. In protected locations such as lagoons, where more stable environmental conditions exist, the breakdown of sea grasses and coastal zone plants creates detritus. The water is more turbid, permitting infaunal colonization of the finer sediments by suspension and deposit-feeding bivalves (Table II). Kay and Switzer (1974) have shown, in Fanning Atoll lagoon, that bivalves are selectively distributed according to the relative turbidity of the water, and Salvat (1967) has shown that the relative dominance of some species within a lagoon system is related to the degree of exchange with the open sea.

1. The Veneridae, the Tellinidae, and Others

Many bivalves have been recorded from coral-associated habitats, but Table II lists those which have been recorded specifically from sands, gravel, and rubble in close proximity to corals in the Indo-Pacific. Those from more offshore habitats are not considered, and the Tridacnidae are treated separately later. Some 12 families are represented, a number of them by only a few genera and species. Some of the more important families are dealt with separately because of their more general association with coral-associated habitats.

The Mesodesmatidae live in clean sands, often associated with surf action (Nielsen, 1976b). Occurring in a family of the Cardiacea, the Hemidonacidae, are species of *Hemidonax* (Ponder et al., 1980). Of particular interest here is *H. donaciformis,* which is apparently adapted to living in coarse coral sand in Australasia. The shell is cardiform, the anterior and posterior ends being about equal in length, and the surface is sculptured with coarse radial ribs. Because of this shell structure, the species is presumed to occur in low-energy environments; other species, being more donaciform, occupy higher energy habitats.

The most widely distributed and diverse group of bivalves of a coral sand system is the Veneridae. Generally speaking, these are solid shelled, shallow burrowers with short, fused siphons, feeding on suspended material. Some species of *Gafrarium* [e.g., *G. tumidum* (Banner and Randall, 1952; Randall et al., 1975; Kay and Switzer, 1974)] are more typical of estuarine shores, being recorded as consistently from mangroves (Taylor, 1968; Morton, 1983b), of which they are probably more characteristic. Others [e.g., *G. pectinatum* and *G. dispar* (Taylor, 1968, 1971b)] occur in cleaner coral sands. *Glycymeris pectiniformis* (Glycymeridae) was recorded from Red Sea coral sands by Mastaller (1978). Species of the Psammobiidae are gravel and rubble inhabitats [e.g., *Asaphis deflorata* (Banner and Randall, 1952)]. *Anisodonta lutea* (Thomassin and Ganelon, 1977) is a member of the Sportellidae; little is known of this family. Species of *Corbula* [e.g., *C. subquadrata* (Corbulidae) (Lewis and Taylor, 1966)], are inequivalve; as they probably lie on the sand surface, the valve inequality may assist in sediment removal (Yonge, 1946). The Tellinidae are also well represented in coral sand habitats. Taylor (1968) recorded *Tellinella crucigera* and *Quadrans gargadia* as common in clean coral sands at Mahé, Seychelles, with five other species also present. They are deep burrowers, smooth and slender, with long siphons projecting to the surface above, from which they collect detritus (Yonge, 1949).

2. The Cardiidae

Corculum cardissa is an Indo-Pacific inhabitant of coral reefs (at least in the adult stage) (Nielsen, 1976b), occupying high intertidal areas of dead coral rubble covered with a little sand. It is byssally attached to coral fragments. This cockle is anteriorly-posteriorly foreshortened, with a corresponding increase in breadth so that the posterior face of the shell, which is covered by greatly expanded pallial tissues, lies flush with the sand surface.

Corculum is of special interest because of the occurrence of zooxanthellae in the greatly expanded mantle tissues and the gills (Kawaguti, 1950, 1968). The shell is somewhat transparent, permitting light penetration to the zooxanthellae below. Although the zooxanthellae seem to propagate there within the blood spaces, there are no signs of their digestion which Kawaguti believed took place in the digestive diverticula.

Other cardiids recorded from coral sands include species of *Cardium, Laevicardium, Ctenocardium,* and *Vasticardium,* but the most consistently recorded species is *Fragum fragum,* which seems to have a very wide distribution throughout the Indian and Pacific Oceans (Taylor, 1968). In Anaa Lagoon, Richard (1982) has shown that *Fragum fragum* has a biomass of 1.4 tons/ha (600^6 individuals in the whole lagoon), which is much more than the combined biomasses of *Arca ventricosa* (43 kg/ha) and *Tridacna maxima* (67 kg/ha) in the Takapoto lagoon (Richard, 1977). Growth is rapid, maximum size being attained in 3 years. The Cardiidae are very shallow burrowers, predominantly of intertidal habitats. Later, their close relatives, the Tridacnidae, will be seen to be the dominant colonizers of the coral reef epifaunal habitat.

3. The Lucinidae

The Lucinidae are an ancient group with unusual feeding methods (Allen, 1958). They occur in a variety of habitats, mostly intertidal, with a wide range from the softest muds of mangroves [e.g., *Phacoides pectinatus* (Morton, 1983b)] to the cleanest coral sands (species of *Codakia* and *Lucina*). In Mahé, Seychelles, Taylor (1968) has shown that the bivalve fauna of *Thalassia* beds is dominated by the lucinoid bivalves *Codakia punctata, C. tigerina, Ctena divergens,* and *Anodontia edentula.*

One of the best-known coral sand lucinaceans is *Fimbria fimbriata* (Allen and Turner, 1970; Morton, 1979c). This bivalve is unusual in that the anterodorsal end of the shell is located at the sand–water interface, the inhalant flow being anterior. The posterior region of the shell is located deeper in the sand with the exhalant siphon projecting to the surface above. There is a posterior inhalant aperture too, but this is used only for the expulsion of pseudofeces. The animal progresses forward with a large foot that is also used to collect food material from the sand. When the foot is withdrawn into the mantle cavity, the material is wiped off onto "pallial palps" (secondarily developed structures replacing diminutive true labial palps) and thence passed to the mouth. In many ways *Fimbria* is very primitive, resembling (for example) the extinct *Babinka* (McAlester, 1965), a supposed early bivalve.

4. The Malleidae

Iredale (1939) recorded *Malleus malleus* as a species characteristic of the Great Barrier Reef coral community. This species occurs vertically embedded in coarse sand, with byssus threads attached to fragments within the substrate. The "hammer head" results from the extraordinary anterior and posterior extensions of the hinge line, which are separated, in this species, by a byssal notch. In the very similar *M. albus,* the shell is stouter and the byssus progressively lost as adult size is achieved. The animal becomes anchored in the substrate exclusively by the anterior and posterior shell extensions (Yonge, 1968). Yonge (1968, 1974) has also investigated *M. regula,* which does not have the hinge extensions; it has a powerful byssal attachment vertically to rocky substrates associated with mud. Yonge (1968) argues that *M. malleus, M. albus,* and *M. regula* are representative of an evolutionary trend in which the notable elongation of the hinge line and the eventual loss of the byssus (in *M. albus*) are associated with a change from an epifaunal to an infaunal life (Fig. 17). Thus the malleids have come, from an epibyssate ancestor, to occupy the cleaner sands of a coral reef system.

M. regula is widely distributed in the Indo-Pacific (Prashad, 1932). Its coun-

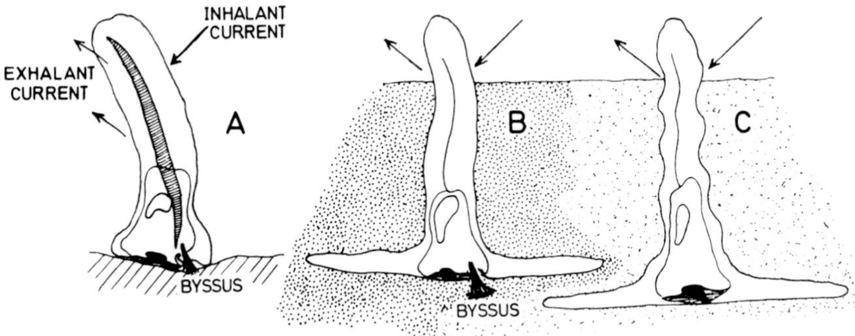

Fig. 17. (**A**) *Malleus regula* byssally attached to hard substrate. (**B**) *M. malleus* with "hammerhead" byssally attached within coarse substrates. (**C**) *M. albus,* unattached in soft mud substrates, anchored only by "hammerhead" in adult. Ctenidia (in A), ligament, byssus, position of solitary adductor, and extent of inner nacreous shell layer are indicated. Arrows show positions of inhalant and exhalant currents. (After Yonge, 1974.)

terpart in the western Atlantic is *M. candeanus* (Boss and Moore, 1967). Although earlier authors (Keen, 1958) alluded to the possible commensal habitat of *M. candeanus* with sponges [as with *Vulsella,* also in sponges (Yonge, 1968; Reid and Porteous, 1980)], Boss and Moore (1967) have shown that the species is byssally attached in the crevices of coral or coral rock. It has been recorded attached to *Agaricia fragilis* and *Colpophyllia natans,* but also to the calcareous alga *Neogonialithon.* Most species occur in shallow water, but some individuals have been taken as deep as 437–546 m. The species appears to be largely restricted to the eastern Caribbean.

In the Indo-Pacific, *M. regula* has the unusual habit of forming intertidal reefs with *Isognomon isognomon.* Yonge (1968) reports such a bed from a very shallow channel between Palau Gaya and Palau Tatagan. Maximum population density of both species was 97/ft^2. Although there was some vertical differential zonation of the two species, Yonge concluded that the separation of apparently otherwise sympatric species resulted from different spawning seasons, thereby reducing larval interspecific competion for space; food availability was not considered to be limiting.

B. The Tridacnidae

Yonge (1980) has recently reviewed much of the literature on the Indo-Pacific Tridacnidae. These bivalves, although evolved from a cardiacean ancestor, live epifaunally on a variety of substrates in the coral reef system. They are perhaps the most characteristic of all coral bivalves. The taxonomy of the Tridacnidae has always been confused, although Rosewater (1965) has monographed the family and resolved most of the difficulties. This author recognized six species in two genera, *Hippopus hippopus* and five species of *Tridacna*. More recently, Rosewater (1982) described a second species of *Hippopus, H. porcellanus,* from the Sulu Archipelago, Philippines. *Tricna gigas* is the largest, with a normal maximum length of about 1 m, although animals with a shell length of 1.5 m have been recorded (Yonge, 1974). *T. derasa* occurs in outer barrier areas and grows to a maximum length of about 0.5 m. *T. squamosa* is recognized by the projecting, broad, leaf-like scales set on five or six strong folds or ribs of the lateral shell surface; it attains a length of some 400 mm. *T. crocea* is a boring giant clam enclosed in a cavity much wider than the aperture. It reaches a maximum length of about 150 mm. *T. maxima* is one of the most common, although not the largest of the giant clams, attaining a maximum length of about 350 mm. It typically lies free on the surface of the reef or sand or is partly embedded in coral.

Rosewater (1965) has produced a distribution map of the six major species (Fig. 18). *T. maxima* and *T. squamosa* are the most common species, with a wide distribution across the Indo-Pacific. *T. maxima* is the only giant clam found in the Gulf of Eliat (Goreau et al., 1973). *T. maxima, T. squamosa,* and *T. crocea* are byssally attached throughout life; *T. crocea* also bores into the soft coral rocks. The distribution of *T. gigas, T. derasa,* and *Hippopus hippopus* is much narrower, and *H. porcellanus* has been recorded only from the Philippines (Rosewater, 1982).

All tridacnids are byssally attached as juveniles, but in *T. gigas* and *T. derasa,* as well as *H. hippopus,* attachment is lost after a certain time when the byssal gland atrophies and the pedal gape closes. Subsequently, these animals maintain themselves by virtue of their great weight. Settlement is in more sheltered areas, *T. squamosa* and *T. gigas* attaching to living corals (Beckvar, 1981), and *Hippopus* coming to rest on sandy substrates where it may dominate. Fankboner

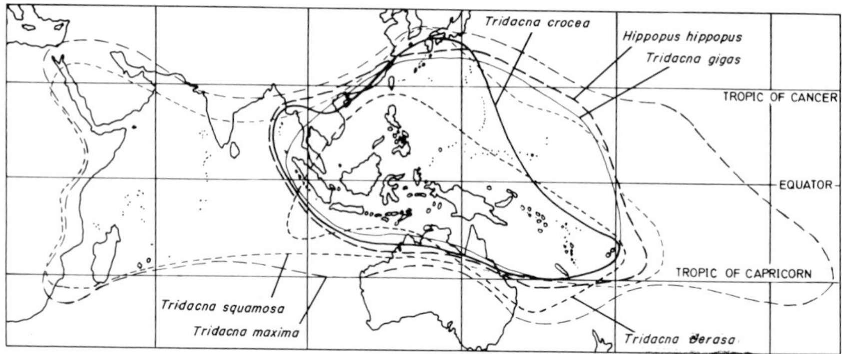

Fig. 18. The distribution of the various species of *Tridacna* and *Hippopus hippopus* in the Indo-Pacific. [From Yonge (1975). Giant clams. *Sci. Am.* **232,** 96–105. Copyright © 1975 by Scientific American, Inc. All rights reserved.]

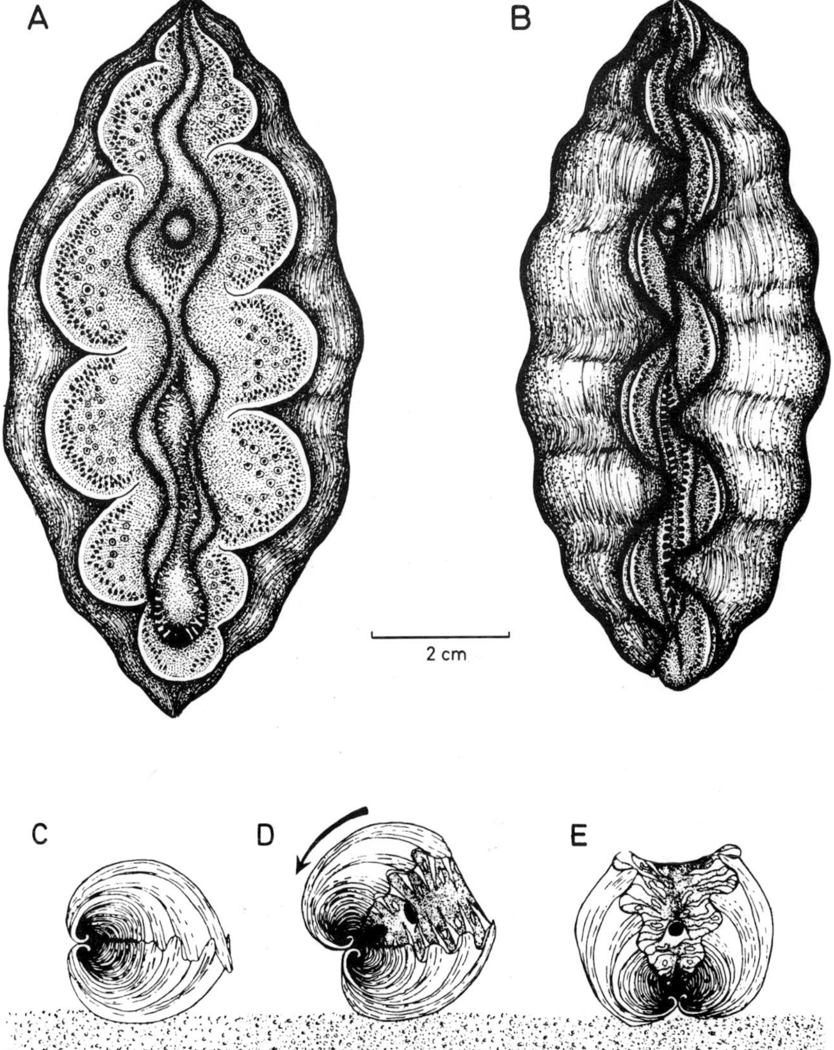

Fig. 19. The siphons of *Tridacna crocea*. (**A**) Fully expanded during the day and (**B**) closed at night. (**C–E**) *Hippopus hippopus*, stepwise sequence of self-righting after displacement possible because of the great thickness of the shell valves in the umbonal region. [C,D,E. After Fankboner (1971b). Reprinted by permission from *Nature* **230,** 579–580. Copyright © 1971 Macmillan Journals Limited.]

(1971b) has shown how this species, when turned on its side by wave action, can right itself. Its remarkable thickness and resultant weight, particularly of the ventral umbones, are responsible for this (Fig. 19C–E).

1. Association with Zooxanthellae

There is strong evidence that the Tridacnidae have evolved their unique form as a result of long association with hermatypic coral reefs. Both possess zooxanthellae, but in the Tridacnidae this acquisition has probably developed by a different method than in the corals. Morton (1978) has suggested that in the former such a relationship was established as a result of the normal defense mechanism to phagocytosed foreign objects.

A considerable volume of literature deals with the subject of the symbiotic relationship of hermatypic corals with dinoflagellate "zooxanthellae" (*Symbiodinium microadriaticum*) (Taylor, 1969, 1973, 1974; Loeblich and Sherley, 1979; Schoenberg and Trench, 1980a,b,c). Taylor (1973) and Trench (1979) have reviewed the occurrence of this alga (and others) with a wide range of other invertebrates, especially coelenterates, including anemones, gorgonians, hydroids, and jellyfishes.

In the Bivalvia, zooxanthellae have been recorded from *Corculum cardissa* (Kawaguti, 1950, 1968), *Fluviolanatus subtorta* (Morton, 1982c), and, most importantly, the various species of the Tridacnidae (Yonge, 1936, 1955, 1975, 1980; Taylor, 1969). These bivalves possess all the typical attributes necessary for the collection, sorting, ingestion, and digestion of filtered food, but they also have symbiotic zooxanthellae in the hypertrophied siphonal tissues. The zooxanthellae are "farmed" by amoebocytes (Yonge, 1936, 1953) as a subsidiary source of food. Fankboner (1971c) believed that this was not "farming" but rather the systematic removal, intracellular digestion, and utilization of degenerate zooxanthellae. (The growth cycle of *S. microadriaticum* has been elucidated by Deane et al., 1979). Goreau et al. (1973) suggested that the Tridacnidae obtain greatest benefits from the photosynthates of their zooxanthellae, although some energy is derived from digestion of old algal cells. It seems fairly clear that tridacnids do obtain nourishment in some way from the zooxanthellae, because it is otherwise anachronistic that clear tropical waters should support dense colonies of the largest bivalve ever to have evolved.

Fankboner (1971c) has also shown that amebocytes cull the zooxanthellae and digest them intracellularly. Some waste material is passed to the kidneys (Morton, 1978), which are appropriately large (Yonge, 1974, 1980) [although the study by Trench et al. (1981) denies kidney excretion of the algae]. Most waste zooxanthellae are passed to the digestive diverticula, where they are expelled from between adjacent digestive cells into the tubule lumina and thence to the stomach. Removal of waste in this way modifies the normal pattern of breakdown of the digestive tubules so that the phases of zooxanthellae release and the times when the tubules ingest filtered food are alternated along a diurnal pattern, the most noticeable manifestation of which is the retraction of the siphons at night (Fig. 19A,B) (Morton, 1978; Yonge, 1980). This observation explained the earlier observations by Yonge (1936) and Mansour (1945) of zooxanthellae in the alimentary tract of tridacnids [Yonge (1974) later asserted that they never occur in the gut]. However, Fankboner and Reid (1981) found zooxanthellae in the stomach of *T. gigas* and concluded that these were of exogenous origin, derived from their mass expulsion by heat-stressed hermatypic corals. Jaap (1979) has monitored a similar event in the Caribbean. Calm weather, high ambient temperatures, and low tides at midday are believed to have caused water temperature elevation sufficient to produce thermal stress, thereby causing expulsion of the zooxanthellae. Ricard and Salvat (1977), however, have shown that zooxanthellae constitute more than 80% of the feces of *T. maxima* even when the clams were maintained in the dark for 10 days. These authors thought filtered material was unimportant in the assimilation of *Tridacna*. Although the latter suggestion does not generally agree with the anatomical structure of the digestive system [a well-developed crystalline style in the stomach and digestive tubules that absorb filtered food (Morton, 1978)], the former observation does indicate that the zooxanthellae in the gut are derived from other parts of the body and are not collected during filtration in zooxanthellae-laden waters. Trench et al. (1981) have shown that the algae do pass from the siphons to the gut where they are voided in the feces. Some defecated zooxanthellae are morphologically intact, photosynthetically active, and capable of being cultured.

The expanded siphonal tissues of tridacnids can have no function other than to expose the zooxanthellae to light (organic salts and CO_2 being available in the blood vascular supply to the mantle). The mantle is also patterned and colored by iridophores (Kawaguti, 1966), which screen the tissues and zooxanthellae from overly intense light and provide optimal lighting conditions for photosynthesis. Jaubert (1977) and Trench et al. (1981) have shown that *Tridacna* has a net oxygen production with a peak at noon. The pallial tissues of *Tridacna*, however, contain numerous "eyes." Yonge (1936) believed that these hyaline organs were lens-like in their construction, but lacked the optic nerve and retina normally associated with a functional photoreceptor. He thus concluded that tridacnid hyaline bodies functioned as light-collecting lenses for internal illumination of zooxanthellae, as the latter are particularly dense around them.

Stasek (1966) disagreed with Yonge's interpretation of these bodies, however, and described an optic nerve, retinal cells with neurites, and a multicellular lens. From these observations, with supporting behavioral evidence of effective photoreception (Stasek, 1965), he contended that the hyaline body was a true eye. Kawaguti and Mabuchi (1969) substantiated many of Stasek's interpretations. Fankboner (1981) has shown that the hyaline organs of *Tridacna* are simple photoreceptors that lack a lens but possess a pigment cup composed of several to many layers of olive-green zooxanthellae. Each eye contains 280–350 polygonal retinal cells which bear numerous ciliary blebs. The blebs give rise to many microvilli that are believed to be the photoreceptive portion of the retinal cell.

Although tridacnid eyes possess directional sensitivity, they are not individually capable of image discrimination because they lack a lens. However, the tridacnid eyes may function cooperatively in small numbers, much like the elements of a primitive compound eye. This study by Fankboner (1981) further suggests that Yonge's original hypothesis may also be credible, because the zooxanthellae surrounding the eye capsule may derive more sunlight than other siphonal tissues.

Jameson (1976) suggested that the acquisition of zooxanthellae in the mantle of juvenile *T. crocea, T. maxima,* and *H. hippopus* occurred at 19–25, 21–40, and 25–27 days, respectively. The smallest juvenile with zooxanthellae was 190, 200, and 210 μm, respectively, for each species. In general, these species acquired zooxanthellae soon after metamorphosis, which corresponds to a time of enhanced growth in all species.

However, Gwyther and Munro (1981) recorded zooxanthellae from the 4-day-old veligers of *T. maxima* and concluded that they had entered the extruded eggs before fertilization, as was originally suggested by Yonge (1936). LaBarbera (1975) had concluded that in *T. maxima* and *T. squamosa* zooxanthellae do not appear until after metamorphosis, and Stephenson (1934) had also failed to detect zooxanthellae in the eggs of *Hippopus hippopus*. This issue has apparently now been resolved by Fitt and Trench (1981) who have shown that neither the eggs nor the trochophore stages of *T. squamosa* possess zooxanthellae. Veligers ingest zooxanthellae, and 2–9 days after metamorphosis, the algal cells move by an unknown mechanism into the siphonal tissues of the host.

Gwyther and Munro (1981) have shown that feeding giant clam veligers with unicellular algal cells did not promote larval growth but did increase survival during metamorphosis. Unfed larvae failed to complete metamorphosis. The similar growth rates of fed and unfed individuals, from 100 to 204 μm in 13 days, indicate that unfed larvae use an energy supply other than particulate organic matter. Of the potential sources of energy, contained zooxanthellae would, if present (and these authors believe they are), account for the similar growth rates observed experimentally, that is, zooxanthellae are very important in larval growth. Fitt and Trench (1981) have shown that survival and growth of veligers and juveniles with zooxanthellae was greater that of those without them. Juvenile *T. squamosa* with zooxanthellae can survive in filtered sea water with light as the sole energy source for more than 10 months, which illustrates the phototropic aspect of the association.

2. *Ecology*

Tridacnids are popular with shell collectors, which has caused depletion of some populations (Rosewater, 1965, 1980). In recent years, Taiwanese commercial fishermen have illegally taken a heavy toll of the giant clams; surveys have shown that clam populations have been decimated, and there has been little return growth (Hardy and Hardy, 1969; Hester and Jones, 1974; Bryan and McConnell, 1976; Pearson, 1977; Hirschberger, 1980).

Hester and Jones (1974) investigated the standing stock of giant clams at Helen Reef in the western Pacific in 1972. High and low estimates of standing stock for five species were made. Low estimates for the stocks were *T. gigas*, 49.8×10^3; *T. derasa*, 32.8×10^3; *T. maxima*, 1.7×10^6; *T. crocea*, 3.7×10^6; and *H. hippopus*, 44.6×10^3. Only one specimen of *T. squamosa* was encountered. On the basis of these data, these authors suggested that a moderate fishery for *T. gigas* and *T. derasa* could be established, but that uncertainties with regard to growth rates and recruitment indicated caution before proceeding. A resurvey of giant clams on Helen Reef in 1975, after foreign vessels had been fishing there, gave the following standing stock estimates: *T. gigas*, 8.6×10^3; *T. derasa*, 12.9×10^3; *T. maxima*, 1.4×10^6; and *H. hippopus*, 47.4×10^3. *T. crocea* was omnipresent (Bryan and McConnell, 1976). Salvat (1972) has shown that *T. maxima* accounts for more than one-half the total annual biomass on Réao atoll.

On reefs, the six tridacnids are separated spatially. *T. crocea* bores into coral blocks; *Hippopus* occurs on sandy areas and occasionally on the reef flat. *T. squamosa* and *T. maxima* are byssally attached on the reef platform, but *T. derasa* and *T. gigas*, because of their large size, are apparently able to live on a variety of habitats. On Helen Reef (Hester and Jones, 1974), large *T. derasa* and *T. gigas* occurred on an extensive sand flat with occasional small coral patches. Large *T. gigas* also occurred along the lagoon edge. Small, medium, and large *T. derasa* and *T. gigas* were found on the coral heads in the lagoon and in association with *Acropora* rubble. The east and south reef flats were hard and relatively bare. On Palau, Hardy and Hardy (1969) found *T. squamosa* and *H. hippopus* together, widely distributed in relatively shallow water both in the lagoon and on the barrier and fringing reefs. *T. squamosa* was more often associated with coral areas of *Acropora* and *H. hippopus* with sandy areas between scattered coral heads. *T. crocea* was highly clumped because of its habit of boring into scattered massive coral heads (*Porites* sp.) and limestone rock. This species was found in the bays, in the lagoon, and on the reefs, most often in shallow water. *T. derasa* was found mostly in the lagoon from the surface down to 20 m, often in association with areas of old *Acropora* rubble. *T. maxima* and *T. gigas* were very seldom found in the bays; sometimes they were in the lagoon, but most often on the barrier and fringing reefs.

In Palau, 178 clams (of all species), totaling 17.17 kg wet body weight, were recorded from an area of 1100 m^2. This is a mean density of 15.6 g/m^2. *T. crocea* was the most abundant, and *T. derasa* was the largest percentage of the total biomass. The relative scarcity of *T. gigas* was attributed to overfishing (Hardy and Hardy, 1969).

Within atolls of the Tuamotus, *T. maxima* may occur in densities as great as

63/m^2; the estimated weight of body tissues/ha is 7 tons and that of shells, 37 tons/ha (Salvat, 1969, 1971). On Michaelmas Reef near Cairns, 120 *T. gigas* and 9 *T. derasa* occur in an area of 150 × 100 m (1.5 ha). Less than 1 km away on the same reef, there are 218 *T. gigas* in approximately 5 ha (Pearson, 1977).

On Tonga, *T. maxima* prefers solid coral limestones or, in many cases, the large rounded heads of *Porites; T. squamosa* lives on sandy or coral rubble areas attached to small pieces of coral; and *T. derasa* is found in deeper areas on sandy or coral rubble areas near reefs (McKoy, 1980). The same author has suggested, using the growth data of McMichael (1974), that on Tonga *T. maxima* lives in excess of 30 years. On the Tongan Islands, *T. derasa, T. squamosa,* and *T. maxima* comprise 0.3–1.0, 1.0–14.1, and 85.6–98% of the population, respectively.

3. Life Span Estimates

Giant clams are protandric consecutive hermaphrodites (Stephenson, 1934; Wada, 1942, 1954; McKoy, 1980). Spawning, as described by Wada (1954) for *T. squamosa, T. derasa,* and *H. hippopus* and by LaBarbera (1975) for *T. squamosa* and *T. maxima,* involves the separate release of sperm and eggs. Sperm are released first, followed by the eggs over a period of 2–3 h (LaBarbera, 1975) to 6 h (Jameson, 1976).

In Fiji, *T. maxima* and *T. squamosa* attain sexual maturity at about 150–200 mm in shell length (LaBarbera, 1975). In Guam, *T. maxima* attains full sexual maturity at a shell length of 110–130 mm.

There is little information on the spawning periodicity of giant clams, but Beckvar (1981) has reviewed the available information derived from laboratory trials. *T. squamosa* spawned from February to March at Eniwetok (Rosewater, 1965), in July in Fiji (LaBarbera, 1974, 1975), and in February and March in Palau (Hardy and Hardy, 1969). This species may have a peak breeding season in winter (Yamaguchi, 1977). In Palau, *H. hippopus* spawned in June (Jameson, 1976), July (Yamaguchi, 1977), and April. The peak breeding period for this species in Australia was the summer months of January to March (Stephenson, 1934). In Palau, *T. derasa* spawned in March and April, and *T. gigas* spawned in July, September, and October (Beckvar, 1981). Beckvar also reports that giant clams may have a lunar-correlated spawning cycle. Five of eight spawnings by *T. gigas, T. derasa,* and *H. hippopus* were on or near the new moon; these spawnings also occurred in the afternoon on a rising tide.

A few hours after fertilization, the developing larvae hatch from the egg as trochophores, which transform into veligers 0.1–0.2 mm in diameter. The larvae of *T. maxima* spend 10–19 days in the plankton (LaBarbera, 1975; Jameson, 1976) before settling. Those of *T. squamosa* settle after 10 days (LaBarbara, 1975), and those of *H. hippopus* after 8–15 days (Jameson, 1976). The juvenile pediveligers of *T. crocea* aggregate on unoccupied coral heads (Hamner, 1978) in protected situations (Jameson, 1976). *T. maxima* larvae will also settle on plastic panels (Gwyther and Munro, 1981).

Life span estimates for tridacnids range from 8 (Pelseneer, 1894) to several hundred years (Comfort, 1957). Growth measurements on a specimen of *T. gigas* near Eniwetok Island indicate a growth rate of 5.1 cm/year. Bonham (1965) used radioautographs to measure strontium-90 deposited in the shells of *T. gigas* at Bikini Atoll after the 1956 and 1958 nuclear weapons tests. He interpreted the regular strontium-90 banding pattern found in the shells as annular growth lines indicating a growth rate of 5 to 8 cm/year, which suggested that a 52-cm clam was in its ninth year. Preliminary growth measurements of tagged clams by Hardy and Hardy (1969) on Palau indicated that growth proceeds slowly. In fact, growth over a 7–11-month period was not great enough to exceed the experimental error in measurement. For example, measurements on 15 tagged *T. squamosa* indicated a growth of 0.7 ± 0.6 cm. These authors argued that growth rate probably varies with size, species, and habitat. Jameson (1976) investigated growth rates in juveniles of *T. crocea, T. maxima,* and *Hippopus hippopus;* these growth curves exhibit similar basic characteristics. The growth rate of the veliger shell is low (1.2 μm/day for *T. crocea,* 2.7 μm/day for *T. maxima,* and 3.5 μm/day for *H. hippopus*). Shell growth rates after settlement and metamorphosis until day 35 (*T. crocea*), day 40 (*T. maxima*), and day 27 (*H. hippopus*) are also low (2.5, 2.3, and 0.9 μm/day, respectively). After day 40 (*T. maxima*) and day 27 (*H. hippopus*), growth rates increase sharply (6.8 and 13.9 μm/day, respectively). Thus, a dramatic burst in growth rate corresponds to the time at which the majority of juveniles acquire zooxanthellae.

Beckvar (1981) has shown that laboratory-reared *T. gigas* reached a mean length of 2.6 cm at 10 months postfertilization; *T. derasa,* 1.1 cm at 5 months; and *T. squamosa,* 6.7 cm at 2 years. Field studies of giant clams 12–25 cm long at Palau predict a high projected growth/year: *T. gigas* grows 8–12 cm/year; *T. derasa* 3–6 cm/year; *H. hippopus* 3–5 cm/year; and *T. squamosa* 2–4 cm/year. Growth rates decrease with increasing length for each species. McMichael (1974) performed a series of measurements on a population of *T. maxima* at One Tree Island, Queensland, in 1966, 1968, and 1969. The measurements indicated that growth is relatively rapid during the early years of life, slowing down with increasing size (as noted by Beckvar, 1981). Based on a composite curve which represents a hypothetical length–age relationship, a clam takes between 10 and 17 years to attain modal size and may not obtain a maximum size for 40 years. During the 14-month period from 1968 to 1969, the mean increment in the length of large specimens (i.e., more than 229 mm long) was slightly less than 1 mm. Newly settled individuals probably grew to a length of 16–80 mm in their first year. The mean increment of clams in the 20–50-mm age class during a 14-month period was 38 mm; the maximum was 63 mm, and the range was 44–107 mm. The considerable variation in growth rate by *T. maxima* at One Tree Island led McMichael to speculate that either some individuals grow much faster than others during the early stages or settlement takes place over a considerable period of time.

Similar results were obtained by Richard (1977) for *T. maxima* on Takapoto Atoll in French Polynesia. A clam 50 mm long was estimated to be 2 years old and would require 9 years to attain 118 mm (95% of the maximum theoretical length). In the lagoon, the *T. maxima* population was estimated at 14 million individuals, 9 million on the reef edge, and 5 million in the central lagoon. Such figures would allow for the theoretical production of 120 tons of meat/year, equivalent to 1.6 kg meat/ha/year. Munro and Gwyther (1982) have shown that growth rate in *T. gigas* is much faster than for *H. hippopus, T. squamosa,* and *T. maxima*. The mariculture potential for *T. gigas* is, thus, high; a weight of 29 kg may be obtained in 6 years.

Beckvar (1981) has shown that the amount of light affects growth. *T. derasa* juveniles reared in 1000-liter tanks in direct sunlight had the most rapid growth, with a mean length of 0.2 cm; those receiving the least sunlight were only 0.13 cm in mean length.

4. Evolution

The Tridacnidae are associated with the Cardiidae in the Cardiacea, but the totally distinctive structure of the Tridacnidae indicates a long and intimate association with coral reefs (Yonge, 1980). The unique form of the Tridacnidae has been explained by Yonge (1936, 1953, 1974, 1975, 1980) (Fig. 20); the siphons are extended along the entire upper surface, and an anticlockwise rotation in the sagittal plane of the mantle shell in relation to the visceropedal mass is involved. The latter is effectively unaltered; foot and byssus still emerge midventrally. Because of the rotation, the dorsal region of the shell and hinge eventually become situated just anterior to the byssal gape. In the course of this 180° rotation the anterior adductor is lost and the posterior is enlarged. Because most modifications are to the mantle-shell, the visceral mass is only slightly modified, although the anus is now located on the upper surface of the adductor thus maintaining its necessary relationship to the exhalant siphon. Forward extension of the siphons (in the Cardiidae, from immediately ventral to the posterior adductor) has occurred in the Tridacnidae, so that they stretch broadly forward above the adductor along the entire upper surface of the valves. Thus, the fleshy lobes of brilliantly colored pallial tissue exposed to the light in the Tridacnidae are, in fact, the siphons; a corresponding development of the orbital muscles effects their retraction, although complete retraction is possible only in *Hippopus*.

Stasek (1961) believes that Yonge's explanation of form in the Tridacnidae is untenable; he has suggested that the equivalent regions of the mantle-shell are intimately associated with equivalent regions of the body in both *Clinocardium* and *Tridacna,* and therefore the mantle-shell of *Tridacna* has not rotated about the body. Instead, Stasek argues, the dissimilarities in form of *Clinocardium* and *Tridacna* are based on differences in the distribution of centers of growth rather than on alterations of functional relationships between the body and the mantle-shell. The former grows primarily in a ventral direction; the latter grows primarily to the posterior. However, all authors agree that the evolution of the Tridacnidae has proceeded from a cardiid stock with a corresponding change from a shallow-burrowing infaunal habit to the epifaunal colonization of hard, coralliphorous substrates. *Hippopus* is somewhat intermediate in that it still occurs on sandy substrates. Yonge (1980) has further suggested that the changes that have occurred to the tridacnid shell are changes occurring in phylogeny but not in ontogeny. But both Stasek (1961) and Rosewater (1980) have shown that during the first few months of postlarval life young *T. elongata* (= *T. squamosa*) and *T. gigas* have a wedge-shaped shell rather than the fan-shaped outline characteristic of older individuals. At 4–5 months of age, the shell shape in *T. gigas* has changed and the angle formed by the hinge line and the ventral margin may exceed 90°. Older individuals measuring from 5.2 to 9.3 cm probably range in age from 1 to 2 years (Rosewater, 1965). The angle of these shells varies from 131° to 150°, and other considerably larger shells do not significantly exceed the larger angle. Shells approximately 1 year of age, therefore, exhibit a mature configuration. The change from a wedge-shaped to a fan-shaped outline thus occurs during the second to fifth month of postlarval life. Larvae of *Tridacna,* although distinctive, thus do resemble those of the related Cardiidae (LaBarbera, 1975; Jameson, 1976). Acquisition of the zooxanthellae [occurring between 19 and 40 days postsettlement (Jameson, 1976)] appears to coincide temporally with the increase in shell angle and with the presumed hypertrophy of the siphonal tissues which house the zooxanthellae.

Kawaguti (1950) believes that the acquisition of zooxanthellae by the Tridacnidae may have accelerated an evolutionary tendency within the Cardiacea to settlement on the surface of coral reefs and the concomitant evolution of a thick shell, which appears to have occurred about the beginning of the Cenozoic (Yonge, 1980).

IX. Predator Defense Mechanisms

An available food resource is usually tapped by a predator of some kind. Relatively little is known of the predators of coral bivalves, although some information is available on predatory gastropods. Taylor (1977) has reviewed the food and habitats of predatory gastropods on coral reefs. Members of the Naticacea, Cymatidae, Muricidae, Melongenidae, Fasciolaridae, Volutidae, and

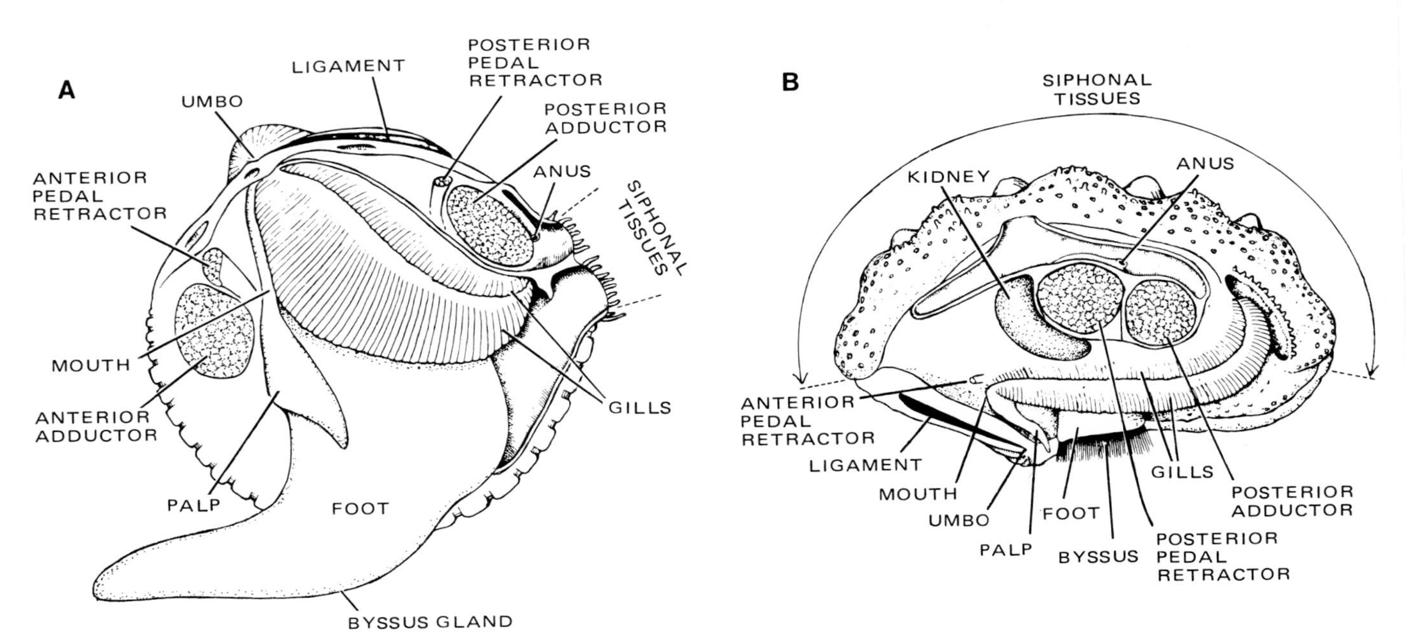
A
LIGAMENT
UMBO
POSTERIOR
PEDAL
RETRACTOR
POSTERIOR
ADDUCTOR
ANUS
SIPHONAL
TISSUES
ANTERIOR
PEDAL
RETRACTOR
MOUTH
ANTERIOR
ADDUCTOR
GILLS
PALP
FOOT
BYSSUS GLAND
B
SIPHONAL
TISSUES
KIDNEY
ANUS
ANTERIOR
PEDAL
RETRACTOR
LIGAMENT
MOUTH
UMBO
PALP
FOOT
BYSSUS
GILLS
POSTERIOR
ADDUCTOR
POSTERIOR
PEDAL
RETRACTOR

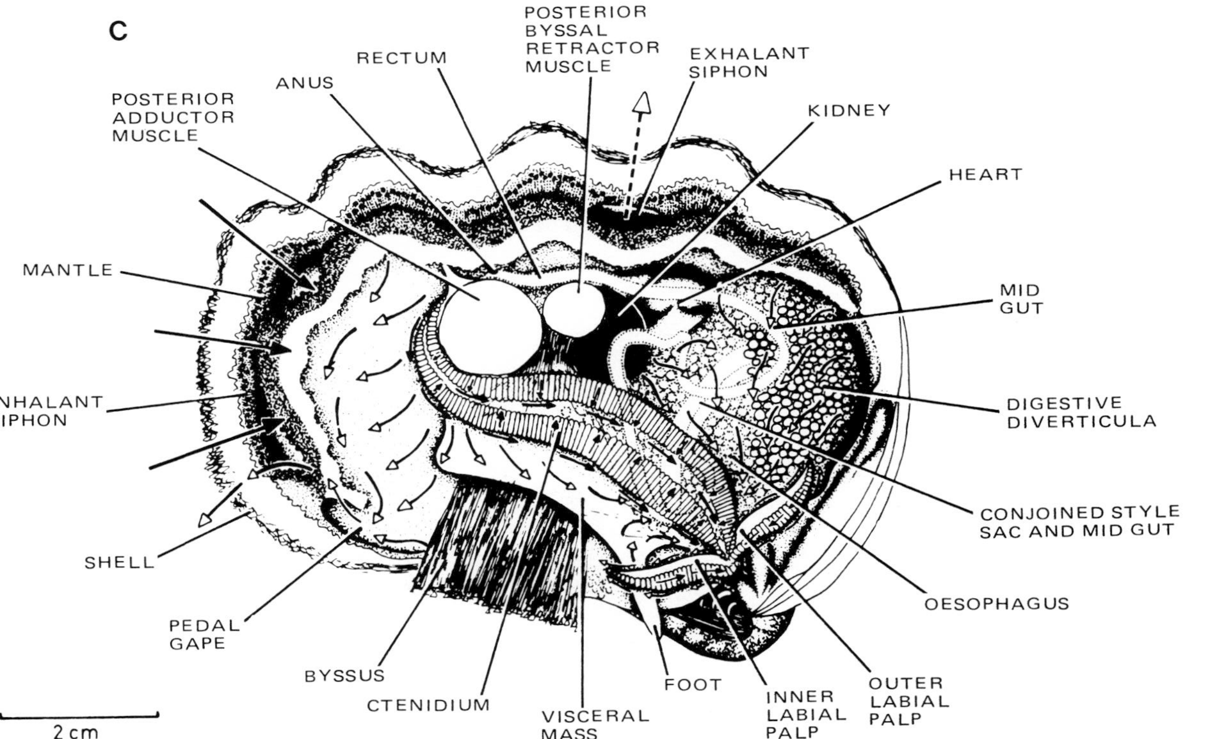

Fig. 20. (**A**) The anatomy of a cockle (Cardiidae), as compared (**B**) with a giant clam (Tridacnidae), showing the great expansion of the siphonal region of the latter. (**C**) The anatomy and ciliary currents of the mantle cavity of *Tridacna crocea* as seen from the right side after removal of the right shell valve and mantle lobe. [A,B from Yonge (1975). Giant clams. *Sci. Am.* **232,** 96–105.]

Olividae are known to feed on bivalves. On the reefs of Oahu, Hawaii, Houbrick and Fretter (1969) have shown that *Cymatium pileare* feeds on *Crassostrea gigas, Isognomon perna,* and *Ostrea sandvichensis; C. muricinum* feeds on the last two species only. Taylor (1976) has investigated the feeding habits of muricacean predators on Aldabra Atoll. *Thais aculeata* feeds to a small extent on *Brachidontes variabilis, Isognomon dentifer,* and *Saccostrea cucullata. T. fusconigra* feeds on *B. variabilis. Purpura rudolphi* feeds on *B. variabilis* and *S. cucullata,* whereas *Morula granulata* also feeds on *B. variabilis, I. dentifer,* a few other bivalves, and *Lithophaga nasuta.* At different localities the various species have subtly different feeding habits, but bivalves are not the major prey of any of the muricaceans.

On Tonga, *Chicoreus ramosus* feeds on *Tridacna maxima* up to a length of 186 mm. A single clam would be eaten in one day, the bivalve being attacked through the dorsal opening between the valves (McKoy, 1980). From Hong Kong, Taylor (1980) has shown that species of *Chicoreus* occasionally feed on *Lithophaga* sp., although the principal diet of this muricid is the oyster *Alectryonella plicatula.* Other muricids, *Ergalatax contracta, Mancinella echinata,* and *Cronia margariticola,* occasionally feed on members of the bivalve epibion, but their principal food is *Balanus trigonus.*

Most coral bivalves, however, possess defense mechanisms. Perhaps the best example is seen in the Tridacnidae; the exhalant siphon can be oriented to squirt a jet of water at predatory fishes (Stasek, 1965). The massive shell confers some protection, although in some the valves cannot be fully apposed because of the swollen edges of the inhalant siphon. Bivalves that are cemented (Chamidae, Spondylidae) or byssally attached (Arcacea) to the bases of dead coral heads are typically thick-shelled, the two valves interlocking closely. The Spondylidae have an array of spines growing out from the upper valve.

In the dead coral borer *Gastrochaena hians,* the siphons are especially extensible and retractable, thereby ensuring that this species can survive coral overgrowth. Consequently, however, the siphonal entrance to the burrow is sufficiently wide to allow predation by gastropods (Carter, 1978). Conversely, the small siphon diameter of *G. ovata* and the siphonal separation of *S. rostrata* exclude gastropod predation, but these species are less capable of preventing coral overgrowth. Thus, defense strategies are seen to be effective against one or the other form of predation, but rarely both.

Dead coral-boring lithophagines typically have thick, ridged shells and are securely attached byssally. In *L. myoforceps,* the posterior edges of the valves are crossed, which may confuse a predator used to a straight valve seal. In other species such as *L. bisulcata,* the regular laying down of a posterior encrustation on the valves may constitute a predator defense mechanism. In living coral borers, the shell is altogether different, being typically thin and smooth. Here protection is from the mass of surrounding polyps, although the secretion from posterior siphon glands may not only inhibit coral attack (Morton and Scott, 1980) but also deter other predators.

In the coral nestlers, other strategies have been adopted to protect the vulnerable posterior regions of the shell. Thus, *Gregariella coralliophaga* produces a posterior byssal cocoon to block the aperture to the burrow and conceal the valve ends (Morton, 1982a). In *Coralliophaga coralliophaga,* each valve has a series of outwardly directed, sharp, elongate ridges. It might be difficult for a predator that either opens or drills the valves (i.e., thaids and muricids), to decide, with so many choices, which are the opposed valve margins (Morton, 1980a). A similar adaptation is seen in *Irus macrophyllus,* which also settles in coral crevices. In *Pedum spondyloideum,* protection is from embedding within overgrowing coral, but also the shell is thick and retraction deeper into the slot against the byssal attachment point is possible.

In coral galleries in Hong Kong, species of the thin-shelled *Limaria* such as *L. fragilis* and *L. hongkongensis* (Morton, 1979a) have tentacles that can autotomize and release a noxious secretion (Gilmour, 1967). They also build a loose byssal nest, but if removed from the gallery, ultimate retreat lies in swimming flight. Leptonaceans of coral galleries typically possess a reflected mantle studded with protuberances and often tentacles that can similarly discharge a secretion, presumably noxious. This is best seen in species of *Galeomma* and *Ephippodonta* (Morton, 1976); the latter is also very compressed so as to fit closely against the gallery wall.

To generalize, it can be seen that exposed epibyssate reef bivalves rely on either a massive shell and/or secure attachment to avoid predation. In solitary coral crevice species or coral borers, the posterior margins of the valves, the only area vulnerable to attack, are disguised or selectively strengthened. In living coral borers and the inhabitants of coral galleries, where associations are much more intimate, the shell is typically reduced and protection is usually the partial responsibility of the host (e.g., the Lithophaginae and the commensal Leptonacea). With a reduced shell, self-protection is typically pallial, and often glands that secrete an inhibitory substance of unknown composition are developed. In many ways, the above sequence of changing predator defense mechanisms also reflects an increasing intimacy with the corals or the coral reef community.

X. Discussion

In the last review of coral-associated molluscs, Hadfield (1976) posed a number of interesting questions, particularly with regard to the methods by which gastropod coral predators avoid nematocyst toxins. The same question can be asked about those bivalves, notably the borers of living corals (i.e., the

Lithophaginae) whose siphons project among the mass of living coral polyps. Morton and Scott (1980) have made some progress in showing how this protection is achieved. These bivalves, which significantly are not dead coral-boring lithophagines, possess glands in the siphons which (these authors postulate) may inhibit nematocyst discharge. Other glands in the mantle prevent coral calcification of the burrow aperture. Reviewing the literature, Hadfield (1976) found a consensus that a mucus desensitizes the nematocysts, inhibiting their discharge. Such a protective device is by no means unusual among molluscan predators of zoantharians (Harris, 1968; Robertson, 1970; Salvini-Plawen, 1972; Hadfield, 1976). Nothing is known of the nature of the secretion produced by *Lithophaga*. Some other bivalves, notably the scallop *Pedum spondyloideum* and the leptonacean *Montacutona compacta,* also live in very intimate association with corals. In the latter bivalve, glands also occur in the mantle around the siphons and may have a function similar to those of *Lithophaga* (Morton, 1980b).

Hadfield (1976) has also asked, How does a settling molluscan larva find the specific coral on which it will live? In the Lithophaginae, an additional question is also pertinent: How does the larva penetrate the coral tissue? Clearly, in the case of wide-spectrum borers, the former question is not particularly important because a wide array of corals is suitable for habitation, but it assumes much greater importance with narrow-spectrum borers, which are, in every sense, commensal with only a few species of corals. In the case of *Fungiacava eilatensis,* which lives in strict association with *Fungia,* it is clear that the larva must have very specific responses to well-defined signals produced by the coral. However, the bivalve occurs only over a very small part of the total range of the coral (Goreau et al., 1969; Yonge, 1974). In this case, it is clear that factors other than the availability of the coral influence settlement in *Fungiacava*. Nothing is known of the larval adaptations that enable *Lithophaga* larvae first to find and penetrate coral tissue and then to commence boring the skeleton.

Our knowledge of this association, especially with regard to the larva, is very poor, and it is hoped that the present interest in live coral-associated lithophagines will be extended to the juvenile phases. But it is also important for other research to continue with the adults because, as this review demonstrates, the taxonomy of the Lithophaginae at both the genus and species levels is by no means defined (Wilson, 1979; Morton, 1982a); the broad question of the degree of intimacy between coral and bivalve cannot be properly answered until taxonomic arguments have been resolved.

Because of the review by Carter (1978), the Gastrochaenidae are far better understood and, it is clear, they attack only dead corals. Other borers (e.g., members of the Petricolidae, Pholadidae, and Clavagellidae) are less well understood; broad research on their boring techniques is required especially with regard to the latter two families. The Lithophaginae are probably the major bivalve element in coral reef erosion. Highsmith (1980) and Scott (1980) have initiated research on quantitative assessment of their numbers, but there are almost no data on the impact of borers (of any kind) upon dead (or living) corals, or on their rates of erosion and production. The only exception is the work of Hamner and Jones (1976) on *Tridacna crocea,* which, at average population densities, produces sediment at the rate of 200 $g/m^2/year$. Barthel (1982) has offered the novel suggestion that the calcium laid down by a lithophagine to smooth its burrow may strengthen and stabilize coral blocks. If this is true, present attitudes about the role of these borers in the destruction of coral reefs may need to be radically reappraised.

The second major component characterizing the bivalve fauna of coral reefs in the Indo-Pacific are giant clams (Tridacnidae). In contrast to the Lithophaginae, their taxonomy has broad agreement (Rosewater, 1965, 1982), but other interesting questions require investigation. The relationship with the zooxanthellae is as yet incompletely defined, and there is debate regarding just when the clams obtain the zooxanthellae, whether after metamorphosis (LaBarbera, 1975; Jameson, 1976) or earlier. Yonge (1936) and Gwyther and Munro (1981) argued that zooxanthellae are present in the fertilized egg, implying maternal inheritance.

Fitt and Trench (1981), however, argued the former for *T. squamosa.* The symbiotic algae were not seen in the fertilized eggs or trochophore. Zooxanthellae introduced to veligers were taken into the stomach via the mouth and within 2 to 9 days after metamorphosis moved into the developing siphonal tissues. Similarly, the role of the tridacnid kidney in zooxanthellae disposal has not been investigated, although the kidney is extraordinarily large (Yonge, 1980) and Morton (1978) has recognized zooxanthellae in it [this is denied by Trench et al. (1981)]. The nutritional importance of the zooxanthellae to the giant clams and how (or if) they assist the latter in obtaining their large size requires attention, although Fitt and Trench (1981) have shown that survival and growth of veligers and juveniles of *T. squamosa* with zooxanthellae was greater than of those lacking them.

No clear unequivocable data on growth rates and length of life in the various tridacnids are available, and this would seem to be crucial to a broad understanding of the likely impact of fisheries upon established populations. Overfishing has caused a marked decline in the number of these animals at many localities (Hardy and Hardy,1969; Hester and Jones, 1974; Bryan and McConnell, 1976; Pearson, 1977; Hirschberger, 1980). A concerted program to cultivate giant clams (Beckvar, 1981) probably would have beneficial applications in restocking depleted reefs and also would provide a welcome income for some island peoples. Also, in this context, various other reef molluscs are receiving more attention as potential mariculture organisms; for example, *Arca ventricosa* and *Fragum fragum* (Richard, 1982).

The remaining bivalves of coral areas can be classified according to microhabitat and are generally of less immediate interest than are the Lithophaginae and Tridacnidae. Some of the more characteristic species of bivalves living with corals, such as *Pedum spondyloideum* (Yonge, 1967b), *Corculum cardissa* (Kawaguti, 1950, 1968), *Limaria fragilis,* and *Montacutona compacta* (Morton, 1979a, 1980b), have been investigated, but again only superficially.

A stratification of the degree of dependence upon, or interdependence with, corals is shown by all associated bivalves. This chapter attempts to characterize some by reference to particular microhabitats, producing a number of categories. Within each category, however, it is likely that there are degrees of interdependence; for example, some species are able to settle and survive on a dead coral base much as if it were a rock. Other species appear to be much more characteristic of coral areas, such as *Arca avellana* and *Barbatia helblingi,* both widely dominant colonizers of corals in the Indo-Pacific (Table I).

The bivalve fauna of coral reef areas is not especially rich in comparison with the gastropods. The reason for this probably lies in the fact that the Bivalvia have evolved mostly in shallow, coastal, continental waters rich in suspended material (Salvat, 1967; Taylor, 1971c), and adaptive radiation into coral areas has been by only a few families involving relatively few genera. Thus, the coral sand bivalves are largely represented by the Lucinidae, Tellinidae, Cardiidae, and Veneridae. The former two families are deposit-feeders, which might be expected to occur in areas more organically rich, whereas the latter two families are typically suspension-feeders, and are more often characteristic of cleaner sands. Although these families—the Lucinidae (Allen, 1958), the Tellinidae (Yonge, 1949), and the Veneridae (Ansell, 1961)—have been investigated at a broad level, a few studies of one or two individuals have shown fascinating adaptations for life in coral sands. Thus, *Fimbria fimbriata* [the Fimbriidae are closely related to the Lucinidae (Morton, 1979c)], feeds on subsurface deposits collected by a probing sticky foot. There may be little in the overlying water for this coral sand bivalve, but it can accumulate nutrients over a wide area from *within* the sediment. Other lucinids, some of which have been reviewed by Allen (1958), may be worthy of greater attention to reveal feeding strategies adapting them to their mode of life in coral sands. *Corculum cardissa,* a member of the family Cardiidae, which is closely related to the Tridacnidae in the Cardiacea, possesses zooxanthellae in the mantle and gills. Clearly, the Tridacnidae have evolved from the cardiid-like ancestor, which argues for a longer and more intimate relationship with coral reef systems than is apparent in the other infaunal bivalves (except possibly the Lucinacea). Similarly, the hammer oysters (Malleidae) occur in a number of reef habitats, and Yonge (1968) has suggested evolution of free-living sand dwellers from an ancestor byssally attached to coral heads. Conversely, the crevice-dwelling *Irus* seems to have evolved in the opposite direction from an infaunal venerid ancestor.

The epibyssate colonizers of dead coral heads are also interesting. The community also appears to be dominated by a small number of families; the Arcidae and Pteriidae are especially important. In estuarine mainland shores, often mangrove-dominated, the epibyssate mode of life is almost exclusively the domain of the Mytilidae, bivalves which have radiated widely into a variety of microhabitats (Morton, 1983b). On coral shores, the Mytilidae are not conspicuous; even the widely recorded *Septifer bilocularis* is more typically an inhabitant of exposed rocky beaches, and *Brachidontes variabilis* is more typically a nestler of rock oyster clusters, although also occurring in large monospecific masses. Their counterparts in coral areas are members of the Arcidae. These more primitive bivalves have radiated widely in this habitat, simpler anterior-posterior filtering currents in the mantle cavity being less capable of dealing with a high sediment load but equipping them adequately for life in cleaner waters.

Members of the Isognomonidae and Pteriidae, on the other hand, occur with equal facility in both coral and estuarine areas, but are typical of the tropics and seem adapted to byssal attachment under an array of conditions. Some pteriids, however, have attained a more intimate (but as yet not properly defined) relationship with deeper-water gorgonian and antipatharian corals, attaching to the axial skeleton by a byssus (R. Robertson, 1963).

A similar argument can be made for the cemented bivalves of coral areas. There is a marked absence of true oysters (Ostreidae). Morton (1983b) has shown that each mangrove area always possesses a distinctive, widely dominant oyster, but this is not true in coral areas in the Indo-Pacific, where there appears to be a number of rather solitary species. The records of *Saccostrea cucullata* by Abbott (1950) and Nielsen (1976b) are probably from sheltered beaches, this species being everywhere, over its very broad range, intertidal. Like the pteriids, some of these oysters are more intimately associated with gorgonians and ceriantipatharians, such as *Lopha frons* and *L. cristigalli* (Masteller, 1978). The cemented habit on corals has been adopted and dominated by members of the Spondylidae (closely related to the byssiferous Pectinidae) and Chamidae, with the Plicatulidae being occasionally represented [e.g., *Plicatula plicata* (Mastaller, 1978; Taylor, 1980)]. The former two families appear to have divided the habitat between them, the Spondylidae being possibly more intimately associated with living corals and the Chamidae always cementing to dead coral bases. These bivalves rarely occur in estuarine areas on continental land masses (Morton, 1983b) and, like the byssate colonizers of corals (i.e., the Arcidae and Pteriidae), have evolved in parallel with scleractinian reef situations.

Some families of byssally attached bivalves, such as the Veneridae, Trapeziidae, and Carditidae, are represented by only single genera and, often, single species, all with a very wide distribution pattern. These bivalves are solitary inhabitants of narrow crevices. Abbott (1950), Taylor (1971c), and Kay (1971) recorded species of *Trapezium* from corals; Morton (1979b) has shown that *Trapezium sublaevigatum,* which can occur in coral crevices, has a wide distribution over a variety of shores and occurs at a variety of depths. This probably also applies to *T. oblongum* and *T. bicarinatum. Coralliophaga,* on the other hand, appears to be specifically associated with coral crevices and has a wide distribution in the Caribbean (Bromley, 1978) and the Indo-Pacific (Solem, 1954). *Gregariella coralliophaga* (Mytilacea) has a similarly wide distribution (Morton, 1982a). Representatives of these families apparently have only rarely produced representatives capable of colonizing this habit successfully, and here they occupy very narrow, specialized niches. *G. coralliophaga* (Lithophaginae) (Morton, 1982a) is also a nestler but is capable of enlarging its burrow. If this bivalve is typical of ancestral conditions in the Lithophaginae, diversification has been from a narrow stock, resulting in the adoption of a boring habit and ultimately leading to a more commensal mode of life with corals; this is the only bivalve lineage to do this, all others boring dead corals.

It thus seems clear that the coral habitat has been adopted by a spectrum of bivalves representing a few (major) families or by a few species of other less widely dominant families. These families, by and large, are not those which colonize more continental, intertidal, and shallow sublittoral habitats. [Compare these faunal lists from coral areas with those from mangrove areas (Morton, 1983b).]

Salvat (1967) and Taylor (1971c) have made major contributions to our understanding of the biogeographic distribution of the coral-associated Mollusca of the Pacific and Indian oceans, respectively. The area of greatest diversity, as with the mangroves and their molluscan fauna (Vermeij, 1974; Morton, 1983b), appears to center on Malaysia, Indonesia, and the Philippines. From this focus, recruitment to island faunas is low, resulting in a reduced diversity often with the local dominance of one or more species; in this dominance bivalves are very important (Salvat, 1967). The best example is, of course, the giant clams; in some lagoons, the number may be considerable. In coral systems a few families have become very important, each possessing a few genera represented by many species, for example, the Pteriidae, Arcidae, Spondylidae, and Chamidae attached to coral bases, and the Lucinidae, Veneridae, Cardiidae, and Tellinidae buried in lagoon sands (Tables I and II). Such a situation is the opposite of mangrove mollusc faunas, in which many families are typically represented by single genera (and often single species) (Vermeij, 1974; Morton, 1983b).

Low familial and generic but high species diversity on coral reefs argues for wide adaptive radiation within this system from narrow stocks [with little recruitment from mainland waters (Salvat, 1967; Taylor, 1971c)]. This is exemplified by the above families but is better demonstrated by the coral-boring Lithophaginae, in which evolution has clearly proceeded from a narrow stock, with subsequent wide adaptive radiation and the adoption of an increasingly sophisticated, interdependent, almost commensal relationship with living corals (Morton, 1982a). Restricted habitats in the coral reef such as crevices and galleries are often occupied by widely distributed single species, each representing single families [e.g., *Coralliophaga coralliophaga* and *Gregariella coralliophaga*, both of which have an Indo-Pacific and a Caribbean distribution (Morton, 1980a, 1981a)]; [as does *Petricola lapicida* (Otter, 1937; Bromley, 1978)].

The marine bivalves of island reefs are relatively depauperate compared to continental reefs. This may result from long isolation, a low nutrient supply, and high temperatures. However, the greater diversity of continental nearshore seas may also result from a diversity of habitats; for example, estuaries created by river systems (within which mangroves flourish) often establish high nutrient levels inshore. Moreover, a possibly greater variability in the degree of exposure to wave action on continental shores enhances this diversity.

References

Abbott, R. T. (1950). The molluscan fauna of the Cocos-Keeling Islands, Indian Ocean. *Bull. Raffles Mus.* **22,** 68–98.

Abbott, R. T. (1960). The genus *Strombus* in the Indo-Pacific. *Indo-Pac. Mollusca* **1**(2), 33–146.

Adams, H. (1868). Further descriptions of new species of shells collected at Mauritius by Geoffrey Nevill, Esq. *Proc. Zool. Soc. London* **36,** 12–14.

Allen, J. A. (1958). The basic form and adaptations to habitat in the Lucinacea (Eulamellibranchia). *Philos. Trans. R. Soc. London, Ser. B* **241,** 421–484.

Allen, J. A., and Turner, J. F. (1970). The morphology of *Fimbria fimbriata* (Linné) (Bivalvia: Lucinidae). *Pac. Sci.* **24,** 147–154.

Amemiya, I. Y., and Ohsima, Y. (1923). Note on the habit of rock-boring Mollusca on the coast of central Japan. *Proc. Imp. Acad. (Tokyo)* **9,** 120–123.

Ansell, A. D. (1961). The functional morphology of the British species of Veneracea (Eulamellibranchia). *J. Mar. Biol. Assoc. U.K.* **41,** 489–515.

Ansell, A. D., and Nair, N. B. (1969). A comparative study of bivalves which bore mainly by mechanical means. *Am. Zool.* **9,** 857–868.

Appukuttan, K. K. (1973). Distribution of coral boring bivalves along Indian coasts. *J. Mar. Biol. Ass. India* **15,** 427–430.

Arakawa, K. Y. (1960). Ecological observations on an aberrant lamellibranch, *Ephippodonta murakamii* Kuroda. *Venus* **21,** 50–60.

Arnaud, P. M., and Thomassin, B. A. (1976). First records and adaptive significance of boring into a free-living scleractinian coral (*Heteropsammia michelini*) by a date mussel (*Lithophaga lessepsiana*). *Veliger* **18,** 367–374.

Austin, A. D., Austin, S. A., and Sale, P. F. (1980). Community structure of the fauna associated with the coral *Pocillopora damicornis* (L.) on the Great Barrier Reef. *Aust. J. Mar. Freshwater Res.* **31,** 163–174.

Bacon, P. R. (1975). Shell form, byssal development and habitat of *Mytella guyanensis* (Lamarck) and *M. falcata* (Orbigny) (Pelecypoda: Mytilidae) in Trinidad, West Indies. *Proc. Malacol. Soc. London* **41,** 511–520.

Banner, A. H., and Randall, J. E. (1952). Preliminary report on a marine biological study of Onotoa Atoll, Gilbert Islands. *Atoll Res. Bull.* **13,** 1–62.

Barthel, K. W. (1982). *Lithophaga obesa* (Philippi) reef-dwelling and cementing pelecypod—a survey of its boring. *Proc. Int. Coral Reef Symp., 4th, 1981* **2,** 649–659.

Beckvar, N. (1981). Cultivation, spawning, and growth of the giant clams *Tridacna gigas, T. derasa,* and *T. squamosa* in Palau, Caroline Islands. *Aquaculture* **24,** 21–30.

Bertram, G. C. L. (1936). Some aspects of the breakdown of coral at Ghardaqa, Red Sea. *Proc. Zool. Soc. London* **106,** 1011–1026.

Bonham, K. (1965). Growth rate of giant clam *Tridacna gigas* at Bikini Atoll as revealed by radio autography. *Science* **149,** 300–302.

Boshoff, P. H. (1965). Pelecypoda of Inhaca Island, Mocambique. *Mem. Inst. Invest. Cient. Mocamb.* **A7,** 65–206.

Boss, K. J. (1965). Symbiotic erycinacean bivalves. *Malacologia* **3,** 183–195.

Boss, K. J. (1967). On the evolution of *Spengleria* (Gastrochaenidae; Bivalvia). *Am. Malacol. Union Annu. Rep.* pp. 15–16.

Boss, K. J. (1971). Critical estimate of the number of Recent Mollusca. *Occas. Pap. Mollusks Mus. Comp. Zool. Harv. Univ.* **3**(40), 81–135.

Boss, K. J., and Moore, D. R. (1967). Notes on *Malleus (Parimalleus) candeanus* (D'Orbigny) (Mollusca: Bivalvia). *Bull. Mar. Sci.* **17,** 85–94.

Bourne, G. C. (1906). Report of *Jousseaumiella,* a new genus of eulamellibranchs commensal with the corals *Heterocyathus* and *Heteropsammia,* collected by Professor Herdman in Ceylon. *Rep. Pearl Oyster Fish. Ceylon* **5,** 243–266.

Bouvier, M. E. L. (1894). A new instance of commensalism: Association of worms of the genus *Aspidosiphon* with madreporarian polyps and a bivalve mollusk. *Ann. Mag. Nat. Hist.* **14,** 312–314.

Bromley, R. G. (1978). Bioerosion of Bermuda reefs. *Palaeogeogr., Palaeoclimatol., Palaeoecol.* **23,** 169–197.

Bryan, P. G., and McConnell, D. B. (1976). Status of giant clam stocks (Tridacnidae) on Helen Reef, Palau, Western Caroline Islands, April 1975. *Mar. Fish. Rev.* **38,** 15–18.

Bullock, R. C., and Boss, K. J. (1971). *Lithophaga aristata* in the shell-plates of chitons (Mollusca). *Breviora* **369,** 1–10.

Cailliaud, F. (1856). Mémoire sur les Mollusques perforants. *Nat. Verh. Bataaf. (Koning. Hollandsche) Maatsch. Wet. Haarlem, Verz. 2* **11,** 1–58.

Carazzi, D. (1892). La perforazione delle rocce calcaree per opera dei Datteri (*Lithodomus dactylus*). *Atti Soc. Ligust. Sci. Nat. Geogr.* **3,** 279–297.

Carazzi, D. (1903). Contributo all'istologia e alla fisiologia dei Lamellibranchi. *Int. Monatsschr. Anat. Physiol.* **20,** 57–90.

Carter, J. G. (1976). The ecology of a tropical Western Atlantic endolithic bivalve community. *Geol. Soc. Am. Abstr.* **8,** 148–149.

Carter, J. G. (1978). Ecology and evolution of the Gastrochaenacea (Mollusca, Bivalvia) with notes on the evolution of the endolithic habit. *Bull. Peabody Mus.* **41,** 1–92.

Carter, J. G., and Aller, R. C. (1975). Calcification in the bivalve periostracum. *Lethaia* **8,** 315–320.

Chavan, A. (1969). Superfamily Leptonacea Gray, 1847. *In* "Treatise on Invertebrate Paleontology" (R. C. Moore, ed.), Part N, 2, Mollusca 6, pp. N518–N537. Geol. Soc. Am. and Univ. of Kansas Press, Lawrence.

Comfort, A. (1957). The duration of life in mollusks. *Proc. Malacol. Soc. London* **32,** 219–241.

Connell, J. H. (1973). Population ecology of reef building corals. *In* "Biology and Geology of Coral Reefs". Vol. 2. Biology. 1. (R. E. Endean and R. A. Jones, eds.). pp. 205–245. Academic Press, New York.

Coomans, H. E. (1969). Biological aspects of mangrove Mollusca in the West Indies. *Malacologia* **9,** 79–84.

Deane, E. M., Duxbury, T., and O'Brien, R. W. (1979). Growth cycle of free-living *Gymnodinium microadriaticum* from *Tridacna maxima. Br. Phycol. J.* **14,** 407–412.

Deshayes, G. P. (1846). Examen anatomique du gastrochène de la Méditerraneé. *C. R. Hebd. Seances Acad. Sci.* **22,** 37–38.

Dudgeon, D., and Morton, B. S. (1982). Coral associated Mollusca of Tolo harbour, Hong Kong. *In* "Proceedings of the First International Marine Biological Workshop: The Marine Flora and Fauna of Hong Kong and Southern China. Hong Kong, 1980" (B. S. Morton and C. K. Tseng, eds.), pp. 627–650. Hong Kong Univ. Press, Hong Kong.

Eldredge, L. G., Tsuda, R. T., Moore, P., Chernin, M., and Neudecker, S. (1977). A natural history of Maug, northern Mariana Islands. *Tech. Rep.—Univ. Guam Mar. Lab.* **43,** 1–87.

Evseev, G. A. (1982). On the ecology of coral-boring bivalve *Lithophaga kuehnelti* Kleeman (Bivalvia: Mytilidae), Great Barrier Reef, Australia. *Proc. Int. Coral Reef Symp., 4th, 1981* **2,** 661–663.

Fankboner, P. V. (1971a). The ciliary currents associated with feeding, digestion and sediment removal in *Adula (Botula) falcata* Gould 1851. *Biol. Bull. (Woods Hole, Mass.)* **140,** 28–45.

Fankboner, P. V. (1971b). Self righting by tridacnid clams. *Nature (London)* **230,** 579–580.

Fankboner, P. V. (1971c). Intracellular digestion of symbiotic zooxanthellae by host amoebocytes in giant clams (Bivalvia: Tridacnidae), with a note on the role of the hypertrophied siphonal epidermis. *Biol. Bull. (Woods Hole, Mass.)* **141,** 222–234.

Fankboner, P. V. (1981). Siphonal eyes of giant clams (Bivalvia: Tridacnidae) and their relationship to adjacent zooxanthellae. *Veliger* **23,** 245–249.

Fankboner, P. V., and Reid, R. G. B. (1981). Mass expulsion of zooxanthellae by heat stressed reef corals: a source of food for giant clams. *Experientia* **37,** 251–252.

Field, L. R. (1949). Sea anemones and corals of Beaufort, North Carolina. *Bull. Duke Univ. Mar. Stn.* **5,** 1–39.

Fishelson, L. (1973). Ecological and biological phenomena influencing coral-species composition on the reef tables at Eilat (Gulf of Aqaba, Red Sea). *Mar. Biol. (Berlin)* **19,** 183–196.

Fitt, W. K., and Trench, R. K. (1981). Spawning, development, and acquisition of zooxanthellae by *Tridacna squamosa* (Mollusca, Bivalvia). *Biol. Bull. (Woods Hole, Mass.)* **161,** 213–235.

Franc, A. (1960). Classe de bivalves. *In* "Traité de Zoologie" (P.-P. Grassé, ed.), Vol. 5(2), pp. 1845–2164, Figs. 1605–1830. Masson, Paris.

Frank, P. W. (1969). Growth rates and longevity of some gastropod molluscs on the coral reef at Heron Island. *Oecologia* **2,** 232–250.

Gardiner, J. S. (1903). The Maldive and Laccadive Groups, with notes on other coral formations in the Indians Ocean. *Fauna Geogr., Maldive Laccadive Archipelagoes* **1,** 333–341.

Gilmour, T. H. J. (1967). The defensive adaptations of *Lima hians* (Mollusca, Bivalvia). *J. Mar. Biol. Assoc. U.K.* **47,** 209–221.

Gohar, H. A. F., and Soliman, G. N. (1963a). On three mytilid species boring in living corals. *Publ. Mar. Biol. Stn., Ghardaqa, Red Sea* **12,** 65–98.

Gohar, H. A. F., and Soliman, G. N. (1963b). On the rock-boring lamellibranch *Rocellaria ruppelli* (Deshayes). *Publ. Mar. Biol. Stn., Ghardaqa, Red Sea* **12,** 145–157.

Gohar, H. A. F., and Soliman, G. N. (1963c). On two mytilids boring in dead coral. *Publ. Mar. Biol. Stn., Ghardaqa, Red Sea* **12,** 205–218.

Goreau, T. F., Goreau, N. I., Soot-Ryen, T., and Yonge, C. M. (1969). On a new commensal mytilid (Mollusca: Bivalvia) opening into the coelenteron of *Fungia scutaria* (Coelenterata). *J. Zool.* **158,** 171–195.

Goreau, T. F., Goreau, N. I., and Yonge, C. M. (1972). On the mode of boring in *Fungiacava eilatensis* (Bivalvia, Mytilidae). *J. Zool.* **166,** 55–60.

Goreau, T. F., Goreau, N. I., and Yonge, C. M. (1973). On the utilization of photosynthetic products from zooxanthellae and of a dissolved amino acid in *Tridacna maxima* f. *elongata* (Roeding) (Mollusca: Bivalvia). *J. Zool.* **169,** 417–454.

Goreau, T. F., Goreau, N. I., Yonge, C. M., and Neumann, Y. (1970). On feeding and nutrition in *Fungiacava eilatensis* (Bivalvia: Mytilidae), a commensal living in fungiid corals. *J. Zool.* **160,** 159–172.

Gwyther, J., and Munro, J. L. (1981). Spawning induction and rearing of larvae of tridacnid clams (Bivalvia: Tridacnidae). *Aquaculture* **24,** 197–217.

Habe, T. (1973). *Halcampella maxima* Hertwig, host of *Nipponomontacuta actinariophila* Yamamoto and Habe. *Venus* **31,** 157.

Hadfield, M. G. (1976). Molluscs associated with living corals. *Micronesica* **12,** 133–148.

Hamner, W. M. (1978). Intraspecific competition in *Tridacna crocea,* a burrowing bivalve. *Oecologia* **34,** 267–281.

Hamner, W. M., and Jones, M. S. (1976). Distribution, burrowing, and growth rates of the clam *Tridacna crocea* on interior reef flats. *Oecologia* **24,** 207–227.

Hardy, J. T., and Hardy, S. A. (1969). Ecology of *Tridacna* in Palau. *Pac. Sci.* **23,** 467–472.

Harris, L. G. (1968). Notes on the biology and distribution of the aeolid nudibranch (Gastropoda), *Phestilla melanobranchia* Bergh, 1874. *Publ. Seto Mar. Biol. Lab.* **16,** 193–198.

Hass, F. (1943). Malacological Notes. (b), the boring of *Lithophaga. Chicago Field Mus. Nat. Hist.* **29,** 1–23.

Hester, F. J., and Jones, E. C. (1974). A survey of giant clams, Tridacnidae, on Helen Reef, a Western Pacific atoll. *Mar. Fish. Rev.* **36,** 17–22.

Highsmith, R. C. (1980). Burrowing by the bivalve mollusc *Lithophaga curta* in the living reef coral *Montipora berryi* and a hypothesis of reciprocal larval recruitment. *Mar. Biol. (Berlin)* **56,** 155–162.

Hirschberger, W. (1980). Tridacnid clam stocks on Helen Reef, Palau, Western Caroline Islands. *Mar. Fish. Rev.* **42,** 8–15.

Hodgkin, N. M. (1962). Limestone boring by the mytilid *Lithophaga. Veliger* **4,** 123–129.

Houbrick, J. R., and Fretter, V. (1969). Some aspects of the functional anatomy and biology of *Cymatium* and *Bursa. Proc. Malacol. Soc. London* **38,** 415–429.

Iredale, T. (1939). Mollusca. Part 1. *Sci. Rep. Great Barrier Reef Exped.* **5,** 209–425.

Jaap, W. C. (1979). Observations on zooxanthellae expulsion at Middle Sambo Reef, Florida Keys. *Bull. Mar. Sci.* **29,** 414–422.

Jaccarini, V., Bannister, W. H., and Micallef, H. (1968). The pallial glands and rock boring in *Lithophaga lithophaga* (Lamellibranchia; Mytilidae). *J. Zool.* **154,** 397–401.

Jameson, S. C. (1976). Early life history of the giant clams *Tridacna crocea* Lamarck, *Tridacna maxima* (Roding), and *Hippopus hippopus* (Linnaeus) *Pac. Sci.* **30,** 219–233.

Jaubert, J. (1977). Height, metabolism and the distribution of *Tridacna maxima* on a South Pacific atoll: Takapoto (French Polynesia). *Proc. Int. Coral Reef Symp., 3rd, 1977* pp. 489–494.

Kawaguti, S. (1950). Observations on the heart shell, *Corculum cardissa* (L.), and its associated zooxanthellae. *Pac. Sci.* **4,** 43–49.

Kawaguti, S. (1966). Electron microscopy on the mantle of the giant clam with special reference to zooxanthellae and iridophores. *Biol. J. Okayama Univ.* **12,** 81–92.

Kawaguti, S. (1968). Electron microscopy on zooxanthellae in the mantle and gill of the heart shell. *Biol. J. Okayama Univ.* **14,** 1–12.

Kawaguti, S., and Mabuchi, K. (1969). Electron microscopy on the eyes of the giant clam. *Biol. J. Okayama Univ.* **15,** 87–100.

Kay, E. A. (1971). The littoral marine mollusks of Fanning Island. *Pac. Sci.* **25,** 260–281.

Kay, E. A., and Switzer, M. F. (1974). Molluscan distribution patterns in Fanning Island Lagoon and a comparison of the mollusks of the lagoon and the seaward reefs. *Pac. Sci.* **28,** 275–295.

Keen, A. M. (1958). "Sea Shells of Tropical West America." Stanford Univ. Press, Stanford, California.

Keen, A. M. (1969). Superfamily Gastrochaenacea Gray, 1840. *In* "Treatise on Invertebrate Paleontology," (R. C. Moore, ed.), Part N, Mollusca 6, pp. N699–N700. Geol. Soc. Am. and Univ. of Kansas Press, Lawrence.

Kleeman, K. (1977). A new species of *Lithophaga* (Bivalvia) from the Great Barrier Reef, Australia. *Veliger* **20,** 151–154.

Kleeman, K. H. (1980a). Korallenbohrende Muschel sert dem Mittleren Lias unverändert. *Beitr. Palaeontol. Oesterr.* **7,** 239–249.

Kleeman, K. H. (1980b). Boring bivalves and their host corals from the Great Barrier Reef. *J. Moll. Stud.* **46,** 13–54.

Knudsen, J. (1970). The systematics and biology of abyssal and hadal Bivalvia. *Galathea Rep.* **11,** 1–241.

Kühnelt, W. (1930). Bohrmuschelstudien. I. *Paleobiologica* **3,** 51–91.

Kühnelt, W. (1933). Uber Anpassungen der Muscheln an ihren Aufenthaltsort. *Biol. Gen.* **9,** 189–200.

Kühnelt, W. (1934). Bohrmuschelstudien. II. *Palaeobiologica* **5,** 371–408.

LaBarbera, M. (1974). Calcification of the first larval shell of *Tridacna squamosa* (Tridacnidae: Bivalvia). *Mar. Biol. (Berlin)* **25,** 233–238.

LaBarbera, M. (1975). Larval and post-larval development of the giant clams *Tridacna maxima* (Roding) and *Tridacna squamosa* Lamarck (Tridacnidae: Bivalvia). *Malacologia* **15,** 69–79.

Ladd, H. S. (1960). Origin of the Pacific Island molluscan fauna. *Am. J. Sci.* **258A,** 137–150.

Lewis, M. S., and Taylor, J. D. (1966). Marine sediments and bottom communities of the Seychelles. *Philos. Trans. R. Soc. London, Ser. A* **259,** 279–290.

List, T. (1902). Die Mytiliden des Golfes von Neapal. *Fauna Flora Golfes Neapal, Monogr.* **27,** 1–312.

Loeblich, A. R., and Sherley, J. L. (1979). Observations on the theca of the motile phase of free-living and symbiotic isolates of *Zooxanthella microadriatica* (Freudenthal) Comb. nov. *J. Mar. Biol. Assoc. U.K.* **59,** 195–205.

Loi, T. N. (1967). Peuplements animausx et végétausx du substrat dur intertidal de la Baie de Nha Trang (Viet Nam). *Inst. Oceanogr. Nha Trang Mem.* **11,** 1–236.

McAlester, A. L. (1965). Evolutionary and systematic implications of a transitional Ordovician lucinoid bivalve. *Malacologia* **3,** 433–439.

McKoy, J. L. (180). Biology, exploitation and management of giant clams (Tridacnidae) in the Kingdom of Tonga. *Fish. Bull., Fish. Div. Tonga* **1,** 1–61.

McMichael, D. F. (1974). Growth rate, population size and mantle coloration in the small giant clam *Tridacna maxima* (Röding), at One Tree island, Capricorn group, Queensland. *Proc. Int. Coral Reef Symp., 2nd, 1974* pp. 241–254.

Maes, V. (1967). The littoral marine mollusks of Cocos-Keeling Islands (Indian Ocean). *Proc. Acad. Nat. Sci. Philadelphia* **119,** 93–217.

Mansour, K. (1945). The zooxanthellae, morphological peculiarities and food and feeding habits of the Tridacnidae with reference to other lamellibranchs. *Proc. Egypt. Acad. Sci.* **1,** 1–11.

Mastaller, M. (1978). The marine molluscan assemblages of Port Sudan, Red Sea. *Zool. Med ed.* **53,** 117–144.

Morton, B. S. (1973). Some aspects of the biology, population dynamics and functional morphology of *Musculista senhausia* Benson (Bivalvia: Mytilacea). *Pac. Sci.* **28,** 19–33.

Morton, B. S. (1976). Secondary brooding of temporary dwarf males in *Ephippodonta (Ephippodontina) oedipus* n.sp. (Bivalvia: Leptonacea). *J. Conchol.* **29,** 31–39.

Morton, B. S. (1978). The diurnal rhythm and the processes of feeding and digestion in *Tridacna crocea* (Bivalvia: Tridacnidae). *J. Zool.* **185,** 371–388.

Morton, B. S. (1979a). A comparison of lip structure and function correlated with other aspects of the functional morphology of *Lima lima, Limaria (Platilimaria) fragilis,* and *Limaria (Platilimaria) hongkongensis* sp. nov. (Bivalvia: Limacea). *Can. J. Zool.* **57,** 728–742.

Morton, B. S. (1979b). Some aspects of the biology and functional morphology of *Trapezium (Neotrapezium) sublaevigatum* (Lamarck) (Bivalvia: Arcticacea). *Pac. Sci.* **33,** 177–194.

Morton, B. S. (1979c). The biology and functional morphology of the coral-sand bivalve *Fimbria fimbriata* (Linnaeus 1758). *Rec. Aust. Mus.* **32,** 389–420.

Morton, B. S. (1980a). Some aspects of the biology and functional morphology of *Coralliophaga (Coralliophaga) coralliophaga* (Gmelin 1791) (Bivalvia: Arcticacea): a coral associated nestler in Hong Kong. *In* "Proceedings of the First International Workshop on the Malacofauna of Hong Kong and Southern China, Hong Kong 1977" (B. S. Morton, ed.), pp. 311–330. Hong Kong Univ. Press, Hong Kong.

Morton, B. S. (1980b). Some aspects of the biology and functional morphology (including the presence of a ligamental lithodesma) of *Montacutona compacta* and *M. olivacea* (Bivalvia: Leptonacea) associated with coelenterates in Hong Kong. *J. Zool., Lond.* **192,** 431–455.

Morton, B. S. (1980c). The biology and some aspects of the functional morphology of *Arcuatula elegans* (Mytilacea: Crenellinae). *In* "Proceedings of the First International Workshop on the Malacofauna of Hong Kong and Southern China, Hong Kong 1977" (B. S. Morton, ed.), pp. 331–345. Hong Kong Univ. Press, Hong Kong.

Morton, B. S. (1982a). The mode of life and functional morphology of *Gregariella coralliophaga* (Gmelin 1791) (Bivalvia: Mytilacea) with a discussion on the evolution of the boring Lithophaginae and adaptive radiation in the Mytilidae. *In* "Proceedings of the First International Marine Biological Workshop: The Marine Flora and Fauna of Hong Kong and Southern China" (B. S. Morton and C. K. Tseng, eds.), pp. 875–895. Hong Kong Univ. Press, Hong Kong.

Morton, B. S. (1982b). Pallial specializations in *Gastrochaena (Cucurbitula) cymbium* Spengler 1783 (Bivalvia: Gastrochaenacea). *In* "Proceedings of the First International Marine Biological Workshop: The Marine Flora and Fauna of Hong Kong and Southern China, Hong Kong, 1980" (B. S. Morton and C. K. Tseng, eds.), pp. 859–873. Hong Kong Univ. Press, Hong Kong.

Morton, B. S. (1982c). The biology, functional morphology and taxonomic status of *Fluviolanatus subtorta* (Bivalvia: Trapeziidae), a heteromyarian bivalve possessing "zooxanthellae." *J. Malacol. Soc. Aust.* **5,** 113–140.

Morton, B. S. (1983a). The biology and functional morphology of the twisted ark *Trisidos semitorta* (Bivalvia: Arcacea) with a discussion on shell 'torsion' in the genus. *Malacologia* **23,** 375–396.

Morton, B. S. (1983b). Mangrove bivalves. *In* The Mollusca" (K. M. Wilbur and W. D. Russell-Hunter, eds.), Vol. 6, pp. 77–140. Academic Press, New York.

Morton, B. S., and Scott, P. J. B. (1980). Morphological and functional specializations of the shell, musculature and pallial glands in the Lithophaginae (Mollusca: Bivalvia). *J. Zool.* **192,** 179–204.

Munro, J. L., and Gwyther, J. (1982). Growth rates and maricultural potential of tridacnid clams. *Proc. Int. Coral Reef Symp., 4th, 1981* **2,** 633–666.

Narchi, W. (1975). Functional morphology of a new *Petricola* (Mollusca, Bivalvia) from the littoral of Sao Paulo, Brazil. *Proc. Malacol. Soc. London* **41,** 451–465.

Nielsen, C. (1976a). Notes on boring bivalves from Phuket, Thailand. *Ophelia* **15,** 141–148.

Nielsen, C. (1976b). An illustrated checklist of bivalves from PMBC beach with a reef flat at Phuket, Thailand. *Phuket Mar. Biol. Cent., Res. Bull.* **9,** 1–7.

Ostergaard, J. M. (1935). Recent and fossil marine molluscs of Tongatabu. *Bull. Bishop Mus., Honolulu* **131,** 1–59.

Otter, G. W. (1937). Rock destroying organisms in relation to coral reefs. *Sci. Rep. Great Barrier Reef Exped.* **1,** 323–352.

Patton, W. K. (1972). Studies on the animal symbionts of the gorgonian coral, *Leptogorgia virgulata* (Lamarck). *Bull. Mar. Sci.* **22,** 419–431.

Patton, W. K. (1975). Animal associates of living coral reefs. *In* "Biology and Geology of Coral reefs" (O. Jones and R. Endean, eds.), Vol. 3, Biol. 2, pp. 1–36. Academic Press, New York.

Pearson, R. G. (1977). Impact of foreign vessels poaching giant clams. *Aust. Fish.* pp. 8–11, 23.

Pease, W. H. (1865). Description of new genera and species of marine shells from the islands of the Central Pacific. *Proc. Zool. Soc. London* **33,** 512–517.

Pelseneer, P. (1894). Introduction à l'étude des mollusques. *Mem. Soc. R. Malacol. Belg.* **27,** 31–243.

Pelseneer, P. (1911). Les lamellibranches de l'expedition du Siboga. Partie Anatomique. *Siboga Exped. Monogr.* **53a,** 1–126.

Pojeta, J., and Palmer, T. J. (1976). The origin of rock boring in mytilacean pelecypods. *Alcheringa* **1,** 167–179.

Ponder, W. F. (1971). *Montacutona ceriantha* n.sp., a commensal leptonid bivalve living with "*Cerianthus.*" *J. Conchyliol.* **109,** 15–25.

Ponder, W. F., Colman, P. H., Yonge, C. M., and Colman, M. H. (1980). The taxonomic position of *Hemidonax* Mörch, 1871 with a review of the genus (Bivalvia: Cardiacea). *J. Malacol. Soc. Aust.* **5,** 41–64.

Prashad, B. (1932). The Lamellibranchia of the Siboga expedition, Systematic Part II. Pelecypoda. *Siboga Exped. Monogr.* **53,** 1–353.

Purchon, R. D. (1954). A note on the biology of the Lamellibranch *Rocellaria (Gastrochaena) cuneiformis* Spengler. *Proc. Zool. Soc. London* **124,** 17–23.

Purchon, R. D. (1955a). The functional morphology of the rock-boring lamellibranch *Petricola pholadiformis* Lamarck. *J. Mar. Biol. Assoc. U.K.* **34,** 257–278.

Purchon, R. D. (1955b). The structure and function of the British Pholadidae (rock-boring Lamellibranchia). *Proc. Zool. Soc. London* **124,** 859–911.

Purchon, R. D. (1955c). A note on the biology of *Tridacna crocea* Lam. *Proc. Malacol. Soc. London* **31,** 95–110.

Randall, R. H., Tsuda, R. T., Jones, R. S., Gawel, M. J., Chase, J. A., and Rechebei, R. (1975). Marine biological survey of the Cocos Barrier reefs and enclosed lagoon. *Tech. Rep.—Univ. Guam Mar. Lab.* **17,** 1–160.

Rees, W. J. (1967). A brief survey of the symbiotic associations Cnidaria with Mollusca. *Proc. Malacol. Soc. London* **37,** 213–231.

Reid, R. G. B., and Porteous, S. (1980). Aspects of the functional morphology and digestive physiology of *Vulsella vulsella* (Linne) and *Crenatula modiolaris* (Lamarck), bivalves associated with sponges. *In* "Proceedings of the First International Workshop on the Malacofauna of Hong Kong and Southern China, Hong Kong, 1977" (B. S. Morton, ed.), pp. 291–310. Hong Kong Univ. Press, Hong Kong.

Ricard, M., and Salvat, B. (1977). Faeces of *Tridacna maxima* (Mollusca–Bivalvia), composition and coral reef importance. *Proc. Int. Coral Reef Symp. 3rd, 1977* pp. 495–501.

Richard, G. (1976). Étude geomorphologique et biologique de L'atoll fermé de Taiaro (Tuamoto, Polynésie Francaise). V. Transport de materiaux et évolution Récente de la faune malacologique lagunaire de Taiaro. *Cah. Pac.* **19,** 265–282.

Richard, G. (1977). Quantitative balance and production of *Tridacna maxima* in the Takapoto Lagoon (French Polynesia). *Proc. Int. Coral Reef Symp., 3rd, 1977* pp. 599–605.

Richard, G. (1982). A first evaluation of the findings on the growth and production of lagoon and reef molluscs in French Polynesia. *Proc. Int. Coral Reef Symp., 4th, 1981* **2,** 637–641.

Richard, G., Salvat, B., and Millous, O. (1979). Mollusques et faune benthique du lagon de Takapoto. *J. Soc. Oceanist.* **35,** 59–68.

Robertson, P. B. (1963). A survey of the marine rock-boring fauna of Southeast Florida. M.Sc. Thesis, University of Miami, Coral Gables.

Robertson, R. (1963). The mollusks of British Honduras. *Proc. Philadelphia Shell Club* **1,** 15–20.

Robertson, R. (1970). Review of the predators and parasites of stony corals, with special reference to symbiotic prosobranch gastropods. *Pac. Sci.* **24,** 43–54.

Rodriguez, G. (1959). The marine communities of Margarita Island, Venezuela. *Bull. Mar. Sci.* **9,** 237–280.

Rosewater, J. (1961). The family Pinnidae in the Indo-Pacific. *Indo-Pac. Mollusca* **1**(4), 175–226.

Rosewater, J. (1965). The family Tridacnidae in the Indo-Pacific. *Indo-Pac. Mollusca* **1**(6), 347–396.

Rosewater, J. (1980). Changes in shell morphology of post larval *Tridacna gigas* Linne (Bivalvia: Heterodonta). *Bull. Am. Malacol. Union* pp. 45–48.

Rosewater, J. (1982). A new species of *Hippopus* (Bivalvia: Tridacnidae). *Nautilus* **96,** 3–6.

Salvat, B. (1967). Importance de la faune malacologique dans les atolls Polynésiens. *Cah. Pac.* **11,** 7–49.

Salvat, B. (1969). Dominance biologique de quelques mollusques dans les atolls fermes (Tuamotu, Polynesia); phénomène recent–conséquences actuelles. *Malacologia* **9,** 187–189.

Salvat, B. (1971). Evaluation quantitative totale de la faune benthique de la bordure lagunaire d'un atoll de Polynesia Française. *C.R. Hebd. Seances Acad. Sci.* **272,** 211–214.

Salvat, B. (1972). La faune benthique de lagon de l'atoll de Reao (Tuamotu, Polynésie). *Cah. Pac.* **16,** 29–110.

Salvini-Plawen, L. V. (1972). Cnidaria as food-sources for marine invertebrates. *Cah. Biol. Mar.* **13,** 385–400.

Schoenberg, D. A., and Trench, R. K. (1980a). Genetic variation in *Symbiodinium (= Gymnodinium) microadriaticum* Freudenthal, and specificity in its symbiosis with marine invertebrates. I. Isoenzyme and soluble protein patterns of axenic cultures of *Symbiodinium microadriaticum. Proc. R. Soc. London, Ser. B* **207,** 405–427.

Schoenberg, D. A., and Trench, R. K. (1980b). II. Morphological variation in *Symbiodinium microadriaticum. Proc. R. Soc. London, Ser. B* **207,** 429–444.

Schoenberg, D. A., and Trench, R. K. (1980c). III. Specificity and infectivity of *Symbiodinium microadriaticum. Proc. R. Soc. London, Ser. B* **207,** 445–460.

Scott, P. J. B. (1980). Associations between scleractinians and coral boring molluscs in Hong Kong. *In* "Proceedings of the First International Workshop on the Malacofauna of Hong Kong and Southern China, Hong Kong, 1977" (B. S. Morton, ed.), pp. 121–128. Hong Kong Univ. Press, Hong Kong.

Smith, B. D. (1978). Field observations of the gastropod and bivalve molluscs of Yap. *Tech. Rep.—Univ. Guam Mar. Lab.* **45,** 72–80.

Solem, A. (1954). Living species of the pelecypod family Trapeziidae. *Proc. Malacol. Soc. London* **31,** 64–84.

Soliman, G. N. (1969). Ecological aspects of some coral-boring gastropods and bivalves of the Northwestern Red Sea. *Am. Zool.* **9,** 887–894.

Soliman, G. N. (1971). On a new clavagellid bivalve from the Red Sea. *Proc. Malacol. Soc. London* **39,** 389–397.

Soliman, G. N. (1973). On the structure and behaviour of the rock-boring bivalve *Rocellaria retzii* (Deshayes) from the Red Sea. *Proc. Malacol. Soc. London* **40,** 313–318.

Soot-Ryen, T. (1969). Superfamily Mytilacea Rafinesque 1815. *In* "Treatise on Invertebrate Paleontology" (R. C. Moore, ed.), Part N, Mollusca 6, pp. N271–N280. Geol. Soc. Am. and Univ. of Kansas Press, Lawrence.

Spry, J. F. (1964). The sea-shells of Dar es Salaam. Pt. II. Pelecypoda (Bivalves). *Tanganyika Notes Rec.* **63,** 1–41.

Stasek, C. R. (1961). The form, growth and evolution of the Tridacnidae (giant clams). *Arch. Zool. Exp. Gen.* **101,** 1–40.

Stasek, C. R. (1965). Behavioural adaptation of the giant clam *Tridacna maxima* to the presence of grazing fishes. *Veliger* **8,** 29–35.

Stasek, C. R. (1966). The eye of the giant clam (*Tridacna maxima*). *Occas. Pap. Calif. Acad. Sci.* **58,** 1–9.

Stephenson, A. (1934). The breeding of reef animals. Part II. Invertebrates other than corals. *Sci. Rep. Great Barrier Reef Exped.* **3,** 247–272.

Stojkovich, J. O. (1977). Survey and species inventory of representative pristine marine communities on Guam. *Tech. Rep.—Univ. Guam Mar. Lab.* **40,** 1–183.

Sumadhiharga, O. K. (1971). A preliminary study on the ecology of the reef coral of Pombo Island. *Mar. Res. Indonesia* **17,** 29–49.

Taylor, D. L. (1969). Identity of zooxanthellae isolated from some Pacific Tridacnidae. *J. Phycol.* **5,** 336–340.

Taylor, D. L. (1973). The cellular interactions of algal–invertebrate symbiosis. *Adv. Mar. Biol.* **11,** 1–56.

Taylor, D. L. (1974). Symbiotic marine algae: taxonomy and biological fitness. *In* "Symbiosis in the Sea" (W. B. Vernberg, ed.), pp. 245–262. Univ. of South Carolina Press, Columbia.

Taylor, J. D. (1968). Coral reef and associated invertebrate communities (mainly molluscan) around Mahe, Seychelles. *Philos. Trans. R. Soc. London Ser. B* **254,** 129–206.

Taylor, J. D. (1971a). Observations on the shallow-water marine fauna. *Atoll Res. Bull.* **149,** 31–39.

Taylor, J. D. (1971b). Marine Mollusca from Diego Garcia. *Atoll. Res. Bull.* **149,** 105–125.

Taylor, J. D. (1971c). Reef associated molluscan assemblages in the western Indian Ocean. *Symp. Zool. Soc. London* **28,** 501–534.

Taylor, J. D. (1976). Habits, abundance and diets of muricacean gastropods at Aldabra Atoll. *J. Linn. Soc. London, Zool.* **59,** 155–193.

Taylor, J. D. (1977). Food and habitats of predatory gastropods on coral reefs. *Rep. Underwater Assoc.* **2** (new ser.), 17–34.

Taylor, J. D. (1978). Habitats and diet of predatory gastropods at Addu Atoll, Maldives. *J. Exp. Mar. Biol. Ecol.* **31,** 83–103.

Taylor, J. D. (1980). Diets and habits of shallow water predatory gastropods around Tolo Channel, Hong Kong. *In* "Proceedings of the First International Workshop on the Malacofauna of Hong Kong and Southern China, Hong Kong, 1977" (B. S. Morton, ed.), pp. 163–180. Hong Kong Univ. Press, Hong Kong.

Thomassin, B. A., and Ganelon, P. (1977). Molluscan assemblages on the boulder tracts of Tulear coral reefs (Madagascar). *Proc. Int. Coral Reef. Symp., 3rd, 1977* pp. 247–252.

Trench, R. K (1979). The cell biology of plant–animal symbiosis. *Annu. Rev. Plant Physiol.* **30,** 485–532.

Trench, R. K., Wethey, D. S., and Porter, J. W. (1981). Observations on the symbiosis with zooxanthellae among the Tridacnidae (Mollusca: Bivalvia). *Biol. Bull. (Woods Hole, Mass.)* **161,** 180–198.

Turner, R. D. (1955). The family Pholadidae in the western Atlantic and eastern Pacific. Pt. 2. Martesiinae, Jouannetiinae and Xylophaginae. *Johnsonia* **3,** 65–160.

Turner, R. D., and Boss, K. J. (1962). The genus *Lithophaga* in the western Atlantic. *Johnsonia* **4,** 81–115.

Vaillant, L. (1865). Recherches sur la faune malacologique de la baie de Suez. *J. Conchyliol.* **13,** 97–127.

Vermeij, G. J. (1974). Molluscs in mangrove swamps: physiognomy, diversity and regional differences. *Syst. Zool.* **22,** 609–624.

Wada, S. K. (1942). Notes on the tridacnid clams in Palau. *Sci. South Seas* **5,** 62–69.

Wada, S. K. (1954). Spawning in the tridacnid clams. *Jpn. J. Zool.* **11,** 273–285.

Waller, T. R. (1972). The Pectinidae (Mollusca: Bivalvia) of Eniwetok Atoll, Marshall Islands. *Veliger* **14,** 221–264.

Warme, J. E., and Marshall, N. F. (1969). Marine borers in calcareous terrigenous rocks of the Pacific coast. *Am. Zool.* **9,** 765–774.

Warmke, G. L., and Abbott, R. T. (1961). "Caribbean Seashells." Livingston Publ., Narbeth Pennsylvania.

Wilson, B. R. (1979). A revision of Queensland lithophagine mussels (Bivalvia: Mytilidae: Lithophaginae). *Rec. Aust. Mus.* **32,** 435–489.

Yamaguchi, M. (1977). Conservation and culture of small giant clams in the tropical Pacific. *Biol. Conserv.* **11,** 13–20.

Yamamoto, G., and Habe, T. (1961). *Nipponomontacuta actinariophila* gen. et sp. nov., a new commensal bivalve of the sea anemone. *Publ. Seto Mar. Biol. Lab.* **9,** 265–266.

Yonge, C. M. (1936). Mode of life feeding, digestion and symbiosis with zooxanthellae in the Tridacnidae. *Sci. Rep. Great Barrier Reef Exped.* **1,** 283–321.

Yonge, C. M. (1946). On the habits and adaptations of *Aloidis (Corbula) gibba. J. Mar. Biol. Assoc. U.K.* **26,** 358–376.

Yonge, C. M. (1949). On the structure and adaptations of the Tellinacea, deposit-feeding Eulamellibranchia. *Philos. Trans. R. Soc. London, Ser. B* **234,** 29–76.

Yonge, C. M. (1953). Mantle chambers and water circulation in the Tridacnidae (Mollusca). *Proc. Zool. Soc. London* **123,** 551–561.

Yonge, C. M. (1955). Adaptation to rock boring in *Botula* and *Lithophaga* (Lamellibranchia, Mytilidae) with a discussion on the evolution of this habit. *Q. J. Microsc. Sci.* **96,** 383–410.

Yonge, C. M. (1958). Observations on *Petricola carditoides* (Conrad). *Proc. Malacol. Soc. London* **33,** 25–31.

Yonge, C. M. (1963). Rock boring organisms. *In* "Mechanisms of Hard Tissue Destruction" (R. F. Sognnaes, ed.), Publ. 75, pp. 1–24. Am. Assoc. Adv. Sci., Washington, D.C.

Yonge, C. M. (1967a). Form, habit and evolution in the Chamidae (Bivalvia) with reference to conditions in the rudists (Hippuritacea). *Philos. Trans. R. Soc. London, Ser. B* **252,** 49–105.

Yonge, C. M. (1967b). Observations on *Pedum spondyloideum* (Chemnitz) Gmelin, a scallop associated with reef building corals. *Proc. Malacol. Soc. London* **37,** 311–323.

Yonge, C. M. (1968). Form and habit in species of *Malleus* (including the 'hammer oysters') with comparative observations on *Isognomon. Biol. Bull. (Woods Hole, Mass.)* **135,** 378–405.

Yonge, C. M. (1969). Function, morphology and evolution within the Carditacea (Bivalvia). *Proc. Malacol. Soc. London* **38,** 493–527.

Yonge, C. M. (1973). Functional morphology with particular reference to hinge and ligament in *Spondylus* and *Plicatula* and a discussion on relations within the superfamily Pectinacea (Mollusca: Bivalvia). *Philos. Trans. R. Soc. London, Ser. B* **267,** 173–208.

Yonge, C. M. (1974). Coral reefs and molluscs. *Trans. R. Soc. Edinburgh* **69,** 147–166.

Yonge, C. M. (1975). Giant clams. *Sci. Am.* **232,** 96–105.

Yonge, C. M. (1977). Form and evolution in the Anomiacea—*Pododesmus* (Monia), *Anomia, Patro, Enigmonia* (Anomiidae); *Placunanomia, Placuna* (Placunidae Fam. Nov.). *Philos. Trans. R. Soc. London, Ser. B* **276,** 453–523.

Yonge, C. M. (1979). Cementation in bivalves. *In* "Pathways in Malacology" (S. van der Spoel, A.c.van Bruggen, and J. Lever, eds.), pp. 83–106. Bohn, Scheltema & Holkema, Utrecht.

Yonge, C. M. (1980). Functional morphology and evolution in the Tridacnidae (Mollusca: Bivalvia: Cardiacea). *Rec. Aust. Mus.* **33,** 735–777.

Yonge, C. M. (1981). On adaptive radiation in the Pectinacea with a description of *Hemipecten forbesianus. Malacologia* **21,** 23–34.

5

Reproductive and Trophic Ecology of Nudibranch Molluscs

CHRISTOPHER D. TODD

Department of Marine Biology
Gatty Marine Laboratory
The University of St. Andrews
St. Andrews, Fife, Scotland

I. Introduction

As a result of the evolutionary loss of the restrictive but nonetheless protective gastropod shell and operculum, the nudibranch opisthobranchs demonstrate perhaps the greatest morphological and ecological diversity of any group of invertebrates of ordinal status. Although primarily a macrofaunal marine epibenthic group, the nudibranchs have also successfully invaded both the pelagic oceanic (see Cheng, 1975, for review) and interstitial benthic meiofaunal habitats (Swedmark, 1964). Loss of the adult shell has necessitated the evolution of elaborate adult defensive mechanisms, although the free-living larval stages have retained the shell. Success of the group primarily resides with this very diversity, although many species may be of at least transient quantitative importance in some benthic communities. It is perhaps their apparent scarcity which has led to their frequent neglect by ecologists.

The Mollusca, Vol. 6
Ecology

ISBN 0-12-751406-6

The bulk of the ecological literature relating to the nudibranch molluscs has been reviewed (Harris, 1973; Todd, 1981), and an outline of the systematics of the order Nudibranchia is included in the latter article. Briefly, the "true nudibranchs" (as opposed to other shell-less opisthobranchs, e.g., order Sacoglossa) comprise four suborders; Dendronotacea, Doridacea, Arminacea, and Aeolidacea. Almost without exception, they are exclusive predators of sessile marine invertebrates. The Dendronotacea are the most primitive (Thompson, 1976), and are characterized by an elongate body form, sheathed retractile rhinophores [anterior cephalic tentacles which may be chemosensory and/or rheosensory (Wolter, 1967; Storch and Welch, 1969)], and a variously elaborated dorsum. Evolutionary loss of the shell and thereby the mantle cavity (with its accompanying ctenidium) has invariably necessitated the development of "secondary gills," or at least epidermal proliferations of the dorsum to provide for gaseous exchange, among these molluscs. Within the Dendronotacea, such gills range from the extensive feather-like structures of *Dendronotus* and *Tritonia* to the simple disc-shaped plates exhibited by *Melibe*.

The Doridacea are typically somewhat flattened with a broad, rounded foot and a similarly shaped dorsum. With the exception of the extremely unusual tropical species *Okadaia elegans* (Baba, 1937), the dorids possess complex secondary gills surrounding the anus in a middorsal position which may be retractile and contractile into a protective pouch. Problems of fouling of the gills and dorsum, and indeed of providing a respiratory flow of water, are largely overcome by the activity of epidermal cilia which cover the entire body surface. The constant secretion of mucus and its movement across the body surface also prevent fouling. Many tropical chromodorids constantly undulate the branchial crown, presumably in order to enhance respiratory exchange.

Within the Doridacea, an evolutionary tendency toward reduction of the mantle and its accompanying calcareous endoskeletal spicules in conjunction with elongation of the body is detectable, (e.g., Chromodoridae) (Fig. 1) and in some cases (e.g., Polyceridae) with the development of dorsal processes resembling the cerata which characterize the Aeolidacea. The Dendronotacea are exclusively coelenterate predators, and the Doridacea prey upon a wide range of encrusting sessile invertebrates, especially sponges, bryozoans, and ascidians.

The suborder Arminacea may require some systematic revision, but currently it encompasses a heterogeneous assemblage of genera, including the dorid-like *Armina* (Fig. 1) and *Phyllidia* (with rows of secondary *gill lamellae* along the dorsal undersurface of the pallial groove) and *Antiopella* (an aeolid-like genus with a middorsal posterior anus).

Evolution of the fourth nudibranch suborder, the Aeolidacea, is closely related to their invariable association with coelenterates as prey organisms; indeed, the occurrence of aeolid–zooxanthellae symbioses (Rudman, 1981a,b) in some tropical species testifies to this. In general the aeolid body is elongate, with a narrow foot and long oral tentacles and rhinophores. Secondary gills are absent, but the *cerata* (finger-like projections of the dorsum that contain the diverticula of the digestive gland) preclude their necessity by virtue of the increased body surface area : volume ratio. Whereas jaws only occasionally accompany the typical molluscan radula in the previous suborders, the Aeolidacea all possess a pair of large chitinous jaws in conjunction with a simple radula.

Figure 1 illustrates the morphological diversity of this opisthobranch order. It is apparent that this is primarily attributable to the respective prey associations of the four suborders. Thus, the aeolids, with their elongate body form, slender foot, and increased buoyancy are primarily associated with erect coelenterates; the dorids, by contrast, with their squat, rounded, frequently heavily spiculate bodies, are typically predators of sessile epibenthic encrusting organisms such as sponges and bryozoans. Furthermore, it is only to be expected that the truly pelagic oceanic (e.g., *Glaucus*) and interstitial meiobenthic (e.g., *Pseudovermis*) nudibranchs should be included in the Aeolidacea, because this suborder lacks endoskeletal spicules and shows the most streamlined profile.

A major feature of the ecology of these molluscs to be emphasized in this chapter is the frequently specific predator–prey associations displayed. A review of the diets of the British nudibranch fauna (Todd, 1981) shows that approximately 50% of the species are associated with one, or perhaps two, prey species. Clearly, the maintenance of such monophagous predator–prey associations depends upon the success of the reproductive propagules in locating and exploiting temporally and spatially transient resources; in this context, it is relevant to note that the majority of nudibranch species reproduce by means of dispersive, pelagic, planktotrophic veliger larvae. In possessing planktonic larvae that will settle only in the presence of the (adult) benthic diet, however, the major problem confronting the benthic adult, notably location of a conspecific mate and copulation, is ameliorated by virtue of this more or less specific "associative settlement." A major objective of this chapter is to emphasize the essential synergism between the reproductive and trophic ecology of these molluscs.

II. Reproductive Ecology and Larval Biology

All nudibranchs are hermaphrodite and oviparous, with internal fertilization resulting from reciprocal exchange of spermatozoa between copulating individuals. It is probable that all species are simultaneous hermaphrodites (Fig. 2), although it has been noted that the appearance of mature spermatozoa slightly precedes that of mature oocytes at least in the dorid *Onchidoris muricata* (Todd, 1978). Insofar as single-species populations of nudibranchs are generally small, discontinuously distributed, and composed of individuals of low motility, the simultaneous hermaphroditism of these molluscs might be considered a good

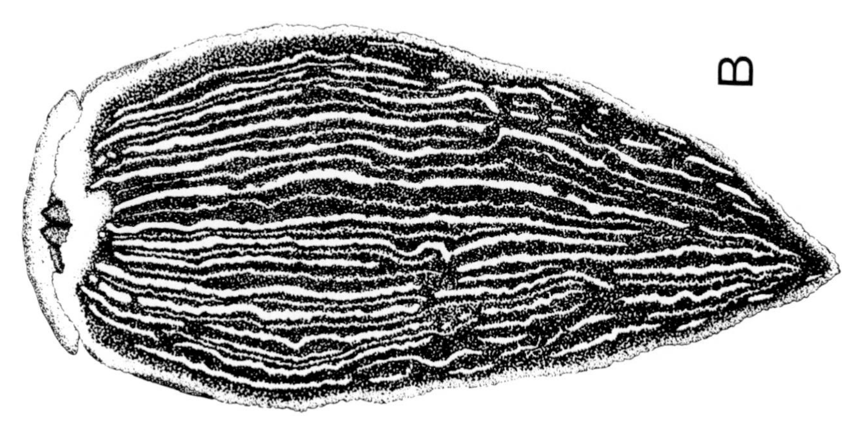
B

A

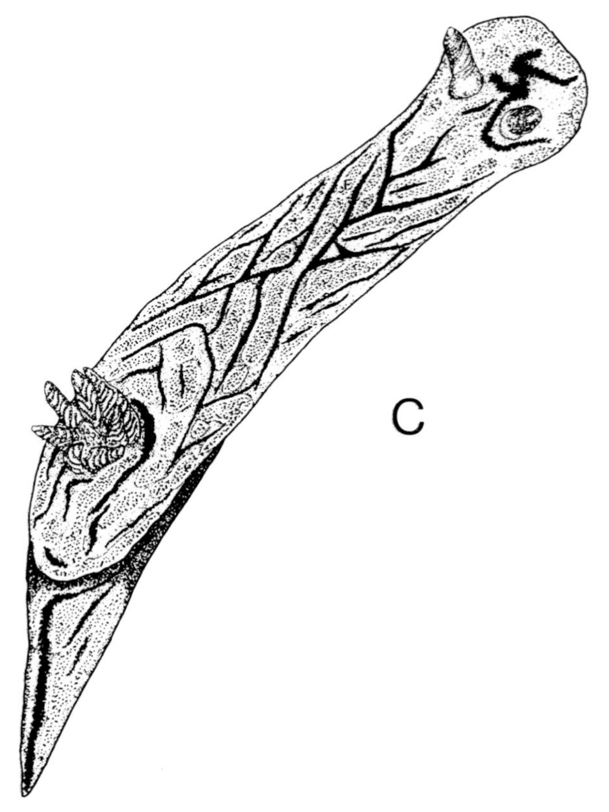

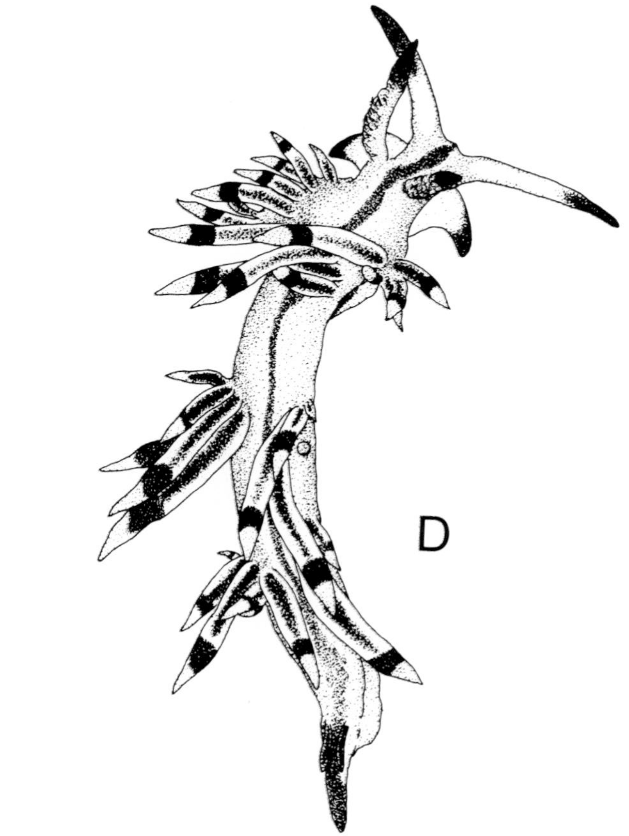

Fig. 1. Examples of the four nudibranch suborders: (**A**) Dendronotacea (40 mm), (**B**) Arminacea (30 mm), (**C**) Doridacea (25 mm), and (**D**) Aeolidacea (20 mm). (A and B from Todd, 1981.)

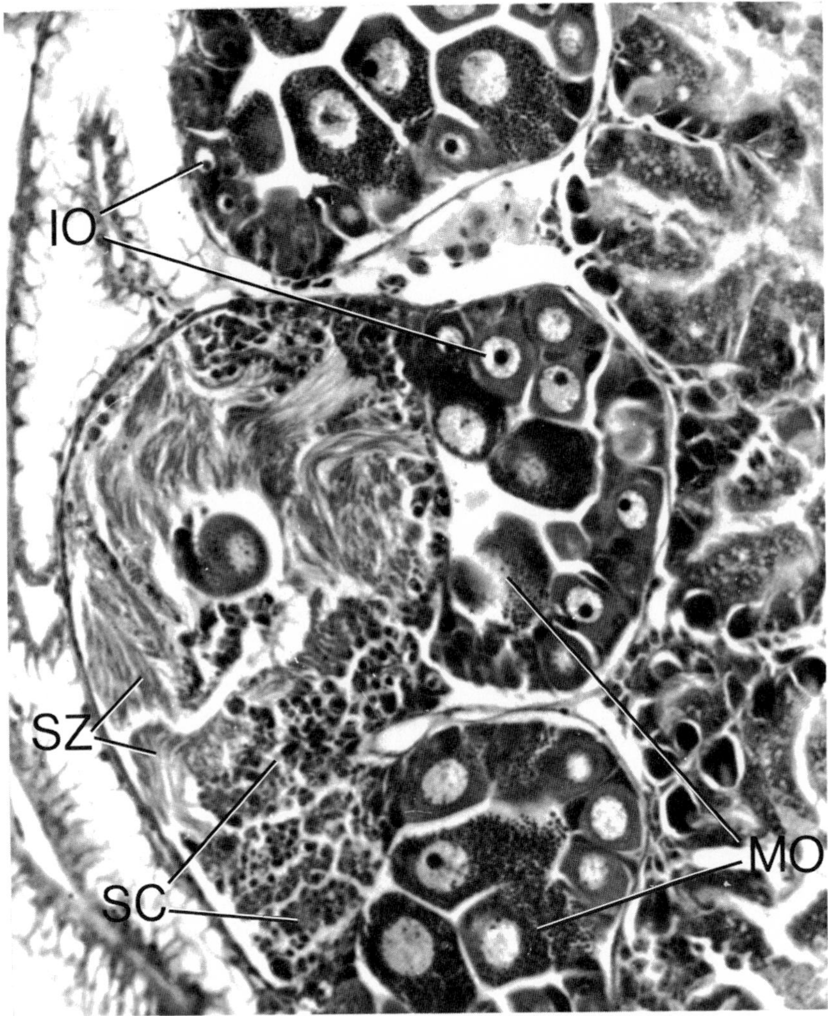

Fig. 2. Vertical longitudinal sections through the posterior ovotestis of the dorid *Onchidoris muricata* (Müller). All stages of oogenesis and spermatogenesis are present: MO (mature oocytes); IO (immature oocytes); SZ (spermatozoa); SC (spermatocytes, primary and secondary). The proximal male (♂) and distal female (♀) parts of the acini are also visible. Scale: 100 μm.

illustration of the low-density model alluded to by Tomlinson (1966) and augmented by Ghiselin (1969) when attempting to explain the evolution of hermaphroditism.

Because the reproductive system may simultaneously contain female gametes at various stages of development (from oogonia to fertilized ova in the egg-

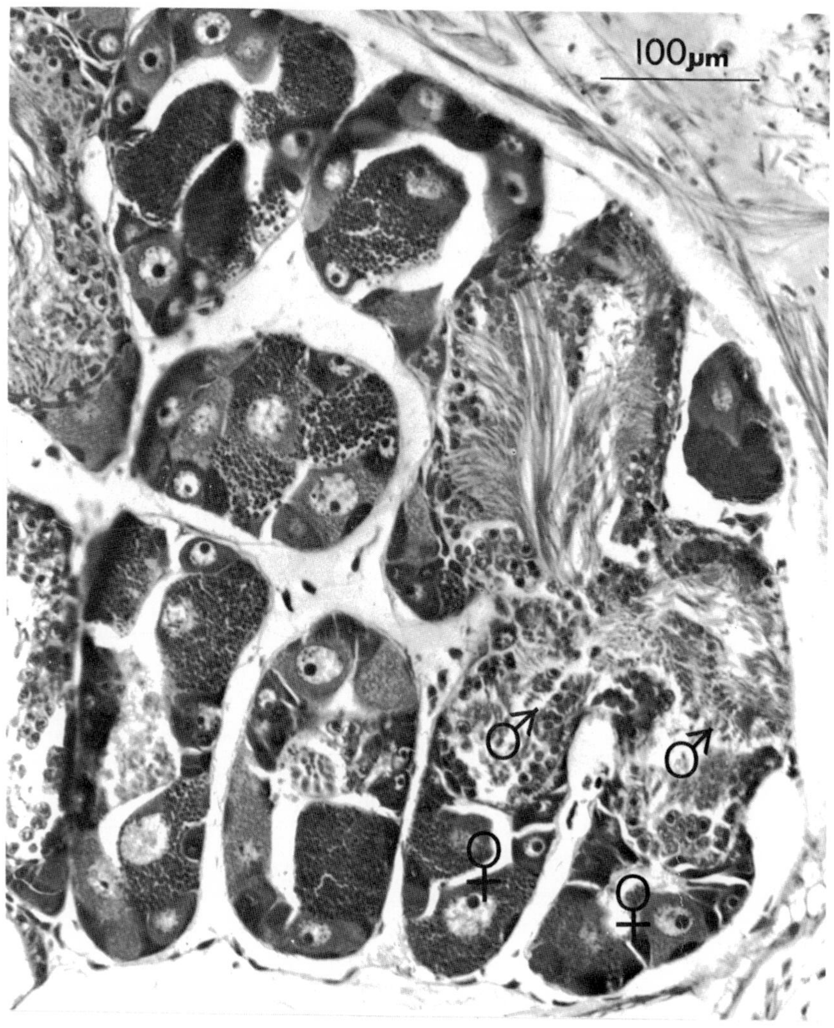

string), autosperms (from spermatogonia to mature spermatozoa), and mature allosperms (''foreign'' sperms), it is no surprise that nudibranchs, to preclude self-fertilization, have ''extraordinarily complex reproductive structures'' (Ghiselin, 1966; Beeman, 1977). Both Thompson (1976) and Beeman (1977) provide exhaustive reviews of the structure and function of nudibranch reproductive systems.

From an ecological standpoint, it is important to consider reproduction in nudibranchs in terms of the duality conferred by their sexuality. For example, the female role of the individual involves both oviposition (of fertilized ova) and the long-term storage of mature allosperms, whereas the male role consists solely of insemination of another mature individual. As is discussed below, natural selection maximizes the individual's female role, although circumstances may prevail in which predominantly male behavior is expedient. In the present context, it is relevant that spawned-out adults are still capable of spermatogenesis (Todd, 1978). This implies that individuals may continue as functional males within the population for a short period prior to death at the completion of reproduction.

In some forms, notably certain aeolid genera, the male and female gametes are formed in separate acini (or lobes) of the ovotestis (Beeman, 1977). More usually, gametogenesis occurs within the same acinus [e.g., *Tritonia hombergi* (Thompson, 1961)], with the female system distal and the male proximal in disposition (Fig. 2). The ovotestis always overlies the digestive gland and the two structures, although distinct, are intimately associated. Copulation occurs with the pair of nudibranchs in a characteristic configuration, facing in opposite directions with their right anterior body surfaces in intimate contact throughout. Copulation may continue for several hours, or even days [*e.g., Adalaria proxima* (C. D. Todd, personal observation)]. Exchange of spermatozoa is mutual and probably simultaneous, with the exception of the arminid *Melibe* (Franc, 1968), in which transfer is unilateral.

Allosperms received during copulation are stored by the nudibranch in a seminal vesicle of one form or another depending upon the species. Thus, in the dendronotid *T. hombergi* for example, allosperms are stored in the *bursa copulatrix* (gametolytic gland) (Thompson, 1961), whereas in the dorid *Archidoris pseudoargus* they are stored in the *receptaculum seminis* (Thompson, 1966).

The consensus of current opinion (Beeman, 1977) is that self-sterility is maintained by spermatozoa that are incapable of fertilization until "capacitated" within the (storing) seminal receptacle (Thompson, 1961). Mature oocytes, on leaving the acinus, are moved along the hermaphroditic duct by ciliary action and are fertilized in a specialized chamber (often the albumin gland). Following fertilization the ova are encapsulated, either singly or in groups, passed through a more or less complex mucus gland, and coated with a gel composed largely of galactogen with some proteinaceous material. The mucous string then passes to the genital aperture, through which it is extruded, and adhered to the substratum (or, more frequently, the adult prey organism) by movements of the foot. Neither parental protection nor brooding of spawn-masses has been recorded. Similarly, there are no recorded instances of asexual reproduction [i.e., polyembryony or parthenogenesis) (Beeman, 1977)].

The form of the spawn-mass is often species specific and falls into one of three categories, as defined by Hurst (1967). Type A spawn-masses (which are characteristic of dorids) are in the form of long ribbons attached along one edge. The egg capsules are frequently distributed in a tight helical coil oriented horizontally to the substratum and firmly embedded within the gel matrix. Type B spawn-masses (characteristic of dendronotids, arminids, and some aeolids) are a hollow, cylindrical, capsule-filled cord attached along one side; the capsules are not embedded in a gel and may be lost if the cord is ruptured. Type D spawn-masses (exclusive to certain aeolid species) are small sac-like structures, sometimes reniform and attached along one side. The egg capsules are irregularly but firmly embedded throughout the matrix. (Type C has no nudibranch representatives.)

As a rule, those species in which embryonic development is the most protracted possess the most durable spawn matrices. For example, *Tritonia plebeia* (spawn-mass type B) hatches in 10 days at 10°C (Thompson, 1967) and has a very fragile cord gel; *T. hombergi* (also spawn-mass type B), which requires 36–38 days to hatch at 9–10°C (Thompson, 1961), has a considerably tougher and more durable cord matrix. This trend is exemplified in its most extreme by the Antarctic dorid *Austrodoris macmurdensis* (Gibson et al., 1970); this species reinforces the matrix with strong chitin-like and other proteinaceous materials. Alternatively, other species with extended embryonic development (e.g., *Adalaria proxima*) provide the eggs with the additional necessary protection by simply increasing the quantity of gel associated with a given number of eggs.

Cleavage of fertilized ova proceeds through the typical molluscan spiral pattern, and gastrulation is by epiboly. The embryo passes through a trochophore-like stage before developing velar lobes and the formation of a coiled gastropod shell. Typically, the hatching stage of nudibranchs is a free-swimming shelled veliger larva. The ciliated bilobed locomotory velum also generates the feeding current (Thompson, 1959). Larval development, which is discussed in detail below, is both morphologically and ecologically the most complex and traumatic stage of the entire life cycle. At metamorphosis the larva is changed in a matter of hours from a shelled, torted, swimming phytoplanktivore to a shell-less, detorted, benthic carnivore. These morphological and physiological transitions occur after the requisite feeding and/or further development in the obligatory pelagic phase and are ultimately possible only on acquisition of the correct biochemical stimuli at the appropriate time. Clearly, this is a most crucial phase of the life-history of these invertebrates.

Hatching occurs by release from the capsule and escape of the veliger through the gel matrix after its disruption by the "swimming" of the larvae. Release of veligers from the capsule may be effected by the action of the velar cilia on the capsule membrane (Perron and Turner, 1977), or by the larva taking the membrane into its mouth and rupturing it (e.g., *Phestilla* spp.) (Harris, 1973, 1975).

Most species possess pelagic planktotrophic larvae, which is generally considered the most primitive mode of development, although some species are lecithotrophic. These latter forms are only briefly pelagic and do not require extrinsic planktonic nutrition to complete their development. Many of those "directly developing" species, which hatch as benthic juveniles, do so by suppressing the pelagic lecithotrophic larva, although a very few species do undergo a specialized embryogenesis involving vestigial larval stages through which they pass entirely within the capsule.

A. Description of Larval Types

Planktotrophic larvae hatch from small (approximately 100 μm) eggs which develop comparatively rapidly (see Todd and Doyle, 1981, for review). Embryonic development time depends largely upon egg size and temperature (and thereby geographical locality); for example, the embryonic (i.e., intracapsular) phase of the dorid *Doridella obscura* (northeastern United States) ranges from 4 to 20+ days at temperatures between 25 and 4°C, respectively (Perron and Turner, 1977). Planktotrophic veligers are comparatively poorly developed, having only a small foot, no propodium, and no eyes (Fig. 3). Feeding, growth, and further development proceed in the plankton, although the relative dietary importance of phytoplankton, bacteria, particulate and/or dissolved organic carbon, and other dissolved nutrients is as yet unknown. Competence to metamorphose is indicated externally by the presence of a well-developed foot and propodium, and of eyes after an obligatory pelagic phase (during which the larva is incapable of metamorphosis) of varying duration depending upon species and temperature. At this juncture, the larva is termed a *pediveliger* and is capable of intermittent crawling and swimming while searching for the correct stimuli to metamorphose.

Lecithotrophic veligers, which develop from intermediate-sized eggs (Todd and Doyle, 1981) of approximately 120–230 μm, are broadly similar to their planktotrophic counterparts except that at hatching they are larger, have a well-

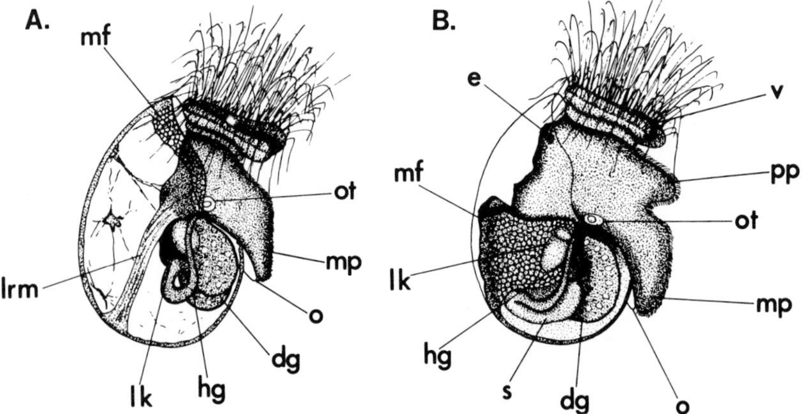

Fig. 3. (**A**) Planktotrophic and (**B**) lecithotrophic veligers of dorid nudibranchs. Both species have type 1 shells (length 166 and 290 μm, respectively); dg, digestive gland; e, eye; hg, hindgut; lk, larval kidney; lrm, larval retractor muscle; mf, mantle fold; mp, metapodium; o, operculum; ot, otocyst (= statocyst); pp, propodium; s, shell; v, veluma. (From Todd, 1981.)

developed foot, propodium, and eyes, and have noticeably large yolk reserves within the digestive gland. The velum is, however, rather poorly developed, and in some species the larva may swim only with difficulty, if at all [e.g., *Cuthona nana* (Rivest, 1978)]. Such larvae need not feed in the plankton to complete development, although many are capable of feeding, and they have an obligatory pelagic phase of a few hours to a few days during which they are again incapable of metamorphosis [e.g., *A. proxima* (Thompson, 1958)]. Apart from anything else, this ensures a limited amount of dispersal away from the adult microhabitat. Once initiated, metamorphosis goes to completion in a matter of hours.

Most directly developing species appear to have simply suppressed the pelagic phase of the lecithotrophic larva, with the consequence that the shell is cast soon after hatching, perhaps after a brief crawling phase. Some species may actually cast the shell within the capsule or spawn-mass gel and exit as benthic juveniles (Roginskaya, 1962a; Schmekel, 1966; Hamatani, 1967; Schönenberger, 1969). However, a few instances have been recorded of species developing from large (approximately 250–400 μm) eggs, undergoing vestigial larval phases within the capsule, and hatching as fully formed benthic juveniles. In the case of the northwestern European dorid *Cadlina laevis,* a vestigial veliger phase is characterized by a very thin transient shell and reduced velar lobes (Thompson, 1967). The metapodial mucus glands, nephrocysts, and larval retractor muscle (all characteristic of normal veligers) are absent, and at no time is the embryo capable of retracting into the shell. Furthermore, an operculum is never formed. During the 20 days (at 10°C) preceding hatching, the mantle fold reflexes over the shell which then degenerates. Hatching of the benthic juvenile occurs after approximately 50 days (at 10°C). The rhinophores and circumanal branchiae are absent at hatching but develop subsequently.

In the case of the northeastern North American aeolid *Coryphella salmonacea,* the uncleaved zygote is 250 μm in diameter and requires 52 days (4–5°C) to develop to hatching (Morse, 1971). As for *Cadlina,* an operculum never develops and the vestigial veliger possesses a shell only transiently. After resorption of the velar lobes, the first cerata appear, and by the time of hatching a fully formed aeolid with rhinophores, oral tentacles, and cerata is established. Feeding may commence immediately.

The classification of nudibranch shells has been extensively reviewed by Todd (1981), who established that two schemes of categorization are necessary to encompass all developmental modes. Thus, *larval* shells are divisible into types 1 and 2 (*sensu* Thompson, 1967), whereas *veliger* shells (which include the vestigial stages of the nonpelagic mode) are separable into types A, B, and C (*sensu* Vestergaard and Thorson, 1938; Thorson, 1946). Type A shells are shallow and cup like and lack whorls (Fig. 4), whereas type B (Thompson type 1) are tightly coiled sinistral shells with a large aperture. Type C (Thompson type 2) includes those that are inflated and egg shaped, with a small aperture and little

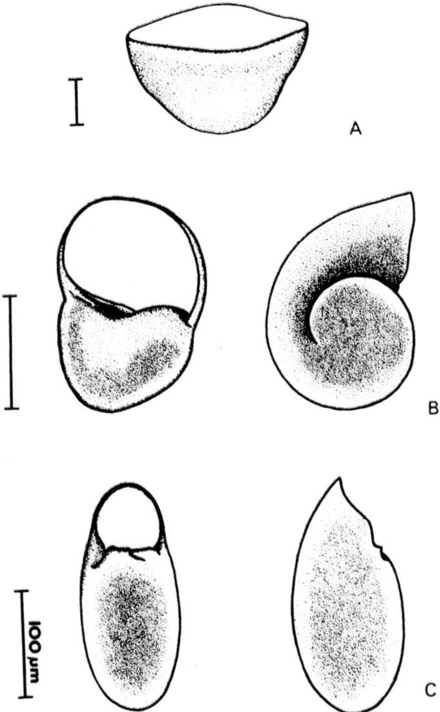

Fig. 4. Nudibranch veliger shells. Types A and B (= Thompson type 1), and C (= Thompson type 2), respectively. (From Todd, 1981.)

asymmetry. Type A shells are exhibited by only two directly developing tropical species, the veligers of which are clearly not adapted to a free-living pelagic existence (Soliman, 1977). Thompson's scheme, applied to larval stages only, permits the classification of all pelagic nudibranch larvae (Table I). Planktotrophic species are divisible into two groupings (short- and long-term) (Table I) according to the duration of the pelagic phase, which may range from a few days to several weeks or months. All short-term planktotrophic species have the inflated type 2 shell. In larvae with type 2 shells, the mantle fold is withdrawn from the shell aperture prior to hatching and hence further shell growth is not possible. However, a relatively large shell volume is available, which is sufficient to accommodate all pelagic somatic growth of the larva. All larvae with type 2 shells, whatever the reproductive strategy, have relatively small velar lobes and a long slender foot, both consequences of the restricted shell aperture. It is to be expected that such larval shells should characterize the Aeolidacea (with their narrow, elongate bodies), and this is indeed the case. For those aeolids which undergo a long-term growth and development phase in the plankton, a type 1 shell (which permits posthatching growth) is the rule.

TABLE I

Shell Type and Reproductive Strategies of Representative Nudibranch Species

Nudibranch species	Larval shell type (Thompson)	Veliger shell type (Thorson)	Suborder	Reference
Short-term planktotrophic species				
Eubranchus cingulatus	2	C	Aeolidacea	Tardy (1970)
Eubranchus exiguus	2	C	Aeolidacea	Fischer (1892)
Phestilla melanobranchia	2	C	Aeolidacea	Harris (1975)
Tergipes tergipes	2	C	Aeolidacea	Tardy (1964)
Long-term planktotrophic species				
Tritonia diomedea	1	B	Dendronotacea	Kempf and Willows (1977)
Doridella obscura	1	B	Doridacea	Perron and Turner (1977)
Onchidoris bilamellata	1	B	Doridacea	Todd (1981)
Rostanga pulchra	1	B	Doridacea	Chia and Koss (1978)
"Aeolid sp."	1	B	Aeolidacea	Tardy (1970)
Facelina coronata	1	B	Aeolidacea	Tardy (1970)
Hermissenda crassicornis	1	B	Aeolidacea	Harrigan and Alkon (1978)
Pelagic lecithotrophic species				
Dendronotus frondosus	1	B	Dendronotacea	Miller (1958)
Lomanotus stauberi	(?)1	(?)B	Dendronotacea	Clark and Goetzfried (1978)
Tritonia hombergi	1	B	Dendronotacea	Thompson (1962)
Adalaria proxima	1	B	Doridacea	Thompson (1958)
Armina tigrina	1	B	Arminacea	Eyster (1981)
Aeolidiella mannarensis	1	B	Aeolidacea	Rao and Alargarswami (1960)
Austraeolis catina	(?)2	(?)C	Aeolidacea	Clark and Goetzfried (1978)
Catriona adyarensis	2	C	Aeolidacea	Rao (1962)
Cuthona gymnota	2	C	Aeolidacea	Clark (1975)
Eubranchus doriae	2	C	Aeolidacea	Tardy (1962a)
Eubranchus farrani	2	C	Aeolidacea	Todd (1981)

(Continued)

TABLE I ***Continued***

Nudibranch species	Larval shell type (Thompson)	Veliger shell type (Thorson)	Suborder	Reference
Favorinus auritulus	2	C	Aeolidacea	Clark and Goetzfried (1978)
Phestilla lugubris	2	C	Aeolidacea	Harris (1975)
Phidiana lynceus	2	C	Aeolidacea	Clark and Goetzfried (1978)
Spurilla neapolitana	2	C	Aeolidacea	Clark and Goetzfried (1978)
Tenellia mediterranea	2	C	Aeolidacea	Vannuci and Hoscoe (1953)
Nonpelagic lecithotrophic species				
Dendrodoris krebsi	(?)1	(?)B	Doridacea	Clark and Goetzfried (1978)
Dermatobranchus striatellus	1	B	Doridacea	Hamatani (1967)
Doridopsis limbata	1	B	Doridacea	Tchang-Si (1931)
Doriopsilla sp.	(?)1	(?)B	Doridacea	Clark and Goetzfried (1978)
Doriopsilla pharpa	(?)1	(?)B	Doridacea	Clark and Goetzfried (1978)
Aeolidiella alderi	2	C	Aeolidacea	Tardy (1962b)
Cuthona granosa	2	C	Aeolidacea	Schmekel (1966)
Cuthona nana	2	C	Aeolidacea	Rivest (1978)
Cuthona pustulata	2	C	Aeolidacea	Roginskaya (1962a,b)
Tenellia pallida	2	C	Aeolidacea	Eyster (1979)
Directly developing species				
Cadlina laevis	"1"	"B"	Doridacea	Thompson (1967)
Casella obsoleta		A	Doridacea	Gohar and Soliman (1967)
Glossodoris sibogae		A	Doridacea	Usuki (1967)
Coryphella salmonacea	"2"	"C"	Aeolidacea	Morse (1971)

Shell growth of long-term planktotrophs may be considerable [e.g., 106–310 μm for *Hermissenda crassicornis* (Harrigan and Alkon, 1978) and 166–440 μm for *Onchidoris bilamellata* (Todd, 1981)], and the velum and shell aperture are correspondingly large. Only at the onset of metamorphosis (i.e., the termination of the obligatory pelagic phase) does the mantle fold retract from the shell aperture and growth cease (Perron and Turner, 1977; Harrigan and Alkon, 1978). For all forms of pelagic veliger, the head, velum, and foot are completely retractile within the protective shell, and the operculum closes the aperture.

B. Growth and Development of Pelagic Larvae

1. Planktotrophic Veligers

Most evidence and data for the growth and development of pelagic nudibranch larvae must necessarily depend upon laboratory studies, which are framed in a broader ecological context by extrapolation to the field situation. An exception is the study by Thiriot-Quievreux (1977) of the apparently unique developmental mode displayed by the northwestern European dorid *Aegires punctilucens*. Her study of *Aegires* is based largely upon plankton tows off Banyuls-sur-Mer and Villefranche-sur-Mer, France. *Aegires* hatches as a planktotrophic veliger with a type 1 shell 220 μm in length and is unusual in two respects: first, the larva possesses eyes and a pair of rhinophores at hatching; second, the velum is exceedingly large (300 μm in breadth). Shelled larvae were rarely collected, but it is clear that at a more advanced stage the mantle commences proliferation and the operculum and shell are rapidly cast in the plankton. The mantle undergoes further growth and develops spiculated tubercles, and the velum also continues to expand. Settlement and metamorphosis are preceded by growth of the foot and subsequent incorporation of the visceral mass during detorsion. Metamorphosis was induced in about a dozen larvae, not by the adult sponge diet, surprisingly, but by algal fragments.

A considerable body of data is now available on the development, growth, and metamorphosis of a wide range of nudibranch species. The most comprehensive accounts are of the aeolids *Phestilla melanobranchia* (Harris, 1975) and *Hermissenda crassicornis* (Harrigan and Alkon, 1978), the dendronotid *Tritonia diomedea* (Kempf and Willows, 1977), and the dorids *Doridella obscura* (Perron and Turner, 1977), *Rostanga pulchra* (Chia and Koss, 1978), *Doridella steinbergae* (Bickell and Chia, 1979), and *Onchidoris bilamellata* (Todd, 1981). The specific details of feeding regimen, temperature of culture, use of antibiotics, and developmental timetable have been comprehensively reviewed by Todd (1981) and need not concern us here, but a brief account of the principal events during ontogeny and metamorphosis will be given.

Newly hatched planktotrophic larvae are strongly photopositive (and geonegative) and rise to the surface of culture vessels. The strength of this response wanes after a few days and is perhaps also affected by an increase in body size and a concomitant reduction in swimming ability.

Growth of the shell usually ceases a few days prior to settlement (Perron and Turner, 1977; Kempf and Willows, 1977; Harrigan and Alkon, 1978; Chia and Koss, 1978; Todd, 1981) and is attributable to retraction of the mantle fold from the shell aperture. The remaining days of the pelagic phase thus appear to be a period of somatic preparation for metamorphosis (competence). Interestingly, Bickell and Chia (1979) noted the cessation of shell growth in *D. steinbergae* as early as day 12 of the 27-day pelagic phase, although somatic growth and development did continue. They interpreted this as a partitioning of larval energy to somatic growth, preparing the larva for benthic life as opposed to obtaining a larger size. Clearly, the determinant of such a bias is the timing of mantle retraction. The significance of this, discussed below, undoubtedly is important in terms of the relative sizes of the postmetamorph nudibranch and the prey organism upon which it must subsequently feed.

Competence to metamorphose is visibly indicated by the development of the foot, notably the propodium, and the appearance of eyes in addition to the retraction of the mantle. At this stage, intermittent swimming and crawling are possible as the larva seeks to acquire the correct stimuli for metamorphosis.

Bickell and Chia (1979) and Bickell et al. (1981) have described the morphogenetic processes involved in growth, development, and metamorphosis in the planktotrophic veliger of *D. steinbergae* in great histological detail. From an ecological standpoint, perhaps the most important feature in nudibranch development (aside from the complexities of converting a torted swimming larva to a detorted benthic juvenile) is the morphogenesis of the alimentary tract, notably the conversion of a (pelagic) phytoplanktivore to a (benthic) carnivore. Thus, for example, the gastric shield, style sac, and intestinal groove of the larva are all destroyed at metamorphosis, and this part of the gut is reorganized to deal with the benthic bryozoan prey. Histochemical studies are still lacking but will presumably demonstrate associated biochemical and physiological adaptations to the corresponding larval and juvenile dietary regimens.

Data defining the extent to which metamorphosis might be delayed in the absence of the correct stimuli are largely absent from the literature, although Bickell and Chia (1979) observed that such a delay may exceed 20 days postcompetence in *D. steinbergae;* the obligatory phase is 27 days at the same temperature. Certainly, the ability to delay metamorphosis will be of paramount importance in maintaining the integrity of specific predator–prey associations, particularly if the adult microhabitat is discontinuously distributed.

Metamorphosis of nudibranch veligers comprises a structured sequence of constructive and destructive events which obviously involve considerable cellular and tissue reorganization. Thompson (1976, p. 90) outlines the broad sequence of events, and many other studies provide specific detailed analyses (e.g., Bonar and Hadfield, 1974; Bickell and Chia, 1979; Bickell et al., 1981). Briefly, resorption of the velar lobes, casting of the shell and operculum, reflexion of the mantle fold, detorsion of the alimentary tract, and incorporation of the visceral mass within the foot occur. The latter stages also imply an elongation of body form followed by the development of rhinophores, oral tentacles, and cerata (in aeolids), in addition to extensive reorganization of the gut preparatory to the carnivorous habit. Once the correct stimuli have been acquired by the crawling pediveliger, the larva becomes effectively committed to the benthic habit. Commencement of resorption of the velum evidently precludes further swimming and is an irreversible process. In the case of the barnacle-eating dorid *O. bilamellata* (Todd, 1981), it was found that larvae permitted to crawl over barnacles for only a few minutes before being isolated in watch glasses underwent subsequent metamorphosis. This indicates that once the correct sequence or combination of stimuli has been acquired, continued barnacle contact is not necessary: that is, a "trigger" mechanism is implied.

Most reports (Bonar and Hadfield, 1974; Perron and Turner, 1977; Kempf and Willows, 1977; Harrigan and Alkon, 1978) indicate that metamorphosis is largely complete within 24–48 h, although benthic feeding may not commence for a further 4–5 days (Kempf and Willows, 1977; Harrigan and Alkon, 1978). Thus, from an ecological viewpoint, it is clearly important to recall that this period of reorganization precludes any feeding activity and hence must be accomplished by utilizing previously acquired reserves.

2. *Lecithotrophic Veligers*

Both larval shell types are represented among the lecithotrophic species, and typically the veliger hatches at an advanced state of development with a pronounced propodium, eyes, radula, and statocysts (Fig. 3). No shell growth occurs in the pelagic phase, which is, by definition, brief. Feeding in the plankton is not necessary for the completion of development, although many species are capable of feeding (Thorson, 1950; Thompson, 1958, 1959; Ockelmann, 1965). Lecithotrophic larvae probably do feed in the obligatory phase and any such nutritional gain will enhance individual survival during and immediately after the (nonfeeding) metamorphic period. In addition, pelagic feeding and the accretion of energy reserves will almost certainly be necessary during any delay phase should the metamorphic stimuli not be acquired immediately on attaining competence.

In lecithotrophic species, as with planktotrophic species, the pelagic phase can be divided into an obligatory period and a competent or searching period. For example, *A. proxima* undergoes an obligatory 24–48 h phase which involves morphological changes in the mantle and the digestive, nervous, and muscular systems preparatory for metamorphosis (Thompson, 1958). During this period the larva is phototactic and swims upward; presumably, this provides limited dispersal away from the adult microhabitat. Similarly, at hatching, *Phestilla lugubris* is not competent of metamorphosis (state A: Bonar and Hadfield, 1974), but within a few hours it develops a propodium and retracts the mantle in readiness for settlement. For *T. hombergi* (Thompson, 1962), the onset of the searching phase (after an obligatory phase of 24–48 h) is indicated only by a switch from positive phototaxis. By contrast, the obligatory period for the aeolid *Eubranchus farrani* was observed to continue for only a few hours, and metamorphosis could be initiated within 10 h of hatching (Todd, 1981).

Settlement behavior and the metamorphic processes are broadly similar for both lecithotrophic and planktotrophic veligers. It should be emphasized, however, that the nutritive reserves present at hatching of a lecithotrophic larva must support the veliger throughout the pelagic phase and the postsettling, nonfeeding period. Data indicating the potential extent of this "delay" phase are largely lacking, although Thompson (1958) reports 14 days beyond the obligatory 2 days for *A. proxima* and 10 days for *T. hombergi* (Thompson, 1962).

3. *Metamorphosis Stimulation and Postlarval Development*

It has become apparent that the major if not only stimulus to metamorphosis for both competent planktotrophic and competent lecithotrophic nudibranch veligers is invariably the presence of the live adult diet, or at least some biochemical component thereof (Thompson, 1958, 1961; Tardy, 1964, 1970; Harris, 1973, 1975; Kempf and Willows, 1977; Perron and Turner, 1977; Chia and Koss, 1978; Harrigan and Alkon, 1978; Rudman, 1979, 1981a; Todd, 1981). It should be stated, however, that there are a number of records of metamorphosis of both planktotrophic (Perron and Turner, 1977; Thiriot-Quievreux, 1977) and lecithotrophic (Rao and Alargarswami, 1960; Rao, 1962) species without the presence of the adult diet or, indeed, any stimulus in most cases. Metamorphosis of nonpelagic lecithotrophic species certainly appears to be a predetermined developmental event not requiring extrinsic stimulus (Roginskaya, 1962a; Tardy, 1962b; Schmekel, 1966; Hamatani, 1967; Schönenberger, 1969; Eyster, 1979); the lecithotrophic veligers of *Cuthona nana* (Harris et al., 1975; Rivest, 1978), although released from the spawn-mass, appeared incapable of swimming and metamorphosed without apparent stimulus within 2 days of hatching.

The tropical aeolid *Phestilla lugubris* preys upon the coral *Dendrophyllia elegans,* and the veligers will settle and metamorphose only on the dead coral skeleton (Harris, 1975); larvae settling on live polyps are consumed. By contrast, Rudman (1979) observed metamorphosis of *C. poritophages* (a predator of the coral *Porites somaliensis*) on the mucous sheath which may cover parts of the live coral; presumably, metamorphs are protected from the prey polyps by the mucus. In this context, it is perhaps relevant to note that Harris (1975) found that a protein component of the mucus of *Dendrophyllia* was all that was necessary to promote metamorphosis in *P. lugubris*. Curiously, although the larvae of the dendronotid *T. diomedea* are not consumed by the adult prey (*Virgularia*), metamorphosis will similarly occur only near, but not actually on, the prey pennatulid (Kempf and Willows, 1977). In a different manner, Thompson (1958) showed that although contact of *A. proxima* pediveligers with the live bryozoan *Electra pilosa* promoted the most successful metamorphosis, some larvae that were screened from this substratum by plankton gauze still underwent subsequent development. Detailed specific analyses are still lacking, but in general it does appear that the trigger to metamorphosis comprises a biochemical "cue" component of the prey organism that is usually acquired by contact chemoreception.

Poecilogony, or "strategy switching," seems to have been established for the aeolids *C. nana* (planktotrophy : nonpelagic lecithotrophy) (Harris et al., 1975; Rivest, 1978) and *Tenellia pallida*. For this latter species Rasmussen (1944) noted both pelagic and nonpelagic lecithotrophy (Denmark), and Eyster (1979) reported both planktotrophy and nonpelagic lecithotrophy within the one population in North Carolina. Significantly, Eyster found that individual adults did not switch development type during their lifetime and that individuals of one type would successfully copulate with the other. Both phenotypes were sympatric, with nonpelagic development recorded for about 20% of field populations. The ecological significance of variability in developmental mode is discussed below.

Only in the case of *D. obscura* has an intermediate diet been observed for the postlarval stage of a nudibranch species. In this case, the postmetamorphs subsisted on detritus for the first 5 days of benthic life before commencing feeding on the definitive bryozoan prey. Most other species commence feeding directly on the definitive prey within a similar period. Undoubtedly, the relative sizes of the postlarval nudibranchs and their respective prey items, and the extent to which nudibranch reproduction can be geared to the seasonal behavior of the prey, are primary ecological determinants of these often specific associations. These factors, among others, are considered in the final section on reproductive ecology.

C. Reproductive Strategies

The considerable variety of reproductive strategies displayed by the nudibranchs, in conjunction with their potential for laboratory culture, provides possibilities of examining some interesting genetic and evolutionary questions. What effect, for example, does the wide range of dispersal capabilities exert on the balance between gene flow and intensity of selection in local populations? The absence of lengthy dispersal in lecithotrophic species presumably increases the likelihood of genetic isolation and the possibilities of local adaptation. On the other hand, we would anticipate that those species which incorporate a highly dispersive long-term planktotrophic larva will possess regionally homogeneous gene pools which, by foregoing the possibilities of local adaptation, will perhaps display rather broader ecological and/or physiological tolerances. The sympatric British dorids *Onchidoris muricata* (planktotrophic) and *A. proxima* (lecithotrophic) perhaps illustrate the point. Throughout its British distribution, *O. muricata* is a uniform pale white-cream color. *A. proxima,* by contrast, is bright yellow on west coasts, but local populations around the coasts of Scotland and down the east coast of England show a progressive cline through pale yellow to cream and finally to pure waxy white. Whether these regional differences are also reflected in physiological and ecological adaptations has yet to be established.

One of the unusual aspects of the reproductive ecology of the opisthobranchs is that in general they constitute a marked exception to "Thorson's rule." Thorson (1950) postulated that invertebrate larval development rates would be broadly correlated with sea temperature, and hence geographical locality or latitude, but that differential patterns of habitability and productivity of the surface waters were superimposed on this. In principle, he predicted a high incidence of direct development in the high Arctic with a decrease in this mode, and a corresponding increase in planktotrophy toward the tropics. For the Southern Hemisphere, he further hypothesized an increase in the incidence of brooding behavior and parental protection of directly developing embryos in the more ancient Antarctic province. Analyses by Clark (1975), Franz (1975), and Clark and Goetzfried (1978) have shown an absence of directly developing species in southern New England (unstable maritime climate) but a considerable proportion of these in the British Isles (stable cold temperate) and Florida (subtropical). In attempting to explain these patterns, these authors have invoked such factors as climatic stability, and seasonality (or its lack) in the production, persistence, availability, and quality of the associate prey organisms. It should be noted that these authors do not differentiate nonpelagic lecithotrophy from true direct development as defined in this chapter. In principle, therefore, we need to explain the increased incidence of pelagic and nonpelagic lecithotrophy at lower latitudes: nonpelagic lecithotrophy and direct development do, however, have broadly similar ecological and genetic implications.

An additional unusual feature, albeit of infrequent occurrence among nudibranchs, is strategy switching within the same species. This has been discussed for *Tenellia pallida* and *Cuthona nana,* and Clark and Goetzfried (1978) report the aeolid *Spurilla neapolitana* switches from 90 μm (pelagic lecithotrophic) to 82 μm (planktotrophic) eggs after 5 days of starvation. This latter case is certainly unusual in that starvation of nudibranchs generally affects only numbers and sizes of spawn-masses because egg size, and thereby developmental mode, is genetically set. Ecologically, the differences between the various developmental modes, and the possibility of switching mode in response to (for example) environmental variables (Clark and Goetzfried, 1978), are extremely important.

Todd and Doyle (1981) showed that there is a clear temperature-independent relationship between embryonic (intracapsular) development time and egg size, between egg size and larval type, and thereby between larval type and development time for a wide range of north temperate nudibranch species (Fig. 5). These development time differences are further amplified, in a different ordination, when the pelagic phase is also included (Fig. 6). These differences may be of adaptive significance should the predator need to gear its reproduction to a qualitative or quantitative seasonal trait in the life history of the prey organism. In deriving the settlement-timing hypothesis, Todd and Doyle (1981) utilized the specific example of *O. bilamellata*. This dorid preys exclusively on acorn barnacles and is particularly associated with *Balanus balanoides*. Settlement of *B. balanoides* is a discrete annual event continuing for only a few weeks with slightly variable timing in any one locality. Because no intermediate diet is inferred for the postmetamorph stages of *O. bilamellata*, in view of the small nudibranch size at settling it is evident that only barnacle cyprids and/or newly metamorphosed spat can be handled at this stage. For *O. bilamellata* there is, therefore, a distinct but temporally variable settlement target. In addition, there is an optimal time for adults to spawn (when population biomass is at a peak), and it is the gap between the optimal time to spawn and the optimal time to settle

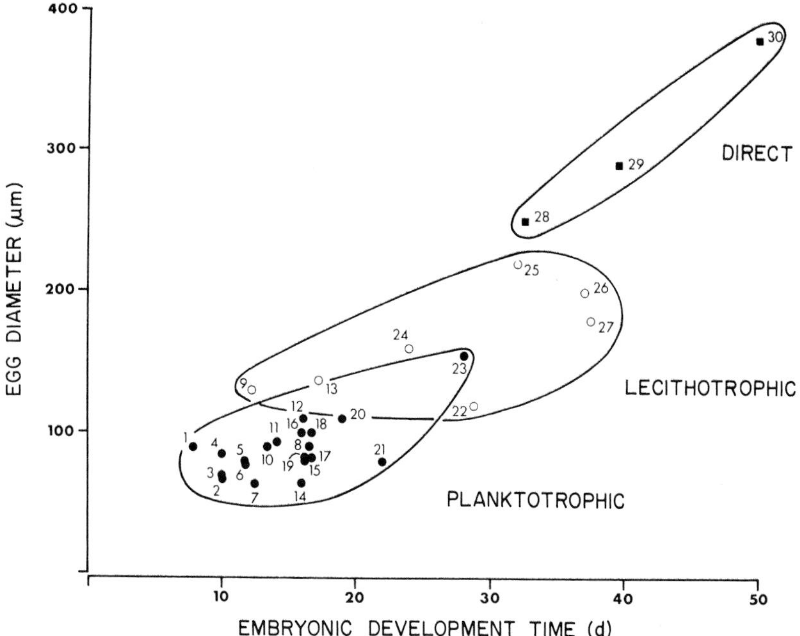

Fig. 5. Interrelationships of egg size, development time, and reproductive strategy for 21 species of North Atlantic nudibranchs. (From Todd and Doyle, 1981.)

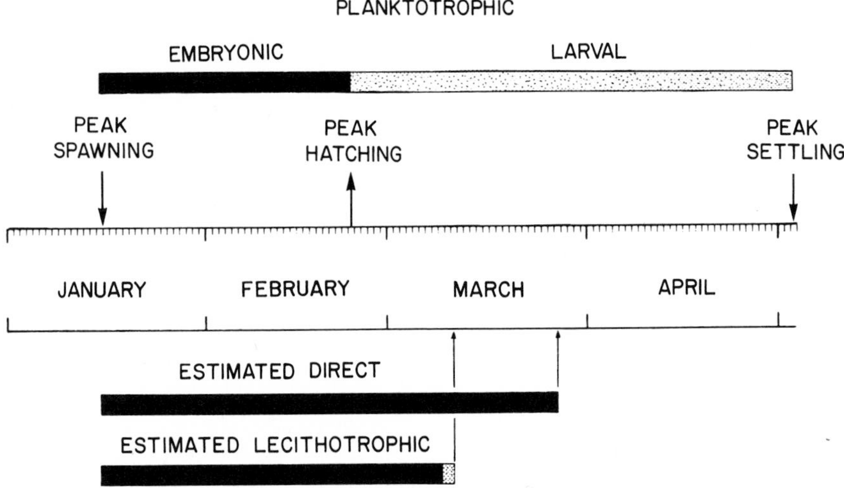

Fig. 6. The egg-to-juvenile period of *Onchidoris bilamellata* in northeastern England (upper bar). The predicted periods for hypothetical direct and lecithotrophic strategies are shown by the lower bars. (From Todd and Doyle, 1981.)

that must be bridged by the embryonic–larval phase. The lecithotrophic strategy confers the shortest egg-to-juvenile period, and the planktotrophic strategy the longest; direct development is intermediate (Fig. 6). Because there is no reason on energetic grounds for *O. bilamellata* to be incapable of supporting either the lecithotrophic or direct strategies, the planktotrophic strategy was considered as the only expedient means of satisfactorily bridging the time gap. This hypothesis does not presume to explain why every marine invertebrate reproduces in its given way; rather, it shows that, in certain cases, total development time may be the deterministic factor within the constraints of energetic and genetic prerequisites.

It is my contention that energetic studies will provide the basis for rationalizing specific differences in selection for different larval strategies. That is, energetic criteria will provide a coarse filter for the reproductive options available, should different absolute energetic thresholds apply for the various strategies (Todd, 1979a; Todd and Doyle, 1981). If the species can "afford" only the "cheapest" strategy, there is no option. If, however, two or more strategies are alternative on energetic grounds, then such genetic considerations as dispersal and local adaptation and the implications of egg-to-juvenile periods may prevail at this finer level of resolution.

Reproductive effort (the spawn calorific output expressed as a percentage of total body calories) varies considerably among species. For example, *O. bilamellata* (planktotrophic) exerts an effort of as much as 168% (Todd, 1979b) and *O. muricata* (planktotrophic) as much as 143%, but *A. proxima* (lecithotrophic) only to 55% (Todd, 1979a). As yet there are no data available for non-pelagic lecithotrophic or directly developing species, and hence the strategy-specific pattern is unconfirmed. Obviously, in order for these exceptionally high levels of reproductive output to be maintained, continued feeding and gametogenesis throughout the spawning period are necessary.

As remarked in Section I, individual fitness is measured in terms of the success of the female role for such hermaphrodites. Recent observations (C. D. Todd, unpublished) have shown that for both *O. muricata* and *A. proxima,* starvation throughout the spawning period results in the expected marked decrease in reproductive effort. Significantly, however, there is an additional increase in the frequency and duration of copulatory behavior. Because feeding is not undertaken during copulation, this must be considered a reproductive "cost" to the parent inasmuch as it is a non-energy-acquiring commitment. This response appears to be an attempt by the individual to maximize its reproductive success as a male in view of its clearly "poor" performance as a female. These preliminary observations underline the necessity of considering nudibranch reproduction in terms of the duality of individual sexuality.

III. Predator–Prey Associations

A. Life Cycle Types

The foregoing review of reproductive biology and larval settlement responses has emphasized the intimacy of the prey associations displayed by these molluscs from the very commencement of the benthic phase of the life cycle. The fact that large-scale migrations by nudibranch populations do not occur (see Miller, 1962; Potts, 1970; Nybakken, 1978 for review) can only increase the importance of larval electivity. The majority of nudibranch species undergo simple annual life cycles with little or no generation overlap and total postspawning mortality. All species are semelparous and there is no survival for a second year, even if access to a mate and copulation are prevented (Todd, 1979b). Some species have an extended biennial life cycle but again semelparity is the rule; there is only one spawning period, at the end of the second year of life. However, many species, especially small aeolids and dorids, do undergo subannual or "ephemeral" life cycles, and egg-to-egg periods may be accordingly brief (Swennen, 1961; Miller, 1962; Nybakken, 1974, 1978; Clark, 1975; Franz, 1975; Todd, 1981). For example, the north temperate aeolid *Tenellia pallida* matures 14–21 days posthatching (Rasmussen, 1944; Eyster, 1979), whereas the small dorid *Doridella obscura* may proceed from egg to egg within only 26 days at field temperatures (Perron and Turner, 1977).

The annual life cycle seems to be the "rule" for nudibranchs, and the subannual and biennial variants are independent derivatives. The interrelationship of both prey and predator seasonal behavior is discussed below, but it may be illustrative to summarize the major features of the population ecology of the annual dorid *Onchidoris bilamellata.* This species is circum-Arctic in distribution, and spawns between December and May in the British Isles with subsequent benthic recruitment in June (Todd, 1978, 1979b, 1981). *Balanus balanoides,* the major prey item, is available throughout the year in the intertidal and itself recruits to the benthic phase over a brief period in May–June. Indeed, it is only at barnacle spatfall that newly metamorphosed *Onchidoris* can exploit *Balanus.* Veligers metamorphosing much before or after cyprid spatfall are confronted by prey which have already grown into a size refuge. Clearly, in the absence of an intermediary diet, *O. bilamellata* could only be annual or biennial in order to key into this resource. Mortality of field populations was shown to be density-dependent, with less than 1% of those nudibranchs initially recruited surviving to spawn 9 months later. Growth rates vary among local populations on the same shore in response to the availability of *Balanus,* with low shore populations exhibiting the greatest densities but also intraspecific competition-induced small size. Mid shore populations are probably controlled more by physical factors (e.g., temperature effects and desiccation), and low shore populations are more biologically controlled. As anticipated for an annual species with a long-term planktotrophic larva, annual fluctuations in recruitment were considerable.

Figure 7 describes the growth, gametogenesis, "de-growth," and postspawning death of a hypothetical annual individual such as *O. bilamellata.* The same features, with an appropriate alteration of the time scale, apply to both subannual and biennial species. Note that spermatogenesis slightly precedes oogenesis and may continue beyond the completion of spawning. Most if not all nudibranchs continue feeding during spawning, which accounts for both the high reproductive effort and the periodic increases in body size during spawning. Prior to the termination of spawning the digestive gland is invariably atrophied and autolyzed, and the metabolic products are output in gametes (Todd, 1978) (Fig. 2).

B. Prey Stability/Availability and Nudibranch Life Cycles

The seasonal prey size restrictions and in fact the barnacle-eating habit of *O. bilamellata* are perhaps exceptions rather than the rule for nudibranchs. Certainly, the sponge grazers, hydroid predators, and bryozoan-associated species are not usually confronted with such problems. With the exception perhaps of *D. obscura* (outlined above), the Bryozoa might be considered ideal prey for nudibranchs because the prey item (polypide) is available in multiples of small unit size and the colonies are annual/polyannual and available year round. Sponge colonies similarly present no small or large size constraints, although

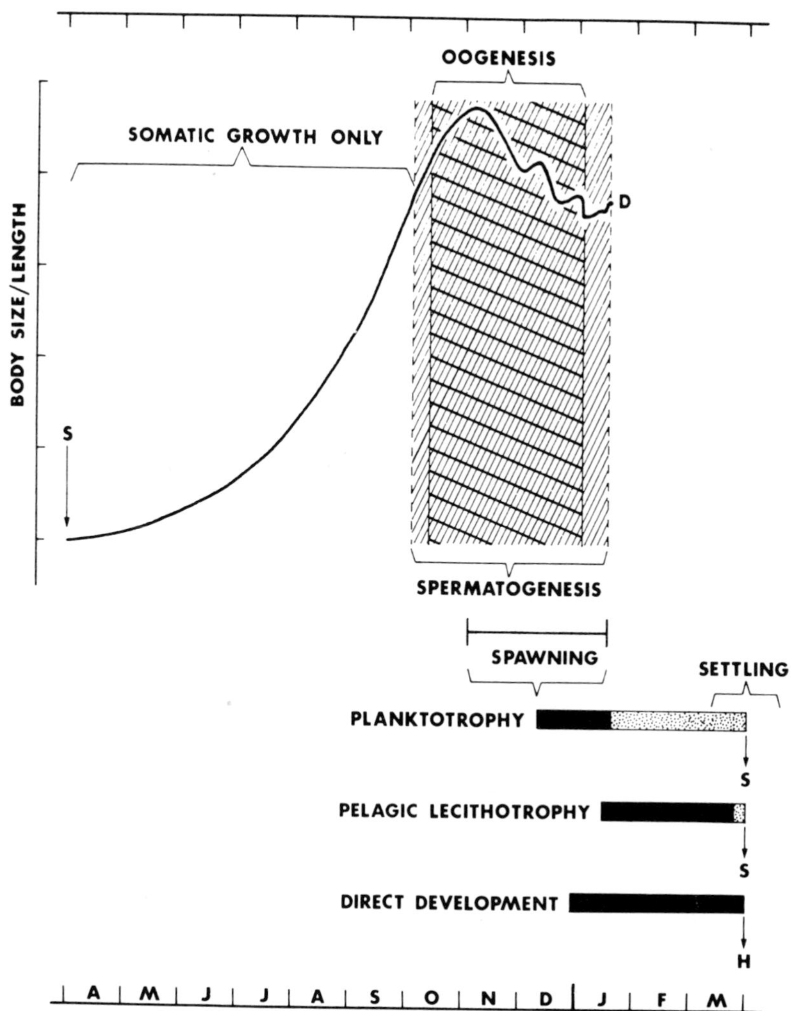

Fig. 7. Summary of the growth, gametogenesis, reproduction, and larval life of a typical annual nudibranch. The relative durations of the three fundamental reproductive strategies are indicated for comparison. See text for further explanation.

probably both solitary and colonial ascidians do so for newly metamorphosed juveniles. Although few size problems (see below) apparently are encountered by the aeolid predators of calyptoblastic hydroids, considerable temporal and spatial difficulties are introduced. Typically, there is a vernal "bloom" of these hydroids in northern temperate waters but the growths of hydroids disappear rapidly in late summer, both naturally and as a result of predation. Such prey populations tend to be thinly scattered or locally absent from shallow water during the winter before redeveloping the following spring. Many aeolid predators behave accordingly in rapidly colonizing such resources, maturing, reproducing, and dying within a few weeks; several generations may pass during the summer (Swennen, 1961; Miller, 1962; Nybakken, 1974, 1978; Clark, 1975). By contrast, the two largest British nudibranchs, *Archidoris pseudoargus* and *Tritonia hombergi,* prey upon temporally and spatially stable and predictable resources, notably the sponge *Halichondria panicea* and the octocorallian *Alcyonium digitatum,* respectively. For these two molluscs there are, therefore, no prey size or availability complications, and consequently extended biennial life cycles are supportable.

In light of the above observations on prey seasonality, availability, and size constraints, it is perhaps permissible to indulge in some speculation on the derivation of the subannual and biennial cycles. Expressed most crudely and simplistically, selection should go toward abbreviating (subannual) or prolonging (biennial) the life cycle, according to the differential interaction of age-specific survivorship and age-specific fecundity. Thus, selection will favor prolongation from the simple annual cycle if the consequent increase in (size/age-specific) fecundity overcompensates the decreased (size/age-specific) survivorship.

Figure 8 presents an array of nudibranch life-cycle strategies supportable by a variety of prey resources. Resource A is a temporally and spatially stable and predictable prey item (e.g., a sponge species) which exerts no size constraints. Selection should favor annual or biennial cycles rather than subannual cycles on the basis of differential fecundities. Whether or not fitness is increased by adopting a biennial strategy over the annual cycle depends upon survivorship–fecundity interactions.

Resource B is either unavailable for a short period [B(i)] (e.g., when the prey species is itself in a pelagic phase) or exerts prey size constraints on settling veligers [B(ii)] (e.g., barnacles, ascidians). Clearly, an annual cycle with both predator and prey occupying the pelagic phase together is the most plausible predator response. An extended cycle is, however, possible if the 1-year-old nudibranchs can tolerate starvation or overcome prey unavailability by switching to another species. Many nudibranchs, both large and small, can survive extended starvation for as long as several months, but presumably only with a concomitant reduction of survivorship and overall fitness in nature.

Resource C(i) is predictably but briefly available in a particular habitat (e.g., intertidal) and present in another (e.g., subtidal) year round. This may apply to many hydroid species which are only transient in shallow water. Such an intertidal prey resource is most profitably exploited by an abbreviated life cycle and colonization at the beginning of the following year from deeper water populations.

Resource C(ii) is temporally and spatially unpredictable, and again an appropriately abbreviated life cycle is expedient. Support of extended cycles is not possible in any one locality, and new resource locations must be sought and exploited by dispersive larvae. The emphasis here is on rapid location, colonization, reproduction, and dispersal, which are especially well illustrated by many small, hydroid-grazing aeolids.

These proposals can only be a crude approximation of the three basic life history strategies displayed by these molluscs but, within the prescribed provisos, I believe they are conceptually valid.

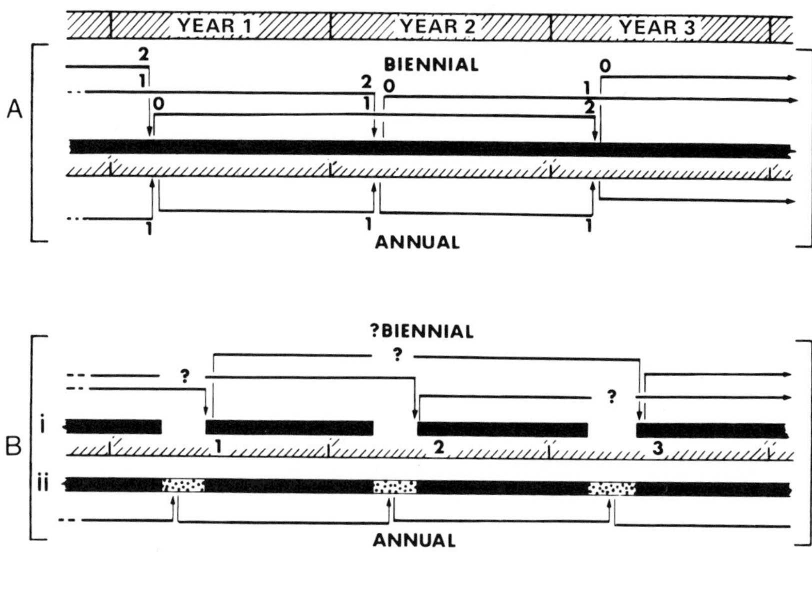

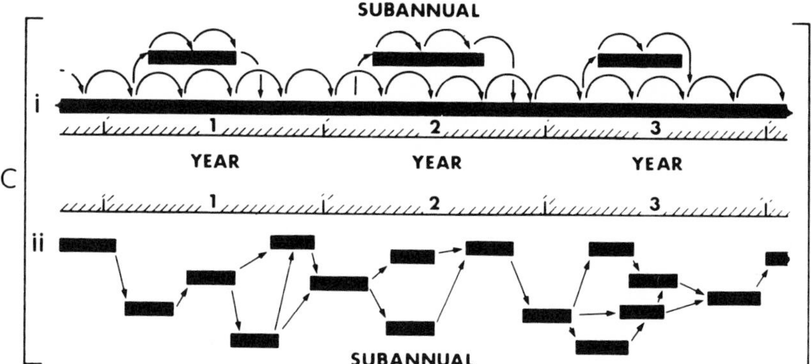

Fig. 8. Conceptual representation of the annual, biennial, and subannual life cycle strategies in relation to resource availability. See text for further explanation.

C. Adaptations to Prey Types

Morphologically and ecologically, the nudibranchs are divisible into four major groupings according to their prey associations: the sponge grazers, bryozoan grazers, hydroid grazers, and a miscellaneous category. These trophic groupings follow an approximate taxonomic ordination inasmuch as the sponge predators are invariably dorids or arminids, the bryozoan grazers predominantly dorids, and those associated with hydroids are almost exclusively aeolids. The miscellaneous grouping is represented by all four suborders, with species preying upon one or more of the following subcategories: ascidians, cirripedes, other coelenterates, corals, other nudibranchs and their eggs, and teleost fish eggs.

From a morphological viewpoint, it is apparent that the bulky, yet flattened, dorids are well adapted to preying upon encrusting sponges and bryozoans, whereas at the other extreme the elongate, buoyant aeolids are well suited to their delicate, erect hydroid prey. Within this broad framework there are, naturally, many exceptions, including aeolids preying on bryozoans (Edmunds, 1975), ascidians (Morse, 1969), and alcyonarians (Rudman, 1981b), and the unique dorid *Okadaia elegans* which not only preys upon serpulid polychaetes but also drills through their calcareous tubes (Baba, 1937). The adaptations and diets of British (and other) nudibranch species have been recently reviewed by Todd (1981), but three subgroupings, the hydroid predators, pleuston associates, and coral grazers, may be profitably appraised specifically.

Tubularia spp. are large, colonial, gymnoblastic hydroids which are the preferred prey of a wide variety of aeolid and other nudibranchs (for reviews, see Thompson, 1964; Todd, 1981). Their widespread sublittoral distribution and year round availability, but perhaps most importantly their remarkable capacity for asexual regeneration of lost polyps (Todd, 1981), suggest *Tubularia* spp. are an ideal nudibranch diet. Such appears true for perhaps 19 British aeolids and dendronotids are particularly associated with *Tubularia* spp. These gymnoblasts, however, possess a virulent nematocyst defense which, although it does not discourage adult nudibranch predators, is effective against juveniles. It is relevant that no one species of nudibranch preys exclusively upon *Tubularia* spp., and Miller (1962) noted the example of juvenile *Dendronotus frondosus* shifting from comparatively harmless calyptoblastic hydroid prey species to *Tubularia* spp. only at an intermediate size. By the same token, *Obelia* spp. grow rapidly, and are widely distributed and abundant in shallow water in addition to having a very wide aperture to the hydrotheca; it is thus perhaps no surprise that many small species (and juveniles of larger species) are commonly associated with this calyptoblastic hydroid.

Cheng (1975) has reviewed the literature pertaining to pleustonic nudibranchs. Among these the aeolids *Glaucus* and *Fiona*, and *Scyllaea*, an elongate dendronotid (as might be anticipated from buoyancy considerations), are, with the exception of *Fiona,* exclusively associated with coelenterates. *Glaucus* spp. float upside down at the sea–air interface and are appropriately camouflaged (blue ventrally, silvery white dorsally). Flotation is effected by gulping air into the stomach (Bieri, 1966; Thompson and McFarlane, 1967; Thompson and Bennett, 1969); the prey consists primarily of the chondrophores *Velella* and *Porpita* and the siphonophore *Physalia.* In common with most hydroid-grazing aeolids, *Glaucus* stores functional nematocysts from the ingested prey tissue, which it may use in its own defense, in special cnidosacs at the cerata tips [see Edmunds (1966) and Todd (1981) for review of defensive mechanisms]. Unlike *Glaucus, Scyllaea* and *Fiona* lack morphological or behavioral adaptations to pleustonic life; *Scyllaea* is cryptically camouflaged on *Sargassum* and preys upon epiphytic hydroids, and *Fiona* is associated with both *Velella* and the goose barnacle *Lepas.* The ability to swim actively, although displayed by a range of species throughout the order (Agersborg, 1925; Willows, 1967; Hurst, 1968; Willows and Hoyle, 1969; Schumacher, 1973; Thompson and Brown, 1976), does not extend to these pleuston associates. Presumably, *Fiona* can transfer to new prey items only on contact; the infrequency of such contacts is perhaps indicated by the ability of this aeolid to grow from 8 mm to maturity within only 4 days (Bayer, 1963).

The sponge-grazing dorids invariably have broad radulae with multiple rows of simple hook-shaped teeth. The bryozoan grazers have somewhat simpler radulae, often with a well-developed central tooth and numerous small laterals in each row. Feeding is commonly by rasping, with the radula ingesting material and passing this to the esophagus; some of the bryozoan and ascidian predators feed suctorially, utilizing a heavily muscular buccal diverticulum. All aeolids possess a pair of large cutting chitinous jaws in association with a simple radula; the radula is invariably a single row of (central) teeth, often with extensive serrations.

The rhinophores of nudibranchs may be simple and tentacular or more or less elaborately laminate, presumably to increase sensory surface area. The rhinophores of probably all species are chemosensory (Wolter, 1967; Storch and Welch, 1969) and in the particular case of the anemone predator, *Aeolidia papillosa,* such prey detection is well documented (Stehouwer, 1952; Braams and Geelen, 1953; Haaften and Verwey, 1960; Robson, 1961; Swennen, 1961; Waters, 1973; Edmunds et al., 1974, 1976; Harris and Howe, 1979; Hall et al., 1983). Presumably, chemoresponses are effective in the marine environment only over short distances of perhaps a few centimeters. Nevertheless, the discriminatory abilities of *A. papillosa,* for example (Hall et al., 1982), testify to their importance in the trophic ecology of these molluscs.

At this juncture, it may be useful to discuss nudibranch defenses briefly. These generally include two lines (Edmunds, 1966; Todd, 1981); the primary defense is commonly crypsis (or aposematism) which purports to evade visual detection, but should this fail, other adaptations may be used. These include behavioral (e.g., swimming), morphological (e.g., autotomy and inedibility), and chemical (e.g., acid or toxin production) mechanisms. Cryptic homochromy, with the prey organism as the substratum, is frequently achieved by disruption of the body outline with tubercles (dorids) or cerata (aeolids), and body coloration matching that of the prey. In many instances, the degree of homochromy is enhanced by the utilization of prey-derived pigments in conjunction with those of the nudibranch. The large, brilliantly colored tropical arminid *Phyllidia varicosa,* which grazes on the sponge *Hymeniacidon,* displays an even more cohesive prey association in utilizing a prey-derived isocyanide toxin in its own defensive secretions (Burreson et al., 1975).

Perhaps the most elaborate illustration of the integrity of some nudibranch associations is provided by the coelenterate-grazing aeolids. The major, although seemingly not necessarily the most efficacious, defensive mechanism of this suborder is the storage and utilization of functional prey nematocysts. Untriggered and/or immature nematocysts pass through the digestive gland unharmed and are maintained within specialized cnidosacs at the ceras tip. Ejection of the sac contents through the terminal cnidopore elicits discharge of the nematocysts, although clearly this can be effective only as a close contact defense. The general unpalatability of aeolids appears to be attributable to synergism between nematocyst discharge and mucous and other defensive secretions.

Finally, I will mention some recent observations of Rudman (1981a) on coral-eating nudibranchs. To date, four species of aeolid and one arminid (Harris, 1975; Rudman, 1981a) have been recorded as obligate predators of scleractinian corals. Of these, Rudman noted that four are sympatric specialists on the coral *Porites somaliensis* and all are well camouflaged on the live prey. All four nudibranch species ingest all the soft parts of the coral tissue, although each has specialized on one or more components thereof. Thus, *Phestilla lugubris* utilizes all components of the prey tissue, and *P. minor* obtains at least a major portion of its nutrition from the prey spirocysts. *Cuthona poritophages* obtains its nutrition largely from the zooxanthellae; the arminid *Pinufius rebus* deploys intact and functional prey zooxanthellae in the tissues of the digestive gland and dorsal body wall in a symbiotic association. [See also Rudman (1981b) on zooxanthellae symbioses in alcyonarian-grazing aeolids.] This assemblage of nudibranchs thus displays the remarkable ecological specificity and diversity of the order Nudibranchia; although all four species have evolved strikingly similar radula teeth and ingest all parts of the coral tissue, each is isolated in its dietary specialization. Furthermore, although all three aeolids have abbreviated pelagic larval phases, their success is manifest by their wide distribution throughout the Indo-West Pacific.

References

Agersborg, H. P. K. (1925). The sensory receptors and the structure of the oral tentacles of the nudibranchiate mollusk *Hermissenda crassicornis* syn. *H. opalescens*. *Acta Zool. (Stockholm)* **6,** 167–182.

Baba, K. (1937). Contributions to the knowledge of a nudibranch *Okadaia elegans* Baba. *Jpn. J. Zool.* **7,** 147–190.

Bayer, F. M. (1963). Observations on pelagic Molluscs associated with the siphonophores *Velella* and *Physalia*. *Bull. Mar. Sci.* **13,** 454–466.

Beeman, R. D. (1977). *In* "Reproduction of Marine Invertebrates" (A. C. Giese and J. S. Pearce, eds.), vol. 4, pp. 115–179. Academic Press, New York.

Bickell, L. R., and Chia, F. S. (1979). Organogenesis and histogenesis in the planktotrophic veliger of *Doridella steinbergae* (Opisthobranchia, Nudibranchia). *Mar. Biol. (Berlin)* **52,** 291–313. [See Chia and Koss (1978)]

Bickell, L. R., Chia, F. S., and Crawford, B. J. (1981). Morphogenesis of the digestive system during metamorphosis of the nudibranch *Doridella steinbergae* (Gastropoda): conversion from phytoplanktivore to carnivore. *Mar. Biol.* **62,** 1–16.

Bieri, R. (1966). Feeding preferences and rates of the snail *Ianthina prolongata,* the barnacle *Lepas ansifera,* the nudibranchs *Glaucus atlanticus* and *Fiona pinnata,* and the food web in the marine neuston. *Publ. Seto Mar. Biol. Lab.* **14,** 161–170.

Bonar, D. B., and Hadfield, M. G. (1974). Metamorphosis of the marine gastropod *Phestilla sibogae* (Nudibranchia: Aeolidacea). I. Light and electron microscope analysis of larval and metamorphic stages. *J. Exp. Mar. Biol. Ecol.* **16,** 227–255.

Braams, W. G., and Geelen, H. F. M. (1953). The preference of some nudibranchs for certain coelenterates. *Archs. Neerl. Zool.* **10,** 241–264.

Burreson, B. J., Scheuer, P. J., Finer, J., and Clardy, J. (1975). 9-Isocyanopupukeanane, a marine invertebrate allomone with a new sesquiterpene skeleton. *J. Am. Chem. Soc.* **97,** 4763–4764.

Cheng, L. (1975). Marine pleuston—animals at the sea–air interface. *Oceanogr. Mar. Biol.* **13,** 181–212.

Chia, F. S., and Koss, R. (1978). Development and metamorphosis of the planktotrophic larvae of *Rostanga pulchra* (Mollusca: Nudibranchia). *Mar. Biol. (Berlin)* **46,** 109–119.

Clark, K. B. (1975). Nudibranch life-cycles in the Northwest Atlantic and their relationship to the ecology of fouling communities. *Helgol. Wiss. Meeresunters.* **27,** 28–69.

Clark, K. B., and Goetzfried, A. (1978). Zoogeographic influences on development patterns of North Atlantic Ascoglossa and Nudibranchia, with a discussion of factors affecting egg size and number. *J. Moll. Stud.* **44,** 283–294.

Edmunds, M. (1966). Protective mechanisms in the Eolidacea (Mollusca: Nudibranchia). *J. Linn. Soc. London, Zool.* **47,** 27–71.

Edmunds, M. (1975). An eolid nudibranch feeding on Bryozoa. *Veliger* **17,** 269–270.

Edmunds, M., Potts, G. W., Swinfen, R., and Waters, V. (1974). The feeding preferences of *Aeolidia papillosa* (L.). *J. Mar. Biol. Assn., U.K.,* **54,** 939–947.

Edmunds, M., Potts, G. W., Swinfen, R., and Waters, V. (1976). Defensive behaviour of sea-anemones in response to predation by the opisthobranch mollusc *Aeolidia papillosa* (L.). *J. Mar. Biol. Assoc. U.K.* **56,** 65–83.

Eyster, L. S. (1979). Reproduction and developmental variability in the opisthobranch *Tenellia pallida*. *Mar. Biol. (Berlin)* **51,** 133–140.

Eyster, L. S. (1981). Observations on the growth, reproduction and feeding of the nudibranch *Armina tigrina*. *J. Moll. Stud.* **47,** 171–180.

Fischer, P. (1892). Recherches sur la morphologie du foie des Gasteropodes. *Bull. Sci. Fr. Belg.* **24,** 260–346.

Franc, A. (1968). *In* "Traite de Zoologie" (P.-P. Grassé, ed.), Vol. II, pp. 608–893, 1079–1082. Masson, Paris.

Franz, D. R. (1975). An ecological interpretation of nudibranch distribution in the Northwest Atlantic. *Veliger* **18,** 79–83.

Ghiselin, M. T. (1966). Reproductive function and the phylogeny of Opisthobranch gastropods. *Malacologia* **3,** 327–378.

Ghiselin, M. T. (1969). The evolution of hermaphroditism among animals. *Q. Rev. Biol.* **44,** 181–208.

Gibson, R., Thompson, T. E., and Robilliard, G. A. (1970). Structure of the spawn of an Antarctic dorid nudibranch *Austrodoris macmurdensis* Odhner. *Proc. Malacol. Soc. Lond.* **39,** 221–225.

Gohar, H. A. F., and Soliman, G. N. (1967). The direct development of the nudibranch *Casella obsoleta* (Ruppell & Leuckart) (with remarks on the metamorphosis). *Publ. Mar. Biol. Stn., Ghardaqa, Red Sea* **14,** 149–166.

Haaften, J. L., and Verwey, L. (1960). The role of water currents in the orientation of marine animals. *Arch. Neerl. Zool.* **13,** 493–499.

Hall, S. J., Todd, C. D., and Gordon, A. (1982). The effects of ingestive conditioning on the prey species selection in *Aeolidia papillosa* (L.) (Mollusca: Nudibranchia). *J. Anim. Ecol.* **51,** 907–921.

Hamatani, I. (1967). Notes on veligers of Japanese opisthobranchs (7). *Publ. Seto Mar. Biol. Lab.* **15,** 121–131.

Harrigan, J., and Alkon, D. L. (1978). Larval rearing, metamorphosis, growth and reproduction of the aeolid nudibranch *Hermissenda crassicornis* (Eschscholtz, 1831) (Gastropoda: Opisthobranchia). *Biol. Bull. (Woods Hole, Mass.)* **154,** 430–439.

Harris, L. G. (1973). Nudibranch associations. *Curr. Top. Comp. Pathobiol.* **2,** 213–315.

Harris, L. G. (1975). Studies on the life-history of two coral-eating nudibranchs of the genus *Phestilla. Biol. Bull. (Woods Hole, Mass.)* **149,** 539–550.

Harris, L. G., and Howe, N. R. (1979). An analysis of the defensive mechanisms observed in the anemone *Anthopleura elegantissima* in response to its nudibranch predator *Aeolidia papillosa. Biol. Bull. (Woods Hole, Mass.)* **157,** 138–152.

Harris, L. G., Wright, L., and Rivest, B. R. (1975). Observations on the occurrence and biology of the aeolid nudibranch *Cuthona nana* in New England waters. *Veliger* **17,** 264–268.

Hurst, A. (1967). The egg masses and veligers of thirty Northeast Pacific opisthobranchs. *Veliger* **9,** 255–288.

Hurst, A. (1968). The feeding mechanism and behaviour of the opisthobranch *Melibe leonina. Symp. Zool. Soc. London* **22,** 151–166.

Kempf, S., and Willows, A. O. D. (1977). Laboratory culture of the nudibranch *Tritonia diomedea* Bergh (Tritoniidae: Opisthobranchia) and some aspects of its behavioral development. *J. Exp. Mar. Biol. Ecol.* **30,** 261–276.

Miller, M. C. (1958). Studies on the nudibranchiate Mollusca of the Isle of Man. Ph.D. Thesis, University of Liverpool, England.

Miller, M. C. (1962). Annual cycles of some Manx nudibranchs, with a discussion of the problem of migration. *J. Anim. Ecol.,* **31,** 545–569.

Morse, M. P. (1969). On the feeding of the nudibranch *Coryphella verrucosa rufibranchialis,* with a discussion of its taxonomy. *Nautilus* **83,** 37–40.

Morse, M. P. (1971). Biology and life-history of the nudibranch mollusc, *Coryphella stimpsoni* (Verrill, 1879). *Biol. Bull. (Woods Hole, Mass.)* **140,** 84–94.

Nybakken, J. (1974). A phenology of the smaller dendronotacean, arminacean and aeolidacean nudibranchs at Asilomar State Beach over a twenty seven month period. *Veliger* **16,** 370–373.

Nybakken, J. (1978). Abundance, diversity and temporal variability in a California nudibranch assemblage. *Mar. Biol. (Berlin)* **45,** 129–146.

Ockelmann, K. W. (1965). Developmental types in marine bivalves and their distribution along the Atlantic coast of Europe. *Proc. Eur. Malacol. Congr., 1st, 1962* pp. 25–35.

Perron, F. E., and Turner, R. D. (1977). Development, metamorphosis, and natural history of the nudibranch *Doridella obscura* Verrill (Corambidae: Opisthobranchia). *J. Exp. Mar. Biol. Ecol.* **27,** 171–185.

Potts, G. W. (1970). The ecology of *Onchidoris fusca* (Nudibranchia). *J. Mar. Biol. Assoc. U.K.* **50,** 269–292.

Rao, K. V. (1961). Development and life-history of a Nudibranchiate gastropod *Cuthona adyarensis* Rao. *J. Mar. Biol. Assoc. India* **3,** 186–197.

Rao, K. V., and Alargarswami, K. (1960). Account of the structure and early development of a new species of a nudibranchiate gastropod, *Eolidina mannarensis. J. Mar. Biol. Assoc. India* **2,** 6–16.

Rasmussen, E. (1944). Faunistic and biological notes on marine invertebrates I. The eggs and larvae of *Branchystomia rissoides* (Hanl.), *Eulimella nitidissima* (Mont.), *Retusa truncatula* (Brug.), and *Embletonia pallida* (Alder & Hancock) (Gastropoda, marina). *Vidensk. Medd. dans. Naturh. Foren.* **107,** 207–233.

Rivest, B. R. (1978). Development of the eolid nudibranch *Cuthona nana* (Alder & Hancock, 1842) and its relationship with a hydroid and hermit crab. *Biol. Bull. (Woods Hole, Mass.)* **154,** 157–175.

Robson, E. A. (1961). The swimming response and its pacemaker system in the anemone *Stomphia coccinea. J. Exp. Biol.* **38,** 685–694.

Roginskaya, I. S. (1962a). The egg masses of nudibranchs of the White Sea. T. Belomorsk. Biol. *Stn. Mosk. Gos. Univ.* **1,** 201–214.

Roginskaya, I. S. (1962b). Reproductive biology and life-cycle of *Cuthona pustulata* (Gastropoda, Nudibranchia). *Dokl. Akad. Nauk SSSR* **146,** 488–491.

Rudman, W. D. (1979). The ecology and anatomy of a new species of aeolid opisthobranch mollusc; a predator of the scleractinian coral *Porites. Zool. J. Linn. Soc.* **65,** 339–350.

Rudman, W. D. (1981a). Further studies on the anatomy and ecology of opisthobranch molluscs feeding on the scleractinian coral *Porites. Zool. J. Linn. Soc.* **71,** 373–412.

Rudman, W. D. (1981b). The anatomy and biology of alcyonarian-feeding aeolid molluscs and their development of symbiosis with zooxanthellae. *Zool. J. Linn. Soc.* **72,** 219–262.

Schmekel, L. (1966). Zwei neue Arten der Familie Cuthonidae Aus dem Golf von Neapel: *Trinchesia granosa* n.sp. und *Trinchesia ocellata* n.sp. (Gastropoda, Opisthobranchia). *Pubbl. Stn. Zool. Napoli* **35,** 13–28.

Schönenberger, N. (1969). Beitrage zur Entwicklung und Morphologie von *Trinchesia granosa* Schmekel (Gastropoda, Opisthobranchia). *Pubbl. Stn. Zool. Napoli* **37,** 236–292.

Schuhmacher, H. (1973). Notes on occurrence, feeding and swimming behaviour of *Notarchus indicus* and *Melibe bucephala* at Elat, Red Sea (Mollusca: Opisthobranchia). *Isr. J. Zool.* **22,** 13–25.

Soliman, G. N. (1977). A discussion of the systems of classification of dorid nudibranch veliger shells and their taxonomic significance. *J. Moll. Stud.* **43,** 12–17.

Stehouwer, E. C. (1952). The preference of the sea-slug *Aeolidia papillosa* for the sea-anemone *Metridium senile. Arch. Neerl. Zool.* **10,** 161–170.

Storch, V., and Welsch, U. (1969). Uber bau und Funktion der Nudibranchier–Rhinophoren. *Z. Zellforsch. Mikrosk. Anat.* **97,** 528–539.

Swedmark, B. (1964). The interstitial fauna of marine sand. *Biol. Rev.* **39,** 1–42.

Swennen, C. (1961). Data on the distribution, reproduction and ecology of the nudibranchiate molluscs occurring in the Netherlands. *Neth. J. Sea Res.* **1,** 191–240.

Tardy, J. P. (1962a). Observations et experiences sur la metamorphose d'*Eolidina alderi* (Gasteropode, Nudibranche). *C. R. Hebd. Seances Acad. Sci.* **254,** 2242–2244.

Tardy, J. P. (1962b). Cycle biologique et metamorphose d'*Eolidina alderi* (Gasteropode, Nudibranche). *C. R. Hebd. Seances Acad. Sci.* **255,** 3250–3252.

Tardy, J. P. (1964). Observation sur le développement de *Tergipes despectus* (Gasteropode, Nudibranche). *C. R. Hebd. Seances Acad. Sci.* **255,** 1635–1637.

Tardy, J. P. (1970). Contribution a l'étude des metamorphoses chez les nudibranches. *Ann. Sci. Nat. Zool. Biol. Anim.* **12,** 299–370.

Tchang-Si, T. (1931). Contribution a l'étude des Mollusques Opisthobranches de la cote provencale. Ph.D. Thesis, Université de Lyons, France (unpublished).

Thiriot-Quievreux, C. (1977). Veligere planctotrophe du doridien *Aegires punctilucens* (D'Orbigny) (Mollusca: Nudibranchia: Notodorididae): description et metamorphose. *J. Exp. Mar. Biol. Ecol.* **26,** 177–190.

Thompson, T. E. (1958). The natural history, embryology, larval biology, and post-larval development of *Adalaria proxima* (Alder & Hancock) (Gastropoda, Opisthobranchia). *Philos. Trans. R. Soc. London, Ser. B* **242,** 1–58.

Thompson, T. E. (1959). Feeding in nudibranch larvae. *J. Mar. Biol. Assoc. U.K.* **38,** 239–248.

Thompson, T. E. (1961). The importance of the larval shell in the classification of the Sacoglossa and the Acoela (Gastropoda, Opisthobranchia). *Proc. Malacol. Soc. London* **34,** 233–238.

Thompson, T. E. (1962). Studies on the ontogeny of *Tritonia hombergi* Cuvier (Gastropoda: Opisthobranchia). *Philos. Trans. R. Soc. London, Ser. B* **245,** 171–218.

Thompson, T. E. (1964). Grazing and the life-cycles of British nudibranchs. *Brit. Ecol. Soc.* Symp. **4,** 275–297.

Thompson, T. E. (1966). Studies on the reproduction of *Archidoris pseudoargus* (Rapp) (Gastropoda, Opisthobranchia). *Philos. Trans. R. Soc. London, Ser. B* **250,** 343–374.

Thompson, T. E. (1967). Direct development in the nudibranch *Cadlina laevis,* with a discussion of developmental processes in Opisthobranchia. *J. Mar. Biol. Assoc. U.K.* **47,** 1–22.

Thompson, T. E. (1976). "Biology of Opisthobranch Molluscs." Ray Society, London.

Thompson, T. E., and Bennett, I. (1969). *Physalia* nematocysts: utilized by molluscs for defence. Science **166,** 1532–1533.

Thompson, T. E., and Brown, G. H. (1976). "A Synopsis of the British Opisthobranch Molluscs: Mollusca: Gastropoda: Keys and Notes for the Identification of the Species." Academic Press, New York.

Thompson, T. E., and McFarlane, I. (1967). Observations on a collection of *Glaucus* from the Gulf of Aden, with a critical review of published records of Glaucidae (Gastropoda, Opisthobranchia). *Proc. Linn. Soc. London* **178,** 107–123.

Thorson, G. (1946). Reproduction and larval development of Danish marine bottom invertebrates, with special reference to planktonic larvae in the Sound (Oresund). *Medd. Komm. Dan. Fisk. Havunders., Ser. Plankton* **4,** 1–523.

Thorson, G. (1950). Reproduction and larval ecology of marine bottom invertebrates. *Biol. Rev. Cambridge Philos. Soc.* **25,** 1–45.

Todd, C. D. (1978). Gonad development of *Onchidoris muricata* (Müller) in relation to size, age and spawning (Gastropoda: Opisthobranchia). *J. Moll. Stud.* **44,** 190–199.

Todd, C. D. (1979a). Reproductive energetics of two species of dorid nudibranchs with planktotrophic and lecithotrophic larval strategies. *Mar. Biol. (Berlin)* **53,** 57–68.

Todd, C. D. (1979b). The population ecology of *Onchidoris bilamellata* (L.) (Gastropoda: Nudibranchia). *J. Exp. Mar. Biol. Ecol.* **41,** 213–255.

Todd, C. D. (1981). Ecology of nudibranch molluscs. *Oceanogr. Mar. Biol.* **19,** 141–234.

Todd, C. D., and Doyle, R. W. (1981). Reproductive strategies of benthic marine invertebrates: a settlement-timing hypothesis. *Mar. Ecol.: Prog. Ser.* **4,** 75–83.

Tomlinson, J. (1966). The advantages of hermaphroditism and parthenogenesis. *J. Theoret. Biol.* **11,** 54–58.

Usuki, I. (1967). The direct development and the single cup-shaped larval shell of a nudibranch *Glossodoris sibogae. Sci. Rep. Niigata Univ., Ser. D* **4,** 75–85.

Vannuci, M., and Hoscoe, K. (1953). Sobre *Embletonia mediterranea* (Costa); Nudibrasquiro da regias lagunar de Cananeia. *Bol. Inst. Oceanogr. (Univ. Sao Paulo)* **4,** 103–120.

Vestergaard, K., and Thorson, G. (1938). Uber den Laich und die Larven von *Duvaucelia plebeia, Polycera quadrilineata, Eubranchus pallidus* und *Limapontia capitata* (Gastropoda, Opisthobranchiata). *Zool. Anz.* **124,** 129–138.

Waters, V. E. (1973). Food-preference of the nudibranch *Aeolidia papillosa,* and the effect of defences of the prey on predation. *Veliger* **15,** 174–192.

Willows, A. O. D. (1967). Behavioral acts elicited by stimulation of single, identifiable brain cells. *Science* **157,** 570–574.

Willows, A. O. D., and Hoyle, G. (1969). Neuronal network triggering a fixed action pattern. *Science* **166,** 1549–1551.

Wolter, H. (1967). Beitrage zur biologie, histologie und sinnesphysiologie (insbesondere der Chemorezeption) einiger Nudibranchier (Mollusca, Opisthobranchia) der Nordsee. *Z. Morphol. Oekol. Tiere* **60,** 275–337.

6

Physiological and Trophic Ecology of Cephalopods

WILLIAM C. SUMMERS

Western Washington University
Bellingham, Washington

I. Introduction

An analysis of trophic ecology or physiological adaptation requires a clear definition of the subject organisms. The class Cephalopoda (phylum Mollusca) is a subject which must be further qualified to evaluate living forms satisfactorily. What is a cephalopod? It is revealing that a practical answer can be supplied by listing squid, octopus, and cuttlefish; accepting reasonable variability, little more is required for the casual observer. The systematic definition given in the second sentence is useful in an academic and probably evolutionary sense, but suffers from circularity and abstraction. What about *Nautilus?* Should it not be included

The Mollusca, Vol. 6
Ecology

ISBN 0-12-751406-6

too? The answer is yes for both practical and systematic reasons, but with the caution that this relict genus is biologically distinct from modern, naked coleoids, and hence is not central to the evaluation of their physiology. Any ascription of degeneracy simply because coleoids lack an external shell and/or an adequate fossil record may be set aside; they happen to represent the class throughout oceanic depths worldwide and are no less faithful to their molluscan roots than the nautiloids or fully extinct forms.

Malacologists have little difficulty relating modern cephalopods to a hypothetical ancestral mollusc, a small, limpet-like creature with a conical shell, a highly ciliated visceral mass, and a muscular foot. This herbivore, it is thought, crept about on the more consolidated substrates of ancient shallow seas, licking up food with a radular tongue. The extent to which this bilateral organism experienced segmentation is overshadowed by major events in the evolution of consistent cephalopod themes: development of centralized nervous systems, closed circulation, jet propulsion, ink glands, buoyancy regulation, large sizes, and general activity patterns. The shell was internalized and lost in most groups; fins were evolved to aid in swimming (and lost again in the incirrate octopods); eight or 10 sucker-bearing arms were perfected for prey capture (60–90 sticky arms in *Nautilus*), and complex adaptations in color and texture modification, as well as bioluminescence, were developed under sensory feedback control.

It is convenient to think of the first "leap" in cephalopod evolution from the ancestral mollusc prototype as occurring by the development of an apical, gas-filled chamber within the shell. This was accomplished by secretion of a calcareous septum by the mantle and then (through redirection of the shell gland chemistry) by the absorption of water and release of respiratory gases into the enclosed space by a narrow, penetrating tube (the siphuncle). The chamber structure has advantages in repetition; 20–40 units are found in modern *Nautilus* and *Spirula,* although the latter bears the shell internally. Thus buoyed, the protocephalopod had some relief from its gravitational tie to the ocean floor, but this would be advantageous only with related changes at about the same time. The shell must have been necessary for protection (from what?); otherwise, it could have been reduced to save weight at a lesser cost than its elaboration. It is also hard to imagine a simple, snail-footed creature benefiting from intermittent contact with the substratum; the foot could neither push off nor effectively catch the animal on its descent. Furthermore, without modification the simple foot would have no advantage when falling onto sessile (plant) food organisms. Its only probable advantage would be for a leaping escape response.

On the other hand, the presence of the septum/chamber strongly suggests a need to regulate the volume enclosed within the shell margin, a requirement if hydrostatic jetting had evolved by rapid withdrawal of the body into the shell.

Possible orientation of the outwash current could be used to dislodge food or cleanse the immediate environment. Gas filling of the chamber(s) may have been an incidental refinement after such an invention, resulting in subsequent modification of the foot and the eventual development of a bouncing predatory mode of life. Note that shell growth is incumbent from the beginning, which suggests an accompanying change in food preference. Further, buoyed and able to leave the bottom at will, the cephalopod line had already invented its distinctive, two-way locomotion and the ability to cover large areas in search of food.

Packard (1972) has described the vast expanse of cephalopod evolution as convergent with fishes and, to a lesser extent, with certain groups of marine reptiles and mammals. The cephalopods came first; working within different genetic limitations they have been in direct competition with vertebrates for so long that the resultant distribution and adaptive groupings bear much intercomparison for their common solutions to the problems inherent in the higher trophic levels of the marine envionment. Indeed, cephalopods and marine fishes have probably shaped the intermediate trophic levels as a competitive unit throughout recent time.

Using workable quantitative units, we can again refer to systematics. Voss (1977, 1979) submits that there are approximately 650 species, 130 genera, and 40 families within the Cephalopoda. The greatest numbers, many monotypic, are found among the oceanic squids, and the most specious are from the better-known genera which live benthic coastal lives. In round numbers, allowing for new discoveries and some revisions in single species or single life stage descriptions, there are probably fewer than 1000 extant species of cephalopods. That this number is small when compared with either extinct forms or modern fishes (about 30,000 species) does not in itself presage a threat of extinction to the class.

As stated above, the genus *Nautilus* is a distinctive remnant of one of the earliest cephalopod lines. All other living cephalopods are categorized as coleoids and might be thought of as decapods and octopods, with the exception of a single species constituting the Vampyromorpha, a specialized deep-water form that generally bears more resemblance to octopods despite a mixture of characteristics. The decapods have several traditional subgroupings, as well as a strong convergence in structures, which defy clear separations. The designation of sepioids and teuthoids is partially instructive, so long as one recognizes that the genus *Spirula* does not easily fit the former and that teuthoids include oceanic (oegopsid) forms and myopsids; the latter share many features with the shell-less sepioids. Octopods also have a major, divergent subgrouping, the deep-water cirrate octopods, which often bear fins and always have deeply webbed arms.

For the purpose of physiological discussion, I offer the following simple taxonomy.

Class Cephalopoda
Subclass Nautiloidea
Genus *Nautilus*

Subclass Coleoidea
Decapods
Genus *Spirula*

Order Sepioidea[1] (cuttlefishes)
Order Teuthoidea[2] (squids)
Octopods
Order Octopoda
Suborder Cirrata
Suborder Incirrata
Order Vampyromorpha
Genus *Vampyroteuthis*

Aside from the great unevenness of this classification, I trust that its functional groupings will have utility. This is best illustrated by examining habitat continuity and adaptations, especially buoyancy-regulating mechanisms.

The treatment of cephalopod ecology is difficult because few specimens have ever been seen alive; still fewer have been maintained for even brief periods under conditions which might permit reasonably normal behavior, and even fewer (a few large-egged octopods and a half-a-dozen benthic sepioids) have been cultivated. Collections of material are disparate and especially limited from locations other than fishing grounds or shallow depths; few specific efforts at cephalopod collections have been made. Finally, the cephalopod habit of mincing food as it is ingested makes stomach analyses relatively difficult, and the identification of partially digested cephalopods in the stomachs of vertebrate predators has often disgusted evaluators to the point of merely recording categorical occurrences.

II. Adaptation and Tolerance

A. Materials and Laboratory Maintenance

Cephalopods are limited to marine conditions and are found in all open waters from shallow inshore waters to great depths. (Their distribution beneath polar ice masses is not known.) They range in adult size from a few centimeters (*Idiosepius*) to well over 10 m (*Architeuthis*), and in location from permanent pel-

[1]Myopsid condition.
[2]Mostly oegopsid condition except for the loliginids, which are myopsids.

agic forms to those which cruise the demersal region, settle regularly, bury in soft sediments, or slink from one hiding place to the next. Although newly hatched cephalopods cannot be called true larval forms (Boletzky, 1974a), some, which vary considerably from the adults, have lengths of about 1 mm. A few millimeters is more common. Thus, the capture of cephalopod material requires a diversity of sampling techniques which often exceeds practical limitations.

Not surprisingly, the usual types of gear employed to capture cephalopods are little modified from those used to harvest fish. The main exception is the substitution of barbless, multihook lures (jiggers) for the baited hooks commonly used for fishing. Trap fishing is apparently restricted to the capture of benthic octopods and *Nautilus*. Mid-water net fishing is equally effective with fish and pelagic cephalopods (that is, not especially productive, but possible). Dragged nets and seines are used mostly for commercial species, notably the aggregating neritic ones. Stomach analyses of sea birds, marine mammals, and a variety of fishes have provided additional information on cephalopod distributions, particularly discoveries of large numbers of squid beaks in whale stomachs. In recent years, especially valuable information has resulted from photographs taken by remotely controlled cameras and from submersibles.

The major faunal regions, which were described by Packard (1972), include coastal waters which are populated by the sepioids (except *Spirula*), the myopsid squids, certain groups of oegopsid squids, and the more typical forms of (incirrate) octopods. The myopsid condition (possession of a closed cornea, a fluid-filled space in front of the eye lens, and a tendency toward a horizontal pupil) occurs in all but the most pelagic of these cephalopods. A transition zone between the continental shelf and deeper zones is probably selected by certain pelagic squids (including *Architeuthis?*) and characteristic octopods. Faunal types are stratified in open waters with epipelagic forms in or near the photic zone, a mesopelagic assemblage, and a bathypelagic zone; the latter two overlap at about 1000 m (during the day), and the last extends to the decapod limit of about 3000 m. *Vampyroteuthis* and the cirrate octopods inhabit the demersal zone to a depth of several thousand meters. The mesopelagic is dominated by adult decapods (mostly bioluminescent) and the bathypelagic by octopods and decapods which have buoyancy adaptations. The faunal condition between the bathypelagic and the abyssal demersal region, where depths permit, is not really known. Nesis (1977) summarizes the ontogenetic, distributional, and migrational aspects of cephalopods into categories suitable for comparative purposes.

In general, the maintenance of wild-caught cephalopods is at a rudimentary stage. These animals are relatively fragile and often are injured as a result of the capture techniques. Furthermore, because they are active animals they often respond negatively to enclosure and handling before they can be delivered to adequate aquarium facilities. The most tractable cephalopods (and definitely the most readily available) are benthic octopods, the sepioids (except *Spirula*), and possibly *Nautilus*.Cuttlefish (*Sepia*) and various shallow-water octopods (*Octopus*) have been kept in aquaria for days or weeks for as long as there have been marine stations, and some of these have been cultivated. *Octopus joubini* has been reared in a closed-system aquarium (Forsythe and Hanlon, 1980). Boletzky (1974b) summarized his cultivation success with four sepiolid species, and Summers and Bergström (1980) cultivated another (*Sepietta oweniana*). J. M. Arnold (personal communication) states that he has extended his posthatching experiments with the sepiolid *Euprymna scolopes* through a complete generation. All of these animals have a strongly benthic orientation, and all have large eggs that hatch directly to produce benthic young; none have planktonic stages or large, active adult forms. Attempts to maintain active and/or small-egged forms have been unsuccessful (Hurley, 1976; Balch et al., 1978; Hanlon et al., 1979).

B. Oceanographic Factors

Because cephalopods exist throughout the world's oceans, they have developed adaptations adequate for most naturally occurring variations in such factors as temperature and salinity. Individual species typically are restricted to narrower ranges and/or water masses although the class is represented worldwide. Nearshore salinity reduction presents an obvious limitation to cephalopods, and shallow, brackish waters usually lack cephalopods even if the freshwater input is only occasional. The northwest coastal Atlantic squid, *Lolliguncula brevis,* reportedly tolerates salinities as low as 17.5‰ (Hendrix, 1980). This is apparently the lowest level that any cephalopod will tolerate; it is equivalent to one-half the concentration of seawater. More commonly, neritic species function best in seawater with salinities no lower than about 25‰. For instance, Vecchione (1981) reported that newly hatched *Loligo pealei* were confined to coastal waters off New Jersey and Virginia at salinities of 31.5 to 34.0‰ especially if noticeably affected by Gulf Stream eddies and storm wind conditions. Hypersaline conditions may also exclude tropical, shallow-water (tide pool) cephalopods (Akimushkin, 1963); however, oxygen and/or temperature restrictions probably come into effect earlier (i.e., are more restrictive).

The temperature preferences of most cephalopods are unknown, although seasonal collections have been correlated with annual temperature cycles and the behavior of healthy animals has been observed in temperature-regulated aquaria. These conditions have seldom been met. For example, I proposed, based on extensive winter trawl sampling off New England (Summers, 1969), that the northwest Atlantic squid *Loligo pealei* was restricted to bottom temperatures of 8°C or higher. These data have been confirmed by more recent work (Lange, 1980), and temperature ranges of maximal trawl catches have been set at 10–13°C (spring) and 10–17°C (autumn). *L. pealei* occurs farther south, where

Whittaker (1978) reported an upper temperature limit of 24°C, and generally less than 20°C. This species also occurs in the Gulf of Mexico; Hixon (1980) placed the upper temperature limit for its capture in trawls there at 22°C, although he did take it at the surface under night lights at 30°C. In the aquarium *L. pealei* becomes inactive at temperatures lower than 8°C; its eggs have been hatched at temperatures ranging from 12 to 23°C (McMahon and Summers, 1971). Although this is a wide temperature range, the species is distinctly limited by temperature and range; inshore–offshore migrations and seasonal activities are clearly fixed by these parameters.

Some attempts have been made to correlate long-term collections of pelagic cephalopods with oceanographic data, or at least to compare catches with known current and water mass conditions. Okutani and McGowan (1969) evaluated decapods taken in systematic 1-m plankton net samplings conducted by the California Cooperative Ocean Fisheries Investigation (CalCOFI) over a 4-year period in the California current system. This information was supplemented by Young (1972), who examined all pelagic cephalopods taken in Isaacs-Kidd trawls in the same area over a later 6-year period. Both studies have contributed to a subsequent study of *L. opalescens* (Recksiek and Frey, 1978), among other investigations, by providing a basis for establishing faunal regions within the oceanographic regime off the coast of California and northwestern Mexico.

More restricted studies have been made possible by examination (long-term) of incidental captures and museum collections such as the taxonomic and biogeographic evaluation of ommastrephid squid in the Pacific Ocean by Wormuth (1976). In this case, comparisons were made with distributions of other marine organisms and physical oceanographic data which suggested that water masses (defined by temperature and salinity) played an indirect role in observed distribution. Competition and predation pressure were noted as probable causes of additional variability for these epipelagic squid, although the oceanic and neritic groupings were generally distinctive. Continuing fisheries investigations (particularly in conjunction with international treaty organizations) have produced the bases for single species evaluations and/or management plans. *L. pealei* has benefited (Lange, 1980) from the International Commission for Northwest Atlantic Fisheries, as has *Octopus dofleini*, among others, from joint halibut studies (Mottet, 1975).

Except for its gas-filled shells (*Nautilus, Sepia,* and *Spirula*), there are no known structures in cephalopods which should be affected by changes in depth or pressure. Even these are limited only by implosion depths, and the three genera are regarded as being limited to nearshore or mesopelagic conditions. Pressures associated with depth certainly after chemical reactions and possibly nervous conduction within cephalopods, but whether the apparently sluggish nature of deep-water cephalopods is brought about by pressure, changes in food availability, or buoyant adaptations cannot be stated at this time.

C. Buoyancy and Locomotion

The theoretical archetypal cephalopod buoyed itself with a gas-filled external shell which was divided into a number of chambers. As stated above, that arrangement persists in *Nautilus* (although probably not in the same arrangement as when it first evolved), as an internal spiral shell in the macroplanktonic *Spirula,* and in a reduced form in the demersal *Sepia.* The last has modified the shell into a number of parallel calcareous plates separated by closely spaced pegs and sealed against the environment (the familiar cuttlebone). All three genera can regulate the relative water/gas volumes, and thus buoyancy, and none is known to pressurize the shell (Packard, 1972). With the exception of *Nautilus,* all cephalopod lines have dispensed with the buoyancy and protection of the shell to create naked, muscular, dense-bodied organisms which either swim actively or support themselves on the bottom or both. The decapods and octopods have secondarily developed buoyancy control mechanisms based on chemical substitutions and the storage of low-density materials in certain groupings. The sepioids apparently either depend upon the shell (*Spirula, Sepia*) or manage without other buoyancy compensation.

Clarke et al. (1979) reviewed the buoyancy mechanisms in squid, noting that of all the specimens examined (26 families), examples were either 2–4% more dense than seawater or approximately neutrally buoyant; none were intermediate. Twelve of the families contained neutrally buoyant species which have special tissues in which ammonium solutions (isosmotic with seawater) are stored in thin-walled reticula, usually in the arms, coelom, or mantle. One family, the Enoploteuthidae, provided both dense and buoyant examples. These ammoniacal squids are abundant and usually associated with the bathypelagic fauna. They are all teuthoids, partly gelatinous in appearance, and are preyed upon by whales. The measured ammonium concentrations varied by a factor of 100 among different parts of the body and were attributed to substitution for sodium ions (Clarke et al., 1979). Another widespread family of teuthoid squid, the Gonatidae, stores fats, apparently for buoyancy control purposes. These are generally accepted as mesopelagic, or coastal transitional zone epipelagic, species. The distribution of buoyant tissues is not uniform and does not seem to relate consistently to the center of gravity; hence, squid probably assume various positions in the water when at rest, and some may change attitude with development.

Another group of cephalopods, the cirrate octopods, also live in deep water and have a gelatinous appearance. These (at least some of them) have tissues in which chloride is substituted for sulfate ions and also assume buoyant neutrality. Photographs of cirrate octopods show them, with deeply webbed arms widespread, near the bottom at abyssal depths (Roper and Young, 1975). *Vampyroteuthis* is also soft bodied and associated with the bottom; however, its buoyancy mechanism apparently has not been reported. The cirrates have fins, and *Vampyroteuthis* has two pairs which it exchanges at a relatively early stage of life.

One further approach to buoyancy must be noted. The female *Argonauta* secretes a secondary shell which it carries externally and into which it places its eggs. This shell traps air and provides some buoyancy for the pelagic octopod (Packard, 1972). Packard (1972) suggests that a related octopod, *Ocythoe,* may have a swim bladder-like structure which provides buoyancy when the animal is heavy with young.

Most of the remaining "dense" cephalopods are demersal and neritic myopsids (sepioids and teuthoids), oegopsids (pelagic teuthoids), or incirrate octopods that are strictly benthic. These have two general body forms, sac-shaped or stiffened and elongated; the latter is associated with rapid and/or constant swimming (teuthoids) and the former is typical of known cephalopod young and the larger octopods. The shell-less sepioids are both small and intermediate in form, but live squid-like lives.

All cephalopods possess some form of jet propulsion in which the reaction from an expelled volume of water is directed by the funnel to move the animal. Water is drawn in around the head through an opening which is sealed off at the beginning of the power stroke like a poppet valve (the connection is reinforced by a pair of snap fastener structures in the decapods). *Nautilus* simply displaces water contained in the open chamber of its shell by withdrawing its head and viscera. The coleoids contract the entire mantle-cavity by muscular action, most of it supplied by circumferential muscles within the mantle wall. The pulse intensity is regulated, under dual nervous control in many cases, including the uniform stimulation provided by the well-known "giant axon" system which functions in escape responses. The jet can be directed in many different ways for swimming or turning, or it may be used by demersal species (e.g., *Sepia*) to uncover food organisms. The bathypelagic cranchiids have a divided mantle-cavity to accommodate their ammoniacal coelom, but the overall effect is about the same. Cephalopod swimming by jet propulsion has been studied in detail (Trueman, 1975); the mantle muscle was investigated by Bone et al. (1981).

The most adept swimmers among the cephalopods (the decapods) all have fins and, as with fishes, the fastest usually have the most aerodynamically shaped and/or smallest fins. These are lateral and paired and operate synchronously except in slow maneuvering. During escape responses, the fins are folded down against the mantle. In addition, the decapods and some octopods have swimming fins or keels on the outboard edges of the arms, which with the arms provide additional control surfaces (at the opposite end from the fins) and may assist in directing the funnel jet in some cases. Thus, decapods have foils on each end (either side of the center of gravity) with a pulse jet "engine" in between. Octopods often lack fins but generally have relatively long arms which trail behind, and they reverse direction less easily. The coordination between fin strokes and jet pulses is very exact in the decapods and is seen best in the dense sepioids, which often have large, rounded fins.

The jet–fin combination has taken some squid literally out of the water. Cole and Gilbert (1970) show a large squid (*Dosidicus gigas,* whose mantle length is about 1.2 m) accelerating out of the water using a continuing jet of water. Ommastrephids and other epipelagic families are frequently reported to leap out of the water, landing on boat decks and hurling themselves considerable distances. As discussed by Packard (1972), these flying squid are not as well-equipped for gliding in air as the flying fishes, but are unique in their mode of launching.

All octopods utilize the arm webs in swimming to some degree. The deep-water forms may depend upon this as a singular form of locomotion because of their extensive webbing. In any case, all octopods studied are adept at walking with the arms and very graceful at negotiating obstacles. The myopsid squids use the arms as props when resting and may shift position by their movement. Sepioids often cover themselves with sediment by the coordinated use of the arms (Boletzky and Boletzky, 1970).

D. Surface-Related Phenomena

With few exceptions (some cirrate octopods), cephalopods have large, image-forming eyes and are highly visual in their behavior. In fact, the eyes are often so elaborated, enlarged, or surrounded by photophores as to suggest that they are the main sensory organs. One family of oegopsids, the histioteuthids (*Calliteuthis*), have unequal eyes; the left is much larger than the right. Young (1977) notes that the retinas are subdivided so that the smaller right eye looks both upward and downward, whereas the large left eye looks solely upward. This squid is ammoniacal and ranges from the surface to at least 1000 m depth. Retinal subdivisions also occur in other deep-water squids and octopods, and probably also in *Vampyroteuthis* (Young, 1977).

Light penetration is attenuated by depth and wavelength is selected to a more monochromatic color. Biological detection and utilization of light probably terminate at about 1000 m in clear seawater and at a much shallower depth in turbid coastal areas. The depth of minimal visibility changes with time of day, season, and atmospheric conditions for any one species and set of water conditions. It is safe to assume that animals living constantly below 1000 m have little or no benefit from sunlight, but most cephalopods at these depths possess well-developed eyes. Were it not for the presence of photophores on these and some other organisms at those depths, eyes would seem to be unnecessary. Indeed, light organs are apparently most common in the mesopelagic, but are also found on *Vampyroteuthis.* Young and Roper (1976) suggest that countershading is the main achievement of bioluminescent midwater squids.

The diel migrations of mesopelagic fauna are directly associated with light levels, and the daily movements of neritic species appear to be similarly keyed to light penetration. Clearly, cephalopods as a group avoid bright illumination, seeking lower levels which provide sufficient visual information about their immediate environment without exposing themselves more than necessary. The direction and color of background illumination change greatly in shallow water, which is certainly related to the spectacular surface texture and color-changing abilities of many cephalopods.

Beside the eyes, cephalopods have photosensitive vesicles associated with the central nervous system (Mauro, 1977). Their role is uncertain, but their effect as light receptors cannot be doubted.

III. Gross Behavior and Habitat Selection

In his review of oceanic squid systematics and ecology, Clarke (1966) states that "the taxonomic tangle prevents any but the most limited analysis of ecological data" and submits that "the most obvious reason for [knowing so little about their ecology] is the great difficulty we have in catching them." His work, the first synoptic attempt with this cephalopod subgrouping in 54 years, has been little modified in systematics since then; however, a brief and improved key to teuthoid families is now readily available (Roper et al., 1969). Probably even less information has been obtained on cirrate octopods and *Vampyroteuthis*. The brightest areas are those related to commercial neritic stocks, in which systematics are more advanced, and trophic data gathered in association with epipelagic or transitional zone fishes. Fortunately, it is no longer necessary for organismic marine biologists to subdivide the sea intellectually into isolated segments.

Roper and Young (1975) suggest that diel vertical migration is mainly restricted to species living in the upper 800 m and is not exhibited by those living below that depth. Further, "apparently very few [pelagic] species live below 1000 m." At the risk of re-creating an "azoic zone," it should be noted that deep (>1000 m) sampling has been relatively infrequent and that our ecological expectation is for a decrease in animal abundance with depth because the singular source of energy input is at the surface. Whether this portends a distinct lower limit for particular groups or certain trophic levels has not been established. Roper and Young (1975) state that familial groupings show some consistency in distributional patterns, but there are major exceptions. They list the following categories for observed patterns of vertical distribution: (1) near-surface dwellers, (2) first-order diel migrators, (3) second-order diel migrators, (4) diel vertical shifters, (5) diel vertical spreaders, (6) nonmigrators, (7) vertical wanderers, (8) species associated with the ocean bottom, and (9) species exhibiting ontogenetic descent. Roper (1977) later compared the effectiveness of three midwater trawls for sampling pelagic cephalopods, concluding that each one presents significant problems in operation and selectivity.

In an areal sense, distribution is thought to be basically cosmopolitan for the bathypelagic and abyssal groups (because of inadequate sampling and relatively uniform environmental conditions). Shallower faunas are thought to be selected by latitudes (temperature) and/or ocean basins. Coastal, especially neritic, species are typically related to relatively small ranges. This last group is somewhat qualified by examples such as the small-egged octopods, which have planktonic young and wide distributions (*Octopus vulgaris, O. dofleini*). Clarke (1977) illustrates a marked decline in the number of species taken with a modern midwater trawl in a series of stations from the tropics to 60°N in the Atlantic Ocean. It appears that the number of families is smaller as well. Perhaps the ultimate in sampling is described in this report; the major theme is an evaluation (to the family level where possible) of squid beaks found in the stomachs of whales. This is reported by region, although the beaks may have been transported some distance within the whales. Sperm whales feed on bathypelagic and benthic squids (mostly ammoniacal species), which Clarke places at the fourth trophic level or higher. Thus, we may feel confident that a trophic structure extends to the depths of the sea, even if our sampling nets have not brought up evidence that confirms it.

Coastal migrations are well established for loliginids (*Loligo vulgaris, L. pealei*), ommastrephids (*Illex illecebrosus, Todarodes pacificus*), and octopods (*Octopus vulgaris, O. dofleini*). (My Swedish colleague Bo Bergström and I have recently also determined that the sepiolid *Sepietta oweniana* moves annually in the North Sea). Typically, depth and proximity to shore change together, and temperature can be used as a descriptive parameter of the selected sites. As noted above, the California current system is inhabited by distinctive epipelagic squids which respond geographically to seasonal changes in the currents. The conditions are widespread and comparable to those of fish migrations (and perhaps related).

The more sedentary cephalopods of the neritic areas (incirrate octopods and sepioids) go to great lengths behaviorally to hide themselves during the day. *Octopus* spp. are notorious for squeezing into narrow spaces or escaping from closed containers (the popular literature does not exaggerate this point). The group has highly developed hiding behavior and minimal physical limitations to prevent access. As stated earlier, the sepioids burrow in soft substrates to avoid detection. Deep-water forms are usually dark colored and do not appear (judging from photographs) to seek cover, but they of course live in permanent darkness. Pelagic cephalopods are cryptic in swimming ability, color, and pattern. When alarmed, cephalopods (even if recently hatched) can release a puff of ink in the exhalant jet. The brown or black ''object'' must confer some survival advantage by diversion because ink glands are ubiquitous in cephalopods (they are absent in *Nautilus*).

IV. Feeding Strategies

A. General Patterns

All cephalopods are carnivores; all possess a biting apparatus resembling an inverted parrot beak, and almost all have some form of radula. The food-capturing mechanism, however, is the crown of sucker-bearing arms and tentacles (where present; none of the octopods has tentacles, but *Vampyroteuthis* has a pair of retractile filaments, and all sepioids and most adult teuthoids have tentacles). *Vampyroteuthis* and the cirrate octopods have cirri in conjunction with the suckers and are extensively webbed. The teuthoids often have stalked suckers, possibly equipped with teeth, and they may have hooks on the tentacles. Thus, for food-gathering purposes, the ''gape'' is limited by the arms and extended forward by the tentacles. The physical mouth is secondary in function and not restrictive in food selection. The octopods and *Nautilus* have a crop; the decapods swallow chunks bitten off by the beak directly to the stomach. Enzymatic breakdown is initiated by salivary glands, and digestion is both incomplete and rapid (Bidder, 1966).

B. Pelagic Forms

Nektonic squids are very impressive predators. As has been observed under lights near the surface at night, ommastrephids move through the illuminated area singly or in formation, often tail first. They may gather in food organisms without altering their progress, feign a strike or reverse direction, attack, and withdraw in very quick sequence. Collectors with squid jiggers, who have witnessed these acts repeatedly, will feel (with luck) a sudden jerk on the line as a squid impales its arms and tentacles on the hooks after discovering, too late, that it has been tricked. Squires (1966) suggested that *Illex* creates a turbulent wake to capture small food organisms when swimming tail first. The head-first approach and strike is more obvious and probably has a consistent pattern in all decapods. It is hard to imagine this strike scaled to the dimensions of a large architeuthid (perhaps 20 m in overall length). Fortunately, most pelagic squids are smaller and some are only macroplanktonic.

The bathypelagic, gelatinous squids have not been observed when feeding. Their morphology is not unique in any sense, which suggests a different food-gathering pattern, except that these animals are apparently relatively immobile and may be oriented other than horizontally. Pelagic octopods are usually small;

lacking tentacles, they must be particularly adept at stalking or (in the more illuminated depths) make good use of other objects to approach their prey.

C. Demersal Forms

It is thought that the cirrate octopods and *Vampyroteuthis* feed on small crustaceans or possibly just small organic particles at or near the bottom. The wide web, multiple (sensory?) cirri, and reduction of the feeding apparatus indicate a relatively simple feeding pattern. Why does *Vampyroteuthis* have filaments? These are often compared with decapod tentacles, but have little in common with them (Young, 1977).

The strike of nekto-benthic decapods is especially well documented, principally because these animals can be observed in aquaria. Messenger (1977) reported a three-phase attack in *Sepia* and correlated this with learning and brain functioning. The cuttlefish are particularly suitable for study because of docility and adaptability to near-surface illumination. In addition, the nekto-benthic forms (mainly sepioids) lack mobility and thus depend more upon the strike than do the more active forms (teuthoids), which can relocate or reorient more rapidly to supplement the strike. As noted above, some pelagic teuthoids lack tentacles as adults (*Lepidoteuthis, Gonatopsis,* octopoteuthids, and grimalditeuthids; see Roper et al., 1969). This represents an opposite extreme in prey capture behavior, although not necessarily the same solution presented by pelagic octopods. Messenger (1977) concluded that the *Sepia* strike is under visual feedback control during its early stages and that the animals improve in ability with age and learning. Should we choose to extend this hypothesis, a correlate idea is that the more generalized strike of small *Sepia,* which is comparable to that of other decapods, requires less learned skill and selectivity. In other words, the prey size and location may determine the type of strike ability which is required for successful feeding. Tactile inputs must also replace some of the visual linkages in attack by deep-sea cephalopods.

D. Benthic Forms

Another major category of feeding behavior is illustrated by neritic octopods. These animals cue visually on distant prey, which they approach by rising to partial arm length and gliding in that direction. The attack is a pounce during which many of the arms and the web are thrown over the prey to immobilize it. A bite and poisoning from the salivary glands are usual in many species (Wells, 1978). The initial prey evaluation is often preceded by a cat-like bobbing of the head, which probably provides parallax information if not a better view of the object.

In addition to this behavior which has been observed in aquaria, shallow-water octopods create midden heaps near their dens that depict their food selection and the results of their feeding. Among these heaps one often finds neatly bored shells of other molluscs (bivalves and gastropods) resulting from another form of feeding, a leisurely boring activity. Apparently, the radula is employed at selected sites to wear a hole through the shell so the octopus can weaken the shellfish and extract the soft body for food (Arnold and Arnold, 1969; Wodinsky, 1969). Middens and boring have not been reported for deep-water octopods, and may not exist because of the lack (or need) of hiding places and the lack of calcareous shellfish at greater depths.

E. Food of Cephalopods

The generalizations and minimal specific information about this aspect afford little useful synthesis but much need for additional work. All cephalopods are carnivores, and there is ample reason to suspect that they are adaptive opportunists at all life stages. In captivity, many have been "trained" to take nonliving food, but they are all attracted by live (moving) prey and appear to prefer it. The general progression is for newly hatched and/or planktonic cephalopods to focus on crustaceans. As they mature, the food preference shifts to larger crustaceans, fish (mainly in the teuthoids), other molluscs (including cannibalism), and a minor, diverse selection of invertebrates (annelids, ctenophores, etc.). As noted above, the feeding behavior is selective for particular food organisms. Neritic octopods thrive on crabs but neglect shrimps and mysids. The sepioids prefer mysids and compete with smaller demersal teuthoids for krill. Pelagic squids generalize on fish and/or other squids; their weakness for jiggers suggests that they are not selective. Mesopelagic decapods are inexplicably woven into the deep scattering layer community, but in some cases they do forage on shallower forms. Finally, the bathypelagic and abyssal cephalopods have occasionally been documented to contain crustacean remains; they are largely unknown.

In an unpublished manuscript on cephalopod diets, Boletzky summarized the literature on this subject, most of it from aquarium data. The listing is extensive and anthropogenic, but has crustaceans as a central theme. I submit that the subject cephalopods are purposely teaching us how adaptable they are. The major areas for work that remain are planktonic stages and the less aquarium-adapted groupings. Prey size is important for selection (LaRoe, 1971), and species changes occur at distinctive life stages (B. Bergström and W. C. Summers, unpublished results on the cultivation of *Sepietta oweniana*). Large-egged sepioids and octopods sometimes maintain lifelong food preferences.

F. Predation on Cephalopods

At this point in our appreciation of marine ecology, the main strategy of higher trophic levels seems to be growth to avoid predation at the cost of having to search further for suitable prey. As Packard (1972) points out, the long coevolution of cephalopods and fish has sharpened their competitiveness and brought about common solutions to the same problems. We can observe that fishes and cephalopods prey on each other and on their own kinds with more regard for access than for systematics. The cephalopods may have an innate advantage in rapid growth (because no hard tissue is generated) and in external handling of prey, but fishes can sustain rapid swimming and may have advantages that result from more complex social structures. Neither clearly outclasses the other in any marine habitat, and they appear to interact at every possible level and location.

Aside from fishes, cephalopods are sought as food by diving birds and, more spectacularly, by a number of whales. Nemoto and Kawamura (1977) single out fin and sei whales as the major (commercial) baleen predators on oceanic squid. It should be noted that these whales seek copepods and euphausiids, which probably are also prey for the squid. The humpback whale also appears in their data as a minor squid predator. Clarke's (1977) report on squid beaks in sperm whales provided an impressive estimate: 110 million tons of oceanic squid are consumed annually by that species of toothed whale alone. Smaller toothed whales are closely associated with aggregations of squid (e.g., the pothead whale and feeding masses of *Illex illecebrosus* in Newfoundland).

Some single-species studies provide information on the cephalopod role in marine ecosystems. Ogi (1980) lists squid as 72.6% of the diet of pelagic feeding, thick-billed murres in the North Pacific (mainly three oegopsid species). The lancetfish *Alepisaurus* has been reported to feed on 34 species from 17 families of cephalopods by Rancurel (1970) and on 18 species from 14 families by Okutani and Kubota (1976). From another point of view, *Loligo opalesceus* was significantly placed within the food web of Monterey Bay, California, with clear evidence that the commercial fishery for this species utilizes only a small part of its abundance (Morejohn et al., 1978).

In terms of trophic position, cephalopods are probably born into the third trophic level and progress one or two steps with age. The calculation is clouded by cephalopods feeding on organisms (e.g., euphausiids) which can themselves change levels and/or on scavengers (crabs, in the case of *Octopus*). They seem never to reach "top carnivore" status because it is known that some vertebrate eats even the largest cephalopods (this has not always been true). This predatory attention, widespread distribution, consistency in form and function, moderate speciation, and a long record of ecological success clearly indicate that the cephalopods have an assured part in the marine environment. As with all competent groups, however, the survival of cephalopods is based on participation and adaptability.

Acknowledgments

I am grateful for the continuous encouragement and open generosity of my teuthological colleagues. Support for my efforts have been derived from Western Washington University (particularly for library services and manuscript preparation), a recent Fulbright Hayes travel grant, and the Swedish Natural Science Research Council. Finally, I respectfully acknowledge the intellectual stimulation of my students, which has had the effect of making some parts of my search both real and personal.

References

Akimushkin, I. I. (1963). "Cephalopods of the Seas of the U.S.S.R." Natl. Sci. Found., Washington, D.C. TT 65-50013, 1965 (Engl. Transl.).

Arnold, J. M., and Arnold, K. O. (1969). Some aspects of hole-boring predation by *Octopus vulgaris*. *Am. Zool.* **9,** 991–996.

Balch, N., Amaratunga, T., and O'Dor, R. K., eds. (1978). Proceedings of the workshop on the squid *Illex illecebrosus*. *Tech. Rep.—Fish. Mar. Serv. (Can.)* **833,** 1–298.

Bidder, A. M. (1966). Feeding and digestion in cephalopods. *In* "Physiology of Mollusca" (K. M. Wilbur and C. M. Yonge, eds.), Vol. 2, pp. 97–124. Academic Press, New York.

Boletzky, S. V. (1974a). The "larvae" of cephalopoda: A review. *Thalassia Jugosl.* **10,** 45–76.

Boletzky, S. V. (1974b). Elevage de céphalopodes en aquarium. *Vie Milieu* **24** (Ser. A), 309–340.

Boletzky, S. V., and Boletzky, M. V. (1970). Das eingraben in sand bei *Sepiola* und *Sepietta* (Mollusca, Cephalopoda). *Rev. Suisse Zool.* **77,** 536–548.

Bone, Q., Pulsford, A., and Chubb, A. D. (1981). Squid mantle muscle. *J. Mar. Biol. Assoc. U.K.* **61,** 327–342.

Clarke, M. R. (1966). A review of the systematics and ecology of oceanic squids. *Adv. Mar. Biol.* **4,** 93–323.

Clarke, M. R. (1977). Beaks, nets and numbers. *Symp. Zool. Soc. London* **38,** 89–126.

Clarke, M. R., Denton, E. J., and Gilpin-Brown, J. B. (1979). On the use of ammonium for buoyancy in squids. *J. Mar. Biol. Assoc. U.K.* **59,** 259–276.

Cole, K. S., and Gilbert, D. L. (1970). Jet propulsion of squid. *Biol. Bull. (Woods Hole, Mass.)* **138,** 245–246.

Forsythe, J. W., and Hanlon, R. T. (1980). A closed marine culture system for rearing *Octopus joubini* and other large-egged octopods. *Lab. Anim.* **14,** 137–142.

Hanlon, R. T., Hixon, R. F., Hulet, W. H., and Won Tack Yang (1979). Rearing experiments on the California Market Squid *Loligo opalescens* Berry 1911. *Veliger* **21,** 428–431.

Hendrix, J. P., Jr. (1980). Osmotic concentration changes in the bay squid *Lolliguncula brevis* (Blainville, 1823) at various salinities. *Proc. Am. Malacol. Union,* 1979, pp. 69–70 (abstr.).

Hixon, R. F. (1980). Growth, reproductive biology, distribution and abundance of three species of loliginid squid (Myopsida, Cephalopoda) in the northwestern Gulf of Mexico. Ph. D. Thesis, University of Miami, Coral Gables, Florida.

Hurley, A. C. (1976). Feeding behavior, food consumption, growth, and respiration of the squid *Loligo opalescens* raised in the laboratory. *Fish. Bull.* **74,** 176–182.

Lange, A. M. T. (1980). The biology and population dynamics of the squids, *Loligo pealei* (LeSueur) and *Illex illecebrosus* (LeSueur), from the northwest Atlantic. M.S. Thesis, University of Washington, Seattle.

LaRoe, E. T. (1971). The culture and maintenance of the loliginid squids *Sepioteuthis sepioidea* and *Doryteuthis plei*. *Mar. Biol. (Berlin)* **9,** 9–25.

McMahon, J. J., and Summers, W. C. (1971). Temperature effects on the developmental rate of squid (*Loligo pealei*) embryos. *Biol. Bull. (Woods Hole, Mass.)* **141,** 561–567.

Mauro, A. (1977). Extra-ocular photoreceptors in cephalopods. *Symp. Zool. Soc. London* **38,** 287–308.

Messenger, J. B. (1977). Prey-capture and learning in the cuttlefish, *Sepia*. *Symp. Zool. Soc. London* **38,** 347–376.

Morejohn, G. V., Harvey, J. T., and Krasnow, L. T. (1978). The importance of *Loligo opalescense* in the food web of marine vertebrates in Monterey Bay, California. *Calif. Fish. Game, Fish. Bull.* **169,** 67–98.

Mottet, M. G. (1975). The fishery biology of *Octopus dofleini* (Wülker). *State Washington, Dep. Fish., Tech. Rep.* **16,** 1–39.

Nemoto, T., and Kawamura, A. (1977). Characteristics of food habits and distribution of baleen whales with special reference to the abundance of North Pacific sei and Bryde's whales. *Rep. Int. Whale Comm., Spec. Issue* **1,** 80–87.

Nesis, K. N. (1977). Vertical distribution of pelagic cephalopods. *Zh. Obshch. Biol.* **38,** 547–558.

Ogi, H. (1980). The pelagic feeding ecology of thick-billed murres in the North Pacific, March–June. *Bull. Fac. Fish., Hokkaido Univ.* **31,** 50–72.

Okutani, T., and Kubota, T. (1976). Cephalopods eaten by lancetfish. *Alepisaurus ferox* Lowe, in Suruga Bay, Japan. *Bull. Tokai Reg. Fish. Res. Lab.* **84,** 1–9.

Okutani, T., and McGowan, J. A. (1969). Systematics, distribution, and abundance of the epipelagic squid (Cephalopoda, Decapoda) larvae of the California current April, 1954–March, 1957. *Bull. Scripps Inst. Oceanogr.* **14,** 1–90.

Packard, A. (1972). Cephalopods and fish: The limits of convergence. *Biol. Rev. Cambridge Philos. Soc.* **47,** 241–307.

Rancurel, P. (1970). Le contenus stomacaux d'*Alepisaurus ferox* dans le sudouest Pacifique (Céphalopodes). *Cah. ORSTOM, Ser. Oceanogr.* **8,** 3–87.

Recksiek, C. W., and Frey, H. W. (1978). Biological, oceanographic, and acoustic aspects of the Market Squid, *Loligo opalescens* Berry. *Calif. Fish Game, Fish Bull.* **169,** 1–185.

Roper, C. F. E. (1977). Comparative capture of pelagic cephalopods by midwater trawls. *Symp. Zool. Soc. London* **38,** 61–88.

Roper, C. F. E., and Young, R. E. (1975). Vertical distribution of pelagic cephalopods. *Smithson. Contrib. Zool.* **209,** 1–51.

Roper, C. F. E., Young, R. E., and Voss, G. L. (1969). An illustrated key to the families of the order Teuthoidea (Cephalopoda). *Smithson. Contrib. Zool.* **13,** 1–32.

Squires, H. J. (1966). Feeding habits of the squid. *Illex illecebrosus. Nature (London)* **211,** 1321.

Summers, W. C. (1969). Winter population of *Loligo pealei* in the mid-Atlantic Bight. *Biol. Bull. (Woods Hole, Mass.)* **137,** 202–216.

Summers, W. C., and Bergström, B. (1980). Cultivation of the sepiolid squid, *Sepietta oweniana,* and its ecological significance. *Am. Zool.* **20,** 1106 (abstr.).

Trueman, E. R. (1975). Swimming by jet propulsion. *In* "The Locomotion of Soft-Bodied Animals," pp. 129–194. Arnold, London.

Vecchione, M. (1981). Aspects of *Loligo pealei* early life history. *J. Shellfish Res.* **1**(2).

Voss, G. L. (1977). Classification of recent cephalopods. *Symp. Zool. Soc. London* **38,** Appendix 2.

Voss, G. L. (1979). Cephalopoda. *In* "Encyclopaedia Britannica," 15th ed., Vol. 3, pp. 1149–1154. Encyclopaedia Britannica, Inc., Chicago, Illinois.

Wells, M. J. (1978). "Octopus, Physiology and Behavior of an Advanced Invertebrate." Chapman & Hall, London.

Whittaker, J. D. (1978). A contribution to the biology of *Loligo pealei* and *Loligo plei* (Cephalopoda, Myopsida) off the southeastern coast of the United States. M.S. Thesis, College of Charleston, Charleston, South Carolina.

Wodinski, J. (1969). Penetration of the shell and feeding on gastropods by *Octopus. Am. Zool.* **9,** 997–1010.

Wormuth, J. H. (1976). The biogeography and numerical taxonomy of the oegopsid squid family Ommastrephidae in the Pacific Ocean. *Bull. Scripps Inst. Oceanogr.* **23,** 1–90.

Young, J. Z. (1977). Brain, behavior and evolution of cephalopods. *Symp. Zool. Soc. London* **38,** 377–434.

Young, R. E. (1972). The systematics and areal distribution of pelagic cephalopods from the seas off Southern California. *Smithson. Contrib. Zool.* **97,** 1–159.

Young, R. E., and Roper, C. F. E. (1976). Bioluminescent countershading in midwater animals: evidence from living squid. *Science* **191,** 1046–1047.

7

Physiological Ecology of Freshwater Bivalves

ALBERT J. BURKY

Department of Biology
University of Dayton
Dayton, Ohio

I. Introduction

The living freshwater bivalve molluscan fauna (using the classification of Vokes, 1980) is represented primarily by three superfamilies: the Unionacea, Corbiculacea, and Dreissenacea. The Unionacea (freshwater mussels or naiades) is dominated by the families Margaritiferidae and Unionidae, the latter of which is composed of a relatively large number of subfamilies, genera, and species. The Corbiculacea includes the Asiatic clam *Corbicula fluminea* (family Cor-

The Mollusca, Vol. 6
Ecology

ISBN 0-12-751406-6

biculidae) and the pisidiids (family Pisidiidae; ICZN Declaration 27, precedence over Sphaeriidae). The pisidiids are commonly known as fingernail, pill, or pea clams and belong primarily to the genera *Sphaerium, Musculium,* and *Pisidium.* The Dreissenacea is notably represented by *Dreissena polymorpha.*

The physiological ecology of freshwater bivalves is historically based on early descriptive work which includes basic taxonomy with notes on habitat, community composition, abundance, and distribution. Modern physiological ecology of freshwater clams has more recently involved paired field and laboratory studies on basic life cycles, growth, reproduction, population dynamics, and energetics. Further, measurable physiological parameters involving metabolism, feeding, and excretion are essential. Comparative information on closely related forms and for intraspecific interpopulation comparisons is critical to the basic understanding of ecological function, as well as the broader concepts of the evolution of both physiological and life-history tactics of freshwater bivalves.

The relative impermanence and spatial discontinuity of most freshwater habitats compared to those of marine and terrestrial environments have been claimed to be related to the evolution of many freshwater faunas, including snails (e.g., Russell-Hunter, 1978, and references therein) because many populations, particularly pulmonate snails, have evolved in relatively small transient bodies of water (Russell-Hunter, 1978). Significantly, many pisidiid clams have evolved in such transient habitats, whereas unionacean clams are generally associated with major drainage systems of relatively longer duration. These characteristics of freshwater habitats have a major impact on considerations of the physiological ecology of clams.

This chapter covers aspects of the ecology of distribution, physiology, population phenomena, and evolution. However, the coverage is in some respects uneven, with an emphasis on seasonal studies of the physiology and population phenomena of natural populations. This is due to the lack of information on some topics or groups of clams, the chapter (Chapter 12) on *Corbicula* by McMahon in this volume, and my own predilection for certain subjects.

II. Distributional Ecology of Freshwater Bivalves

A. General Occurrence and Distribution

The Unionacea and Corbiculacea constitute the two major superfamilies of freshwater bivalves which have representatives of worldwide distribution (Fuller, 1974). The Asiatic clam *Corbicula fluminea* (= *C. manilensis*) was introduced into North America (West Coast) in the late 1800s and has since spread across the continent to become a major pest/pollutant/weed (see McMahon, 1982; Chapter 12, this volume). Further, *Dreissena polymorpha* (Dreissenacea) has similarly become a pest in Britain, Europe, and Asia (see Section IV,A). The reproductive similarities of *C. fluminea* and *D. polymorpha* make them subjects of major concern. With man's help, these species have the potential of near worldwide distribution. However, the occurrence of pisidiid clams (a small number of genera and species; see Section V) in habitats ranging from temporary ponds to swamp woodlands (with only a few millimeters of standing water) to permanent ponds, lakes, streams, and major rivers make them the most cosmopolitan, abundant (Section IV), and widely distributed family of freshwater bivalves. Characteristics of hermaphroditism and viviparity (Section IV,A,B), along with physiological adaptations (Section III,A,B), efficient dispersal (Section II,D), and evolutionary aspects of life history (Sections IV,A,B and V), have contributed to their continuing success. Although unionacean clams are widely distributed (a large number of genera and species; Section V), peculiarities of a life cycle with a parasitic link have placed this diverse group with many endemic forms in peril (Section IV,A,B). Consequently, there are many extinct and endangered species (Stansbery, 1970, 1971) as a result of anthropogenic changes in hydrography (Fuller, 1974; Section II,C) and water quality (Fuller, 1974; Section II,B). Apart from the topics of water chemistry, trophic conditions, hydrography, substrates, and dispersal, the distribution of freshwater bivalves involves geological (Van der Schalie, 1945; Sepkoski and Rex, 1974; Cvancara, 1976; Section V) and recent (Carr and Hiltunen, 1965; Sinclair, 1971; Chapter 12, this volume; Sections I,D, IV,A, and V) time and geography. There are numerous studies of the geographic distribution of bivalves within specific lakes, rivers, or regions; however, only a very few of these references are provided (Boycott, 1936; Wolff, 1968, 1970; Cvancara, 1970; Kuiper and Wolff, 1970; Feldmann, 1971; Clarke, 1973; Sepkoski and Rex, 1974; Bishop and Hewitt, 1976; Wu, 1978; Brandauer and Wu, 1978; Gordon et al., 1979; Ökland, 1979; Brice and Lewis, 1979; Mackie et al., 1980) because the major emphasis of this chapter is physiological ecology.

B. Dissolved Ions and Trophic Conditions

One of the most important ions for freshwater molluscs is Ca^{2+}. According to Boycott (1936), most molluscs in Britain occur in waters with >20 ppm Ca^{2+}, with only a few living at a lower concentration. Harman (1969) found some mussel species at a water hardness as low as 21 ppm (as $CaCO_3$, = 8.4 ppm Ca^{2+}) in New York State. However, Fuller (1974) points out that there are no published values that determine the presence or absence of a given mussel species. Dussart (1976) found that in northwestern England, *Sphaerium corneum* and species of *Pisidium* occur in waters over a range of 6.7 to 70.3 ppm Ca^{2+}, but that they are most dense in harder waters. For South Norway, Ökland (1971) reported up to seven *Pisidium* species in soft waters with 1.5–10.5 ppm Ca0

(1.07–7.46 ppm Ca^{2+}). Further, Hinz (1976) proposes that apart from Ca^{2+} availability, acid rain is responsible for lower densities of *Pisidium* species in South Norway compared to North Norway. Also, the importance of shells in the calcium budget of aquatic systems has been discussed by Aldridge and McMahon (1978) for *C. fluminea* and by Green (1980) for *Anodonta grandis,* and the relationship of water hardness to shell secretion is discussed in Section III,D.

The trophic condition of the habitat is obviously important for fresh-water bivalves because they are primarily suspension feeders. Green (1971) found that the best prediction of bivalve distribution was one based on trophic rather than physical or chemical definitions of the niche. Further, Clarke (1979) claimed that the presence and abundance of certain pisidiid clams appear to be good indicators of trophic lake stages. Apart from considerations of water, the trophic condition of the substrate is also important. Harrison and Rankin (1978) showed that the greatest densities of *Pisidium punctiferum* occurred where there was the greatest percentage of organic material in the substrate. This would undoubtedly be important for deposit feeding (Sections II,C, III,B, and IV,D).

A growing concern is the effect of pollution on aquatic organisms (Ingram, 1957; Fuller, 1974). Little has changed since Ingram (1957) discussed the use of freshwater snails and clams as indicators of pollution. He pointed out that unionacean clams are generally associated with conditions more characteristic of unpolluted water and are thus a negative index of pollution. However, those pisidiid clams which are often associated with domestic sewage in septic areas (e.g., *Musculium transversum* and *Pisidium idahoense*) are also common in clean, unpolluted water and are, therefore, not indicator organisms of organic pollution. However, Paparo and Sparks (1977) showed that ciliary function (on gills of *M. transversum*) is inhibited at the higher potassium concentrations found in certain rivers and proposed the use of ciliary function as an approach to the assessment of water quality. Also, Mackie (1978a) proposed that assessments of reproductive output (natality) in pisidiids can be used to assess the effects of salt and organic pollution because significant effects are seen in *M. securis*.

C. Hydrography and Substrates

The consideration of hydrography and substrates also involves distribution (Section II,A), dissolved ions (Section II,B), dispersal (Section II,D), and aspects of depth–density relationships (Section IV,C). One of the more complete studies of freshwater bivalves in relation to a specific hydrographic region is a series of papers dealing with the estuarine delta area of the Rhine, Meuse, and Scheldt rivers in the southwestern part of the Netherlands (Wolff, 1968, 1969, 1970; Kuiper and Wolff, 1970). This series of papers examined ecological habitat descriptions for bivalve species (Unionacea, Pisidiidae, Dreissenacea) and provided predictions for the sequence of bivalve colonization after impoundment in relation to subsequent changes from brackish to fresh water. Among the freshwater bivalves, species of mussels (Unionacea) are the most greatly affected by anthropogenic alterations of the waterways. Fuller (1974) presents a lengthy and well-documented account of the influences of channelization, impoundment, siltation, and the mussel industry; and pollution from wood products, organic eutrophication, acid mine wastes, pesticides, radionuclides, etc. In general, most mussel species require relatively clean, unpolluted conditions. However, the most crucial effects of channelization, impoundment, and associated silt concern disruptions of the life cycle (Fuller, 1974; Section IV,A,B). This involves problems of a reduction in number or the elimination of host fish for glochidal larvae, or the loss of appropriate substrates and/or conditions for juvenile survival.

Although substrate characteristics are important to bivalve distribution and density, they are also critical components of hydrographic considerations. Changes in substrate characteristics can cause problems for juvenile (Negus, 1966) and adult mussels (e.g., Harman, 1972; Fuller, 1974; Ghent et al., 1978) as well as for pisidiids (Rogers, 1976). It has been shown that when species of *Pisidium* occupy the same square meter of substrate there can be both vertical and horizontal distributions based on variations in the coarseness and organic nature of the substrate (Meier-Brook, 1969; Hinz and Scheil, 1976). Meier-Brook also demonstrated that these tiny clams can remain buried while drawing water through the substrate for purposes of respiration and feeding (Section III,B). Also, it has been shown that although juvenile *Sphaerium striatinum* have a preference for mud, larger individuals show no preference for mud, sand, or sandy mud (Gale, 1973). Gale (1973) states that substrate does not appear to be a major factor in the distribution of *S. striatinum. S. transversum,* however, prefers mud to sandy mud or sand in the laboratory although it occurs in equal densities on all substrates in the field (Gale, 1971). *S. transversum* was also observed as deep as 16 cm in the substrate of Pool 19 of the Mississippi River, and juveniles can burrow to 24 cm (in experimental conditions) in sandy mud (Gale, 1976). Further, Rogers (1976) demonstrated that *S. striatinum* and *S. transversum* can survive burial by various substrates (mud, sandy mud, sand), a potential field condition caused by dredging activities for channelization. However, it is unclear how far these clams can burrow up to the surface. Currents (flood conditions) may be important to uncover deeply buried clams so that they can complete their life cycle (Gale, 1976; Rogers, 1976). These studies at Pool 19 (Gale, 1971, 1976; Rogers, 1976) are important because these small clams are a major food source for wildlife (e.g., fish and ducks).

D. Dispersal

It is well known that passive dispersal by aerial means is common for freshwater molluscs. There are numerous notes, reports, and reviews which document these mechanisms for unionacean and corbiculacean bivalves, and references can be found in Rees (1965), Thompson and Sparkes (1977), and Mackie (1979a). Apparently, wind (waterspouts) can transport unionacean bivalves (Kew, 1893). Man has probably also been a major contributing factor in the dispersal of *C. fluminea* (Thompson and Sparks, 1977). Apart from this, most aerial dispersal is recorded in relation to tiny clams clamped on the legs of insects and the legs and feathers of birds (Rees, 1965; Mackie, 1979a). Mackie (1979a) examined the survival time of adults and fully developed larvae (extramarsupial larvae, ready for birth) of six species of pisidiid clams to aerial exposure. Some adults of all six species survived for 30 minutes, and 50% or more of the unborn larvae survived this treatment for 1 hour. Mackie concludes that the most likely survivors of aerial transport are the extramarsupial larvae or newborn clams. Also, there has been some controversy about survival of clams in the guts of waterfowl. Thompson and Sparks (1977) investigated the survival of *C. fluminea* in the gut of the lesser scaup duck, *Aythya affinis,* and concluded that dispersal in the gut of ducks is not a likely mechanism because there was 100% mortality of clams and their veliger larvae. However, Mackie (1979a) fed pisidiids to mallards, *Anas platyrhynchos platyrhynchos,* and collected clams from regurgitated material (food which was only stored in the crop) in pans below their cages. Of the many clams recovered, one *Musculium securis* and two *Sphaerium occidentale* survived and reproduced in laboratory culture. Thus, it is possible that ducks can disperse pisidiids over short distances in their crops. Only one of these hermaphroditic and viviparous clams is necessary to establish a new population. Clarke (1973) points out that after each glacial event much of Canada was invaded by numerous molluscs. Obviously, colonization involves passive dispersal. One only needs to consider *C. fluminea* as an example of rapid colonization and dispersal. The first established population was recorded on the west coast of North America in the 1920s, and today there are numerous populations along the east coast (Thompson and Sparks, 1977; McMahon, 1982; Chapter 12, this volume).

Although unionacean clams are apparently dispersed by the same mechanisms as those described for pisidiids, the likelihood of survival and reproductive success of a dispersed individual is undoubtedly less. Unionaceans are comparatively longer lived, and have both dioecious and hermaphroditic species in addition to a parasitic link of the glochidial larvae on vertebrates (mostly fish). Some mussel species have a limited number of hosts, whereas others utilize many (Fuller, 1974; Section IV,A,B). The parasitic larval stage on fish provides an excellent means of dispersal within a watershed, but even a waterfall or a dam apparently can become a formidable barrier. These clams have been traditionally considered as restricted to their native watershed. The nature of less efficient dispersal and/or low success upon arrival in a new habitat has been the basis of the use of mussels to trace the geological confluence of drainage systems (Van der Schalie, 1945). Isolation and a low level of colonization success makes it possible to treat the rivers of the North American East and Gulf coasts as biogeographic islands (Sepkoski and Rex, 1974).

III. Physiological Adaptations, Acclimation, and Environmental Stress

Consideration of physiological adaptations requires a brief synopsis of some basic relationships: First, there is the relationship between metabolism and body size; second, there is the response of metabolism to immediate changes in temperature or oxygen availability; third, there are long-term adaptational responses to seasonal temperature, oxygen concentration, or absence of water; and fourth, there are sequential influences of changing life-history commitments to growth and reproduction. First, the relationship between metabolism for a whole clam [(e.g., $M = \mu lO_2/(\text{clam}\cdot h)$] in relation to body size (e.g., W = mg dry tissue/clam) can be expressed by $M = kW^b$, where k and b are constants. This power curve equation gives a linear relationship on a double logarithmic plot where k equals the y intercept for the M of a 1.0-mg clam and b is the slope of the line. Although textbook dogma reiterates the concept that $b = 0.75$ for animals (Zeuthen, 1947, 1953; Hemmingsen, 1950), investigations on clam respiration and filtration show important seasonal changes. When $b = 1.0$, the metabolism per unit weight of tissue (*weight specific metabolism*) is the same for animals of all sizes. When b is >1.0 or <1.0, the metabolism per unit weight of tissue is relatively lower or greater, respectively, for clams of smaller size. Second, the magnitude of immediate response to temperature can be expressed by Q_{10}. This is the ratio of rates or velocity constants 10° apart (e.g., $Q_{10} = 2.0$, a doubling of the rate for each 10° increase). Important seasonal responses to higher temperatures where metabolism may increase, remain unchanged (temperature insensitivity), or decrease would correspond to Q_{10} values of >1.0, 1.0, or <1.0, respectively. Immediate responses to lower oxygen concentrations evoke oxygen conformity (dependence) where metabolic rate decreases as oxygen concentration decreases, or oxygen independence (regulation) where metabolism is maintained at a constant rate down to some low critical level of oxygen availability. Third, long-term responses to changes in seasonal oxygen availability can cause changes from oxygen dependence to independence and the reverse; and changes in water availability can bring about dormancy in some clams when water is absent. Further, patterns of acclimation to temperature under controlled laboratory conditions (Precht et al., 1973) can potentially produce five distinct patterns where the value of k (in $M = kW^b$) changes. The responses of natural populations involves many adaptational factors, and although one can interpret patterns in relation to the standard definitions of Precht et al. (1973), one is really looking at what has been classified as acclimatization. Fourth, changing commitments to life-history demands of growth and reproduction are dealt with in detail in other sections of this chapter. However, these energetic commitments are usually reflected in all aspects of the seasonal responses in physiology.

It has been pointed out that there is a need for more seasonal studies that should be interpreted not only in terms of physical and chemical factors but also in relation to population energetics in a broader ecological–environmental context (Burky, 1971; McMahon, 1973; Burky and Burky, 1976; Russell-Hunter, 1978). I will, therefore, concentrate more on studies of natural populations.

A. Respiration

There are many reports on respiration in freshwater clams; however, most of these deal directly with responses to temperature, anaerobiosis, and drying under controlled laboratory conditions. Other studies provide information on respiratory variation for field-adapted clams during one season (usually summer) and estimate other seasonal levels of metabolism from limited data on Q_{10} and the size rate relationship ($M = kW^b$). Most of these reports provide valuable but limited estimates because the states of growth, reproduction, or life cycle are often ignored or unknown. However, others have used these types of data to aid in the construction of seasonal energy balance sheets for population energetics (see Section IV,D). Although comprehensive studies on life cycle and habitat characteristics are seldom coupled with respiratory physiology, this type of paired investigation is exactly what is needed if we are to understand the complexity of adaptive seasonal responses. The studies of Burky and Burky (1976) on *Pisidium walkeri,* Hornbach (1980) and Hornbach et al. (1983a) on *S. striatinum,* Way et al. (1981) and Buchwalder (1983) on *Musculium partumeium,* and Alexander (1982) and Alexander and Burky (1982) on *M. lacustre* provide this type of information. Apart from these studies, students of respiratory physiology are directed to the following references: Culbreth (1941), Dance (1958), Segal (1961), Berg et al. (1962), Alimov (1965a, 1975), Berg and Jónasson (1965), Lukacsovics (1966a,b), Collins (1967), Salánki and Lukacsovics (1967), Tudorancea and Florescu (1968b, 1970), Nagabhushanam and Lomte (1970), Johnson and Brinkhurst (1971), Lomte and Nagabhushanam (1971), Bayne (1971), Jónasson (1972), Badman and Chin (1973), Hochachka et al. (1973), Badman (1974, 1975), Zs-Nagy (1974), Dietz (1974), De Zwaan and Wijsman (1976), Holopainen and Ranta (1977a,b), McMahon (1979a,b), Waite and Neufeld (1977), and Bleck and Heitkamp (1980).

Seasonal studies of respiration on freshwater clams using Clark-type oxygen electrodes (Clark, 1956) in conjunction with acrylic chamber units (Burky, 1977) have been carried out (Burky and Burky, 1976; Way et al., 1981). This technique can provide quick measurements of initial metabolic rates and also allow the assessment of responses to oxygen depletion over time (Waite and Neufeld, 1977; McMahon, 1979a). Unfortunately the mechanical vibrations of stirring disturb some clams and can render the use of electrodes inappropriate. Hornbach et al. (1983a) used the standard Winkler titration (Strickland and Parsons, 1965; American Public Health Service, 1975) to assess seasonal metabolic adaptations in *S. striatinum.* In seasonal studies of this type, a large number of values on field-acclimated clams representing a size series are simultaneously assessed at two or more temperatures. Hornback et al. (1983a) sealed clams in 40-ml glass-stoppered dropping bottles for maintenance at each experimental temperature. At the end of each experiment, the water was fixed (Winkler method: bottles with and without clams) and the corresponding aliquots titrated for oxygen content. Under the caustic conditions necessary to fix Winkler samples, the clams simply close their shell valves. However, they must be immediately removed from the bottles and washed, and subsequently muffled to determine the ash-free dry weight (equivalent to the tissue dry weight). D. J. Hornbach (*personal communications,* 1982) claims that the approach works on operculate gastropods, and it may be applicable to pulmonate snails as well. This method facilitates the collection of the large amounts of data necessary for seasonal studies and has been successfully employed in studies on *S. striatinum* (Hornbach et al., 1983a), *M. lacustre* (Alexander, 1982; Alexander and Burky, 1982), and *M. partumeium* (Buchwalder, 1983).

General studies on metabolism suggest that the effect of body size or *b* value (from $M = kW^b$) is constant over a broad range of conditions. Consequently, only values of absolute metabolism (*k* values) are usually reported. Further, Alimov (1975) concludes that *b* values of 0.713 and 0.895 for the Unionidae and Pisidiidae (Sphaeriidae) are constant. Alimov combined these values with those of other freshwater clams (e.g., 0.63 for *D. polymorpha*) and claimed that the *b* value is 0.72 for all freshwater clams or 0.73 for freshwater and marine forms combined. Alimov apparently supports the dogma of a *b* value equal to 0.75, but many studies do not support this view. Burky and Burky (1976) and Way et al. (1981) claim that *b* values for *P. walkeri* and *M. partumeium,* respectively, are not significantly different from 1.0. It is possible that the small range of body size for *P. walkeri,* along with developing broods in these hermaphroditic and viviparous clams, has an influence. Values of *b* range from 0.25 to 0.85 for four North American (Johnson and Brinkhurst, 1971) and from 0.58 to 1.049 for four European (Holopainen and Ranta, 1977a,b) species of *Pisidium;* they range from 0.53 to 1.45 for *S. striatinum* (Hornbach et al., 1983a) and from 0.57 to 1.34 for *M. lacustre* (Alexander, 1982; Alexander and Burky, 1982).

The *b* value varies seasonally, with values >1.0 in the winter and <1.0 in the summer, respectively, for *S. striatinum* (Hornbach et al., 1983a). This indicates that larger clams have both a greater absolute metabolism and a higher weight-specific metabolism in the winter. Values of *b* for *M. lacustre* (Alexander, 1982;

Alexander and Burky, 1982) decline before birth periods, suggesting that smaller clams may be growing faster; the *b* values increase during birth periods, indicating that the presence of larger embryos in brood sacs may have an influence. Also, Holopainen and Ranta (1977b) have shown that the *b* value is positively correlated with changes in temperature in littoral and profundal species of *Pisidium* (i.e., as temperature increases, the *b* value increases). However, Hornbach et al. (1983a, for *S. striatinum*) and Alexander and Burky (1982, for *M. lacustre*) have shown that in general field-adapted clams show an inverse relationship between *b* value and changing temperature (i.e., as temperature increases, the *b* value decreases). These studies (Hornbach, et al., 1983a; Alexander, 1982; Alexander and Burky, 1982) show an inverse seasonal relationship between *b* and *k* values, so that as *k* increases, the increase is proportionately greater for smaller clams (i.e., *b* decreases). It also follows that when metabolism (*k*) reaches a maximum, the absolute increase is greater although it is proportionately less for larger clams. Maximum changes in *k* and *b* values appear to be associated with peak reproduction in the studies on both *S. striatinum* and *M. lacustre*. Further, Hornbach et al. (1983a) claim that lower *b* values for metabolism, in conjunction with *b* values of 1.0 for rates of filter feeding, suggest an increase in the relative efficiency of energy allocation for reproduction.

Seasonally, Q_{10} values can range from <1.0 (Burky and Burky, 1976; Hornbach et al., 1983a; Alexander, 1982) to 9.76 (Johnson and Brinkhurst, 1971). In studies on species of *Pisidium,* Johnson and Brinkhurst (1971) and Burky and Burky (1976) observed that the Q_{10} of metabolism is lower in the winter than in the summer. This presumably dampens the response to short-term elevations in temperature during the winter. Also, Alexander (1982) reports elevated Q_{10} values during reproduction in *M. lacustre,* with higher values for smaller clams, presumably for the rapid growth of smaller clams and their developing broods in response to increases in temperature. Johnson and Brinkhurst (1971), Burky and Burky (1976), Hornbach et al. (1983a), and Alexander (1982) all point out that different seasons and biological conditions may affect patterns of metabolism as well as Q_{10}. In Hornbach's study on *S. striatinum,* Q_{10} values were greatest during the winter, and smaller clams have greater Q_{10} values during the summer as well as seasonal variations in *b* value. Hornbach argues that this situation allows smaller clams to take advantage of short warm periods in the winter, when an equivalent response would bring larger clams to a premature reproductive condition at an inappropriate time. Further, Q_{10} values of <1.0 during the summer for smaller clams prevent inappropriate growth at higher temperatures (i.e., as temperature increases, metabolism decreases), whereas larger clams with a Q_{10} of 1.0 show temperature insensitivity for purposes of reproductive stability. It is significant that Waite and Neufeld (1977) demonstrated temperature insensitivity (Q_{10} = 1.0) for *S. simile* over a range of 15° to 25°C. However, when excised gill and mantle tissues were examined, the temperature insensitivity was destroyed. This suggests that control mechanisms for temperature responses operate at the whole organism level.

Many bivalves can tolerate the high summer ambient temperature characteristic of some freshwater habitats and show appropriate adaptations in Q_{10}, as discussed above. There are also a number of other studies which approach temperature adaptations by investigating upper tolerance limits for clams conditioned at a series of temperatures. When oxygen consumption is reduced above a specific temperature range, McMahon (1979a) refers to this reduction as *respiratory impairment* for *C. fluminea*. Certainly, there is impairment for all organisms above some critical temperature. However, within limits, reduced metabolism at higher temperatures can be an adaptational response to dampen potential effects of temperature (e.g., Q_{10} values <1.0). This is particularly true when this reduction in metabolism corresponds to survival during a period of seasonally high temperatures. *Corbicula fluminea* apparently survives such hot periods in some populations (Aldridge and McMahon, 1978). *C. fluminea* can tolerate exposure to 34°C (30°C acclimated) (Mattice and Dye, 1976), and *Parreysia corrugata* can tolerate 39.5°C (33°C acclimated) (Nagabhushanam and Lomte, 1970). However, oxygen consumption is severely impaired above an acclimation temperature of 25° to 30°C for *C. fluminea* (McMahon, 1979a), whereas *P. corrugata* shows no such impairment (Lomte and Nagabhushanam, 1971). Such comparisons by McMahon (1979a) may not be valid without more information on the chemical and physical nature of normal habitats. Also, a continued rise in metabolism as temperature increases could represent poorly developed adaptations (in some clams) for coping with high temperatures and could lead to a permanent impairment of life itself. As pointed out earlier, metabolic interpretations require integration of all aspects of an organism's biology.

Responses to low oxygen availability is another topic of interest because bivalves occur under conditions ranging from oxygen saturation to complete anoxia. In *S. simile* there is a relatively broad range of oxygen independence down to an oxygen tension of 16–32 mm Hg when metabolism collapses (Waite and Neufeld, 1977). This oxygen-independent metabolism is most marked at 30°C, with some amount of oxygen-dependent metabolism at 25°, 20°, and 15°C. Good regulation is also shown by *Pisidium casertanum* (Berg et al., 1962). This is the complete opposite of the oxygen-dependent response reported for *Parreysia corrugata* (Lomte and Nagabhushanam, 1971), *C. fluminea* (McMahon, 1979a), and *M. partumeium* and *M. lacustre* (Buchwalder, 1983), and many unionacean (e.g., Lukacsovics, 1966a,b; Culbreth, 1941; Dietz, 1974; Lomte and Nagabhushanam, 1971; Tudorancea and Florescu, 1970) and corbiculacean (e.g., Juday, 1908; Berg et al., 1962; Thomas, 1963, 1965; Berg and Jónasson, 1965; Jónasson, 1972; McMahon, 1979a,b; Way et al., 1980) clams can tolerate varying periods of anoxic conditions. This ranges from 3 days for *Anodonta cygnaea* (Lukacsovics, 1966a,b), to 4–5 days for *Ligumia subrostrata* (Dietz, 1974), to 7–14 days for *C. fluminea* (McMahon, 1979b), to 14 days for *Pisidium idahoense* (Juday, 1908), to 30 days for *Anodonta hallenbeckii* (Culbreth, 1941), to >6 weeks for *M. partumeium* (D. G. Conover and A. J. Burky, unpublished data). Interestingly, most of these time periods represent resistance adaptations to carry the clams over until oxygen becomes available. For *M. partumeium* this is not the case because most often growth and reproduction take place under anaerobic conditions (Thomas, 1963, 1965; Way et al., 1980). Significantly, the greatest oxygen-dependent response in metabolism is during seasonal periods of low oxygen availability (Buchwalder, 1983).

B. Feeding

Filter feeding in bivalves is achieved by passing water (propelled by cilia) between gill filaments and straining fine particulate material from suspension for subsequent ingestion. The general structure and function of the bivalve gill for both feeding and respiration can be found in a number of general texts (Russell-Hunter, 1979; Barnes, 1980). More specific references provide background information on the type, size, and concentration of natural suspended foods as well as gill porosity, control of filtration rate, and responses to specific conditions (Jørgensen, 1966, 1975; Winter, 1978). The rate at which bivalves pump water can be estimated using both direct and indirect approaches. Direct approaches either collect or measure the water flow through the mantle cavity or calculate this from observations on the cross section and velocity of the exhalant current (Jørgensen, 1966). The indirect approach of measuring filtration rates by the clearance of particles from suspension is well known (Coughlan, 1969; Jørgensen, 1975; Burky and Benjamin, 1979), where particle suspensions such as colloidal graphite or clay, cultures of algae, and latex beads have been used. This allows the calculation of clearance rates or the volume of water depleted of particles over time. Actual or relative particle concentrations have been assessed photometrically, with radioactive labels, and by the use of electronic counters. Clearance rates approach actual rates as particle retention becomes more efficient. Such clearance or water transport rates are sometimes referred to as *filtration rates* (FR), expressed for whole clams [e.g., FR = ml H_2O/(clams·h)] or per unit weight [e.g., FR/W = ml H_2O/(g dry tissue·h)], and can be related to the size–rate equation ($FR = kW^b$). However, clearance rates can provide information on the actual particle retention only if accurate particle concentrations are known. Burky and Benjamin (1979) adapted the use of uniform-size latex beads for measuring filtration rates on tiny pisidiid clams in 20 ml of water. This adaptation facilitates the simultaneous measuring of the large number of values necessary for seasonal studies on natural populations (Benjamin, 1978; Benjamin and Burky, 1978; Hornbach, 1980; Hornbach et al., 1983b.

Jørgensen (1975) discusses a number of papers dealing with both the rate of water transport and gill porosity. Gill function is probably under some neural control because filaments receive both serotonergic and dopaminergic nerve fibers, forming a presumed excitatory-inhibitory mechanism for the control of the beat of lateral cilia. The laterofrontal cilia presumably beat independently of gill innervation, but experimentally serotonin can alter the angle of beat. This would then change porosity within limits of the distance between filaments. There is still much controversy over the efficiency of particle retention, but Jørgensen (1975) states that generally there is a low gill porosity in undisturbed bivalves, with little difference in retention in the 1–200 μm range.

Two general rate responses to particle concentration are reported. First, there is evidence that water clearance rates vary inversely with the concentration of suspended material (Monakov, 1972; Jørgensen, 1975; Winter, 1978, and references therein); second, there is evidence that some bivalves maintain feeding currents at a constant rate which is independent of the suspension concentration over a broad range of concentrations (Ali, 1970; Jørgensen, 1966, 1975). It appears that there are a number of potential strategies which could produce distinct patterns of responses in regard to particle concentration and the rate of water transport. Therefore, understanding each strategy demands knowledge of many aspects of seasonal ecology and physiology for each organism. In freshwater clams, an inverse relationship between clearance rate and particle concentration has been reported for *Dreissena polymorpha* (Mikheev, 1966), *S. suecicum* (Alimov and Bulion, 1972), *S. striatinum* (Hornbach, 1980; Hornbach et al., 1983b), and *M. partumeium* (Conover et al., 1981). For some clams, the inverse relationship can be described by $\text{FR} = A + B(R)^{\text{conc}}$ where FR is the water transport (ml H_2O/h) for a standard-size clam (e.g., 1.0 mg dry tissue); A is the asymptotic or minimum value of FR; B is the difference between A and the value of FR when the particle concentration is zero; R is the ratio of successive differences along the curve; and conc is the particle concentration in milligrams per liter. The concentration response is described as $\text{FR} = 0.568 + 9.676(0.902)^{\text{conc}}$ for *S. striatinum* at 20°C (Hornbach, 1980; Hornbach et al., 1983b) and $\text{FR} = 0.5 + 5(0.9)^{\text{conc}}$ for *M. partumeium* at 20°C (Conover et al., 1981). If one multiplies FR by the particle concentration, the rate of particle retention can be predicted for each concentration. This illustrates that at extremely low concentrations the rate of particle removal is low. As the concentration increases, the rate of removal increases rapidly to a peak level, after which the rate either plateaus or shows some decrease. At even higher concentrations, the rate increases again as the suspension concentration increases. In reality, pseudofeces production begins after the initial peak removal rate. It follows that actual ingestion is the amount of suspension filtered minus the pseudofeces production. The ingestion rate potentially can level off or decrease at higher concentrations of suspension (Winter, 1978; Hornbach et al., 1983b; Burky, unpublished data). Mattice (1979) and Conover et al. (1981) report that filtration rates are independent of particle concentration for *C. fluminea* and *M. lacustre,* respectively.

Although Conover et al. (1981) report an inverse relationship for *M. partumeium* under aerobic conditions, the rate of filtration is uniformly low and is independent of particle concentration under anaerobic conditions. These rates were measured during the summer, when *M. partumeium* is normally dormant under anaerobic conditions in its natural habitat. Further, Jørgensen (1975) contends that ingestion rates vary directly with the particle concentration in natural waters where particle concentrations are below levels at which an effect could be produced. This direct relationship between ingestion rate and concentration would also apply to clams such as *C. fluminea* and *M. lacustre*.

Also, it is known that many clams do not continuously filter water because shell valves may remain closed for various periods of time (Benedens and Hinz, 1980; Salanki and Vero, 1969). Obviously, variations in rates of water transport due to the concentration of natural suspensions, oxygen availability, and diurnal activity should be integrated to provide actual values of daily ingestion.

For freshwater clams, only the studies on *M. partumeium* (Benjamin, 1978; Benjamin and Burky, 1978; Burky and Benjamin, 1979; Conover et al., 1981) and *S. striatinum* (Hornbach, 1980; Hornbach et al., 1983b) provide seasonal data on filtration dynamics in relation to the energy budget, life cycle, and ecology of natural populations. These studies examine seasonal changes in both b and k values (from $\mathrm{FR} = kW^b$), and temperature effects (Q_{10} values for *M. partumeium*), as well as the response to particle concentration. Although the b value is claimed to be constant in species of *Sphaerium* (b = 0.6; Alimov, 1965b) both *M. partumeium* (b values range about 0.4–1.2) and *S. striatinum* (b values range about 0.5–1.0) show important seasonal changes. In general, both of these species show an inverse seasonal relationship between b and k values (i.e., as b decreases, k increases). For *M. partumeium* absolute metabolism (k value) peaks in the spring and fall, corresponding to periods of peak reproduction, and although the absolute clearance rates are higher for larger clams, the weight-specific rate (FR/W) is lower. Smaller clams feed faster and presumably grow faster as embryos continue to develop in their brood sacs. There is a similar relationship between peak reproduction and peak k values in *S. striatinum*. However, the changes in k and b values take place over longer periods of time for *S. striatinum*. For *S. striatinum*, reproduction is spread over a longer period of time and is less synchronous than in *M. partumeium* (see Section IV,B, IV,D, and V). For *M. partumeium*, the periods of peak k values do not correspond to periods of peak concentrations of natural suspended material or to periods of peak temperature. Also, temperature sensitivity changes seasonally for *M. partumeium*, with Q_{10} values between 2 and 3 in the winter and about 1.0 (i.e., temperature insensitivity) in the summer. The lack of exact correspondence among peak reproduction (and corresponding changes in b and k values), seasonal changes in Q_{10}, seasonal changes in pond temperature, and seasonal changes in the concentration of suspended particles suggests that other factors (e.g., photoperiod; see Section IV,A) as well as a complex integration of all reported adaptations are involved in the control of clearance rates. It must be emphasized that filtration studies on *M. partumeium* and *S. striatinum* are part of broader attempts to construct energy balance sheets for natural populations. However, studies on both of these clams report that they cannot meet their energy needs from predictions of filter feeding and thus deposit feeding probably is involved. It has been suggested that deposit feeding is probably important for a number of freshwater pisidiid clams (Monokov, 1972; Mitropolskii, 1966a,b; Benjamin, 1978; Benjamin and Burky, 1978; Hornbach, 1980; Hornbach et al., 1983b). Species of *Pisidium* apparently pump water into the mantle cavity along the foot, as well as through normal inhalant areas (Mitropolskii, 1966a,b; Meier-Brook, 1969; Monokov, 1972). This suggests a direct uptake of detritus associated with cilia along the foot. Deposit feeding and filter feeding by these tiny clams are ecologically important if one considers conditions of high population density (Hinz and Scheil, 1972; Section IV,C). Other mechanisms for obtaining nutrition in freshwater clams involve the direct absorption of dissolved organics in *Pisidium casertanum* (Efford and Tsumura, 1973) and uptake resulting from the symbiotic relationship of zoochlorellae and *Anodonta* (Pardy, 1980).

C. Ionic and Osmotic Factors

Like other organisms living in fresh water, the bivalves face problems of continuous solute loss and water gain and must maintain body fluid concentrations above those in the surrounding water through regulatory mechanisms. Interestingly, freshwater bivalves regulate their body fluids at a lower concentration than any other freshwater organism (Robertson, 1964; Potts, 1967, 1968; Prosser, 1973; Dietz, 1977; Matsushima, 1980; McCorkle and Dietz, 1980; and references in each of these works). Most of the published research concerns mussels (Unionacea), *D. polymorpha,* and corbiculid clams. There does not appear to be any published work on the pisidiids.

Apparently, the lowest blood concentration is reported for *A. cygnaea* (Potts, 1954), with an overall osmotic concentration of about 22 m*M*/liter (as NaCl), which is about 4–5% of the concentration of sea water. Other freshwater animals have higher concentrations, with values generally between 50 and 100 m*M*/liter (except some snails) for soft-bodied invertebrates and between 100 and 200 m*M*/liter for arthropods and teleosts (Prosser, 1973). In general, the blood ion concentration for freshwater mussels is approximately 15–20 m*M* Na^+, 0.5 m*M* K^+, 4–8 m*M* Cl^-, and 11–14 m*M* HCO_3^-. Apparently, *C. fluminea* is different, because NaCl is the predominant salt (about 29 m*M* Na^+ and 25 m*M* Cl^-), with <3 m*M* HCO_3^- (Dietz, 1977; McCorkle and Dietz, 1980). These clams maintain their ion concentration through active uptake across exposed epithelia including the gut. Na^+ and Cl^- transport are independent; Na^+ is exchanged for NH_4^+ or H^+ and Cl^- is exchanged for HCO_3^- or OH^- (Deitz, 1977, and references therein). Also, the rate of Na^+ transport for *C. fluminea* is apparently greater than that of other freshwater bivalves and most other freshwater animals (McCorkle and Dietz, 1980). It has been pointed out that not only do freshwater mussels have well-developed transport systems, but the mechanisms involved are indistinguishable in capacity from those of other freshwater animals. Further, freshwater clams have well-developed pericardial glands which presumably are for the production of large-volume fluid elimination (Martin and Harrison, 1966). Nevertheless, Dietz (1977) argues that the low concentration of blood solutes is not a deficiency of the ion transport mechanism, but instead may be due to a limited capacity of the bivalve kidney for water excretion. A limited kidney function would be reflected in the presumed loss of ability for volume regulation in their evolution from marine to freshwater conditions (Gainey and Greenberg, 1977).

Apart from the generalizations already discussed, it should be pointed out that free amino acids are important in the osmoregulation of blood and cells for *Ligumia subrostrata* (Hanson and Dietz, 1976) and that blood concentration increases under conditions of dehydration and stress (Hanson and Dietz, 1976; Dietz, 1977). Also, a number of studies have examined heavy metal uptake (Smith and Green, 1975; and references in Fuller, 1974, and Merlini et al., 1978), a component of increasing concern in water pollution. Further, an ion of major importance for bivalves is Ca^{2+} because shell secretion is critical. Picken (1937) demonstrated that *A. cygnea* can maintain higher blood concentrations of Ca^{2+} and HCO_3^- in harder water. However, the ionic composition and concentration of the extrapallial fluid are probably more important for shell secretion (references in Wilbur, 1964, and Pietrzak et al., 1976), but the amount of $CaCO_3$ allocated to shells can be independent of environmental concentrations of Ca^{2+} (see Section III,D).

D. Shells

In clams the shell valves provide mechanical, chemical, and biological protection and give support for soft body parts. The shells of many clams are composed of three basic structural layers: the outer periostracum, which is primarily composed of proteinaceous material; a middle prismatic layer, which is made up mostly of calcium carbonate in an organic matrix; and an inner nacreous layer (absent in corbiculacean clams), which is mostly calcium carbonate (Wilbur, 1964; Denis, 1979).

Natural populations of freshwater molluscs show extreme variations in the pattern of shell secretion which may or may not be related to water hardness and/or trophic conditions. This is particularly true for freshwater pulmonate snails (Russell-Hunter et al., 1981, and references therein). There is considerably less information on the nature of shell secretion of freshwater clams in relation to water chemistry. Agrell (1949) reported that in general the eutrophic waters had greater concentrations of dissolved calcium. He showed that for populations of *Unio tumidus,* shell weight decreases as trophic conditions and dissolved calcium increase, and that the shell weights of *U. pictorum* and *A. piscinalis* increase as trophic conditions and dissolved calcium increase. Agrell argued that both available calcium and trophic conditions are important for shell secretion and that these species were most abundant in habitats where they were able to secrete the heaviest shells. Green (1972) analyzed numerous environmental factors in 19 Canadian lakes and concluded that stability of the overall water chemistry is important for the physiological processes involved in $CaCO_3$ deposition for the production of thicker shells in *Lampsilis radiata* and *A. grandis.* His data suggest that high NaCl is a limiting factor in unionid distribution. Like the report on *U. tumidus* (Agrell, 1949), Burky et al. (1979) reported an inverse relationship for six populations of *S. striatinum* between the amount of shell calcification and water hardness. Burky et al. (1979) also reported that there is an inverse relationship between the amount of shell $CaCO_3$ and shell organics interspecifically for *S. striatinum, S. simile, P. walkeri, M. partumeium* and *M. transversum.* This relationship also holds intraspecifically for populations of *S. striatinum, S. simile,* and *M. partumeium.* Further, values converted from data on *C. fluminea* (from Aldridge and McMahon, 1978) and data on *M. lacustre* (Alexander, 1982) closely agree with this relationship. It can be argued that these findings (inverse relationships between the amount of shell calcification and shell organics) provide convincing evidence that pisidiid shells are consistently built for some base level of adaptive need for mechanical protection. That is, the mechanical integrity of the shell is provided by structural proteins and crystalline $CaCO_3$, and these bivalves do not wastefully incorporate structural components into their shells. This suggests that pisidiid clams have a common strategy for shell secretion which would presumably be under genetic control. Mackie (1978c) showed that within the Corbiculacea there is little or no variation in shell structure and crystalline type (orthorhombic aragonite) with environmental conditions. One would expect such basic structural similarities if selection for shells of basic mechanical strength is real. Energetically, it would seem most economical for clams to build the most calcareous shells in the hardest waters (this pattern holds for *U. pictorum* and *A. piscinalis*) (Agrell, 1949) or the most proteinaceous shells under the most eutrophic conditions. However, in terms of water hardness this relationship does not hold for some clams because there is no interspecific relationship between shell calcium (or weight) and water hardness in pisidiids, and there is an inverse relationship for *S. striatinum* (Burky et al., 1979) and *U. tumidus* (Agrell, 1949). In terms of trophic conditions, a weak argument can be made for *M. partumeium* because it occurs in the most eutrophic habitats (Burky et al., 1979) and produces the thinnest shells (Mackie, 1978c;

Burky et al., 1979) with the highest proportions of shell organics (Burky et al., 1979). Obviously, more work is needed on the relationship of shells to environmental conditions. Such studies should include many populations of each species (for taxonomic relationships) over a broad range of habitat types.

Another interesting aspect of pisidiid shells is the presence of punctal canals (Rosso, 1954; Mackie, 1978c; Robertson and Coney, 1979). These are minute canals extending from the inside of the shell to the outer periostracum. Studies using scanning electron microscopy (Robertson and Coney, 1979) show that for *M. securis* the punctal canals do not penetrate the periostracum. However, Rosso (1954) assumed that mantle tissue in these canals penetrates the periostracum. Robertson and Coney (1979) speculate that although the periostracum provides a structural separation between the environment and mantle tissue, these canals may provide a route for gas exchange and waste removal when the valves are closed. They also propose that for species inhabiting temporary ponds, such as *M. securis* and *M. partumeium,* these canals may provide a mechanism for sensing water conditions as well. Although shells of *Musculium* are translucent, these punctal canal areas could have concentrations of photoreceptors. It has been proposed that *M. partumeium* might respond to photoperiod cues (Way et al., 1980; Hornbach et al., 1980c; Conover and Burky, 1981).

IV. Life Histories, Population Dynamics, and Population Energetics

A. Life-Cycle Patterns

There is a considerable amount of information on the nature of the life-history patterns for temperate unionacean clams or freshwater mussels. Fuller (1974) provides an excellent documentation of basic information for these clams, which there is no need to duplicate. However, he points out that a general account of their normal life cycle reveals important biological aspects of their ecology. It is well known that mussels have a required period of parasitism for their glochidial larvae on vertebrate hosts (usually fish). The consequences of this parasitic phase are ultimately the most vulnerable aspect in their life history (Stein, 1971; Fuller, 1974) and are one reason why they are sensitive to anthropogenic influences on fish populations as well as hydrology.

A short summary of the general life-cycle pattern is provided: male mussels (dioecious and hermaphroditic species) shed sperm [as packets or volvocoid bodies (Edgar, 1965)], which enter the female on inhalant currents; fertilized eggs are incubated in the marsupial portion of the gill, and the miniature bivalve (glochidia) stage are released in large numbers through the exhalant aperture [except for most Lampsilinae (Kraemer, 1970; Fuller, 1974)]. The glochidia are frequently suspended on mucuous threads or lace (Yokley, 1972) to facilitate host infection by surface contact or through feeding. The glochidia are not free-swimming, and although some are found alive in the plankton, this is apparently not an important mode of host infection and/or dispersal (Fuller, 1974, and references therein). Glochidia encyst [the cysts are formed by the growth of host epithelium (Heard and Hendrix, 1964; Lukacsovics and Labos, 1965)] on fins, under scales, or on softer tissues such as gills, depending on the fish host species. Glochidia grow and metamorphose, and eventually aid in the rupture of their cysts for release (Arey, 1932). Juvenile mussels must find a suitable stable substrate (Negus, 1966) for continued growth, after which time they may migrate to other substrates. Apparently, anything that disrupts the parasitic link in the life cycle imperils these clams. The association is particularly fragile for those mussel species capable of infecting a limited number of host species, whereas the more successful and abundant mussels such as *Amblema plicata, Anodonta grandis,* and *Megalonaias gigantes* can infect a large number of fish species (Fuller, 1974). Fuller (1974) provides a comprehensive list of glochidial hosts. Given successful recruitment to the adult population, mussels can be long lived (years), with an iteroparous pattern of reproducing many times over their life span.

Since the late eighteenth century, the freshwater dreissenacean clam *D. polymorpha* has apparently spread throughout Great Britain, Europe, and parts of Asia (Morton, 1969c) and shares an infamous pest distinction with *C. fluminea.* Like *C. fluminea, D. polymorpha* is oviparous, with a free-living veliger larva. *D. polymorpha* exists in dioecious populations (Walz, 1978b) or those which also have a small number of hermaphroditic individuals (Antheunisse, 1963). These clams can live for 5 years (Morton, 1969c) and are iteroparous, reaching a maximum size of about 5 cm (Morton, 1969c). For further insight into the biology of *D. polymorpha,* see the works of Wiktor (1963), Stánczykowska (1963, 1964), Morton (1969a,b,c,), and Walz (1978a–f, 1979).

The freshwater corbiculacean bivalves are hermaphroditic and relatively much smaller (maximum shell length about 4 cm) and shorter lived than most unionacean clams. *Corbicula fluminea* is oviparous, semelparous (Heinsohn, 1958), or iteroparous (Morton, 1977; Aldridge and McMahon, 1978) and has a lifespan of 14–17 months in a Texas lake (Aldridge and McMahon, 1978), 3+ years in a Hong Kong reservoir (Morton, 1977), and possibly 5–7 years in a Tennessee population (Sinclair and Isom, 1963). More details on *C. fluminea* can be found in Chapter 12 by McMahon, this volume. Studies on the hermaphroditic and viviparous (ovoviviparous, according to Mackie, 1978b) pisidiid clams have revealed a number of life-cycle patterns which are influenced by seasonal, physical, chemical, and trophic factors as well as adaptations for habitats ranging from deep profundal lake sediments to large rivers and streams to marshes with little standing water to small permanent or temporary ponds. There is evidence for what may be common self-fertilization in this group (Thomas, 1959; Hornbach et al., 1980a,b; McLeod et al., 1981). As might be expected from the range of habitats, there is considerable variation between interspecific and intraspecific populations in life-cycle pattern (most data are essentially on temperate populations). Notably, the classification of life-cycle patterns adapted by Russell-Hunter (1964, 1978) for freshwater snails (mostly pulmonates) generally applies to these tiny clams and undoubtedly reflects some similarities in selection for life in transient aquatic habitats, despite the obvious differences in life style between most snails and clams. Russell-Hunter identifies two biennial patterns with reproduction in 2 successive years or a single reproduction after about 22 months; two simple annual patterns with either a single spring or a single late summer reproductive period; two patterns involving two generations per year, either with complete semelparous replacement of sequential generations or with both generations overwintering to contribute to the next year's breeding; and, lastly, three successive generations per year (a pattern not documented for freshwater clams). Further, for pisidiid clams there are populations with continuous reproduction throughout the year coupled with major birth period(s); some pisidiids can live for years. The plasticity of life-cycle responses is well documented for freshwater snails (Russell-Hunter, 1978, and references therein). Similarly, many pisidiid clams illustrate a wide variation in life-cycle pattern. This is particularly true for cosmopolitan species such as *P. casertanum* and *S. striatinum*. Also, some pisidiid clams have life-cycle adaptations for temporary aquatic habitats, with some fixed responses to conditions and cues. Studies on life-cycle pattern and reproduction include reports on *Eupera* (Mackie and Huggins, 1976); *Musculium* (Thomas, 1963, 1965; Mitropolskii, 1965; Mackie et al., 1976a,b; Mackie, 1979b; Gale, 1977; Heard, 1977; Hornbach et al., 1980c; McKee and Mackie, 1981; Way et al., 1980); *Pisidium* (Heard, 1965; Thut, 1969; Gillespie, 1969; Ladle and Baron, 1969; Meir-Brook, 1970; Jónasson, 1972; Danneel and Hinz, 1976; Holopainen, 1978, 1979, 1980; Holopainen and Hanski, 1979; Mackie, 1979b; Hamill et al., 1979; Heitkamp, 1980; Burky et al., 1981; Way and Wissing, 1982); and *Sphaerium* (Monk, 1928; Foster, 1932; Mitropolskii, 1966a; Alimov, 1967; Collins, 1967; Avolizi, 1970, 1976; Zumoff, 1973; Heard, 1977; Mackie, 1979b; Hornbach, 1980; McKee and Mackie, 1981; Hornbach et al., 1982).

Although *P. casertanum* and *S. striatinum* are the most cosmopolitan pisidiid species; this distinction of abundance has led to only a relatively few studies of life history. In distinct natural populations of *P. casertanum* there is a wide range of patterns: a simple annual pattern in permanent (Thut, 1969) and temporary (Mackie, 1979b) habitats; two generations per year with complete replacement in a temporary pond (Mackie, 1979b); a 30–33-month life span for a low-temperature stream population with a single reproductive period after 24 months of growth (Burky et al., 1981); populations with 3-year life spans and one or two reproductive periods (Holopainen, 1979); and data suggesting a life span of 5 years in a population under conditions of generally low temperature, oxygen concentration, and productivity (Jónasson, 1972). For *S. striatinum* there is a pattern for the recruitment of two generations per year, with a life span of about 1 year for a pond (Foster, 1932) and a stream population (Hornbach, 1980; Hornbach et al., 1982). Avolizi (1970, 1976) reported a life span of 18 to 24 months in a lake population with two periods of reproduction. Further, peak reproduction for *S. striatinum* occurred in spring–summer and fall in the studies of Avolizi (1970, 1976) and Hornbach et al. (1982), but in the summer and winter in Foster's study; Heard (1977) reported a stream population with a single period of summer reproduction. Although reproduction in *S. striatinum* is identified for specific periods, there may be continuous births at a low level between periods of major recruitment (Hornbach, 1980; Hornbach et al., 1982). Continuous birth with two major periods of recruitment was also reported for *S. simile* by Zumoff (1973).

Habitat notes on species of pisidiid clams (Herrington, 1962; Burch, 1975a; Mackie et al., 1980) reveal that *M. lacustre, M. partumeium, M. securis, S. occidentale,* and *P. casertanum* are often found in temporary aquatic habitats in North America. The requirement for submersion with a water-filled mantle cavity for filter feeding necessitates that survival during dry periods involves resistance adaptations without energy intake. It is therefore not surprising that some life-cycle adaptations for temporary habitats persist in many populations under conditions of permanent water. As the aquatic habitat dries, both juvenile and adult *S. occidentale* burrow into the substrate to wait for the return of water. Under these conditions, embryos continue to develop in the brood sacs of adults (McKee and Mackie, 1981). *M. partumeium* and *M. securis* complete adult growth and reproduction before the habitat dries. Apparently, only juveniles are capable of surviving a dry period in the substrate. *M. securis* has a life span of about 12 months, with one birth period per year in a temporary pond (univoltine and semelparous), but two birth periods per year (bivoltine and iteroparous) in a stream and in a permanent pond (Mackie et al., 1976a,b). *M. partumeium* similarly lives about 12 months, with a single birth period per year in temporary ponds (Thomas, 1963, 1965; Hornbach et al., 1980c; Way et al., 1980) and two birth periods per year in permanent ponds (Hornbach et al., 1980c; Way et al., 1980). Although *M. securis* and *M. partumeium* are univoltine in temporary ponds, they switch to a bivoltine pattern in the occasional year when these ponds remain full. Although temporary pond populations respond to wet years with the production of an extra generation in the fall, newborn clams of permanent pond populations are dormant (no growth) during the summer, thereby mimicking populations in temporary habitats. It has been suggested that this pattern evolved in the temporary habitat (Mackie et al., 1976a,b; Mackie, 1979b; Hornbach et al., 1980c; McKee and Mackie, 1981; Way et al., 1980; McLeod et al., 1981).

Synchrony for the initiation of births and the dormancy of newborns during the summer in both temporary and permanent habitats of *M. partumeium* is claimed to be an evolutionary strategy for survival in temporary or potentially temporary ponds, and photoperiod may provide the appropriate cues for birth, dormancy, and growth (Hornbach et al., 1980c; Conover et al., 1981; Way et al., 1980; McLeod et al., 1981). *P. casertanum* (Mackie, 1979b) and *M. lacustre* (Mitropolskii, 1965; Mackie, 1979b; Alexander, 1982) show similar patterns of life-cycle adaptations for temporary habitats under permanent conditions. Further, studies on *S. occidentale* (Collins, 1967; McKee and Mackie, 1980), *P. personatum* and *P. obtusale* (Bleck and Heitkamp, 1980; Heitkamp, 1980), and *P. casertanum* and *P. personatum* (Danneel and Hinz, 1974) have concerned resistance adaptations to desiccation under xeric conditions. Herrington (1962) speculated that *S. occidentale* may require a dry period because it is found only in temporary ponds. However, McKee and Mackie (1981) have demonstrated that this species grows and reproduces under permanent aquatic conditions in the laboratory and that a dry period is not necessarily a requirement. It appears that species of *Musculium* may require a dormant period after birth during the summer in both temporary and permanent habitats (Mackie et al., 1976a,b; Mackie, 1979b; Hornbach et al., 1980c; Way et al., 1980; McKee and Mackie, 1981; Alexander, 1982).

B. Reproductive Tactics

Discussions of larval biology in molluscs can be found in a number of publications (Okada, 1935a,b,c; Bondesen, 1950; Thorson, 1950; Russell-Hunter et al., 1972; Mackie et al., 1974; Heard, 1977; Ellis, 1978; Mackie, 1979b; Morton, 1979). Thorson (1950) recognizes three types of free-living larvae in the Mollusca. These larvae fall into two functional categories: those with a prolonged larval life span of two to three months which involve feeding and dispersal, and those with short larval histories of less than a week which primarily involve dispersal. In nonmarine molluscs, selection has generally favored a few eggs of large size (except for unioacean, corbiculid, and dreissenacean clams) and the reduction of the temporal extent of immature growth, as well as the suppression of free-living larval stages. This has led to the hatching of miniature adults in pulmonate (Bondesen, 1950; Russell Hunter, 1953; Burky, 1971, McMahon, 1975) and ampullariid (Burky, 1974) snails. Presumably this strategy enhances larval survival with some independence from adverse environmental conditions. However, the freshwater viviparid gastropods (Browne, 1978; Vail, 1978) and pisidiid clams have maximized this strategy not only by the reduction of larval stages, but by internal development and a small number of relatively large newborns (miniature adults). Viviparity allows recruitment to populations during most seasons without the need for a period of free-living larval growth or development within exposed eggs.

Oviparity is the rule for unionacean, corbiculid, and dreissenacean clams. The production of free-living larval stages is coupled with high annual selection ratios for *C. fluminea* [>20,000 : 1 (Aldridge and McMahon, 1978)] and *D. polymorpha* [as much as 1,000,000 : 1 for large individuals (Walz, 1978b)]. A high selection ratio and oviparity has been a very successful strategy, as evidenced by the rapid spread of *C. fluminea* in North America and *D. polymorpha* throughout Great Britain, Europe, and Asia. Their reproductive success has become a problem due to spat fall and subsequent growth in industrial water systems. Although oviparity with free-living larvae is an excellent dispersal mechanism, the parasitic link of the glochidal larvae of unionacean clams [selection ratio >5,000,000 : 1 (Ellis, 1978)] is tied to host (mostly fish) movement, which limits both larval survival and dispersal (in most instances) to the confluence of each system.

Viviparity in pisidiid clams has been dealt with by many authors (Monk, 1928; Okada, 1935a,b,c, 1936; Thomas, 1959; Avolizi, 1970, 1976; Heard, 1964, 1965, 1977; Zumoff, 1973; Mackie *et al.*, 1976a,b, 1978; Mackie, 1979b; Meier-Brook, 1977; Hornbach *et al.*, 1980c, 1982; McKee and Mackie, 1981; Alexander, 1982). Unlike most other molluscs, these clams have remarkably low fecundity. The range for annual selection ratio is 6 : 1 to 24 : 1 for species of *Sphaerium* (Avolizi, 1970, 1976; Heard, 1977, and references therein; Hornbach et al., 1982); 2 : 1 to 136 : 1 for species of *Musculium* (Thomas, 1959; Mackie et al., 1976a,b, 1978; Mackie, 1979b; Heard, 1977; Hornbach et al., 1980c; Alexander, 1982); and 3.3 : 1 to 6.7 : 1 for species of *Pisidium* (C. M. Way, unpublished data). Values on brood or litter size can also be found in most of the above studies. However, actual litter size or the number of developing embryos does not necessarily reflect the number of births per adult (annual selection ratio). Brood mortality is high, ranging from about 50% in *M. partumeium* (Hornbach et al., 1980c) to 97.5% in *S. striatinum* (Hornbach et al., 1982).

A few studies on population natality are worth discussing. Mackie et al. (1978) have shown that both intrapopulation and interpopulation variations in the number of embryos per adult *M. securis* can occur, and that both population density and interspecific competition can affect natality. Also, Mackie et al. (1976b) claimed that for *M. securis*, interpopulation variations in reproductive output are better explained by differences in embryonic mortality than in the actual number of embryos produced. However, Hornbach et al. (1980c) have demonstrated that for *M. partumeium* there are not only intra- and interpopulation differences in mortality but also in the actual number of embryos produced. Further, the studies of Meier-Brook (1977), Mackie (1979b), Hornbach et al. (1980c, 1982), and Alexander (1982) show that embryonic mortality can take place at different stages of development. It is further suggested that smaller embryos in embryonic sacs may serve as a nutrient source for larger developmental stages and are at least a partial explanation for embryonic mortality (Meier-Brook, 1977; Hornbach et al., 1980c; Alexander, 1982).

The dynamics of reproduction are further complicated by the relationship between adult and embryonic growth. In *M. partumeium* (Hornbach et al., 1980c; Way et al., 1980), newborns have a shell length of about 1.4 mm. By the time they reach a shell length of 2.3 mm, small developing embryos are present. However, adults are incapable of giving birth until they have reached a shell length of about 4.0 mm. This relationship is also true for other pisidiids. In *M. partumeium,* not only do new embryos begin developing as adults continue to grow, but developmental stages are remarkably synchronous in adults of each generation cohort (Hornbach et al., 1980c). This is apparently also true for *M. lacustre* (Alexander, 1982). As embryos reach birth size, they are retained until what appears to be a controlled and synchronous initiation of births. This is undoubtedly tied to the evolution of life-cycle patterns in temporary habitats for these forms. The effects on the energetics of reproductive effort must be varied because embryonic development to minimum birth size in *M. partumeium* ranges from 1 to 9 months depending on generation and/or population [similar variations occur in *M. lacustre* (Alexander, 1982)]. Conversely, selection appears to have favored a pattern of asynchronous development and birth in a stream population of *S. striatinum* (Hornbach et al., 1982), i.e., adults of similar size do not necessarily give birth at the same time even though most births are usually confined to two prolonged periods. Not only is the minimum size at birth in this stream population (Hornbach et al., 1982) slightly larger (4.0 versus 3.6 mm) than in a lake population (Avolizi, 1970, 1976), the annual selection ratio is lower (10.49 : 1 versus 12 : 1). Asynchronous characteristics of stream reproduction, along with larger newborns, probably represent selection for harsh conditions and can undoubtedly buffer such populations against unpredictable events such as flooding, low water, substrate shifts, and temperature extremes. Synchronous and asynchronous reproductive patterns in pisidiid clams have also been discussed by Heard (1965, 1977).

C. Population Densities and Dynamics

Most of the information on the numbers of freshwater clams involves reports of population densities at one or more times during the year. However, few of these studies attempt to collect extensive sequences of data for assessments of the dynamics of survivorship (life tables: Pearl and Miner, 1935; Deevey, 1947). Other aspects of population dynamics are covered in Sections IV,A,B, and D.

In their introduction, Davis and Fuller (1981) point out that the unionacean bivalves of North America constitute one of the most diverse groups of living macroinvertebrates in freshwater systems, and one of the richest known localities is a section of the Tennessee River (known as Mussel Shoals) which contained about 70 species (Conrad, 1834). Apparently, sympatric species (packed valve to valve) paved the river shoals; this bit of history must represent one of the most dense records of unionacean communities. Such abundant mussel populations apparently serve as prey for raccoon, muskrat, and other mammals (evidence from midden heaps; Stansbery, 1973). Apart from this, freshwater mussels are reportedly influenced by depth and type of substrate. In general, mussels are found in waters less than 10 m deep (Negus, 1966; Cvancara, 1972; Tudorancea, 1972; Haukioja and Hakala, 1974; Krzyżanek, 1976), with many species apparently most abundant in the shallower zones. In the study of Haukioja and Hakala (1974), *U. pictorum* was limited to shallower waters (maximum density of 10 clams/m^2 at 0–2 m depth) and *Pseudanodonta complanata* to deeper zones (maximum density of 10 clams/m^2 at 2–3 m depth); *A. picinalis* was found at all depths to 4–5 m (most abundant at 1 m depth, 53 clams/m^2). In this study, the maximum density was 161.4 clams/m^2 for four species at 1 m depth. Negus (1966) also found that densities vary with depth, that the maximum density of four species was at depths of 2–3 m, and that the average density over all depths ranged from 0.40 to 11.7 clams/m^2 for *A. minima* and *A. anatina,* respectively. In another study, Cvancara (1972) similarly found that for three species total density decreased from 54 clams/m^2 at 1 m to 0.6 clams/m^2 at 9 m. Also, the type of substrate has been regarded as being both important (Harman, 1972; Ghent et al., 1978) and unimportant (Cvancara, 1972) to the distribution and densities of mussels. Negus (1966) also pointed out the importance of different substrates for the distribution of juvenile and adult mussels.

Population densities for *Dreissena polymorpha* can approach 2000 clams/m^2 (Stánczykowska, 1976), whereas that of *C. fluminea* can range from 5 to 1452 clams/m^2 (Diaz, 1974) to about 6000 clams/m^2 (Heinsohn, 1958). Aldridge and McMahon (1978) report 17.7–94.6 clams/m^2 for a Texas population and provide the dynamics of change in density for each cohort. However, the most impressive densities for freshwater bivalves are those of pisidiid clams, with *M. transversum* exceeding 100,000 clams/m^2 in Keokuk Pool (pool 19) of the Mississippi River (Gale, 1973, 1975). Apparently, pisidiid clams are often prominent or dominant in the benthos in numbers and/or biomass (Humphries, 1936; Carr and Hiltunen, 1965; Ransom and Prophet, 1974; Gale, 1975; Avolizi, 1976; Healey, 1977, 1978; Eckblad et al., 1977; Thorp and Bergey, 1981; Way et al., 1980; Hornbach, 1980; Hornbach et al., 1982; Alexander, 1982). Further, the importance of the abundance of pisidiid clams in the food chain is evident from their position as primary consumers or collectors (Gale and Lowe, 1971; Section III,B) and as prey for insects (Foote, 1976), fish (Eyerdam, 1968; Jude, 1973), and waterfowl (Thompson, 1973).

The distribution and densities of these pisidiid clams vary with depth. One of the most extensive studies reported on 11 species of *Pisidium* in four Canadian lakes (Healey, 1977). *P. conventus* was the most abundant species (maximum, 2000 clams/m^2) both seasonally and with increasing depth, but *P. casertanum* was restricted to shallower or mid-depth zones (usually <100 clams/m^2). Most of the other species were usually found in shallower areas. Apparently, *P. conventus* is a profundal species with densities of approximately 400, 1000, 2000, 2500, and 2900 clams/m^2 (Healey, 1978; Holopainen, 1978; Healey, 1977; Hamilton, 1971; Holopainen and Hanski, 1979, respectively), and populations recorded to depths of >160 m (Russell Hunter and Slack, 1958). *Pisidium casertanum* is a species of shallower water, with densities of >200 clams/m^2 (Thut, 1969), >5000 clams/m^2 (Jónasson, 1972; Holopainen and Hanski, 1979), and >6000 clams/m^2 (Gillespie, 1969). In *P. walkeri* (A. J. Burky, unpublished data), average densities decrease from 2078 to 910, 200, and zero clams/m^2 at depths of 0–1, 1–2, 2–3 and 3+ m, respectively, in a small pond. For *S. striatinum* and *S. simile,* densities range from 89 to 543 clams/m^2 for lake populations (Avolizi, 1976), whereas densities of up to 899 (maximum average, 417) clams/m^2 have been reported for a stream population of *S. striatinum* (Hornbach, 1980; Hornbach et al., 1982). Heitkamp (1980) reported densities of 1800 and 21,500 clams/m^2 for temporary pond populations of *P. personatum* and *P. obtusale,* respectively. In temporary (univoltine) and permanent (bivoltine) pond populations of *M. partumeium* (Way et al., 1980), average maximum densities are 839 and 4217 clams/m^2, respectively. The average maximum density of a bivoltine population of *M. lacustre* in a permanent habitat is 3420 clams/m^2 (Alexander, 1982). In the studies of Way et al. (1980) and Alexander (1982), a Type III survivorship pattern is reported for all generations of these viviparous clams (Pearl and Miner, 1935; Deevey, 1947). It is likely that all freshwater clams illustrate some variation of a Type III pattern because larval mortality must be even higher in oviparous species (Unionacea, Dreissenacea, Corbiculidae).

D. Comparative Energetics

Since the work of Avolizi (1970, 1976) on *S. striatinum* and *S. simile,* there have been relatively few studies on the bioenergetic turnover of total organic carbon and total nitrogen of tissue and shell for freshwater clams. In some respects this is not surprising, because many ecologists anticipate difficulties in regular population sampling and identification of tiny pisidiid clams or in the inherent problems of working with an endangered unionacean species with complications of a parasitic link in the life cycle. However, there are many advantages in working with pisidiid populations because they are often very abundant as a prominent component of the benthic biomass. In general, these clams are relatively easy to study from sequential population samples, and along with analyses of energy content, their life cycle as well as the energetics of growth and reproduction can be approached. The application of biomass analysis as total organic carbon (Russell-Hunter et al., 1968) or total nitrogen (using a Coleman model-29 semiautomatic nitrogen analyzer, which is based on a modified micro-

Dumas method as described by Gustin, 1960), has been employed in a number of studies on gastropods (Russell-Hunter, 1978, and references therein) and on corbiculacean clams. The studies on clams include work on *C. fluminea* (Aldridge and McMahon, 1978), *S. striatinum* (Avolizi, 1970, 1976; Hornbach, 1980), *M. partumeium* (A. J. Burky, unpublished data), *M. lacustre* (Alexander, 1982), and *P. walkeri* (A. J. Burky, unpublished data). Recently, calorific values for pisidiid clams have been published (Wissing et al., 1982); however, analyses employing bomb calorimetry are not as easily applied to population studies as are measurements on total organic carbon or total nitrogen. Normally, studies involve sequential generation samples in which size series based on shell length are analyzed for energy content (most often as total organic carbon) and related to a natural size frequency distribution of 100 clams (unit population). These values are derived for each date from regressions of energy content to body size and combined with the frequency (expressed as a percentage) of each size category. These values are summed to provide the energy content for a natural size distribution of 100 clams. The unit population values are an excellent measure for making comparisons of the ''average'' (this is not a true average) population unit in terms of growth from one date to the next over each cohort life span. These energy values for natural size distributions of 100 animals provide a convenient measure which can be combined with the appropriate density, survivorship, or fecundity values for the assessment of standing crop biomass or productivity rates. Such unit values have been employed in the above-mentioned studies on corbiculacean clams and in studies on gastropods (Burky, 1969, 1971; Russell-Hunter, 1978, and references therein).

In population studies, the symbols for energy relationships which have been clearly set out by Russell-Hunter (1970) have been applied to studies on snails (Russell-Hunter, 1978, and references therein) and clams (Hornbach, 1980; Alexander, 1982) by many workers. The amount of energy uptake as filtered material is total ingestion (TI). Based on TI, the following relationships hold for the partitioning of energy. Total assimilation (TA) and that which is not assimilated (NA = energy egested as feces) make up the TI. Growth (G) and reproduction (R) make up the nonrespired assimilation (N-RA) or production. The total assimilation is composed of respired assimilation (RA) and N-RA. Thus, TI = TA + NA, TA = RA + N-RA, and N-RA = G + R form the basis for energy flow through an individual clam, cohort, or population. From energy balance sheets, ratios for the efficiency of the conversion or partitioning of energy are often expressed as a percentage and the following relationships are provided: TA : TI represents assimilation or gut efficiency; N-RA : TA represents production efficiency; and R : N-RA represents the efficiency for the partitioning of production between reproduction and growth. Russell-Hunter (1970) has also set out a method for the computation of total biomass production based on information from successive population data on unit population biomass, mortality, and standing crop biomass and has discussed the application of this approach to studies on snails (Russell-Hunter, 1978). This approach has also been applied in some clam studies.

Before considering actual productivity values, assimilation efficiencies (TA : TI) must be considered. However, there are relatively few studies in which there is complete information on all aspects of bioenergetics, including TI. Furthermore, the approaches to measuring filter feeding in bivalves vary (Section III,B), providing a number of noncomplementary studies. Nonetheless, the following comparisons on ingestion are provided. For *S. suecicum* (Alimov and Bulion, 1972), it is claimed that all energy needs can be met by filter feeding. On the other hand, Efford and Tsumura (1973) have shown that the direct uptake of dissolved organics can supplement ingestion by providing 4% of the total energy requirements for *P. casertanum*. The studies of *M. partumeium* (Benjamin, 1978; Benjamin and Burky, 1978) and *S. striatinum* (Hornbach, 1980; Hornbach et al., 1983b) have shown that filter feeding can account for approximately 100% and 24% of the TA, respectively, if all filtered material is ingested and assimilated at 100% efficiency (TA : TI). Obviously, assimilation is less (perhaps 60%), suggesting either that filtration rates are in error or that ingestion is supplemented. Both Hornbach et al. (1983a) and A. J. Burky (unpublished data) claim that their data provide reasonable estimates of filtration rates and that *S. striatinum* and *M. partumeium,* along with *S. corneum* (Mitropolskii, 1966a,b), probably supplement their energy intake through deposit feeding. This could involve ciliary tracts on the foot, direct detrital uptake from the substrate through the general inhalant area or siphon, or directing the exhalant current against rich particulate substrates to suspend organic material. Obviously, future studies should assess both filter feeding and deposit feeding.

The best studies on the population energetics of corbiculacean clams have involved assessments of biomass as organic carbon, as discussed above (Avolizi, 1970, 1976; Aldridge and McMahon, 1978; Hornbach, 1980; Alexander, 1982; A. J. Burky, unpublished data; C. M. Way, unpublished data). However, a number of other studies have involved assessments of biomass based on dry tissue weight in corbiculacean clams (Teal, 1957; Gillespie, 1969; Alimov, 1970; Jónasson, 1972; Lévèque, 1973a,b; Qadri et al., 1974; Holopainen, 1978; Holopainen and Hanski, 1979; Hamill et al., 1979). Such dry weight values can be converted to total organic carbon by using values from Russell-Hunter et al. (1968) and have been referred to in discussions by Aldridge and McMahon (1978), and Hornbach (1980). Studies on the energetics of *D. polymorpha* (Stánczykowska, 1976; Walz, 1978a–f, 1979) and unionacean (Negus, 1966; Tudorancea and Florescu, 1968a,b, 1969; Magnin and Stanczykowska, 1971; Cameron et al., 1979; Huebner, 1980; Strayer et al., 1981) clams have also been reported.

Annual productivity values (N-RA) are 1963 mg C/(m^2·yr) for *C. africana* (Lévèque, 1973a,b); 10,433 mg C/(m^2·yr) for *C. fluminea* (Aldridge and McMahon, 1978); 6939 and 1661 mg C/(m^2·yr) for permanent and temporary pond populations, respectively, of *M. partumeium* (A. J. Burky, unpublished data); 20 and 160 mg C/(m^2·yr) for populations of *M. securis* (Qadri et al., 1974); 1111 mg C/(m^2·yr) for *M. lacustre* (Alexander, 1982); 4–448 mg C/(m^2·yr) for *P. casertanum* from numerous populations (Jónasson, 1972; Holopainen, 1978; Hamill et al., 1979); 1094 and 3846 mg C/(m^2·yr) for *P. compressum* (Gillespie, 1969; C. M. Way, unpublished data); 22 mg C/(m^2·yr) for *P. conventus* (Holopainen and Hanski, 1979); 19 mg C/(m^2·yr) *P. crassum* (Alimov, 1970); 1343 mg C/(m^2·yr) for *S. suecicum* (Alimov, 1970); 1744 mg C/(m^2·yr) for *P. variabile* (C. M. Way, unpublished data); 1679 mg C/(m^2·yr) for *S. simile* (Avolizi, 1970, 1976); and 4818 and 5573 mg C/(m^2·yr) for *S. striatinum* (Avolizi, 1970, 1976; Hornbach, 1980).

Apart from these absolute values, the turnover ratio of annual production to biomass (N-RA : average standing crop biomass) can be used as a more meaningful relative measure of productivity (Waters, 1977; Russell-Hunter, 1978), and the turnover time (365 days divided by the turnover ratio) can be used comparatively as the average time a population takes to produce biomass equivalent to the average standing crop biomass. Therefore, high turnover ratios and short turnover times are relative indices of high productivity. In this regard, *M. partumeium* (A. J. Burky, unpublished data) has the highest reported productivity, with a turnover ratio of 13.5 and a turnover time of 27 days for the fall generation in a permanent pond. The turnover ratios are 11.4 (32 days) and 7.2 (50.7 days) for the combined generations in a permanent pond and the single generation in a temporary pond, respectively. A turnover ratio of 7.9 (46.5 days) is reported for a lake population of *M. lacustre* (a species common in temporary habitats; Alexander, 1982). The only other high turnover ratio is 9.2 (39.7 days) for a stream population of *S. striatinum* (Hornbach, 1980). High productivity may be associated with adaptations for harsh habitats. Temporary ponds and streams make rigorous demands, requiring rapid productivity as a result of distinct selection pressures (Section IV,B). Turnover ratios between 4.0 and 5.0 (i.e., turnover times of 91 and 73 days) are reported for populations of *C. fluminea* (Aldridge and McMahon, 1978), *P. compressum* (Gillespie, 1969), and *S. striatinum* (lake population; Avolizi, 1970, 1976). Turnover ratios of 3.6 and 4.2 (turnover times of 102 and 88 days) are reported for *P. variabile* and *P. compressum,* respectively (C. M. Way, unpublished data). There is a turnover ratio of 1.0 (365 days) for *P. conventus* (Holopainen and Hanski, 1979) and a lower ratio of 0.2 (5 years) for *P. casertanum* (Jónasson, 1972). Intuitively, a ratio of 0.2 suggests a slow-growing, long-lived, iteroparous organism in a stable habitat of low temperature and/or low productivity with potential limitations of oxygen availability. In reality, these general types of conditions are reported by Jónasson (1972), who suggests that a life span of several years may be involved for *P. casertanum* because it can be long lived (Meir-Brook, 1970; Thut, 1969; Burky et al., 1981). Turnover ratios of <1 are reported for *D. polymorpha* (Stánzykowska, 1976) and are probably common for unionid clams, which in general are oviparous with life spans of several years (Magnin and Stánczykowska, 1971). Despite a low turnover ratio, *P. casertanum* (Jónasson, 1972) and other pisidiid clams can be an important component of the productivity of zoobenthic communities (Section IV,C). Jónasson (1972) reports that *P. casertanum* (ratio, 0.2) can make up 26.5% of the average standing crop biomass, and Avolizi (1970, 1976) attributes 33% of zoobenthic productivity to *S. striatinum* (ratio, 4.5) in a lake population. It has also been pointed out that high turnover ratios for a stream population of *S. striatinum* [9.2 (Hornbach, 1980)] and population generations of *M. partumeium* [13.5 (A. J. Burky, unpublished data)] are indicative of a capacity for significant contribution to secondary production.

In the evaluation of energy budgets, the total assimilated energy (TA) must be considered. The relative efficiency of the conversion of TA to production (N-RA) can be expressed (N-RA : TA) as a percentage: Production efficiencies range from 10% for *C. africana* (Lévèque, 1973a,b) and *M. lacustre* (Alexander, 1982) to 79% for *P. conventus* (Holopainen and Hanski, 1979). The highest production efficiencies are reported for *M. partumeium* [44–62% (A. J. Burky, unpublished data)]; *S. striatinum* [61% (Hornbach, 1980)]; *P. compressum* [68% (C. M. Way, unpublished data]; *C. fluminea* [71% (Aldridge and McMahon, 1978)]; and *P. conventus* [79% (Holopainen and Hanski, 1979)]. Apart from overall production efficiency, the conversion of N-RA to growth (G : N-RA) and reproduction (R : N-RA) can provide a meaningful index of reproductive effort. Also, high production efficiencies (N-RA:TA) do not necessarily correspond to high reproductive efficiencies because *S. striatinum* and *C. fluminea* have reproductive efficiencies of 4 and 15%, respectively (Hornbach, 1980; Aldridge and McMahon, 1978).

Of particular note are permanent and temporary pond populations of *M. partumeium* (A. J. Burky, unpublished data). Spring and fall generations in a permanent pond have production efficiencies of 58 and 44% but reproductive efficiencies of 35 and 17%, respectively; in contrast, a temporary pond generation has a production efficiency of 62% and a reproductive efficiency of 18%. These are the highest reproductive efficiencies reported for pisidiid clams, with the exception of 19.5% for *M. lacustre* (Alexander, 1982) and 25% for *P. conventus* (Holopainen and Hanski, 1979). Further, it can be argued that high efficiency values for *M. partumeium* and *M. lacustre* are further evidence of selective specialization for life in ephemeral habitats. Browne and Russell-Hunter (1978) compared values of R : N-RA for a variety of freshwater and marine molluscs and hypothesized that viviparous species have a lower reproductive efficiency (mean for two species = 5.25%) than oviparous species (mean for 14 species = 24.24%). A reproductive efficiency of 4.3% (Hornbach, 1980) and 6.0% (Avolizi, 1976) for *A. striatinum* and 5.7 and 7.2% for *P. variabile* and *P. compressum*, respectively (C. M. Way, unpublished data), further support this idea. However, reproductive efficiencies for viviparous corbiculacean clams

(Pisidiidae) are 35, 17, and 18% for *M. partumeium* (A. J. Burky, unpublished data), 19.0, 19.5, and 19.3% for *M. lacustre* (Alexander, 1982), 25% for *P. conventus* (Holopainen and Hanski, 1979), and 14% for *P. casertanum* (Holopainen, 1978). These values are in general equivalent to the average ratio for oviparous species (24.24%); the 35% value is greater than all but 4 of the 23 values reported for oviparous molluscs by Browne and Russell-Hunter (1978). Even within the Corbiculacea, values for the oviparous *C. fluminea* range from 13.5 to 24.4% (Aldridge and McMahon, 1978; Browne and Russell-Hunter, 1978). The values for corbiculacean clams do not support the ideas of Browne and Russell-Hunter (1978) on reproductive effort in relation to oviparity and viviparity. Further, comparisons between iteroparous and semelparous species by Browne and Russell-Hunter (1978) support the idea of Williams (1966a,b) that reproductive effort is greatest in semelparous species. However, examination of iteroparous and semelparous generations of *M. partumeium* (or considering *M. partumeium* to be an iteroparous species) does not support this contention. This lack of concordance does not necessarily negate the theoretical considerations, but it does underline the inherent dangers of generalization. *M. partumeium* probably evolved in temporary pond habitats with viviparity functioning as a major characteristic in the evolution of its life-history pattern. The specialized efficiency of viviparity with the iteroparous pattern is obviously advantageous to its strategy.

V. Evolutionary Strategies: An Ecological Perspective

The lamellibranch bivalves are basically uniform in structure, yet selection has produced some dichotomous strategies for two major freshwater groups (Unionacea and Pisidiidae). Considerations of the evolution of freshwater clams must take into account a number of factors which cut across both time and scale of ecologically based selection. The ecology of the evolution of freshwater clams includes differences in both the geological duration and scale of distinct freshwater habitats; the long-term duration of certain regular seasonal cycles involving temperature and the availability of oxygen, food, and water; and life-history patterns including the characteristics of reproduction and genetics.

The duration and size of freshwater habitats, which have been the subject of numerous authors, are discussed at length by Russell-Hunter (1978, and references therein) in relation to the evolution of freshwater pulmonate snails. Russell-Hunter points out that most larger freshwater lakes are relatively transitory in geological time, with life expectancies of 10^3 to 10^4 years, and that some ponds may persist for as long as 10^2 years. However, some ancient lakes (e.g., Baikal and Tanganyika) have existed for at least 10^5 years. By comparison, these time durations fall short of the age of some terrestrial ($>10^6$ years) and marine ($>10^9$

years) habitats. Only the few ancient lakes can be claimed to have provided the long time scale and space necessary for the selection of highly specialized and often endemic species (Russell-Hunter, 1978, and references therein). Russell-Hunter also reemphasizes that most pulmonate snail populations not only live in but also evolved in the smaller transient bodies of fresh water. In general, there is a relatively small number of pulmonate species of wide distribution that have a wide range of intraspecific phenotypic plasticity in many aspects of their biology.

Similarly, pisidiid clams live in and evolved in the smaller and more transient freshwater habitats; they also exhibit a wide range of intraspecific phenotypic plasticity but have a low degree of genotypic variability (Hornbach et al., 1980a,b; McLeod et al., 1981). Pisidiid clams are also represented by a relatively small number of widely distributed species which are dependent on passive dispersal from one transient habitat to the next (Sections II,D and IV,B). There are four genera and perhaps 39+ species of these clams in North America (Herrington, 1962; Burch, 1975a; Mackie et al., 1980). Conversely, species of unionacean clams have evolved in aquatic systems which have a longer duration and a larger scale. They are associated with major river drainage systems which can be considered to have a large surface area (and substrate area), and although the exact drainage pattern can and does change (transient lakes and ponds are also formed and destroyed along the flow of these systems), the long-term confluence of such systems is associated with the persistence of continental land masses (e.g., $>10^6$ years; Mississippi River; Amazon–Orinoco River system). This long-term, large-scale isolation has allowed for the selection of a relatively large number of genera with highly specialized (e.g., parasitic link in the life cycle) and often endemic species. Aspects of the importance of river confluence and isolation have been considered in relation to unionacean clams by Van der Schalie (1945) and Sepkoski and Rex (1974), as well as in Section II,D. In contrast to the pisidiids, the unionaceans of North America include approximately 46–50 genera and 225–227 species (Heard and Guckert, 1970; Burch, 1975b). More details and references on the evolution of unionacean clams can be found in the publications of Davis and Fuller (1981) and Davis et al. (1981).

Apart from the scale and duration of aquatic habitats, it is necessary to consider seasonal cycles in temperature and the availability of oxygen, food, water, etc. Both capacity and resistance adaptations to variations in temperature, oxygen, and food availability have been discussed (Section III). These adaptations reflect the long-term regularity of the sequence of winter–spring–summer–fall or wet–dry seasons. There must also be the ability to respond to unexpected variations in these sequences and still obtain food by filtration with a respiratory/feeding organ. These adaptational responses are the manifestation of variations in phenotypic plasticity. Although temporary ponds are often considered to be unstable habitats (relative to the presence of water), the relative regularity of annual drying represents a "stable" climatic–physical–biological set of circumstances over a requisite period of geological time.

The long-term influence of regular dry seasons (stability) in a tropical savannah was discussed by Burky, 1974). The harsh aspect of a dry habitat, along with the regular predictability of the event, has coupled some elements of plasticity and genetic adaptations along with fixed responses for *M. partumeium* (see earlier discussions). The level of genetic heterozygosity is relatively low in *Sphaerium* spp. [0.38% (Hornbach et al., 1980b)] and *C. fluminea* [0.49% (McLeod and Sailstad, 1980, and references therein)], and 0 and 6.14% for permanent and temporary pond populations, respectively, of *M. partumeium* (McLeod et al., 1981). These values are lower than those reported as ranging from 9.7 to 25% [mean, 15.1% (Selander and Kaufman, 1973)] for most other invertebrates; Davis et al. (1981) report values of 3 to >12% for some unionacean clams. Low intraspecific genetic variability for *S. striatinum* and low heterozygosity for four species of *Sphaerium* and for *C. fluminea* may indicate that these clams self-fertilize and/or that there is selection against heterozygotes [at least in *Sphaerium* (Hornbach et al., 1980b)]. Perhaps heterozygotes are selected against in the brood sacs (high mortality) of these viviparous clams (Section IV,B). In *M. partumeium* (McLeod et al., 1981), a temporary pond population has a greater degree of genetic variability (22.72% polymorphism and 6.14% heterozygosity) than a permanent pond population (no polymorphism and heterozygosity). McLeod et al. (1981), state that "this presumably allows the population to remain adapted to the harsh conditions and also possibly subdivide the habitat so that differing monomorphic genetic lines become, in effect, specialists" in the temporary habitats. This population elects quality rather than quantity, producing fewer energy-rich young for survival over the dry period. This adaptation appears to be partly phenotypic because this population adds a second generation in the odd year that the pond remains full. This is similar to the pattern in the permanent pond except that only 23% of the temporary pond population is committed to this second reproduction. The permanent pond population of *M. partumeium* employs the opposite approach. Although dormant over the summer, nearly all individuals become active for the production of a fall generation. Genetic monomorphism and the lack of heterozygosity in this population suggest that the genetic composition has been streamlined through a strategy of self-fertilization.

Selection for life in variable habitats has evoked speculation about the application of the deterministic *r*- and *K*-selection theory to life history tactics (Cole, 1954; Pianka, 1970; Heard, 1977; Stearns, 1976, 1977, 1980; Calow, 1978; Mackie et al., 1978; Hornbach et al., 1980a,c; Way et al., 1980; McLeod et al., 1981. Heard (1977) concludes that at the generic level *S. occidentale* and species of *Musculium* are "*r* strategists" because they produce many young of low energy content and live in variable habitats. In contrast, other species of *Sphaerium* are *K* strategists, occupying more stable habitats and producing few offspring of high energy content. At the intraspecific level, Mackie et al. (1978) claim that populations of *M. securis* exhibit *r*-selection characteristics in litter size, but McKee and Mackie (1981) now believe that another theory (bet hedging) applies to *M. securis* as well as *S. occidentale*. Further, intraspecific comparisons of temporary and permanent pond populations of *M. partumeium* (Hornbach et al., 1980c; Way et al., 1980; McLeod et al., 1981) indicate that most characteristics of a temporary pond population are *K* in nature. *K* strategy in a variable *r* habitat can be expected according to the stochastic bet-hedging theory (Stearns, 1976, 1977) if juvenile mortality is more variable than adult mortality in the variable habitat. In this regard, Hornbach et al. (1980c), Way et al. (1980), and McLeod et al. (1981) point out that the data on *M. partumeium* seem to support a bet-hedging strategy rather than one of *r* and *K* selection.

The application of either theory (*r* and *K* selection or bet hedging) is difficult because population characteristics for *M. partumeium* do not completely fit any theory for the evolution of life-history tactics. This involves some potential conceptual problems. First, temporary ponds or recurring dry seasons represent a set of predictable and, therefore, stable climatic conditions overriding any variable designation based on the presence of water (instability in streams may also represent predictably stable habitat conditions, as discussed earlier). Second, life in temporary aquatic habitats may represent selective specialization which makes comparisons based on current theories seem paradoxical (see Section IV,D). Third, the application of current theories may not be valid even if there is an apparent fit. Stearns (1980) also questions the application of any theory on life-history tactics at the intraspecific level. The reality of the first two conceptual problems has been discussed elsewhere in this chapter. The third is probably an overstatement. In reality, models with an infinite number of feedback loops could possibly provide a best fit; however, each population potentially could have its own strategy (and/or theory). In fact, theoretical generalizations (e.g., *r* and *K* selection, bet hedging) may be inappropriate or inadequate when applied to most populations, although they help provide an important conceptual framework for making comparisons.

Obviously, contributions to our knowledge of the evolutionary process involve studies on physiological adaptations as well as the overall efficiencies of bioenergetic adaptations. General and more meaningful insight into this process should involve ecological studies with complete sets of data at least comparable to those on *M. partumeium* (Benjamin, 1978; Burky and Benjamin, 1979; Burky et al., 1979; Hornbach et al., 1980a,c; Way et al., 1980, 1981; Conover and Burky, 1981; Conover et al., 1981; McLeod et al., 1981; Buchwalder, 1983); *S. striatinum* (Avolizi, 1970, 1976; Hornbach, 1980; Hornbach et al., 1982, 1983a,b); and *M. lacustre* (Alexander, 1982; Alexander and Burky, 1982; Buchwalder,

1983) on everything from life cycle to population dynamics, ingestion, assimilation efficiency, production efficiency, reproduction efficiency, and genetics in both time and extent for inter- and intraspecific interpopulation comparisons.

Acknowledgments

I would like to thank Joanne Maynard of the University of Dayton Library and Erin Davis for help with reference material; and Ann E. Feldmann, Kimie Payne, and Lynda F. Routley for clerical assistance during the preparation of the manuscript. I would also like to acknowledge the importance of my early training under and continued association with Dr. W. D. Russell-Hunter; former research associates from my laboratory for their interest, dedication, and discussions on some of the topics in this chapter (Dr. Daniel J. Hornbach, Richard B. Benjamin, Dr. Carl M. Way); Jeffrey P. Alexander and Stephen H. Buchwalder for reading the manuscript; and my wife, Kathleen A. Burky, for her involvement at all stages of manuscript preparation.

References

Agrell, I. (1949). The shell morphology of some Swedish Unionides as affected by ecological conditions. *Ark. Zool.* **41A,** 1–30.

Aldridge, D. W., and McMahon, R. F. (1978). Growth, fecundity and bioenergetics in a natural population of the Asiatic freshwater clam, *Corbicula manilensis* Philippi, from North Central Texas. *J. Moll. Stud.* **44,** 49–70.

Alexander, J. P. (1982). Energetics of a lake population of the freshwater clam, *Musculium lacustre* Müller (Bivalvia: Pisidiidae), with special reference to seasonal patterns of growth, reproduction and metabolism. Master's Thesis, University of Dayton, Dayton, Ohio.

Alexander, J. P., and Burky, A. J. (1982). Seasonal effect of body size and temperature on respiration of the freshwater clam, *Musculium lacustre. Ohio J. Sci.* **82,** Abstr. Suppl., 98.

Ali, R. M. (1970). The influence of suspension density and temperature on the filtration rate of *Hiatella arctica. Mar. Biol. (Berlin)* **6,** 291–302.

Alimov, A. E., (1965a). Oxygen consumption of the freshwater mollusc *Sphaerium corneum. Zool. Zh.* **44,** 1558–1562.

Alimov, A. E. (1965b). The filtrational ability of mollusks belonging to the genus *Sphaerium* (Scopuli). *Dokl. Biol. Sci.* (*Engl. Transl.*) **164,** 195–197.

Alimov, A. F. (1967). Peculiarities of the life cycle and growth of *Sphaerium corneum* (L.). *Zool. Zh.* **46,** 192–199.

Alimov, A. F. (1970). Potok energii cherez populyatsiyu mollyuskov (Na primere Sphaeriidae). *Gidrobiol. Zh.* **6,** 63–71.

Alimov, A. F. (1975). The rate of metabolism in freshwater bivalve mollusks. *Sov. J. Ecol. (Engl. Transl.)* **6,** 6–13.

Alimov, A. F., and Bulion V. V. (1972). Filtration activity of mollusks *Sphaerium suecicum* Clessin at different concentrations of suspended substances. *Zh. Obshch. Biol.* **33,** 97–104.

American Public Health Association (1975). "Standard Methods for the Examination of Water and Wastewater," 14th ed. APHA, Washington, D.C.

Antheunisse, L. J. (1963). Neurosecretory phenomena in the zebra mussel *Dreissena polymorpha* Pallas. *Arch. Neerl. Zool.* **15,** 237–314.

Arey, L. B. (1932). The formation and structure of the glochidial cyst. *Biol. Bull.* (*Woods Hole, Mass.*) **62,** 212–221.

Avolizi, R. J. (1970). Biomass turnover in natural populations of viviparous sphaeriid clams: Interspecific and intraspecific comparisons of growth, fecundity, mortality and biomass production. Ph.D. Dissertation, Syracuse University, Syracuse, New York (University Microfilms, Ann Arbor, Michigan) (*Diss. Abstr.* **32,** 1918-B, Order No. 71–23, 429).

Avolizi, R. J. (1976). Biomass turnover in populations of viviparous sphaeriid clams: Comparisons of growth, fecundity, mortality and biomass production. *Hydrobiologia* **51,** 163–180.

Badman, D. G. (1974). Changes in activity in a fresh-water clam in response to oxygen concentration. *Comp. Biochem. Physiol. A* **47,** 1265–1271.

Badman, D. G. (1975). Filtration of neutral red by freshwater clams in aerobic and hypoxic conditions. *Comp. Biochem. Physiol. A* **51,** 741–744.

Badman, D. G., and Chin, S. L. (1973). Metabolic responses of the freshwater bivalve, *Pleurobema coccineum* (Conrad), to anaerobic conditions. *Comp. Biochem. Physiol. B* **44,** 27–32.

Barnes, R. D. (1980). "Invertebrate Zoology," 4th ed. Saunders, Philadelphia, Pennsylvania.

Bayne, B. L. (1971). Oxygen consumption by three species of lamellibranch mollusc in declining ambient oxygen tension. *Comp. Biochem. Physiol. A* **40,** 307–313.

Benedens, H. G., and Hinz, W. (1980). Zur Tagesperiodizität der Filtrationsleistung von *Dreissena polymorpha* und *Sphaerium corneum* (Bivalvia). *Hydrobiologia* **69,** 45–48.

Benjamin, R. B. (1978). The seasonal filtration pattern of the freshwater clam, *Musculium partumeium* (Say). Master's Thesis, University of Dayton, Dayton, Ohio.

Benjamin, R. B., and Burky, A. J. (1978). Filtration dynamics in the sphaeriid clam, *Musculium partumeium* (Say). *Bull. Am. Malacol. Union, Inc.* **44,** 60.

Berg, K., and Jónasson, P. M. (1965). Oxygen consumption of profundal lake animals at low oxygen content of the water. *Hydrobiologia* **26,** 131–143.

Berg, K., Jónasson, P. M., and Ockelmann, K. W. (1962). The respiration of some animals from the profundal zone of a lake. *Hydrobiologia* **19,** 1–39.

Bishop, M. J., and Hewitt, S. J. (1976). Assemblages of *Pisidium* species (Bivalvia, Sphaeriidae) from localities in eastern England. *Freshwater Biol.* **6,** 177–182.

Bleck, V., and Heitkamp, U. (1980). Ökophysiologische Untersuchungen an *Pisidium personatum* Malm, 1855 und *Pisidium obtusale* (Lamarck, 1818) (Bivalvia, Sphaeriidae). *Zool. Anz.* **205,** 162–180.

Bondesen, P. (1950). A comparative morphological–biological analysis of the egg capsules of freshwater pulmonate gastropods. *Nat. Jutlandica* **3,** 1–208.

Boycott, A. E. (1936). The habitats of freshwater Mollusca in Britain. *J. Anim. Ecol.* **5,** 116–186.

Brandauer, N., and Wu, S.-K. (1978). Part 2. The freshwater mussels (Family Unionidae). *In* "Natural History Inventory of Colorado," Vol. 2, The Bivalvia of Colorado, pp. 41–60. University of Colorado Museum, Boulder, Colorado.

Brice, J. R., and Lewis, R. B. (1979). Mapping of mussel (Unionidae) communities in large streams. *Am. Midl. Nat.* **101,** 454–455.

Browne, R. A. (1978). Growth, mortality, fecundity, biomass and productivity of four lake populations of the prosobranch snail, *Viviparus georgianus. Ecology* **59,** 742–750.

Browne, R. A., and Russell-Hunter, W. D. (1978). Reproductive effort in molluscs. *Oecologia* **37,** 23–27.

Buchwalder, S. H. (1983). Seasonal respiratory adaptations to temperature and oxygen availability in the freshwater clams *Musculium partumeium* (Say) and *Musculium lacustre* (Müller). Master's Thesis, University of Dayton, Dayton, Ohio.

Burch, J. B. (1975a). "Freshwater Sphaeriacean Clams (Mollusca: Pelecypoda) of North America." Malacological Publ., Hamburg, Michigan.

Burch, J. B. (1975b). "Freshwater Unionacean Clams (Mollusca: Pelecypoda) of North America." Malacological Publ., Hamburg, Michigan.

Burky, A. J. (1969). Biomass turnover, energy balance, and interpopulation variation in the stream limpet, *Ferrissia rivularis* (Say), with special reference to respiration, growth, and fecundity.

Ph.D. Dissertation, Syracuse University, Syracuse, New York (University Microfilms, Ann Arbor, Michigan [*Diss. Abstr.* **30**(8), 3917-B, Order No. 70-1942].

Burky, A. J. (1971). Biomass turnover, respiration, and interpopulation variation in the stream limpet, *Ferrissia rivularis* (Say). *Ecol. Monogr.* **41,** 235–251.

Burky, A. J. (1974). Growth and biomass production of an amphibious snail, *Pomacea urceus* (Müller) from the Venezuelan savannah. *Proc. Malacol. Soc. London* **41,** 127–143.

Burky, A. J. (1977). Respiration chambers for measuring oxygen consumption of small aquatic molluscs with Clark-type polarographic electrodes. *Malacol. Rev.* **10,** 71–72.

Burky, A. J., and Benjamin, R. B. (1979). An accurate microassay for measuring filtration rates of small invertebrates using latex beads. *Comp. Biochem. Physiol. A* **63,** 483–484.

Burky, A. J., and Burky, K. A. (1976). Seasonal respiratory variation and acclimation in the pea clam, *Pisidium walkeri* Sterki. *Comp. Biochem. Physiol. A* **55,** 109–114.

Burky, A. J., Benjamin, M. A., Catalano, D. M., and Hornbach, D. J. (1979). The ratio of calcareous and organic shell components of freshwater sphaeriid clams in relation to water hardness and trophic conditions. *J. Moll. Stud.* **45,** 312–321.

Burky, A. J., Hornbach, D. J., and Way, C. M. (1981). Growth of *Pisidium casertanum* (Poli) in West Central Ohio. *Ohio J. Sci.* **81,** 41–44.

Calow, P. (1978). The evolution of life-cycle strategies in freshwater gastropods. *Malacologia* **17,** 351–364.

Cameron, C. J., Cameron, I. F., and Paterson, C. G. (1979). Contribution of organic shell matter to biomass estimates of unionid bivalves. *Can. J. Zool.* **57,** 1666–1669.

Carr, J. F., and Hiltunen, J. K. (1965). Changes in the bottom fauna of Western Lake Erie from 1930 to 1961. *Limnol. Oceanogr.* **10,** 551–569.

Clark, L. C., Jr. (1956). Monitor and control of blood and tissue oxygen tensions. *Trans. Am. Soc. Artif. Intern. Organs* **2,** 41–57.

Clarke, A. H. (1973). The freshwater molluscs of the Canadian interior basin. *Malacologia* **13,** 1–509.

Clarke, A. H. (1979). Sphaeriidae as indicators of trophic lake stages. *Nautilus* **94,** 178–184.

Cole, L. C. (1954). The population consequences of life history phenomena. *Q. Rev. Biol.* **29,** 103–137.

Collins, T. W. (1967). Oxygen uptake, shell morphology and dessication of the fingernail clam, *Sphaerium occidentale* Prime. Ph.D. Dissertation, University of Minnesota, Minneapolis (*Diss. Abstr.* **28B,** 5238, Order No. 68-07294).

Conover, D. G., and Burky, A. J. (1981). Photoperiod as a cue for growth in the freshwater sphaeriid clam, *Musculium partumeium* (Say). *Ohio J. Sci.* **81,** Abstr. Suppl., 110.

Conover, D. G., Detrick, R. J., Alexander, J. P., and Burky, A. J. (1981). Comparison of filtration rates as a function of suspended particle concentration in the freshwater sphaeriid clams, *Musculium partumeium* (Say) and *Musculium lacustre* (Müller). *Ohio J. Sci.* **81,** Abstr. Suppl., 110.

Conrad, T. A. (1834). "New Fresh Water Shells of the United States with Colored Illustrations, and a Monograph of the Genus *Anculotus* of Say; also a Synopsis of the American Naiades." Philadelphia, Pennsylvania (citation from Davis and Fuller, 1981).

Coughlan, J. (1969). The estimation of filtering rate from the clearance of suspensions. *Mar. Biol (Berlin)* **2,** 356–358.

Culbreth, S. E. (1941). The role of tissues in the anaerobic metabolism of the mussel *Anodonta hallenbeckii* Lea. *Biol. Bull.* (*Woods Hole, Mass.*) **80,** 79–85.

Cvancara, A. M. (1970). Mussels (Unionidae) of the Red River Valley in North Dakota and Minnesota, U.S.A. *Malacologia* **10,** 57–92.

Cvancara, A. M. (1972). Lake mussel distribution as determined with Scuba. *Ecology* **53,** 154–157.

Cvancara, A. M. (1976). Aquatic mollusks in North Dakota during the last 12000 years. *Can. J. Zool.* **54,** 1688–1693.

Dance, S. P. (1958). Drought resistance in an African freshwater bivalve. *J. Conchol.* **24,** 281–282.

Danneel, I., and Hinz, W. (1974). Trockenresistenz dreier *Pisidium*-Populationen (Bivalvia) in Abhangigkeit von der relativen Luftfeuchtigkeit. *Hydrobiologia* **45,** 39–43.

Danneel, I., and Hinz, W. (1976). The biology of *Pisidium amnicum* O. F. Müller (Bivalvia). *Arch. Hydrobiol.* **77,** 213–225.

Davis, G. M., and Fuller, S. L. H. (1981). Genetic relationships among recent Unionacea (Bivalvia) of North America. *Malacologia* **20,** 217–253.

Davis, G. M., Heard, W. H., Fuller, S. L. H., and Hesterman, C. (1981). Molecular genetics and speciation in *Elliptio* and its relationships to other taxa of North American Unionidae (Bivalvia). *Biol. J. Linn. Soc.* **15,** 131–150.

Deevey, E. S. (1947). Life tables for natural populations of animals. *Q. Rev. Biol.* **22,** 283–314.

Denis, A. (1979). L'utilisation du microscope électronique à balayage dans l'examen des tissus minéralisés chez les lamellibranches. *Malacologia* **18,** 19–21.

De Zwaan, A., and Wijsam, T. C. M. (1976). Anaerobic metabolism in bivalvia (Mollusca). Characteristics of anaerobic metabolism. *Comp. Biochem. Physiol. B* **54B,** 313–324.

Diaz, R. J. (1974). Asiatic clam, *Corbicula manilensis,* in the tidal James River, Virginia. *Chesapeake Sci.* **15,** 118–120.

Dietz, T. H. (1974). Body fluid composition and aerial oxygen consumption in the freshwater mussel, *Ligumia subrostrata* (Say): Effects of dehydration and anoxic stress. *Biol. Bull. (Woods Hole, Mass.)* **147,** 560–572.

Dietz, T. H. (1977). Solute and water movement in freshwater bivalve mollusks (Pelecypoda; Unionidae; Corbiculidae; Margaritiferidae). *In* "Water Relations in Membrane Transport in Plants and Animals" (A. M. Jungreis, T. K. Hodges, A. Kleinzeller, and S. G. Schultz, eds.), pp. 111–119. Academic Press, New York.

Dussart, G. B. J. (1976). The ecology of freshwater molluscs in North West England in relation to water chemistry. *J. Moll. Stud.* **42,** 181–198.

Eckblad, J. W., Peterson, N. L., Ostlie, K., and Temte, A. (1977). The morphometry, benthos and sedimentation rates of a floodplain lake in Pool 9 of the Upper Mississippi River. *Am. Midl. Nat.* **97,** 433–443.

Edgar, A. L. (1965). Observations on the sperm of the pelecypod *Anodontoides ferussacianus* (Lea). *Trans. Am. Microsc. Soc.* **84,** 228–230.

Efford, I. E., and Tsumura, K. (1973). Uptake of dissolved glucose and glycine by *Pisidium,* a freshwater bivalve. *Can. J. Zool.* **51,** 825–832.

Ellis, A. E. (1978). "British Freshwater Bivalve Mollusca." Academic Press, New York.

Eyerdam, W. J. (1968). Fresh-water mollusks eaten by trout and other fish. *Nautilus* **81,** 103–104.

Feldmann, R. (1971). Die Kleinmuscheln (Sphaeriidae) des mitterleren Ruhrtales. *Decheniana* **123,** 27–47.

Foote, B. A. (1976). Biology and larval feeding habits of three species of *Renocera* (Diptera: Sciomyzidae) that prey on fingernail clams (Mollusca: Sphaeriidae). *Ann. Entomol. Soc. Am.* **69,** 121–133.

Foster, T. D. (1932). Observations on the life history of a fingernail shell of the genus *Sphaerium. J. Morphol.* **53,** 473–497.

Fuller, S. L. H. (1974). Clams and mussels (Mollusca: Bivalvia). *In* "Pollution Ecology of Freshwater Invertebrates" (C. W. Hart and S. L. H. Fuller, eds.), pp. 215–274. Academic Press, New York.

Gainey, L. F., Jr., and Greenberg, M. J. (1977). Physiological basis of the species abundance-salinity relationship in Molluscs: A speculation. *Mar. Biol. (Berlin)* **40,** 41–49.

Gale, W. F. (1971). An experiment to determine substrate preference of the fingernail clam, *Sphaerium transversum* (Say). *Ecology* **52,** 367–370.

Gale, W. F. (1973). Substrate preference of the fingernail clam, *Sphaerium striatinum* (Lamarck) (Sphaeriidae). *Southwest. Nat.* **18,** 31–37.

Gale, W. F. (1975). Bottom fauna of a segment of Pool 19, Mississippi River, near Fort Madison, Iowa, 1967–1968. *Iowa State J. Res.* **49,** 353–372.

Gale, W. F. (1976). Vertical distribution and burrowing behavior of the fingernail clam, *Sphaerium transversum. Malacologia* **15,** 401–409.

Gale, W. F. (1977). Growth of the fingernail clam, *Sphaerium transversum* (Say) in field and laboratory experiments. *Nautilus* **91,** 8–12.

Gale, W. F., and Lowe, R. L. (1971). Phytoplankton ingestion by the fingernail clam, *Sphaerium transversum* (Say), in Pool 19, Mississippi River. *Ecology* **52,** 507–513.

Ghent, A. W., Singer, R., and Johnson-Singer, L. (1978). Depth distributions determined with SCUBA, and associated studies of the freshwater unionid clams, *Elliptio complanata* and *Anodonta grandis* in Lake Bernard, Ontario. *Can. J. Zool.* **56,** 1654–1663.

Gillespie, D. M. (1969). Population studies of four species of molluscs in the Madison River, Yellowstone National Park. *Limnol. Oceanogr.* **14,** 101–114.

Gordon, M. E., Brown, A. V., and Kraemer, L. R. (1979). Mollusca of the Illinois River, Arkansas. *Proc. Arkansas Acad. Sci.* **33,** 35–37.

Green, R. H. (1971). A multivariate statistical approach to the Hutchinsonian Niche: bivalve molluscs of central Canada. *Ecology* **52,** 543–556.

Green, R. H. (1972). Distribution and morphological variation of *Lampsilis radiata* (Pelecypoda, Unionidae) in some central Canadian Lakes: A multivariate statistical approach. *J. Fish. Res. Board Can.* **29,** 1565–1570.

Green, R. H. (1980). Role of a unionide clam population in the calcium budget of a small Arctic Lake. *Can. J. Fish. Aquat. Sci.* **37,** 219–224.

Gustin, G. M. (1960). A simple, rapid automatic micro-Dumas apparatus for nitrogen determination. *Microchem. J.* **4,** 43–54.

Hamill, S. E., Qadri, S. U., and Mackie, G. L. (1979). Production and turnover ratio of *Pisidium Casertanum* (Pelecypoda: Sphaeriidae) in the Ottawa River near Ottawa-Hull, Canada. *Hydrobiologia* **62,** 225–230.

Hamilton, A. L. (1971). Zoobenthos of fifteen lakes in the Experiment Lakes Area, northwestern Ontario. *J. Fish. Res. Board Can.* **28,** 257–263.

Hanson, J., and Dietz, T. (1976). The role of free amino acids in cellular osmo-regulation in the freshwater bivalve *Ligumia subrostrata* (Say). *Can. J. Zool.* **54,** 1927–1931.

Harman, W. N. (1969). The effect of changing pH on the Unionidae. *Nautilus* **83,** 69–70.

Harman, W. N. (1972). Benthic substrates: their effect on fresh-water Mollusca. *Ecology* **53,** 271–277.

Harrison, A. D., and Rankin, J. J. (1978). Hydrobiological studies of Eastern Lesser Antillean Islands. III. St. Vincent: Fresh-water Molluska—their distribution, population dynamics and biology. *Arch. Hydrobiol., Suppl.* **54,** 123–188.

Haukioja, E., and Hakala, T. (1974). Vertical distribution of freshwater mussels (Pelecypoda, Unionidae) in southwestern Finland. *Ann. Zool. Fenn.* **11,** 127–130.

Healey, M. C. (1977). Experimental cropping of lakes: 4. Benthic communities. *Fish. Mar. Serv., Div. Tech. Rep.* **711,** 1–47.

Healey, M. C. (1978). Sphaeriid mollusc populations of eight lakes near Yellowknife, Northwest Territories. *Can. Field Nat.* **92,** 242–251.

Heard, W. H. (1964). Litter size in the Sphaeriidae. *Nautilus* **78,** 47–49.

Heard, W. H. (1965). Comparative life histories of North American pill clams (Sphaeriidae: *Pisidium*). *Malacologia* **2,** 381–411.

Heard, W. H. (1977). Reproduction of fingernail clams (Sphaeriidae: *Sphaerium* and *Musculium*). *Malacologia* **16,** 421–455.

Heard, W. H., and Guckert, R. H. (1970). A re-evaluation of the recent Unionacea (Pelecypoda) of North America. *Malacologia* **10,** 333–355.

Heard, W. H., and Hendrix, S. S. (1964). Behavior of unionid glochidia. *Am. Malacol. Union, Annu. Rep. Bull.* **29,** 2–4.

Heinsohn, G. E. (1958). Life history and ecology of the freshwater clam *Corbicula fluminea.* Master's Thesis, University of California at Santa Barbara (citation from Aldridge and McMahon, 1978).

Heitkamp, U. (1980). Populationsdynamik von *Pisidium personatum* Malm, 1855 und *Pisidium obtusale* (Lamarck, 1818) (Mollusca, Sphaeriidae) in einem periodischen Tumpel. *Zool. Anz.* **205,** 280–308.

Hemmingsen, A. M. (1950). The relation of standard (basal) energy metabolism to total fresh weight of living organisms. *Rep. Steno. Mem. Hosp. Nord. Insulinlab.* **4,** 7–58.

Herrington, H. B. (1962). A revision of the Sphaeriidae of North America (Mollusca: Pelecypoda). *Misc. Publ. Mus. Zool. Univ. Mich.* **118,** 1–74.

Hinz, W. (1976). Siedlungsdichten limnischer Mollusken in Nordskaninavien und in Südnorwegen. *Norw. J. Zool.* **24,** 205–223.

Hinz, W., and Scheil, H.-G. (1972). Zur Filtrationsleistung von *Dreissena, Sphaerium* und *Pisidium* (Eulamellibranchiata). *Oecologia* **11,** 45–54.

Hinz, W., and Scheil, H.-G. (1976). Substratwahlversuche an *Pisidium casertanum* und *Pisidium amnicum* (Bivalvia). *Basteria* **40,** 89–100.

Hochachka, P. W., Fields, J., and Mustafa, T. (1973). Animal life without oxygen: basic biochemical mechanisms. *Am. Zool.* **13,** 543–555.

Holopainen, I. J. (1978). Ecology of *Pisidium* (Bivalva, Sphaeriidae) populations in an oligotrophic and mesohumic lake. *Publ. Univ. Joensuu, Ser. B2* **9,** 1–12.

Holopainen, I. J. (1979). Population dynamics and production of *Pisidium species* (Bivalvia, Sphaeriidae) in the oligotrophic and mesohumic Lake Pääjärvi, Southern Finland. *Arch. Hydrobiol., Suppl.* **54,** 466–508.

Holopainen, I. J. (1980). Growth of two *Pisidium* (Bivalvia, Sphaeriidae) species in the laboratory. *Oecologia* **45,** 104–108.

Holopainen, I. J., and Hanski, I. (1979). Annual energy flow in populations of two *Pisidium* species (Bivalvia, Sphaeriidae), with discussion on possible competition between them. *Arch. Hydrobiol.* **86,** 338–354.

Holopainen, I. J., and Ranta, E. (1977a). Carbon dioxide output in the respiration of three *Pisidium* species (Bivalvia, Sphaeriidae). *Oecologia* **30,** 1–8.

Holopainen, I. J., and Ranta, E. (1977b). Respiration of *Pisidium amnicum* (Bivalvia) measured by infrared gas analysis. *Oikos* **28,** 196–200.

Hornbach, D. J. (1980). Population energetics of the freshwater clam, *Sphaerium striatinum.* Lamarck (Bivalvia; Sphaeriidae), with special reference to seasonal patterns of growth, reproduction, filter-feeding and metabolism. Ph.D. Dissertation, Miami University, Oxford, Ohio (University Microfilms: Ann Arbor, Michigan) (*Diss. Abstr.* **41,** 2514B-2515B, Order No. 8108485).

Hornbach, D. J., McLeod, M. J., and Guttman, S. I. (1980a). On the validity of the genus *Musculium* (Bivalvia: Sphaeriidae): Electrophoretic evidence. *Can. J. Zool.* **58,** 1703–1707.

Hornbach, D. J., McLeod, M. J., Guttman, S. I., and Seilkop, S. K. (1980b). Genetic and morphological variation in the freshwater clam, *Sphaerium* (Bivalvia: Sphaeriidae). *J. Moll. Stud.* **46,** 158–170.

Hornbach, D. J., Way, C. M., and Burky, A. J. (1980c). Reproductive strategies in the freshwater sphaeriid clam, *Musculium partumeium* (Say), from a permanent and a temporary pond. *Oecologia* **44,** 164–170.

Hornbach, D. J., Wissing, T. E., and Burky, A. J. (1982). Life-history characteristics of a stream population of the freshwater clam, *Sphaerium striatinum* Lamarck (Bivalvia: Pisidiidae). *Can. J. Zool.* **60,** 249–260.

Hornbach, D. J., Wissing, T. E. and Burky, A. J. (1983a). Seasonal variation in the metabolic rates and Q_{10}-values of the fingernail clam, *Sphaerium striatinum* Lamark. *Comp. Biochem. Physiol. A (in press).*

Hornbach, D. J., Way, C. M., Wissing, T. E., and Burky, A. J. (1983b). Effects of particle concentration and season on the filtration rates of the freshwater clam, *Sphaerium striatinum* Lamark (Bivalvia: Pisidiidae). *Hydrobiologia* (in press).

Huebner, J. D. (1980). Seasonal variation in two species of unionid clams from Manitoba, Canada: biomass. *Can. J. Zool.* **58,** 1980–1983.

Humphries, C. F. (1936). An investigation of the profundal and sublittoral fauna of Windermere. *J. Anim. Ecol.* **5,** 29–52.

Ingram, W. M. (1957). The use and value of biological indicators of pollution: freshwater clams and snails. *In* "Biological Problems in Water Pollution" (C. M. Tarzwell, ed.), pp. 94–135. Robert A. Taft Sanit. Eng. Cent., Cincinnati, Ohio.

Johnson, M. G., and Brinkhurst, R. O. (1971). Production of benthic macroinvertebrates of Bay of Quinte and Lake Ontario. *J. Fish. Res. Board Can.* **28,** 1699–1714.

Jónasson, P. M. (1972). Ecology and production of the profundal benthos in relation to phytoplankton in Lake Esrom. *Oikos, Suppl.* **14,** 1–148.

Jørgensen, C. B. (1966). "Biology of Suspension Feeding." Pergamon, Oxford.

Jørgensen, C. B. (1975). Comparative physiology of suspension feeding. *Annu. Rev. Physiol.* **37,** 57–79.

Juday, C. (1908). Some aquatic invertebrates that live under anaerobic conditions. *Wis. Acad. Sci. Arts, Lett.* **16,** 10–16.

Jude, D. J. (1973). Food and feeding habits of Gizzard Shad in Pool 19, Mississippi River. *Trans. Am. Fish. Soc.* **102,** 378–383.

Kew, H. W. (1893). "The Dispersal of Shells." Kegan Paul, London (citation from Rees, 1965).

Kraemer, L. R. (1970). The mantle flap in three species of *Lampsilis* (Pelecypoda: Unionidae). *Malacologia* **10,** 225–282.

Krzyżanek, E. (1976). Preliminary investigations on bivalves (Bivalvia) of the dam reservoir Goczalkowice. *Acta Hydrobiol.* **18,** 61–73.

Kuiper, J. G. J., and Wolff, W. J. (1970). The mollusca of the estuarine region of the rivers Rhine, Meuse, and Scheldt in relation to the hydrography of the area. III. The genus *Pisidium. Basteria* **34,** 1–42.

Ladle, M., and Baron, F. (1969). Studies on three species of *Pisidium* (Mollusca: Bivalvia) from a chalk stream. *J. Anim. Ecol.* **38,** 407–413.

Lévèque, C. (1973a). Dynamiques des peuplements biologie, et estimation de la production des mollusques benthiques du Lac Tchad. *Cah. ORSTOM, Ser. Hydrobiol.* **7,** 117–147.

Lévèque, C. (1973b). Bilans énergétiques des populations naturalles de mollusques benthiques du Lac Tchad. *Cah. ORSTOM, Ser. Hydrobiol.* **7,** 151–165.

Lomte, V. S., and Nagabhushanam, R. (1971). Studies on the respiration of the freshwater mussel *Parreysia corrugata. Hydrobiologia* **38,** 239–246.

Lukacsovics, F. (1966a). Hypoxial examination of *Anodonta cygnaea* L. on the O_2-consumption of gill-tissues and the relation between body dimensions and respiration of the gill tissues. *Ann. Inst. Biol. (Tihany) Hung. Acad. Sci.* **33,** 79–94.

Lukacsovics, F. (1966b). Ciliary activity examinations after anoxybiosis on the isolated gills of *Anodonta cygnaea* L. *Ann. Inst. Biol. (Tihany) Hung. Acad. Sci.* **33,** 95–102.

Lukacsovics, F., and Labos, E. (1965). Chemo-ecological relationship between some fish species in Lake Balaton and glochidia of *Anodonta cygnea* L. *Ann. Inst. Biol. (Tihany) Hung. Acad. Sci.* **32,** 37–54.

McCorkle, S., and Dietz, T. H. (1980). Sodium transport in the freshwater Asiatic clam *Corbicula fluminea. Biol. Bull. (Woods Hole, Mass.)* **159,** 325–336.

McKee, P. M., and Mackie, G. L. (1980). Desiccation resistance in *Sphaerium occidentale* and *Musculium securis* (Bivalvia: Sphaeriidae) from a temporary pond. *Can. J. Zool.* **58,** 1693–1696.

McKee, P. M., and Mackie, G. L. (1981). Life history adaptations of the fingernail clams *Sphaerium occidentale* and *Musculium securis* to ephemeral habitats. *Can. J. Zool.* **59,** 2219–2229.

Mackie, G. L. (1978a). Effects of pollutants on natality of *Musculium securis* (Bivalvia: Pisidiidae). *Nautilus* **92,** 25–33.

Mackie, G. L. (1978b). Are sphaeriid clams ovoviviparous or viviparous? *Nautilus* **92,** 145–146.

Mackie, G. L. (1978c). Shell structure in freshwater Sphaeriacea (Bivalvia: Heterodonta). *Can. J. Zool.* **56,** 1–6.

Mackie, G. L. (1979a). Dispersal mechanisms in sphaeriidae (Mollusca: Bivalvia). *Bull. Am. Malacol. Union, Inc.* **45,** 17–21.

Mackie, G. L. (1979b). Growth dynamics in natural populations of Sphaeriidae clams (Sphaerium, Musculium, Pisidium). *Can. J. Zool.* **57,** 441–456.

Mackie, G. L., and Huggins, D. G. (1976). Biological notes on *Eupera cubensis* (Bivalvia: Sphaeriidae) from Kansas. *J. Fish. Res. Board Can.* **33,** 1652–1656.

Mackie, G. L., Qadri, S. U., and Clarke, A. H. (1974). Development of brood sacs in *Musculium securis* bivalvia: Sphaeriidae. *Nautilus* **88,** 109–111.

Mackie, G. L., Qadri, S. U., and Clarke, A. H. (1976a). Intraspecific variations in growth, birth periods, and longevity of *Musculium securis* (Bivalvia: Sphaeriidae) near Ottawa, Canada. *Malacologia* **15,** 433–446.

Mackie, G. L., Qadri, S. U., and Clarke, A. H. (1976b). Reproductive habits of four populations of *Musculium securis* (Bivalvia: Sphaeriidae) near Ottawa, Canada. *Nautilus* **90,** 76–86.

Mackie, G. L., Qadri, S. U., and Reed, R. M. (1978). Significance of litter size in *Musculium securis* (Bivalvia: Sphaeriidae). *Ecology* **59,** 1069–1074.

Mackie, G. L., White, D. S., and Zdeba, T. W. (1980). "A Guide to Freshwater Mollusks of the Laurentian Great Lakes with Special Emphasis on the Genus *Pisidium.*" U.S. Environ. Prot. Agency, Environ. Res. Lab., Duluth, Minnesota.

McLeod, M. J., and Sailstad, D. M. (1980). An electrophoretic study of *Corbicula fluminea* (Bivalvia: Corbiculacea) in the Catawba River. *Bull. Am. Malacol. Union, Inc.* **46,** 17–19.

McLeod, M. J., Hornbach, D. J., Guttman, S. I., Way, C. M., and Burky, A. J. (1981). Environmental heterogeneity, genetic polymorphism and reproductive strategies. *Am. Nat.* **118,** 129–134.

McMahon, R. F. (1973). Respiratory variation and acclimation in the freshwater limpet, *Laevapex fuscus. Biol. Bull.* (*Woods Hole, Mass.*) **145,** 492–508.

McMahon, R. F. (1975). Growth, reproduction and bioenergetic variation in three natural populations of a freshwater limpet *Laevapex fuscus* (C. B. Adams). *Proc. Malacal. Soc. London,* **41,** 331–351.

McMahon, R. F. (1979a). Response to temperature and hypoxia in the oxygen consumption of the introduced Asiatic freshwater clam *Corbicula fluminea* (Müller). *Comp. Biochem. Physiol. A* **63,** 383–388.

McMahon, R. F. (1979b). Tolerance of aerial exposure in the Asiatic freshwater clam, *Corbicula fluminea* (Müller). *Proc.—Int. Corbicula Symp., 1st, 1977,* pp. 227–241.

McMahon, R. F. (1982). The occurrence and spread of the introduced Asiatic freshwater clam, *Corbicula fluminea* (Müller), in North America: 1924–1982. *Nautilus* **96,** 134–141.

Magnin, E., and Stańczykowska, A. (1971). Quelques données sur la croissance, la biomasse et la production annuelle de trois mollusques Unionidae de la région de Montréal. *Can. J. Zool.* **49,** 491–497.

Martin, A. W., and Harrison, F. M. (1966). Excretion. *In* "Physiology of Mollusca" (K. M. Wilbur and C. M. Yonge, eds.), Vol. 2, pp. 353–386. Academic Press, New York.

Matsushima, O. (1980). The efflux and tissue content of ninhydrin positive substances in brackish and freshwater clams, *Corbicula japonica* and *C. leana,* with special reference to osmotic responses. *Annot. Zool. Jpn.* **53,** 77–88.

Mattice, J. S. (1979). Interactions of *Corbicula* sp. with power plants. *Proc.—Int. Corbicula Symp., 1st, 1977,* pp. 119–138.

Mattice, J. S., and Dye, L. L. (1976). Thermal tolerance of the adult Asiatic clam. *ERDA Symp. Ser.* **40,** 130–135.

Meier-Brook, C. (1969). Substrate relations in some *Pisidium* species (Eulamellibranchiata: Sphaeriidae). *Malacologia* **9,** 121–125.

Meier-Brook, C. (1970). Investigations on the biology of some *Pisidium* species (Mollusca; Eulamellibranchiata; Sphaeriidae). *Arch. Hydrobiol., Suppl.* **38,** 73–150.

Meier-Brook, C. (1977). Intramarsupial suppression of fetal development in sphaeriid clams. *Malacol. Rev.* **10,** 53–58.

Merlini, M., Cadario, G., and Oregioni, B. (1978). The unionid mussel as a biogeochemical indicator of metal pollution. *Environ. Biogeochem. Geomicrobiol. Proc. Int. Symp., 3rd, 1977,* Vol. 3, pp. 955–965.

Mikheev, V. P. (1966). On rate of water filtration by *Dreissena. Trans., Inst. Biol. Inland Waters, Acad. Sci. USSR* **12,** 134–138 (citation from Monakov, 1972).

Mitropolskii, V. I. (1965). Observations on the life cycle, growth rate and tolerance of drying in *Musculium lacustre* (Müller) (Lamellibranchiata). *Tr., Inst. Biol. Vnutr. Vod., Akad. Nauk SSSR* **8,** 118–124.

Mitropolskii, V. I. (1966a). Notes on life-cycle and nutrition of *Sphaerium corneum* L. (Mollusca, Lamellibranchia). *Tr., Inst. Biol. Vnutr. Vod., Akad. Nauk SSSR* **12,** 125–128.

Mitropolskii, V. I. (1966b). On mechanisms of filtration and nutrition of sphaeriids (Mollusca, Lamellibranchia). *Tr., Inst. Biol. Vnutr. Vod., Akad. Nauk SSSR* **12,** 129–133.

Monokov, A. V. (1972). Review of studies on feeding of aquatic invertebrates conducted at the Institute of Biology of Inland Waters, Academy of Sciences, USSR. *J. Fish. Res. Board Can.* **29,** 363–383.

Monk, C. R. (1928). The anatomy and life-history of a freshwater mollusk of the genus *Sphaerium. J. Morphol. Physiol.* **45,** 473–503.

Morton, B. S. (1969a). Studies on the biology of *Dreissena polymorpha* Pall. 1. General anatomy and morphology. *Proc. Malacol. Soc. London* **38,** 301–321.

Morton, B. S. (1969b). Studies on the biology of *Dreissena polymorpha* Pall. 2. Correlation of the rhythms of adductor activity, feeding, digestion and excretion. *Proc. Malacol. Soc. London* **38,** 401–414.

Morton, B. S. (1969c). Studies on the biology of *Dreissena polymorpha* Pall. III. Population dynamics. *Proc. Malacol. Soc. London* **38,** 471–482.

Morton, B. (1977). The population dynamics of *Corbicula fluminea* (Bivalvia: Corbiculacea) in Plover Cove Reservoir, Hong Kong. *J. Zool.* **181,** 21–42.

Morton, B. (1979). Freshwater fouling bivalves. *Proc.—Int. Corbicula Symp., 1st, 1977,* pp. 1–14.

Nagabhushanam, R., and Lomte, V. S. (1970). Effects of temperature on the heat tolerance of the freshwater mussel. *Parreysia corrugata. Broteria, Ser. Trimest.: Cienc. Nat.* **39,** 211–220.

Negus, C. L. (1966). A quantitative study of growth and production of unionid mussels in the river Thames at Reading. *J. Anim. Ecol.* **35,** 513–532.

Okada, K. (1935a). Some notes on *Musculium heterodon* (Pilsbry), a freshwater bivalve. I. The genital system and the gametogenesis. *Sci. Rep. Tohoku Imp. Univ., Ser. 4* **9,** 315–328.

Okada, K. (1935b). Some notes on *Musculium heterodon* (Pilsbry), a freshwater bivalve. II. The gill, the breeding habits and the marsupial sac. *Sci. Rep. Tohoku Imp. Univ., Ser. 4* **9,** 373–391.

Okada, K. (1935c). Some notes on *Musculium heterodon* (Pilsbry), a freshwater bivalve. III. Fertilization and segmentation. *Sci. Rep. Tohoku Imp. Univ., Ser. 4* **10,** 467–483.

Okada, K. (1936). Some notes on *Sphaerium japonicum biwaense* Morei, a freshwater bivalve. IV. Gastrula and fetal larva. *Sci. Rep. Tohoku Imp. Univ., Ser. 4* **11,** 49–68.

Ökland, K. A. (1971). On the ecology of Sphaeriidae in a high mountain area in South Norway. *Norw. J. Zool.* **19,** 133–143.

Ökland, K. A. (1979). Sphaeriidae of Norway: A project for studying ecological requirements and geographical distribution. *Malacologia* **18,** 223–226.

Paparo, A. A., and Sparks, R. E. (1977). Rapid assessment of water quality using the fingernail clam, *Musculium transversum. ASTM Spec. Tech. Publ.* **607,** 96–109.

Pardy, R. L. (1980). Symbiotic algae and C incorporation in the freshwater clam, *Anodonta. Biol. Bull. (Woods Hole, Mass.)* **158,** 349–355.

Pearl, R., and Miner, J. R. (1935). Experimental studies on the duration of life. XIV. The comparative mortality of certain lower organisms. *Q. Rev. Biol.* **10,** 60–79.

Pianka, E. R. (1970). On *r*- and K-selection. *Am. Nat.* **104,** 592–597.

Picken, L. E. R. (1937). The mechanism of urine formation in invertebrates. II. The excretory mechanism in certain Mollusca. *J. Exp. Biol.* **14,** 20–34.

Pietrzak, J. E., Bates, J. M., and Scott, R. M. (1976). Constituents of unionid extrapallial fluid. II. pH and metal ion composition. *Hydrobiologia* **50,** 89–93.

Potts, W. T. W. (1954). The inorganic composition of the blood of *Mytilus edulis* and *Anodonta cygnea. J. Exp. Biol.* **31,** 376–385.

Potts, W. T. W. (1967). Excretion in the molluscs. *Biol. Rev. Cambridge Philos. Soc.* **42,** 1–41.

Potts, W. T. W. (1968). Aspects of excretion in the molluscs. *Symp. Zool. Soc. London* **22,** 187–192.

Precht, H., Christophersen, J., Hensel, H., and Larcher, W. (1973). "Temperature and Life" Springer-Verlag, Berlin and New York.

Prosser, C. L., ed. (1973). "Comparative Animal Physiology," 3rd ed. Saunders, Philadelphia, Pennsylvania.

Qadri, S. U., Mackie, G. L., Hamill, S. E., and Clair, T. A. (1974). Macrobenthos: Standing crop biomass, production and distribution. *In* "Distribution and Transport of Pollutants in Flowing Water Ecosystems," Interim Rep. No. 2, pp. 15-1 to 15-38. University of Ottawa, National Research Council of Canada.

Ransom, J. D., and Prophet, C. W. (1974). Species diversity and relative abundance of benthic macroinvertebrates of Cedar Creek Basin Kansas USA. *Am. Midl. Nat.* **92,** 217–222.

Rees, W. J. (1965). The aerial dispersal of Mollusca. *Proc. Malacal. Soc. London* **36,** 269–282.

Robertson, J. D. (1964). Osmotic and ionic regulation. *In* "Physiology of Mollusca" (K. M. Wilbur and C. M. Yonge, eds.), Vol. 1, pp. 283–311. Academic Press, New York.

Robertson, J. L., and Coney, C. C. (1979). Punctal canals in the shell of *Musculium securis* (Bivalvia: Pisidiidae). *Malacol. Rev.* **12,** 37–40.

Rogers, G. E., (1976). Vertical burrowing and survival of sphaeriid clams under added substrates in Pool 19, Mississippi River. *Iowa State J. Res.* **51,** 1–12.

Rosso, S. W. (1954). A study of the shell structure and mantle epithelium of *Musculium transversum* (Say). *J. Wash. Acad. Sci.* **44,** 329–332.

Russell Hunter, W. (1953). On the growth of the fresh-water limpet, *Ancylus fluviatilis* Müller. *Proc. Zool. Soc. London* **123,** 623–636.

Russell Hunter, W. (1964). Physiological aspects of ecology in nonmarine molluscs. *In* "Physiology of Mollusca" (K. M. Wilbur and C. M. Yonge, eds.), Vol. 1, pp. 83–126. Academic Press, New York.

Russell-Hunter, W. D. (1970). "Aquatic Productivity." Macmillan, New York.

Russell-Hunter, W. D. (1978). Ecology of freshwater pulmonates. *In* "Pulmonates" (V. Fretter and J. Peak, eds.), 2, pp. 335–383. Academic Press, New York.

Russell-Hunter, W. D. (1979). "A Life of Invertebrates." Macmillan, New York.

Russell Hunter, W., and Slack, H. D. (1958). *Pisidium conventus* Clessin in Loch Lomond. *J. Conchol.* **24,** 245–247.

Russell-Hunter, W. D., Meadows, R. T., Apley, M. L., and Burky, A. J. (1968). On the use of a 'wet-oxidation' method for estimates of total organic carbon in mollusc growth studies. *Proc. Malacol. Soc. London* **38,** 1–11.

Russell-Hunter, W. D., Apley, M. L., and Hunter, R. D. (1972). Early life-history of *Melampus* and the significance of semilunar synchrony. *Biol. Bull.* (*Woods Hole, Mass*) **143,** 623–656.

Russell-Hunter, W. D., Burky, A. J., and Hunter, R. D. (1981). Interpopulation variation in calcarous and proteinaceous shell components in the stream limpet, *Ferrissia rivularis. Malacologia* **20,** 255–266.

Salánki, J., and Lukacsovics, F. (1967). Filtration and oxygen consumption related to periodic activity of freshwater mussel *(Anodonta cygnea). Ann. Inst. Biol.* (*Tihany*) *Hung. Acad. Sci.* **34,** 85–98.

Salánki, J., and Vero, M. (1969). Diurnal rhythm of activity in freshwater mussel (*Anodonta cygnea* L.) under natural conditions. *Ann. Inst. Biol.* (*Tihany*) *Hung. Acad. Sci.* **36,** 95–107.

Segal, E. (1961). Acclimation in molluscs. *Am. Zool.* **1,** 235–244.

Selander, R. K., and Kaufman, D. W. (1973). Genic variability and strategies of adaptation in animals. *Proc. Natl. Acad. Sci. U.S.A.* **70,** 1875–1877.

Sepkoski, J. J., Jr., and Rex, M. A. (1974). Distribution of freshwater mussels: Coastal rivers as biogeographic islands. *Syst. Zool.* **23,** 165–188.

Sinclair, R. M. (1971). *Corbicula* variation and *Dreissena* parallels. *Biologist* **53,** 153–159.

Sinclair, R. M., and Isom, B. G. (1963). "Further Studies on the Introduced Asiatic Clam *Corbicula* in Tennessee." Tennessee Department of Public Health, Nashville.

Smith, A. L., and Green, R. H. (1975). Uptake of mercury by freshwater clams (Family Unionidae). *J. Fish. Res. Board Can.* **32,** 1297–1304.

Stańczykowska, A. (1963). Analysis of the age of *Dreissena polymorpha* Pall. in the Masurian Lakes. *Bull. Acad. Pol. Sci., Ser. Sci. Biol.* **9,** 29–33.

Stańczykowska, A. (1964). On the relationship between abundance, aggregations and "condition" of *Dreissena polymorpha* Pall. in 36 Masurian Lakes. *Ekol. Pol., Ser. A* **12,** 653–690.

Stańczykowska, A. (1976). Biomass and production of *Dreissena polymorpha* (Pall.) in some Masurian Lakes. *Ekol. Pol.* **24,** 103–112.

Stansbery, D. H. (1970). Eastern freshwater mollusks, The Mississippi and St. Lawrence River system, pp. 9–21. *Malacologia* **10,** 1–56.

Stansbery, D. H. (1971). Rare and endangered mollusks in eastern United States. *In* "Proceedings of a Symposium on Rare and Endangered Mollusks (Naiads) of the U.S." (S. E. Jorgensen and R. W. Sharp, eds.), pp. 5–18. U.S. Dept. of the Interior, Fish Wildl. Serv., Bur. Sport Fish. Wildl., Twin Cities, Minnesota.

Stansbery, D. H. (1973). A preliminary report on the naiad fauna of the Clinch River in the southern Appalachian Mountains of Virginia and Tennessee (Mollusca: Bivalvia: Unionida). *Am. Malacol. Union, Annu. Rep. Bull.* **38,** 20–22.

Stearns, S. C. (1976). Life-history tactics: A review of the ideas. *Q. Rev. Biol.* **51,** 3–47.

Stearns, S. C. (1977). The evolution of life history traits: A critique of the theory and a review of the data. *Annu. Rev. Ecol. Syst.* **8,** 145–171.

Stearns, S. C. (1980). A new view of life-history evolution. *Oikos* **35,** 266–281.

Stein, C. B. (1971). Naiad life cycles: their significance in the conservation of the fauna. *In* "Proceedings of a Symposium on Rare and Endangered Mollusks (Naiads) of the U.S." (S. E. Jorgensen and R. W. Sharp, eds.), pp. 19–25. U.S. Dept. of the Interior, Fish Wildl. Serv., Bur. Sport Fish. Wildl., Twin Cities, Minnesota.

Strayer, D. L., Cole, J. J., Likens, G. E., and Buso, D. C. (1981). Biomass and annual production of the freshwater mussel *Elliptio companata* in an oligotrophic softwater lake. *Freshwater Biol.* **11,** 435–440.

Strickland, J. D. H., and Parsons, T. R., (1965). A manual of sea water analysis. *Bull., Fish. Res. Board Can.* **125,** 1–203.

Teal, J. M. (1957). Community metabolism in a temperate cold spring. *Ecol. Monogr.* **27,** 283–302.

Thomas, G. J. (1959). Self-fertilization and production of young in a sphaeriid clam. *Nautilus* **72,** 131–141.

Thomas, G. J. (1963). Study of a population of sphaeriid clams in a temporary pond. *Nautilus* **77,** 37–43.

Thomas, G. J. (1965). Growth in one species of sphaeriid clam. *Nautilus* **79,** 47–54.

Thompson, C. M., and Sparks, R. E. (1977). Improbability of dispersal of adult Asiatic clams, *Corbicula manilenses,* via the intestinal tract of migratory waterfowl. *Am. Midl. Nat.* **98,** 219–223.

Thompson, D. (1973). Feeding ecology of diving ducks on Keokuk Pool, Mississippi River. *J. Wildl. Manage.* **37,** 367–381.

Thorp, J. H., and Bergey, E. A. (1981). Field experiments on responses of a freshwater, benthic macroinvertebrate community to vertebrate predators. *Ecology* **62,** 365–375.

Thorson, G. (1950). Reproductive and larval ecology of marine bottom invertebrates. *Biol. Rev. Cambridge Philos. Soc.* **25,** 1–45.

Thut, R. N. (1969). A study of the profundal bottom fauna of Lake Washington. *Ecol. Monogr.* **39,** 79–100.

Tudorancea, C. L. (1972). Studies on Unionidae populations from the Crapina-Jijila complex of pools (Danube zone liable to inundation). *Hydrobiologia* **39,** 527–561.

Tudorancea, C. L., and Florescu, M. (1968a). Cu privire la fluxul energetic al populatiei de *Unio pictorum* din Balta Crapina. *An. Univ. Bucuresti, Ser. Stiint. Nat., Biol.* **17,** 233–243.

Tudorancea, C. L., and Florescu, M. (1968b). Considerations concerning the production and energetics of *Unio tumidus* Philipsson population from the Crapina marsh. *Trav. Mus. Hist. Nat. "Grigore Antipa"* **8,** 395–409.

Tudorancea, C. L., and Florescu, M. (1969). Aspecte ale productiei si energeticii populatiei de *Anodonta piscinalis* Nilsson din-Balta Crapinan' (zona inundabilă a dunăril. *Acad. Repub. Pop. Rom., Stud. Cercet. Biol., Ser. Zool.* **21,** 43–55.

Tudorancea, C., and Florescu M. (1970). Intensitatea schimburilor respiratorii si a activitatii de mineralizare in conditii naturale la unele specii de unioide din zona inundabila a Dunarii. *Acad. Repub. Pop. Rom., Stud. Cercet., Biol., Ser. Zool.* **22,** 377–384.

Vail, V. A. (1978). Seasonal reproductive patterns in 3 viviparid gastropods. *Malacologia* **17,** 73–97.

Van der Schalie, H. (1945). The value of mussel distribution in tracing stream confluence. *Mich. Acad. Sci. Arts Lett. Pap.* **30,** 355–373.

Vokes, H. E. (1980). "Genera of the Bivalvia: A Systematic and Bibliographic Catalogue" (revised and updated). Paleontol. Res. Inst., Ithaca, New York.

Waite, J., and Neufeld, G. (1977). Oxygen consumption by *Sphaerium simile. Comp. Biochem. Physiol. A* **57,** 373–375.

Walz, N. (1978a). The energy balance of the freshwater mussel *Dreissena polymorpha* Pallas in laboratory experiments and in Lake Constance. I. Pattern of activity feeding and assimilation efficiency. *Arch. Hydrobiol., Suppl.* **55,** 83–105.

Walz, N. (1978b). The energy balance of the freshwater mussel *Dreissena polymorpha* Pallas in laboratory experiments and in Lake Constance. II. Reproduction. *Arch. Hydrobiol., Suppl.* **55,** 106–119.

Walz, N. (1978c). The energy balance of the freshwater mussel *Dreissena polymorpha* Pallas in laboratory experiments and in Lake Constance. III. Growth under standard conditions. *Arch. Hydrobiol., Suppl.* **55,** 121–141.

Walz, N. (1978d). The energy balance of the freshwater mussel *Dreissena polymorpha* Pallas in laboratory experiments and in Lake Constance. IV. Growth in Lake Constance. *Arch. Hydrobiol., Suppl.* **5,** 142–156.

Walz, N. (1978e). Growth rates of *Dreissena polymorpha* Pallas under laboratory and field conditions. *Verh.—Int. Ver. Theor. Angew. Limnol.* **20,** 2427–2430.

Walz, N. (1978f). The production significance of the *Dreissena* population in the nutrient cycle in Lake Constance. *Arch. Hydrobiol.* **82,** 482–499.

Walz, N. (1979). The energy balance of the freshwater mussel *Dreissena polymorpha* Pallas in laboratory experiments and in Lake Constance. V. Seasonal and nutritional changes in biochemical composition. *Arch. Hydrobiol., Suppl.* **55,** 235–254.

Waters, T. F. (1977). Secondary production in inland waters. *Adv. Ecol. Res.* **10,** 91–164.

Way, C. M. and Wissing, T. E. (1982). Environmental heterogeneity and life history variability in the freshwater clams, *Pisidium variabile* (Prime) and *Pisidium compressum* (Prime) (Bivalvia: Pisidiidae). *Can. J. Zool.* **60,** 2841–2851.

Way, C. M., Hornbach, D. J., and Burky, A. J. (1980). Comparative life history tactics of the sphaeriid clam, *Musculium partumeium* (Say), from a permanent and a temporary pond. *Am. Midl. Nat.* **104,** 319–327.

Way, C. M., Hornbach, D. J., and Burky, A. J. (1981). Seasonal metabolism of the sphaeriid clam, *Musculium partumeium,* from a permanent and a temporary pond. *Nautilus* **95,** 55–58.

Wiktor, J. (1963). Research on the ecology of *Dreissena polymorpha* Pall. in the Szczecin Lagoon. *Ekol. Pol., Ser. A* **11,** 275–280.

Wilbur, K. M. (1964). Shell formation and regeneration. *In* "Physiology of Mollusca" (K. M. Wilbur and C. M. Yonge, eds.), Vol. 1, pp. 243–282. Academic Press, New York.

Williams, G. C. (1966a). Natural selection, the costs of reproduction, and a refinement of Lack's Principle. *Am. Nat.* **100,** 687–690.

Williams, G. C. (1966b). "Adaptations and Natural Selection." Princeton Univ. Press, Princeton, New Jersey.

Winter, T. E. (1978). A review on the knowledge of suspension-feeding in lamellibranchiate bivalves, with special reference to artificial aquaculture systems. *Aquaculture* **13,** 1–33.

Wissing, T. E., Hornbach, D. J., Smith, M. S., Way, C. M., and Alexander, J. P. (1982). Caloric contents of sphaeriacean clams (Bivalvia: Heterodonta) from freshwater habitats in the United States and Canada. *J. Moll. Stud.* **48,** 80–83.

Wolff, W. J. (1968). The Mollusca of the estuarine region of the rivers Rhine, Meuse and Scheldt in relation to the hydrography of the area. I. The Unionidae. *Basteria* **32,** 13–47.

Wolff, W. J. (1969). The Mollusca of the estuarine region of the rivers Rhine, Meuse, and Scheldt in relation to the hydrography of the area. II. The Dreissenidae. *Basteria* **33,** 93–103.

Wolff, W. J. (1970). The Mollusca of the esturine region of the rivers Rhine, Meuse and Scheldt in relation to the hydrography of the area. IV. The genus *Sphaerium. Basteria* **34,** 75–90.

Wu, S.-K. (1978). Part 1. The fingernail and pill clams (Family Sphaeriidae). *In* "Natural History Inventory of Colorado," Vol. 2, The Bivalvia of Colorado, pp. 1–39. University of Colorado Museum, Boulder.

Yokley, P., Jr. (1972). Life history of *Pleurobema cordatum* (Rafinesque 1820) (Bivalvia: Unionacea). *Malacologia* **11,** 351–364.

Zeuthen, E. (1947). Body size and metabolic rate in the animal kingdom with special regard to the marine micro-fauna. *C.R. Trav. Lab. Carlsberg, Ser. Chim.* **26,** 17–161.

Zeuthen, E. (1953). Oxygen uptake as related to body size in organisms. *Q. Rev. Biol.* **28,** 1–12.

Zs.-Nagy, I. (1974). Some quantitative aspects of oxygen consumption and anaerobic metabolism of molluscan tissue—a review. *Comp. Biochem. Physiol. A* **49,** 399–405.

Zumoff, C. H. (1973). The reproductive cycle of *Sphaerium simile. Biol. Bull.* (*Woods Hole, Mass.*) **144,** 212–228.

8

Physiological Ecology of Freshwater Prosobranchs

D. W. ALDRIDGE

Department of Biology
North Carolina Agricultural and Technical
State University
Greensboro, North Carolina

The Mollusca, Vol. 6
Ecology

ISBN 0-12-751406-6

I. Introduction

The freshwater prosobranchs are a diverse group which present the ecologist with a rich array of adaptations to the freshwater environment. Many species constitute the dominant members of their communities and are thus very important ecologically. Other species are intermediate hosts of parasitic flukes which infect humans and domestic animals. Because all freshwater prosobranchs are derived from aquatic stock, many of their physiological adaptations differ markedly from those seen in cooccurring pulmonates, which are derived from terrestrial stock. It is hoped that this chapter will spark further interest in the physiological ecology of this fascinating group of animals.

Somewhat apart from detailed studies of internal physiological processes, physiological ecology deals generally with vital functions occurring at the intimate animal–environment interface. Environmental factors impinging upon the animal include biotic and abiotic aspects. Prominent biotic factors include competition, predation, and symbiosis; important abiotic factors are nutrients, oxygen, temperature, ions, water, and seasonality. In freshwater environments, abiotic factors tend to overshadow biotic ones, especially in temperate climates, and this chapter will be largely concerned with abiotic factors relevant to the physiological ecology of freshwater and some estuarine prosobranchs.

II. Taxonomy

The freshwater prosobranchs are a heterogeneous mixture of at least 5–10 major animal stocks. The taxonomy of many of the families is uncertain, and genera commonly have lengthy synonymy lists. As this state of affairs does not facilitate ecological or physiological research on these interesting animals, it is hoped that the family and generic levels of classification will soon stabilize.

The list which follows is meant to be a practical guide to freshwater prosobranchs, not a critical classification scheme. It is intended to provide the reader with a convenient reference to relationships among genera mentioned in the text. Further, it is admittedly a lumped classification which, in general, follows Thiele's (1931) scheme.

Subclass Prosobranchia
- Order Archaeogastropoda
  - Superfamily Neritacea
    - Family Neritidae
      - *Theodoxus*
- Order Mesogastropoda
  - Superfamily Cyclophoracea
    - Family Cyclophoridae

Family Viviparidae
Bellamya, Campeloma, Lioplax
Viviparus
Family Ampullariidae (Pilidae)
Ampullarius, Marisa, Pila, Pomacea
Superfamily Valvatacea
Family Valvatidae
Valvata
Superfamily Rissoacea
Family Hydrobiidae
Amnicola, Aroapyrgus, Bithynia, Bulimus, Hydrobia, Oncomelania, Pomatiopsis, Potamopyrgus
Family Micromelaniidae
Superfamily Cerithiacea
Family Melaniidae (Thiaridae)
Cleopatra, Melania, Melanoides, Melanopsis
Family Pleuroceridae
Goniobasis, Leptoxis, Pleurocera

III. Feeding

The majority of freshwater prosobranchs are herbivores that feed on bacteria, protists, and algae. All of these animals are, to a greater or lesser extent, radular grazers on the *Aufwuchs;* most utilize this food resource almost exclusively. *Aufwuchs* can be defined as the scum flora of diatoms, single-celled and filamentous algae, bacteria, fungi, protozoans, and other associated microscopic plants and invertebrates, all attached to submerged hard surfaces including macrophytes, rocks, wood, and miscellaneous debris. Many larger freshwater prosobranchs will graze macrophytes directly. A large number of prosobranch species feed, at least occasionally, on detritus, the decomposed remains of macrophytes.

Among the animals ingesting large amounts of predominantly inorganic sediments during the course of feeding, one may distinguish between deposit feeders and epipsammic browsers. Deposit feeders ingest relatively large amounts of sediment during the feeding process; this material, with its accompanying nutritive constituents, is passed through the gut to extract available nutrients. Many freshwater snails will resort to this mode of feeding at least occasionally to obtain unicellular algae, protists, and bacteria. Deposit feeding occurs most often on sediments that are fine-grained relative to the body size of the animal (Fenchel and Kofoed, 1976; Lopez and Kofoed, 1980). Epipsammic browsing involves taking one or a few sediment particles into the buccal cavity, where organic material adherent to the particles is removed and ingested, and the particles are regurgitated. In contrast to deposit-feeding, this mode is restricted to situations in which particle size is large relative to animal body size (Fenchel and Kofoed, 1976; Lopez and Kofoed, 1980).

Many species of freshwater prosobranchs in the families Viviparidae and Hydrobiidae have become adapted for filter-feeding using the ctenidial gill. The importance of filter-feeding in such forms depends upon life cycle stage and the availability of suspended food (Tashiro and Colman, 1982).

Most small snails feed at least occasionally while hanging at the air–water interface by surface tension or by virtue of their own buoyancy. A few larger snails (ampullariids) also feed on surface film material while clinging to convenient hard substrates.

Although prosobranchs generally feed readily on microscopic animals, a few feed on larger ones. Several large freshwater prosobranchs are opportunistic and occasionally dedicated predators of smaller snails and snail spawn.

The relative importance of these many different feeding mechanisms in freshwater prosobranchs needs much more study. Specific modes aside, general feeding activity usually cycles on both seasonal and diel bases. As is typical for most temperate animals, prosobranchs found in freshwater usually cease (Russell-Hunter, 1964) or perhaps sharply curtail feeding in winter (van Cleave and Altringer, 1937; Medcof, 1940). In the tropics, feeding activity typically cycles seasonally in response to temperature extremes and/or water availability (Berry, 1975), although some tropical and cave habitats may offer stable trophic conditions where feeding may continue uniformly during the year.

Diel feeding cycles of freshwater prosobranchs have not been well documented, but available evidence suggests that the phenomenon is widespread. Feeding activity in *Goniobasis clavaeformis* peaks at night (Malone and Nelson, 1969); in *Oncomelania quadrasi* (Dazo and Moreno, 1962), *G. livescens,* and *Pleurocera acuta* (Dazo, 1965), it peaks during the day. These data should be borne in mind by investigators evaluating ingestion rates during time intervals of less than 24 h.

A. Food Quality

Nitrogenous food availability appears to be the principal limiting factor in most environments (White, 1978), and food availability is an important determinant of snail distribution patterns (Horst and Costa, 1971). A convenient index of nitrogen content and hence of food quality is the weight-to-weight carbon : nitrogen (C : N) ratio of the food material. A lower C : N ratio for food (when total carbon biomass availability is above a minimum value) favors higher growth rates and fecundities (McMahon et al., 1974; Aldridge, 1980). For the relatively nitrogen-rich *Aufwuchs,* C : N ratios ranging from 3.7 : 1 to 10.1 : 1 have been reported (McMahon et al., 1974). In contrast, macrophytes, which are not widely fed upon by prosobranchs, have considerably higher C:N ratios: 17:1 [as reported by McMahon et al. (1974)]. In the case of detritus as a food source, it should be noted that bacterial growth on the material is important in improving its quality (i.e., lowering the C:N ratio) (Hunter, 1976).

Another useful index of food quality is its content of organic carbon and nitrogen per unit dry mass. Aldridge (1982) noted this as an important determinant of growth and reproduction in *Leptoxis carinata.*

B. Food Localization, Ingestion, and Egestion

The pulmonate limpet *Ancylus fluviatilis,* which feeds on the quite homogeneously distributed *Aufwuchs,* shows a foraging style characterized by random movement and contact chemoreception for distinguishing suitable food (Calow, 1973). Conversely, animals feeding on food items that have a patchier distribution appear to utilize distant chemoreception. The pulmonate snail *Planorbis contortus,* a detritivore, exhibits such distant chemoreception (Calow, 1973), as does the prosobranch *Viviparus georgianus,* which can detect diatom clusters at distances of 10 cm (Duch, 1976). The prosobranch *Pomacea paludosa* can detect food on surface film several centimeters away (McClary, 1964). Food localization in other prosobranchs is presumably by similar combinations of chemoreception and random movement or foraging, but further experimental evidence to that effect would be most welcome.

Once a suitable food has been located, ingestion may commence. However, certain factors may limit the availability of such food. For *Hydrobia ulvae,* attachment of diatoms and bacteria to sediment particles reduces their usefulness as food for the snail (Lopez and Levinton, 1978). Particle ingestion in some snails appears to be a quite specialized process: the four common estuarine hydrobiids in northern Europe (*H. ulvae, H. ventrosa, H. neglecta,* and *Potamopyrgus jenkinsi*) seem to be adapted to feed on particles of different size (Lopez and Kofoed, 1980).

Ingestion rates for *H. ventrosa* are not greatly affected by the nature of the diet (Kofoed, 1975b), and maximum ingestion rates for *H. neglecta, H. ulvae,* and *H. ventrosa* show an optimal response to salinity and temperature (Hylleberg, 1975). Similarly, ingestion rates for *G. clavaeformis* show an optimal response to temperature (Elwood and Goldstein, 1975). Mass specific ingestion rates for *Pila globosa* are an inverse function of live mass (Haniffa and Pandian, 1974); this is probably related to the fact that the gut cross-sectional area : mass ratio declines with increasing mass.

Food passage times and egestion rates have not been well studied in prosobranchs. At 15°C, the average residence time for food in the gut of *G. clavaeformis* (Malone and Nelson, 1969) and *Leptoxis carinata* (Aldridge, 1980) is 2 h.

For *H. ventrosa,* egestion rates are inversely related to ingested particle size (Lopez and Levinton, 1978).

C. General Food Habits

General food habit data for the group are summarized in Table I; some special feeding strategies merit more detailed discussion.

In prosobranchs, surface film feeding is widespread among ampullariids (McClary, 1964), and juvenile pleurocerids have been observed floating on the surface film (Dazo, 1965). Among the ampullariids, Cheesman (1956) noted that young *Pomacea canaliculata* ingest surface film by using their foot as a funnel trap to concentrate proteinaceous material. If food is present on the surface film, *P. paludosa* will commence ciliary feeding, using its foot as a funnel trap (Johnson, 1952; McClary, 1964), although apparently no ciliary tracts within the funnel are particularly specialized for food transport (McClary, 1964). McClary (1964) further observed that *P. paludosa* may accomplish this surface feeding while either attached to a solid substrate at the air–water interface or hanging from the surface film itself. Movement to the surface under high density conditions has been observed in *H. ulvae* (Barnes, 1981) and *H. ventrosa* (Levinton, 1979).

It appears that *Marisa cornuarietis* predates its own egg masses (Demian and Lufty, 1965a), and it is well established that this species also predates the adults, spat, and eggs of *Bulinus truncatus* (Demian and Lufty, 1965a) and *Lymnaea cailludi* (Demian and Lufty, 1966) as well as the spat and eggs of *Biomphalaria alexandrina;* however, it seems that as a result of their narrow shell aperture adults of this last species cannot be effectively predated (Demian and Lufty, 1965b).

A detailed study of *Marisa* predation on *Bulinus* (Demian and Lufty, 1966) revealed a number of interesting points. At constant densities of *Bulinus,* the rate of *Bulinus* predated per *Marisa* was an inverse function of *Marisa* density. Further, if the ratio of predator to prey is kept constant, there is little effect of density on predation rates either per predator or per prey individual. The predation rate shows an optimal response to temperature and is reduced in the presence of vegetable matter. Whether this latter effect results from provision of an alternate food resource for the predator or from provision of cover for the prey is unclear. A similar vegetation effect has been noted for *Marisa* predation on *Biomphalaria glabrata* (Jobin et al., 1973). Experience in *Bulinus* predation by a *Marisa* individual had a positive effect on future rates of *Bulinus* predation by that individual. Further, such predation experience reduced the efficacy of vegetation "protection." These observations suggest that *Marisa* would be very suitable for studies involving optimal foraging theory.

TABLE I

Diet of Freshwater Prosobranchs

Species	Diet (references)
Bithynia leachi	Detritus, diatoms (Fretter and Graham, 1962)
Bithynia tentaculata	Filamentous green algae and diatoms (Lilly, 1953)
	Prefers unicellular green algae to other algae (Reavell, 1980)
	Occasional detritivore (Wesenberg-Lund, 1939, via Fretter and Graham, 1962)
	Gut contents: 95% detritus, 5% algae (Reavell, 1980)
	Feces: mineral particles, diatoms (mostly), detritus, plant fragments (Ladle and Griffiths, 1980)
Campeloma decisum	Detritus (Chamberlain, 1958)
Goniobasis livescens	Green and red algae, desmids (Dazo, 1965)
	Feces: sand, diatoms, desmids (Goodrich, 1945)
Oncomelania hupensis quadrasi	Detritus, microscopic animals, algae (Garcia, 1972)
Oncomelania nosophora	Soil, diatoms, algae (Komiya, 1961)
Oncomelania quadrasi	Green algae, diatoms, to a lesser degree desmids, euglenoids (Dazo and Moreno, 1962)
Pila globosa	Macrophytic vegetation: *Chara, Hydrilla, Ceratophyllum* (Haniffa, 1980)
Pleurocera acuta	Green and red algae, demids (Dazo, 1965)
Potamopyrgus jenkinsi	Detritus (Fretter and Graham, 1962)
	Gut contents: 90% detritus, 7% sand grains, 1% algae (Reavell, 1980)
	Feces: mineral particles, diatoms, detritus, plant fragments (Ladle and Griffiths, 1980)
Theodoxus fluviatilis	Feces: mineral particles, diatoms, detritus (Ladle and Griffiths, 1980)
Valvata piscinalis	Winter feeding: detritus, diatoms, desmids (Fretter and Graham, 1962)
	Gut contents: detritus (Reavell, 1980)
	Feces: diatoms, detritus, mineral particles, plant fragments (Ladle and Griffiths, 1980)
Viviparus georgianus	Filamentous green algae, diatoms (Duch, 1976)
Viviparus viviparus	Gut contents: 90% detritus, 10% algae (Reavell, 1980)

Pomacea adults have been observed predating *Biomphalaria glabrata* spat and eggs, and the presence or absence of vegetation appears to have no major effect on the egg predation (Paulinyi and Paulini, 1972).

IV. Bioenergetics

Partitioning of assimilated material into maintenance, growth, and reproduction is a complex process affected by such factors as age, size, sexuality, reproductive state, temperature, food quality, and season. For uniformity, the following relationships and nomenclature (mostly proposed by Russell-Hunter, 1970) will be used:

1. Assimilation = ingestion − egestion
2. Assimilation = respired assimilation (maintenance) + nonrespired assimilation
3. Nonrespired assimilation = growth + reproduction
4. Assimilation efficiency = assimilation/ingestion
5. Net growth efficiency = nonrespired assimilation/assimilation

Resources most commonly measured include dry biomass, organic carbon and/or nitrogen, and energy (in joules or calories).

A. Assimilation Efficiency

Carbon-based assimilation efficiencies for prosobranchs feeding on preferred foods are quite high, generally between 70 and 95% (Calow and Calow, 1975; Haniffa, 1980), although they may vary considerably with diet. For example, some diet-specific assimilation efficiencies for *H. ventrosa* are bacteria and hay, 56%, and hay, 34% (Kofoed, 1975b); bacteria, 75%; diatoms, 60–71%; *Oscillatoria,* 50%; and *Chroococcus,* 8% (Kofoed, 1975a). For this species, which usually is a deposit feeder, no clear relationship between ingested particle size and assimilation efficiency exists (Lopez and Levinton, 1978).

On a diet of the macrophyte *Ceratophyllum,* the assimilation efficiency of *Pila globosa,* which was 73–88%, varied inversely with the ingestion rate (Vivekanandan et al., 1974). There is a strong correlation between assimilation efficiency and gut extract cellulase-like activity in 13 species of freshwater snails, and such activity tends to be higher in freshwater prosobranchs than in cooccurring pulmonates (Calow and Calow, 1975).

The phosphorus-based assimilation efficiency for *G. clavaeformis* is about 40% and is relatively temperature independent (Elwood and Goldstein, 1975).

B. Net Growth Efficiency

In a study of *L. carinata,* Aldridge (1982) found that net growth efficiencies for protein carbon ranged from 20 to 26% and were inversely related to the proteinaceous percentage of total assimilated carbon. Net growth efficiencies for nonprotein carbon in these animals were 1–5%. Further, growth rates and net growth efficiencies declined as animals aged, and males of the species had lower protein and nonprotein net growth efficiencies than did females. For three tropical species of freshwater prosobranchs, average net production efficiencies were 17.0, 26.0, and 32.3% for *Cleopatra bulimoides, Melania tuberculata,* and *Bellamya unicolor,* respectively (Leveque, 1973).

On a diet of the macrophyte *Ceratophyllum, P. globosa* had a net growth efficiency of 2 to 14% (Vivekanandan et al., 1974), and when fed a mixed diet of *Chara, Hydrilla,* and *Ceratophyllum,* it showed an average net growth efficiency of 10%. Further evidence that net growth efficiency is diet dependent was presented by Kofoed (1975b), in his study of *H. ventrosa;* carbon-based net growth efficiencies of 33, 34, and 15% were found for bacteria, bacteria and hay, and hay diets, respectively.

C. Partitioning

Aldridge (1982), in a study of *L. carinata,* and Tashiro (1980), of *Bithynia tentaculata,* found that the respired assimilation of protein and nonprotein carbon was dependent on the absolute and relative availability of these two classes in the food and on the changing energy demands of the snail. Such partitioning flexibility appears to be highly adaptive for snails living in markedly periodic environments where nutrient availability and metabolic requirements may vary considerably over time. The $N:O_2$ ratio, an index of protein carbon use in respired assimilation, varied over the life cycle of *L. carinata* in response to season, trophic input, life cycle events, and sexual dimorphism. *L. carinata* individuals lose some (10–25%) biomass while overwintering.

A 2-day fast in *M. cornuarietis* led to an increase in protein catabolism and a 20% reduction in respiratory oxygen uptake (Horne, 1979). Starvation in snails has been observed to depress the respiratory quotient as well (Yanagisawa and Komiya, 1961).

Subpartitioning of nonrespired assimilation has not been well studied, although Kofoed (1975b) reported that mucus production in *H. ventrosa* accounts for 9% of assimilated carbon, and many workers have collected data on the percentage of life span biomass growth laid as eggs (i.e., reproductive effort) (see Chapters 11 and 15, this volume).

D. Growth

Snail growth rates generally decline with age, and in *Oncomelania* (Chi and Wagner, 1957) and *Leptoxis* (Aldridge, 1982), growth rates level off following attainment of sexual maturity. Intraspecific differential growth and maturation rates appear to be an important isolating mechanism among sympatric races of *Lithoglyphopsis aperta* (Davis et al., 1976). Mass specific growth rates for *P. globosa* are an inverse function of snail mass (Haniffa, 1980). Finally, high densities appear to stunt growth in *Potamopyrgus jenkinsi* individuals (Lumbye and Lumbye, 1965). All of these observations probably have more general application among freshwater prosobranchs.

Sexually dimorphic growth is a common phenomenon in the group, and many workers have reported sexually dimorphic size in these snails (Browne, 1978; Vail, 1978; Aldridge, 1982).

Seasonal effects on growth are well documented. Young (1975) reported that *B. tentaculata* in southern England did not grow during the winter. In upstate New York, this species shows no overwinter growth; during this time individuals become detached from the substrate (Mattice, 1972). In New York. *B. tentaculata* grows actively from May through October (Mattice, 1972). *Valvata piscinalis* also shows no winter growth in southern England, although *Viviparus viviparus* grows there throughout the year (Young, 1975), a fact which should temper broad generalizations about the trophic state of overwintering snails.

V. Respiration

Freshwater prosobranchs are essentially aerobic animals, and the term *respiration* will therefore be used to indicate respiratory oxygen uptake. Most freshwater prosobranchs are mesogastropods, possessing within their mantle-cavities a single unipectinate gill (*ctenidium*) which is the principal site of respiratory oxygen uptake. A few freshwater archaeogastropods (Neritacea) have bipectinate ctenidia. In addition to this aquatic gill, many freshwater and amphibious prosobranchs have modified all or part of their mantle-cavity as a lung for aerial respiration.

During aquatic respiration of mesogastropod forms, a maneuverable, inwardly rolled extension of the mantle edge, the inhalant siphon, ensures delivery of clean water into the mantle-cavity. In many tropical mesogastropods, the inhalant siphon is used as a snorkel for air breathing. Oxygen uptake across the gill is certainly augmented to various degrees by cutaneous uptake.

Although no freshwater prosobranch has been reported to be active under anoxic conditions, most are capable of tolerating such conditions for short periods (von Brand et al., 1950), and metabolism during aestivation is apparently anaerobic in some forms, although not in all.

Three important variables affect availability of oxygen to aquatic animals: temperature, salinity, and the oxygen demand of the environment relative to available oxygen supplies. Oxygen solubility is an inverse function of temperature and salinity, and high environmental oxygen demand, especially in poorly aerated waters, will reduce oxygen tension.

Because of their dependence on an aquatic gill, freshwater prosobranchs typically require better oxygenated water than their more tolerant cousins, the air-breathing pulmonates (Boycott, 1936; Palmieri et al., 1980). However, many freshwater prosobranchs, particularly in the tropics, have evolved a more amphibious respiratory strategy as an apparent adaptation to periodically lowered oxygen availability in their normal aquatic environment: the mantle-cavity lung alluded to earlier.

A. Oxygen Uptake Rates

As is true for many animals, mass specific oxygen uptake rates for snails appear to be an inverse function of individual snail mass (Yanagisawa and Komiya, 1961; Dos Santos and Mendes, 1981). Oxygen uptake rates conform to environmental oxygen tension in some snails, such as *B. tentaculata* (Berg and Ockelmann, 1959; Berg, 1961), *B. leachi* (Berg, 1961), *Marisa cornuarietis* (Akerlund, 1974), *Theodoxus fluviatilis,* and *Potamopyrgus jenkinsi* (Lumbye, 1958), and *Viviparus* (Studier and Pace, 1978). Other snails maintain fairly constant oxygen uptake rates against changing environmental oxygen tension. Included in this latter group are *Goniobasis cochliaris* and *G. cahawbensis* (Hawkins and Ultsch, 1979), and *Valvata piscinalis* (Berg and Ockelmann, 1959; Berg, 1961).

Temperature is perhaps the single most important factor affecting oxygen uptake rates. These rates generally increase with increasing temperature up to some critical limit (typically 35–40°C), followed by a rate plateau and decline as temperature increases beyond the snail's ability to cope (Lumbye, 1958; Berg and Ockelmann, 1959; Buckingham and Freed, 1976; Freiberg and Hazelwood, 1977). For *M. cornuarietis,* higher Q_{10} values for oxygen uptake have been reported for juveniles than for adults (Akerlund, 1969).

In addition to temperature, a number of other factors also affect oxygen uptake rates in these animals. For example, the response to starvation is quite variable. Some snails respond with a depressed oxygen uptake rate, such as *B. leachi* (Berg and Ockelmann, 1959), *M. cornuarietis* (Akerlund, 1969), *Oncomelania nosophora* (Yanagisawa and Komiya, 1961), and *P. jenkinsi* (Lumbye and Lumbye, 1965). On the other hand, Berg and Ockelmann (1959) reported that starvation had no marked effect on oxygen uptake rates in *B. tentaculata* or *V. piscinalis*. Studier and Pace (1978) reported that oxygen uptake rates decline in starved *Viviparus georgianus* females but not in males.

Environmental osmolarity also has a variable effect on oxygen uptake rates in these snails. In *P. jenkinsi,* both oxygen uptake rates and activity show a maximum response to salinity at 28‰ (Duncan, 1966; Klekowski and Duncan, 1966), and Lumbye (1958) reported that individuals from freshwater populations of this species had lower oxygen uptake rates than did individuals from brackish-water populations. However, for *Theodoxus fluviatilis,* such rates are about the same for individuals from freshwater and brackish-water populations (Lumbye, 1958). Juvenile *P. jenkinsi* have higher oxygen uptake rates in freshwater than they do in brackish water (Skoog, 1973). Oxygen uptake rates in *V. georgianus* are an inverse function of environmental NaCl concentrations (Studier and Pace, 1978).

Sexually dimorphic oxygen uptake rates are well established for *V. georgianus,* in which species males have higher mass specific oxygen uptake rates than females (Fitch, 1975; Buckingham and Freed, 1976; Fitch, 1976), although Q_{10} values for oxygen uptake rates are higher for females than for males (Buckingham and Freed, 1976). Aldridge (1982) found sexually dimorphic oxygen uptake rates in *L. carinata,* with males showing higher rates than females; two other species, *Pomacea paludosa* and *P. lineata,* are reported to show no such sexual dimorphism (Freiberg and Hazelwood, 1977; Dos Santos and Mendes, 1981).

Less obvious factors may also affect oxygen uptake in snails. For example, uptake rates in *V. georgianus* showed an optimal response to pH (Buckingham and Freed, 1976), whereas substrate type had no apparent effect (Studier and Pace, 1978). Various temporal patterns of oxygen uptake have been reported. Not surprisingly, mass specific oxygen uptake rates for many snails vary seasonally (Berg, 1961; Aldridge, 1980). A related observation in *V. georgianus* (Fitch, 1976) showed that oxygen uptake rates in this species are an inverse function of the percentage of the day spent under illumination, although there was no apparent diel pattern to these rates.

B. Facultative Air Breathing

The ampullariid *P. paludosa,* as are many other cyclophoraceans, is a facultative air breather, and oxygen uptake rates are reported to be the same whether the snail is breathing air or water so long as the temperature remains below 30°C (Freiberg and Hazelwood, 1977). Above this temperature, aerial rates decline sharply, probably due to desiccation. Similar data have been reported for *P. lineata* (Dos Santos and Mendes, 1981).

Pomacea will commence air-breathing whenever chemical irritants are added to the water, aquatic oxygen tensions decline, water temperature increases, or starvation conditions exist (McClary, 1964) (Section III,C). Such air inspiration results in buoyancy, and the snails are often found floating in poorly oxygenated waters (Burky and Burky, 1977). It appears that gas retention in the mantle-cavity may be a general buoyancy mechanism in other prosobranchs as well (Thomas, 1975; D. W. Aldridge, personal observations). In the less amphibious ampullariid *M. cornuarietis,* aerial respiration does occur (Demian, 1965), although oxygen uptake rates during air-breathing are lower than those during water-breathing (Freiberg and Hazelwood, 1977).

C. Response to Hypoxia and Anoxia

Hypoxic and anoxic tolerance in prosobranchs is quite variable. Operculates are generally more resistant to anoxic conditions than are pulmonates (von Brand et al., 1950). In their study of anoxic survival, more than 95% of the tested individuals of four operculate species (*Melanoides tuberculata, Oncomelania nosophora, Pomatiopsis lapidaria,* and *Goniobasis livescens*) survived for 24 h, and 84% of the *M. tuberculata* individuals were still alive after 84 h. During the period of anoxia, these four operculate snails produced very little lactate.

Among other prosobranchs, *Pomacea urceus* (Burky et al., 1972) and *Marisa cornuarietis* (Horne, 1979) are anoxia intolerant, and the latter species must repay an oxygen debt following brief anoxic exposure (Akerlund, 1974). *G. cochliaris* and *G. cahawbensis* tolerate hypoxic episodes lasting as long as 6 days (Hawkins and Ultsch, 1979).

VI. Nitrogen Excretion

Nitrogenous compounds excreted by freshwater prosobranchs include ammonia, urea, uric acid, and amino acids. The relative and absolute contributions of each of these to the total excreted nitrogen in an individual snail may be influenced by qualitative and quantitative changes in diet, sexuality, life cycle stage, osmotic and ionic stress, and the more general need to conserve water under the very different environmental circumstances of activity, reproduction, and aestivation of its occurrence (Little, 1981).

Most aquatic and amphibious freshwater prosobranchs appear to be primarily ammonotelic when active, although significant, highly variable amounts of urea may also be excreted (Haggag and Fouad, 1968; Shylaja and Alexander, 1975a; Aldridge, 1982). Some forms, such as *Pila* (Reddy and Swami, 1975; Shylaja and Alexander, 1975a; Chaturvedi and Agarwall, 1979) and *Pomacea* (Little, 1968; Burky et al., 1972), are distinctly uricotelic during aestivation, when water conservation is critical. Embryos of forms that lay terrestrial eggs, such as *P.*

paludosa (Sloan, 1964), also appear to be uricotelic. It may also be that viviparous aquatic forms, particularly when young, are more ureotelic than their oviparous relatives because of the need to reduce the potentially toxic effects of protein catabolism upon young and parents during rapid juvenile growth.

VII. Reproduction

Although valvatids are hermaphroditic (Heard, 1963), sexes are separate in all other freshwater prosobranchs that are not parthenogenetic. Parthenogenesis, usually associated with viviparity, is found in many prosobranch families including the Viviparidae, Hydrobiidae, and Melaniidae (Boycott, 1919; van Cleave and Altringer, 1937; Medcof, 1940; Chamberlain, 1958; Winterbourn, 1970; Berry and Kadri, 1974; Muley, 1977).

As in all gastropods, gametes are produced by a single gonad. Fertilization is internal in all freshwater prosobranchs and usually involves a penis or spermatophore, although for many species, such as the pleurocerids, the exact method of sperm transfer is unclear (Jones and Branson, 1964).

A peculiar feature of many prosobranch families is the presence of dimorphic sperm. The functional significance of this sperm dimorphism is not clear, but it may be related to the extended sperm storage seen in many female freshwater prosobranchs (Otori et al., 1954; Chi and Wagner, 1957).

The female reproductive tract of freshwater prosobranchs is complex, having specialized areas for receiving and storing sperm and for fertilization, and specialized glands for adding nutrients to the eggs (Fretter and Graham, 1962). Following ovulation, eggs are typically first invested with a layer of albumin, and after fertilization, other energy stores and a protective capsule are added. The capsule may contain, depending upon the species, one or many eggs. Reproductive modes for female freshwater prosobranchs include oviparity, ovoviviparity, and viviparity.

A. Oviparity

In addition to the more general egg capsule attributes mentioned earlier, many oviparous freshwater and amphibious prosobranchs surround these capsules with soil and sand grains: *Pomatias* (Creek, 1951), *Pomatiopsis* (van der Schalie and Walter, 1957), *Oncomelania* (Abbott, 1946), some pleurocerids (van Cleave, 1932; Dazo, 1965), *Lithoglyphopsis* (Davis et al., 1976), and others. Also, in some cyclophoraceans that lay terrestrial eggs (e.g., *Pila, Pomacea*), the eggs are surrounded with a calcified egg capsule (Snyder and Snyder, 1971; Meenakshi et al., 1974; Kushlan, 1978; Tompa, 1980).

Oviparous freshwater prosobranchs typically "lay minute eggs in obscure locations" (Berry and Kadri, 1974). Egg diameters ranging from 0.2 to 1 mm have been reported (Jewell, 1931; van Cleave, 1932; Abbott, 1946; van der Schalie and Walter, 1957; Fretter and Graham, 1962; Dazo, 1965; Lassen, 1979; Aldridge, 1982), although most cyclophoracean genera (e.g., *Ampullarius, Pomacea, Pila*) lay larger eggs. These larger eggs may have been a factor in the development of ovoviviparity in this latter group. The aforementioned tiny eggs, laid singly, may be an antipredator strategy. Chi et al. (1971) found that *M. cornuarietis,* an avid predator on the egg-masses of pulmonates, was not an effective predator on the small eggs of *Oncomelania.*

In forms that lay egg-masses, there is considerable intra- and interspecific variation in the number of eggs per egg-mass and the number of egg-masses produced per laying female. There is also considerable intraspecific variation in egg size and composition (C : N ratio) (Aldridge, 1982). In *Oncomelania,* spat size and probable survivorship correlate with egg size (Chi and Wagner, 1957).

Most freshwater prosobranchs lay their eggs in wet or at least moist places, although some, such as *Pomacea* (Andrews, 1933; Sloan, 1964; Snyder, 1967; Burky et al., 1972) and *Pila* (Prashad, 1932), oviposit calcareous-shelled eggs under fairly dry conditions. In the case of the many prosobranchs which lay aquatic eggs, such as *Bithynia, Amnicola,* pleurocerids, *Hydrobia, Valvata,* and others, most are oviposited on hard substrates such as vegetation, rock, and organic debris (Cleland, 1954; Fretter and Graham, 1962; Heard, 1963; Sloan, 1964; Horst and Costa, 1975; Skoog, 1976; Aldridge, 1982). Eggs in these species are typically intolerant to desiccation. Of the amphibious forms, *Oncomelania, Pomatiopsis,* and others many oviposit in moist soil or organic debris at or near the waterline (Abbott, 1946; McMullen, 1947; Creek, 1951; Ritchie, 1955; Wagner and Wong, 1956; Dundee, 1957; van der Schalie and Walter, 1957). Such semiterrestrial eggs are somewhat resistant to desiccation, but moisture is still critical for successful embryo development (van der Schalie and Walter, 1957).

Among the taxons laying truly terrestrial eggs are many cyclophoracean genera that lay large, essentially cleidoic eggs above the waterline, usually on emergent vegetation. Two possible reasons for adopting a terrestrial or semiterrestrial ovipositing strategy have been suggested. The most likely hypothesis was proposed by Boettger (1942), who suggested that such forms oviposit out of water to avoid hypoxic or potentially hypoxic conditions for their eggs, conditions common in tropical aquatic environments. It has also been proposed that such ovipositing is a predator avoidance strategy (Clench, 1959).

Among a few oviparous forms, one individual within an egg capsule may develop more quickly than its siblings or potential siblings, and subsequently will feed on them. This adaptation is common in marine snails (Fretter and Graham, 1962), although relatively rare in freshwater species; here it is limited to the neritid genus *Theodoxus,* whose egg capsules clearly contain many nurse eggs which provide food for the single earliest hatching embryo (Bondesen, 1940). An analogous opportunistic strategy is seen in some other freshwater species, in which the siblings hatching earliest feed on portions of the egg-mass prior to beginning more normal foraging activities (Burky, 1973, 1974).

The eggs of freshwater prosobranchs typically have longer development times than those of cooccurring pulmonates. Hatching times range from 15 to 50 days, with an average time of 1 month under natural conditions (Abbott, 1946; McMullen, 1947; Chi and Wagner, 1957; Dundee, 1957; Chi and Wagner, 1962; Dazo, 1965; Lang and Dronen, 1970; Mattice, 1972; Burky, 1973; Aldridge, 1982). In some cases, the young aestivate for considerable periods after hatching (Burky, 1974). Factors affecting egg development time include egg size, temperature, and salinity. Development times for *Oncomelania* (Chi and Wagner, 1957) and *L. carinata* (Aldridge, 1982) are inversely related to egg size and are often also inversely related to environmental temperature (Otori et al., 1956; Fish and Fish, 1977; Lassen and Clark, 1979). In estuarine forms, development times usually show an optimal response to salinity (Fish and Fish, 1977).

B. Numerical Fecundity

Numerical fecundity is a function of female size, which shows considerable intraspecific variation in *L. carinata* (Aldridge, 1982) and probably in other species as well. Intraspecific variation in numerical fecundity has also been observed in *Hydrobia* (Lassen, 1979) and *Oncomelania* (Chi and Wagner, 1957).

A number of environmental factors appear to play a role in the initiation and maintenance of ovipositing. Factors in the initiation of egg laying include high water levels (*Oncomelania;* Abbott, 1946) and a critical temperature and/or day length (Russell-Hunter, 1964). Important modifiers of numerical fecundity include salinity (Lassen and Clark, 1979), temperature (Wagner and Wong, 1956; Lassen and Clark, 1979; Aldridge, 1982), interspecific competition (Lassen and Clark, 1979), egg density (Aldridge, 1982), and diet (Skoog, 1978; Tashiro, 1980). Particularly apparent in subtropical and tropical regions is the reduction in egg laying by high midsummer temperatures (Komiya, 1961). Ovipositing by *L. carinata* females is depressed by high egg densities (40%) in the enviornment during the second half of the normal egg-laying period (at about the time when eggs begin to hatch) (Aldridge, 1982). In *B. tentaculata,* poor food quality (high C : N ratio) depresses the fecundity of 2-year-old females but not of 3-year-olds (last year of life) (Tashiro, 1980, 1982).

Temporal ovipositing patterns have not been well studied. Egg-laying intensity cycles during reproduction in some species, such as *Oncomelania formosana* (Roth and Wagner, 1960; Chi and Wagner, 1962) and *V. piscinalis* (Frömming,

1956), although not in others, such as *L. carinata* (Aldridge, 1982). Diel cycles of egg laying have been studied infrequently but probably occur commonly; Snyder and Snyder (1971) reported that *P. paludosa* more often lays at night.

C. Ovoviviparity and Viviparity

Ovoviviparity, and perhaps viviparity, are widespread among freshwater prosobranchs. True viviparity has been difficult to establish in these snails, and more research on the energetic relations of offspring and parents would be helpful. All members of the Viviparidae and Melaniidae are either ovoviviparous or viviparous, as are many hydrobiid genera including *Potamopyrgus* (Winterbourn, 1970) and *Aroapyrgus* (Malek et al., 1975).

Ovoviviparous and viviparous snails typically differ from oviparous ones in having larger eggs, reduced numerical fecundities, and longer embryonic development times. In temperate regions, embryonic development times are 9–12 months for *Viviparus georgianus* (Browne, 1978; Vail, 1978), 10–12 months for *Lioplax pilsbryi,* and 8–10 months for *Campeloma geniculum* (Vail, 1978).

Melania scabra broods its young internally (Muley, 1978b), and *Aroapyrgus costaricensis* hatch and are active within the brood pouch prior to release (Malek et al., 1975). *Melanoides tuberculata* eggs are about 60 μm in diameter, and young are released from the brood pouch at a shell length of 3 mm (Berry and Kadri, 1974). Release of young by *P. jenkinsi* varies with temperature (Frenzel, 1979). Juvenile emergence from *M. tuberculata* females occurs most commonly between nightfall and midnight (Berry and Kadri, 1974), and the viviparid *Filopaludina sumatrensis* also reportedly releases its young at night (Berry, 1974).

D. Patterns of Reproduction

Most freshwater prosobranchs, both temperate and tropical, show periodic patterns of reproduction. Those dwelling in temperate regions usually lay their eggs in spring and early summer. Genera with species having such habits include *Amnicola* (Horst and Costa, 1975), *Bithynia* (Young, 1975; Tashiro, 1980), *Hydrobia* (Lassen and Clark, 1979), *Oncomelania* (McMullen et al., 1951), *Pomatiopsis* (van der Schalie and Walter, 1957), *Potamopyrgus* (Michaut, 1968; Winterbourn, 1970), *Campeloma* (van Cleave and Altringer, 1937; Chamberlain, 1958; Vail, 1978), *Lioplax* (Vail, 1978), *Viviparus* (van Cleave and Lederer, 1932; Young, 1975; Vail, 1978), *Pomatias* (Creek, 1951), *Valvata* (Heard, 1963; Young, 1975), *Leptoxis* (Aldridge, 1982), *Pleurocera* (van Cleave, 1932), and *Goniobasis* (Dazo, 1965). Some of these spring and early summer egg layers may also oviposit in the fall, for example *Amnicola limosa* (Horst and Costa, 1975) and *Potamopyrgus jenkinsi* (Frenzel, 1980). The rarity of fall egg laying in most freshwater prosobranchs probably results from their small spat size, which would limit overwinter survivorship. Reproductive periodicities in snails living in the tropics seem to be adapted for the avoidance of ovipositing during dry or very hot periods. Tropical species displaying such periodicities include *Oncomelania quadrasi* (McMullen, 1947), *Pomacea urceus* (Burky, 1973, 1974), and *Siamopaludina martensi* and *Filopaludina sumatrensis* (Berry, 1974).

Some species living under fairly constant tropical or troglodytic conditions may breed throughout the year; examples include *Bellamya unicolor* (Leveque, 1973), *Melanoides tuberculata* (Leveque, 1973; Berry, 1975), and *Melania scabra* (Muley, 1977, 1978a).

E. Copulation

In contrast to patterns of ovipositing and spat release, seasonal aspects of copulatory activity and success have received only slight attention for the obvious reason that such behavior is often rather obscure and difficult to evaluate in the field. However, observations indicate that most species copulate just prior to and during ovipositing, and many species appear to copulate at other times as well (Dundee, 1957; Komiya, 1961; Burky, 1974). Sexually dimorphic survivorship patterns and skewed sex ratios are known for many freshwater prosobranch populations, as is the female's ability to store sperm; further observations on copulating behavior are badly needed to evaluate the adaptive significance of the temporally dimorphic life cycles of males and females in many species (van der Schalie and Dundee, 1955; Dundee, 1957; Dazo, 1965; Berry, 1974; Burky, 1974; Browne, 1978; Aldridge, 1982).

F. Life Cycle Patterns

Some female freshwater prosobranchs life cycles are annual (McMullen, 1947; McMullen et al., 1951; van der Schalie and Dundee, 1955; Horst and Costa, 1975; Young, 1975; Frenzel, 1980). More commonly, such life cycles are biennial or triennial and the females are iteroparous (Medcof, 1940; Dazo, 1965; Stiven and Walton, 1967; Houp, 1970; Jobin, 1970; Burky, 1974; Young, 1975; Browne, 1978; Dussart, 1979; Haniffa, 1980; Tashiro, 1980), although some north temperate forms are semelparous (Aldridge, 1982). Typically, freshwater prosobranchs have longer life cycles than cooccurring pulmonates. A more detailed account of life cycle patterns and evolution can be found in Chapter 15, this volume.

VIII. Fluid and Electrolyte Relations

Freshwater prosobranchs possess a single coelomoduct kidney (Fretter and Graham, 1962). Most of the fluid received by the kidney (if not all) comes from the pericardial sac which surrounds the prosobranch heart. Fluid from the pericardial sac is generally considered to be an ultrafiltrate of blood, with filtration being accomplished across the wall of the ventricle during its contraction (Fretter and Graham, 1962; Little, 1981). It is clear that this initial ultrafiltrate is substantially modified by secretory and absorptive activities in the kidney (Fretter and Graham, 1962). Of course, not all fluid and electrolyte interchange between animal and environment occurs at the kidney, but contributions by the gill, mantle-cavity, gut, and general body surface have not been well studied.

A. Osmoregulation

Bedford (1972) examined the osmoregulatory capabilities of the prosobranch *Melanopsis trifasciata* and found that in waters exceeding 100 mOsm the snail was essentially isosmotic with its environment. At lower environmental osmolarities all the way down to freshwater, the animal maintained its internal osmolarity at about 100 mOsm. Body water content and the size of the extracellular compartment remained quite stable under all test conditions. In this estuarine species, Na^+ and Cl^- are the most abundant blood ions, and there is some indication that nitrogenous compounds may be involved in osmoregulation (Bedford, 1971). A similar involvement of nitrogenous compounds, specifically free amino acids, has been reported for *Pila virens* (Shylaja and Alexander, 1975b), and the phenomenon is generally widespread among invertebrates and some vertebrate groups.

Salinity adaptation in *P. jenkinsi* usually occurred within 1 day (Todd, 1964; Skoog, 1973), although the species is intolerant to rapid changes in salinity (Duncan, 1967). Blood osmolarity in *P. jenkinsi* was slightly lower in freshwater than in brackish-water individuals (Duncan, 1967). *V. viviparus* produces a hypotonic urine when in fresh water (Little, 1965), whereas *Potamopyrgus* produces an isotonic or only slightly hypotonic one at all environmental salinities tested (Todd, 1964). *Pomacea lineata* and *P. depressa* produce hypotonic urine when active (Little, 1968).

Urine flow has not been well studied in prosobranchs. In *Viviparus malleatus*, blood volume is 46% of wet tissue weight and 9.2% of the blood volume is excreted each hour (Monk and Stewart, 1966).

Freshwater *Theodoxus fluviatilis* individuals are more tolerant of freshwater than are individuals from brackish-water populations, and vice versa (Kangas and Skoog, 1978). Whether this is an acclimation phenomenon or reflects some genetic differences is unclear.

B. Ion Regulation

Studies on ion regulation within the group have centered mostly on Ca^{2+}, which is important in shell deposition and seems to play a crucial role in general fluid and electrolyte tolerance as well as other physiological processes in vertebrates and invertebrates alike (Little, 1981). The gill of *Viviparus bengalensis* sequesters Ca^{2+} (Gupta, 1977), and similar observations have been made for other freshwater prosobranchs (Fretter and Graham, 1962).

M. cornuarietis obtains sufficient Ca^{2+} from the environment to meet 60% of its shell growth demands (Meier-Brook, 1978). Some freshwater prosobranchs, such as *Valvata piscinalis* (Aho, 1978), are rather tolerant of low environmental Ca^{2+} concentrations, although among prosobranchs generally there is broad interspecific variation in environmental Ca^{2+} levels tolerated (Shoup, 1943).

Other specific ion relations, including the regulation of internal acid–base status, have not been well studied. Several authors (Avery, 1946; Ritchie, 1955; Komiya, 1961) have noted that *Oncomelania* species survive well in slightly acid waters, and Aho (1978) suggested that various chemical factors have their most profound effects on snail community composition when total water hardness conditions are low.

C. Desiccation Resistance

Among the amphibious freshwater prosobranchs, desiccation resistance has been well studied in *Oncomelania*. Species within this genus are truly amphibious (Ritchie, 1955), although the young are aquatic (McMullen et al., 1951; Komiya and Hashimoto, 1958). Adults of *O. nosophora* are resistant to desiccation, but this resistance declines as the temperature rises (Komiya and Hashimoto, 1958; Komiya, 1961), and adults die when 40% of their live body mass has been lost by desiccation (Komiya, 1961). Desiccation rates in *O. nosophora* are sexually dimorphic; males losing water faster than females, probably because of the male's smaller body size (Komiya and Hashimoto, 1958). Desiccation resistance in this species varies intraspecifically, with snails from drier habitats being more resistant to desiccation than those from wetter habitats (Komiya and Iijima, 1958). *O. quadrasi* adults are mostly amphibious, but return to water at high temperatures (Avery, 1946). Eggs of this species can resist moderate desiccation for about 12 h (Abbott, 1946).

Among other prosobranch genera, young *Pomatiopsis cincinnatiensis* are less resistant to desiccation than are adults and consequently prefer moister habitats (van der Schalie and Getz, 1961). For the primitive neritid prosobranch *T. fluviatilis*, survivorship at any given relative humidity is an inverse function of temperature (Skoog, 1976).

IX. Adaptations to Climatic Variation

In coping with the long-term variability of freshwater environments, animals have evolved various strategies. Acclimation, migration, and aestivation are three such strategies which may be adopted by animals depending upon their relative ability to cope with or escape from environmental stress.

Temperature is an important abiotic variable in the freshwater environment. Many snails show distinct temperature preferences (van der Schalie and Getz, 1963; Ross and Ultsch, 1980), and tolerance to temperature extremes is an intra- and interspecific variable (van der Schalie and Getz, 1963; Skoog, 1976; Chitramvong et al., 1981). At least one snail, *Pomacea urceus,* has been reported to cool itself (5–10°C below ambient) by evaporation when heat stressed (Burky et al., 1972). This manner of handling heat stress is available only to terrestrial snails. Aquatic snails, in coping with temperature fluctuations, must either tolerate or acclimate. The temperature acclimation of various responses has been demonstrated in a number of freshwater prosobranchs (Yasuraoka, 1961; Skoog, 1976; Ross and Ultsch, 1980). Acclimation responses to variables other than temperature, which are widespread, are discussed in other sections of this chapter.

Migration is a common strategy for avoiding adverse environmental conditions such as winter ice and high temperatures. Many north temperate species migrate to deeper water in winter (Medcof, 1940; Goodrich, 1945; Bovbjerg, 1952; Bickel, 1966; Horst and Costa, 1975; Young, 1975) and during periods of fluctuating water levels (Medcof, 1940).

Aestivation, in the broadest sense, is a widespread phenomenon among molluscs (Boss, 1974) which involves lowered metabolic demands along with physical isolation from the environment to cope *in situ* with adverse environmental conditions such as temperature extremes and desiccation. It is a very common adaptation among freshwater prosobranchs (Ramanan, 1903; Annandale et al., 1919; Cort, 1919; Medcof, 1940; McMullen et al., 1951; van der Schalie and Dundee, 1955; Dundee, 1957; Komiya, 1961; Vinogradov, 1962; Meenakshi, 1964; Little, 1968; Burky et al., 1972).

A number of detailed studies on aestivating cyclophoraceans (*Pomacea* and *Pila*) have been reported. Biomass losses during aestivation can be considerable (35–60%); most of the loss is water (Burky et al., 1972; Burky, 1973, 1974; Haniffa, 1978). Thus, it is not surprising that temperature and humidity strongly influence aestivation survivorship (Haniffa, 1978; Horne, 1979). As further evidence that water stress is significant during aestivation, Little (1968) reported that *Pomacea lineata* and *P. depressa* blood and urine osmolarities may double during aestivation, and Reddy et al. (1974) observed that the normally ammonotelic *Pila globosa* shifted to uricotelism during the process.

Glycogen provides most of the energy used by *P. globosa* and *P. virens* during aestivation (Meenakshi, 1956; Reddy and Swami, 1976); *M. cornuarietis* also utilizes primarily carbohydrate during aestivation (Horne, 1979). Oxygen uptake rates in aestivating snails are reduced to a fraction of their active levels (Coles, 1968; Burky et al., 1972; Rao et al., 1972; Haniffa, 1978).

Metabolism during aestivation is reported to be aerobic in *P. ovata* (Coles,

1968) and *Pomacea urceus* (Burky et al., 1972), and anaerobic in *Pila virens* (Meenakshi, 1956, 1957) and *P. globosa* (Krishnamoorthy and Brahmanandam, 1970). Reports on lactate production during anaerobiotic aestivation are variable (Reddy and Swami, 1976).

X. Future Studies

Freshwater prosobranchs are choice candidates for a variety of experimental studies; for example, (1) the effects of food quality and quantity on growth, maintenance, and reproduction; (2) the effects of biotic and abiotic factors on propagule size and numerical fecundity; (3) preferred temperatures and moisture conditions in amphibious forms; (4) acclimation and acclimatization responses to important environmental factors other than temperature; (5) the adaptive value of the various nitrogen excretion strategies; (6) physiological mechanisms underlying the "oblomovic" responses to extreme environmental conditions in poikilothermic invertebrates; (7) adaptive significance of sexual dimorphism; and finally (8) the adaptive value of such strategies as semelparity versus iteroparity, oviparity versus ovoviviparity versus viviparity, and parthenogenesis. Further, because these animals are so widespread and in rather intimate contact with their environments, they are often sensitive indicators of environmental quality (Harman, 1974).

Pursuant to the above suggestions, this author would like to take issue with the increasingly endothermic bias in the field of physiological ecology. Poikilotherms are interesting animals with flexible and highly adaptable physiologies which, when considered in light of their worldwide abundance and often intimate relationships with their environments, make their study crucial to our understanding of life on Earth. The freshwater prosobranchs offer many opportunities for such studies.

Acknowledgments

I would like to thank my brother, James B. Aldridge, for assistance with the manuscript, especially the sections on fluid and electrolyte relations and respiration. Thanks also to Therese M. Allen-Aldridge for help with the bibliography, and to the Duke University libraries for access to much of the reference material cited here.

References

Abbott, R. (1946). The egg and breeding habits of *Oncomelania quadrasi*, the schistosomiasis snail of the Philippines. *Occas. Pap. Mollusks, Harv. Univ. Mus. Comp. Zool.* **1**, 41–48.

Aho, J. (1978). Freshwater snail populations and the equilibrium theory of island biogeography. II. Relative importance of chemical and spatial variables. *Ann. Zool. Fenn.* **15,** 155–164.

Akerlund, G. (1969). Oxygen consumption of the ampullariid snail *Marisa cornuarietis* L. in relation to body weight and temperature. *Oikos* **20,** 529–533.

Akerlund, G. (1974). Oxygen consumption in relation to environmental oxygen concentrations in the ampullariid snail *Marisa cornuarietis* (L.). *Comp. Biochem. Physiol. A* **47,** 1065–1075.

Aldridge, D. W. (1980). Life cycle, reproductive tactics, and bioenergetics in the freshwater prosobranch snail *Spirodon carinata* (Bruguiere) in upstate New York. Ph.D. Dissertation, Syracuse University, Syracuse, New York.

Aldridge, D. W. (1982). Reproductive tactics in relation to life-cycle bioenergetics in three natural populations of the freshwater snail, *Leptoxis carinata. Ecology* **63,** 196–208.

Andrews, E. A. (1933). Eggs of *Ampullaria* in Jamaica. *Nautilus* **46,** 93–97.

Annandale, N., Prashad, B., and Kemp, S. W. (1919). Report on the aquatic fauna of the Seistan with subsidiary studies. The Mollusca of the inland waters of Baluchistan and Seistan with a note on the liver-fluke of sheep in Seistan. *Rec. Indian Mus.* **18,** 17–63.

Avery, J. L. (1946). The habitat of the snail host of *Schistosoma japonicum* in the Philippines. *Science* **104,** 5.

Barnes, R. S. K. (1981). Factors affecting climbing in the coastal gastropod, *Hydrobia ulvae. J. Mar. Biol. Assoc. U. K.* **61,** 301–306.

Bedford, J. J. (1971). Osmoregulation in *Melanopsis trifasciata* Gray 1843. 3. The intracellular nitrogenous compounds. *Comp. Biochem. Physiol. A* **40A,** 899–910.

Bedford, J. J. (1972). Osmoregulation in *Melanopsis trifasciata.* 2. The osmotic pressure and the principal ions of the hemocoelic fluid. *Physiol. Zool.* **45,** 261–269.

Berg, K. (1961). On the oxygen consumption of some freshwater snails. *Verh.—Int. Ver. Theor. Angew. Limnol.* **14,** 1019–1022.

Berg, K., and Ockelmann, K. W. (1959). The respiration of freshwater snails. *J. Exp. Biol.* **36,** 690–708.

Berry, A. J. (1974). Reproductive condition in two Malayan freshwater viviparid gastropods. *J. Zool.* **174,** 357–367.

Berry, A. J. (1975). Patterns of breeding activity in west Malaysian gastropod molluscs. *Malays. J. Sci.* **3,** 49–59.

Berry, A. J., and Kadri, A. B. H. (1974). Reproduction in the Malayan freshwater cerithiacean gastropod *Melanoides tuberculata. J. Zool.* **172,** 369–381.

Bickel, D. (1966). Stranded *Campeloma. Nautilus* **79,** 106–107.

Boettger, C. R. (1942). Vertreter der Schneckenfamilie Ampullariidae als Aquarienbewohner in Deutschland. *Wochenschr. Aquar. Terrarienkd.* pp. 77–80.

Bondesen, P. (1940). Preliminary investigations into the development of *Neritina fluviatilis* L. in brackish and fresh water. *Vidensk. Medd. Dank. Naturh. Foren.* **104,** 283–318.

Boss, K. J. (1974). Oblomovism in the Mollusca. *Trans. Am. Microsc. Soc.* **93,** 460–481.

Bovbjerg, R. V. (1952). Ecological aspects of dispersal of the snail *Campeloma decisum. Ecology* **33,** 169–176.

Boycott, A. E. (1919). Parthenogenesis in *Paludestrina jenkinsi. J. Conchol.* **16,** 54.

Boycott, A. E. (1936). The habitats of freshwater Mollusca in Britain. *J. Anim. Ecol.* **5,** 116–186.

Browne, R. A. (1978). Growth, mortality, fecundity, biomass and productivity of four lake populations of the prosobranch snail, *Viviparus georgianus. Ecology* **59,** 742–750.

Buckingham, M. J., and Freed, D. E. (1976). Oxygen consumption in the prosobranch snail *Viviparus contectoides* (Mollusca:Gastropoda). 2. Effect of temperature and pH. *Comp. Biochem. Physiol. A* **53A,** 249–252.

Burky, A. J. (1973). Organic content of eggs and juveniles of an amphibious snail, *Pomacea urceus* (Müller), from the Venezualan savannah and their ecological significance. *Malacol. Rev.* **6,** 59.

Burky, A. J. (1974). Growth and biomass production of an amphibian snail, *Pomacea urceus* (Müller), from the Venezuelan savannah. *Proc. Malacol. Soc. London* **41,** 127–143.

Burky, A. J., Pacheco, J., and Pereyra, E. (1972). Temperature, water, and respiratory regimes of an amphibious snail, *Pomacea urceus* (Müller), from the Venezuelan savannah. *Biol. Bull. (Woods Hole, Mass.)* **143,** 304–316.

Burky, K. A., and Burky, A. J. (1977). Buoyancy changes as related to respiratory behavior in an amphibious snail, *Pomacea urceus* (Mueller), from Venezuela. *Nautilus* **91,** 97–104.

Calow, P. (1973). Field observations and laboratory experiments on the general food requirements of two species of freshwater snail, *Planorbis contortus* (Linn.) and *Ancylus fluviatilis* Müll. *Proc. Malacol. Soc. London* **40,** 483–489.

Calow, P., and Calow, L. J. (1975). Cellulase activity and niche separation in freshwater gastropods. *Nature (London)* **255,** 478–480.

Chamberlain, N. A. (1958). Life history studies of *Campeloma decisum. Nautilus* **72,** 22–29.

Chaturvedi, M. L., and Agarwall, R. A. (1979). Excretion, accumulation and site of synthesis of urea in the snail, *Pila globosa,* during active and dormant periods. *J. Physiol. (Paris)* **75,** 233–238.

Cheesman, D. F. (1956). The snail's foot as a Langmuir trough. *Nature (London)* **178,** 987–988.

Chi, L. W., and Wagner, E. D. (1957). Studies on reproduction and growth of *Oncomelania quadrasi, O. nosophora,* and *O. formosana,* snail hosts of *Schistosoma japonicum. Am. J. Trop. Med. Hyg.* **6,** 949–959.

Chi, L. W., and Wagner, E. D. (1962). Oviposition observations in *Oncomelania formosana. Trans. Am. Microsc. Soc.* **81,** 244–246.

Chi, L. W., Winkler, L. R., and Colvin, R. (1971). Predation of *Marisa cornuarietis* on *Oncomelania formosana* eggs under laboratory conditions. *Veliger* **14,** 184–186.

Chitramvong, E. S., Upatham, E. S., and Sukhapanth, N. (1981). Effects of some physico-chemical factors on the survival of *Bithynia siamensis siamensis, Radix rubiginosa,* and *Indoplanorbis exustus. Malacol. Rev.* **14,** 43–48.

Cleland, D. M. (1954). A study of the habits of *Valvata piscinalis* (Müller) and the structure and function of the alimentary canal and reproductive system. *Proc. Malacol. Soc. London* **30,** 167–203.

Clench, W. J. (1959). Mollusca. *In* "Freshwater-water Biology" (H. B. Ward and G. C. Whipple, eds.), pp. 1117–1160. Wiley, New York.

Coles, G. C. (1968). The termination of aestivation in the large fresh-water snail *Pila ovata* (Ampullariidae). 1. Changes in oxygen uptake. *Comp. Biochem. Physiol.* **25,** 517–522.

Cort, W. (1919). On the resistance to desiccation of the intermediate host of *Schistosoma japonica* Katsuroda. *J. Parasitol.* **6,** 84–88.

Creek, G. (1951). The reproductive system and embryology of the snail *Pomatias elegans. Proc. Zool. Soc. London* **121,** 599–640.

Davis, G. M., Kitikoon, V., and Temcharoen, P. (1976). Monograph of "*Lithoglyphopsis*" *aperta,* the snail host of Mekong River schistosomiasis. *Malacologia* **15,** 241–287.

Dazo, B. C. (1965). The morphology and natural history of *Pleurocera acuta* and *Goniobasis livescens* (Gastropoda: Cerithiacea: Pleuroceridae). *Malacologia* **3,** 1–80.

Dazo, B. C., and Moreno, R. G. (1962). Studies on the food and feeding habits of *Oncomelania quadrasi,* the snail intermediate host of *Schistosoma japonicum* in the Philippines. *Trans. Am. Microsc. Soc.* **81,** 341–347.

Demian, E. S. (1965). The respiratory system and the mechanism of respiration in *Marisa cornuarietis* (L.). *Ark. Zool.* **17,** 539–560.

Demian, E. S., and Lufty, R. G. (1965a). Predatory activity of *Marisa cornuarietis* against *Bulinus truncatus,* the transmitter of urinary schistosomiasis. *Ann. Trop. Med. Parasitol.* **59,** 331–336.

Demian, E. S., and Lufty, R. G. (1965b). Predatory activity of *Marisa cornuarietis* against *Biomphalaria alexandrina* under laboratory conditions. *Ann. Trop. Med. Parasitol.* **59,** 337–339.

Demian, E. S., and Lufty, R. G. (1966). Factors affecting the predation of *Marisa cornuarietis* on *Bulinus truncatus, Biomphalaria alexandrina* and *Lymnaea cailludi. Oikos* **17,** 212–230.

Dos Santos, C. A. Z., and Mendes, E. G. (1981). Oxygen consumption of the amphibious snail *Pomacea lineata;* influence of weight, sex and environments. *Comp. Biochem. Physiol. A* **69A,** 595–598.

Duch, T. M. (1976). Aspects of the feeding habits of *Viviparus georgianus. Nautilus* **90,** 7–10.

Duncan, A. (1966). The oxygen consumption of *Potamopyrgus jenkinsi* (Smith) (Prosobranchiata) in different temperatures and salinities. *Verh.—Int. Ver. Theor. Angew. Limnol.* **16,** 1739–1751.

Duncan, A. (1967). Osmotic balance in *Potamopyrgus jenkinsi* (Smith) from two Polish populations. *Pol. Arch. Hydrobiol.* **14,** 1–10.

Dundee, D. S. (1957). Aspects of the biology of *Pomatiopsis lapidaria* (Say). *Misc. Publ. Mus. Zool., Univ. Mich.* **100,** 1–37.

Dussart, G. B. J. (1979). Life cycles and distribution of the aquatic gastropod mollusks *Bithynia tentaculata, Gyraulus albus, Planorbus planorbis,* and *Lymnaea peregra* in relation to water chemistry. *Hydrobiologia* **67,** 223–240.

Elwood, J. W., and Goldstein, R. A. (1975). Effects of temperature on food ingestion rate and absorption, retention, and equilibrium burden of phosphorus in an aquatic snail, *Goniobasis clavaeformis* Lea. *Freshwater Biol.* **5,** 397–406.

Fenchel, T., and Kofoed, L. H. (1976). Evidence for exploitive interspecific competition in mud snails (Hydrobiidae). *Oikos* **27,** 367–376.

Fish, J. D., and Fish, S. (1977). The effects of temperature and salinity on the embryonic development of *Hydrobia ulvae* (Pennant). *J. Mar. Biol. Assoc. U. K.* **57,** 213–218.

Fitch, D. D. (1975). Oxygen consumption in the prosobranch snail *Viviparus contectoides* (Mollusca:Gastropoda). I. Effects of weight and activity. *Comp. Biochem. Physiol. A* **51A,** 815–820.

Fitch, D. D. (1976). Oxygen consumption in the prosobranch snail *Viviparus contectoides* (Mollusca:Gastropoda). III. Effects of light. *Comp. Biochem. Physiol. A* **54A,** 253–257.

Freiberg, M. W., and Hazelwood, D. H. (1977). Oxygen consumption of two amphibious snails: *Pomacea paludosa* and *Marisa cornuarietis* (Prosobranchia:Ampullariidae). *Malacologia* **16,** 541–548.

Frenzel, P. (1979). Biology and population dynamics of *Potamopyrgus jenkinsi* (Gastropoda:Prosobranchia) in the littoral of Lake Constance, West Germany. *Arch. Hydrobiol.* **85,** 448–464.

Frenzel, P. (1980). Production of *Potamopyrgus jenkinsi* (Gastropoda:Prosobranchia) in Lake Constance. *Hydrobiologia* **74,** 141–144.

Fretter, V., and Graham, A. (1962). "British Prosobranch Molluscs." Ray Society, London.

Frömming, E. (1956). "Biologie der mitteleuropäische Süsswasserschnecken." Duncker & Humbolt, Berlin.

Garcia, R. G. (1972). Tolerance of *Oncomelania hupensis quadrasi* to varying concentrations of dissolved oxygen and organic pollution. *Bull. W.H.O.* **47,** 59–70.

Goodrich, C. (1945). *Goniobasis livescens* of Michigan. *Misc. Publ. Mus. Zool., Univ. Mich.* **64,** 1–26.

Gupta, A. S. (1977). Observations on the gill of *Viviparus bengalensis* in relation to calcium uptake and storage. *Acta Zool. (Stockholm)* **58,** 129–134.

Haggag, G., and Fouad, Y. (1968). Comparative study of nitrogenous excretion in terrestrial and fresh-water gastropods. *Z. Vergl. Physiol.* **57,** 428–431.

Haniffa, M. A. (1978). Energy loss in an aestivating population of the tropical snail *Pila globosa. Hydrobiologia* **61,** 169–182.

Haniffa, M. A. (1980). Studies on lifespan, growth increments and energy budget of the freshwater snail, *Pila globosa. Proc.—Indian Acad. Sci. [Ser.]: Anim. Sci.* **89,** 275–286.

Haniffa, M. A., and Pandian, T. J. (1974). Effect of body weight on feeding rate and radula size in the freshwater snail *Pila globosa. Veliger* **16,** 415–418.

Harman, W. N. (1974). Snails (Mollusca: Gastropoda). *In* "Pollution Ecology of Freshwater Invertebrates" (C. W. Hart and S. L. H. Fuller, eds.), pp. 275–312. Academic Press, New York.

Hawkins, M. J., and Ultsch, G. R. (1979). Oxygen consumption in 2 species of freshwater snails (*Goniobasis*): Effects of temperature and ambient oxygen tension. *Comp. Biochem. Physiol. A* **63A,** 369–372.

Heard, W. H. (1963). Reproductive features of *Valvata. Nautilus* **77,** 64–68.

Horne, F. R. (1979). Comparative aspects of an aestivating metabolism in the gastropod, *Marisa cornuarietis. Comp. Biochem. Physiol. A* **64A,** 309–312.

Horst, T. J., and Costa, R. R. (1971). Distribution patterns of five selected gastropod species from McCargo Lake. *Nautilus* **85,** 38–43.

Horst, T. J., and Costa, R. R. (1975). Seasonal migration and density patterns of the freshwater snail *Amnicola limosa. Nautilus* **89,** 56–59.

Houp, K. H. (1970). Population dynamics of *Pleurocera acuta* in a central Kentucky limestone stream. *Am. Midl. Nat.* **83,** 81–88.

Hunter, R. D. (1976). Changes in carbon and nitrogen content during decomposition of three macrophytes in freshwater and marine environments. *Hydrobiologia* **51,** 119–128.

Hylleberg, J. (1975). The effect of salinity and temperature on egestion in mud snails (Gastropoda:Hydrobiidae). A study on niche overlap. *Oecologia* **21,** 279–289.

Jewell, D. D. (1931). Observations on reproduction in the snail *Goniobasis. Nautilus* **44,** 115–119.

Jobin, W. R. (1970). Population dynamics of aquatic snails in three farm ponds of Puerto Rico. *Am. J. Trop. Med. Hyg.* **19,** 1038–1048.

Jobin, W. R., Ferguson, F. F., and Berrios-Duran, L. A. (1973). Effect of *Marisa cornuarietis* on populations of *Biomphalaria glabrata* in farm ponds of Puerto Rico. *Am. J. Trop. Med. Hyg.* **22,** 278–284.

Johnson, B. M. (1952). Ciliary feeding in *Pomacea paludosa. Nautilus* **66,** 1–5.

Jones, W. C., and Branson, B. A. (1964). The radula, genital system, and external morphology in *Mudalia potosiensis* (Lea) 1841 (Gastroposa: Prosobranchiata: Pleuroceridae) with life history notes. *Trans. Am. Microsc. Soc.* **83,** 41–62.

Kangas, P., and Skoog, G. (1978). Salinity tolerance of *Theodoxus fluviatilis* (Mollusca, Gastropoda) from freshwater and from different salinity regimes in the Baltic Sea. *Estuarine Coastal Mar.Sci.* **6,** 409–416.

Klekowski, R. Z., and Duncan, A. (1966). The oxygen consumption in saline water of young *Potamopyrgus jenkinsi* (Smith) (Prosobranchiata). *Verh.—Int. Ver. Theor. Angew. Limnol.* **16,** 1753–1760.

Kofoed, L. H. (1975a). The feeding biology of *Hydrobia ventrosa* (Montague). I. The assimilation of different components of the food. *J. Exp. Mar. Biol. Ecol.* **19,** 233–241.

Kofoed, L. H. (1975b). The feeding biology of *Hydrobia ventrosa* (Montague). II. Allocation of the components of the carbon budget and the significance of the secretion of dissolved organic material. *J. Exp. Mar. Biol. Ecol.* **19,** 243–256.

Komiya, Y. (1961). The ecology of *Oncomelania nosophora:* A review. *Jpn. J. Med. Sci. Biol.* **14,** 1–9.

Komiya, Y., and Hashimoto, I. (1958). The survival of *Oncomelania nosophora,* the vector snail of *Schistosoma japonicum,* under dried condition and their water loss. *Jpn. J. Med. Sci. Biol.* **11,** 339–346.

Komiya, Y., and Iijima, T. (1958). The local difference of the resistance of *Oncomelania nosophora,* the vector snail of *Schistosoma japonicum,* to dessication. *Jpn. J. Med.Sci. Biol.* **11,** 455–459.

Krishnamoorthy, R. V., and Brahmanandam, V. (1970). Hepatopancreatic lipolytic activity in the aestivated pond snail *Pila globosa*. *Indian J. Exp. Biol.* **8,** 145–146.

Kushlan, J. A. (1978). Predation on apple snail eggs (*Pomacea*). *Nautilus* **92,** 57–58.

Ladle, M., and Griffiths, B. S. (1980). A study on the feces of some chalk stream invertebrates. *Hydrobiologia* **74,** 161–172.

Lang, B. Z., and Dronen, N. O. (1970). Eggs and attachment sites for egg capsules of *Valvata lewisi*. *Nautilus* **84,** 9–12.

Lassen, H. H. (1979). Reproductive effort in Danish mud snails (Hydrobiidae). *Oecologia* **40,** 365–370.

Lassen, H. H., and Clark, M. E. (1979). Comparative fecundity in 3 Danish mud snails (Hydrobiidae). *Ophelia* **18,** 171–178.

Leveque, C. (1973). Dynamique des peuplements biologie, et estimation de la production des mollusques benthiques du lac Tchad. *Cah. ORSTOM, Ser. Hydrobiol.* **1,** 117–147.

Levinton, J. S. (1979). The effect of density upon deposit-feeding populations: Movement, feeding, and floating of *Hydrobia ventrosa* (Gastropoda:Prosobranchia). *Oecologia* **43,** 27–40.

Lilly, M. M. (1953). The mode of life and the structure and functioning of the reproductive ducts of *Bithynia tentaculata* (L.). *Proc. Malacol. Soc. London* **30,** 87–110.

Little, C. (1965). The formation of urine by the prosobranch gastropod mollusc *Viviparus viviparus* Linn. *J. Exp. Biol.* **43,** 39–54.

Little, C. (1968). Aestivation and ionic regulation in two species of *Pomacea* (Gastropoda:Prosobranchia). *J. Exp. Biol.* **48,** 569–585.

Little, C. (1981). Osmoregulation and excretion in prosobranch gastropods. Part I. Physiology and biochemistry. *J. Moll. Stud.* **47,** 221–247.

Lopez, G. R., and Kofoed, L. H. (1980). Epipsammic browsing and deposit-feeding in mud snails (Hydrobiidae). *J. Mar. Res.* **38,** 585–600.

Lopez, G. R., and Levinton, J. S. (1978). The availability of microorganisms attached to sediment particles as food for *Hydrobia ventrosa* Montague (Gastropoda:Prosobranchia). *Oecologia* **32,** 263–276.

Lumbye, J. (1958). The oxygen consumption of *Theodoxus fluviatilis* (L.) and *Potamopyrgus jenkinsi* (Smith) in brackish and fresh water. *Hydrobiologia* **10,** 245–262.

Lumbye, J., and Lumbye, L. E. (1965). The oxygen consumption of *Potamopyrgus jenkinsi* (Smith). *Hydrobiologia* **25,** 489–500.

McClary, A. (1964). Surface inspiration and ciliary feeding in *Pomacea paludosa* (Prosobranchia:Mesogastropoda:Ampullariidae). *Malacologia* **2,** 87–104.

McMahon, R. F., Hunter, R. D., and Russell-Hunter, W. D. (1974). Variation in *Aufwuchs* at six freshwater habitats in terms of carbon biomass and of carbon : nitrogen ratio. *Hydrobiologia* **45,** 391–404.

McMullen, D. B. (1947). The control of schistosomiasis. I. Observations on the habits, ecology and life cycle of *Oncomelania quadrasi*, the intermediate host of *Schistosoma japonicum* in the Philippine Islands. *Am. J. Hyg.* **45,** 259–273.

McMullen, D. B., Komiyama, S., and Endo-Habaski, T. (1951). Observations on the habits, ecology and life cycle of *Oncomelania nosophora*, the molluscan intermediate host of *Schistosoma* in Japan. *Am. J. Hyg.* **54,** 402–415.

Malek, G. A., Brenes,R., and Rojas, G. (1975). *Aroapyrgus costaricensis*, hydrobiid snail host of paragonimiasis in Costa Rica. *J. Parasitol.* **61,** 355–359.

Malone, C. R., and Nelson, D. J. (1969). Feeding rates of freshwater snails (*Goniobasis clavaeformis*) determined with cobalt^{60}. *Ecology* **50,** 728–730.

Mattice, J. S. (1972). Production of a natural population of *Bithynia tentaculata* L. (Gastropoda, Mollusca). *Ekol. Pol.* **20,** 525–539.

Medcof, J. C. (1940). On the life cycle and other aspects of the snail *Campeloma* in the Speed River. *Can. J. Res., Sect. D* **18,** 165–172.

Meenakshi, V. R. (1956). Physiology of hibernation of the apple snail *Pila virens* (Lamarck). *Curr. Sci.* **25,** 321–322.

Meenakshi, V. R. (1957). Anaerobiosis in the South Indian apple snail *Pila virens* (Lamarck) during aestivation. *J. Zool. Soc. India* **9,** 62–71.

Meenakshi, V. R. (1964). Aestivation in the Indian apple snail *Pila*. I. Adaptation in natural and experimental conditions. *Comp. Biochem. Physiol.* **11,** 379–386.

Meenakshi, V. R., Blackwelder, P. L., and Watabe, N. (1974). Studies on the formation of calcified egg-capsules of ampullarid snails. *Calcif. Tissue Res.* **16,** 283–291.

Meier-Brook, C. (1978). Calcium uptake by *Marisa cornuarietis* (Gastropoda; Ampullariidae), a predator of schistosome-bearing snails. *Arch. Hydrobiol.* **82,** 449–464.

Michaut, P. (1968). Données biologiques sur un Gastéropode prosobranche récemment introduit en Côte-d'Or, *Potamopyrgus jenkinsi*. *Hydrobiologia* **32,** 513–527.

Monk, C., and Stewart, D. M. (1966). Urine formation in a freshwater snail, *Viviparus malleatus*. *Am. J. Physiol.* **210,** 647–651.

Muley, E. V. (1977). Studies on the breeding habits and development of the brood-pouch of a viviparous prosobranch; *Melania scabra*. *Hydrobiologia* **54,** 181–186.

Muley, E. V. (1978a). Embryology and development of a freshwater prosobranch, *Melania scabra*. *Hydrobiologia* **58,** 89–92.

Muley, E. V. (1978b). Studies on growth indices of the freshwater prosobranch, *Melania scabra*. *Hydrobiologia* **58,** 137–144.

Otori, Y., Sandiford, C. S., and Ritchie, L. S. (1954). Persistence of egg-laying by *Oncomelania nosophora* females following isolation from males. *J. Parasitol.* **40,** 43.

Otori, Y., Ritchie, L. S., and Hunter, G. W. (1956). The incubation period of the eggs of *Oncomelania nosophora*. *Am. J. Trop. Med. Hyg.* **5,** 559–561.

Palmieri, M. D., Palmieri, J. R., and Sullivan, J. T. (1980). A chemical analysis of the habitat of nine commonly occurring Malaysia freshwater snails. *Malay. Nat. J.* **34,** 39–45.

Paulinyi, H. M., and Paulini, E. (1972). Laboratory observations on the biological control of *Biomphalaria glabrata* by a species of *Pomacea* (Ampullariidae). *Bull. W.H.O.* **46,** 243–247.

Prashad, B. (1932). *Pila* (the apple snail). *Indian Zool. Mem.* **4,** 1–83.

Ramanan, V. V. (1903). On the respiratory and locomotory habits of *Ampullaria globosa* Swainson. *J. Malacol.* **10,** 107–113.

Rao, D. B., Venkatasubbaiah, M. C., Reddy, R. S., Raju, A. N., Rao, P. V., and Swami, K. S. (1972). Metabolism of brooding young from aestivating adults of the banded pond snail *Viviparus bengalensis*. *Malacologia* **11,** 281–286.

Reavell, P. E. (1980). A study of the diets of some British freshwater gastropods. *J. Conchol.* **30,** 253–271.

Reddy, Y. S., and Swami, K. S. (1975). Uric acid synthesis during aestivation in *Pila globosa*. *Curr. Sci.* **44,** 235–236.

Reddy, Y. S., and Swami, K. S. (1976). Some metabolic effects of aestivation on glycolysis in *Pila globosa* (Swainson). *Indian J. Exp. Biol.* **14,** 191–193.

Reddy, Y. S., Rao, P. V., and Swami, K. S. (1974). Probable significance of urea and uric acid accumulation during aestivation in the gastropod, *Pila globosa* (Swainson). *Indian J. Exp. Biol.* **12,** 454–456.

Ritchie, L. S. (1955). The biology and control of the amphibious snails that serve as intermediate hosts for *Schistosoma japonicum* in Japan. *Am. J. Trop. Med. Hyg.* **4,** 426–441.

Ross, M. J., and Ultsch, G. R. (1980). Temperature and substrate influences on habitat selection in 2 pleurocerid snails (*Goniobasis cahawbensis* and *Goniobasis carinifera*). *Am. Midl. Nat.* **103,** 209–217.

Roth, A. A., and Wagner, E. D. (1960). Notes on the production of eggs in *Oncomelania nosophora* and *O. formosana*. *Nautilus* **73,** 147–151.

Russell Hunter, W. (1964). Physiological aspects of ecology in nonmarine molluscs. *In* ''Physiology of Mollusca'' (K. M. Wilbur and C. M. Yonge, eds.), Vol. 1, pp. 83–126. Academic Press, New York.

Russell-Hunter, W. D. (1970). ''Aquatic Productivity.'' Macmillan, New York.

Shoup, C. S. (1943). Distribution of fresh water gastropods in relation to total alkalinity of streams. *Nautilus* **56,** 130–134.

Shylaja, R., and Alexander, K. M. (1975a). Studies on the physiology of excretion in the freshwater prosobranch *Pila virens:* Part I. Pattern of excretion in the normal and post aestivating *P. virens. Indian J. Exp. Biol.* **13,** 363–365.

Shylaja, R., and Alexander, K. M. (1975b). Studies on the physiology of excretion in the freshwater prosobranch *Pila virens.* II. Effect of osmotic stress on excretion. *Indian J. Exp. Biol.* **13,** 366–368.

Skoog, G. (1973). Salinity reactions of two freshwater snails from brackish water. *Oikos, Suppl.* **15,** 253–260.

Skoog, G. (1976). Effects of acclimatization and physiological state on the tolerance to high temperatures and reactions to dessication of *Theodoxus fluviatilis* and *Lymnaea peregra. Oikos* **27,** 50–56.

Skoog, G. (1978). Influence of natural food items on growth and egg production in brackish water populations of *Lymnaea peregra* and *Theodoxus fluviatilis* (Mollusca). *Oikos* **31,** 290–298.

Sloan, W. C. (1964). The accumulation of nitrogenous compounds in terrestrial and aquatic eggs of prosobranch snails. *Biol. Bull. (Woods Hole, Mass.)* **126,** 302–306.

Snyder, N. F. R. (1967). An alarm reaction of aquatic gastropods to intraspecific extract. *Mem.—N.Y., Agric. Exp. Stn. (Ithaca)* **403.**

Snyder, N. F. R., and Snyder, H. A. (1971). Defenses of the Florida apple snail *Pomacea paludosa. Behaviour* **40,** 175–215.

Stiven, A. E., and Walton, C. R. (1967). Age and shell growth in the freshwater snail, *Goniobasis proxima* (Say). *Am. Midl. Nat.* **78,** 207–214.

Studier, E. H., and Pace, G. L. (1978). Oxygen consumption in the prosobranch snail *Viviparus contectoides* (Mollusca; Gastropoda). IV. Effects of dissolved oxygen level, starvation, density, symbiotic algae, substrate composition and osmotic pressure. *Comp. Biochem. Physiol. A* **59A,** 199–204.

Tashiro, J. S. (1980). Bioenergetic background to reproductive partitioning in an iteroparous population of the freshwater prosobranch *Bithynia tentaculata.* Ph.D. Dissertation, Syracuse University, Syracuse, New York.

Tashiro, J. S. (1982). Grazing in *Bithynia tentaculata:* Age-specific bioenergetic patterns in reproductive partitioning of ingested Carbon and Nitrogen. *Am. Midl. Nat.* **197,** 133–150.

Tashiro, J. S., and Colman, S. D. (1982). Filter feeding in the freshwater prosobranch snail *Bithynia tentaculata:* Bioenergetic partitioning of Ingested Carbon and Nitrogen. *Am. Midl. Nat.* **197,** 114–132.

Thiele, J. (1931). ''Handbuch der Systematischen Weichtierkunde,'' Vol. I. Fischer, Jena.

Thomas, K. J. (1975). Biological control of *Salvina* by the snail *Pila globosa* Swainson. *Biol. J. Linn. Soc.* **7,** 243–247.

Todd, M. E. (1964). Osmotic balance in *Hydrobia ulvae* and *Potamopyrgus jenkinsi* (Gastropoda:Hydrobiidae). *J. Exp. Biol.* **41,** 665–677.

Tompa, A. S. (1980). Studies on the reproductive biology of gastropods. Part III. Calcium provision and the evolution of terrestrial eggs among gastropods. *J. Conchol.* **30,** 145–154.

Vail, V. A. (1978). Seasonal reproductive patterns in 3 viviparid gastropods. *Malacologia* **17,** 73–98.

van Cleave, H. J. (1932). Studies on snails of the genus *Pleurocera.* I. The eggs and egg laying habits. *Nautilus* **46,** 29–34.

van Cleave, H. J., and Altringer, D. A. (1937). Studies on the life cycle of *Campeloma rufum*, a freshwater snail. *Am. Nat.* **71,** 167–184.

van Cleave, H. J., and Lederer, L. G. (1932). Studies on the life cycle of the snail *Viviparus contectoides*. *J. Morphol.* **53,** 499–522.

van der Schalie, H., and Dundee, D. S. (1955). The distribution ecology and life history of *Pomatiopsis cincinnatiensis* (Lea), an amphibious operculate snail. *Trans. Am. Microsc. Soc.* **74,** 119–133.

van der Schalie, H., and Getz, L. L. (1961). Comparisons of adult and young *Pomatiopsis cincinnatiensis* (Lea) in respect to moisture requirements. *Trans. Am. Microsc. Soc.* **80,** 211–220.

van der Schalie, H., and Getz, L. L. (1963). Comparison of temperature and moisture responses of the snail genera *Pomatiopsis* and *Oncomelania*. *Ecology* **44,** 73–83.

van der Schalie, H., and Walter, H. J. (1957). The egg-laying habits of *Pomatiopsis cincinnatiensis* (Lea). *Trans. Am. Microsc. Soc.* **76,** 404–422.

Vinogradov, L. I. (1962). On the biology of the mollusc *Bithynia leachi* (Shepard, 1823). *Zool. Zh.* **41,** 464–465.

Vivekanandan, E., Haniffa, M. A., Randian, T. J., and Raghuraman, R. (1974). Studies on energy transformation in the freshwater snail *Pila globosa*. *Freshwater Biol.* **4,** 275–280.

von Brand, T., Baerstein, H. D., and Mehlman, B. (1950). Studies on the anaerobic metabolism and the aerobic carbohydrate consumption of some freshwater snails. *Biol. Bull. (Woods Hole, Mass.)* **98,** 266–276.

Wagner, E. D., and Wong, L. W. (1956). Some factors influencing egg laying in *Oncomelania nosophora* and *Oncomelania quadrasi*, intermediate hosts of *Schistosoma japonicum*. *Am. J. Trop. Med. Hyg.* **5,** 544–552.

White, T. C. R. (1978). The importance of a relative shortage of food in animal ecology. *Oecologia* **33,** 71–86.

Winterbourn, M. J. (1970). Population studies on the New Zealand freshwater gastropod, *Potamopyrgus antipodarum* (Gray). *Proc. Malacol. Soc. London* **39,** 139–149.

Yanagisawa, T., and Komiya, Y. (1961). Physiological studies on *Oncomelania nosophora* the vector snail *Schistosoma japonicum*. On the oxygen uptake and effect of starvation upon it. *Jpn. J. Med. Sci. Biol.* **14,** 69–76.

Yasuraoka, K. (1961). Seasonal variation in the temperature-extruding response relation of *Oncomelania nosophora*, the vector snail of *Schistosoma japonicum* in Japan. *Jpn. J. Med.Sci. Biol.* **14,** 195–200.

Young, M. R. (1975). The life cycles of six species of freshwater molluscs in the Worcester-Birmingham canal. *Proc. Malacol. Soc. London* **41,** 533–548.

9

Physiological Ecology of Freshwater Pulmonates

ROBERT F. McMAHON

Section of Comparative Physiology
Department of Biology
The University of Texas at Arlington
Arlington, Texas

The Mollusca, Vol. 6
Ecology

ISBN 0-12-751406-6

I. Introduction

A. Conceptual Framework

Physiological (or *environmental*) *ecology* has almost as many definitions as there are treatises on the subject (see Bligh et al., 1976; Calow and Townsend, 1981; Phillips, 1975; Hill, 1976, for discussions of physiological ecology). Thus, it is imperative that any discussion of the physiological ecology of freshwater pulmonate snails be preceded by the author's working definition of his subject. For the purposes of this chapter, physiological ecology has been broadly defined as the study of the functioning of an organism in relation to its environment, including both the physical and biotic factors that comprise that environment. If physiological traits are assumed to be genetically based phenotypes, then they should be as subject to natural selection as are, more obviously, morphological traits. As such, those genes associated with physiological phenotypes that bestow the greatest *fitness* allow individuals that carry them to make relatively greater contributions to the gene pools of successive generations. Indeed, it can be quite reasonably argued that physiological traits generally have much higher selective values than most morphological traits, a fact that accounts for the extensive formation of physiological races in freshwater pulmonate snails (Russell-Hunter, 1964, 1978; McMahon and Payne, 1980). Therefore, it is the environment and the selective forces associated with it that have shaped each species' or population's particular set of physiological adaptations. Variations in function between or within species will have a historical, evolutionary significance but will not necessarily represent the best possible adaptation at any particular instant. Environments change through time, and adaptational response to such changes, by its very nature, lags behind them (Bligh et al., 1976). Such physiological adaptations are assumed to endow organisms with a high capacity for homeostatic control over their internal environments, allowing maintenance of cellular conditions within the rather narrow range required for both efficient function and life, although the external environmental conditions vary over a much wider range. As argued by Bligh et al. (1976), such definitions for physiological ecology appear to include the whole of biochemical, cellular, and organismal physiology, as well as a large proportion of the realm of ecology. Indeed, the discipline becomes even more complex when physiological ecology is considered from an energetic, life-history, and evolutionary point of view, i.e., how organisms ingest, assimilate, and utilize energy resources to optimize the physiological processes which affect gene transmission (Browne and Russell-Hunter, 1978; Calow and Townsend, 1981; Chapter 11, this volume).

This chapter attempts to describe the physiological ecology of freshwater pulmonate snails, a subject extensive enough to warrant an entire volume if considered in detail. Fortunately, the biology, ecology, and physiology of freshwater pulmonates have been admirably reviewed (Russell-Hunter, 1964, 1978; Wilbur and Yonge, 1964, 1966; Fretter and Peake, 1975, 1978a; Florkin and Scheer, 1972). Such previous reviews make this author's task much easier. This chapter will attempt to cover in detail only those aspects of the subject on which new information has been published or which appear to require further discussion in the light of new information. The approach, perhaps more physiologically oriented than that of previous reviewers, is essentially that of a comparative physiologist trying to understand the adaptive significance of the physiological traits of freshwater basommatophoran snails in terms of homeostasis, capacity adaptation, energetic efficiency, and adaptational fitness (the causal basis of a selective advantage; see Calow and Townsend, 1981). The chapter will center on the evolution of physiological adaptations, such as those involved in oxygen consumption, response to temperature, osmoregulation, response to hypoxia, and desiccation tolerance, which have the greatest interaction with environmental conditions.

B. Evolution of the Basommatophora

The adaptive significance of the physiological traits of modern freshwater basommatophoran snails can be properly understood only in light of their evolutionary history within the class Pulmonata. The pulmonates appear to have arisen in the Devonian from a prosobranch, monotocardian archeogastropod from which also diverged the Cephalaspidea (Bullomorpha), the probable stem group of the Opisthobranchia. Pulmonates arose either from an early cephalaspid stock or from an archeogastropod prosobranch ancestor common to both groups (Morton, 1955). The line leading to the pulmonates was one whose evolutionary history was characterized by the continual invasion of successively higher and more aerial levels of estuaries or intertidal shores (Morton, 1955; Ellis, 1926), eventually leading to a totally terrestrial, air-breathing, and reproductive schesis. This evolutionary pattern is today recapitulated in the basommatophoran family Ellobiidae, whose representatives range from such amphibious, high estuarine, and intertidal forms as *Melampus bidentatus* Say, which spend a vast majority of their time (90–95%) in air (Apley, 1967, 1970; Russell-Hunter et al., 1972; Price, 1980; McMahon and Russell-Hunter, 1981), resembling higher land snails in their behavior and physiology (Price, 1980; McMahon and Russell-Hunter, 1981) although retaining a series of physiological adaptations to aquatic life (McMahon and Russell-Hunter, 1981) and an aquatic reproductive schesis involving a planktonic veliger larva (Apley, 1967, 1970; Russell-Hunter et al., 1972), to the genera *Carychium* (Müller) and *Zospeum* Bourg., which live in moist terrestrial environments and have evolved a tough, leathery, nearly cleidoic, terrestrial egg (larval stages are suppressed) from which hatch miniature adults ready to take up life in an aerial environment (Morton, 1955; Hubendick,

1978; Watson and Verdcourt, 1953; McMahon and Russell-Hunter, 1981). The earliest pulmonates appear to have been terrestrial, having lost the characteristic molluscan gill (ctenidium) and evolved the mantle-cavity as a highly vascularized diffusion lung with a narrow, muscular external opening (the pneumostome) through which gas exchange occurs by diffusion down respective concentration gradients (Morton, 1955; Purchon, 1977; Russell-Hunter, 1964, 1978; Ghiretti and Ghiretti-Magaldi, 1975; Morton and Yonge, 1964; Fretter, 1975). This terrestrial habit was also associated with evolution of a terrestrial reproductive schesis (Duncan, 1975) and loss of external epithelial ciliation (Russell-Hunter, 1964, 1978; Machin, 1975).

Present-day pulmonates are divided into two orders: (1) the Stylommatophora, the higher land snails which are the culmination of terrestrial adaptation within the Pulmonata (Morton and Yonge, 1964; Fretter, 1975; Purchon, 1977; Solem, 1978), and (2) the Basommatophora, the aquatic pulmonates, characterized by a single pair of noninvaginable tentacles with basal eyes (Morton and Yonge, 1964). This group, which consists primarily of aquatic forms and is undoubtedly an artificial assemblage (Morton, 1955; Hubendick, 1978), includes primitive amphibious families such as the Ellobiidae, Otinidae, Amphibolidae (which, unlike other pulmonates, retain the operculum), Siphonariidae, and Trimusculidae, all of which inhabit high estuarine or intertidal environments and, with the rare exception of a few terrestrial genera such as *Carychium* and *Zoespeum,* have never evolved a completely terrestrial schesis (Hubendick, 1978; Morton and Yonge, 1964); and the more advanced hygrophilous families, including the more primitive related families Chilinidae, Acroloxidae, and Latiidae and the more advanced and related families Lymaeidae, Physidae, Planorbidae, and Ancylidae (the latter two were recently combined by Hubendick, 1978, as the family Ancyloplanorbidae). The advanced hygrophilous families of freshwater basommatophorans are a monophyletic group which comprise the higher, truly successful, freshwater pulmonates (Hubendick, 1978). The freshwater pulmonates, which appear to have reevolved an aquatic habit from a more terrestrial ancestor, represent an evolutionary continuum from primitive, nearly terrestrial, occasionally amphibious species living on the shores of lakes, ponds, and streams, such as *Lymaea truncatula* (Müller) (Walton and Jones, 1926), *L. palustris* (Müller) (Hunter, 1972, 1975a), and *L. bulimoides* (Haldeman) (R. F. McMahon, unpublished observations), through intermediately adapted forms such as the Physidae and some of the Lymnaeidae and Planorbidae, which are aquatic but retain a primarily aerial, pulmonary mode of gas exchange (Russell-Hunter, 1964, 1978; Hubendick, 1978), to advanced, purely aquatic groups such as the freshwater limpets and some of the smaller Planorbidae which remain continually submerged, have evolved neomorphic gills, and have an entirely aquatic mode of oxygen consumption (Russell-Hunter, 1964, 1978; Ghiretti and Ghiretti-Magaldi, 1975; Purchon, 1977).

Thus, unlike the vast majority of higher organisms, the freshwater basommatophorans have had an evolutionary history marked by increasingly greater adaptation to an aquatic schesis rather than a more terrestrial one, an evolutionary progression believed to be still incomplete because freshwater basommatophorans retain many of the traits of more terrestrial pulmonate species, including the lack of surface ciliation particularly on the neomorphic gills, which appears to be nonadaptive in the more nearly aquatic species (Russell-Hunter, 1964, 1978). The physiological adaptations of basommatophoran snails to fresh water can best be understood in relation to their evolutionary progression toward a more continuously aquatic habit. The early ancestors of the group were, no doubt, terrestrial, living in the moist environments provided by stream, pond, and lake margins, and marshes, as the primitive terrestrial ellobiid genus *Carychium* does today. Members of the group have evolved to become increasingly amphibious, most likely to take advantage of the rich food resources provided by the highly productive *Aufwuchs* or periphyton (mixed algal, fungal, and bacterial slime communities growing on hard surfaces) and abundant detritus available in shallow, eutrophic, nearshore waters (Calow, 1975a; McMahon et al., 1974; Hunter, 1976; Russell-Hunter, 1978). The vast majority of such marginal freshwater environments are characterized by a relatively high degree of short-term water level fluctuation associated with changes in rainfall and subsequent runoff which periodically cause the exposure of sublittoral areas to air (Russell-Hunter, 1953a; Cheatum, 1934; Clampitt, 1970, 1974). As such, shallow, nearshore environments are not generally suitable for freshwater prosobranch gastropods, which retain the ctenidium or gill and are totally aquatic in their oxygen consumption and relatively intolerant of desiccation (Russell-Hunter, 1964; Aldridge, 1983, see Chapter 8). Indeed, the vast majority of freshwater prosobranchs occur only in larger more stable bodies of fresh water, comprising the majority of gastropod fauna below 4 m depth (Russell-Hunter, 1978; Payne, 1979; Aldridge, 1980, 1982; Chapter 8, this volume). Due to their terrestrial origin, most freshwater basommatophorans have retained a high tolerance of prolonged aerial exposure (Machin, 1975) and an aerial mode of respiration which is highly adaptive in their shallow, eutrophic, microenvironments where exposure to air, desiccation, and hypoxic conditions can be frequent. In such environments, freshwater pulmonates are highly successful. They make up a large proportion of the benthic animal biomass of the margins of larger lakes and rivers, but are particularly well adapted and successful in smaller, more variable aquatic habitats such as ponds, marshes, streams, and ditches (Russell-Hunter, 1978).

Such shallow, eutrophic habitats are extremely unstable and variable, presenting a range of environmental extremes which is far greater than that associated with the more stable deeper water habitats occupied by the majority of freshwater prosobranchs. In readapting to fresh waters, basommatophoran pulmonates had to evolve not only adjustments of physiological functions such as those associated with osmoregulation, gas exchange, reproduction, and buoyancy, but also extensive capacity and compensatory adaptations to allow survival under the high levels of stressful environmental variation associated with their shallow freshwater microhabitats. Such adaptations were required to allow efficient functioning under the high levels of temperature fluctuation [water temperatures can rise several degrees above ambient air temperatures due to insolation (McMahon and Payne, 1980)], under periodic or continual hypoxia, particularly in eutrophic environments (McMahon, 1972, 1973), and under periodic and occasionally long-term exposure to air and subsequent desiccation (Cheatum, 1934; Clampit, 1970, 1974; Jokinen, 1978; Russell Hunter, 1953a, 1957; Brown, 1982). Freshwater pulmonates have high capacities both for tolerating environmental extremes and for physiological compensation or acclimation to short-term and longer term environmental variations. Such plasticity is characteristic of freshwater pulmonates as a whole (Russell-Hunter, 1978), and its contribution to the evolution of this group is a major thesis of this chapter.

C. Interspecific and Intraspecific Variation

Freshwater basommatophoran pulmonates include a small number of families (six to nine, depending on the classification scheme), each with relatively few genera (Hubendick, 1978; Fretter and Peake, 1978b). These genera, although depauparate in number of species, tend to have cosmopolitan, worldwide distributions and to be ubiquitous, with highly euryecic individual species inhabiting a broad range of habitats (Russell-Hunter, 1964, 1978; Hubendick, 1978). The vast majority of freshwater habitats, particularly those of smaller extent preferred by freshwater pulmonates, are characterized by temporal instability and considerable isolation. Most lakes and streams remain in existence approximately 10^3 to 10^4 years, whereas ponds generally exist for only 10^2 years (Russell-Hunter, 1964, 1978). The vast majority of freshwater habitats are limited in extent and highly isolated from one another (Hubendick, 1952, 1954, 1962a; Russell-Hunter, 1952, 1957, 1961a,b, 1964, 1978; Russell-Hunter et al., 1967, 1970, 1981; Lassen, 1975) with little interpopulation migration even between closely adjacent populations of organisms, because pulmonate snails do not have desiccation-resistant aerial dispersal stages in their life cycles (Russell-Hunter, 1964, 1978; Russell-Hunter et al., 1967, 1970, 1981; McMahon and Payne, 1980). It has been speculated that species populations such as freshwater pulmonates inhabiting temporal, freshwater environments are not isolated long enough to allow formation of new species before destruction or alteration of habitat causes population extinction (Russell-Hunter, 1952, 1957, 1961a,b, 1964, 1978; Hubendick, 1952, 1954, 1962a; Lassen, 1975); therefore, a great radiation of species in freshwater basommatophorans such as has occurred in stylommatophoran pulmonate snails in more temporally stable habitats has been prevented. Indeed, extraordinary speciation seems to have occurred in certain pulmonate stocks isolated in very old lakes which have been stable long enough ($>10^6$ years) for speciation to occur (Russell-Hunter, 1964, 1978; Boss, 1978).

Although interspecific variation and species diversity and numbers are low in basommatophoran pulmonates as a result of the temporal instability of their preferred shallow-water habitats, the intraspecific variation within this group is extraordinarily high (Russell-Hunter, 1952, 1957, 1961a,b, 1964, 1978; Russell-Hunter et al., 1967, 1970, 1981; Hubendick, 1952, 1954, 1962a; Berrie, 1959; Burky, 1971; McMahon, 1975a,b, 1976a,b; McMahon and Payne, 1980; Ibarra and McMahon, 1981; Hunter, 1975a,b; Nickerson, 1972), both in morphological traits such as the shape of the shell and radular teeth, and in physiological traits such as temperature tolerance, reproductive rate, metabolic efficiency, calcium metabolism, growth rate, and actuarial bioenergetics. Although much of this variation has an environmentally induced component, such as trophic effects on shell height indices [(aperture length: shell length, or aperture width: shell length) (Nickerson, 1972; Hunter, 1975b)] or rates of growth and fecundity and actuarial bioenergetics (Burky, 1971; Hunter, 1975a; McMahon, 1975a,b, 1976a; Eversole, 1978; Romano, 1980; Chapter 11, this volume), a large component of intraspecific, interpopulation variation in freshwater pulmonates appears to have a genetic basis which cannot be eliminated by acclimation in the laboratory (Forbes and Crampton, 1942; McMahon, 1976b; McMahon and Payne, 1980; Ibarra and McMahon, 1981) or by reciprocal field transfers of individuals between two different environments (Burky, 1969, 1971; Hunter, 1972, 1975a,b; McMahon, 1972, 1975a; Nickerson, 1972; Eversole, 1974; Romano, 1980).

The high degree of observable plasticity in freshwater pulmonates appears to be correlated with correspondingly high levels of enzyme polymorphism. An electrophoretic study of seven enzyme systems involving 9 loci in 16 populations of *Ferrissia rivularis* (Say) in upstate New York showed that 5 of the loci were uniformly polymorphic across all 16 populations and that 2 were polymorphic in several populations, with the level of polymorphism in individual populations ranging between 56 and 78%; the mean heterozygosity was 27% per average individual, an extraordinarily high level of isozyme variation when compared to most natural animal populations (Nickerson, 1972).

Such high levels of intraspecific, interpopulation variation are, in part, a result of founder effect and subsequent genetic drift, because new populations of freshwater pulmonates are usually established by very small numbers of individuals (Russell-Hunter, 1964, 1978) or even by the self-fertilization of a single hermaphroditic individual. Founder effect and genetic drift appear to be the causes of such geographically random interpopulation variation of *selectively neutral* traits in freshwater pulmonates as the rate of mineral deposition in the shell of *F. rivularis* (Nickerson, 1972; Russell-Hunter et al., 1967, 1970, 1981), interpopulation variation in shell aperture shape (Nickerson, 1972; Ibarra and

McMahon, 1981), or interpopulation variation in the shape of radular teeth in *Lymnaea peregra* (Müller) (Berrie, 1959) or *L. palustris* (Hunter, 1972, 1975b). Interpopulation variation in other genetically controlled characteristics of higher selective value, particularly physiological traits, appears to be at least partially the result of genetic adaptation to selective pressures specific to the microenvironment of each population (Forbes and Crampton, 1942; Russell Hunter, 1961a,b; Burky, 1969, 1971; Hunter, 1972, 1975b; McMahon, 1972, 1975a, 1976b; McMahon and Payne, 1980; Ibarra and McMahon, 1981; Eversole, 1974; Romano, 1980).

Such genetic adaptation and the evolution of "physiological races" can take place in the relatively short periods during which the majority of freshwater habitats exist (10^2–10^4 years) because resident pulmonate populations experience a high degree of reproductive isolation and have relatively high fecundities (Russell-Hunter, 1964, 1978; Russell-Hunter and Buckley, 1983; McMahon, 1976b; Eversole, 1974; Chapter 11, this volume) which allow them to tolerate high levels of selective pressure (McMahon, 1976b). For example, McMahon (1976b) reports that only 54 generations were required to produce a thermally resistant physiological race of *Physa virgata* Gould under the selective pressures exerted by artificially elevated ambient water temperatures maintained by thermal effluents from a steam-electric generating station.

This high degree of physiological plasticity in terms of both resistance and compensatory responses within individuals and the high degree of natural phenotypic and genotypic physiological variation maintained within and between populations of single species is considered to be highly adaptive in freshwater pulmonates (Russell-Hunter, 1964, 1978). It allows individuals to invade, survive in, and be successful in small, shallow, temporal, and highly variable aquatic habitats; provides the genetic variation required for short-term adaptation to specific microenvironmental conditions; and allows populations to survive severe changes in microhabitats, all of which are characteristic of the invasive nature of freshwater pulmonates (Lassen, 1975). As microenvironmental regimes vary widely among isolated freshwater habitats, such physiological race formation and, therefore, selection for high levels of genetic variability, are the rule rather than the exception among freshwater pulmonates. For these reasons, descriptions of the ecology, physiology, or life-history traits of pulmonate species based only on studies of a single population may be highly unreliable, particularly in investigations of resistance adaptations to environmental extremes such as those of thermal tolerance (McMahon, 1976b; McMahon and Payne, 1980; Ibarra and McMahon, 1981). Although it requires greater effort, a comparative interpopulation approach to studies of freshwater pulmonate ecology and physiology appears to provide far better estimates of the physiological variability and capacity for adaptation of this remarkable group of molluscs (Forbes and Crampton, 1942; Russell Hunter, 1953b,c, 1961a,b; Russell-Hunter et al.,

1967, 1970, 1981; Burky, 1971; Hunter, 1975b; McMahon, 1975a, 1976b; McMahon and Payne, 1980; Nickerson, 1972; Eversole, 1974; Romano, 1980).

II. Respiratory Adaptations to Aquatic Life

A. Pulmonary versus Cutaneous Respiration

The hygrophilous basommatophorans display varying degrees of readaptation to aquatic life. The majority of lymnaeids are generally characterized as truly amphibious. Living at the margins of fresh waters and experiencing only occasional submergence, they rarely extend to depths below 0.25 m (Russell Hunter, 1953a,c, 1957; Hunter, 1972, 1975a). Some species, such as *Lymnaea truncatula* (Walton and Jones, 1926) and *L. palustris* (Jokinen, 1978), are more nearly terrestrial, and others, such as *L. stagnalis* (L.), have evolved a more aquatic schesis (Boag and Bentz, 1980; Jones, 1961). The marginal habit of lymnaeids induces populations to make extensive migrations up and down the shore with changing water levels, even when these involve movements to unfavorable environments (Russell Hunter, 1953a).

The family Physidae includes species somewhat more aquatic than are lymnaeids. Unlike amphibious lymnaeids, the majority of physids rarely leave the water but are still restricted to shallow, nearshore margins of aquatic habitats. They generally extend to greater depths than lymnaeids, being found at depths of 1 m both in Europe [*Physa fontinalis* (L.) (Russell Hunter, 1953c, 1957)] and in North America (*P. gyrina* Say and *P. integra* Haldeman) (Clampitt, 1970, 1974).

The planorbids (subfamily Planorbinae) (Hubendick, 1978) have several distinct adaptations to aquatic life. This group is rarely amphibious and generally penetrates to greater depths (>1 m) than do either lymnaeids or physids (Cheatum, 1934; Jones, 1961; Rowan, 1966; Clampitt, 1972; Russell-Hunter, 1978).

In Texan fresh waters, these depth distribution differences among the major basommatophoran groups are clearly evident among the most common freshwater species. The lymnaeids *L. bulminoides* and *Pseudosuccinea columella* (Say) are usually found emerged at the edge of the water or submerged just below the water surface (<1–3 cm in depth). *Physa virgata* is always found near the shore below the water surface to depths of 10 to 15 cm, and the planorbid *Helisoma trivolvis* (Say), rarely found near the water surface, can occur many meters from the shore and its range extends to depths greater than 1 m (R. F. McMahon, unpublished observations).

The freshwater limpets of the families Acroloxidae and Ancyloplanorbidae (Ancylidae) (Hubendick, 1978) represent the most advanced freshwater pulmonates. In these groups the mantle-cavity is reduced to a rudimentary, vestigial structure, and respiratory exchange is purely cutaneous, allowing a totally aquatic habit (Basch, 1963; Hubendick, 1962b, 1964, 1970; Russell-Hunter, 1953b, 1964, 1978).

This evolutionary trend toward an increasingly aquatic mode of life in hygrophilous basommatophoran snails is reflected in their respiratory structures. Lymnaeids have a large mantle-cavity lung in which air is renewed by periodically returning to the water surface and opening the pneumostome into the external atmosphere. During exposure to air, lost lung volume is restored; O_2 diffuses into and CO_2 out of the lung down their respective concentration gradients (Russell-Hunter, 1964, 1978; Jones, 1961). In smaller pulmonates such as *L. peregra* and *Physa fontinalis,* diffusion is sufficient to completely equilibrate pulmonary PO_2 and PCO_2 values with the external atmosphere (Russell Hunter, 1953c), but in larger species such as *L. stagnalis* and *Planorbarius corneus* (L.) diffusional gas exchange is incomplete, with pulmonary PO_2 values lower and PCO_2 values higher than atmospheric values (Jones, 1961). Such diving behavior is reported to occur at mean intervals of from 18 min in *L. peregra* (Russell-Hunter, 1953c) to 34 min in *L. stagnalis* (Jones, 1961) at medium PO_2 values near air saturation. In *L. stagnalis* pulmonary oxygen consumption was shown to be equal to cutaneous oxygen consumption at air saturation levels and to increase rapidly in importance relative to cutaneous uptake at PO_2 levels below 140 torr (Jones, 1961).

Physids, although still highly dependent on pulmonary gas exchange, have evolved a rudimentary neomorphic gill consisting of finger-like lobes of the mantle edge which are held reflected back over the external shell surface. These mantle projections greatly increase the epithelial surface area available for aquatic cutaneous respiratory gas exchange compared to similar-sized lymnaeids which do not have such gills (Russell Hunter, 1953c), which may partially account for the more aquatic habit of this group.

The Planorbinae, although retaining pulmonary gas exchange in their larger species, have evolved a highly vascularized evagination of the rectal portion of the mantle (Purchon, 1977) which is protruded into the water and serves as an accessory neomorphic gill (Ghiretti and Ghiretti-Magaldi, 1975; Russell-Hunter, 1978). The presence of this gill allows the larger species of this group, such as *P. corneus,* to maintain more than 50% of total oxygen uptake via cutaneous exchange at PO_2 levels as low as 40 torr (Jones, 1961, 1964a) compared to 140 torr for species without gills, such as *L. stagnalis* (Jones, 1961). Such decreased dependence on pulmonary respiration allows planorbids to maintain periods of submergence that are more extended than those of lymnaeids (Russell Hunter, 1953c; Jones, 1961, 1964a,b). The trend toward aquatic life in planorbids reaches its zenith in smaller species that have become completely aquatic; these fill the mantle-cavity with water and maintain a purely cutaneous mode of respiratory oxygen consumption (Russell-Hunter, 1978).

It can be reasonably argued that the ancylid freshwater limpets, now placed in the family Ancyloplanorbidae by Hubendick (1978), represent the most advanced freshwater basommatophorans. They have lost nearly all traces of the pulmonary cavity and, to an even greater degree than the closely related Planorbinae, have developed well-vascularized, extensive neomorphic gills from evaginations of rectal mantle lobes between the foot and roof of the mantle on the left side of the body (Basch, 1963; Hubendick, 1964, 1970; Purchon, 1977; Russell-Hunter, 1978). These gills provide an extensive surface area for aquatic gas exchange, allowing freshwater limpets to be entirely cutaneous and therefore totally aquatic in their respiration. The morphology of the neomorphic gills of freshwater limpets is strongly correlated with the degree of hypoxia each species experiences in its own microhabitat. More primitive species such as *Ancylus fluviatilus* Müller or *Ferrissia rivularis,* which live on the exposed surfaces of boulders in fast-flowing, well-oxygenated lotic habitats (Russell Hunter, 1953b; Burky, 1971), have small, single-lobed gills with relatively few convolutions and correspondingly small surface areas (Basch, 1963; Hubendick, 1964). In contrast, *Hebetancylus excentricus* (Morelet), which is also lotic but inhabits more hypoxic microhabitats on the underside of rocks where it is surrounded by the substratum (McMahon, 1976a), has a more extensive bilobed gill that is more convoluted and therefore provides greater surface area than that of *Ferrissia* or *Ancylus* (Basch, 1963; Hubendick, 1964). Some species, such as *Laevapex fuscus* (C. B. Adams), inhabit very hypoxic microhabitats below mud and silt substrata on the sides of stones and boulders or emergent vegetation in impounded, often stagnant waters (McMahon, 1972, 1973, 1975a, 1976a) and have the most extensive, highly convoluted bilobed gills of all the freshwater limpets (Basch, 1959, 1963; Hubendick, 1964). It is interesting to note that a group of freshwater limpets represented by *Acroloxus lacustris* (L.) (family Acroloxidae) (Hubendick, 1978) also occur on macrophytes and boulders in impounded, stagnant waters, but only above the bottom substratum in more oxygen-rich open waters (R. F. McMahon, unpublished observations), and have a correspondingly smaller, single-lobed gill that is far less extensively convoluted than that of *L. fuscus* (Basch, 1963; Hubendick, 1962b, 1964).

The gills of the freshwater limpets are believed to have arisen from a lobe of the rectum (Purchon, 1977; Russell-Hunter, 1964, 1978). These gills contain isolated cells bearing tufts of cilia (Russell-Hunter, 1964, 1978), perhaps originally derived from the ciliated epithelium lining the rectal lumen. In lotic species such as *A. fluviatilis* these ciliated cells are sparsely distributed over the gill surface and appear to function in the movements of mucus (gill cleansing) rather than in gill ventilation (Russell-Hunter, 1978, also personal communication).

However, in *L. fuscus,* greater numbers of ciliated cells are found on the gill surface, and their cilia are remarkably long. Small particles placed on the leading (anterior) edge of the gill of *L. fuscus* are swept rapidly to the posterior (trailing) edge of the gill, indicating that gill ciliation may also have evolved a ventilatory function in species from more hypoxic microhabitats.

B. Buoyancy

Although physid and lymnaeid snails are all divers dependent to some extent on pulmonary gas exchange, several populations of *Physa fontinalis* and *Lymnaea peregra* from deeper lake waters and from isolated shoals where excursions to the surface are impossible, have been reported to have water-filled mantle-cavities, surviving throughout the entire life cycle on purely cutaneous respiration (Russell-Hunter, 1953c, 1961b, 1964, 1978). Such populations immediately initiate pulmonary respiration when brought to the water surface and appear to survive only in well-oxygenated waters (Russell Hunter, 1953c) where cutaneous respiration can be maintained at high levels (Jones, 1961). Their existence indicates that maintenance of large volumes of gas in the mantle-cavity of basommatophoran pulmonates may have a functional significance other than that of gas exchange.

Mantle-cavity gases appear to be renewed long before they are depleted of oxygen stores. *Planorbarius corneus* increases pulmonary air volume by only 4.2–15.6% at the end of each dive when lung PO_2 levels are still relatively high (Jones, 1961). Corresponding values for *L. stagnalis* and *Biomphalaria sudanica* Martens were 4–16% and 6–26%, respectively (Jones, 1961, 1964b). Such renewals of mantle-cavity gases occur over intervals much shorter than the extended periods of submergence that can be tolerated by these species (Henderson, 1961, 1963; Jones, 1961, 1964b). Because the restored volume of gas in the lung after surfacing is always approximately the same regardless of the length of the dive (Henderson, 1961, 1963), it has been suggested that the pulmonary gas bubble also functions in buoyancy, providing support for the dense shell (Russell Hunter, 1953c; Jones, 1961, 1964a; Henderson, 1961, 1963). Therefore, the termination of a dive may be related more to the necessity to readjust buoyancy than to the renewal of pulmonary oxygen stores. Henderson (1961, 1963) demonstrated that both *L. stagnalis* and *P. corneus* were positively buoyant at the initiation of a dive and became negatively buoyant within approximately 1 h of submergence, a period close to the normal dive duration of these two species (Jones, 1961), as mantle gas volume decreased. These results indicate that return to the surface may be involved with maintenance of positive buoyancy. Correspondingly, Jones (1961, 1964b) showed that increasing pressure by 1 atm and therefore reducing gas bubble volume and buoyancy by 50%, immediately initiated surfacing behavior and pulmonary air renewal in *P. corneus* and *L. stagnalis,* confirming earlier similar observations on freshwater pulmonates by Pretch (1939).

Jones (1961) also observed that dive duration decreased significantly when *L. stagnalis* and *P. corneus* were allowed to surface in atmospheres deficient in oxygen (PO_2 < 160 torr) (also see Janse, 1981). The quickest return to the surface was observed when snails were allowed to surface in anoxic atmospheres of pure nitrogen. Under such conditions snails return to the surface while still highly buoyant (Henderson, 1961, 1963), indicating that the frequency of pulmonary air renewal may be bimodally controlled in basommatophoran pulmonates in response to both hydrostatic cues and the oxygen concentration of pulmonary gases.

C. Physical Gills

Although oxygen is taken up directly from pulmonary air stores in freshwater pulmonates, as indicated by reduction of mantle cavity PO_2 values below those of the surrounding medium (Russell Hunter, 1953c; Jones, 1961, 1964a,b), the fact that the volume of gas lost from the bubble can be far greater than that accounted for by oxygen uptake alone (Henderson, 1961, 1963; Jones, 1961, 1964a,b) indicates that it may function as a *physical gill.* In physical gills, respiratory gases diffuse down their respective concentration gradients across the gas–water interface exposed at the pneumostome. As artificial increases in oxygen concentration of the medium cause increases in the pulmonary gas volume of submerged snails, inward diffusion of O_2 into the mantle-cavity can occur (Henderson, 1961, 1963). The extent of diffusive pulmonary oxygen uptake is extremely difficult to determine, but it may play an important role in the total oxygen consumption of at least some freshwater pulmonates (see data for *P. corneus* and *L. stagnalis* in Henderson, 1961, 1963, and Jones, 1961). In contrast, other species such as *Physa fontinalis* keep the pneumostome closed during a dive and apparently do not utilize the mantle-cavity gas bubble as a physical gill (Russell Hunter, 1953c).

D. Respiratory Pigments

The vast majority of gastropods utilize hemocyanin, a copper-based respiratory pigment circulated in solution in the blood as the major oxygen-transport molecule (Ghiretti, 1966a,b; Ghiretti and Ghiretti-Magaldi, 1972). However, some basommatophoran pulmonates are exceptions. Lymnaeids (represented by *L. stagnalis*) appear to have no circulating respiratory pigments (Ghiretti and

Ghiretti-Magaldi, 1972, 1975; Jones, 1961), and planorbids have a hemoglobin of high molecular weight dissolved in hemolymph as the major respiratory pigment (Read, 1966; Russell-Hunter, 1964, 1978; Ghiretti and Ghiretti-Magaldi, 1972, 1975; Jones, 1961, 1964a,b).

The lack of hemocyanin in lymnaeids may reflect their largely aerial respiratory schesis. They maintain higher oxygen concentrations in the pulmonary gas bubble, have higher pulmonary exchange volumes on surfacing than do other basommatophorans (Russell Hunter, 1953c; Jones, 1961, 1964a,b; Henderson, 1961, 1963), and remain above or just below the water surface where mantle-cavity gases are readily renewed. Under such conditions, lymnaeids are generally exposed to external PO_2 levels at or near air saturation, allowing sufficient oxygen to meet tissue metabolic demands to be transported entirely in solution in the hemolymph. One manifestation of the lack of blood respiratory pigments in lymnaeids is that in waters of even moderately low PO_2 (80–100 torr) an extremely high proportion of oxygen consumption is pulmonary, although under the same conditions cutaneous respiration accounts for only 50% of total oxygen consumption in planorbids (Jones, 1961, 1964a,b).

The independently evolved blood hemoglobin of the subfamily Planorbinae, as investigated in *Planorbarius corneus,* has a very low p_{50}, ranging between 2.7 and 7.0 torr (depending on pH) at 20°C (Zaaijer and Wolvekamp, 1958; Read, 1966) and, as is characteristic of most hemoglobins, it has a normal bohr shift (Jones, 1964a) and an increase in oxygen affinity with decreasing temperature (Zaaijer and Wolvekamp, 1958; Read, 1966). The very high affinity of planorbid hemoglobins for oxygen indicates that they would not function in oxygen transport at PO_2 levels near normal air saturation (Ghiretti and Ghiretti-Magaldi, 1972, 1975). *In vitro* measurements of the oxygen saturation and dissociation of hemoglobin in *P. corneus* demonstrated that the pigment functions in oxygen transport when the PO_2 of the external medium in 20–65 torr (Jones, 1964a), levels of hypoxia not uncommon in the lentic (standing-water) habitats of many members of the Planorbinae (Boycott, 1936). Jones (1961, 1964a,b) claims that hemoglobin allows *P. corneus* and *B. sudanica* to make better use of available pulmonary oxygen stores, reducing pulmonary PO_2 to 21 torr (*P. corneus*) or 18 torr (*B. sudanica*) during a dive, whereas *L. stagnalis* lowers pulmonary PO_2 only to 65 torr (Jones, 1961, 1964b). It is notable that the operating range of the hemoglobin of *P. corneus* (20–65 torr) corresponds almost exactly to the above reported difference in the ability of *P. corneus* and the pigmentless *L. stagnalis* to utilize pulmonary O_2 stores during a dive (Jones, 1961, 1964a,b). This increased pulmonary efficiency may partially account (along with the well-developed neomorphic gill) for the ability of planorbid snails to remain submerged longer and to extend to greater depths than either lymnaeids or physids.

The advantage conferred by the presence of hemoglobin to cutaneous gas exchange in planorbids is less obvious than that of pulmonary exchange. However, planorbids are the only freshwater pulmonates which routinely crawl over mud and silt sediments of deeper waters (Jones, 1961, 1964b; Rowan, 1966; Clampitt, 1972; Boerger, 1975), a microhabitat that can occasionally become quite hypoxic (Jones, 1961, 1964b). Physids and lymnaeids are almost always associated with hard surfaces such as macrophytic plants, boulders, and other debris in shallow waters (Russell-Hunter, 1953a,c, 1957, 1961a,b, 1964, 1978; Clampitt, 1970, 1974; Boag and Bentz, 1980; McMahon, 1975b, 1976b), microhabitats which are far less susceptible to hypoxic conditions than those of planorbids (Jones, 1961). In hypoxic waters a hemoglobin of high oxygen affinity confers a distinct advantage on planorbids, allowing maintenance of relatively high levels of cutaneous oxygen uptake in relation to pulmonary oxygen consumption and reducing the demands on pulmonary oxygen stores. Such has been demonstrated for *P. corneus*, but under the same levels of hypoxia the oxygen uptake of *L. stagnalis* becomes almost entirely pulmonary (Jones, 1961, 1964a).

The planorbids are also the only basommatophorans (with the exception of certain freshwater limpets) (McMahon, 1972, 1975, 1976a) reported to burrow routinely in mud and silt sediments. *Helisoma trivolvis* and *P. corneus* are both reported to burrow into and remain beneath sediments for several hours (Jones, 1961; Boerger, 1975), a period long enough to severely deplete mantle-cavity oxygen and gas stores (Henderson, 1961, 1963; Jones, 1961). Overwintering individuals of *H. trivolvis* also burrow into and remain in muds and silts for several months (Rowan, 1966; Clampitt, 1972; Boerger, 1975). Such burrowing behavior is not reported in overwintering lymnaeids or physids, which either undergo terrestrial estivation (Daniels and Armitage, 1969; Jokinen, 1978; Brown, 1982), remain active overwinter (Cheatum, 1935; DeWitt, 1955; McMahon, 1975b, 1976b), or migrate to deeper waters where they cling to exposed hard surfaces (Cheatum, 1934; Russell Hunter, 1953a, 1961a,b; Clampitt, 1970, 1974; Boag and Bentz, 1980), all of which represent movement to overwintering microhabitats in which oxygen concentrations are high. While planorbids are buried in muds or silts for periods of several hours and, certainly, for overwintering periods of several months, their pulmonary oxygen and gas stores would be decimated (Jones, 1961, 1964a,b; Henderson, 1961, 1963), causing a shift to a purely cutaneous mode of gas exchange. As muds and silts in the standing waters preferred by planorbids (Boycott, 1936) can be hypoxic (Jones, 1961, 1964a), the presence of a hemoglobin with a high affinity for oxygen in this group appears to be highly adaptive because it greatly enhances the ability of planorbids to consume O_2 cutaneously during long periods of burial in hypoxic substrata.

III. Osmotic and Ionic Regulation in a Hypoosmotic Environment

A. Osmotic Concentration of the Body Fluids and Osmoregulation

Osmoregulation in molluscs has been the subject of several reviews (Robertson, 1964; Martin and Harrison, 1966; Schoffeniels and Gilles, 1972), including an extensive review of water relations in pulmonates (Machin, 1975). Unlike osmoconforming intertidal and estuarine basommatophora whose body fluid osmotic concentrations reflect those of the external environment (Robertson, 1964; Machin, 1975), freshwater pulmonates, as do all freshwater animals, maintain their body fluid and tissue osmolarities above those of the environment (hyperosmotic regulation). Fresh waters are generally considered to have osmolarities below 30 mOsm/kg H_2O and more usually range from 1 to 10 mOsm/kg H_2O (Robertson, 1964; Prosser, 1973a,b). All freshwater pulmonates have hemolymph osmolarities well above those of even the most concentrated fresh waters (Machin, 1975), so that an osmotic gradient is continually maintained between their hyperosmotic body fluids and tissues and the dilute, freshwater environment from which they gain water by osmosis and to which they lose solutes (primarily ions) by diffusion. Therefore, in order to regulate body fluid and tissue concentrations at hyperosmotic concentrations, freshwater pulmonates must constantly excrete excess water and actively regain lost ions against their respective osmotic and ionic gradients (Robertson, 1964; Prosser, 1973a,b; Machin, 1975).

Unfortunately, although there is a relatively extensive literature on the water relations of terrestrial pulmonates, there are few such studies on freshwater species (Machin, 1975). Freshwater pulmonates, because of their small size, soft bodies, and ability to retract deeply into the shell, present difficulties in drawing appropriately sized blood and urine samples, and therefore make poor subjects for osmoregulation investigations. For this reason, most investigations of osmoregulatory physiology in freshwater pulmonates have been carried out on *L. stagnalis,* the largest of the common freshwater pulmonates in northern temperate regions (Robertson, 1964; van Aardt, 1968; Greenaway, 1970; Machin, 1975).

The body fluid osmolarity of most basommatophoran pulmonates ranges from 124 to 150 mOsm/kg H_2O, levels somewhat less than those of terrestrial pulmonates (142–360 mOsm/kg H_2O) (Machin, 1975). Such decreases in body fluid osmolarity are adaptive in freshwater animals, reducing osmotic and ionic gradients between the tissues and fresh water and therefore decreasing water gain and solute loss rates (Prosser, 1973a,b). Curiously, freshwater pulmonates generally maintain higher body fluid osmolarities than freshwater prosobranch snails, for which reported values range from 74 to 113 mOsm/kg H_2O (van Aardt, 1968; Machin, 1975). Freshwater prosobranchs retain the ctenidium (gill), and therefore have greater relative surface areas over which water and solute exchange can occur than do gill-less basommatophorans. Thus, the apparent further reduction of body fluid osmolarity in freshwater prosobranchs may be an adaptation to life in fresh water that reduces ion and water exchange to controllable levels.

Despite being more concentrated than freshwater prosobranchs and bivalves (Robertson, 1964; Prosser, 1973a,b), freshwater pulmonates are still relatively "dilute" organisms compared to freshwater crustaceans and teleost fish (Prosser, 1973b), presumably because they have never evolved impermeable outer coverings or epithelia. Indeed, the body wall of freshwater pulmonates remains highly permeable to both water (van Aardt, 1968) and ions (Greenaway, 1970).

In order to remain hyperosmotic to their medium, freshwater pulmonates must void the body fluids of excess water entering the body by osmosis. The major organ of such excretion is the kidney or coelomoduct. The single coelomoduct of freshwater pulmonates lies in the dorsal mantle wall of the pulmonary cavity. Under hydrostatic pressure, excess water is filtered from the hemolymph through the thin walls of the heart atrium into the pericardial cavity surrounding the heart (Robertson, 1964; Martin and Harrison, 1966; van Aardt, 1968; Machin, 1975). In *L. stagnalis* the osmolarities of the hemolymph and pericardial fluid are essentially equal (van Aardt, 1968), providing circumstantial evidence for such an ultrafiltration mechanism. The filtrate enters the coelomoduct through an opening in the pericardial wall to a short renopericardial canal connected to the tubular elements of the coelomoduct. The excretory fluid leaves the kidney by way of a ureter which lies in the left side of the mantle-cavity and opens near the pneumostome (Martin and Harrison, 1966; van Aardt, 1968; Purchon, 1977; Machin, 1975). When *L. stagnalis* was injected with inulin, a polysaccharide that is small enough to pass through the atrial filter but is not actively secreted into or reabsorbed from the coelomoduct, the concentration of inulin in the pericardial fluid was essentially that of hemolymph (an indication of hemolymph ultrafiltration); however, inulin was slightly more concentrated in the urine than in the hemolymph (inulin urine conc/inulin hemolymph conc = 1.11), evidence for passive tubular reabsorption of water following active solute reabsorption in the coelomoduct tubule (van Aardt, 1968).

Such active tubular reabsorption of solutes is also manifested by a decrease in the osmolarity of the urine of *L. stagnalis* in relation to the pericardial fluid. The average osmotic concentrations of the hemolymph and pericardial fluid are nearly equal at 127.0 and 123.1 mOsm/kg H_2O, respectively, but that of the excretory fluid was less at 96.8 mOsm/kg H_2O (van Aardt, 1968). Decreased excretory fluid osmolarity indicates that passive water reabsorption is incomplete. Such production of a dilute excretory fluid in hyperosmotic regulators is considered to be highly adaptive, because it conserves solute loss (particularly ions) during excretion of excess water (Prosser, 1973a,b). Similar dilute excretory fluids are produced by freshwater prosobranchs, but both freshwater pulmonates and prosobranchs produce relatively concentrated excretory fluids (75% of hemolymph concentration) (Robertson, 1964; van Aardt, 1968). Therefore, the excretory fluid represents a site of considerable solute loss which must be restored by active uptake of solutes from the environment. Freshwater teleost fish and crustaceans (crayfish) produce excretory fluids much more dilute (generally 10% of hemolymph concentrations) than do freshwater pulmonates and therefore are generally more efficient hyperosmotic regulators (Prosser, 1973a,b).

The external body wall of freshwater pulmonates apparently is highly permeable to water. In fresh water (6–8 mOsm/kg H_2O), *L. stagnalis* has a total water efflux of 7% of total body water per minute (63 μl H_2O/g wet tissue weight/h) or a turnover of the total body water every 14.3 min (van Aardt, 1968). Net water influx in *L. stagnalis* was estimated to be 0.145 μl $H_2O/(g \cdot min)$ (which was well within the average excretory fluid production rate of 1.60 μl $H_2O/(g \cdot min)$ or a release of 10% of total body water or 19.8% of total hemolymph fluid volume per hour (van Aardt, 1968).

L. stagnalis is a less strict hyperosmotic regulator than are certain other freshwater animals (Prosser, 1973a,b), as body fluid osmolarity declines significantly with decreasing medium osmotic concentrations below the isosmotic point (127 mOsm/kg H_2O) (van Aardt, 1968). Although they are incapable of hypoosmotic regulation (as are most freshwater organisms), freshwater pulmonates can tolerate relatively high levels of environmental salinity in certain microhabitats. *L. stagnalis, L. peregra, L. palustris, Physa fontinalis,* and *Planorbis vortex* (L.) were all reported to exist at salinities of 11% (320 mOsm/kg H_2O) in the Baltic (Jaeckel, 1950). Such increasing osmotic concentrations greatly reduce excretory fluid flow in *L. stagnalis,* values of 1.02 μl $H_2O/(g \cdot h)$ occurring at 127 mOsm/kg H_2O (isosomotic point) and minimal levels of 0.19 μl $H_2O/(g \cdot h)$ at 249 mOsm/kg H_2O. Body fluid osmolarity approaches that of the medium at these concentrations, greatly reducing net water uptake (van Aardt, 1968). Such reduction in water uptake and excretory fluid production in medium osmolarities above the isosmotic point may account for the atrophied coelomoducts reported in populations of freshwater pulmonates living in brackish water (Hubendick, 1948).

B. Ionic Regulation of the Body Fluids

There are even fewer extensive studies on the mechanisms of ionic regulation in freshwater pulmonates than on osmoregulatory mechanisms, probably because of the technical difficulties associated with performing these experiments on small, soft-bodied animals. Ionic regulation in molluscs has been reviewed by Robertson (1964) and in pulmonates specifically by Machin (1975). Again, *L. stagnalis,* because of its large size and ease of laboratory maintenance, appears to be the "species of choice" in these experiments. In hyperosmotically regulating freshwater pulmonates, ions and other solutes are lost to the environment over the general body surface down their respective concentration gradients and with the release of large quantities of relatively concentrated excretory fluid (Robertson, 1964; van Aardt, 1968; Machin, 1975) (see Section III,A). The external epithelium of freshwater pulmonates does not appear to be as permeable to Na^+ as it is to water. In fresh water, the average sodium flux in *L. stagnalis* with a hemolymph concentration of 57.0 m*M* Na/liter was 0.132 μ*M* Na/g wet tissue weight/h, or a turnover time of 0.35% of total hemolymph Na per hour, total hemolymph Na being turned over every 285 h (Greenaway, 1970). The site of active Na^+ uptake appears to be the external epithelium, with a significant proportion of Na^+ transport (~40%) occurring by exchange diffusion for other cations (Greenaway, 1970). An electrical potential difference of 16.4 mV (blood negative) was found to be maintained between the hemolymph and the external medium. The magnitude of this potential difference was not sufficiently large to account for the total gradient of ionic concentration between the hemolymph and the medium, indicating that blood Na^+ concentrations were maintained by active Na^+ uptake from the medium (Greenaway, 1970).

The k_m (half-saturation point) for sodium influx in *L. stagnalis* occurs at a medium concentration 0.25 m*M* Na/liter, indicating that the Na^+ transport system of this species has a relatively high affinity for Na compared to that of other freshwater animals (Greenaway, 1970; Prosser, 1973b). This high affinity for Na may compensate for its proportionately higher epithelial and excretory ion loss rates (van Aardt, 1968; Greenaway, 1970). As with most freshwater animals, maintenance of *L. stagnalis* individuals in ion-depleted media (which causes diffusional loss of hemolymph Na and reduction of hemolymph Na concentration) stimulated net sodium influx rates but did not affect the relative proportion of Na influx accounted for by exchange diffusion (Greenaway, 1970) as has been reported for some freshwater bivalves (McCorkle and Dietz, 1980). The stimulus for increased active uptake of sodium in *L. stagnalis* seems to be either a relatively large loss of sodium or a relatively small decrease in hemolymph Na concentration (Greenaway, 1970). When forced to retract suddenly into the shell, specimens of *L. stagnalis* release hemolymph from the mantle-cavity hemal pore, reducing hemolymph volume by 35%. After such release of hemolymph, net Na^+ uptake in this species reached maximal levels (higher than those stimulated by salt depletion), indicating that Na^+ uptake may be intimately involved in volume regulation in pulmonate snails (Greenaway, 1970).

C. Calcium Regulation

In common with all shelled molluscs, freshwater pulmonates maintain a continual net uptake of calcium from their environment which is sequestered as $CaCO_3$ (calcium carbonate), the major mineral component of the shell. In fresh waters, calcium is the major dissolved cation, but even here it occurs at relatively low concentrations, ranging from 1–3 mg Ca/kg H_2O to concentrations greater than 400 mg Ca/kg H_2O (Russell Hunter et al., 1967; McMahon, 1972; Hunter, 1975b; Hunter and Lull, 1977). Whether pulmonates live in soft or hard waters, uptake of Ca from the external environment and subsequent precipitation against an electrochemical gradient as crystalline $CaCO_3$ in the shell require relatively large metabolic expenditures (Russell-Hunter, 1978; Russell-Hunter et al., 1967, 1981). The process therefore requires special consideration in any general discussion of the ion regulation of freshwater pulmonates.

It has long been recognized that the vast majority of freshwater basommatophoran snails are highly calciphilic. More than 95% of the species in this group occur in waters with Ca concentrations greater than 3 mg/kg H_2O and more than 45% of the species occurring only in relatively hard waters (Ca >25 mg/kg H_2O) (Boycott, 1936; McKillop and Harrison, 1972; Russell-Hunter, 1964, 1978; Russell-Hunter et al., 1967, 1981). However, species tolerant of relatively soft waters are also common but usually are more successful in waters of higher Ca concentration (Young, 1975a). Indeed, many species of freshwater pulmonates are remarkably euryoecic with regard to environmental Ca concentration. Some environmental Ca concentration ranges reported for freshwater pulmonates are: 4.6–67.6 mg/kg H_2O for *Ferrissia rivularis* (Russell-Hunter et al., 1967, 1981); 3–66 mg/kg H_2O for *Helisoma anceps* (Menke); 31–152 mg/kg H_2O for *Physa gyrina;* 25–177 mg/kg H_2O for *P. intergra* Haldeman (Hunter and Lull, 1977); 30–280 mg/kg H_2O for *Lymnaea palustris* (Hunter, 1972, 1975b); 66–400 mg/kg H_2O for *Laevapex fuscus* (McMahon, 1972); and 6.9–93 mg/kg H_2O for *Biomphalaria pfeifferi* (Krauss) (Harrison et al., 1970). In laboratory studies, environmental Ca concentration has been shown to have significant effects on the life history traits of freshwater pulmonates. Growth rate, survivorship, and fecundity rates all decline with reduction in Ca concentrations below optimal environmental values (Harrison et al., 1970; Thomas et al., 1974). In contrast, Ca concentrations above optimal levels for fecundity were found to increase survivorship in *B. pfeifferi* (Harrison et al., 1970).

Several studies have shown that freshwater pulmonates can absorb the calcium required for shell growth directly from the external medium (van der Borght, 1963; van der Borght and van Puymbroeck, 1964, 1966; Greenaway, 1971; see Wilbur, 1964, 1972, for reviews of shell formation in molluscs). However, a significant proportion of shell Ca may also be absorbed from ingested material when Ca is taken up across the digestive tract epithelium (van der Borght and van Puymbroeck, 1966; Young, 1975b).

Radioactive tracer studies (utilizing ^{45}Ca) have demonstrated that Ca^{2+} was absorbed directly from the external medium by *Lymnaea stagnalis*. In this species the external epithelium appeared to be highly selectively permeable to Ca ion, and the Ca^{2+} uptake mechanism followed standard Michaelis-Menten kinetics, having a k_m at a medium Ca concentration of 0.3 mM/liter (12 mg/kg H_2O) and approaching saturation at 1.0–1.5 mM/liter (40–60 mg/kg H_2O), indicating that *L. stagnalis* has a relatively high affinity for Ca^{2+}, as it does for Na^+ (k_m Na = 2.5 mM/liter) (Greenaway, 1970, 1971). The presence of an inwardly directed active transport of Ca^{2+} was demonstrated in *L. stagnalis* only at medium Ca concentrations below 0.5 mM/liter (20 mg/kg H_2O). Above this concentration, Ca^{2+} uptake occurred either in the absence of an electrochemical gradient or under one favorable for Ca^{2+} absorption (Greenaway, 1971). This lower limit of 0.5 mM/liter for passive Ca^{2+} uptake corresponds exactly with the reported minimum hardness of 0.5 mM/liter reported for natural populations of *L. stagnalis* in Great Britain (Boycott, 1936). Little or no reduction in the rate of Ca^{2+} loss occurred when specimens of *L. stagnalis* were adapted to concentrations of Ca as low as 0.065 mM/liter (2.6 mg/kg H_2O). The mean minimum Ca concentration for maintenance of Ca hemolymph and tissue balance (no net loss of Ca) was 0.062 mM/liter (2.5 mg/kg H_2O) (Greenaway, 1971), a value far below the lower natural Ca concentration limit (0.5 mM/liter) for this species (Boycott, 1936). The apparent ability of freshwater pulmonates to maintain Ca balance at such low Ca concentrations may account for the ability of hard-water species to survive and reproduce when artificially transferred to soft waters (Young, 1975a). Calcium depletion of individuals of *L. stagnalis* did not increase net Ca^{2+} uptake rates or Ca^{2+} influx, which occur for Na^+ net uptake and influx in salt-depleted specimens (Greenaway, 1970, 1971). Calcium depletion, in fact, was not associated with significant decreases in hemolymph or tissue Ca concentrations in this species, despite a significant loss of Ca^{2+} to the medium. This indicates that the shell of freshwater pulmonate snails acts as a calcium buffer or store from which lost hemolymph or tissue Ca can be released during periods of calcium depletion caused by short-term decreases in external Ca concentration such as caused by flooding (Greenaway, 1971).

As the electrochemical gradient appears to be neutral or even favorable for Ca^{2+} absorption in the medium to hard waters preferred by freshwater pulmonates, most species apparently do not expend large amounts of metabolic energy in absorbing calcium from the external medium (Greenaway, 1971). However, as net Ca uptake is eventually deposited in the shell against a considerable electrochemical gradient, shell growth must require considerable amounts of energy, and the relative amounts of $CaCO_3$ in the shells of standardized individuals must reflect the amounts of metabolic energy required to form them (Russell-Hunter, 1978; Russell-Hunter et al., 1967, 1981; Hunter and Lull, 1977). Comparisons of variations in population shell $CaCO_3$ content with environmental Ca concentrations should thus provide information regarding interpopulation variations in shell Ca metabolism among isolated populations of freshwater pulmonates (Russell-Hunter, 1978; Russell-Hunter et al., 1967, 1981). Russell-Hunter et al. (1981) have identified four different patterns of shell Ca variation among species of freshwater pulmonates. The first is a direct relationship between Ca concentration and shell calcium, which occurs in *L. peregra* (Russell-Hunter, 1978; Russell-Hunter et al., 1967, 1981) and *Laevapex fuscus* (McMahon, 1972, 1975b), in which snails appear to expend approximately the same amount of energy on shell formation in a wide variety of environments, Ca availability therefore reflects the Ca deposition rate; i.e., snails from environments with high Ca concentrations have shells with higher Ca contents, and vice versa. Such increased deposition of shell Ca with increased Ca concentration of the external environment has been demonstrated in controlled laboratory experiments for several species of freshwater basommatophorans (Harrison et al., 1970; Thomas et al., 1974; Young, 1975b). The second pattern is that of apparent regulation, in which shell Ca content is maintained within very narrow limits (reduced interpopulation variation) over a wide range of environmental Ca concentrations, as has been described for *P. gyrina* (Hunter and Lull, 1977). The third pattern is one in which interpopulation variations in shell Ca concentration are associated with specific microenvironmental conditions, as described for *Helisoma anceps* (Menke) and *P. integra* (Hunter and Lull, 1977). In the fourth pattern, shell Ca appears to have no distinct relationship with either environmental Ca concentrations or trophic conditions, as reported for populations of the freshwater limpet *Ferrissia rivularis* (Nickerson, 1972; Russell-Hunter et al., 1967, 1981) and of *Lymnaea palustris* (Hunter, 1972, 1975b). Such random, apparently nonadaptive variation is hypothesized to result from genetically discrete, isolated physiological races which have genetically fixed rates of shell Ca deposition (Russell-Hunter et al., 1967, 1981). Founder effects and subsequent genetic drift are presumed to be responsible for the fixation of alleles specifically associated with different rates of shell Ca metabolism in isolated populations, which explains their observed interpopulation variation in shell Ca content. Such physiological races appear to be associated with temporally unstable freshwater habitats which do not exist long enough for selective forces to produce phenotypes with shell Ca deposition rates appropriate to their respective microenvironmental Ca concentrations (Nickerson, 1972; Hunter, 1972, 1975b; Russell-Hunter et al., 1967, 1981). In species with such a random pattern of shell Ca variation, transfer of individuals between environments of different Ca concentration has been shown to *not* affect the characteristic shell Ca contents of transferred individuals. These shell contents remain essentially similar to those characterizing their populations of origin, indicating that such species have a relatively rigid genetic control of shell mineralization that is independent of the effects of environmental Ca concentrations (Nickerson, 1972; Hunter, 1972, 1975b).

In spite of a relatively large literature on the intraspecific, interpopulation variation of shell Ca, the exact relationships between microenvironmental variation and shell Ca variation remain unclear. Although epithelial absorption of Ca has been demonstrated in several species of freshwater pulmonates, laboratory investigations of Ca uptake utilizing ^{45}Ca as a radioactive tracer have demonstrated that a high proportion of shell Ca may also come from ingested food. The contribution of food Ca to the Ca content of the shell has been estimated to be 20% in *L. stagnalis* (van der Borght and van Puymbroeck, 1966). More recently, much higher values of food Ca contribution to shell Ca have been reported for *L. peregra* (46% in hard water, 70% in soft water) and *Planorbarius corneus* (46% in hard water, 79% in soft water) (Young, 1975b). It has been estimated that Ca uptake from the digestive tract of *L. stagnalis* is extremely efficient, 95% of all ingested Ca being absorbed across the gut epithelium (van der Borght and van Puymbroeck, 1966). In these investigations of alimentary Ca absorption, snails were fed dried lettuce, which may have had a Ca content very different from that of the *Aufwuchs* (periphyton) or detritus normally fed upon by natural populations of freshwater pulmonates. In fact, minerals have been shown to comprise a large and highly variable component of the *Aufwuchs* communities on which many freshwater pulmonates graze (McMahon et al., 1974). As such, even larger proportions of shell Ca may be derived from ingested materials in natural populations than has been estimated for freshwater pulmonates in the laboratory. Absorption of shell Ca from the external medium may be only of secondary importance, being utilized as a source of Ca uptake primarily during periods of starvation (such as overwintering) when ingested intake of Ca is very low or nonexistent. If shell Ca in some freshwater pulmonate species is derived primarily from ingested food and substrata, then variations in shell Ca levels may be more highly correlated with *Aufwuchs* or detrital Ca levels than with water Ca hardness. Unfortunately, there are no published reports on the effects of food Ca content on shell formation or shell Ca content in natural populations of freshwater pulmonates.

Certainly, some microenvironmental forces seem to be able to influence shell Ca levels in at least some species of pulmonate snails. Hunter and Lull (1977) described variations in shell Ca for 17 isolated populations of *P. integra* and 18 populations of *H. anceps*. The shell Ca content of these two species showed great interpopulation variation which was correlated with neither the Ca hardness nor the geographical distribution of their respective microenvironments. Curiously, however, in the seven localities where they co-occurred there was a high positive correlation between the shell Ca content of both species. This was taken as circumstantial evidence either that (1) the alleles governing shell Ca deposition in both species were subject to selective pressures associated with unidentified environmental variation, such as the degree of exposure to wave action, or that (2) the observed plasticity of shell Ca content was an *ecophenotypic* characteristic resulting from such environmental influences as trophic conditions, operating directly on individual shell Ca deposition rates and, therefore, not a result of selective forces operating on the population gene pools (Hunter and Lull, 1977).

One form of interpopulation ecophenotypic variation in shell Ca content not yet considered by investigators is the effect that growth rates may have on shell formation. Hunter and Lull (1977) demonstrated that the shell Ca contents of *P. integra* and *H. anceps* populations, in freshwater habitats where they co-occurred, were highly correlated not only with each other but also with the trophic conditions of their environments. In oligotrophic environments (characterized by low primary productivity), shell Ca/dry tissue weight ratios were two to three times greater than those from eutrophic environments (characterized by high primary productivity). Growth rates in freshwater pulmonates are highly variable (Russell-Hunter, 1964, 1978) and highly correlated with environmental trophic conditions (Russell-Hunter, 1964, 1978; McMahon et al., 1974), populations from eutrophic environments generally displaying higher growth rates than those from more oligotrophic habitats. Therefore, the data of Hunter and Lull (1977) can also be interpreted as indicating that shell Ca content is inversely related to growth rate in the two species studied. In such a model, the tissue growth of fast-growing populations can be seen as outstripping shell Ca deposition (which may be operating at genetically fixed levels) and therefore producing thinner shells with the low shell Ca/tissue dry weight ratios characteristic of populations from more eutrophic habitats. In contrast, slower growing populations expand the perimeter of the shell at a lower rate, allowing relatively longer periods for shell Ca deposition and thus producing thicker shells with the high Ca/dry tissue ratios characteristic of populations from more oligotrophic environments. The interpopulation variation in shell Ca of the freshwater limpet *Laevapex fuscus* displays just this sort of pattern with growth rate; the slowest growing population had the heaviest shells with the highest Ca content, whereas the fastest growing population had the lightest shells with the lowest Ca content and the populations with intermediate growth rates had shells with intermediate levels of Ca (McMahon, 1972, 1975a). Similarly, of three populations of *L palustris,* that with the slowest growth rate and only one generation per year had a much higher shell Ca/dry tissue weight ratio than the other two populations which had higher growth rates and two generations per year (Hunter, 1972, 1975a,b).

Certainly, the relationship of trophic condition and population growth rates to patterns of intraspecific, interpopulation variation in the shell mineral component of freshwater plumonates warrants further investigation before the genetic and/or environmental bases of such variation can be understood. However, it is also important to recognize that in any one species variations in shell components are probably influenced by a combination of endogenous and exogenous factors whose degree of expression involves much interspecific variation. Certainly, trophic conditions, growth rate, Ca availability, and other environmentally influenced variables appear to have little effect on the observed shell Ca variation in populations of *F. rivularis*. In this species, the vast majority of interpopulation differences appear to have a purely genetic basis resulting in the formation of isolated and occasionally nonadaptive physiological races with respect to shell Ca deposition. At the other end of the continuum are species such as *L. peregra, L. fuscus, P. integra,* and *H. anceps,* in which the vast majority of interpopulation variation in shell Ca appears to be nongenetic and ecophenotypic, induced by specific conditions associated with each population microhabitat.

IV. Nitrogen Excretion

Nitrogenous wastes from the catabolism of amino acids and other nitrogenous organic compounds are generally excreted by animals as ammonia (NH_3 or NH_4^+), urea, or purines (uric acid, xanthine, and quanine) (Campbell, 1973). Although it is now generally accepted that no one organism excretes only one of these waste products exclusively (Potts, 1967), normally, one of the three forms of nitrogenous wastes predominates in any one species. Thus, species which excrete ammonia as the major nitrogenous waste are referred to as *ammonotelic,* those in which urea predominates are *ureotelic,* and those in which uric acid or other purines predominate are *uricotelic* or *purinotelic* (Campbell, 1973; Campbell and Bishop, 1970; Machin, 1975). An apparent correlation often occurs between the pattern of nitrogenous excretion and the availability of water in the microhabitat (Campbell, 1973; Campbell and Bishop, 1970).

Species from aquatic habitats are primarily ammonotelic. Ammonia is highly soluble in water, and the external epithelial surface of aquatic species is readily permeable to it (Campbell, 1973; Campbell and Bishop, 1970). Because the concentration of ammonia in natural environments is extremely low, aquatic organisms can maintain low body concentrations of ammonia by simple diffusion from body fluids to the external medium over the external epithelium. In contrast, for terrestrial organisms, excretion of nitrogenous wastes as ammonia in an aqueous excretory fluid requires a prohibitive loss of body water stores [because ammonia is toxic above 0.05 mg NH_3/liter (Machin, 1975), it cannot be concentrated much in excretory fluids)], and release of nitrogenous wastes as gaseous ammonia directly to the atmosphere involves maintenance of tissue and body fluid concentrations at toxic levels (Campbell, 1973). Therefore, most terrestrial organisms have evolved biochemical detoxification pathways to convert ammonia to either urea or purines as uric acid, forms of nitrogenous excretions energetically less efficient but far less toxic than ammonia that either can be retained in relatively higher concentrations in the body or excretory fluids (urea) or are nearly insoluble in water and can be stored or excreted as a solid crystalline waste product with little loss of body water (uric acid and other purines) (Campbell, 1973).

Nitrogen excretion in molluscs has been the subject of several reviews (Potts, 1967; Florkin, 1966; Florkin and Bricteux-Grégoire, 1972; Campbell and Bishop, 1970) and, more recently, in pulmonates specifically (Campbell et al., 1972; Machin, 1975). Marine and terrestrial molluscs follow the patterns of nitrogenous excretion described above. Marine gastropods and bivalves are primarily ammonotelic (Florkin and Bricteux-Grégoire, 1972; Duerr, 1968), and terrestrial pulmonates (order Stylommatophora) are almost completely purinotelic, whether active or inactive (aestivating) (Florkin, 1966; Forkin and Bricteux-Grégoire, 1972; Campbell et al., 1972; Machin, 1975). Freshwater pulmonates have presumably reevolved an aquatic mode of life from that of a more nearly terrestrial ancestral stock, and because living species represent a gradient of evolutionary readaptation to aquatic life from amphibious, nearly terrestrial species (some lymnaeids) to a purely aquatic schesis (some planorbids and all freshwater limpets), the evolutionary implications of the patterns of nitrogen excretion in this group have long been of interest to investigators. Needham (1938) proposed that the high uric acid content of the coelomoducts of some freshwater pulmonates, such as *L. stagnalis,* reflected an incomplete readaptation to aquatic life. However, such retention of uricotelism has more recently been suggested as having little adaptive significance for the group as a whole (Duerr, 1967).

Comparative descriptions of nitrogen excretion patterns in freshwater pulmonates are complicated by the fact that retention of significant amounts of uric acid in the coelomoduct and tissues appears to be characteristic of gastropods in general. Of 20 species of marine prosobranchs, many of which were subtidal and purely aquatic, all were found to retain measurable levels of uric acid in their tissues and excretory organs (Needham, 1938; Duerr, 1967). Some species had concentrations comparable to those of freshwater pulmonates (Needham, 1938; Duerr, 1967). Curiously, even freshwater prosobranchs whose entire evolutionary history appears to have been aquatic retain body loads of uric acid that are comparable to those of freshwater pulmonates and even of some terrestrial species (Needham, 1938). However, recent detailed examinations of nitrogen excretion in relation to oxygen consumption rates and nitrogen content of ingested food suggest that three species of freshwater prosobranchs, *Goniobasis livescens* (Menke), *Bithynia tentaculata* (L.), and *Leptoxis* (= *Spirodon*) *carinata* (Bruguiere), are almost entirely ammonotelic in their nitrogen excretion (Payne, 1979; Aldridge, 1980; Tashiro, 1982; Tashiro and Colman, 1982).

Freshwater pulmonates such as *L. stagnalis,* although they are primarily ammonotelic, retain considerable amounts of uric acid in the coelomoduct as do freshwater prosobranchs. Earlier suggestions that freshwater pulmonates were primarily uricotelic, with the apparent excretion of ammonia and urea being an artifact produced by microbial decomposition of fecal material (Duerr, 1966, 1967), have been discounted by more recent experiments in which bacterial contamination of the medium was carefully controlled by antibiotics (Bayne and Friedl, 1968; Friedl, 1974; Becker and Schmale, 1978). Such studies have shown that in *L. stagnalis* ammonia accounts for 52% of excreted nitrogen, urea 28%, unidentified volatile nitrogenous compounds 4%, and nonvolatile nitrogenous compounds (the fraction that could include uric acid or other purines) only 13% of the total nitrogen excretion (Bayne and Friedl, 1968; Friedl, 1974). For *Biomphalaria glabrata,* nitrogen excretion has similarly been reported to be 80% ammonia nitrogen and 20% urea nitrogen (Becker and Schmale, 1978), which was associated with correspondingly high hemolymph concentrations of these two compounds but uric acid could not be detected in the hemolymph (Becker and Schmale, 1975).

Freshwater pulmonates appear to be ideally suited to excrete ammonia because of the high permeability of their external epithelium to water and ions (van Aardt, 1968; Greenaway, 1970, 1971) and the proportionately large fluid volumes excreted by the coelomoducts (van Aardt, 1968; Machin, 1975). Certainly, the high levels of ammonia present in the mucus of *L. truncatula* (0.011 μM NH_3/mg fresh mucus) are indicative of cutaneous excretion of ammonia by diffusion in freshwater pulmonates (Wilson, 1968).

Curiously, although urea appears to play a significant role in the nitrogen excretion of most freshwater pulmonates (Bayne and Friedl, 1968; Friedl, 1974; Becker and Schmale, 1978; Newman and Thomas, 1975), attempts to demonstrate the presence of all the enzymes required for a complete, classic urea cycle in the tissues of freshwater pulmonates have failed both in *L. stagnalis* (Friedl and Bayne, 1966) and in *B. glabrata* (Schmale and Becker, 1977). However, the latter investigators report that inability to detect certain key enzymes for urea synthesis may have resulted from the interference of other enzymes in the crude homogenate of digestive diverticula tissue utilized in such studies (Schmale and Becker, 1977).

If the nitrogen excretion of freshwater pulmonates is primarily ammonotelic and ureotelic, one must question why some (perhaps most) species retain the ability to produce and store uric acid (and perhaps other purines) in the coelomoduct. The long evolutionary history of pulmonates in fresh water [possibly extending from the Jurassic (Morton, 1955)] would seem to preclude the retention of such a nonadaptive and energetically inefficient mode of nitrogenous excretion. Therefore, retention of biosynthetic pathways for both urea and uric acid by freshwater pulmonates must be understood in terms of their selective advantage to present-day species in their respective microenvironments. Indeed, uric acid excretion must be of high selective value to freshwater pulmonates, because the capacity to synthesize and store uric acid occurs in developing embryos as soon as the organs begin to form (Conway et al., 1969).

As described earlier, most freshwater pulmonates inhabit the shallow, eutrophic, nearshore waters of small, unstable aquatic habitats (see Section I,B). In such environments, both short- and long-term exposure to air by receding water levels can occur frequently and often seasonally. Certainly, many basommatophorans are amphibious and are routinely exposed to air during periodic excursions between aquatic and terrestrial environments (See Section II,A). When exposed to air, an ammonotelic schesis is highly disadvantageous because the high solubility of ammonia requires maintenance of hemolymph ammonia concentrations at deleteriously high levels to allow adequate diffusion of ammonia as a gas into the atmosphere (such gaseous diffusion appears to be impossible in estivating snails that withdraw deeply into the shell to conserve evaporative water loss, because it reduces the aerial exposure of permeable surfaces required for gas release). Alternatively, excretion in solution would require the release of very large volumes of water, which would result in the rapid desiccation of aerially exposed individuals. Therefore, urea and uric acid cycles may be retained in freshwater pulmonates as an adaptation to periodic exposure in air. In air, synthesis of relatively nontoxic urea or uric acid allows the accumulation of large quantities of nitrogenous wastes in the hemolymph and coelomoduct, while maintaining ammonia concentrations below toxic levels. Not surprisingly, earlier studies showed that the body loads of uric acid in freshwater pulmonates generally correspond to the degree of aquatic readaptation. Studies of amphibious species such as *Lymnaea stagnalis* and *L. Palustris* have reported body loads of 3.5–6.5 mg uric acid/g dry tissue weight; semiaquatic, shallow-water, diving species such as *Physa parkeri,* Currier 2.8 mg/g; more nearly aquatic species such as *Planorbarius corneus,* 1.5–2.0 mg/g; and the totally aquatic freshwater limpet *Ancylus fluviatilus,* 0.6 mg/g (Needham, 1938; Duerr, 1967). However, notable exceptions to such generalizations do occur; for example, the highly amphibious *L. peregra* has a uric acid body load of only 0.2 mg/g (Needham, 1938).

There is increasing evidence that the patterns of nitrogen excretion in freshwater pulmonate species are far more plastic than had been previously suspected. High levels of protein catabolism resulting from starvation or schistosome infection induced increased urea and uric acid synthesis in *B. glabrata.* After only 5 days of starvation, hemolymph ammonia concentration in this species increased by 41% (fed = 79 μg NH_3/100 ml; starved = 112 μg NH_3/100 ml), but hemolymph urea showed an extraordinary 16-fold increase in concentration (fed = 63.5 μg urea/100 ml; starved = 1017 μg urea/100 ml) (Becker and Schmale, 1975). Increases in hemolymph urea concentrations in starved individuals were shown to correspond to increases in the specific activity of selected urea cycle enzymes (Schmale and Becker, 1977). In *B. glabrata* total nitrogen excretion increased by 62 or 79% over fed controls in individuals starved for 5 days and infected with schistosomes, respectively. The proportions of excreted nitrogenous waste products were also affected by starvation or schistosome infection, with urea excretion increasing from 20% of total excreted nitrogen in fed control snails to 60 and 56% of total nitrogen excretion in starved and schistosome-infected individuals, respectively. In contrast, the rates of ammonia excretion were not affected by such treatments (Becker and Schmale, 1978). Such stimulation of urea production and excretion by starvation may render invalid the results of previous studies of nitrogen excretion in freshwater pulmonates in which short-term isolation and starvation (>24 h) preceded determination of hemolymph levels or excretion rates of nitrogenous wastes. Such large increases in urea production and excretion during starvation appear to be a detoxification mechanism which prevents accumulation of ammonia in the hemolymph to deleterious levels during periods when the majority of metabolic demand is supported by protein catabolism (Becker and Schmale, 1975, 1978; Schmale and Becker, 1977).

Similar adaptive plasticity in nitrogenous excretion has also been reported for freshwater pulmonates exposed in air. Maintenance of *B. glabrata* in air (95% relative humidity) caused no significant increase in hemolymph ammonia concentration (controls = 281 μM NH_3/liter; in air = 232 μM NH_3/liter), but resulted in a massive 155-fold increase in hemolymph urea concentration (controls = 167 μM/liter; in air = 25,870 μM/liter) and a 38-fold increase in urea synthesis rates (Schmale and Becker, 1977; Becker and Schmale, 1978). Increases in urea hemolymph concentration and rates of urea synthesis were correlated with a 2.4- to 32-fold increase in the specific activity of selected urea cycle enzymes in aerially exposed specimens (Schmale and Becker, 1977). Such extreme increases in hemolymph urea concentrations appear to be highly adaptive in aerially exposed snails which cannot efficiently excrete their nitrogenous wastes as ammonia. Conversion of metabolically produced ammonia to urea is a homeostatic mechanism that allows retention of large quantities of nitrogenous wastes in the hemolymph, while ammonia remains well below toxic levels. On return to an aquatic medium, accumulated urea could be quickly excreted via the coelomoduct. This is reported to occur in land pulmonates on arousal from estivation (Horne, 1971), allowing a return to normal ammonotelic excretion within a relatively short time.

Nitrogenous excretion has been studied over extensive periods of aerial estivation in the freshwater lymnaeid *Bakerilymnaea cockerelli* (Pilsbry and Ferriss) [= *B. techella* (Haldeman)]. This species inhabits temporary ponds and small lakes that retain water for only a few months of the year (Taylor, 1981). As its aquatic habitats dry, *B. cockerelli* burrows into the substratum and spends June to November or December in a hot, xeric soil environment, reemerging in the winter when the pond once again floods (Newman and Thomas, 1975). In water this species is primarily ammonotelic, its body loads of ammonia nitrogen accounting for 58% of total waste nitrogen (18% urea nitrogen, 24% uric acid nitrogen) (Newman and Thomas, 1975). After estivating in dry soil for 30 days, the total body waste nitrogen (as NH_3, urea, and uric acid) increased 15-fold over controls, with 74% of nitrogen as urea nitrogen (13% ammonia nitrogen, 13% uric acid nitrogen). This indicated a decided shift toward ureotelism in response to short-term aerial exposure (Newman and Thomas, 1975), which is very similar to the response observed in aerially exposed specimens of *B. glabrata* (Becker and Schmale, 1978) and apparently has the same adaptive value. However, longer periods of aerial estivation (97–174 days) in *B. cockerelli* were associated with a decided shift toward uricotelism with 49–58% of retained excretory nitrogen as uric acid nitrogen (17% ammonia nitrogen, 25–34% urea nitrogen) (Newman and Thomas, 1975). Such accumulation of uric acid in *B. cockerelli* during long-term estivation (>30 days), similar to that of estivating styommatophorans (Machin, 1975), prevents the long-term concentration of potentially toxic nitrogenous wastes as ammonia and urea in the hemolymph by sequestration of waste nitrogen in the lumen of the excretory organ as crystallized uric acid or other purines. It is interesting that in aerially exposed specimens of *B. cockerelli* urea accumulation precedes that of uric acid at least for the first 30 days. This adaptation may allow a quick return to normal function after short periods of aerial exposure between rainfalls. Uricotelism, in contrast, appears to be reserved for longer periods of aerial estivation associated with the seasonal drying up of this species' temporary pond habitats (Newman and Thomas, 1975).

In summary, it would appear that the vast majority of freshwater pulmonates excrete ammonia as the primary nitrogenous waste product when normally submerged. However, urea and uric acid synthesis appear to be retained as an adaptation that allows detoxification of ammonia during periods of exposure in air, which can frequently occur in the variable, shallow-water, nearshore habitats characteristic of the pulmonates. Although the degree of uricotelism displayed by any one group of freshwater pulmonates seems to be associated with their degree of readaptation to aquatic life (the amphibious lymnaeid species retain relatively high levels of uric acid, and the totally aquatic freshwater limpets appear to produce and store minimal levels of uric acid) (Needham, 1938; Duerr, 1967), there are many apparent exceptions to such a generalization (Needham, 1938). It is more likely that the degree to which urea and uric acid synthetic cycles are retained in pulmonates may be associated more with the extent of aerial exposure experienced by a particular species in its microhabitat than with broad phylogenetic relationships. Lymnaeids, physids, and planorbids include species specialized for life in temporary freshwater habitats that subject them to seasonal extended exposures to air (Boycott, 1936; Brown, 1982). It is among these species that retention of synthetic pathways for urea and uric acid should be maximal (e.g., *B. cockerelli*). Conversely, these pathways should be much less fully developed in more permanently aquatic species. Certainly, urea and uric acid synthesis should be least developed in the totally aquatic freshwater limpets. Unfortunately, no systematic comparative surveys of the nitrogen metabolism and excretion of a wide variety of freshwater pulmonates in both water and air have appeared in the literature. Research on this question would do much to clarify the environmental and evolutionary relationships associated with the wide variations observed in the patterns of nitrogen excretion of freshwater pulmonates.

V. Temperature

A. Temperature Variation in Freshwater Habitats

Freshwater pulmonates inhabit small, shallow, marginal, freshwater habitats (see Section I,B) that are subject to much greater extremes of diurnal and seasonal temperature variation than are the microhabitats of animals such as freshwater prosobranchs which inhabit the deeper areas of larger lakes and rivers. The thermal regimes of such shallow-water habitats generally reflect ambient air temperatures (see the data of Burky, 1969; Eversole, 1974; Hunter, 1972; McMahon, 1972, 1975a,b, 1976a,b; Nickerson, 1972; Romano, 1980). Typical seasonal temperature variation may be as great as 0–37°C in some Texas fresh waters inhabited by pulmonates (R. F. McMahon, unpublished observations). Seasonal temperature ranges reported for species of freshwater pulmonates include: *Ferrissia rivularis,* 0–28°C (Nickerson, 1972); *Laevapex fuscus,* 0–31°C (McMahon, 1972, 1976a); *Hebetancylus excentricus* (Morelet), 10–31°C (McMahon, 1976a); *Lymnaea stagnalis,* 0–30°C (Brown, 1979); *L. palustris,* 1–28°C (Brown, 1979); *Physa virgata* Gould, 8–39°C (McMahon, 1975a); *P. gyrina,* 0–33°C (Sankurathri and Holmes, 1976); *P. integra,* 0–31°C (Brown, 1979); *P. hawnii* Lea, up to 36°C (Daniels and Armitage, 1969); *Biomphalaria pfeifferi* and *Bulinus globosus* (Morelet), 7–28°C (Appleton, 1976a); and *Helisoma trivolvis,* 0.5–28°C (Eversole, 1978). Maximum ambient temperatures can reach even higher levels in tropical regions, where swamps inhabited by pulmonates can have an annual temperature range of 10 to 42°C (Russell-Hunter, 1964). The environmental temperature ranges reported for freshwater pulmonates are generally higher than those reported for freshwater prosobranchs which characteristically inhabit larger, more stable bodies of water. The reported annual ranges for two species of freshwater prosobranchs in upstate New York were 1–26°C for three river populations of *Leptoxis carinata* (Aldridge, 1980) and 1–28°C for river and lake populations of *Goniobasis livescens* (Payne, 1979).

Wide diurnal temperature fluctuations are characteristic of the shallow freshwater environments preferred by basommatophorans because they remain relatively close to air temperatures during dark hours or overcast days but often rise to more than 5°C above air temperature on clear, sunny days as a result of insolation (McMahon, 1975b, 1976b; McMahon and Payne, 1980). Diurnal water temperature variations on bright, sunny days can be quite extraordinary, with ranges of up to 15°C occurring occasionally in Texas (R. F. McMahon, unpublished observations). In 1 day of observation, Russell Hunter (1953a) reported diurnal temperature fluctuations ranging between 2 and 8°C in six closely adjacent Scottish freshwater habitats, and diurnal variations as great as 11°C have been reported for ponds in Great Britain (Boycott, 1936). From 2 years of continuous temperature records, Appleton (1976a) reported *monthly mean* diurnal temperature spans ranging from 3.0 to 7.5°C in a South African stream inhabited by *Biomphalaria pfeifferi* and *Bulinus globosus*. Such wide mean monthly values indicate extreme diurnal temperature fluctuations in this environment.

Freshwater pulmonate snails have evolved a high capacity to tolerate and compensate for the widely fluctuating diurnal and seasonal temperature regimes characteristic of their shallow-water habitats. The remainder of this section discusses the adaptive physiology of freshwater pulmonates with regard to their resistance, capacity, behavioral, and life history adaptations to temperature.

B. Resistance Adaptations to Temperature

Thermal resistance adaptations involve the survival limits of the whole organism and can be described as the range of tolerated temperatures associated with a specific set of thermal conditions (Bullard, 1964; Pohl, 1976). The extent of such thermal resistance adaptation can be determined by measuring the thermal tolerance of individuals acclimated to a range of temperatures representative of the variation in the thermal regimes of their natural microhabitats (Pohl, 1976).

As an adaptation to the high levels of temperature variation associated with their habitats, freshwater pulmonates have evolved high tolerance of both short-term and seasonal temperature fluctuations. Certainly, most northern temperate zone species are highly cold tolerant. Most species overwinter in streams or below the ice in lakes and ponds where temperatures remain near 0°C for several months (Boerger, 1975; Boycott, 1936; Burky, 1969, 1971; Cheatum, 1934; Clampitt, 1970, 1974; Eversole, 1974; Hunter, 1972, 1975a; McMahon, 1972, 1973; Sankurathri and Holmes, 1976). Species of overwintering freshwater plumonates have even been observed to remain sluggishly active, locomoting under thin ice at water temperatures near 0°C. These include *P. gyrina* (Cheatum, 1934; DeWitt, 1955), *P. sayii* Walker, *H. trivolvis, Lymnaea columella* (Say), *Gyralus deflectus* (Say) (Cheatum, 1934), *L. peregra,* and *Planorbarius corneus* (Boycott, 1936). Other species have been reported to remain immobile on hard substrata at low winter temperatures, e.g., *L. stagnalis* (Boag and Bentz, 1980), *F. rivularis* (Burky, 1969), *Acroloxus lacustris,* and *Ancylus fluviatilis* (R. F. McMahon, unpublished observations), or to overwinter burrowed into the substratum, e.g., *H. trivolvis* (Boerger, 1975; Eversole, 1974; Rowan, 1966) and *Laevapex fuscus* (McMahon, 1972, 1973, 1975a).

Freshwater pulmonates are also highly tolerant of sudden decreases in temperature. Laboratory studies have shown that individuals acclimated to temperatures in the range of 20 to 25°C tolerated sudden decreases to less than 5°C in species such as *Lymnea stagnalis* (Harrison, 1977a,b; Boag and Bentz, 1980; van der Schalie and Berry, 1973), *H. trivolvis* (Sheanon and Trama, 1972), *Physa hawnii* (Daniels and Armitage, 1969), *A. fluviatilis, Planorbis contortus* (L.) (Calow, 1975b), *Physa virgata,* and *Laevapex fuscus* (R. F. McMahon, unpublished observations).

Perhaps of greater environmental significance are the very high upper lethal temperatures reported for freshwater pulmonates. Some of the tolerance levels to high temperatures characteristic of freshwater basommatophorans lie near the maximum reported values for any multicellular animal in natural aquatic environments (Russell-Hunter, 1964, 1978) (excepting only those species restricted to hot springs). Freshwater pulmonates have been reported to survive and reproduce in waters receiving natural (hot springs) or artificial (industrial) thermal effluents that can attain very high midsummer temperatures. Such populations include *P. virgata* at 39°C (McMahon, 1975b, 1976b); *P. anatina* at 40°C (Beames and Lindeborg, 1967); *P. gyrina* at 35–38.7°C (Brues, 1928; Agersborg, 1932); and *Lymnaea peregra* at 45°C (Issel, 1908). The high tolerance of freshwater pulmonates for elevated ambient temperatures may partially explain their success in colonizing areas receiving heated effluents from industrial installations (McMahon, 1975b; Mann, 1965; Sankurathri and Holmes, 1976).

Laboratory studies have also shown that freshwater pulmonates have extremely high tolerance to acute increases in temperature. Reported upper thermal limits of freshwater pulmonates include: *P. anatina,* 41–43°C (Beames and Lindeborg, 1967); *P. integra* and *P. gyrina* acclimated to 24°C, greater than 40°C (Clampitt, 1970); and *L. acuminata* acclimated to 32°C, equal to 40.1°C (Nagabhushanam and Azmatunnisa, 1976). When measured as *temperature of heat coma,* a reversible condition associated with loss of nervous integration and characterized by cessation of locomotion, ventral curling of the foot, and an inability to cling to hard surfaces (McMahon, 1976b; McMahon and Payne, 1980), the thermal tolerance of *L. peregra* acclimated to 17°C was 40°C (Skoog, 1976) and that of *P. virgata* acclimated to 30°C (McMahon, 1976b) was 44°C. McMahon (1976b) reported that the lethal temperature of *P. virgata* was 3–4°C above the heat coma temperature, yielding a short-term upper lethal limit of 47–48°C for this species.

There have been comparatively few studies on the effects of temperature acclimation on thermal tolerance levels in freshwater pulmonates. The period required for complete acclimation to increased temperature is relatively short, estimated at approximately 6 days in *P. virgata* (McMahon and Payne, 1980) and 9 days in *L. acuminata* (Nagabhushanam and Azmatunnisa, 1976). As is common in ectothermic animals, cold acclimation takes somewhat longer than warm acclimation in pulmonates (Nagabhushanam and Azmatunnisa, 1976).

The response of temperature tolerance limits to temperature acclimation in freshwater pulmonates ranges from a 1.3–1.4°C increase in thermal tolerance limit per 5°C increase in acclimation temperature (1.3–1.4°C/5°C) in *P. virgata* acclimated to 10, 20, and 30°C (McMahon, 1976b) to 2.4°C/5°C in *L. acuminata* acclimated to 10 and 32°C (Nagabhushanam and Azmatunnisa, 1976). Therefore, thermal tolerance levels will be 10–20°C greater than acclimation temperature within the ambient temperature range of most freshwater pulmonates. It is interesting that the acclimatory response to temperature in field populations of *P. virgata* whose thermal tolerances were recorded over an entire annual cycle was 2.0–2.5°C/5°C, which was higher than laboratory values (1.3–1.4°C/5°C) (McMahon, 1976b). This suggested that variables other than ambient temperature may affect thermal tolerance in natural populations of freshwater pulmonates. Maintenance of upper thermal limits at such high levels above ambient temperatures may be an adaptation to the wide diurnal temperature fluctuations characteristic of the shallow habitats of freshwater pulmonates, in which maximal midday temperatures may be 15°C greater than minimal evening temperatures. In contrast, the upper thermal limits of freshwater prosobranchs which inhabit waters with more stable thermal regimes are generally lower than those of freshwater pulmonates (Skoog, 1976).

There has been only one extensive study of the long-term survival of elevated temperatures in freshwater pulmonates (van der Schalie and Berry, 1973). In these tests, survival rates of six North American species were monitored at a variety of constant temperatures over periods ranging from 77 to 121 days. The results indicated that species with a more northerly distribution, such as *L. emarginata* (Say) and *L. stagnalis,* had long-term upper lethal limits of about 28–30°C whereas species endemic to warmer climates and with more extensive southern distributions in North America, such as *Helisoma campanulatum* (Say), *H. anceps, H. trivolvis,* and *P. gyrina,* were able to survive at temperatures above 30°C (32–33°C) (van der Schalie and Berry, 1973). The same study revealed that the long-term thermal tolerance of the freshwater prosobranch *Amnicola limosa* (Say), which is distributed throughout the eastern half of the United States, was lower than that of all six pulmonates tested (van der Schalie and Berry, 1973). This is another apparent indication that freshwater pulmonates, as an adaptation to the extremely unstable thermal regimens of their shallow-water habitats have evolved higher capacities for thermal tolerance than have prosobranchs.

There is some evidence that the thermal tolerance of isolated populations of freshwater pulmonates may become genetically adapted to local microenvironmental thermal regimes. For example, Beames and Lindeborg (1967) have reported a hot springs population of *P. anatina* that has an upper thermal limit (43°C) which was 2°C greater than that of a nearby river population (41°C). Similarly, McMahon (1976b) has demonstrated that individuals from a population of *P. virgata* isolated in a portion of a lake receiving thermal effluents for 18 years (54 generations) had a thermal tolerance averaging 2°C greater than that of snails from another isolated population in the same lake but unaffected by thermal effluents, when both were acclimated to 10, 20, or 30°C. Such physiologically distinct thermal races also appear to occur in natural populations of freshwater pulmonates. After acclimation to 20°C, the mean thermal tolerance (measured as mean heat coma temperatures) of 17 isolated Texan populations of *P. virgata* was found to vary between 35.6°C and 40.1°C (McMahon and Payne, 1980). Of 136 possible paired comparisons among the 17 populations, 78 (58%) were significantly different at the 90% probability level or greater ($p \leq .01$). There was a significant correlation between the thermal tolerance of these populations and their geographic distribution. Thermal tolerance increased 1.8°C over a 100.6-km east-to-west cline, which was associated with increasing maximum summer temperatures in the more westerly distributed habitats of this species in Texas (McMahon and Payne, 1980). More recently, thermal tolerances were determined similarly for 31 additional populations of *P. virgata* over a much larger 612 km north by 924 km west area including portions of Texas, Oklahoma, and Arkansas (Ibarra and McMahon, 1981). After acclimation to 20°C, mean heat coma temperature varied greatly among populations, ranging from 38.7 to 42.6°C, with 217 of 496 (43.8%) possible paired comparisons of population mean values being significantly different at the 90% probability level or greater (J. A. Ibarra and R. F. McMahon, unpublished observations). Mean thermal tolerance values increased significantly with the southerly and westerly distribution of populations from lotic (natural) habitats. These tolerance values were apparently correlated with clines of increasing temperature in these habitats to both the west and the south in the area where the populations were sampled. The positive relationship between thermal tolerance levels and the thermal regime of the microhabitats of these reproductively isolated populations of *P. virgata* (McMahon and Payne, 1980; Ibarra and McMahon, 1981) is an indication that high environmental temperatures can cause the selection of resistant phenotypes or genetically and physiologically distinct thermal races in more thermally stressed environments (McMahon and Payne, 1980).

C. Capacity Adaptations to Temperature

Capacity adaptations to temperature involve the adjustment of rate functions in relation to environmental temperature fluctuations (Bullard, 1964; Pohl, 1976).

Such regulation is generally related to the ability of an organism to compensate metabolically for changes in ambient temperature. Perhaps the rate function of ectothermic animals most directly related to metabolic rate and most sensitive to temperature is the oxygen consumption rate; therefore, the respiratory response of freshwater pulmonates to temperature is an important consideration in any discussion of their physiological ecology. In all freshwater pulmonates that have been examined, oxygen uptake rates increase with temperature until individuals are thermally stressed. Thereafter, uptake rates decline rapidly (Beames and Lindeborgh, 1967; Berg, 1952; Berg and Ockelmann, 1959; Berg et al., 1958; Burky, 1971; Calow, 1975b; Cheatum, 1934; Daniels and Armitage, 1969; McMahon, 1973; Sheanon and Trama, 1972).

Q_{10} is the factor by which metabolic rates increase over a 10°C increase in temperature. Prosser (1973c) notes that for ectotherms the Q_{10} values of most biological functions fall between 2.0 and 2.5 and that values below 2.0 are indicative of the increasing independence of metabolic rate from temperature (or of increasing temperature regulation of metabolic rates). The Q_{10} values corresponding to the respiratory response to temperature in a number of freshwater pulmonate and prosobranch species are displayed in Table I. It is notable that Q_{10} values of less than 2.0, indicative of some temperature regulation of metabolic rate, have been reported to occur over at least a portion of the normal ambient temperature range of 6 (of 10) pulmonate species, but for none of 7 species of freshwater prosobranchs, with the single exception of male individuals of *Viviparus contectoides* (W. G. Binney) (Buckingham and Freed, 1976). The apparent ability to partially regulate metabolic rates with acute temperature fluctuation in some species of freshwater pulmonates may be an adaptation that allows maintenance of relatively stable and efficient metabolic rates under the high diurnal ambient temperature fluctuations characteristic of their shallow habitats. This adaptation is similar to that displayed by certain species of intertidal gastropods and bivalves (McMahon and Russell-Hunter, 1977; McMahon and Wilson, 1981), including the high estuarine basommatophoran *Malampus bidentatus* (McMahon and Russell-Hunter, 1981). Lack of such regulation in freshwater prosobranchs (Table I) may be associated with their apparent preference for larger, deeper, and therefore more temperature-stable freshwater habitats.

The vast majority of freshwater pulmonates appear to be able to compensate or acclimate their metabolic rate to longer term, seasonal ambient temperature fluctuations when examined either in natural field populations or in controlled laboratory experiments. This compensation involves elevation of oxygen consumption rates in cold-acclimated individuals over those of warm-acclimated snails so that cold-acclimated individuals have higher metabolic or oxygen consumption rates than warm-acclimated snails at any one test temperature. These acclimatory responses to long-term temperature fluctuation are considered to be highly adap-

TABLE I

Q_{10} Values for Oxygen Consumption Rate in Response to Temperature Change in Freshwater Pulmonate and Prosobranch Snails

Species	Temperature range (°C)	Q_{10}	Reference
Pulmonates			
Lymnaea peregra	11–18	2.04	Berg and Ockelmann (1959)
Physa hawnii	10–15	1.02–3.21	Daniels and Armitage (1969)
	15–20	1.09–3.02	
	20–25	1.33–2.73	
Physa anatina	25–35	1.76–2.16	Beames and Lindeborg (1967)
Physa fontinalis	11–18	2.50	Berg and Ockelmann (1959)
Myxas glutinosa	11–18	2.50	Berg and Ockelmann (1959)
Helisoma trivolvis	5–15	3.14	Sheanon and Trama (1972)
	15–25	1.37	
Biomphalaria glabrata	15–25	1.84	Meakins (1980)
	25–35	1.57	
Ancylus fluviatilis	4–10	1.71–1.77	Calow (1975b)
	10–18	1.97	
	2–31	2.43	Berg (1952)
	7–15	2.23	Berg et al. (1958)
Acroloxus lacustris	6–31	2.13	Berg (1952)
Ferrissia rivularis	10–20	2.0–3.3	Burky (1971)
Laevapex fuscus	10–20	1.26–2.38	McMahon (1973)
Prosobranchs			
Bithynia leachi	11–18	2.41	Berg and Ockelmann (1959)
Bithynia tentaculata	11–18	2.13	Berg and Ockelmann (1959)
Potamopyrgus jenkinsi	12–29	2.32–2.40	Lumbye (1958)
	15–25	2.00–2.21	Duncan (1966)
Valvata piscinalis	13–21	2.08	Berg and Ockelmann (1959)
Theodoxus fluviatilis	12–29	2.07–3.01	Lumbye (1958)
Viviparus contectoides (males)	12–22	2.53	Buckingham and Freed (1976)
	17–27	1.91	
	22–32	1.61	
Vivaparus contectoides (females)	17–27	3.96	Buckingham and Freed (1976)
	22–32	4.28	
Goniobasis livescens	2–15	5.06	Payne (1979)
	15–22	2.89	

tive, because they allow maintenance of relatively stable metabolic rates (presumably near the most efficient levels) in widely fluctuating seasonal temperature regimes (so that the metabolic rates of cold-acclimated and warm-acclimated individuals are relatively similar at their respective acclimation temperatures) (Prosser, 1973c). Temperature acclimation also affects the short-term respiratory response to temperature, as reflected by changes in Q_{10} values (Table I).

Among freshwater pulmonates, such temperature acclimation of oxygen consumption rates or of other rate functions such as heartbeat (Harrison, 1977a,b) is never complete (i.e., showing absolute compensation, with the oxygen consumption of cold- and warm-acclimated individuals being equal at their respective acclimation temperatures) (Prosser, 1973c). Such incomplete respiratory acclimation to temperature is characteristic of most ectothermic animals (Prosser, 1973c; Pohl, 1976). Normal respiratory compensation for seasonal temperature fluctuations has been reported for *Planorbis contortus* (Calow, 1975b), *Physa hawnii* (Daniels and Armitage, 1969), *P. anatina* (Beames and Lindeborg, 1967), *B. glabrata* (Meakins, 1980), *Lymnaea peregra* and *L. palustris* (Berg and Ockelmann, 1959), and in the heartbeat rate of *L. stagnalis* (Harrison, 1977a).

Among freshwater pulmonates, the freshwater limpets of the family Ancyloplanorbidae (previously placed in a separate family, the Ancylidae) and the family Acroloxidae display the physiological phenomenon of *reverse* or *inverse* respiratory acclimation to temperature; when reverse-acclimated, warm-conditioned snails have oxygen consumption rates higher than those of cold-conditioned individuals at any one test temperature, rather than lower rates (the more normal condition) (Prosser, 1973c; McMahon and Russell-Hunter, 1981). Such reverse acclimation has been reported in four species of freshwater limpets: *Ancylus fluviatilis* (Berg, 1951, 1952, 1953; Berg and Ockelmann, 1959; Berg et al., 1958; Calow, 1975b), *Ferrissia rivularis* (Burky, 1969, 1971), and *Laevapex fuscus* (McMahon, 1972, 1973), all members of the family Ancyloplanorbidae, and *Acroloxus lacustris* of the family Acroloxidae (Berg, 1951, 1952, 1953; Berg and Ockelmann, 1959). Among molluscs other than freshwater limpets, reverse acclimation has been reported for the intertidal limpet *Patella vulgata* L. (Davies, 1965), the terrestrial stylommatophoran pulmonate *Helix pomatia* L. (Blazka, 1954), the primitive high salt marsh basommatophoran *Melampus bidentatus* (McMahon and Russell-Hunter, 1981), and the intertidal bivalves *Tellina tenuis* De Costa and *Cerastoderma edule* (L.) (McMahon and Wilson, 1981). Reverse respiratory acclimation to temperature has also been reported in barnacles (Barnes et al., 1963), freshwater gammarids (Krog, 1954), goldfish (Roberts, 1960, 1966), a hylid frog (Packard, 1972), and lizards (Davies et al., 1981).

As has been suggested by McMahon and Russell-Hunter (1981), reverse acclimation is an adaptation that appears to have been independently evolved in a variety of ectothermic species and phyletic groups in response to a wide range of selective pressures; therefore, it cannot be described in terms of a single unified adaptive advantage in all the species in which it occurs. A number of hypotheses regarding the adaptive significance of reverse acclimation have been proposed. They can be summarized as follows:

1. Reduction of oxygen demand in overwintering individuals is an adaptation to hypoxic overwintering microenvironments (Krog, 1954; Roberts, 1960, 1966).
2. Metabolic rate is dependent on growth rate, a higher metabolic rate in warm-conditioned individuals being due to rapid summer growth (Parry, 1978).
3. The resultant decrease in metabolic rate conserves overwintering energy stores in inactive, nonfeeding individuals (Burky, 1969, 1971; McMahon and Russell-Hunter, 1981; McMahon and Wilson, 1981; Packard, 1972).
4. Reverse acclimation is a manifestation of a more general, temperature-mediated respiratory compensation to hypoxic overwintering habitats involving a shift toward greater regulation of oxygen consumption in cold-conditioned individuals (McMahon, 1972, 1973).
5. The increased oxygen consumption of warm-acclimated individuals is caused by the onset of reproductive activity (Berg et al., 1958).
6. Reverse acclimation is an adaptation to aquatic environments characterized by low, rapidly fluctuating temperatures (Calow, 1975b).

In most species of molluscs, reverse acclimation certainly appears to be associated with overwintering estivation in which metabolic demands are reduced (as are growth rates; see Parry, 1978) to conserve overwintering energy stores at times when food is not available [as has been claimed for *P. vulgata* (Davies, 1965), *H. pomatia* (Blazka, 1954), intertidal bivalves (McMahon and Wilson, 1981), and *M. bidentatus* (McMahon and Russell-Hunter, 1981)]. However, the adaptive value of reverse respiratory acclimation is not as obvious in the freshwater limpets as it is in other molluscan species. Indeed, there are almost as many hypotheses concerning the adaptive significance of reverse acclimation in freshwater limpets as there are studies of the phenomenon. Berg et al. (1958) proposed that reverse acclimation was linked to increases in metabolic rate associated with the onset of reproductive behavior in the European stream limpet *Ancylus fluviatis* and the European pond limpet *Acroloxus lacustris*. However, later detailed studies of both laboratory- and field-acclimated specimens of *Ancylus fluviatilis* and *L. fuscus* indicated that oxygen consumption rates increased in warm-acclimated individuals well before spawning was initiated (Calow, 1975b; McMahon, 1972, 1973). Russell-Hunter and McMahon (1976) have also demonstrated that spermiogenesis occurs in winter months in *L. fuscus,* whereas oogenesis is initiated in the late spring and early summer, long after metabolic rates have begun to increase as a result of reverse acclimation to rising ambient water temperatures (McMahon, 1973). Further, maximal levels of reverse acclimation are achieved in July and August (periods of maximal water temperatures), when most *L. fuscus* populations consist of immature, nonreproductive individuals of the spring generation (McMahon, 1972, 1973, 1975a; Russell-Hunter and McMahon, 1976).

McMahon (1972, 1973) has demonstrated that reverse respiratory acclimation is associated with a temperature-induced, seasonal migratory behavior in the North American pond limpet *L. fuscus*. This species occurs on hard surfaces (boulders, macrophytic aquatic plants, and debris) in well-oxygenated microhabitats just below the sediment surface during the summer months, but migrates down the stems of macrophytes or the sides of boulders into deeper sediments (to avoid freezing as ice forms), where it overwinters in quite hypoxic conditions (McMahon, 1972, 1973, 1975a). Decreasing temperatures in autumn, which appear to trigger this migration, are also associated with a radical shift from a respiratory schesis in summer-conditioned individuals of *L. fuscus* characterized by oxygen conformity in response to declining oxygen tensions to one of strict oxygen regulation in winter-conditioned specimens. As a result of this shift toward oxygen regulation or independence, winter-conditioned specimens of *L. fuscus* have a lower rate of oxygen consumption than summer-conditioned specimens at full air saturation with oxygen (the pattern associated with reverse acclimation), but have much higher levels of oxygen consumption than summer-conditioned snails in the low oxygen tensions characteristic of their hypoxic overwintering habitats (a normal pattern of respiratory temperature acclimation) (McMahon, 1972, 1973). Therefore, reverse acclimation in *L. fuscus* appears to be of little real adaptive significance, but is instead an apparent manifestation of a more general temperature-induced compensatory respiratory response to the hypoxic microhabitats of overwintering individuals. These shifts between oxygen conformity and regulation of oxygen consumption appear to be mediated by environmental temperature alone, because they can be reproduced in the laboratory by cold-acclimating or warm-acclimating individuals of *L. fuscus* maintained under constant aeration and constant light (McMahon, 1972, 1973).

Although the value of reverse acclimation as a compensatory mechanism for hypoxic overwintering microenvironments in *L. fuscus* appears clear, the adaptive significance of reverse acclimation in species such as *A. fluviatilis* or the North American stream limpet *F. rivularis,* which inhabit lotic environments characterized by high levels of oxygenation, remains somewhat uncertain. Burky (1969, 1971) has claimed that reverse acclimation in *F. rivularis* is associated with conservation of energy stores in estivating, nonfeeding, overwintering individuals. However, reduction of oxygen uptake in species inhabiting more hypoxic overwintering environments could cause a partial dependence on glycolysis to meet metabolic demands and would accelerate catabolism of energy stores rather than conserve them (McMahon, 1972, 1973). Calow (1975b) has proposed that reverse acclimation is an adaptation to the low, rapidly fluctuating temperature regimes of the small lotic habitats of *A. fluviatilis,* which causes Q_{10} values for respiratory response to increase with temperature in acclimated specimens and allows *A. fluviatilis* to maintain high levels of activity during short-term increases in ambient temperature. This hypothesis, although suitable for lotic species such as *A. fluviatilis* and *F. rivularis,* appears to have little relevance for the adaptive significance of reverse acclimation in pond species such as *L. fuscus* and *A. lacustris,* whose impounded, lentic marsh and standing-water habitats are generally characterized by high ambient temperatures (Hubendick, 1962b; McMahon, 1972, 1973, 1975a).

Although reverse respiratory acclimation has, no doubt, been independently evolved in a number of mollusc species in response to different selective pressures, it seems unrealistic to assume that it could have so many different adaptive functions in such a highly monophyletic group as the freshwater limpets of the family Ancyloplanorbidae (excluding the family Acroloxidae) (Hubendick, 1978). Reverse respiratory acclimation to temperature in this group requires further intensive investigation, including detailed examinations of the overwintering behavior and microenvironments of both lentic and lotic species of freshwater limpets, before the adaptive and evolutionary significance of reverse or inverse acclimation (termed "paradoxical" by Pretch et al., 1973) in this group can be completely understood.

D. Behavioral Responses and Adaptations to Temperature

Freshwater pulmonates display a variety of behavioral responses to temperature. These adaptations involve shifts in the period of submergence and the frequency of renewal of pulmonary air stores at the air–water interface, temperature selection behavior, variations in the extent of locomotor activity, and longer-term migratory activity associated with seasonal temperature fluctuations.

It has long been known that temperature directly affects the diving behavior of freshwater pulmonates. The frequency of return to the surface to renew pallial air stores increases with temperature, presumably as a result of the increasing dependence on pulmonary oxygen consumption to meet temperature-induced elevations in oxygen demand (Cheatum, 1934). The reduced oxygen tensions associated with elevated ambient temperatures have also been shown to increase the frequency of surfacing behavior (Jones, 1961, 1964b; Janse, 1981). Such temperature-induced increases in the frequency of return to the surface may partially account for the reported restriction of pulmonates to very shallow, nearshore habitats during warm summer months when reduced dive duration limits snails to excursions to only the most minimal depths (Cheatum, 1934; Clampitt, 1970, 1974; Russell Hunter, 1953a) even when such this respiratory limitation causes migration away from food resources as water levels rise (Russell Hunter, 1953a).

Distinct diurnal periodicity in behavior patterns has been reported for several species of freshwater pulmonates, in which activity, duration of submergence, preferred temperatures, and oviposition have all been shown to increase during dark hours (Appleton, 1976b; Kavaliers, 1980, 1981; McDonald, 1973). Increases in locomotory activity and duration of submergence during dark hours have been reported for *Bulinus globosus* (Morelet) (Appleton, 1976b), *Lymnaea stagnalis* (McDonald, 1973), and *Helisoma trivolvis* (Kaveliers, 1981). Diurnal and crepuscular activity rhythms (activity greatest during dark hours in isolated individuals and in light–dark or dark–light transitions in groups of individuals) in *H. trivolvis* were demonstrated both to be entrainable to artificial light–dark cycles and to be free running under conditions of constant light or dark, an indication of the endogenous nature of these rhythms in this species (Kaveliers, 1981). McDonald (1973) has demonstrated that under constant temperature (20°C), the durations of submergence, activity, and feeding all increase during dark periods in *L. stagnalis*.

Although light–dark periods appear to be associated with submergence and activity patterns in freshwater pulmonates, daily fluctuations in temperature may also be involved in the maintenance of diurnal behavioral rhythms. Cheatum (1934) demonstrated in nine species of freshwater pulmonates that short-term decreases in temperature reduce surfacing frequency and stimulate excursions to greater depths. It has been suggested that decreases in metabolic demand at lower temperatures allow freshwater pulmonates to support a greater proportion of their oxygen consumption by cutaneous respiration, which, in turn, allows longer periods of submergence and penetration to greater depths (Russell Hunter, 1953a). Therefore, decreases in the ambient temperature of natural habitats during dark hours (see Section V,A) may allow freshwater pulmonates to move to and feed at greater depths during the evening and, along with increased oviposition during dark hours, may partially explain the observation that freshly laid egg-masses of shallow-water species such as *P. virgata* are often found in abundance at greater depths and distances from the shore than is the adult population during daylight hours (R. F. McMahon, unpublished observations).

Pulmonates also appear to select temperatures higher than the acclimation temperature when placed in a thermal gradient (Clampitt, 1970; Sodeman and Dowda, 1974). In addition, a circadian rhythm of temperature selection has been observed in *H. trivolvis,* which moves toward warmer temperatures in a thermal gradient during dark hours (light = 17–18°C, dark = 21–22°C) (Kavaliers, 1980). Therefore, diurnal migrations and activity patterns in freshwater pulmonates could also be partially accounted for by movements toward rapidly warming, shallow, nearshore waters during the day and migration away from the shore to warmer, deeper waters at night when shallow, nearshore waters cool rapidly.

Pulmonates are also known to make annual migrations to deeper waters to overwinter, presumably an adaptation that allows them to avoid being frozen when ice covers their shallow-water, summer habitats (Boerger, 1975; Boag and Bentz, 1980; Cheatum, 1934; Clampitt, 1970, 1974; Rowan, 1966; Russell Hunter, 1953a), although Cheatum (1934) reports that many species of freshwater pulmonates are capable of surviving for extended periods while frozen in ice. In a special case, the less mobile freshwater limpet *Laevapex fuscus* migrates deeply into the substratum, down the sides of boulders and stems of aquatic macrophytes to overwinter (McMahon, 1972, 1973). Such seasonal migrations appear to be almost universally induced by temperature. Laboratory experiments have shown that many species of freshwater pulmonates respond to declining temperatures by moving to greater depths. Boag and Bentz (1980) showed that at temperatures less than 9°C, *Lymnea stagnalis* displayed a distinct preference for greater depths which reached a maximum at temperatures below 5°C, the snails becoming immobile in deeper waters at temperatures of 2°C or less. It has been proposed that seasonal movements toward greater depths may be initiated when environmental temperatures fall to levels at which metabolic demands can be completely supported by cutaneous respiration (Russell Hunter, 1953a). Thus, freed from the necessity of returning to the surface, snails move to greater depths. Similarly, increases in oxygen demands with increasing temperatures may also be responsible for the initiation of spring migrations to the water's edge, allowing a return to pulmonary respiration (Russell Hunter, 1953a). In at least one species of freshwater pulmonate (*Laevapex fuscus*), such temperature-induced seasonal migrations are associated with extensive compensatory adjustments in the respiratory responses to temperature and hypoxia (McMahon, 1972, 1973; see section V,C). Further studies on the physiological basis of the diurnal and seasonal behavioral rhythms of freshwater pulmonates could provide valuable new information about the behavioral physiology of this group.

E. Life History Adaptations to Temperature

Temperature has been shown to affect directly life history traits such as growth rate, age of maturity, and fecundity levels in freshwater pulmonates. Growth rates have been shown to be directly related to temperature in natural populations (Burky, 1969, 1971; Clampitt, 1970; DeWitt, 1955; Eversole, 1974; Hunter, 1972, 1975a; McMahon, 1972a, 1975a,b, 1976a; Nickerson, 1972; Romano, 1980; Russell Hunter, 1953b, 1961a,b; Streit, 1975). Growth rates are minimal or nonexistent during cold winter months and increase to a maximum in the summer, with peak rates generally corresponding to periods of maximum ambient temperature. Laboratory studies have confirmed that growth rates rise with increasing temperatures in most freshwater pulmonates (Brown, 1979; Calow, 1973; Imhof, 1973; Prinsloo and van Eeden, 1973; van der Schalie and Berry, 1973). However, certain species of *Lymnaea* appear to be exceptions to this generalization, as their growth rates are reported to reach maximum levels at rather low temperatures (15–18°C) (Mattice, 1976; Prinsloo and van Eeden, 1973). Such results correspond directly to the elevated growth rates reported for natural populations of freshwater pulmonates in habitats where ambient temperatures are maintained at artifically high levels by thermal effluents (McMahon, 1975b; Sankurathri and Holmes, 1976). Age of maturity and first reproduction are also reported to decrease with increasing temperature in freshwater pulmonates (Brown, 1979; Calow, 1973).

In most freshwater pulmonates studied, the initiation of oogenesis, copulation, and oviposition appears to be stimulated by a spring increase in temperature above a critical lower limit. Some critical spawning temperatures reported for freshwater pulmonates include: *Ferrissia rivularis,* 10°C (Burky, 1971); *Ancylus fluviatilis,* 7–13°C (Geldiay, 1956; McMahon, 1980; Streit, 1975); *Physa gyrina,* 7–12°C (Clampitt, 1970; DeWitt, 1955; Duncan, 1959); *P. fontinalis,* 7–8°C (deWit, 1955); *P. virgata,* 13°C (McMahon, 1975b); *Lymnaea palustris,* 15°C (Hunter, 1975a); *L. emarginata,* >13°C (van der Schalie and Berry, 1973); *Helisoma trivolvis,* 10°C (Eversole, 1978); *H. anceps,* >13.5°C (van der Schalie and Berry, 1973); *Laevapex fuscus,* 16–18°C (McMahon, 1975a, 1976a); *Hebetancylus excentricus,* 15°C (McMahon, 1976a); and *Acroloxus lacustris,* 14–15°C (McMahon, 1980). Above the lower critical limit for spawning, fecundity levels increase with temperature in most freshwater pulmonate species (Brown, 1979; Mattice, 1976; van der Schalie and Berry, 1973; Streit, 1975).

Laboratory investigations have also shown that the initiation of spawning in most freshwater pulmonates is independent of the length of the photoperiod (DeWitt, 1967; Imhof, 1973). McMahon (1972, 1975a) demonstrated that specimens of *L. fuscus* kept in constant light initiated oviposition at the same temperature (18°C) as did natural field populations. *Physa pomilia* Conrad and *Pseudosuccinea columella* became reproductively active at the same time whether reared under normal light–dark cycles or in constant darkness (DeWitt, 1967). Jenner (1951), however, has suggested that a minimal light–dark cycle of 13.5 light : 10.5 dark is required for oviposition in *Lymnaea stagnalis.* Although generally not associated with the initiation of spawning, length of photoperiod appears to affect fucundity in certain species of freshwater pulmonates; however, the effects of day length are different in various species. Increased day length stimulated spawning in *L. stagnalis* and *L. peregra* but did not affect fecundity in *Planorbarius corneus* or *Planorbis planorbis* (L.) (Imhof, 1973). In contrast, the egg production in *Physa pomilia* was 3–30-fold greater in specimens kept in constant darkness as opposed to those exposed to a normal photoperiod (DeWitt, 1967). Although such data indicate that day length may affect fecundity rates in some species of freshwater pulmonates, temperature appears to be the primary environmental stimulus for both the initiation and cessation of spawning. Temperature control of spawning in freshwater pulmonates has been demonstrated by an observed short-term cessation in oviposition of *F. rivularis* as ambient temperatures fell below the lower limit for spawning of 10°C for a short time in the middle of the spawning period of two natural populations (Burky, 1969, 1971), and by a report of continual, year-round spawning in a population of *P. gyrina* in a habitat receiving thermal effluents in which ambient temperatures remained above the lower critical limit for spawning throughout the year (Sankurathri and Holmes, 1976).

Northern temperate species of freshwater pulmonates generally spawn at lower ambient temperatures than do freshwater prosobranchs. Although many freshwater pulmonates spawn at temperatures below 15°C, the majority of freshwater prosobranchs require higher temperatures for oviposition (Aldridge, 1980, 1982; McMahon, 1980; Payne, 1979). Low spawning temperatures allow freshwater pulmonates to commence oviposition as soon as water temperatures begin to rise in the spring (Russell-Hunter, 1964, 1978). Early spawning leads to early hatching and a long period of summer growth, with some populations growing fast enough to have a second reproductive period in the late summer or fall (Russell-Hunter, 1964, 1978). Therefore, early spawning allows a high proportion of the population to overwinter as large, mature, relatively cold-resistant individuals that are ready to reproduce the following spring. Such early reproduction conforms to the general pattern of life history traits of freshwater pulmonates, whose semelparity (one reproductive effort in a lifetime), short life spans, early age of maturity (they are generally annuals), high growth rates, and high reproductive capacities (Russell-Hunter, 1964, 1978) are considered to maximize the production and survival of offspring in highly variable habitats such as the shallow, marginal environments of basommatophorans, where chances for survival to a second reproductive period are low (Stearns, 1976, 1977). Conversely, the longer life spans (>2 years), delayed maturity, lower growth rates, lower fecundity, and interoparity (more than one reproductive effort in a lifetime) of freshwater prosobranchs (Aldridge, 1980, 1982; McMahon, 1980; Payne, 1979) are generally considered to be adaptations that maximize production and survival of offspring in stable habitats such as the large, deep-water environments of most prosobranchs, where the chances of survival to future reproductive periods are relatively high (Stearns, 1976, 1977).

In light of the general tendency for reproduction at low ambient temperatures in freshwater basommatophorans, it is interesting that one group of freshwater pulmonates, the freshwater limpets inhabiting standing or impounded waters, have relatively high spawning temperatures compared to other species of freshwater pulmonates, specifically, closely related stream and river species of freshwater limpets. Thus, the North American pond limpet *L. fuscus* initiates spawning at 16–18°C (McMahon, 1972, 1975a, 1976a), whereas the North American stream limpet *F. rivularis* spawns at temperatures above 10°C (Burky, 1969, 1971). In Europe a similar pattern occurs with the pond species *Acroloxus lacustris* spawning at temperatures above 14–15°C (McMahon, 1980), whereas the stream species *Ancylus fluviatilis* oviposits at temperatures as low as 7°C (McMahon, 1980; Streit, 1975). The characteristically high spawning temperatures of pond limpets may be an adaptation preventing premature spawning during the occasional periods of elevated temperatures that can occur in shallow, standing-water habitats during unusually warm periods in winter months. If pond species retained the very low spawning temperatures of their stream relatives, it is likely that oviposition could be initiated during transient, midwinter warm spells, causing the loss of both valuable overwintering energy stores and reproductive effort when the temperature subsequently returns to normal winter levels that are incapable of supporting embryonic development. Accordingly, evolution of relatively elevated spawning temperatures appears highly adaptive, as it precludes the possibility of oviposition in pond limpets during periods when water temperature could subsequently fall below optimal levels for embryonic development and juvenile survivorship.

VI. Adaptations to Low Environmental Oxygen Concentration

A. Resistance Adaptations to Anoxia

In the shallow, eutrophic, standing-water habitats of many freshwater pulmonates, exposure to hypoxic conditions is frequent (Jones, 1961, 1964b; Boycott, 1936; Russell-Hunter, 1964, 1978). Species such as *Planorbarius corneus, Helisoma trivolvis, Biomphalaria sudanica,* and *Laevapex fuscus* are reported to crawl over or burrow into reducing bottom muds (Boerger, 1975; Clampitt, 1972; Jones, 1961, 1964b; McMahon, 1973, 1975a; Rowan, 1966), where oxygen tensions can be extremely low. When exposed to waters of low oxygen tension, some species of freshwater pulmonates generally increase their dependency on pulmonary respiration, returning more frequently to the surface to renew pulmonary air stores and reducing both the depth and the duration of underwater activity (Janse, 1981; Jones, 1961, 1964b). The ability to respire in air and the diving habit are adaptations which presumably allowed basommatophorans to penetrate shallow, hypoxic, standing-water habitats during the course of their evolution. However, such responses to hypoxia can also result in the restriction of populations of freshwater pulmonates to the air–water interface during warm summer months, when low environmental oxygen tensions are commonly encountered (Russell Hunter, 1953a). The level of dependency on pulmonary respiration with declining oxygen tension appears to be associated with the degree of aquatic readaptation and is, therefore, somewhat species-specific. Thus, the pulmonary oxygen consumption rate of *Lymnaea stagnalis* (which is a highly aerial, amphibious species) surpasses cutaneous uptake rates at medium oxygen concentrations below 140 torr, a partial pressure nearly equivalent to full air saturation (159 torr), whereas in the more aquatic *P. corneus,* pulmonary oxygen uptake does not exceed cutaneous uptake until medium oxygen tensions fall below 40 torr (Jones, 1961).

Freshwater pulmonates can also be exposed to long periods of hypoxia in conditions that preclude pulmonary respiration, as occurs when ice cover prevents oxygen diffusion into overwintering habitats (Krog, 1954; McMahon, 1972, 1973) or when the drying out of temporary habitats induces burrowing into hypoxic bottom sediments (Cheatum, 1934; Olivier, 1956; Olivier and Barbosa, 1955a,b, 1956; see Section VII). It is not surprising, then, that the majority of freshwater pulmonates have a relatively high capacity for facultative anaerobic metabolism and are relatively tolerant of anoxia. Of 15 species of freshwater pulmonates, all were found to tolerate anoxia for periods greater than 6 h, with some species of Planorbinae and freshwater limpets able to tolerate anoxia for well over 3 days (Berg, 1952; von Brand et al., 1950; McMahon, 1972, 1973). Resistance adaptations to anoxia are not uniformly distributed among the major groups of hygrophilous basommatophorans. The more aerial Lymnaeidae and Physidae are far less tolerant of anoxia (<16–24 h) than are the more aquatic Planorbinae (>24–64 h) (von Brand et al., 1950). The totally aquatic freshwater limpets have the highest toleration of anoxia, lotic forms such as *A. fluviatilus* surviving up to 4 days and lentic species such as *L. fuscus* and *A. lacustris* surviving up to 4–11 days of anaerobiosis (Berg, 1952; McMahon, 1972, 1973). Like the Planorbinae and freshwater limpets, freshwater prosobranchs generally have high tolerances to anoxia (von Brand et al., 1950). Of 13 species of freshwater pulmonates, all were found to produce lactic acid during anoxia, but it was produced in larger amounts by the hypoxia-intolerant lymnaeids and physids, accounting for 5–15% of the carbohydrates metabolized in the 6 species tested. Lactic acid was utilized as an anaerobic end product to a much lesser extent by the more resistant planorbids, in which it accounted for only 1–7% of the carbohydrates metabolized in the majority of species tested (von Brand et al., 1950). Further, although lactic acid was excreted to the medium by all tested species, lymnaeids and physids were found to retain significant amounts in their tissues, whereas none could be detected in planorbids (von Brand et al., 1950).

Lactic acid was found to be the major anaerobic end product in only 2 of 13 tested species of freshwater pulmonates, *Lymnaea stagnalis* and *L. natalensis* Krauss. Instead, the major end products of anaerobic metabolism were the volatile acids, acetic acid and propionate, which were produced in a ratio of 2 : 1 (Mehlman and von Brand, 1951). Of 5 species tested, only 1 showed a large increase in carbohydrate metabolism during anoxia, indicating that most species of freshwater pulmonates are able to reduce metabolic demands to conserve the stored energy utilized during periods when metabolic energy must be produced by relatively inefficient anaerobic metabolic pathways (von Brand et al., 1950).

The production of volatile acids such as propionate as the primary anaerobic waste products indicates the presence of alternative anaerobic metabolic pathways in freshwater pulmonates as is common in other facultatively anaerobic marine molluscs and freshwater invertebrates. In these alternative anerobic metabolic pathways, the ATP yield is greater than that of classic vertebrate glycolytic pathways and the major metabolic end products are substances such as propionate, succinate, and alanine which are less toxic than lactic acid. These end products are not retained in the body fluids but are generally excreted to the external medium (Chen and Awapara, 1969a,b; Hochachka and Mustafa, 1972; Hochachka et al., 1973; Livingstone, 1978; Malanga and Aiello, 1972; McManus and James, 1975a,b,c; Stokes and Awapara, 1968; Wieser, 1980; de Zwaan and Zandee, 1972; de Zwaan and van Marrewijk, 1973; de Zwaan and Wijsman, 1976, and references therein).

It is apparent that there is a relationship between resistance adaptation to anoxia in species of freshwater pulmonates and their phylogenetic origins within the hygrophilous basommatophora. More primitive lymnaeids and physids, which rely to a great extent on pulmonary respiration (Jones, 1961; Russell Hunter, 1953c) and are characteristically amphibious or, if aquatic, inhabitats of shallow, nearshore waters, appear to depend on increased pulmonary respiration to tolerate short-term exposures to hypoxic waters (Jones, 1961). Thus, they are rarely exposed to hypoxic conditions and are relatively intolerant of anoxic stress, under which they produce relatively high levels of lactic acid and allow anaerobic metabolic wastes to accumulate in their tissues and hemolymph to toxic levels (von Brand et al., 1950). In contrast, the more advanced planorbids are highly aquatic and penetrate to greater depths; therefore, they are much more likely to be exposed to extended periods of hypoxia. As a result of adaptation to a more aquatic schesis, planorbids tend to be highly tolerant of anoxia, to suppress glycolysis in favor of alternative, more efficient anaerobic pathways, and to excrete anaerobic waste products rather than allow them to accumulate to deleterious levels in the hemolymph and tissues (von Brand et al., 1950; Mehlman and von Brand, 1951). Presumably, these adaptations occur to an even greater extent in the totally aquatic freshwater limpets, which appear to be the most tolerant of anoxia of all freshwater pulmonates (Berg, 1952, McMahon, 1972, 1973).

B. Capacity Adaptations to Hypoxia

Most species of freshwater pulmonates exhibit considerable regulation of oxygen consumption rates with declining ambient oxygen concentration from air saturation (oxygen independency). The degree of oxygen regulation in different species can be compared by expressing oxygen uptake rates ($\dot{V}O_2$) as a fraction of the rate at nearly full air saturation (PO_2 = 159 torr) and fitting the derived values of $\dot{V}O_2$ versus PO_2 in torr units to a quadratic equation. The resulting quadratic coefficient (b_2) is a predictor of the degree of regulatory (independence) of oxygen uptake from oxygen concentration. Expressed as b_2 (10^3), it becomes increasingly negative as the degree of regulation increases. Values approaching zero indicate nonregulation (oxygen dependence), and values approaching −0.1 indicate nearly perfect regulation (oxygen independency) (Mangum and Van Winkle, 1973). Previously published data on the respiratory response to hypoxia in freshwater pulmonate and prosobranch snails were subjected to this analysis; and the results, including b_2 (10^3) values, are presented in Table II. Perhaps somewhat surprisingly, freshwater pulmonates appear to be relatively good regulators of oxygen uptake rates. Of nine pulmonate species for which data are available, eight displayed some degree of oxygen regulation [b_2 (10^3) < -0.02], and five showed good to nearly perfect regulation of oxygen uptake rates [b_2 (10^3) < -0.07] (Table II). Only one species of freshwater pulmonate, *A. lacustris,* was an oxygen conformer [b_2 (10^3) > -0.02] incapable of regulating oxygen consumption with declining oxygen tensions (Berg, 1952). In contrast, freshwater prosobranchs appear to be generally poorer regulators of oxygen consumption, with four of six tested species being oxygen conformers [b_2 (10^3) > -0.02] (Table II) and only two species showing moderate to good regulation of oxygen consumption (Table II). Data for one species, *Viviparus contectoides* (Studier and Pace, 1978), could not be subjected to quadratic analysis in its published form. The adaptational significance of the high capacity of freshwater pulmonates to regulate oxygen consumption may lie in the relatively high levels of hypoxia that can occur in their shallow, often eutrophic habitats (Russell-Hunter, 1964, 1978). Most prosobranchs are more generally associated with large, open, deeper aquatic habitats in which severe hypoxia is rarely encountered (Russell Hunter, 1964; Aldridge, 1983; see Chapter 8, this volume). However, environmental correlations with the respiratory response to hypoxia must be addressed warily, because some species of molluscs, including the freshwater limpet *Laevapex fuscus* (McMahon, 1972, 1973), can greatly modify their respiratory response to hypoxia in response to short-term fluctuations in ambient temperature (McMahon, 1972, 1973; McMahon and Wilson, 1981) and oxygen concentration (Bayne, 1975; Bayne and Livingstone, 1977; McMahon, 1972, 1973; McMahon and Russell-Hunter, 1978, 1981; Shumway, 1981).

The oxygen uptake rates of freshwater pulmonates generally increase after periods of anoxia. In four of five tested species, postanoxic oxygen consumption rates remained elevated for 3 to 7 h, a pattern characteristic of classic oxygen debt payment (von Brand and Mehlman, 1953). Repayment of oxygen debt in these species was incomplete (von Brand and Mehlman, 1953), which is not surprising in light of the high capacity of freshwater pulmonates to excrete anaerobic wastes (von Brand et al., 1950; Mehlman and von Brand, 1951).

TABLE II

Quadratic Coefficients [b_2 (10^3)], Relating the Degree of Regulation of Oxygen Consumption to Declining Oxygen Concentration for Freshwater Pulmonate and Prosobranch Gastropods[a]

Species	Temperature (°C)	Special conditions	b_2 (10^3)	Degree of regulation	Reference
Pulmonates					
Lymnaea peregra	18		−0.0537	Moderate	Berg and Ockelmann (1959)
Lymnaea palustris	18		−0.0348	Poor	Berg and Ockelmann (1959)
Myxas glutinosa	18		−0.0871	Nearly perfect	Berg and Ockelmann (1959)
Lymnaea auricularia	18		−0.0792	Good	Berg and Ockelmann (1959)
Physa fontinalis	18		−0.0758	Good	Berg and Ockelmann (1959)
Biomphalaria glabrata	30	Before anoxia	−0.0702	Good	von Brand and Mehlman (1953)
		After 16 h of anoxia	−0.0803	Good	
Ancylus fluviatilis	16		−0.0970	Nearly perfect	Berg (1952)
Acroloxus lacustris	16		+0.0143	Nonregulator	Berg (1952)
Laevapex fuscus	20	Before anoxia	−0.0639	Moderate	McMahon (1973)
		After 37 h of anoxia	−0.0485[b]	Nearly perfect[b]	
		After 80 h of anoxia	−0.0485[b]	Nearly perfect[b]	
Prosobranchs					
Theodoxus fluviatilis	17–22	Fresh water	−0.0150	Nonregulator	Lumbye (1958)
		Brackish water	−0.0098	Nonregulator	
Potamopyrgus jenkinsi	17–22	Fresh water	+0.0308	Nonregulator	Lumbye (1958)
		Brackish water	−0.0083	Nonregulator	
Bithynia tentaculata	18		−0.0108	Nonregulator	Berg and Ockelmann (1959)
Bithynia leachi	18		−0.0401	Poor	Berg and Ockelmann (1959)
Valvata piscinalis	18		−0.0628	Moderate	Berg and Ockelmann (1959)

[a] b_2 (10^3) values relate to the degree of regulation of oxygen consumption as follows: < -0.085, nearly perfect regulation; −0.070 to −0.0850, good regulation; −0.050 to −0.0699, moderate regulation; −0.020 to −0.0499, poor regulation; and > -0.020, nonregulation (Mangum and Van Winkle, 1973).

[b] The degree of regulation is beyond the range of applicability to a quadratic model. Therefore, even though these b_2 (10^3) values are low, they represent nearly perfect regulation (Mangum and Van Winkle, 1973).

In their studies of the postanoxic oxygen consumption of freshwater pulmonates, von Brand and Mehlman (1953) observed a second, somewhat anomalous pattern of respiratory response to anoxia in which the respiration rate of postanoxic individuals at first remained low for 1 to 3 h after return to full air saturation and subsequently rose and remained well above preanoxic levels for periods which were far longer than could be accounted for by oxygen debt payment. Similar respiratory responses to short-term (5 h) exposure to anoxia have been reported for several species of marine prosobranch snails (McMahon and Russell-Hunter, 1978) and for the high salt marsh estuarine basommatophoran gastropod *Melampus bidentatus* (McMahon and Russell-Hunter, 1981). In these species, this pattern of delayed respiratory response to anoxia was not associated with oxygen debt payment, but rather with a compensatory change in the respiratory behavior of posthypoxic individuals, in which oxygen uptake rates were both greatly elevated at all oxygen tensions and decidedly shifted toward strict regulation or independency of oxygen consumption (McMahon and Russell-Hunter, 1978, 1981). These compensatory adjustments in posthypoxic individuals are highly adaptive, allowing hypoxically stressed snails to maintain oxygen consumption rates close to normal prestress levels at full air saturation under very hypoxic conditions ($PO_2 < 15$–30 torr) (McMahon, 1972, 1973; McMahon and Russell-Hunter, 1978, 1981).

Such acclimatory adjustment of the respiratory response to hypoxic stress was first described in the North American pond limpet *L. fuscus* (McMahon, 1972, 1973). This species inhabits the portions of leaves and stems of aquatic macrophytes and of boulders that extend below the substratum, a microenvironment in its standing-water habitats in which it can be exposed to extended periods of hypoxia (McMahon, 1972, 1973). After a 37-h period of hypoxia, the oxygen consumption rates of specimens of *L. fuscus* were double those of prehypoxic specimens over the full range of oxygen tensions below air saturation, and had shifted from moderate regulation in prehypoxic individuals to extreme oxygen independency (Table II). Posthypoxic individuals maintained oxygen consumption rates equal to or greater than those of prehypoxic individuals at full air saturation ($PO_2 = 159$ torr) at PO_2 levels as low as 6.9 torr (4.3% of full air saturation with oxygen) (McMahon, 1972, 1973). When subjected to a longer, 80-h period of hypoxia, the oxygen uptake rates of posthypoxic snails were 5.5-fold greater than those of prehypoxic controls at full air saturation. Again, the pattern was characterized by a shift to extreme regulation of oxygen uptake rates, so that the oxygen uptake of posthypoxic specimens was equal to or greater than that of prehypoxic individuals in full air saturation at PO_2 levels as low as 4.3 torr (2.7% of full air saturation) (McMahon, 1972, 1973). Such elevated posthypoxic oxygen uptake rates and high regulation of oxygen consumption were retained in posthypoxic individuals for more than 135 h after return to air-saturated water. This period was far too long to be accounted for by repayment of an oxygen debt (McMahon, 1972, 1973), because the majority of anaerobic wastes appear to be excreted rather than metabolized in freshwater pulmonates (von Brand et al., 1950; Mehlman and von Brand, 1951). Such respiratory compensation for hypoxia, involving progressively higher rates of oxygen uptake at all oxygen tensions and shifts to progressively higher levels of oxygen regulation in response to periods of hypoxic stress of increasing duration, has subsequently been shown to occur in the primitive estuarine basommatophoran *Amphibola crenata* (Gmelin) (Shumway, 1981), and the intertidal mussel *Mytilus edulis* L. (Bayne, 1975; Bayne and Livingstone, 1977).

As pointed out by McMahon and Russell-Hunter (1978), aquatic habitats are rarely completely anoxic, but are frequently hypoxic. Although most freshwater pulmonates have the capacity for facultative anaerobiosis (see Section VI,A), the ATP yield from anaerobic metabolism (even from the relatively efficient alternative anaerobic pathways which occur in molluscs) is far less than that produced by aerobic metabolism and oxidative phosphorylation (Hohachka et al., 1973; de Zwaan and Wijsman, 1976). Therefore, the ability to compensate for hypoxia by increasing the oxygen uptake rate and shifting to greater regulation of oxygen consumption is of the highest adaptive value, allowing snails to remain fully aerobic at efficient metabolic rates and to maintain normal levels of activity including feeding, and preventing partial or complete dependence on energetically inefficient anaerobic metabolism in all but the lowest environmental oxygen tensions (McMahon, 1972, 1973; McMahon and Russell-Hunter, 1978, 1981). Consequently, the greatly elevated oxygen consumption rates of posthypoxic individuals at oxygen tensions near full air saturation are merely one manifestation of the more general acclimatory increase in respiration rates at lower oxygen tensions to efficient prehypoxic levels (McMahon, 1972, 1973; McMahon and Russell-Hunter, 1978, 1981).

It is rather difficult to account for such extensive compensatory adjustments of respiratory response to hypoxia in *L. fuscus* and other molluscs in terms of adjustments in circulation rate and gill perfusion to ventilation ratios because these changes should be able to occur instantaneously and should not result in the observed greatly elevated respiration rates at high ambient oxygen tensions. Instead, such compensation may represent more fundamental physiological changes at the tissue or cellular level in posthypoxic individuals. This hypothesis has recently been addressed and partially confirmed by stereological electron microscopic examinations of the mitochondria of pedal muscle cells of specimens of *L. fuscus* before, immediately after, and 24 h after return to fully air-saturated water following a 24-h period of hypoxia. After 24 h of hypoxia, the mitochondrial volume of pedal muscles increased fivefold and the cristae surface area threefold over those of prehypoxic controls. These changes were associated with both a decided increase in the oxygen consumption rate and a shift to strict regulation of oxygen consumption compared to prehypoxic control specimens.

Twenty-four hours after return to fully air-saturated water, the mitochondrial volume of posthypoxic specimens remained 200% and cristae surface area 35% greater than those of prehypoxic control specimens, associated with a retention of a high degree of oxygen regulation (R. F. McMahon and W. R. Fagerberg, unpublished observations). Similar cellular responses have been reported in terrestrial vertebrates after much longer exposures to hypoxia at high altitudes (Monge and Whittembury, 1976). In *L. fuscus* these cellular changes in mitochondrial volume and internal membrane surface area are presumed to be adaptations which would increase the aerobic capacity of cells and tissues, allowing the maintenance of relatively stable aerobic metabolic rates under the reduced tissue PO_2 levels associated with a decline in ambient oxygen tensions. Such hypoxia-induced enhancement of cellular aerobic capacity in *L. fuscus* and other molluscs displaying respiratory compensation for low environmental oxygen concentrations could also account for the greatly elevated metabolic rates observed on a return from hypoxia to fully aerobic conditions.

VII. Aerial Exposure and Desiccation Resistance

Many species of freshwater pulmonates are amphibious. Other species inhabit temporary fresh waters, in which they are exposed to air for extended periods of time (Boycott, 1936; Brown, 1979, 1982; Cheatum, 1934; Clampitt, 1970; Jokinen, 1978; Newman and Thomas, 1975; Olivier and Barbosa, 1955a,b; Richardot, 1977a). Also, even species inhabiting permanent fresh waters can be exposed to air by abrupt decreases in water level which leave them stranded onshore (Russell Hunter, 1953a). Freshwater pulmonates normally associated with permanent habitats can survive and reproduce in temporary fresh waters (Boycott, 1936; Clampitt, 1970; Olivier and Barbosa, 1955a,b). The high degree of plasticity in the life history traits of some species is an apparent adaptation to life in temporary habitats (Brown, 1979, 1982).

Freshwater pulmonates are generally highly tolerant of short-term aerial exposure and high levels of evaporative water loss, a probable adaptation to their shallow-water, marginal environments and diving habit. When exposed in air, *B. glabrata* withstands a 70% reduction of its total body water (von Brand et al., 1957). This level of desiccation, which is not tolerated by most terrestrial stylommatophorans (Machin, 1975) is similar to that of the high salt marsh basommotophoran *M. bidentatus,* which tolerates a 78.5% loss of its body water (Price, 1980). Such high tolerances of desiccation allow freshwater pulmonates to survive extended periods of aerial exposure in high relative humidities (RH) (for a review of desiccation tolerance in pulmonates, see Machin, 1975). Species inhabiting temporary ponds have been reported to survive for 3 to 7 months without water (Brown, 1979; Jokinen, 1978; Newmann and Thomas, 1975;

Olivier and Barbosa, 1955a,b). As these habitats dry up, pulmonates move to dense vegetation or into matted aquatic vegetation, or burrow into the moist bottom substrata (Cheatum, 1934; Daniels and Armitage, 1969; DeWitt, 1955; Jokinen, 1978; Leonard, 1959; Olivier, 1956; Olivier and Barbosa, 1955a,b, 1956); some species have been reported to burrow to depths as great as 7 cm (Cheatum, 1934).

Both Cheatum (1934) and Olivier and Barbosa (1955a,b) report that survivorship is greatly reduced in specimens of freshwater pulmonates exposed to air directly on the substratum surface in natural habitats. However, in laboratory experiments, position within the substratum did not appear to affect survival in aerially exposed specimens of *B. glabrata* or *Tropicorbis centimetralis* (Lutz) (Olivier, 1956; Olivier and Barbosa, 1956), indicating that in natural environments convection and elevated temperatures may be responsible for increased levels of evaporative water loss and reduced survival in specimens exposed directly to the atmosphere at the substratum surface. Therefore, selection of moist, cool microenvironments by burrowing into bottom substrata or organic debris or by moving to dense covers of vegetation is a highly adaptive behavior in aerially exposed basommatophorans. It appears to be an active process in at least one species, *Lymnea palustris,* which has been observed to seek sites appropriate for estivation before its habitats dry up (Jokinen 1978).

Little is known of estivation in freshwater limpets, the most aquatic freshwater pulmonates, although several species have been reported to inhabit temporary fresh waters (Basch, 1963; Richardot, 1977a; Turner, 1978). In air, limpets such as *Laevapex fuscus* appear to tolerate very high levels of desiccation, but they survive only short periods (1–2 days) of aerial exposure due to the high evaporative loss of body water associated with an open patelliform shell. However, in a Texas population of *L. fuscus,* individuals survived 2 months of aerial exposure as a result of receding water levels by crawling down the bases of aquatic macrophytes deeply beneath the moist substratum, reappearing after the pond refilled with water in the autumn (R. F. McMahon, unpublished observations). Some limpets from standing, temporary waters, such as *Ferrissia wautieri* (Mirolli), *F. fragelis* (Tryon), and *Hebetancylus excentricus,* secrete a thin, calcaerous septum that partially closes the aperture of the patelliform shell (Basch, 1963; Richardot, 1977a; Turner, 1978). The septum is thought to reduce the rate of evaporative water loss by decreasing the surface area of epithelial tissues exposed directly to air, and has been observed to be secreted prior to the drying up of the pond and ditch habitats of these species (Basch, 1963; Boss, 1974; Richardot, 1977a,b). As septate shells are produced in populations before the habitat completely dries up, the environmental stimulus for septum formation appears not to be aerial exposure per se. Instead, the increases in ambient temperature and organic content of the water, declining oxygen concentrations, and longer photoperiods associated with the summer decline in the level of temporary fresh waters appear to be the primary stimuli for septum formation in limpets such as *F. wautieri* (Richardot, 1977a,b). Initiation of septum secretion in response to such environmental cues is highly adaptive, allowing septate shells to be formed in advance of aerial exposure (aerially exposed specimens would desiccate before septum formation could be completed). It may also explain the anomalous formation of septate shells in overwintering populations of *H. excentricus* in permanent waters (Turner, 1978), because the environmental conditions required to induce septum formation can also occur in permanent aquatic habitats.

The interspecific variation in aerial exposure tolerance of freshwater pulmonates is extraordinarily high. Table III displays survivorship values for aerial exposure in a number of freshwater pulmonate species. Fifty percent survivorship of aerial exposure can be as low as 18 h in *L. peregra* at 18°C and 80% RH to as high as 220–230 days in *B. glabrata* and *T. centimetralis* (Olivier, 1956; Olivier and Barbosa, 1956). Unlike certain other adaptations such as gill structure, response to hypoxia and nitrogenous excretion, whose interspecific variation appears to have a partially phyletic basis associated with the degree of aquatic readaptation in the major freshwater basommatophoran families and subfamilies, those involving resistance to aerial exposure and survival of long periods of estivation occur in some species of all the major hygrophilous basommatophoran groups (lymnaeids, physids, planorbids, and freshwater limpets) (Table III). Cheatum (1934) has pointed out that in freshwater pulmonates, resistance to aerial exposure is generally greater in species which can inhabit temporary aquatic environments than in those restricted to more permanent waters. Temporary ponds can support many species of freshwater pulmonates, including species of lymnaeids and physids which live close to the surface and are dependent on pulmonary respiration, as well as more aquatic species of planorbids and freshwater limpets (Basch, 1963; Boycott, 1936; Brown, 1979, 1982; DeWitt, 1955; Richardot, 1977a,b). When their temporary habitats are covered with water, pulmonates occupy a wide variety of microhabitats that are generally associated with the degree of aquatic adaptation and depth of penetration of each species (see Section II). However, the ability to survive long periods of aerial exposure apparently has been evolved independently in those species of each major group of hygrophilous basommatophorans that have penetrated temporary freshwater habitats during the course of their evolution, regardless of their degree of readaptation to an aquatic schesis.

When estivating in air, freshwater pulmonates generally retract deeply into the shell. Although the epiphragm formation characteristic of estivating stylommatophorans does not generally occur in freshwater pulmonates, some species obstruct the shell aperture with dried, foamy layers of mucus which reduce convective water loss (Machin, 1975). As in stylommatophorans, the external epithelium of freshwater pulmonates may function to regulate evaporative water

TABLE III

Resistance to Exposure in Air of Freshwater Pulmonate Snails

Species	Conditions of exposure in air	Temperature (°C)	Relative humidity (%)	Days of exposure	% surviving	Reference
Lymnaea peregra	Adults in air	5	80	2.2	70	Skoog (1976)
		15	80	1.5	50	
		18	80	0.8	50	
	Juveniles in air	18	80	0.6	50	
Lymnaea bulimoides	In soil	Field ambient	—	45	50	Cheatum (1934)
Lymnaea emarginata	In soil	Field ambient	—	62	30	Cheatum (1934)
Lymnaea megasoma	In soil	Field ambient	—	62	40	Cheatum (1934)
Lymnaea palustris	in soil	Field ambient	—	62	21	Cheatum (1934)
Lymnaea stagnalis	In soil	Field ambient	—	62	31	Cheatum (1934)
Bakerilymnaea cockerelli	In soil	Room ambient	—	>174	—	Newman and Thomas (1975)
Physa integra	In air	Room ambient	—	3	44	Clampitt (1970)
				6	28	
				12	14	
Physa gyrina	In air	Room ambient	—	3	83	Clampitt (1970)
				6	73	
				12	51	

Physa sayii	In soil	Field ambient	—	62	27	Cheatum (1934)
Physa parkeri	In soil	Field ambient	—	62	15	Cheatum (1934)
Helisoma trivolvis	In soil	Field ambient	—	62	53	Cheatum (1934)
Helisoma campanulatum	In soil	Field ambient	—	62	25	Cheatum (1934)
Helisoma antrosum	In soil	Field ambient	—	62	50	Cheatum (1934)
Biomphalaria glabrata	In air	27	96	64	50	von Brand et al. (1957)
			85	25	50	
			74	8.2	50	
			57	4.7	50	
			30	3.5	50	
			15	2.8	50	
	In air	27–30	90–95	220	50	Olivier (1956)
			85–90	126	50	
			70–80	61	50	
	In dry soil	25–30	75–90	30	90	Olivier and Barbosa (1956)
				60	73	
				90	64	
				120	23	
				154	6	
Tropicorbis centimetalis	In dry soil	25–30	75–90	210	81	Olivier and Barbosa (1956)
				270	36	
				365	7	

loss (Machin, 1968, 1972, 1974, 1975). The rate of evaporative water loss and survivorship of aerial exposure in freshwater pulmonates are dependent on RH and temperature (Table III). Evaporative water loss is accelerated with both decreased RH and increased temperature (von Brand et al., 1957; Skoog, 1976). which explains the preference of estivating species for moist, cool microenvironments. The rate of oxygen consumption in aerially exposed estivating individuals of *B. glabrata* declines to less than 20% that of the normal aquatic rate, and is less than that of starved snails in water (von Brand et al., 1957), presumably an adaptation that conserves energy stores. Therefore, aerially estivating individuals of *B. glabrata* survive longer than starved individuals in water (von Brand et al., 1957). After 128 days in air at 96% RH, *B. glabrata* metabolizes 50–60% of the body organic matter, the primary metabolite being body protein (von Brand et al., 1957). A similar reduction in body protein has been reported in starving specimens of *H. trivolvis* in water (Russell-Hunter and Eversole, 1976). Lactic acid accumulation, indicative of anaerobic metabolism, occurs only in the early stages of aerial estivation in *B. glabrata*. Lactic acid disappears from the hemolymph of this species within the first 30 days of estivation; thereafter, catabolism of energy during estivation is totally aerobic (von Brand et al., 1957).

Intraspecific interpopulation differences in resistance to aerial exposure also appear to occur among isolated species populations of freshwater pulmonates. Olivier (1956) observed that individuals of *B. glabrata* drawn from populations in temporary habitats survived much longer periods of aerial exposure than did individuals of the same species drawn from populations inhabiting permanent fresh waters. The existence of such distinct physiological races with regard to tolerance of aerial exposure indicates that there may be a selection for resistant phenotypes in temporary freshwater habitats, resulting in populations characterized by high proportions of desiccation-tolerant individuals.

Juveniles appear to be less resistant to desiccation than are larger adult freshwater pulmonates (Skoog, 1976), their higher surface-to-volume ratios and thinner shells resulting in higher rates of evaporative water loss. However, in natural populations of *L. palustris,* apparently only juveniles estivate when their temporary ponds disappear in summer (Jokinen, 1978). The eggs of freshwater pulmonates are far more intolerant of aerial exposure and desiccation than are adults. Egg masses of *B. glabrata* exposed to air at 25–27°C for 60 min in 25–27% RH were completely inviable, whereas only 4% of eggs hatched after 3 h in 74–78% RH and 11% hatched after 3 days in 90–100% RH (Chernin and Adler, 1967), conditions which estivating adults survived for 1 to 8 months (von Brand et al., 1957; Olivier, 1956; Olivier and Barbosa, 1956) (Table III). Similarly, the eggs and developing embryos of *L. peregra* were shown to be inviable after a 7.5-h exposure in air at 80% RH at 18°C, whereas adults tolerated the same conditions for up to 35 h (Skoog, 1976). The relatively low tolerance of eggs and embryos to aerial exposure may account for the observation that the egg masses of freshwater pulmonates are often oviposited at greater depths than are those over which the adult population is normally distributed. Skoog (1976) reported that 65–66% of the egg masses of two populations of *L. peregra* occurred at depths between 10 and 70 cm, whereas the vast majority of adults of this species are reported to remain at the water's edge in summer (Russell-Hunter, 1953a). Similar distributions of egg masses to greater depths than adults have been observed in *P. virgata* (R. F. McMahon, unpublished observations). Differences in adult and egg mass distribution may result from migration to and oviposition at greater depths during dark hours by freshwater pulmonates (see Section V,D) which have the obvious adaptive advantage of insulating desiccation-prone eggs and embryos from the deleterious effects of aerial exposure associated with short-term water level variations in the shallow, nearshore habitats of adults.

During the early stages of aerial estivation, freshwater pulmonates have been reported to shift from ammonia to urea production as the major nitrogenous waste product (see Section IV). Urea appears to be retained in the body fluids of estivating snails. Hemolymph urea concentrations increased to 155 times those of aquatic controls after 5 days of aerial exposure in *B. glabrata* (Becker and Schmale, 1978; Schmale and Becker, 1977) and to 63 times those of controls in *Bakerilymnaea cockerelli* after 30 days of exposure in air (Newman and Thomas, 1975). In some species, such as *B. cockerelli,* long-term estivation involves a further shift from urea to uric acid production, but hemolymph urea concentrations are still maintained well above those of aquatic individuals throughout the aerial estivation period (Newman and Thomas, 1975). Maintenance of high levels of urea production and hemolymph concentration during prolonged aerial estivation in freshwater pulmonates has the obvious advantage of detoxifying the ammonia released by protein catabolism, the apparent primary energy source in estivating snails (von Brand et al., 1957). It may also function to radically increase body fluid osmotic concentration, which, in turn, lowers the vapor pressure of the hemolymph and therefore functions to reduce the rate of evaporative water loss (Becker and Schamale, 1975; Campbell et al., 1972; Horne, 1971; Newman and Thomas, 1975).

VIII. Summary

The major groups of the hygrophilous freshwater Basommatophora appear to recapitulate the evolutionary history of freshwater pulmonates from a primitive, nearly terrestrial ancestral stock which gave rise to present-day species through the evolution of intermediate forms progressively more adapted to aquatic life. These various stages of aquatic readaptation are represented in the modern, extant groups of freshwater basommatophorans. The Lymnaeidae retain a primitive amphibious habit and remain highly dependent on pulmonary respiration. The Physidae, although primarily aquatic, remain close to the air–water interface and are also highly dependent on pulmonary oxygen consumption. The Planorbinae are less dependent on pulmonary gas exchange, having developed a relatively extensive neomorphic gill. They penetrate to greater depths than either lymnaeids or physids, some smaller species of planorbids having developed an entirely aquatic schesis. The freshwater limpets (mostly members of the family Ancyloplanorbidae) have lost the mantle-cavity lung and have evolved an entirely aquatic mode of life, including the development of a complex neomorphic gill.

Although most of the general physiological characteristics of freshwater pulmonates, including retention of urea and uric acid synthetic pathways of nitrogen metabolism, and high tolerance of temperature, hypoxia, aerial exposure, and desiccation, are adaptations related to their preferred eutrophic, shallow-water, nearshore, small, and often temporary freshwater habitats, much of the interspecific variation in the physiological ecology of freshwater pulmonates is associated with the degree of aquatic readaptation displayed by individual species. If it is accepted (with some notable exceptions in each group) that the series Lymnaeidae-Physidae-Planorbinae-freshwater limpets of the family Ancyloplanorbidae represents increasing levels of adaptation to aquatic life, then the physiological adaptations of species from each of these groups can be related to such an evolutionary continuum. Thus, the above series is associated with (1) increasing dependence on aquatic respiration and suppression of pulmonary gas exchange, (2) progressive development of neomorphic gills, (3) decreasing dependence on ureotelic and uricotelic nitrogen excretion, with a corresponding development of ammonotelism, and (4) an increasing capacity to tolerate anoxic and hypoxic conditions, reaching its zenith in totally aquatic freshwater limpets, which are unable to resort to pulmonary respiration during prolonged oxygen deprivation and which have instead evolved extensive respiratory adaptations to compensate for both short-term and seasonal exposures to hypoxia (McMahon, 1972, 1973). Also, in the more aquatic Ancyloplanorbidae there is an apparent shift toward the utilization of more efficient alternative anaerobic metabolic pathways and toward the evolution of a greatly enhanced capacity to excrete toxic anaerobic waste products to the external medium, rather than retaining them in the tissues and hemolymph as in lymnaeids and physids (von Brand et al., 1950; Mehlman and von Brand, 1951).

In contrast, much of the functional significance of many of the adaptations described in this chapter for freshwater pulmonates cannot be related specifically to the degree of aquatic readaptation. Instead, such adaptations occur irregularly throughout each functional series. Thus, the calcium metabolism, temperature tolerance, capacity for desiccation and aerial exposure, respiratory response to temperature, and spawning temperatures of freshwater pulmonates appear to form an anacoluthic or discontinuous progression of adaptations across the evolutionary continuum from a nearly terrestrial to a purely aquatic mode of life represented in the major living groups of aquatic Basommatophora. Although some of this apparently anacoluthic interspecific variation may be accounted for by the high levels of intraspecific, interpopulation variation, and physiological plasticity characteristic of many species of freshwater pulmonates (see Section I,C), most of such interspecific physiological variation appears to be associated with the adaptation of each species to the microenvironmental characteristics of its specific niche. Thus, temperature tolerance appears to be related to environmental thermal regimes, and is distributed in clines both between populations of a single species (Ibarra and McMahon, 1981; McMahon and Payne, 1980) and between species from different latitudes (van der Schalie and Berry, 1973). Oviposition temperatures also seem to have similar interspecific distributions related to the microenvironmental temperature regimes of specific habitats (lotic versus lentic) or to the seasonal temperature variation associated with the latitudinal distribution of a species (van der Schalie and Berry, 1973). The respiratory response to temperature and the capacity for temperature acclimation of respiration are closely associated with the extent of short-term and seasonal temperature variations associated with each species microhabitat. Finally, the level of desiccation and aerial exposure tolerance is apparently associated more with the degree of each species' adaptations to temporary aquatic habitats than with phylogenetic origin. Each major basommatophoran group includes species capable of inhabiting temporary fresh waters and characterized by far greater tolerance of exposure in air compared to species restricted to more permanent habitats.

The author has not only intended that this chapter provide a comprehensive review of recent advances in the study of the physiological ecology of freshwater basommatophoran gastropods, but also that some of the observations and hypotheses presented, many of which were purposely highly speculative, will stimulate further investigations of the adaptive physiology of this remarkable group of organisms. Many years of experience have resulted in familiarity with the advantages of freshwater pulmonates as research animals, not the least of which are their ease of collection and accessibility in the field and their adaptability to laboratory experimentation. A single freshwater pond, stream, river, or lake may often contain 5 to 10 or even more species of basommatophoran snails, each representing a specific suite of adaptations to the physicochemical and biotic characteristics of a unique niche. The importance of such systems to comparative studies is obvious and, if well utilized in the future, could greatly increase our understanding of the comparative ecology and physiology of not only freshwater pulmonates but of aquatic organisms in general.

Acknowledgements

I wish to thank Della Tyson, Lori McDowell, and Colette O'Byrne-McMahon for their assistance with the preparation of this chapter. I also wish to express my gratitude to Dr. W. D. Russell-Hunter, the editor of this volume, for his patience and consideration. It was he who first introduced me to the freshwater pulmonates as experimental animals, and his advice and assistance, both as a graduate advisor and as a research colleague, have proved invaluable in my studies of the physiological ecology of aquatic molluscs. My research on the physiological ecology of freshwater pulmonates during the period in which this chapter was written was supported by a National Science Foundation Minority Graduate Fellowship (to Juan Ibarra) and by a grant from Organized Research Funds from The University of Texas at Arlington.

References

Agersborg, H. P. K. (1932). The relation of temperature to continuous reproduction in the pulmonate snail *Physa gyrina* Say. *Nautilus* **45,** 121–123.

Aldridge, D. W. (1980). Life cycle, reproductive tactics, and bioenergetics in the freshwater prosobranch snail *Spirodon carinata* (Bruguiere) in upstate New York. Ph.D. Dissertation, Syracuse University, Syracuse, New York.

Aldridge, D. W. (1982). Reproductive tactics in relation to life-cycle bioenergetics in three natural populations of the freshwater snail, *Leptoxis carinata. Ecology* **63,** 196–208.

Aldridge, D. W. (1983). Physiological ecology of freshwater prosobranchs. *In* The Mollusca, (W. D. Russell-Hunter, ed.) Vol. 6, pp. 329–358. Academic Press, New York.

Apley, M. L. (1967). Field and experimental studies on pattern and control of reproduction in *Melampus bidentatus* (Say). Ph.D. Dissertation, Syracuse University, Syracuse, New York.

Apley, M. L. (1970). Field studies on life history, gonadal cycle and reproductive periodicity in *Melampus bidentatus* (Pulmonata: Ellobiidae). *Malacologia* **10,** 381–397.

Appleton, C. C. (1976a). Observations on the thermal regime of a stream in the Eastern Transvaal, with reference to certain aquatic pulmonata. *S. Afr. J. Sci.* **72,** 20–23.

Appleton, C. C. (1976b). Evidence for a circadian rhythm in the freshwater snail *Bulimus* (*Physopsis*) *globosus* (Morelet). *S. Afr. J. Sci.* **72,** 310–311.

Barnes, H., Barnes, M., and Finlayson, D. M. (1963). The seasonal changes in body weight, biochemical composition, and oxygen uptake of two common arctic cirripedes, *Balanus balanoides* and *B. balanus. J. Mar. Biol. Assoc. U.K.* **43,** 185–211.

Basch, P. F. (1959). The anatomy of *Laevapex fuscus,* a freshwater limpet (Gastropoda: Pulmonata). *Misc. Publ.—Mus. Zool. Univ. Mich.* **108,** 1–56.

Basch, P. F. (1963). A review of the recent freshwater limpet snails of North America. *Bull. Mus. Comp. Zool.* **129,** 399–461.

Bayne, B. L. (1975). Aspects of the physiological condition in *Mytilus edulis* L., with special reference to the effects of oxygen tension and salinity. *Proc. Eur. Mar. Biol. Symp., 9th, 1975* pp. 213–238.

Bayne, B. L., and Livingstone, D. R. (1977). Responses of *Mytilus edulis* L. to low oxygen tension: Acclimation of the rate of oxygen consumption. *J. Comp. Physiol.* **114,** 129–142.

Bayne, R. A., and Friedl, F. E. (1968). The production of externally measurable ammonia and urea in the snail, *Lymnaea stagnalis jugularis* Say. *Comp. Biochem. Physiol.* **25,** 711–717.

Beames, C. G., Jr., and Lindeborg, R. G. (1967). Temperature adaptation in the snail *Physa anatina. Proc. Okla. Acad. Sci.* **48,** 12–14.

Becker, W., and Schmale, H. (1975). The nitrogenous products of degradation—ammonia, urea and uric acid—in the hemolymph of the snail *Biomphalaria glabrata. Comp. Biochem. Physiol. A* **51A,** 407–411.

Becker, W., and Schmale, H. (1978). The ammonia and urea excretion of *Biomphalaria glabrata* under different physiological conditions: Starvation, infection with *Schistosoma mansoni,* dry keeping. *Comp. Biochem. Physiol.* **59B,** 75–79.

Berg, K. (1951). On the respiration of some molluscs from running and stagnant water. *Ann. Biol.* **27,** 329–335.

Berg, K. (1952). On the oxygen consumption of Ancylidae (Gastropoda) from an ecological point of view. *Hydrobiologia* **4,** 225–267.

Berg, K. (1953). The problem of respiratory acclimatization illustrated by experiments with *Ancylus fluviatilis* (Gastropoda). *Hydrobiologia* **5,** 331–350.

Berg K., and Ockelmann, K. W. (1959). The respiration of freshwater snails. *J. Exp. Biol.* **36,** 690–708.

Berg, K., Lumbye, J., and Ockelmann, K. W. (1958). Seasonal and experimental variations of the oxygen consumption of the limpet *Ancylus fluviatilis* (O. F. Müller). *J. Exp. Biol.* **35,** 43–73.

Berrie, A. D. (1959). Variation in the radula of the freshwater snail *Lymnaea peregra* (Müller) from Northwestern Europe. *Ark. Zool.* **12,** 391–404.

Blazka, P. (1954). Temperatur adaptation des gesamt metabolis mus bei der weinbergschnecke *Helix pomatia* L. *Zool. Jahrb., Abt. Allg. Zool. Physiol. Tiere* **65,** 130–138.

Bligh, J., Cloudsley-Thompson, J. L., and MacDonald, A. G. (1976). "Environmental Physiology of Animals." Wiley, New York.

Boag, D. A., and Bentz, J. A. (1980). The relationship between simulated seasonal temperatures and depth distributions in the freshwater pulmonate, *Lymnaea stagnalis. Can. J. Zool.* **58,** 198–201.

Boerger, H. (1975). Movement and burrowing of *Helisoma trivolvis* (Say) (Gastropoda, Planorbidae) in a small pond. *Can. J. Zool.* **53,** 456–464.

Boss, K. J. (1974). Oblomovism in the Mollusca. *Trans. Am. Microsc. Soc.* **93,** 460–481.

Boss, K. J. (1978). On the evolution of gastropods in ancient lakes. *In* "Pulmonates" (V. Fretter and J. Peake, eds.), Vol. 2A, pp. 385–428. Academic Press, New York.

Boycott, A. E. (1936). The habitats of freshwater Mollusca in Britain. *J. Anim. Ecol.* **5,** 116–186.

Brown, K. M. (1979). The adaptive demography of four freshwater pulmonate snails. *Evolution* **33,** 417–432.

Brown, K. M. (1982). Resource overlap and competition in pond snails: An experimental analysis. *Ecology* **63,** 412–422.

Browne, R. A., and Russell-Hunter, W. D. (1978). Reproductive effort in molluscs. *Oecologia* **37,** 23–27.

Brues, C. T. (1928). Studies on the fauna of hot springs in the Western United States and the biology of thermophilous animals. *Proc. Am. Acad. Arts Sci.* **63,** 139–228.

Buckingham, M. J., and Freed, D. E. (1976). Oxygen consumption in the prosobranch snail *Viviparus contectoides* (Mollusca: Gastropoda).—II. Effects of temperature and pH. *Comp. Biochem. Physiol.* **53A,** 249–252.

Bullard, R. W. (1964). Animals in aquatic environments: Annelids and molluscs. *In* "Handbook of Physiology" (D. B. Dill, E. F. Adolph, and C. G. Wilber, eds.), Sect. 4, pp. 683–695. Am. Physiol. Soc., Washington, D.C.

Burky, A. J. (1969). Biomass turnover, energy balance and interpopulation variation in the stream limpet, *Ferrissia rivularis* (Say), with special reference to respiration growth and fecundity. Ph.D. Dissertation, Syracuse University, Syracuse, New York.

Burky, A. J. (1971). Biomass turnover, respiration and interpopulation variation in the stream limpet, *Ferrissia rivularis* (Say). *Ecol. Monogr.* **41,** 235–251.

Calow, P. (1973). On the regulatory nature of individual growth: Some observations from freshwater snails. *J. Zool.* **170,** 415–428.

Calow, P. (1975a). The feeding strategies of two freshwater gastropods, *Ancylus fluviatilus* Müll and *Planorbis contortus* Linn. (Pulmonata), in terms of ingestion rates and absorption efficiencies. *Oecologia* **20,** 33–49.

Calow, P. (1975b). The respiratory strategies of two species of freshwater gastropods (*Ancylus fluviatilis* Müll and *Planorbis contortus* Linn.) in relation to temperature, oxygen concentration, body size, and season. *Physiol. Zool.* **48,** 114–129.

Calow, P., and Townsend, C. R. (1981). Energetics, ecology and evolution. *In* "Physiological Ecology: An Evolutionary Approach to Resource Use" (C. R. Townsend and P. Calow, eds.), pp. 3–19. Blackwell, Oxford.

Campbell, J. W. (1973). Nitrogen excretion. *In* "Comparative Animal Physiology" (C. L. Prosser, ed.), 3rd ed., pp. 279–316. Saunders, Philadelphia, Pennsylvania.

Campbell, J. W., and Bishop, S. H. (1970). Nitrogen metabolism in molluscs. *In* "Comparative Biochemistry of Nitrogen Metabolism" (J. W. Campbell, ed.), Vol. 1, pp. 103–206. Academic Press, New York.

Campbell, J. W., Drotman, R. B., McDonald, J. A., and Tramell, P. R. (1972). Nitrogen metabolism in terrestrial invertebrates. *In* "Nitrogen Metabolism and the Environment" (J. W. Campbell and L. Goldstein, eds.), pp. 1–54. Academic Press, New York.

Cheatum, E. P. (1934). Limnological investigations on respiration, annual migratory cycle and other related phenomena in fresh water pulmonate snails. *Trans. Am. Microsc. Soc.* **53,** 348–407.

Chen, C., and Awapara, J. (1969a). Intracellular distribution of enzymes catalyzing succinate production from glucose in *Rangia* mantle. *Comp. Biochem. Physiol.* **30,** 727–737.

Chen, C., and Awapara, J. (1969b). Effect of oxygen on the end products of glycolysis in *Rangia cuneata. Comp. Biochem. Physiol.* **31,** 395–401.

Chernin, E., and Adler, V. L. (1967). Effects of desiccation on eggs of *Australorbis glabratus. Ann. Trop. Med. Parasitol.* **61,** 11–14.

Clampitt, P. T. (1970). Comparative ecology of the snails *Physa gyrina* and *Physa integra* (Basommatophora: Physidae). *Malacologia* **10,** 113–151.

Clampitt, P. T. (1972). Seasonal migrations and other movements in Douglas Lake pulmonate snails (Abstract). *Malacol. Rev.* **5,** 11–12.

Clampitt, P. T. (1974). Seasonal migratory cycle and related movements of the fresh water pulmonate snail, *Physa integra. Am. Midl. Nat.* **92,** 275–300.

Conway, A. F., Black, R. E., and Morrill, J. B. (1969). Uric acid synthesis in embroys of the pulmonate pond snail, *Limnaea palustris:* Evidence for a unique pathway. *Comp. Biochem. Physiol.* **30,** 793–802.

Daniels, J. M., and Armitage, K. B. (1969). Temperature acclimation and oxygen consumption in *Physa hawnii* Lea (Gastropoda: Pulmonata). *Hydrobiologia* **33,** 1–13.

Davies, P. M. C., Patterson, J. W., and Bennett, E. L. (1981). Metabolic coping strategies in cold tolerant reptiles. *J. Therm. Biol.* **6,** 321–330.

Davies, P. S. (1965). Environmental acclimation in the limpet, *Patella vulgata* L. *Nature* (*London*) **205,** 924.

deWit, W. F. (1955). The life-cycle and some other biological details of the fresh-water snail *Physa fontinalis* (L.). *Basteria* **19,** 35–73.

DeWitt, R. M. (1955). The ecology and life history of the pond snail *Physa gyrina. Ecology* **36,** 40–44.

DeWitt, R. M. (1967). Stimulation of egg production in a physid and a lymnaeid. *Malacologia* **5,** 445–453.

de Zwaan, A., and van Marrewijk, W. J. A. (1973). Anaerobic glucose degradation in the sea mussel *Mytilus edulis* L. *Comp. Biochem. Physiol.* **44B,** 429–439.

de Zwaan, A., and Wijsman, T. C. M. (1976). Anaerobic metabolism in Bivalvia (Mollusca). Characteristics of anaerobic metabolism. *Comp. Biochem. Physiol.* **54B,** 313–324.

de Zwaan, A., and Zandee, D. I. (1972). The utilization of glycogen and accumulation of some intermediates during anaerobiosis in *Mytilus edulis* L. *Comp. Biochem. Physiol.* **43B,** 47–54.

Duerr, F. G. (1966). Nitrogen excretion in the fresh water pulmonate snail *Lymnaea stagnalis appressa* (Say). *Physiologist* **9,** 172.

Duerr, F. G. (1967). The uric acid content of several species of prosobranch and pulmonate snails as related to nitrogen excretion. *Comp. Biochem. Physiol.* **22,** 333–340.

Duerr, F. G. (1968). Excretion of ammonia and urea in seven species of marine prosobranch snails. *Comp. Biochem. Physiol.* **26,** 1051–1059.

Duncan, A. (1966). The oxygen consumption of *Potamopyrgus jenkinsi* (Smith) (Prosobranchiata) in different temperatures and salinities. *Verh. Int. Ver. Theor. Limnol. Angew* **16,** 1739–1751.

Duncan, C. J. (1959). The life cycle and ecology of the freshwater snail *Physa fontinalis* (L.). *J. Anim. Ecol.* **28,** 97–117.

Duncan, C. J. (1975). Reproduction. *In* "Pulmonates" (V. Fretter and J. Peake, eds.), Vol. 1, pp. 309–365. Academic Press, New York.

Ellis, R. A. (1926). "British Snails: A Guide to the Non-Marine Gastropoda of Great Britain and Ireland Pleistocene to Recent." Oxford Univ. Press, London and New York.

Eversole, A. G. (1974). Fecundity in the snail *Helisoma trivolvis:* Experimental, bioenergetic and field studies. Ph.D. Dissertation, Syracuse University, Syracuse, New York.

Eversole, A. G. (1978). Life-cycles, growth and population bioenergetics in the snail *Helisoma trivolvis* (Say). *J. Moll. Stud.* **44,** 208–218.

Florkin, M. (1966). Nitrogen metabolism. *In* "Physiology of Mollusca" (K. M. Wilbur and C. M. Yonge, eds.), Vol. 2, pp. 309–351. Academic Press, New York.

Florkin, M., and Bricteux-Grégoire, S. (1972). Nitrogen metabolism in mollusks. *In* "Chemical Zoology" (M. Florkin and B. T. Scheer, eds.), Vol. 7, pp. 301–348. Academic Press, New York.

Florkin, M., and Scheer, B. T., eds. (1972). "Chemical Zoology," Vol. 7. Academic Press, New York.

Forbes, G. S., and Crampton, H. E. (1942). The differentiation of geographical groups in *Lymnaea palustris. Biol. Bull. (Woods Hole, Mass.)* **82,** 26–46.

Fretter, V. (1975). Introduction. *In* "Pulmonates" (V. Fretter and J. Peake, eds.), Vol. 1, pp. xi–xxix. Academic Press, New York.

Fretter, V., and Peake, J., eds. (1975). "Pulmonates," Vol. 1. Academic Press, New York.

Fretter, V., and Peake, J., eds. (1978a). "Pulmonates," Vol. 2A. Academic Press, New York.

Fretter, V., and Peake, J. (1978b). Appendix. *In* "Pulmonates" (V. Fretter and J. Peake, eds.), Vol. 2A, pp. 527–534. Academic Press, New York.

Friedl, F. E. (1974). Nitrogen excretion by the freshwater pulmonate snail, *Lymnaea stagnalis jugularis* Say. *Comp. Biochem. Physiol. A* **49,** 617–622.

Friedl, F. E., and Bayne, R. A. (1966). Ureogenesis in the snail *Lymnaea stagnalis jugularis. Comp. Biochem. Physiol.* **17,** 1167–1173.

Geldiay, R. (1956). Studies on local populations of the freshwater limpet *Ancylus fluviatilis* Müller. *J. Anim. Ecol.* **25,** 389–402.

Ghiretti, F. (1966a). Respiration. *In* "Physiology of Mollusca" (K. M. Wilbur and C. M. Yonge, eds.), Vol. 2, pp. 175–208. Academic Press, New York.

Ghiretti, F. (1966b). Molluscan hemocyanins. *In* "Physiology of Mollusca" (K. M. Wilbur and C. M. Yonge, eds.), Vol. 2, pp. 233–248. Academic Press, New York.

Ghiretti, F., and Ghiretti-Magaldi, A. (1972). Respiratory proteins in mollusks. *In* "Chemical Zoology" (M. Florkin and B. T. Scheer, eds.), Vol. 7, pp. 201–217. Academic Press, New York.

Ghiretti, F., and Ghiretti-Magaldi, A. (1975). Respiration. *In* "Pulmonates" (V. Fretter and J. Peake, eds.), Vol. 1, pp. 33–52. Academic Press, New York.

Greenaway, P. (1970). Sodium regulation in the freshwater mollusc *Lymnaea stagnalis* (L.) (Gastropoda: Pulmonata). *J. Exp. Biol.* **53,** 147–163.

Greenaway, P. (1971). Calcium regulation in the freshwater mollusc, *Lymnaea stagnalis* (L.) (Gastropoda: Pulmonata). I. The effect of internal and external calcium concentration. *J. Exp. Biol.* **54,** 199–214.

Harrison, A. D., Williams, N. V., and Greig, G. (1970). Studies on the effects of calcium bicarbonate concentrations on the biology of *Biomphalaria pfeifferi* (Krauss) (Gastropoda: Pulmonata). *Hydrobiologia* **36,** 317–327.

Harrison, P. T. C. (1977a). Seasonal changes in the heart rate of the freshwater pulmonate *Lymnaea stagnalis* (L.). *Comp. Biochem. Physiol.* **58A,** 37–41.

Harrison, P. T. C. (1977b). Laboratory induced changes in the heart rate of *Lymnaea stagnalis* (L.). *Comp. Biochem. Physiol.* **58A,** 43–46.

Henderson, A. E. (1961). Studies on the respiratory and hydrostatic functions of the mantle cavity in two freshwater pulmonate snails. Ph.D. Dissertation, University of Glasgow.

Henderson, A. E. (1963). On the underwater weights of freshwater snails. *Z. Vergl. Physiol.* **46,** 467–490.

Hill, R. W. (1976). "Comparative Physiology of Animals: An Environmental Approach." Harper, New York.

Hochachka, P. W., and Mustafa, T. (1972). Invertebrate facultative anaerobiosis. *Science* **178,** 1056–1060.

Hochachka, P. W., Fields, J., and Mustafa, T. (1973). Animal life without oxygen: Basic biochemical mechanisms. *Am. Zool.* **13,** 543–555.

Horne, F. R. (1971). Accumulation of urea by a pulmonate snail during aestivation. *Comp. Biochem. Physiol.* **38A,** 565–570.

Hubendick, B. (1948). Sur les variations de la taille du rein chez *Lymnaea limosa* (L.). *J. Conchyliol.* **88,** 1–10.

Hubendick, B. (1952). On the evolution of the so-called thalassoid molluscs of Lake Tanganyika. *Ark. Zool.* **3,** 319–323.

Hubendick, B. (1954). Viewpoints on species discrimination with special attention to medically important snails. *Proc. Malacol. Soc. London* **31,** 6–11.

Hubendick, B. (1962a). Aspects of the diversity of the freshwater fauna. *Oikos* **13,** 249–261.

Hubendick, B. (1962b). Studies on *Acroloxus* (Moll. Basomm.). *Medd. Göeteborgs Mus. Zool. Avd.* **133,** 1–68.

Hubendick, B. (1964). Studies on Anclidae, the subgroups. *Medd. Göeteborgs Mus. Zool. Avd.* **137,** 1–72.

Hubendick, B. (1970). Studies on Ancylidae, the palearctic and oriental species and formgroups. *Acta Reg. Soc. Sci. Litt. Gothoburgensis, Zool.* **5,** 1–52.

Hubendick, B. (1978). Systematics and comparative morphology of the Basommatophora. *In* "Pulmonates" (V. Fretter, and J. Peake, eds.), Vol. 2A, pp. 1–47. Academic Press, New York.

Hunter, R. D. (1972). Energy budgets and physiological variation in natural populations of the freshwater pulmonate, *Lymnaea palustris*. Ph.D. Dissertation, Syracuse University, Syracuse, New York.

Hunter, R. D. (1975a). Growth, fecundity, and bioenergetics in three populations of *Lymnaea palustris* in Upstate New York. *Ecology* **56,** 50–63.

Hunter, R. D. (1975b). Variation in populations of *Lymnaea palustris* in Upstate New York. *Am. Midl. Nat.* **94,** 401–420.

Hunter, R. D. (1976). Changes in carbon and nitrogen content during decomposition of three macrophytes in freshwater and marine environments. *Hydrobiologia* **51,** 119–128.

Hunter, R. D., and Lull, W. W. (1977). Physiologic and environmental factors influencing the calcium-to-tissue ratio in populations of three species of freshwater pulmonate snails. *Oecologia* **29,** 205–218.

Ibarra, J. A., and McMahon, R. F. (1981). Interpopulation variation in the shell morphometrics and thermal tolerance of *Physa virgata* Gould (Mollusca: Pulmonata). *Am. Zool.* **21,** 1020.

Imhof, G. (1973). Der einfluss von temperatur und photoperiode auf den lebenszyklus einiger süsswasser pulmonaten. *Malacologia* **14,** 393–395.

Issel, R. (1908). Sulla biologia termale. *Int. Rev. Gesamten Hydrobiol. Hydrogr.* **1,** 29–36.

Jaeckel, S. (1950). Die mollusken der chlei. *Arch. Hydrobiol.* **44,** 214–270.

Janse, C. (1981). The effect of oxygen on gravity orientation in the pulmonate snail *Lymnaea stagnalis. J. Comp. Physiol.* **142,** 51–59.

Jenner, C. E. (1951). Photoperiodism in the fresh-water pulmonate snail, *Lymnaea palustris.* Ph.D. Dissertation, Harvard University, Cambridge, Massachusetts.

Jokinen, E. H. (1978). The aestivation pattern of a population of *Lymnaea elodes* (Say) (Gastropoda: Lymnaeidae). *Am. Midl. Nat.* **100,** 43–53.

Jones, J. D. (1961). Aspects of respiration in *Planorbis corneus* L. and *Lymnaea stagnalis* L. (Gastropoda: Pulmonata). *Comp. Biochem. Physiol.* **4,** 1–29.

Jones, J. D. (1964a). The role of haemoglobin in the aquatic pulmonate, *Planorbis corneus. Comp. Biochem. Physiol.* **12,** 283–295.

Jones, J. D. (1964b). Respiratory gas exchange in the aquatic pulmonate, *Biomphalaria sudanica. Comp. Biochem. Physiol.* **12,** 297–310.

Kavaliers, M. (1980). A circadian rhythm of behavioral thermoregulation in a freshwater gastropod, *Helisoma trivolis. Can. J. Zool.* **58,** 2152–2155.

Kavaliers, M. (1981). Circadian and ultradian activity rhythms of a freshwater gastropod, *Helisoma trivolis:* The effects of social factors and eye removal. *Behav. Neural Biol.* **32,** 350–363.

Krog, H. (1954). Temperature sensitivity of freshwater gammarids. *Biol. Bull.* (*Woods Hole, Mass.*) **107,** 397–410.

Lassen, H. H. (1975). The diversity of freshwater snails in view of the equilibrium theory of island biogeography. *Oecologia* **19,** 1–8.

Leonard, A. B. (1959). Handbook of gastropods in Kansas. *Misc. Publ. Nat. Hist. Mus. Univ. Kansas.* No. 20, 1–224.

Livingstone, D. R. (1978). Anaerobic metabolism in the posterior adductor muscle of the common mussel *Mytilus edulis* L. in response to altered oxygen tension and temperatures. *Physiol. Zool.* **51,** 131–139.

Lumbye, J. (1958). The oxygen consumption of *Theodoxus fluviatilis* (L.) and *Potamopyrgus jenkinsi* (Smith) in brackish and fresh water. *Hydrobiologia* **10,** 245–262.

McCorkle, S., and Dietz, T. H. (1980). Sodium transport in the freshwater Asiatic clam *Corbicula fluminea. Biol. Bull.* (*Woods Hole, Mass.*) **159,** 325–336.

McDonald, S. C. (1973). Activity patterns of *Lymnaea stagnalis* (L.) in relation to temperature conditions: A preliminary study. *Malacologia* **14,** 395–396.

Machin, J. (1968). The permeability of the epiphragm of terrestrial snails to water loss. *Biol. Bull.* (*Woods Hole, Mass.*) **134,** 87–95.

Machin, J. (1972). Water exchange in the mantle of a terrestrial snail during periods of reduced evaporative loss. *J. Exp. Biol.* **57,** 103–111.

Machin, J. (1974). Osmotic gradients across snail epidermis: Evidence for a water barrier. *Science* **183,** 759–760.

Machin, J. (1975). Water relationships. *In* "Pulmonates" (V. Fretter and J. Peake, eds.), Vol. 1, pp. 105–163. Academic Press, New York.

McKillop, W. B., and Harrison, A. D. (1972). Distribution of aquatic gastropods across an interface between the Canadian Shield and limestone formations. *Can. J. Zool.* **50,** 1433–1445.

McMahon, R. F. (1972). Interpopulation variation and respiratory acclimation in the bioenergetics of *Laevapex fuscus.* Ph.D. Dissertation, Syracuse University, Syracuse, New York.

McMahon, R. F. (1973). Respiratory variation and acclimation in the freshwater limpet, *Laevapex fuscus. Biol. Bull.* (*Woods Hole, Mass.*) **145,** 492–508.

McMahon, R. F. (1975a). Growth, reproduction and bioenergetic variation in three natural populations of a fresh water limpet *Laevapex fuscus* (C. B. Adams). *Proc. Malacol. Soc. London* **41,** 331–351.

McMahon, R. F. (1975b). Effects of artifically elevated water temperatures on the growth, reproduction and life cycle of a natural population of *Physa virgata* Gould. *Ecology* **56,** 1167–1175.

McMahon, R. F. (1976a). Growth, reproduction and life cycle in six Texan populations of two species of fresh-water limpets. *Am. Midl. Nat.* **95,** 174–185.

McMahon, R. F. (1976b). Effluent-induced interpopulation variation in the thermal tolerance of *Physa virgata* Gould. *Comp. Biochem. Physiol.* **55A,** 23–28.

McMahon, R. F. (1980). Life-cycles of four species of freshwater snails from Ireland. *Am. Zool.* **20,** 927.

McMahon, R. F., and Payne, B. S. (1980). Variation of thermal limits in populations of *Physa virgata* Gould (Mollusca: Pulmonata). *Am. Midl. Nat.* **103,** 218–230.

McMahon, R. F., and Russell-Hunter, W. D. (1977). Temperature relations of aerial and aquatic respiration in six littoral snails in relation to their vertical zonation. *Biol. Bull.* (*Woods Hole, Mass.*) **152,** 182–198.

McMahon, R. F., and Russell-Hunter, W. D. (1978). Respiratory responses to low oxygen stress in marine littoral and sublittoral snails. *Physiol. Zool.* **51,** 408–424.

McMahon, R. F., and Russell-Hunter, W. D. (1981). The effects of physical variables and acclimation on survival and oxygen consumption in the high littoral salt-marsh snail, *Melampus bidentatus* Say. *Biol. Bull.* (*Woods Hole, Mass.*) **161,** 246–269.

McMahon, R. F., and Wilson, J. G. (1981). Seasonal respiratory responses to temperature and hypoxia in relation to burrowing depth in three intertidal bivalves. *J. Therm. Biol.* **6,** 267–277.

McMahon, R. F., Hunter, R. D., and Russell-Hunter, W. D. (1974). Variation in Aufwuchs at six freshwater habitats in terms of carbon biomass and of carbon:nitrogen ratio. *Hydrobiologia* **45,** 391–404.

McManus, D. P., and James, B. L. (1975a). Tricarboxylic acid cycle enzymes in the digestive gland of *Littorina saxatilis rudis* (Manton) and the daughter sporocysts of *Microphallus similis* (Jag) (Digenea: Microphallidae). *Comp. Biochem. Physiol.* **50B,** 491–495.

McManus, D. P., and James, B. L. (1975b). Anaerobic glucose metabolism of *Littorina saxatilis rudis* (Manton) and in the daughter sporocysts of *Microphallus similis* (Jag) (Digenea: Microphallidae). *Comp. Biochem. Physiol.* **51B,** 293–297.

McManus, D. P., and James, B. L. (1975c). Pyruvate kinases in the digestive gland of *Littorina saxatilis rudis* (Manton) and in the daughter sporocysts of *Microphallus similis* (Jag) (Digenea: Microphallidae). *Comp. Biochem. Physiol.* **51B,** 299–306.

Malanga, C. J., and Aiello, E. L. (1972). Succinate metabolism in the gills of the mussels *Modiolus demissus* and *Mytilus edulis*. *Comp. Biochem. Physiol.* **43B,** 795–806.

Mangum, C., and Van Winkle, W. (1973). Responses of aquatic invertebrates in declining oxygen conditions. *Am. Zool.* **13,** 529–541.

Mann, K. H. (1965). Heated effluents and their effects on the invertebrate fauna of rivers. *Proc. Soc. Water Treat. Exam.* **14,** 45–53.

Martin, A. W., and Harrison, F. M. (1966). Excretion. *In* "Physiology of Mollusca" (K. M. Wilbur, and C. M. Yonge, eds.), Vol. 2, pp. 353–386. Academic Press, New York.

Mattice, J. S. (1976). Effect of temperature on growth, mortality, reproduction, and production of adult snails. *ERDA Symp. Ser.* **40,** 73–80.

Meakins, R. H. (1980). Studies on the physiology of the snail *Biomphalaria glabrata* (Say): Effects of body size, temperature and parasitism by sporocysts of *Schistosoma mansoni* Sambon upon respiration. *Comp. Biochem. Physiol.* **66A,** 137–140.

Mehlman, B., and von Brand, T. (1951). Further studies on the anaerobic metabolism of some fresh water snails. *Biol. Bull.* (*Woods Hole, Mass.*) **100,** 199–205.

Monge, C., and Whittembury, J. (1976). High altitude adaptations in the whole animal. *In* "Environmental Physiology of Animals" (J. Bligh, J. L. Cloudsley-Thompson, and A. G. MacDonald, eds.), pp. 289–308. Wiley, New York.

Morton, J. E. (1955). The evolution of the Ellobiidae with a discussion on the origin of the Pulmonata. *Proc. Zool. Soc. London,* **125,** 127–168.

Morton, J. E., and Yonge, C. M. (1964). Classification and structure of the Mollusca. *In* "Physiology of Mollusca" (K. M. Wilbur and C. M. Yonge, eds.), Vol. 1, pp. 1–58. Academic Press, New York.

Nagabhushanam, R., and Azmatunnisa, Q. (1976). Temperature relations of the freshwater pulmonate, *Lymnae acuminata. J. Anim. Morphol. Physiol.* **23,** 52–59.

Needham, J. (1938). Contributions of chemical physiology to the problem of reversibility in evolution. *Biol. Rev. Cambridge Philos. Soc.* **13,** 225–251.

Newman, K. C., and Thomas, R. E. (1975). Ammonia, urea and uric acid levels in active and estivating snails, *Bakerilymnae cockerelli. Comp. Biochem. Physiol.* **50A,** 109–112.

Nickerson, R. P. (1972). A survey of enzyme and shell variation in 16 populations of the stream limpet, *Ferressia rivularis* (Say). Ph.D. Dissertation, Syracuse University, Syracuse, New York.

Olivier, L. (1956). Observations on vectors of *Schistosomiasis mansoni* kept out of water in the laboratory. I. *J. Parasitol.* **42,** 137–146.

Olivier, L., and Barbosa, F. S. (1955a). Seasonal studies on *Australorbis glabratus* Say from two localities in Eastern Pernambuco, Brazil. *Publ. Avulsas Inst. Aggeu Magalhaes, Recife, Braz.* **4,** 79–103.

Olivier, L., and Barbosa, F. S. (1955b). Seasonal studies on *Tropicorbis centimetralis* in Northeastern Brazil. *Publ. Avulsas Inst. Aggeu Magalhaes, Recife, Braz.* **4,** 105–115.

Olivier, L., and Barbosa, F. S. (1956). Observations on vectors of *Schistomiasis mansoni* kept out of water in the laboratory. II. *J. Parasitol* **42,** 277–286.

Packard, G. C. (1972). Inverse compensation for temperature in oxygen consumption of the hylid frog *Pseudacris triseriata. Physiol. Zool.* **45,** 270–275.

Parry, G. D. (1978). Effects of growth and temperature acclimation on metabolic rate in the limpet, *Cellana tramoserica* (Gastropoda: Patellidae). *J. Anim. Ecol.* **47,** 351–368.

Payne, B. S. (1979). Bioenergetic budgeting of carbon and nitrogen in the life-histories of three lake populations of the prosobranch snail, *Goniobasis livescens.* Ph.D. Dissertation, Syracuse University, Syracuse, New York.

Phillips, J. G. (1975). "Environmental Physiology." Wiley, New York.

Pohl, H. (1976). Thermal adaptation in the whole animal. *In* "Environmental Physiology of Animals" (J. Bligh, J. L. Cloudsley-Thompson, and A. G. MacDonald, eds.), pp. 261–286. Wiley, New York.

Potts, W. T. W. (1967). Excretion in molluscs. *Biol. Rev. Cambridge Philos. Soc.* **42,** 1–41.

Precht, H. (1939). Die lüngenatmung der süsswasser pulmonaten (zugleich ein beitrag zur temperaturabhangigheit der atmung). *Z. Vergl. Physiol.* **26,** 696–739.

Precht, H., Christophersen, J., Hensel, H., and Larcher, W. (1973). "Temperature and Life." Springer-Verlag, Berlin and New York.

Price, C. H. (1980). Water relations and physiological ecology of the salt marsh snail, *Melampus bidentatus* Say. *J. Exp. Mar. Biol. Ecol.* **45,** 51–67.

Prinsloo, J. F., and van Eeden, J. A. (1973). The influence of temperature on the growth rate of *Bulinus* (*Bulinus*) *tropicus* (Krauss) and *Lymnaea natalensis* Krauss (Mollusca: Basommatophora). *Malacologia* **14,** 81–88.

Prosser, L. C. (1973a). Water: Osmotic balance, hormonal regulation. *In* "Comparative Animal Physiology" (L. C. Prosser, ed.), 3rd ed., pp. 1–78. Saunders, Philadelphia, Pennsylvania.

Prosser, L. C. (1973b). Inorganic ions. *In* "Comparative Animal Physiology" (L. C. Prosser, ed.), 3rd ed., pp. 79–110. Saunders, Philadelphia, Pennsylvania.

Prosser, L. C. (1973c). Temperature. *In* "Comparative Animal Physiology" (C. L. Prosser, ed.), 3rd ed., pp. 362–788. Saunders, Philadelphia, Pennsylvania.

Purchon, R. D. (1977). "The Biology of the Mollusca." Pergamon, Oxford.

Read, K. R. H. (1966). Molluscan hemoglobin and myoglobin. *In* "Physiology of Mollusca" (K. M. Wilbur and C. M. Yonge, eds.), Vol. 2, pp. 209–232. Academic Press, New York.

Richardot, M. (1977a). Ecological factors inducing estivation in the freshwater limpet *Ferrissia wautieri* (Basommatophora: Ancylidae). I. Oxygen content, organic matter content and pH of the water. *Malacol. Rev.* **10,** 7–13.

Richardot, M. (1977b). Ecological factors inducing estivation in the freshwater limpet *Ferrissia wautieri* (Basommatophora: Ancylidae). II. Photoperiod, light intensity and water temperature. *Malacol. Rev.* **10,** 15–30.

Roberts, J. L. (1960). The influence of photoperiod on thermal acclimation. *Zool. Anz.* **24,** 73–78.

Roberts, J. L. (1966). Systematic versus cellular acclimation to temperature by poikilotherms. *Helgol. Wiss. Meeresunters.* **14,** 451–465.

Robertson, J. D. (1964). Osmotic and ionic regulation. *In* "Physiology of Mollusca" (K. M. Wilbur and C. M. Yonge, eds.), Vol. 1, pp. 283–311. Academic Press, New York.

Romano, F. A., III (1980). Bioenergetics and neurosecretory controls in univoltine and bivoltine populations of *Ferrissia rivularis* (Say). Ph.D. Dissertation, Syracuse University, Syracuse, New York.

Rowan, W. B. (1966). Autumn migration of *Helisoma trivolvis* in Montana. *Nautilus* **79,** 108–109.

Russell Hunter, W. (1952). The adaptations of freshwater gastropoda. *Glasgow Nat.* **16,** 84–85.

Russell Hunter, W. (1953a). On migrations of *Lymnaea peregra* (Müller) on the shores of Loch Lomond. *Proc. R. Soc. Edinburgh. Sect. B: Biol.* **65,** 84–105.

Russell Hunter, W. (1953b). On the growth of the freshwater limpet, *Ancylus fluviatilis* Müller. *Proc. Zool. Soc. London* **123,** 623–636.

Russell Hunter, W. (1953c). The condition of the mantle cavity in two pulmonate snails living in Loch Lomand. *Proc. R. Soc. London, Ser. B:* **65,** 143–165.

Russell Hunter, W. (1957). Studies on freshwater snails at Loch Lomond. *Glasgow Univ. Publ. Stud. Loch Lomond* **1,** 56–95.

Russell Hunter, W. (1961a). Annual variations in growth and density in natural populations of freshwater snails in the west of Scotland. *Proc. Zool. Soc. London* **136,** 219–253.

Russell Hunter, W. (1961b). Life cycles of four freshwater snails in limited populations in Loch Lomond with a discussion of infraspecific variation. *Proc. Zool. Soc. London* **137,** 135–171.

Russell Hunter, W. (1964). Physiological aspects of ecology in nonmarine molluscs. *In* "Physiology of Mollusca" (K. M. Wilbur and C. M. Yonge, eds.), Vol. 1, pp. 83–126. Academic Press, New York.

Russell-Hunter, W. D. (1978). Ecology of freshwater pulmonates. *In* "Pulmonates" (V. Fretter and J. Peake, eds.), Vol. 2A, pp. 334–383. Academic Press, New York.

Russell-Hunter, W. D. and Buckley, D. E. (1983). Actuarial bioenergetics of nonmarine molluscan productivity. *In* "The Mollusca" (W. D. Russell-Hunter, ed.), Vol. 6, pp. 463–503. Academic Press, San Diego.

Russell-Hunter, W. D., and Eversole, A. G. (1976). Evidence for tissue degrowth in starved freshwater pulmonate snails (*Helisoma trivolvis*) from tissue, carbon and nitrogen analysis. *Comp. Biochem. Physiol.* **54A,** 447–453.

Russell-Hunter, W. D., and McMahon, R. F. (1976). Evidence for functional protandry in a freshwater basommatophoran limpet, *Laevapex fuscus. Trans. Am. Microsc. Soc.* **95,** 174–182.

Russell Hunter, W., Apley, M. L., Burky, A. J., and Meadows, R. T. (1967). Interpopulation variations in calcium metabolism in the stream limpet, *Ferrissia rivularis* (Say). *Science* **155,** 338–340.

Russell-Hunter, W. D., Burky, A. J., and Hunter, R. D. (1970). Interpopulation variation in shell components in the stream limpet, *Ferrissia rivularis. Biol. Bull.* (*Woods Hole, Mass.*) **139,** 402.

Russell-Hunter, W. D., Apley, M. L., and Hunter, R. D. (1972). Early life-history of *Melampus* and the significance of semilunar synchrony. *Biol. Bull. (Woods Hole, Mass.)* **143,** 623–656.

Russell-Hunter, W. D., Burky, A. J., and Hunter, R. D. (1981). Interpopulation variation in calcareous and proteinaceous shell components in the stream limpet, *Ferressia rivularis. Malacologia* **20,** 255–266.

Sankurathri, C. S., and Holmes, J. C. (1976). Effects of thermal effluents on the population dynamics of *Physa gyrina* Say (Mollusca: Gastropoda) at Lake Wabamun, Alberta. *Can. J. Zool.* **54,** 582–590.

Schmale, H., and Becker, W. (1977). Studies on the urea cycle of *Biomphalaria glabrata* during normal feeding activity, in starvation and with infection of *Schistosoma mansoni. Comp. Biochem. Physiol.* **58B,** 321–330.

Schoffeniels, E., and Gilles, R. (1972). Ion regulation and osmoregulation in Mollusca. *In* "Chemical Zoology" (M. Florkin and B. T. Scheer, eds.), Vol. 7, pp. 393–420. Academic Press, New York.

Sheanon, M. J., and Trama, F. B. (1972). Influence of phenol and temperature on the respiration of a freshwater snail *Helisoma trivolvis. Hydrobiologia* **40,** 321–328.

Shumway, S. E. (1981). Factors affecting oxygen consumption in the marine pulmonate *Amphibola crenata* (Gmelin, 1791). *Biol. Bull. (Woods Hole, Mass.)* **160,** 332–347.

Skoog, G. (1976). Effects of acclimatization and physiological state on the tolerance to high temperatures and reactions to desiccation of *Theodoxus fluviatilis* and *Lymnaea peregra. Oikos* **27,** 50–56.

Sodeman, W. A., Jr., and Dowda, M. C. (1974). Behavioral responses of *Biomphalaria glabrata. Physiol. Zool.* **47,** 198–206.

Solem, A. (1978). Classification of the land mollusca. *In* "Pulmonates" (V. Fretter and J. Peake, eds.), Vol. 2A, pp. 49–97. Academic Press, New York.

Stearns, S. C. (1976). Life-history tactics: A review of the ideas. *Q. Rev. Biol.* **51,** 3–47.

Stearns, S. C. (1977). The evolution of life history traits: A critique of the theory and a review of the data. *Annu. Rev. Ecol. Syst.* **8,** 145–171.

Stokes, T. M., and Awapara, J. (1968). Alanine and succinate as end-products of glucose degradation in the clam *Rangia cuneata. Comp. Biochem. Physiol.* **25,** 883–892.

Streit, V. B. (1975). Experimentelle untersuchungen zum stoffhaushalt von *Ancylus fluviatilis* (Gastropoda—Basommatophora) 1. Ingestion, assimilation, wachstum und eiablage. *Arch. Hydrobiol.* **4,** 458–514.

Studier, E. H., and Pace, G. L. (1978). Oxygen consumption in the prosobranch snail *Viviparous contectoides* (Mollusca: Gastropoda)—IV. Effects of dissolved oxygen level, starvation, density, symbiotic algae, substrate composition and osmotic pressure. *Comp. Biochem. Physiol.* **59A,** 199–203.

Tashiro, J. S. (1982). Grazing in *Bithynia tentaculata:* Age-specific bioenergetic patterns in reproductive partitioning of ingested carbon and nitrogen. *Am. Midl. Nat.* **107,** 133–150.

Tashiro, J. S., and Colman, S. D. (1982). Filter-feeding in the freshwater prosobranch snail *Bithynia tentaculata:* Bioenergetic partitioning of ingested carbon and nitrogen. *Am. Midl. Nat.* **107,** 114–132.

Taylor, D. W. (1981). Freshwater mollusks of California: A distributional checklist. *Calif. Fish Game* **67,** 140–163.

Thomas, J. D., Benjamin, M., Lough, A., and Aram, R. H. (1974). The effects of calcium in the external environment on the growth and natality rates of *Biomphalaria glabrata* (Say). *J. Anim. Ecol.* **43,** 839–860.

Turner, H. M. (1978). *Hebetancylus excentricus* (Morelet) (Pulmonata: Ancylidae) in Louisiana and a report of septum formation. *Nautilus* **92,** 83–85.

van Aardt, W. J. (1968). Quantitative aspects of the water balance in *Lymnaea stagnalis* (L.). *Neth. J. Zool.* **18,** 253–312.

van der Borght, O. (1963). In- and out-fluxes of calcium ions in freshwater gastropods. *Arch. Int. Physiol. Biochim.* **71,** 46–50.

van der Borght, O., and van Puymbroeck. S. (1964). Active transport of alkaline earth ions as physiological base of the accumulation of some radio-nucleotides in freshwater molluscs. *Nature (London)* **204,** 533–535.

van der Borght, O., and van Puymbroeck, S. (1966). Calcium metabolism in a freshwater mollusc: Quantitative importance of water and food as supply for calcium during growth. *Nature (London)* **210,** 791–793.

van der Schalie, H., and Berry, E. G. (1973). The effects of temperature on growth and reproduction of aquatic snails. *Sterkiana* **50,** 1–92.

von Brand, T., and Mehlman, B. (1953). Relations between pre- and post- anaerobic oxygen consumption and oxygen tension in some fresh water snails. *Biol. Bull.* (*Woods Hole, Mass.*) **104,** 301–312.

von Brand, T., Baernstein, H. D., and Mehlman, B. (1950). Studies on the anaerobic metabolism and the aerobic carbohydrate consumption of some fresh water snails. *Biol. Bull.* (*Woods Hole, Mass.*) **98,** 266–276.

von Brand, T., McMahon, P., and Nolan, M. O. (1957). Physiological observations on starvation and desiccation of the snail *Australorbis glabratus*. *Biol. Bull.* (*Woods Hole, Mass.*) **113,** 89–102.

Walton, C. L., and Jones, W. W. (1926). Further observations on the life-history of *Limnaea truncatula*. *Parasitology* **18,** 144–147.

Watson, H., and Verdcourt, B. (1953). The two British species of *Carychium*. *J. Conchol.* **23,** 306–324.

Wieser, W. (1980). Metabolic end products in three species of marine gastropods. *J. Mar. Biol. Assoc. U.K.* **60,** 175–180.

Wilbur, K. M. (1964). Shell formation and regeneration. *In* "Physiology of Mollusca" (K. M. Wilbur and C. M. Yonge, eds.), Vol. 1, pp. 243–282. Academic Press, New York.

Wilbur, K. M. (1972). Shell formation in mollusks. *In* "Chemical Zoology" (M. Florkin and B. T. Scheer, eds.), Vol. 7, pp. 103–145. Academic Press, New York.

Wilbur, K. M., and Yonge, C. M., eds. (1964). "Physiology of the Mollusca," Vol. 1. Academic Press, New York.

Wilbur, K. M., and Yonge, C. M., eds. (1966). "Physiology of Mollusca," Vol. 2. Academic Press, New York.

Wilson, R. A. (1968). An investigation into the mucus produced by *Lymnaea truncatula*, the snail host of *Fasciola hepatica*. *Comp. Biochem. Physiol.* **24,** 629–633.

Young, J. O. (1975a). Preliminary field and laboratory studies on the survival and spawning of several species of Gastropoda in calcium-poor and calcium-rich waters. *Proc. Malacol. Soc. London* **41,** 429–437.

Young, J. O. (1975b). A laboratory study, using 45Ca tracer, on the source of calcium during growth in two freshwater species of Gastropoda. *Proc. Malacol. Soc. London* **41,** 439–445.

Zaaijer, J. J. P., and Wolvekamp, H. P. (1958). Some experiments on the haemoglobin-oxygen equilibrium in the blood of the ramshorn (*Planorbis corneus* L.). *Acta Physiol. Pharmacol. Neerl.* **7,** 56–77.

10

Physiological Ecology of Land Snails and Slugs

WAYNE A. RIDDLE

Department of Biological Sciences
Illinois State University
Normal, Illinois

I. Introduction

The environmental physiology of land molluscs has been treated in a number of general works (Hyman, 1967; Morton, 1967; Runham and Hunter, 1970; Prosser, 1973; Crawford, 1981). Other reviews relevant to the present chapter have focused on the physiological ecology of nonmarine molluscs (Russell-Hunter, 1964) and the biochemical ecology of mollusca (Gilles, 1972). This chapter examines the physiological, and to some extent the behavioral, adaptations of land snails and slugs to terrestrial environments. Terrestrial habitats, unlike aquatic and marine habitats, limit the availability of free water and are

The Mollusca, Vol. 6
Ecology

ISBN 0-12-751406-6

characterized by daily and seasonal variations in temperature not experienced in more thermally stable aquatic and marine biotopes. Despite the abiotic constraints of terrestrial habitats, pulmonates are successful land animals, occupying some of the most severe habitats on earth. Their success is largely attributable to physiological and behavioral adaptations to problems of water balance and environmental temperature, as well as to the problems of osmotic and ionic regulation and nitrogen excretion. Also important, and closely interrelated with water balance and thermal relations, are adaptive alterations in the respiratory metabolism of land pulmonates. Finally, regulation of locomotor activity during daily and seasonal time periods ensures that essential activities of land snails and slugs occur under the most favorable environmental conditions.

II. Water Relations

A. Water Uptake

Water uptake by dehydrated land snails and slugs potentially occurs through the integument, by drinking or through feeding. Available evidence indicates that rehydration is fairly rapid. In dehydrated *Otala lactea* and *Helix aspersa* allowed to crawl over a wet surface, hemolymph dilution was observed almost immediately and full rehydration was evident within 2–3 h (Burton, 1966). Similar rates of rehydration have been noted in *Anguispira alternata* (Riddle, 1981a). In *H. aspersa,* Machin (1964a) observed a weight increase of 30%/h in snails immersed in water. In a single dehydrated specimen of *Limax maximus* immersed in tap water to a region below the pneumostome, Dainton (1954a) found nearly complete rehydration over a 2-h period. She concluded that most water absorption in slugs occurs through the body surface. Water uptake rates show substantial variation (Machin, 1975), which is probably attributable to variation in the extent of prior dehydration and hemolymph osmolality, as well as to the duration and extent of contact with free water. The hygroscopic properties of superficial mucus, as well as the surface anatomy of the integument, facilitate spreading of mucus and potentially increase the surface area available for water absorption. Machin (1964a) found the mucus of *H. aspersa* to have an osmotic concentration equivalent to a 7.4 g/liter NaCl solution (~ 240 mOsm/kg). The osmotic gradient existing between mucus and free water in the environment appears to be an important determinant of water uptake. Osmotic uptake of vapor water by integumentary mucus, although theoretically possible at a near-saturated humidity, probably is relatively unimportant under natural conditions. In *H. aspersa,* a weight increase of only 5% per week was found in saturated air (Machin, 1964a). Water gain was observed only under saturated humidity conditions in *Euparypha pisana* (Lazaridou-Dimitriadou and Daguzan, 1978).

Water uptake by drinking in either land snails or slugs remains to be established. Buccal mass movements observed by Künkel (1916) in slugs moving over a moist surface were interpreted as indications of drinking. Pusswald (1948) argued that water uptake by both drinking and integumentary absorption occurs in snails. In contrast, oral water uptake was never observed by Dainton (1954a) in the slugs she studied.

Water uptake in feeding has been demonstrated but may be of limited significance to the water balance of land pulmonates. An analysis by Machin (1975) of the results of Pallant (1970) on the slug *Agriolimax reticulatus* and of Mason (1970) on several species of woodland snails showed net water gains in animals that were provided with natural diets. Evidence that water ingested in feeding alone is not sufficient to maintain water balance in slugs comes from work by Hunter (1968), who found that animals provided with food under saturated air conditions continued to lose weight. Blinn (1964) found that in the snails *Mesodon thyroidus* and *Allogona profunda,* water available from feeding on lettuce under near-saturated conditions was not sufficient to maintain body hydration.

B. Water Loss

Machin (1964a) has shown that integumentary water loss in active, extended *H. aspersa* is comparable to that from a free water surface. He estimated that the osmotic permeability of the body wall of *H. aspersa* was too low to provide water to superficial mucus at a rate required to prevent drying of the integument at low humidities. Continuous mucous gland activity was considered necessary to provide water at a sufficient rate to prevent surface drying of the integument at lower humidities. Mucus glands of the foot and body wall are surrounded by rather extensive blood sinuses (Campion, 1961). Machin (1964a) considered that hemocoelic pressure was normally sufficient to cause rather continuous secretion from these glands under normal humidity conditions. At greater rates of water loss, localized muscular undulations increased mucus secretion to rapidly drying areas. Detailed discussions of the physics of integumentary evaporation in active snails can be found in Machin (1964b,c) and as part of a broad treatment of the physiology of the integument of molluscs (Machin, 1977).

Work on the water relations of estivating snails has focused primarily on the effectiveness of the shell, mantle epithelium, and epiphragm as barriers to evaporative water loss. Understandably, the shell constitutes a major barrier to water loss in inactive animals. Measurement of shell permeability to water has required the use of water-filled shells in which the aperture has been sealed. Using this technique, Warburg (1965) attributed nearly one-half of the total water loss in a xeric-adapted snail, *Sinumelon remissum,* to loss through the shell. Much smaller contributions of water loss through the shell were found by Cameron (1970a) in helicid snails from England. He determined that this loss accounted for less than 5% of the total water loss. He found that water loss through the shell did not differ significantly among *Cepaea nemoralis, C. hortensis,* and *Arianta arbustorum.*

Machin (1967) has argued that differences in total water loss rates between dormant snails from xeric and mesic habitats are attributable to species differences in aperture area, shell thickness, and epiphragm thickness rather than to physiological differences. Using a percentage index of relative aperture size (aperture area × 100/aperture area + shell area), he found a relationship between relative aperture area and habitat. Among the three species of snails examined, the lowest relative aperture area was found in the desert snail, *Sphincterochila boissieri* (4.3%), and the highest (12.1%) in the mesic *H. aspersa. Otala lactea,* considered to occupy a semiarid habitat, had an intermediate value of 7.8%. A comparable association between aperture size and habitat was found by Gebhardt-Dunkel (1953). Greater shell thickness, in addition to reduced aperture size, may be important in reducing water loss. Machin (1967) found that shells of *S. boissieri* were approximately twice as thick as those of *O. lactea* and *H. aspersa.* Evidence of an association between shell thickness and habitat dryness within single species has emerged from work on certain helicid snails in England and Israel. In adult *H. pomatia* taken from localities in southern England which differed in rainfall, Pollard (1975a) found thicker shells in individuals from populations in areas of low rainfall. He suggested that a thicker shell in these areas may result in a longer life span and enable snails to survive a number of poor breeding seasons. Laboratory rearing of *H. pomatia* from eggs indicated that population differences in shell thickness were at least partly genetic. Bar (1978) found a clear trend of increasing shell thickness, although not of shell size, in *Theba pisana* along a gradient of increasing habitat aridity in Israel. Bar argued that the observed smaller variation in shell thickness in populations from the most arid habitats was evidence of the result of strong selection for this trait. Intraspecific variation in shell thickness in land snails has also been attributed to differences in the availability of calcium carbonate in the environment. Oldham (1928) found that providing chalk to *A. arbustorum* had little effect on shell size but a substantial influence on shell mass. Murray (1966) noted significant differences in shell thickness in *C. nemoralis* taken from seaside and inland populations, and considered the greater thickness of shells from the seaside population was caused by additional calcium carbonate in the form of marine shells washed up from the sea.

Substantial reductions in water loss rates of land snails accompany withdrawal into the shell at the onset of estivation. Water loss rates in fully inactive snails can be at least three orders of magnitude lower than in active animals (Machin, 1975). Such vast differences in water loss rates cannot be explained solely on the basis of reduced exposure of the moist integument during withdrawal, but rather reflect regulation of evaporation from the mantle surface of inactive animals.

Investigations ultimately resulting in a proposed mechanism of water loss regulation in dormant land snails followed from the observations of Machin (1965), showing that a withdrawn snail forming an epiphragm had a water loss rate some 45 times greater than that of an inactive one. Estimates of mantle permeability confirmed the existence of cutaneous regulation in *H. aspersa* (Machin, 1966). Evidence was subsequently presented supporting the proposed existence of a barrier to evaporative water movement located beneath a superficial hygroscopic layer of mantle tissue in *O. lactea* (Machin, 1972). Analysis of frozen sections of the mantle epithelium of inactive *O. lactea* showed that a steep osmotic gradient was maintained between the surface of epithelial cells and the base of microvilli located approximately 2 μm beneath the cell surface (Machin, 1974). An examination of the ultrastructure of the mantle of dormant *O. lactea* revealed abundant lamellate vesicles in the apical cytoplasm (Newell and Machin, 1976; Appleton and Newell, 1977). These vesicles were far less abundant in snails examined following arousal from dormancy. It was tentatively concluded that lamellate vesicles and their contents constituted the permeability barrier previously envisioned. Appleton et al. (1979) showed that the osmotic gradients in apical cytoplasm of dormant *O. lactea* were caused in part by potassium and chloride ion concentrations, and further argued that lamellate vesicles and possibly horizontally arranged microtubules constituted the permeability barrier. It is not yet clear if the formation of the vesicles precedes or follows the establishment of the osmotic gradient (Newell and Appleton, 1979). Intriguing questions remain concerning the mechanisms of formation of these vesicles as well as the energetics of maintaining (or abolishing) the barrier and the osmotic gradient associated with it.

Water loss from the mantle of many species of land snails is reduced by the formation of a mucus epiphragm across the shell aperture. Although some evidence suggests that the epiphragm may be a largely insignificant barrier to evaporation from a moist mantle surface (Riddle, 1975), its importance in further reducing water loss in dormant "regulating" snails is unquestionable. Epiphragm thickness may vary considerably between species. Machin (1967) found that epiphragms were approximately six times thicker in the desert-adapted *S. boissieri* than in *H. aspersa*. Thicker epiphragms, along with thicker shells and reduced relative aperture size, undoubtedly constitute important adaptations for survival under xeric conditions. Machin (1975) presented data indicating that reductions in total water loss attributable to the epiphragm were substantial in *H. aspersa* (41%), *O. lactea* (37%), and *S. boissieri* (31%). Van der Laan (1975) noted that epiphragms of dormant *Helminthoglypta arrosa* reduced total water loss by 50–60%, depending on the duration of inactivity. Survival time in two species of camaenid tree snails exposed to desiccation was found to be positively correlated with the proportion of exposure time that an epiphragm was present (Heatwole and Heatwole, 1978). Machin (1968) determined that epiphragm permeability was comparable to that of a waterproof cellulose film and varied inversely with its thickness and its water content. Water content was shown to be determined by the average vapor pressure existing across the epiphragm. Completely dried epiphragms, unlike fresh mucus, resisted absorption of liquid water (Machin, 1968) but did absorb water vapor (Warburg, 1965).

The positioning of epiphragms and the environmental conditions favoring their formation have been examined by Rokitka and Herreid (1975a,b). They found that in *O. lactea* the total number of epiphragms secreted depended largely on the duration of dormancy. As the number of epiphragms increased, the distance between the mantle tissue and the nearest membrane decreased. Snails were more likely to become dormant and form epiphragms at low relative humidities and low temperatures. Arousal from dormancy in *O. lactea* is favored by lower temperatures and higher relative humidities (Herreid and Rokitka, 1976). Horne (1973a) noted that the semi-desert snail *Bulimulus dealbatus* tended to produce more epiphragms, with a maximum of seven, during prolonged exposure to low humidity. The desert snail *S. boissieri* may produce as many as five epiphragms (Yom-Tov, 1971a). The effect of producing additional epiphragms on water loss has not been examined. Any advantage of multiple epiphragms in reducing water loss during prolonged dormancy would have to be weighed against the disadvantage of additional water loss associated with mucous gland activity required in epiphragm formation.

Epiphragms may be particularly significant in minimizing respiratory water loss. Evidence for such a function comes from work by Machin (1975) in which the weight of a single dormant specimen of *O. lactea* was continuously recorded. That record showed a sharp weight loss following the release of humid pulmonary air and a subsequent weight gain. The weight gain observed was apparently caused by the absorption of water vapor by the dehydrated mantle surface. This absorption would likely be less effective in the absence of an epiphragm because humid air should be rapidly dispersed.

Evaporative water loss from estivating land snails is potentially influenced by ambient temperature, relative humidity, and air movement, as well as by size and shell morphology. Estivating snails are also sensitive to physical disturbances which can initiate mucus secretion, opening of the pneumostome, and protrusion of the foot (Ross, 1979). Such disturbances can greatly affect water loss rates (Schmidt-Nielsen et al., 1971; Machin, 1975). In a study of three species of Australian land snails, Warburg (1965) found that water loss rates in dry air were strongly affected by temperature. Temperature effects were also noted in other species (Warburg, 1972). In the earlier paper, Warburg reported that water loss at a constant saturation deficit was influenced by temperature in two of the three species examined. This observation was significant because it suggested that temperature itself, rather than saturation deficit, influenced total water loss. Although direct temperature effects on mantle and shell permeability may exist, greater rates of respiratory water loss at higher temperatures may also have contributed to the observed temperature effects. Warburg's use of higher humidities at higher temperatures to maintain a constant saturation deficit may have further enhanced respiratory water loss. Higher respiration rates have been associated not only with higher temperature (see Section V) but also with higher ambient relative humidity (Riddle, 1975, 1977; Herreid, 1977). Although difficult to determine directly, differences in the contribution of respiratory water loss to total water loss might explain some of the differences in water loss rates noted in snails from different habitats. There is some evidence of lower respiration rates during dormancy in snails from xeric habitats (Riddle, 1977). Consistent with the findings of Warburg were those of Heatwole and Heatwole (1978), which indicated that water loss rates were from 2 to 10 times higher at 30°C than at 20°C depending on the species of snail examined. Water loss rates, however, were not affected by relative humidity in four of the five species tested.

The greater surface area : volume status and thinner shells of small snails place them at a potential disadvantage with respect to water loss. In addition, small snails may expose a larger aperture area per unit mass than larger animals (Riddle, 1975). Despite a significant inverse relationship found between size and per unit mass water loss rates in desert snails killed within shells, water loss rates of estivating animals showed no relationship to size (Riddle, 1975). In contrast, Heatwole and Heatwole (1978) found an inverse size dependence on water loss in living animals. Cameron (1970a) suggested that size differences in *Cepaea* may have contributed to intraspecific and interspecific variations in survival during dehydration.

Much of our understanding of the water relations of land snails has been based on laboratory work investigating the effectiveness of structural and physiological barriers to transpiratory water loss. Although these studies have contributed a great deal, they have been understandably limited in their ability to comment on the water relations of animals under natural conditions or to explain fully observed habitat distributions. Deserving of greater emphasis in the future are studies which take into consideration the constraints on water balance of particular habitats, as well as the adaptiveness of behavioral responses made by animals in those habitats. Using such an approach, Cameron (1970a) investigated the water relations of the helicid snails *A. arbustorum, C. hortensis,* and *C. nemoralis* in England. Field observations (Cameron, 1969) indicated that *A. arbustorum* occupied habitats considered the most moist, followed by *C. hortensis* and *C. nemoralis*. Under low humidity conditions, *A. arbustorum* had greater water loss rates and survived less well than the *Cepaea* species. *A. arbustorum* remained active longer and formed epiphragms less readily than did the *Cepaea* species. Cameron argued that the behavioral responses of each species reflected selection maximizing activity under favorable conditions balanced against selection minimizing the risk of death by dehydration. He further proposed that in snails naturally subjected to long dry periods, rapid withdrawal and epiphragm formation were particularly adaptive responses, as was rapid emergence in response to available moisture. In snails occupying damp habitats, maintaining activity for a longer period might be favored despite high water loss rates, because dry periods probably would be of short duration. In a very thorough and broadly comparative ecological study by Heatwole and Heatwole (1978), the higher water loss rates during dormancy found in snails taken from more mesic upland habitats in Puerto Rico were at least partly attributable to the absence of epiphragm formation in upland species and the inability of one species to withdraw fully into its shell.

C. Water Balance in Eggs

Eggs of terrestrial pulmonates show little resistance to water loss, although they do appear capable of withstanding substantial dehydration. Bayne (1968a) found water loss rates of snail and slug eggs to be typically greater than 1%/min at 20°C and 50% relative humidity and to be fairly constant to at least 50% weight loss. Species differences in water loss rates were slight. Bayne (1968b) believed that the calcareous granules found in eggs of some snails and slugs may confer structural support but were of little significance in reducing water loss. Bayne (1969) found that embryos of the slug *Agriolimax reticulatus* were able to survive 60–80% water loss, and that previous dehydration had little effect on viability after eggs were allowed to develop under moist conditions. Tolerated water losses as high as 85% in embryos and 75% in eggs of *Limax flavus* have been reported (Carmichael and Rivers, 1932). Tolerance to dehydration in the eggs of desert snails appears to be comparable to that of slugs. Yom-Tov (1971a) found that eggs of the desert snail *Trochoidea seetzeni* could survive a weight loss of 71%. Apparently nothing is known about the rate of rehydration occurring in eggs following a period of dehydration or about water exchange between the embryo and extraembryonic material; osmotic uptake of soil moisture by eggs seems probable considering the appreciable osmotic concentrations that have been found in the perivitelline fluid in eggs of two desert snails (Yom-Tov, 1971a).

The poor resistance to desiccation of eggs understandably requires that oviposition occurs in suitably moist microhabitats. Egg laying is avoided under dry conditions in slugs (Carrick, 1942; Arias and Crowell, 1963) and in snails (Wolda, 1965; Pollard, 1975b). Environmental water availability has been shown to affect fecundity, the rate of embryogenesis, and hatching success in *Helminthoglypta arrosa* (Van der Laan, 1980). Potts (1975) indicated that in *Helix aspersa* hatching required 3–4 weeks of warm, moist weather during which time the soil became neither dehydrated nor saturated with water. Potts found that in certain populations of *H. aspersa* studied, unsuccessful nesting and oviposition due to unfavorable weather conditions were fairly common. In the desert snails *S. boissieri* and *T. seetzeni*, eggs are deposited in excavated chambers only after rainfalls of 3 mm or more (Yom-Tov, 1971a).

D. Pallial Water

The water content values presented in Table I attest to the wide variation in body hydration tolerated by land snails and slugs. It can be noted that in many

TABLE I

Water Content of Land Snails and Slugs

Species	Water content (mg H_2O/mg shell-free dry weight) Mean	Range	Reference[a]
Slugs			
Arion ater	7.0	—	Pusswald (1948)
Arion ater	6.3	4.1–11.2	Martin et al. (1958)
Snails			
Achatina fulica	6.4	4.7–12.2	Martin et al. (1958)
Cepaea nemoralis	7.5	—	Trams et al. (1965)
Helix aperta	—	3.3–7.0	Burton (1966)
Helix pomatia	—	3.8–10.2	Burton (1964)
Helix pomatia (active)	5.1	4.1–7.4	Von Brand (1931)
Helix pomatia (inactive)	4.3	3.8–5.5	Von Brand (1931)
Otala lactea	—	4.9–10.1	Burton (1966)
Arianta arbustorum	5.6	4.3–9.0[b]	Stöver (1973a,b)
Cepaea nemoralis	—	4.0–5.3[c]	Jaremovic and Rollo (1979)
Sphinterochila boissieri	4.3	—	Schmidt-Nielsen et al. (1971)
Rabdotus schiedeanus	—	5.5–10.5[d]	Riddle (1975)
Anguispira alternata	—	3.2–7.0	Riddle (1981b)
Anguispira alternata	—	4.3–5.7[e]	Riddle (1981c)
Anguispira alternata	—	5.9–7.9[f]	Riddle, 1981b)

[a] Values from work published before 1973 calculated from percentage water content values given in Machin (1975).

[b] Values for animals collected following rainfall, in winter, and after maximum hydration, respectively.

[c] High and low values in field animals taken at the soil surface and on vegetation, respectively.

[d] Following moderate dehydration of fully hydrated animals.

[e] Winter–summer means of field animals.

[f] With mantle (pallial) water.

species the water content of dormant, dehydrated animals may be one-half or less than that of fully hydrated animals. Even under conditions in which free water is provided, substantial rhythmic fluctuations in water content may occur in snails and slugs (Howes and Wells, 1934a,b). In some snails, retention of pallial fluid in the mantle-cavity may contribute to the high water content of hydrated animals (Blinn, 1964; Riddle, 1981b). Blinn (1964) considered the observation of copious water discharge by Künkel (1916) in slugs to indicate pallial fluid expulsion. As an external body water source, pallial fluid has a significant potential influence on the duration of locomotor activity and on the ionic and osmotic concentration of body fluids. Blinn found that when free water was provided, the polygyrid snails *Mesodon thyroidus* and *Allogona profunda* displayed consistent changes in body hydration and locomotor activity associated with variations in the volume of pallial fluid. Smith (1981) determined that in the polygyrid *Triodopsis albolabris,* pallial fluid loss closely corresponded with total water loss during dehydration. He noted that during exposure to dehydration animals retaining fluid remained active for longer periods and had hemolymph osmotic concentrations far below those of animals forced to expel fluid. A similar influence of pallial fluid retention on hemolymph concentration was evident in *Anguispira alternata* (Riddle, 1981b). The source of pallial fluid has not been conclusively determined, but a urinary origin has been suggested (Blinn, 1964; Machin, 1975). Smith (1981) noted that pallial fluid osmolality in *T. albolabris* ranged from 38 to 58 mOsm/kg, roughly 100 mOsm/kg below that of the hemolymph. Considering the osmotic gradient between hemolymph and pallial fluid, osmotic movement of water to the hemolymph may occur readily. Based on the observations of Kerkut and Taylor (1956) showing that spontaneous electrical activity of the slug pedal ganglion decreased with increasing osmolality, Smith speculated that hemolymph absorption of pallial water during dehydration could prolong locomotor activity by delaying increases in hemolymph osmolality. Evidence of a strong association between locomotor activity and hemolymph osmolality comes from work by Wieser and Schuster (1975) on *H. pomatia* which showed that no active snail had a hemolymph osmolality greater than 190 mOsm/kg and that no inactive animal had one below 175 mOsm/kg.

III. Osmotic and Ionic Regulation

Osmotic and ionic regulation in molluscs has been thoroughly reviewed by Robertson (1964), Potts and Parry (1964), Schoeffeniels and Gilles (1972), and Machin (1975). Ionic regulation in land pulmonates has been examined in a series of papers by Burton (1965a,b,c, 1966, 1968a,b, 1969, 1971) and by others (Trams et al., 1965; Meincke, 1972; Stöver, 1973a; Matsumoto et al., 1974) and will not be discussed further here. Machin (1975) has presented hemolymph osmolality values for a wide variety of terrestrial pulmonates. Despite the volume of this data, the only generalization that clearly emerges is that osmotic concentration varies widely with changes in body hydration. No association has yet been established between body fluid concentration and either habitat dryness or taxonomic position in land pulmonates. However, Rumsey (1972) noted that among the terrestrial prosobranchs of the family Pomatiasidae, species from drier habitats tended to have higher hemolymph osmolalities.

The great reductions in water content tolerated by many land snails and slugs during dehydration (Table I) in addition to the rapid replenishment of body water with rehydration suggest that land pulmonates are well adapted to the potential stresses of osmotic concentration and dilution of body fluids. Basic problems associated with changes in body hydration are (1) effects on hemolymph and tissue fluid concentration, (2) exchanges of water and solutes between the hemolymph and intracellular water compartments, and (3) the influence of nitrogenous metabolites on body fluid concentrations. Considering the first problem, Stöver (1973a) found that hemolymph osmolality increased with reductions in body water content and blood volume in *Arianta arbustorum*. Little (1968) found that hemolymph osmolality may nearly double during estivation in the terrestrial prosobranchs *Pomacea lineata* and *P. depressa,* but noted that osmotic concentrations increased less than predicted, which suggested some regulation. In the woodland snail *Anguispira alternata,* hemolymph osmolality caused by solutes other than urea was not regulated during dehydration but changed predictably with variations in body hydration (Riddle, 1981b). In considering the problems of water and solute exchanges between hemolymph and tissue, it is evident from the work of Burton (1964) and Stöver (1973a) that the hemolymph functions as a water reservoir during dehydration, substantially minimizing changes in tissue water content. Burton found that variations in blood volume in *H. pomatia* accounted for most of the variation in body water content. Stöver noted a direct correlation between blood volume and total water content in *Arianta arbustorum*. Water loss does result in some dehydration of tissues, particularly glandular ones (Pusswald, 1948; Gebhardt-Dunkel, 1953; Burton, 1965b). Water content of muscle and hepatopancreas decreases significantly in *H. pomatia,* with reductions in body water content associated with winter dormancy (Perseca et al., 1977). The distribution of certain organic solutes and their influence on tissue water content has been examined by Wieser and Schuster (1975). They found that the water content of foot tissue in *H. pomatia* varied less than predicted with changes in hemolymph osmolality. Foot tissue concentrations of some organic solutes increased with increasing hemolymph osmolality but were approximately compensated in their effect by others withdrawn from the extracellular fluid. Consequently, the net effect on intracellular osmolality was negligible. Questions remain concerning the identity and behavior of possible additional osmotic effectors that contribute to the regulation of tissue water content. The final problem, that of dealing with the accumulation of nitrogenous metabolites during inactive periods, may be important in some snails. Hemolymph urea concentrations as high as 440 m*M* have been reported in the snail *Strophocheilus oblongus* (DeJorge et al., 1965). The level observed in *Anguispira alternata* after prolonged dormancy was 175 m*M*, which accounted for more than one-third of the total hemolymph osmolality (Riddle, 1981b). Hemolymph urea concentrations increase steadily during dormancy and are reduced following rehydration (Horne, 1971, 1973a; Chaturvedi and Agarwal, 1979; Riddle, 1981b). Hemolymph urea concentrations in *H. pomatia,* although comparatively low, were substantially greater than those of the foot tissue (Wieser and Schuster, 1975). Similarly, Tramell and Campbell (1972) noted that in *S. oblongus,* hemolymph urea concentrations exceeded tissue concentrations. On the basis of available information, it appears that land snails, and perhaps slugs, do not regulate hemolymph osmolality with changes in water content but are nonetheless capable of exchanging solutes in such a way as to minimize changes in tissue water content. Accumulation of urea, in those species where it does occur, appears simply to be tolerated.

IV. Nitrogen Excretion

Nitrogen excretion in molluscs has been treated in a number of reviews (Florkin, 1966; Martin and Harrison, 1966; Potts, 1967, 1968; Campbell and Bishop, 1970; Campbell et al., 1972; Florkin and Bricteux-Grégoire, 1972). Evidence presented in these reviews has led to the generalization that land snails are purinotelic, with excretory nitrogen appearing as uric acid, guanine, and xanthine. Results of experiments on *H. pomatia* by Jezewska et al. (1963) presented by Florkin and Bricteux-Grégoire (1972), on *Otala lactea* (Speeg and Campbell, 1968a), and on *S. oblongus* discussed in Campbell and Bishop (1970), clearly indicate the accumulation of purines in these species during estivation and excretion during active periods. Renal ammonia excretion in *Eremina desertorum* reported by Haggag and Fouad (1968) has been questioned, as have earlier reports of substantial urea excretion (Campbell and Bishop, 1970). The criticism of work reporting urea in the excreta of snails has been based on the possibility that at least some of the urea found was derived from microbial degradation of uric acid. This consideration also applies to the work of Horne (1971) and that of Riddle (1981b), who considered the reduction of high levels of tissue and hemolymph urea developed prior to rehydration was attributable to excretion, because the latter had collected excreta under conditions that were not bacteriostatic. Urea excretion by *S. oblongus* as reported by DeJorge et al. (1969) may also be questioned on the same grounds.

Despite the lack of unequivocal evidence of urea excretion, there can be no question that ornithine–urea cycle enzymes are significant in the nitrogen metabolism of land snails. Campbell and Bishop (1970), Campbell et al. (1972), Horne and Barnes (1970), and Horne and Boonkoom (1970) have shown these enzymes to be present in a variety of species. Although these enzymes are present in a number of land snails, they are responsible for significant urea production in only some species. In the semidesert snail *Bulimulus dealbatus,* urea biosynthesis and accumulation during dormancy have been clearly demonstrated (Horne, 1971, 1973a,b); whole body urea content was found to be low or undetectable during active periods, but increased to a greater extent than uric acid during dormancy. Horne (1973b) argued that an increase in the activity of urea cycle enzymes and a decrease in urease activity were responsible for accelerated urea biosynthesis during dormancy. The alleviation of ammonia toxicity was considered to be the major adaptive role of urea biosynthesis. In the South American land snail *S. oblongus,* appreciable tissue and hemolymph urea levels have been noted (DeJorge and Peterson, 1970). In contrast, Tramell and Campbell (1970) found no urea in kidney contents in that species, noting that purines accounted for more than 70% of total excretory nitrogen. Tramell and Campbell (1972) found extremely high urea levels in some dormant *S. oblongus,* but levels were undetectable in others. Evidence of ureolytic activity, possibly provided by urease, was noted.

Purine and urea synthesis and excretion have been found in slugs. Jezewska (1969) considered slugs to be purinotelic, based on the uric acid, guanine, and xanthine found in the excreta of *Deroceras agreste* and *Limax maximus.* In *L. flavus,* Horne (1977a) determined that urea and purines accounted for 59 and 41%, respectively, of total excretory nitrogen. Of the purines produced, uric acid predominated, with lesser contributions made by guanine and xanthine. Considering that microbial action on the excreta was inhibited by an antibiotic and that uric acid decomposition was shown to be minimal, urea excretion seems to be substantiated. In *L. flavus,* Horne (1977b, 1979) found that proteins were predominantly utilized during starvation, and that urea was synthesized faster in starving slugs than in well-fed ones. Unlike many land snails, *L. flavus* did not have an active urease.

Extrarenal ammonia excretion may be significant in some land snails. Speeg and Campbell (1968b) found that in *O. lactea* urea was synthesized but subsequently degraded by urease to form ammonia. In their view, ammonium ions increased the pH of the extrapallial fluid adjacent to the shell and thereby facilitated calcium carbonate deposition in the shell. Campbell and Boyen (1974) proposed a mechanism in which the formation of ammonium ion from ammonia, and ultimately calcium carbonate deposition, were controlled by carbonic anhydrase activity in the mantle. Loest (1979a) noted periodic bursts of ammonia volatilization in 11 species of terrestrial pulmonate and prosobranch snails. Slugs, he found, absorb ammonia rather than volatilize it, suggesting a means of retaining endogenous ammonia, possibly to be used for amino acid synthesis (Loest, 1979b). Urease or adenosine deaminase, possibly responsible for ammonia production, were found in the mantle tissue of a number of species of snails and slugs.

The adaptive significance of the ornithine–urea cycle in land pulmonates can be viewed in a number of ways. From the foregoing discussion, urea formation may be adaptive in some species, avoiding ammonia toxicity and providing a substrate for the ammonia production considered important in shell formation. Another view (Florkin and Bricteux-Grégoire, 1972) considers the cycle significant, not in the production of urea but in the formation of arginine to be used in both protein and phosphoarginine synthesis. Campbell et al. (1972) consider it unlikely that urea formation is involved in ammonia detoxification, but argue that urea accumulation is significant in reducing evaporative water loss in snails by decreasing the vapor pressure of body fluids. This possibility seems unlikely because even the highest levels of hemolymph urea concentrations reported (440 m*M*) could have only minimal effect in reducing the vapor pressure gradient existing between the animal and subsaturated air (Machin, 1975; Riddle, 1981a). Further, because evaporation is primarily regulated at the surface of mantle tissue in land snails, the osmolality of body fluids could have no direct effect on transpiration from mantle tissue. However, retention of urea may be significant in water uptake. Considering the evidence of osmotic absorption of water through the integument (Machin, 1975), high hemolymph urea concentrations developed during dormancy might facilitate osmotic movement of water between tissues and hemolymph on rehydration. Finally, there is also some evidence that urea may minimize the dilution of hemolymph during rehydration (Riddle, 1981b).

The advantages of purionotelic excretion in land snails and slugs by removing nitrogen with reduced urinary water loss seem apparent. Nonetheless, this advantage must be balanced against the greater carbon loss per nitrogen atom incurred with purine excretion than with urea. The accumulation of the comparatively insoluble purines rather than urea during dormant periods would minimize increases in the osmolality of body fluids, but otherwise would seem to confer no particular advantage over urea. The possible role of urates in providing an "ion sink," as proposed in arthropods (Mullins and Cochran, 1974), has not been examined in land pulmonates.

V. Respiratory Metabolism

As terrestrial ectotherms, snails and slugs may experience appreciable daily and seasonal variations in body temperature. To the extent that temperature exerts an effect on the rate of respiration, it also affects the rate of energy expenditure and influences the energy resources available for growth, reproduction, and other activities. An understanding of metabolic adaptations to temperature appears to be particularly relevant to studies of population energetics (Mason, 1971; Richardson, 1975). One can argue that those physiological and behavioral responses which reduce metabolic rates at temperature and moisture conditions unfavorable for activity are clearly adaptive, estivation in land snails being particularly significant. In a dormant condition, pulmonary ventilation rate declines (Nopp, 1974) and lung volume decreases (Krogh, 1941). Barnhart and Arp (1980) note that substantial changes in internal PO_2 and PCO_2 levels occur during dormancy in *O. lactea*. The decline in metabolic rate with estivation may vary among species, but it can be substantial (Nopp, 1974; Herreid, 1977; Riddle, 1977). Body water content exerts an influence on metabolic rate in both active and dormant snails (Wells, 1944; Wieser and Fritz, 1971), as does ambient relative humidity (Riddle, 1975, 1977; Herreid, 1977). Dormant snails may show temporal variations in metabolic rate because of bursts of respiratory activity. In *Sphincterocheila boissieri* these bursts were found to be irregular in periodicity (Schmidt-Nielsen et al., 1971), whereas in *Eobania* (*Otala*) *vermiculata* circadian periodicity was evident (Kratochvil, 1976). Simulated day–night changes in artificial illumination did not influence respiration in the dormant desert snail *Rabdotus schiedeanus* (Riddle, 1977).

Thermal acclimation of metabolic rate has been demonstrated in some, but not all, terrestrial pulmonates thus far examined. In *L. flavus,* prior exposure to lower temperatures resulted in higher metabolic rates (Segal, 1961). Similarly, Roy (1963) found an average depression in metabolic rate of 1–1.5%/°C increase in acclimation temperature in *Arion circumscriptus. L. maximus* and *Philomycus carolinianus* showed acclimatory changes in metabolism but differed in the time required for full acclimation (Rising and Armitage, 1969). Seasonal variations in acclimatory response were also evident. In snails, thermal acclimation in metabolic rate has been found in *Stenotrema leai* (Armitage and Stinson, 1980). In contrast, acclimation was not evident in *Arianta arbustorum* (Nopp, 1965; Wieser et al., 1970; Wieser and Fritz, 1971) or in the woodland snails *Hygromia striolata* and *Discus rotundatus* (Mason, 1971). Similarly, respiration rates were not affected by different prior temperature exposures in *R. schiedeanus* or *Helix aspersa* (Riddle, 1977). Seasonal changes in metabolism– temperature relations caused by natural acclimitization have been found in *S. leai* (Barnhart and Armitage, 1979) and in the rate of metabolism in *A. arbustorum* (Wieser et al., 1970; Wieser and Fritz, 1971). In the former species no distinct cold season depression in metabolism was evident, whereas in the latter, and also in *Helix pomatia* (Blâzka, 1955) and *Cepaea vindobondensis* (Müller, 1943), reductions in metabolism accompanied winter dormancy.

Metabolism–temperature (M-T) relations of snails and slugs have been extensively examined. M-T curves in slugs show strong temperature dependence, with a general trend toward lower respiratory Q_{10} values at higher temperatures, particularly in animals previously adapted to higher temperatures (Segal, 1961; Rising and Armitage, 1969). Presently, there is no evidence that slugs make rapid compensatory reductions in metabolism in response to increasing temperatures. Rather, the available evidence suggests that slugs make slow acclimatory adjustments to temperature changes over days or weeks (Rising and Armitage, 1969). M-T relations in snails, however, may be quite different. Evidence has accumulated from work on a number of species that metabolic rate in snails may be virtually insensitive to temperature over substantial temperature ranges (Nopp, 1965; Grainger, 1969; Wieser et al., 1970; Steigen, 1979). Strong temperature effects are apparent, however, during dormancy in some species (Blâzka, 1955; Schmidt-Nielsen et al., 1971). M-T relations of dormant animals may also be affected in a rather complex manner by changes in ambient humidity (Riddle, 1977).

Considerable effort has been directed at elucidating the biochemical basis of metabolic compensation in snails. Nopp and Farahat (1967) found that temperature influenced metabolic rate in isolated tissues in a manner comparable to that observed in whole animals, and that estivation as well as nutrition influenced tissue metabolism. Changes in acetylcholine esterase activity in *A. arbustorum* with season, water content, and respiration have been noted by Wieser and Fritz (1971). There has been an interest in certain enzymes, notably lactic acid dehydrogenase (LDH), and their roles in metabolic adaptation (Wieser, 1977; Storey, 1977; Gill, 1978; Wieser and Wright, 1978, 1979; Long et al., 1979).

Making estimates of the metabolic rates of land snails which accurately reflect energy expenditure under natural conditions is understandably difficult. Ideally, such estimates should take into account the proportion of time snails are dormant, as well as the temperature and perhaps humidity that prevail during dormant periods. In addition, factors such as water content, nutrition, reproductive state, season, and possible rhythmic respiratory activity would need to be considered. Finally, emerging evidence that land snails may respire anaerobically at certain times (Oudejans and Van der Horst, 1974; Van der Horst, 1974; Kratochovil, 1976; Storey, 1977; Wieser, 1978; Wieser and Wright, 1978) may further complicate studies of energetics by suggesting the use of direct calorimetry rather than respirometry in determining energy metabolism.

VI. Thermal Relations

A. Adaptations to High Temperatures

Slugs and some land snails occupy fairly protected habitats in which daily and seasonal extremes in air temperature are largely avoided. Many land snails, however, endure conditions unfavorable for activity by estivating in very exposed situations on vegetation or other objects. Although these estivation sites may permit avoidance of ground predators (Van der Laan, 1975), they may also confer protection from high temperatures at the soil surface. Pomeroy (1968) found that body temperatures of estivating *Helicella virgata* fully exposed to the sun could exceed ambient temperatures by as much as 10°C, but noted that temperature differences decreased with increasing distance from the soil surface. McQuaid et al. (1979) found a similar relationship between body temperature and distance from the ground. They further noted that snails tended to estivate with apertures directed upward on vertical surfaces in the coolest compass directions. Interestingly, *H. virgata* in Australia, unlike *Theba pisana* in South Africa, estivates on vertical surfaces in compass directions which would be more appropriate in the Northern Hemisphere, from which they, like *T. pisana,* were introduced. A model of heat transfer constructed for *T. pisana* suggested that heat from the substrate was conducted to the body of the snail and to ambient air via the shell surface and cooler airspace in the basal whorl. The upward orientation of the airspace was considered to be adaptive in influencing heat exchange between the insolated shell and the body. In *Sphincteochila boissieri,* a desert snail which estivates on the soil surface, the basal whorl airspace clearly provides crucial insulation between the body and the hot soil surface (Schmidt-Nielsen et al., 1971, 1972).

Upper lethal temperatures of land snails show some relationship to habitat temperature. Hogben and Kirk (1944) found that *Helix pomatia* from England survived at 44°C in saturated air for at least 1 h. *A. arbustorum* taken in the summer from a habitat at 2500 m in Austria and acclimated to 5°C were unable to survive for 1 h at 40°C (Grainger, 1969). Exposure to 45°C proved fatal within 1 h to *H. aspersa* acclimated to 13°C (Grainger, 1975). Heatwole and Heatwole (1978) found upper lethal temperatures of 42 to 46°C in snails from Puerto Rico and some indication of higher lethal limits in animals from warmer lowland populations. The desert snail *R. schiedeanus* survived at 45°C for 4 h (Riddle, 1975). L_{temp} 50 levels for 20-min and 3.5-h exposures of semidesert *T. pisana* were 49 and 44°C, respectively (McQuaid et al., 1979). The highest upper lethal temperatures thus far recorded (50–55°C) have been those of the desert snails *S. boissieri* and *Trochoidea seetzeni* (Schmidt-Nielsen et al., 1971, 1972; Yom-Tov, 1971b).

Grainger (1969, 1975) has shown that marked disturbances in Na^+ and K^+ concentrations, possibly leading to a blockage of neuromuscular transmission, occur with heat death in *A. arbustorum* and *H. aspersa.* Changes in certain inorganic and organic substances at extremely high and low temperatures have been examined in *H. pomatia* (Meincke, 1975). It should be emphasized that using the criteria of response to tactile stimuli and normal locomotor activity to establish survival following high-temperature exposure, it is not possible to evaluate the long-term effects of high temperature on the fitness of surviving animals. Surprisingly, no information seems to be available on the effects of prior thermal acclimation or seasonal acclimitization on upper lethal limits.

Shell color influences the thermal relations of snails exposed to solar radiation. Clearly adaptive are the highly reflective light or chalky white shells of desert snails (Schmidt-Nielsen et al., 1971; Yom-Tov, 1971b). Heatwole and Heatwole (1978) found lighter shell color in snails from more exposed and generally warmer habitats in Puerto Rico. Heller (1981) found no evidence that *Theba pisana* had darker coloration in northern, more mesic habitats in Israel. He argued that dark, mottled shells were cryptic and helped reduce predation. In remaining shielded by vegetation, *T. pisana* avoided overheating by solar radiation. In *C. nemoralis,* Emberton and Bradbury (1963) found greater transmission of light through shells of the yellow compared to the brown morphs, but were unable to measure the amount of reflected light and therefore absorbed energy in the color morphs. In live *C. nemoralis* and in mercury-filled shells exposed to sunlight, higher internal temperatures were associated with darker shell color (Heath, 1975). Garcia (1977) also noted variation in temperature with color phenotype in *C. nemoralis,* as Jones (1973a) did in *C. vindobondensis.*

The significance of color and banding polymorphism on the thermal relations of *Cepaea* has been important in arguments concerning climatic selection in land snails (see the review by Jones et al., 1977; also Chapter 15, this volume). In *C. nemoralis* there is clear evidence of greater frequencies of yellow-shelled morphs in southern populations in Europe (Jones et al., 1977). In *C. vindobondensis* fully pigmented snails were more common in valley frost hollows than on surrounding hillsides in Yugoslavia (Jones, 1973a). Similarly, the brown morph of *Cepacea* species in England may be more common in frost hollows prone to ponding of cold air (Cain and Currey, 1963a; Cain, 1968), although no strict association between the brown morph and the frost hollow habitat has been established (Cain and Currey, 1963b; Cameron, 1969). Jones (1973a,b) has suggested that on cool mornings, darker shelled *C. vindobondensis* and *C. nemoralis* may warm to temperatures suitable for activity more rapidly than lighter shelled animals. Observations of greater heat-associated mortality among brown morphs of *C. nemoralis* under field conditions in summer (Richardson, 1974) suggest the disadvantage of dark shell color. Clearly needed, in the view of Jones et al. (1977), are further studies on the effects of thermal physiology on the various aspects of fitness in *Cepaea* morphs.

B. Adaptations to Low Temperatures

Physiological and behavioral aspects of overwintering and cold hardiness in land molluscs have received comparatively little attention. Segal (1961) found poor survival in *L. flavus* at subfreezing temperatures and showed that prior low temperature acclimation did not improve cold survival. Mellanby (1961) noted that chill–coma temperatures of slugs differed among species but were not influenced by thermal acclimation. Getz (1959) showed that freezing for 5 h at −8°C was lethal to *Arion circumscriptus* and *Deroceras reticulatum* but not to *D. laeve*. Similar exposure to −10°C was lethal to all three species. Frozen *D. laeve* collected outdoors in late winter appeared to be normal on thawing, suggesting that this slug tolerates freezing under natural conditions. Snails appear to be readily killed by freezing but avoid freezing by supercooling. In *Arianta arbustorum* the supercooling point (the temperature at which lethal ice formation occurs) varied with body water content and with the site of initial freezing (Stöver, 1973b); tissue freezing was associated with a rise in K^+ and a lowering of Na^+ concentration in the hemolymph. In *Anguispira alternata* gut clearance and reduced water content in the fall were considered to be essential preparations for overwintering (Riddle, 1981c). In *A. alternata* the mean supercooling point decreased to about −16°C in February and was influenced by thermal acclimation. Cryoprotective polyhydric alcohols (glycerol, sorbitol, and mannitol) were not detected in the hemolymph of *A. alternata* collected in winter. The thick, distinctive winter epiphragms found in some species (Blinn, 1963; Barnhart, 1979; Riddle, 1981c) are probably important in preventing debris from contacting the mantle tissue, which might initiate freezing in supercooled animals (Stöver, 1973b). Blinn (1963) found no significant natural winter mortality in *Allogona profunda* and *Mesodon thyroidus,* whereas Carney (1966) noted appreciable winter mortality in *A. ptychophora.* In *Arianta arbustorum,* Terhivuo (1978) noted higher winter mortality in younger animals. The last three investigators all noted a predominantly upward apertural orientation in snails during overwintering. Oosterhoff (1977) found that winter survival was better in faster than in slower growing juvenile *C. nemoralis.*

Physiological adjustments to overwintering in land snails may include a depression in metabolic rate (see the previous discussion) and a reduction in water content (Table I). Changes in concentrations of hemolymph solutes, in enzyme activities, and in tissue metabolism may also occur (Wieser and Fritz, 1971; Stöver, 1973a; Meincke, 1974; Van der Horst et al., 1974; Perseca et al., 1977; Wieser and Wright, 1979). Physiological responses to the anoxic conditions that may exist in the spring in water-logged hibernacula have been examined in *H. pomatia* (Wieser, 1978, 1981). Low temperatures and short photoperiods are undoubtedly important in the preparation for, and initiation of, winter dormancy (Lind, 1968; Tischler, 1974; Jeppesen and Nygard, 1976; Jeppesen, 1977). Behavioral responses prior to overwintering in the form of autumnal homing to specific hibernacula have been noted in certain species of snails (Edelstam and Palmer, 1950; Blinn, 1963; Pollard, 1975b). The significance of chemoreception in the homing behavior of land pulmonates has been examined (Gelperin, 1974; Cook, 1979a,b, 1980).

VII. Locomotor Activity Patterns and Photoperiodism

Locomotor activity in land pulmonates is often restricted to daily periods of the most favorable light, moisture, and temperature conditions. Over a number of years, considerable effort has been made to evaluate the significance of various external environmental factors in influencing locomotor activity. Emphasis has been placed on aspects of the endogenous control of locomotor activity. Daily patterns of locomotor activity in slugs under natural conditions have been extensively described (Barnes and Weil, 1944, 1945; White, 1959; Webley, 1964, 1970; Hunter, 1968; Daxl, 1969; Crawford-Sidebotham, 1972; Baker, 1973) and in snails (Blinn, 1963; Henne, 1963; Pomeroy, 1969; Cameron, 1970b; Bailey, 1975). A number of investigators have examined the effects of changes in light intensity, relative humidity, and temperature in intitating locomotor activity and controlling its duration. For slugs, Dainton (1943, 1954a,b) argued that temperature changes, rather than changes in light intensity or humidity, were most significant in initiating activity. Karlin (1961) also considered temperature changes to be important but included food supply, humidity, free water availability, and light intensity as significant in slugs. Newell (1966, 1968) emphasized the importance of light intensity in influencing activity in *Agriolimax reticulatus*. Consistent with that emphasis, Lewis (1969a) noted that in *Arion ater* natural light conditions rather than temperature changes were important, and found that slugs readily synchronized their activity with artificial light cycles. In the snail *Polygyra* (*Triodopsis*) *albolabris* kept at constant temperature and light conditions, locomotor activity was distributed fairly uniformly over the 24-h period but was restricted largely to dark periods when a light–dark cycle was provided (Henne, 1963). Predominantly dark-phase locomotor activity was noted in *C. nemoralis, C. hortensis,* and *A. arbustorum,* although activity persisted for some time following the change in illumination to "light on" (Cameron, 1970b). In that study, all three species become more nocturnal at higher temperatures. Bailey (1975) found that seasonal activity of *H. aspersa* in outdoor enclosures was controlled most strongly by soil moisture and that daytime activity was restricted to periods of rainfall. He noted that outdoor activity began before sunset, reached a maximum 2–6 h later, and declined during the evening. A morning peak of activity occurred about 6 h after sunset. The duration of the active period in *H. aspersa* and the timing of evening and morning activity peaks appeared to be related to the time of sunset and were independent of the time of sunrise. Laboratory experiments confirmed the importance of sunset in timing activity. Tercafs (1960) found that cave-dwelling *Oxychilus cellaris* snails retained a predominantly nocturnal activity pattern during constant natural darkness comparable to that of epigean animals exposed to natural photoperiods.

Evidence for endogenous control of locomotor activity in slugs has come from work by Lewis (1969b), who found that *Arion ater* exhibited a locomotor activity rhythm of a roughly circadian period under constant darkness. In *L. maximus* a locomotor activity rhythm persisted under constant darkness with a temperature-compensated period of 23.6–24.6 h (Sokolove et al., 1977). Circadian locomotor activity of *L. maximus* and *L. flavus* has been shown to persist in constant darkness or light and to be phase shifted by the light–dark cycle. Interestingly, the eyes of these slugs are not necessary for the maintenance of rhythmicity or for entrainment to light cycles (Beiswanger et al., 1981). In these animals the eyes do provide photic input, but do not constitute an endogenous circadian oscillator or clock comparable to that envisioned by Jacklet (1969) in the eye of the sea slug *Aplysia*. In recent work on *H. aspersa*, Bailey (1981) demonstrated endogenous circadian rhythmicity in locomotor activity. In examining circannual activity rhythms, he found that at essentially constant temperature conditions, locomotor activity and feeding decreased with shorter simulated day lengths. Bailey argued that the initiation of winter dormancy in *H. aspersa* is controlled by a circannual rhythm entrained under natural conditions by photoperiod cues. Jeppesen (1977) found that the effect on hibernation in *H. pomatia* of short light pulses applied during long subjective night periods depended upon that time during the circadian period at which pulses were administered. He argued that *H. pomatia* displayed a photoperiodic control of hibernation based on a cyclic mechanism of time measurement. Photoperiodism in *H. pomatia* does appear to be consistent with a number of models based on work with other animals, and future work on the various aspects of photoperiodism in land molluscs will undoubtedly prove to be most fruitful.

Other physiological and biochemical processes in land pulmonates show rhythmicity or are influenced by photoperiod. Reproductive maturation and activity appear to be influenced by photoperiod in slugs (Sokolove and McCrone, 1978) and in snails (Bailey, 1981). Diurnal fluctuations in acetylcholine content and acetylcholine esterase activity in the snail *Cryptozona ligulata* may have been partly responsible for the diurnal rhythmicity in spontaneous electrical activity observed in that species (Reddy et al., 1978). Diel variations in total protein and carbohydrate content have been found in the slug *Laevicaulis alte* (Pavan Kumar et al., 1981), but their physiological significance seems unclear.

VIII. Concluding Remarks

Rather than attempt to draw any general conclusions from the foregoing discussion, it would seem more useful at this point to make some specific comments concerning future research. In the area of water relations, it would be premature to conclude that differences in water loss rates in dormant land snails from xeric and mesic habitats are exclusively attributable to structural or behavioral rather than physiological differences. Clearly needed, in the view of this writer, is further work assessing the relative contributions of water loss through the mantle epithelium, shell, and pneumostome in land snails from different habitats. Along this line, careful work investigating the effects of periodic pneumostome opening and closing on both gas exchange and water balance could prove most fruitful. The mechanisms by which external humidity changes are perceived by dormant land snails and influence respiratory metabolism deserve investigation. Concerning excretion, it is evident that land pulmonates as a group synthesize and void a variety of nitrogenous end products. It is presently impossible to associate particular patterns of nitrogen excretion, if indeed they do exist, with specific taxonomic groups or habitats. Here broadly comparative work on closely related species from different habitats or on single species having fairly wide habitat distributions would be pertinent. Finally, the paucity of information presently existing on the overwintering physiology and cold hardiness of land pulmonates should encourage further research, the underlying mechanisms of low temperature adaptation in general and of cold hardiness in particular being clearly worthy of intensive investigation.

References

Appleton, T. C., and Newell, P. F. (1977). X-ray microanalysis of freeze-dried ultrathin frozen sections of a regulating epithelium from the snail *Otala. Nature (London)* **266,** 854–855.

Appleton, T. C., Newell, P. F., and Machin, J. (1979). Ionic gradients within mantle-collar epithelial cells of the land snail *Otala lactea. Cell Tissue Res.* **199,** 83–97.

Arias, R. O., and Crowell, H. H. (1963). A contribution to the biology of the grey garden slug. *Bull. South. Calif. Acad. Sci.* **62,** 83–97.

Armitage, K. B., and Stinson, D. (1980). Metabolic acclimation to temperature in a terrestrial snail. *Comp. Biochem. Physiol. A* **67,** 135–139.

Bailey, S. E. R. (1975). The seasonal and daily patterns of locomotor activity in the snail *Helix aspersa* Müller, and their relation to environmental variables. *Proc. Malacol. Soc. London* **41,** 415–428.

Bailey, S. E. R. (1981). Circannual and circadian rhythms in the snail *Helix aspersa* Müller and the photoperiodic control of annual activity and reproduction. *J. Comp. Physiol.* **142,** 89–94.

Baker, A. N. (1973). Factors contributing towards the initiation of slug activity in the field. *Proc. Malacol. Soc. London* **40,** 329–333.

Bar, Z. (1978). Variation and natural selection in shell thickness of *Theba pisana* along climatic gradients in Israel. *J. Moll. Stud.* **44,** 322–326.

Barnes, H. F., and Weil, J. W. E. (1944). Slugs in gardens: their numbers, activities and distribution. Part I. *J. Anim. Ecol.* **13,** 140–175.

Barnes, H. F., and Weil, J. W. E. (1945). Slugs in gardens: their numbers, activities and distribution. Part II. *J. Anim. Ecol.* **14,** 71–105.

Barnhart, M. C. (1979). Notes on the winter epiphragm of *Pupoides albilabris. Veliger* **21,** 400–401.

Barnhart, M. C., and Armitage, K. B. (1979). Seasonal changes in the temperature affects on oxygen consumption of a terrestrial snail. *Comp. Biochem. Physiol. A* **63,** 539–541.

Barnhart, M. C., and Arp, A. (1980). Low O_2 and high CO_2 pressures in dormant land snails: Relation to respiratory gas exchange and hemocyanin function. *Am. Zool.* **20,** 871.

Bayne, C. J. (1968a). A study of the desiccation of egg capsules of eight gastropod species. *J. Zool.* **155,** 401–411.

Bayne, C. J. (1968b). Histochemical studies on the egg capsules of eight gastropod molluscs. *Proc. Malacol. Soc. London* **38,** 199–212.

Bayne, C. J. (1969). Survival of the embryos of the grey field slug *Agriolimax reticulatus,* following desiccation of the egg. *Malacologia* **9,** 391–401.

Beiswanger, C. M., Sokolove, P. G., and Prior, D. J. (1981), Extraocular photoentrainment of the circadian locomotor rhythm of the garden slug *Limax. J. Exp. Zool.* **216,** 13–24.

Blâzka, P. (1955). Temperaturadaptation des Gesamtmetabolismus bei der weinbergschnecke *Helix pomatia* L. *Zool. Jahrb. Abt. Allg. Zool. Physiol. Tiere* **65,** 430–438.

Blinn, W. C. (1963). Ecology of the land snails *Mesodon thyroidus* and *Allogona profunda. Ecology* **44,** 498–505.

Blinn, W. C. (1964). Water in the mantle cavity of land snails. *Physiol. Zool.* **37,** 329–337.

Burton, R. F. (1964). Variations in the volume and concentration of the blood of the snail, *Helix pomatia* L., in relation to the water content of the body. *Can. J. Zool.* **42,** 1085–1097.

Burton, R. F. (1965a). Sodium, potassium and magnesium in the blood of the snail, *Helix pomatia* L. *Physiol. Zool.* **38,** 335–342.

Burton, R. F. (1965b). Variations in the water and mineral contents of some organs of the snail, *Helix pomatia* L. *Can. J. Zool.* **43,** 771–779.

Burton, R. F. (1965c). Relationships between the cation contents of slime and blood in the snail *Helix pomatia* L. *Comp. Biochem. Physiol.* **15,** 339–345.

Burton, R. F. (1966). Aspects of ionic regulation in certain terrestrial pulmonata. *Comp. Biochem. Physiol.* **17,** 1007–1018.

Burton, R. F. (1968a). Ionic regulation in the snail, *Helix aspersa. Comp. Biochem. Physiol.* **25,** 501–508.

Burton, R. F. (1968b). Ionic balance in the blood of pulmonata. *Comp. Biochem. Physiol.* **25,** 509–516.

Burton, R. F. (1969). Buffers in the blood of the snail, *Helix pomatia* L. *Comp. Biochem. Physiol.* **29,** 919–930.

Burton, R. F. (1971). Natural variations in cation levels in three species of land snails (Pulmonata: Helicidae). *Comp. Biochem. Physiol.* **39,** 267–275.

Cain, A. J. (1968). Studies on *Cepaea.* V. Sand dune populations of *Cepaea nemoralis* (L.). *Philos. Trans. R. Soc. London, Ser. B* **253,** 499–517.

Cain, A. J., and Currey, J. D. (1963a). Area effects in *Cepaea. Philos. Trans. R. Soc. London, Ser. B* **246,** 1–81.

Cain, A. J., and Currey, J. D. (1963b). Area effects in *Cepaea* on the Larkhill Artillery Ranges, Salisbury Plain. *J. Linn. Soc. London, Zool.* **45,** 1–15.

Cameron, R. A. D. (1969). The distribution and variation of three species of land snail near Rickmansworth, Hertfordshire. *J. Linn. Soc. London, Zool.* **48,** 83–111.

Cameron, R. A. D. (1970a). The survival, weight loss and behaviour of three species of land snail in conditions of low humidity. *J. Zool.* **160,** 143–157.

Cameron, R. A. D. (1970b). The effect of temperature on the activity of three species of helicid snail (Mollusca: Gastropoda). *J. Zool.* **162,** 303–315.

Campbell, J. W., and Bishop, S. H. (1970). Nitrogen metabolism in molluscs. *In* "Comparative Biochemistry of Nitrogen Metabolism" (J. W. Campbell, ed.), Vol. 1, pp. 103–206. Academic Press, New York.

Campbell, J. W., and Boyan, B. D. (1974). On the acid–base balance of Gastropod molluscs. *In*

"The Mechanisms of Mineralization in the Invertebrates and Plants" (N. Watabe and A. M. Wilbur, eds.), pp. 109–133. Univ. of South Carolina Press, Columbia.

Campbell, J. W., Drotman, R. B., McDonald, J. A., and Tramell, P. R. (1972). Nitrogen metabolism in terrestrial invertebrates. *In* "Nitrogen Metabolism and the Environment" (J. W. Campbell and L. Goldstein, eds.), pp. 1–54. Academic Press, New York.

Campion, M. (1961). The structure and function of cutaneous glands in *Helix aspersa*. *Q. J. Microsc. Sci.* **102,** 195–216.

Carmichael, E. B., and Rivers, T. D. (1932). The effect of dehydration on the hatchability of *Limax flavus* eggs. *Ecology* **13,** 375–380.

Carney, W. P. (1966). Mortality and apertural orientation in *Allogona ptychophora* during winter hibernation in Montana. *Nautilus* **79,** 134–136.

Carrick, R. (1942). The life history and development of *Agriolimax agrestis* L., the grey field slug. *Trans. R. Soc. Edinburgh* **59,** 563–597.

Chaturvedi, M. L., and Agarwal, R. A. (1979). Excretion, accumulation and site of synthesis of urea in the snail, *Pila globosa,* during active and dormant periods. *J. Physiol.* (*Paris*) **75,** 233–238.

Cook, A. (1979a). Homing by the slug *Limax pseudoflavus*. *Anim. Behav.* **27,** 545–552.

Cook, A. (1979b). Homing in gastropods. *Malacologia* **18,** 315–318.

Cook, A. (1980). Field studies of homing in the pulmonate slug *Limax pseudoflavus*. *J. Moll. Stud.* **46,** 100–105.

Crawford, C. S. (1981). "Biology of Desert Invertebrates." Springer-Verlag, Berlin and New York.

Crawford-Sidebotham, T. J. (1972). The influence of weather upon the activity of slugs. *Oecologia* **9,** 141–154.

Dainton, B. H. (1943). Effect of air currents, light, humidity and temperature on slugs. *Nature* (*London*) **151,** 25.

Dainton, B. H. (1954a). The activity of slugs. I. The induction of activity by changing temperatures. *J. Exp. Biol.* **31,** 165–187.

Dainton, B. H. (1954b). The activity of slugs. II. The effect of light and air currents. *J. Exp. Biol.* **31,** 188–197.

Daxl, R. (1969). Beobachtungen zur diurnalen und saisonellen Aktivitat einiger Nacktschnekenarten. *Z. Angew. Zool.* **56,** 357–370.

DeJorge, F. B., and Peterson, J. A. (1970). Urea and uric acid contents in the hepatopancreas, kidney and lung of active and dormant snails, *Strophocheilus* and *Thaumastus* (Pulmonata, Mollusca). *Comp. Biochem. Physiol.* **35,** 211–219.

DeJorge, F. B., Cintra, A. B. U., Haeser, P. E., and Sawaya, P. (1965). Biochemical studies on the snail *Strophocheilus oblongus musculus* (Becquaert). *Comp. Biochem. Physiol.* **14,** 35–42.

DeJorge, F. B., Peterson, J. A., and Ditadi, A. S. F. (1969). Variations in nitrogenous compounds in the urine of *Strophocheilus* (Pulmonata, Mollusca) with different diets. *Experientia* **25,** 614–615.

Edelstam, D., and Palmer, C. (1950). Homing behaviour in gastropodes. *Oikos* **2,** 259–270.

Emberton, L. R. B., and Bradbury, S. (1963). Transmission of light through shells of *Cepaea nemoralis* L. *Proc. Malacol. Soc. London* **35,** 211–219.

Florkin, M. (1966). Nitrogen metabolism. *In* "Physiology of Mollusca" (K. M. Wilbur and C. M. Yonge, eds.), Vol. 2, pp. 309–351. Academic Press, New York.

Florkin, M., and Bricteux-Grégoire, S. (1972). Nitrogen metabolism in mollusks. *In* "Chemical Zoology" (M. Florkin and B. T. Scheer, eds.), Vol. 7, pp. 301–348. Academic Press, New York.

Garcia, M. C. L. R. (1977). Ecophysiology of the snail *Cepaea nemoralis* (L.) (Helicidae): Some consequences of sunlight. *Arch. Zool. Exp. Gen.* **118,** 495–514.

Gebhardt-Dunkel, E. (1953). Die Trockenresistenz bei Gehauseschnecken. *Zool. Jahrb.* **64,** 235–266.

Gelperin, A. (1974). Olfactory basis of homing behavior in the giant garden slug, *Limax maximus*. *Proc. Natl. Acad. Sci. U.S.A.* **71,** 966–970.

Getz, L. L. (1959). Notes on the ecology of slugs: *Arion circumscriptus, Deroceras reticulatum,* and *D. laeve*. *Am. Midl. Nat.* **61,** 485–498.

Gill, P. D. (1978). Non-genetic variation in isoenzymes of lactate dehydrogenase of *Cepaea nemoralis*. *Comp. Biochem. Physiol. B* **59,** 271–276.

Gilles, R. (1972). Biochemical ecology of Mollusca. *In* "Chemical Zoology" (M. Florkin and B. T. Scheer, eds.), Vol. 7, pp. 467–499. Academic Press, New York.

Grainger, J. N. R. (1969). Heat death in *Arianta arbustorum*. *Comp. Biochem. Physiol.* **29,** 665–670.

Grainger, J. N. R. (1975). Mechanism of death at high temperatures in *Helix* and *Patella*. *J. Therm. Biol.* **1,** 11–13.

Haggag, G., and Fouad, Y. (1968). Comparative study of nitrogenous excretion in terrestrial and freshwater gastropods. *Z. Vergl. Physiol.* **57,** 428–431.

Heath, D. J. (1975). Colour, sunlight and internal temperatures in the land-snail *Cepaea nemoralis* (L.). *Oecologia* **19,** 29–38.

Heatwole, H., and Heatwole, A. (1978). Ecology of the Puerto Rican Camaenid Tree-Snails. *Malacologia* **17,** 241–315.

Heller, J. (1981). Visual versus climatic selection of shell banding in the landsnail *Theba pisana* in Israel. *J. Zool.* **194,** 85–101.

Henne, F. C. (1963). Effect of light and temperature on locomotor activity of *Polygyra albolabris* (Say). *Bios* **34,** 129–133. (Published in Cedar Rapids, Iowa.)

Herreid, C. F., II (1977). Metabolism of land snails (*Otala lactea*) during dormancy, arousal and activity. *Comp. Biochem. Physiol. A* **56,** 211–215.

Herreid, C. F., II, and Rokitka, M. A. (1976). Environmental stimuli for arousal from dormancy in the land snail *Otala lactea* (Müller). *Physiol. Zool.* **49,** 181–190.

Hogben, L., and Kirk, R. L. (1944). Studies on temperature regulation. I. The Pulmonata and Oligochaeta. *Proc. R. Soc. London, Ser. B* **132,** 239–252.

Horne, F. R. (1971). Accumulation of urea by a pulmonate snail during aestivation. *Comp. Biochem. Physiol. A* **38,** 565–570.

Horne, F. R. (1973a). The utilization of foodstuffs and urea production by a land snail during estivation. *Biol. Bull.* (*Woods Hole, Mass.*) **144,** 321–330.

Horne, F. R. (1973b). Urea metabolism in an estivating terrestrial snail *Bulimulus dealbatus*. *Am. J. Physiol.* **224,** 781–787.

Horne, F. R. (1977a). Ureotelism in the slug, *Limax flavus* Linne. *J. Exp. Zool.* **199,** 227–232.

Horne, F. R. (1977b). Regulation of urea biosynthesis in the slug, *Limax flavus* Linné. *Comp. Biochem. Physiol. B* **56,** 63–69.

Horne, F. R. (1979). Comparative aspects of estivating metabolism in the gastropod, *Marisa*. *Comp. Biochem. Physiol. A* **64,** 309–311.

Horne, F. R. and Barnes, G. (1970). Reevaluation of urea biosynthesis in prosobranch and pulmonate snails. *Z. Vergl. Physiol.* **69,** 452–457.

Horne, F. R., and Boonkoom, V. (1970). The distribution of the ornithine cycle enzymes in twelve gastropods. *Comp. Biochem. Physiol.* **32,** 141–153.

Howes, N. H., and Wells, G. P. (1934a). The water relations of snails and slugs. I. Weight rhythms in *Helix pomatia* L. *J. Exp. Biol.* **11,** 327–343.

Howes, N. H., and Wells, G. P. (1934b). The water relations of snails and slugs. II. Weight rhythms in *Arion ater* L. and *Limax flavus* L. *J. Exp. Biol.* **11,** 344–351.

Hunter, P. J. (1968). Studies on slugs of arable ground. III. Feeding Habits. *Malacologia* **6,** 391–399.

Hyman, L. H. (1967). "The Invertebrates," Vol. 6. McGraw-Hill, New York.

Jacklet, J. W. (1969). Circadian rhythm of optic nerve impulses recorded in darkness from the isolated eye of *Aplysia*. *Science* **164,** 562–563.

Jaremovic, R., and Rollo, C. D. (1979). Tree climbing by the snail *Cepaea nemoralis* (L.): A possible method for regulating temperature and hydration. *Can. J. Zool.* **57,** 1010–1014.

Jeppesen, L. L. (1977). Photoperiodic control of hibernation in *Helix pomatia* L. (Gastropoda: Pulmonata). *Behav. Processes* **2,** 373–382.

Jeppesen, L. L., and Nygard, K. (1976). The influence of photoperiod, temperature and internal factors on the hibernation of *Helix pomatia* L. (Gastropoda, Pulmonata). *Vidensk. Medd. Dan. Naturhist. Foren.* **139,** 7–20.

Jezewska, M. M. (1969). The nephridial excretion of guanine, xanthine and uric acid in slugs (Limacidae) and snails (Helicidae). *Acta Biochim. Pol.* **16,** 313–320.

Jezewska, M. M., Gorzkowski, B., and Heller, J. (1963). Nitrogen compounds in the snail *Helix pomatia*. *Acta Biochim. Pol.* **10,** 55–65.

Jones, J. S. (1973a). Ecological genetics and natural selection in molluscs. *Science* **182,** 546–552.

Jones, J. S. (1973b). The genetic structure of a southern peripheral population of the snail *Cepaea nemoralis*. *Proc. R. Soc. London, Ser. B* **183,** 371–384.

Jones, J. S., Leith, B. H., and Rawlings, P. (1977). Polymorphism in *Cepaea:* A problem with too many solutions? *Annu. Rev. Ecol. Syst.* **8,** 109–143.

Karlin, E. J. (1961). Temperature and light as factors affecting the locomotor activity of slugs. *Nautilus* **74,** 125–130.

Kerkut, G. A., and Taylor, B. J. R. (1956). The sensitivity of the slug pedal ganglion to osmotic pressure changes. *J. Exp. Biol.* **33,** 493–501.

Kratochvil, H. (1976). Langfristige messungen des Sauerstoffverbrauches und der Herzschlagrate an trockenschlafenden Landpulmonaten. *Zool. Anz.* **196,** 289–317.

Krogh, A. (1941). "Comparative Physiology of Respiratory Mechanisms." Univ. of Pennsylvania Press, Philadelphia.

Künkel, K. (1916). "Zur Biologie der Lungenschnecken." Carl Winters, Heidelburg.

Lazaridou-Dimitriadou, M., and Daguzan, J. (1978). Study of water balance and its evolution in relation to temperature and relative humidity in *Euparypha pisana* (Mueller) (Gastropoda, Pulmonata, Stylommatophora). *Arch. Zool. Exp. Gen.* **119,** 549–564.

Lewis, R. D. (1969a). Studies on the locomotor activity of the slug *Arion ater* (Linnaeus). I. Humidity, temperature and light reactions. *Malacologia* **7,** 295–306.

Lewis, R. D. (1969b). Studies on the locomotor activity of the slug *Arion ater* (Linnaeus). II. Locomotor activity rhythms. *Malacologia* **7,** 307–312.

Lind, H. (1968). Hibernating behaviour of *Helix pomatia* L. (Gastropoda, Pulmonata). *Vidensk. Medd. Dan. Naturhist. Foren.* **131,** 129–151.

Little, C. (1968). Aestivation and ionic regulation in two species of *Pomacea* (Gastropoda, Prosobranchia). *J. Exp. Biol.* **48,** 569–585.

Loest, R. A. (1979a). Ammonia volatilization and absorption by terrestrial gastropods: A comparison between shelled and shell-less species. *Physiol. Zool.* **52,** 461–469.

Loest, R. A. (1979b). Ammonia-forming enzymes and calcium-carbonate deposition in terrestrial pulmonates. *Physiol. Zool.* **52,** 470–483.

Long, G. L., Ellington, W. R., and Duda, T. F. (1979). Comparative enzymology and physiological role of D-lactate dehydrogenase from the foot muscle of two gastropod molluscs. *J. Exp. Zool.* **207,** 237–248.

Machin, J. (1964a). The evaporation of water from *Helix aspersa*. I. Nature of the evaporating surface. *J. Exp. Biol.* **41,** 759–769.

Machin, J. (1964b). The evaporation of water from *Helix aspersa*. II. Measurement of air flow and the diffusion of water vapour. *J. Exp. Biol.* **41,** 771–781.

Machin, J. (1964c). The evaporation of water from *Helix aspersa*. III. The application of evaporation formulae. *J. Exp. Biol.* **41,** 783–792.

Machin, J. (1965). Cutaneous regulation of evaporative water loss in the common garden snail *Helix aspersa*. *Naturwissenschaften* **52,** 18.

Machin, J. (1966). The evaporation of water from *Helix aspersa*. IV. Loss from the mantle of the inactive snail. *J. Exp. Biol.* **45,** 269–278.

Machin, J. (1967). Structural adaptation for reducing water-loss in three species of terrestrial snail. *J. Zool.* **152,** 55–65.

Machin, J. (1968). The permeability of the epiphragm of terrestrial snails to water vapor. *Biol. Bull (Woods Hole, Mass.)* **134,** 87–95.

Machin, J. (1972). Water exchange in the mantle of a terrestrial snail during periods of reduced evaporative loss. *J. Exp. Biol.* **57,** 103–111.

Machin, J. (1974). Osmotic gradients across snail epidermis: Evidence for a water barrier. *Science* **183,** 759–760.

Machin, J. (1975). Water relationships. *In* "Pulmonates" (V. Fretter and J. Peake, eds.), Vol. 1, pp. 105–163. Academic Press, New York.

Machin, J. (1977). Role of integument in molluscs. *In* "Transport of Ions and Water in Animals" (B. L. Gupta, R. B. Moreton, J. L. Oschman, and B. J. Wall, eds.), pp. 735–759. Academic Press, New York.

McQuaid, C. D., Branch, G. M., and Frost, P. G. H. (1979). Aestivation behaviour and thermal relations of the pulmonate *Theba pisana* in a semi-arid environment. *J. Therm. Biol.* **4,** 47–55.

Martin, A. W., and Harrison, F. M. (1966). *In* "Physiology of Mollusca" (K. M. Wilbur and C. M. Yonge, eds.), Vol. 2, pp. 353–386. Academic Press, New York.

Martin, A. W., Harrison, F. M., Huston, M. J., and Stewart, D. M. (1958). The blood volumes of some representative molluscs. *J. Exp. Biol.* **35,** 260–279.

Mason, C. F. (1970). Food, feeding rates and assimilation in woodland snails. *Oecologia* **4,** 358–374.

Mason, C. F. (1971). Respiration rates and population metabolism of woodland snails. *Oecologia* **7,** 80–94.

Matsumoto, M., Morimasa, T., Takeuchi, H., Mori, A., Kohsaka, M., Kobayashi, J., and Mori, F. (1974). Amounts of inorganic ions and free amino acids in hemolymph and ganglion of the African giant snail, *Achatina fulica* Ferrusac. *Comp. Biochem. Physiol. A* **48,** 465–470.

Meincke, K. F. (1972). Osmotischer Druck und ionale Zusammensetzung der Hämolymph winterschlafender *Helix pomatia* bei konstanter und sich zyklisch ändernder Temperatur. *Z. Vergl. Physiol.* **76,** 226–232.

Meincke, K. F. (1974). Die Wirkung der Temperatur auf den Stoffhaushalt von *Helix pomatia* im Herbst. *J. Comp. Physiol.* **88,** 103–112.

Meincke, K. F. (1975). The influence of extreme temperatures on metabolic substances in hemolymph and foot muscle of *Helix pomatia*. *Comp. Biochem. Physiol. A* **51,** 373–376.

Mellanby, K. (1961). Slugs at low temperatures. *Nature (London)* **189,** 944.

Morton, J. E. (1967). "Molluscs." Hutchinson, London.

Müller, G. (1943). Untersuchungen über die Gesetzlichkeit des Wachstums. X. Weiteres zur Frage der Abhangigkeit der Atmung von der Korpergrosse. *Biol. Zentralbl.* **63,** 446–453.

Mullins, D. E., and Cochran, D. G. (1974). Nitrogen metabolism in the American cockroach: An examination of whole body and fat body regulation of cations in response to nitrogen balance. *J. Exp. Biol.* **61,** 557–570.

Murray, J. (1966). *Cepaea nemoralis* in the Isles of Scilly. *Proc. Malacol. Soc. London* **37,** 167–181.

Newell, P. F. (1966). The nocturnal behaviour of slugs. *Med. Biol. Illust.* **16,** 146–159.

Newell, P. F. (1968). The measurement of light and temperature as factors controlling surface activity of the slug *Agriolimax reticulatus* (Müller). *In* "The Measurement of Environmental Factors in Terrestrial Ecology" (R. M. Wadsworth, ed.), pp. 141–146. Blackwell, Oxford.

Newell, P. F., and Appleton, T. C. (1979). Aestivating snails—The physiology of water regulation in the mantle of the terrestrial pulmonate *Otala lactea*. *Malacologia* **18,** 575–581.

Newell, P. F., and Machin, J. (1976). Water regulation in aestivating snails. *Cell Tissue Res.* **173,** 417–421.

Nopp, H. (1965). Temperaturbezogene regulationen des saurstoffverbrauchs und der herzschlagrate bei einigen landpulmonaten. *Z. Vergl. Physiol.* **50,** 641–659.

Nopp, H. (1974). Physiologische Aspecte des Trockenschlafes der Landschnecken. *Sitzungsber.—Oesterr. Akad. Wiss., Math.-Naturwiss. Kl., Abt. 1***182,** 1–75.

Nopp, H., and Farahat, A. Z. (1967). Temperatur und Zellstoffwechel bei Heliciden. *Z. Vergl. Physiol.* **55,** 103–118.

Oldham, C. (1928). The influence of lime on the shell of *Arianta arbustorum* (L.). *Proc. Malacol. Soc. London* **18,** 143–144.

Oosterhoff, L. M. (1977). Variation in growth rate as an ecological factor in the landsnail *Cepaea nemoralis* (L.). *Neth. J. Zool.* **27,** 1–132.

Oudejans, R., and Van der Horst, D. (1974). Biosynthesis of fatty acids in the pulmonate land snail, *Cepaea nemoralis*. *Comp. Biochem. Physiol. B* **47,** 139–147.

Pallant, D. (1970). A quantitative study of feeding in woodland by the grey field slug (*Agriolimax reticulatus* (Müller)). *Proc. Malacol. Soc.* **39,** 83–87.

Pavan Kumar, T., Ramamurthi, R., and Babu, K. S. (1981). Circadian fluctuations in total protein and carbohydrate content in the slug *Laevicaulis alte* (Ferussac, 1821). *Biol. Bull. (Woods Hole, Mass.)* **160,** 114–122.

Perseca, T., Dordea, M., Irimies, E., and Marosan, V. (1977). Seasonal variations of some biochemical parameters in *Helix pomatia*. *Stud. Univ. Babes-Bolyai*[*Ser.*]*Biol.* **2,** 56–61.

Pollard, E. (1975a). Differences in shell thickness in adult *Helix pomatia* L. from a number of localities in southern England. *Oecologia* **21,** 85–92.

Pollard, E. (1975b). Aspects of the ecology of *Helix pomatia* L. *J. Anim. Ecol.* **44,** 305–329.

Pomeroy, D. E. (1968). Dormancy in the land snail, *Helicella virgata* (Pulmonata: Helicidae). *Aust. J. Zool.* **16,** 857–869.

Pomeroy, D. E. (1969). Some aspects of the ecology of the land snail, *Helicella virgata*, in south Australia. *Aust. J. Zool.* **17,** 495–514.

Potts, D. C. (1975). Persistence and extinction of local populations of the garden snail *Helix aspersa* in unfavorable environments. *Oecologia* **21,** 313–334.

Potts, W. T. W. (1967). Excretion in the molluscs. *Biol. Rev. Cambridge Philos. Soc.* **42,** 1–41.

Potts, W. T. W. (1968). Aspects of excretion in the molluscs. *Symp. Zool. Soc. London* **22,** 187–192.

Potts, W. T. W., and Parry, G. (1964). "Osmotic and Ionic Regulation in Animals." Pergamon, Oxford.

Prosser, C. L. (1973). "Comparative Animal Physiology," Vol. 1. Saunders, Philadelphia, Pennsylvania.

Pusswald, A. W. (1948). Beitrage zum wasserhaushalt der pulmonaten. *Z. Vergl. Physiol.* **31,** 227–248.

Reddy, G., Rajarami, G., Pavan Kumar, T., Murali Mohan, P., and Babu, K. S. (1978). Diurnal variations in physiological activities in the garden snail, *Cryptozona ligulata*. *J. Comp. Physiol.* **125,** 59–66.

Richardson, A. M. M. (1974). Differential climatic selection in natural population of land snail *Cepaea nemoralis*. *Nature (London)* **247,** 572–573.

Richardson, A. M. M. (1975). Energy flux in a natural population of the land snail, *Cepaea nemoralis* L. *Oecologia* **19,** 141–164.

Riddle, W. A. (1975). Water relations and humidity-related metabolism of the desert snail *Rabdotus schiedeanus* (Pfeiffer) (Helicidae). *Comp. Biochem. Physiol. A* **51,** 579–583.

Riddle, W. A., (1977). Comparative respiratory physiology of a desert snail *Rabdotus schiedeanus,* and a garden snail, *Helix aspersa. Comp. Biochem. Physiol. A.* **56,** 369–373.

Riddle, W. A. (1981a). Cuticle water activity and water content of beetles and scorpions from xeric and mesic habitats. *Comp. Biochem. Physiol. A* **68,** 231–235.

Riddle, W. A. (1981b). Hemolymph osmoregulation and urea retention in the woodland snail, *Anguispira alternata* (Say) (Endodontidae). *Comp. Biochem. Physiol. A* **69,** 493–498.

Riddle, W. A. (1981c). Cold hardiness in the woodland snail, *Anguispira alternata* (Say) (Endodontidae). *J. Therm. Biol.* **6,** 117–120.

Rising, T. L., and Armitage, K. B. (1969). Acclimation to temperature by the terrestrial gastropods, *Limax maximus* and *Philomycus carolinianus:* Oxygen consumption and temperature preference. *Comp. Biochem. Physiol.* **30,** 1091–1114.

Robertson, J. D. (1964). Osmotic and ionic regulation. *In* ''Physiology of Mollusca'' (K. M. Wilbur and C. M. Yonge, eds.), Vol. 1, pp. 283–311. Academic Press, New York.

Rokitka, M. A., and Herreid, C. F., II (1975a). Position of epiphragms in the land snail *Otala lactea* (Müller). *Nautilus* **89,** 23–26.

Rokitka, M. A., and Herreid, C. F., II (1975b). Formation of epiphragms by the land snail *Otala lactea* (Müller) under various environmental conditions. *Nautilus* **89,** 27–32.

Ross, R. J. (1979). The effects of mechanical disturbances on the behaviour of inactive terrestrial snails. *J. Moll. Stud.* **45,** 35–38.

Roy, A. (1963). Étude de l'acclimation thermique chez la limace *Arion circumscriptus. Can. J. Zool.* **41,** 671–698.

Rumsey, T. J. (1972). Osmotic and ionic regulation in a terrestrial snail, *Pomatias elegans* (Gastropoda, Prosobranchia) with a note on some tropical Pomatiasidae. *J. Exp. Biol.* **57,** 205–215.

Runham, N. W., and Hunter, P. J. (1970). ''Terrestrial Slugs.'' Hutchinson University Library, London.

Russell-Hunter, W. (1964). Physiological aspects of ecology in nonmarine molluscs. *In* ''Physiology of Mollusca'' (K. M. Wilbur and C. M. Yonge, eds.), Vol. 1, pp. 83–126. Academic Press, New York.

Schmidt-Nielsen, K., Taylor, C. R., and Shkolnik, A. (1971). Desert snails: Problems of heat, water and food. *J. Exp. Biol.* **55,** 385–398.

Schmidt-Nielsen, K., Taylor, C. R., and Shkolnik, A. (1972). Desert snails: Problems of survival. *Symp. Zool. Soc. London* **31,** 1–13.

Schoffeniels, E., and Gilles, R. (1972). Ionoregulation and osmoregulation in mollusca. *In* ''Chemical Zoology'' (M. Florkin and B. T. S. Scheer, eds.), Vol. 7, pp. 393–420. Academic Press, New York.

Segal, E. (1961). Acclimation in molluscs. *Am. Zool.* **1,** 235–244.

Smith, L. H. (1981). Quantified aspects of pallial fluid and its affect on the duration of locomotor activity in the terrestrial gastropod *Triodopsis albolabris. Physiol. Zool.* **54,** 407–414.

Sokolove, P. G., and McCrone, E. J. (1978). Reproductive maturation in the slug, *Limax maximus,* and the effects of artificial photoperiod. *J. Comp. Physiol.* **125,** 317–325.

Sokolove, P. G., Beiswanger, C. M., Prior, D. J., and Gelperin, A. (1977). A circadian rhythm in the locomotor behaviour of the giant garden slug *Limax maximus. J. Exp. Biol.* **66,** 47–64.

Speeg, K. V., Jr., and Campbell, J. W. (1968a). Purine biosynthesis and excretion in *Otala* (=*Helix*) *lactea:* An evaluation of the nitrogen excretory potential. *Comp. Biochem. Physiol.* **26,** 579–595.

Speeg, K. V., Jr., and Campbell, J. W. (1968b). Formation and volatilization of ammonia gas by terrestrial snails. *Am. J. Physiol.* **214,** 1392–1402.

Steigen, A. L. (1979). Temperature effects on energy metabolism in banded and unbanded morphs of the snail *Cepaea hortensis* Müll. *Oecologia* **141,** 163–173.

Storey, K. B. (1977). Lactate dehydrogenase in tissue extracts of the land snail, *Helix aspersa:* Unique adaptation of LDH subunits in a faculative anaerobe. *Comp. Biochem. Physiol. B* **56,** 181–187.

Stöver, H. (1973a). Über den Wasser-und Elektrolythaushalt von *Arianta arbustorum* (L.). *J. Comp. Physiol.* **83,** 51–61.

Stöver, H. (1973b). Cold resistance and freezing in *Arianta arbustorum* L. (Pulmonata). *In* "Effects of Temperature on Ectothermic Organisms" (W. Wieser, ed.), pp. 281–290. Springer-Verlag, Berlin and New York.

Tercafs, R. R. (1960). Comparaison entre les induvidus épigés et cavernicoles d'un mollusque gasteropode troglophile *Oxychilus cellarius* Mull. Affinitiés éthologiques pour le milieu souterrain. *Soc. R. Zool. Belg.* **91,** 85–116.

Terhivuo, J. (1978). Growth, reproduction and hibernation of *Arianta arbustorum* (L.) (Gastropoda, Helicidae) in southern Finland. *Ann. Zool. Fenn.* **15,** 8–16.

Tischler, W. (1974). Aussenfaktoren als Auslöser tageszeitlichen Verhaltens und begrenzter Dormanz bei *Helix pomatia* L. *Zool. Anz.* **193,** 251–255.

Tramell, P. R., and Campbell, J. W. (1970). Nitrogenous excretory products of the giant South American land snail, *Strophocheilus oblongus. Comp. Biochem. Physiol.* **32,** 569–571.

Tramell, P. R., and Campbell, J. W. (1972). Arginine and urea metabolism in the South American land snail, *Strophocheilus oblongus. Comp. Biochem. Physiol. B* **42,** 439–449.

Trams, E. G., Lauter, C. J., Bourke, R. S., and Tower, D. B. (1965). Composition of *Cepaea nemoralis* hemolymph and tissue extracts. *Comp. Biochem. Physiol.* **14,** 399–404.

Van der Horst, D. J. (1974). *In vivo* biosynthesis of fatty acids in the pulmonate land snail *Cepaea nemoralis* (L.) under anoxic conditions. *Comp. Biochem. Physiol. B* **47,** 181–187.

Van der Horst, D. J., Oudejans, R. C. H. M., Meijers, J. A., and Testerink, G. (1974). Fatty acid metabolism in hibernating *Cepaea nemoralis* (Mollusca: Pulmonata). *J. Comp. Physiol.* **91,** 247–256.

Van der Laan, K. L. (1975). Aestivation in the land snail *Helminthoglypta arrosa* (Binney). *Veliger* **17,** 360–368.

Van der Laan, K. L. (1980). Terrestrial pulmonate reproduction: Seasonal and annual variation and environmental factors in *Helminthoglypta arrosa* (Pulmonata: Helicidae). *Veliger* **23,** 48–54.

von Brand, T. (1931). Der Jahreszyklus im stoffbestand der weinbergschnecke (*Helix pomatia*). *Z. Vergl. Physiol.* **14,** 200–264.

Warburg, M. R. (1965). On the water economy of some Australian land snails. *Proc. Malacol. Soc. London* **36,** 297–305.

Warburg, M. R. (1972). On the physiological ecology of the Israeli clausiliidae, a relic group of land snails. *Trans. Conn. Acad. Arts Sci.* **44,** 379–395.

Webley, D. (1964). Slug activity in relation to weather. *Ann. Appl. Biol.* **53,** 407–414.

Webley, D. (1970). Observations on the effects of distribution and numbers of slug pellets on the catch of slugs. *Ann. Appl. Biol.* **66,** 347–352.

Wells, G. P. (1944). The water relations of snails and slugs. III. Factors determining activity in *Helix pomatia* L. *J. Exp. Biol.* **20,** 79–87.

White, A. R. (1959). Observations on slug activity in a northumberland garden. *Plant Pathol.* **8,** 62–68.

Wieser, W. (1977). Slow, fast- and medium fast responses of ectotherms to temperature changes: A new mechanism of metabolic compensation in *Helix pomatia? J. Therm. Biol.* **2,** 197–202.

Wieser, W. (1978). The initial stage of anaerobic metabolism in the snail, *Helix pomatia*. *FEBS Lett.* **95,** 375–378.

Wieser, W. (1981). Responses of *Helix pomatia* to anoxia: Changes of solute activity and other properties of the haemolymph. *J. Comp. Physiol.* **141,** 503–509.

Wieser, W., and Fritz, H., (1971). Seasonal changes of metabolism in *Arianta arbustorum* (Gastropoda; Pulmonata): The cholinesterase of the blood. *Comp. Biochem. Physiol. A* **39,** 63–73.

Wieser, W., and Schuster, M. (1975). The relationship between water content, activity, and free amino acids in *Helix pomatia* L. *J. Comp. Physiol.* **98,** 169–181.

Wieser, W., and Wright, E. (1978). D-lactate formation, D-LDH activity and glycolytic potential of *Helix pomatia* L. *J. Comp. Physiol.* **126,** 249–255.

Wieser, W., and Wright, E. (1979). The effects of season and temperature on D-lactate dehydrogenase, pyruvate kinase and arginine kinase in the foot of *Helix pomatia* L. *Hoppe-Seyler's Z. Physiol. Chem.* **360,** 533–542.

Wieser, W., Fritz, H., and Reichel, K. (1970). Jahreszeitliche Steuerung der Atmung von *Arianta arbustorum* (Gastropoda). *Z. Vergl. Physiol.* **70,** 62–79.

Wolda, H. (1965). The effect of drought on egg production in *Cepaea nemoralis* (L.). *Arch. Neerl. Zool.* **16,** 387–399.

Yom-Tov, Y. (1971a). The biology of two desert snails *Trochoidea* (*Xerocrassa*) *seetzeni* and *Sphincterochila boissieri*. *Isr. J. Zool.* **20,** 231–248.

Yom-Tov. Y. (1971b). Body temperature and light reflectance in two desert snails. *Proc. Malacol. Soc. London* **39,** 319–326.

11

Actuarial Bioenergetics of Nonmarine Molluscan Productivity

W. D. RUSSELL-HUNTER AND DANIEL E. BUCKLEY

Department of Biology, Syracuse University
Syracuse, New York

The Mollusca, Vol. 6
Ecology

ISBN 0-12-751406-6

I. Introduction

The classic distributional studies of molluscan ecology are being reinforced by modern research on physiological ecology and population dynamics. Investigations in both of these fields are enhanced when quantitative questions are posed, and substantive or experimental data sought, in bioenergetic units. Sound assessment of the evolutionary implications of ecological processes now requires a similar basis.

In energetic terms, natural populations of organisms (like individual living organisms) are open systems which approximate a steady state. A balance is maintained between the energy gained from the environment and the necessary catabolic energy expenditures that constitute life combined with a surplus available for partitioning into growth and reproduction. Fiscal and economic analogs are obvious and appropriate. Our designation of certain studies in molluscan ecology as being *actuarial bioenergetics* began by analogy with the older commercial meaning (now becoming obsolete in this more general sense) of actuarial as pertaining to the keeping of accounts for a public corporation (Russell-Hunter and Eversole, 1976; Russell-Hunter, 1978; Browne and Russell-Hunter, 1978; Romano, 1980). However, one increasingly important aspect of these ecological studies has moved closer to the special usage of *actuarial* (which is current in commerce) as referring to computation of insurance and other premiums, which involves interaction of interest rates with certain statistical probabilities (e.g., those of differential mortality).

Ultimately, effective bioenergetic analyses must be concerned with dynamic rather than static units, because life is process rather than structure. Although our usual measures are of the standing crop energy content (or biomass), our assessments of bioenergetic partitioning (e.g., between growth and reproduction) need to be stated in terms of differential energy flux (or fractional productivity rates). As in other areas of biology, valuable data are being gained from both experimental and comparative studies of molluscan bioenergetics. Comparative studies of separate natural populations of the same molluscan species have revealed considerable infraspecific interpopulation differences in growth rates and fecundity (Russell Hunter, 1961b, 1964). Interpopulation differences in life-cycle pattern can be demonstrated (Russell-Hunter, 1964, 1978; Calow, 1978, 1979; also Chapter 15, this volume). Obviously, such comparative studies aid in the substantive quantification of "trade-offs" between immediate fecundity and future reproduction, which are conceptually so important in the current debates on theories of life-cycle evolution (Cody, 1966; Williams, 1966a,b; Charnov and Krebs, 1973; Taylor et al., 1974; Browne and Russell-Hunter, 1978; Schaffer, 1979; Calow, 1978, 1979). In addition to this theoretical significance, investigations of bioenergetics in natural populations of molluscs can be of considerable value to applied ecology (Aldridge and McMahon, 1978; Russell-Hunter, 1970,

1978; McMahon, 1975; Maitland, 1965; Russell-Hunter et al., 1981). Examples used in this chapter have been arbitrarily restricted to nonmarine molluscs, because comparable modern studies on inshore marine molluscs (Widdows and Bayne, 1971; Thompson and Bayne, 1974; Bayne et al., 1976, 1977; Widdows, 1978; Bayne and Widdows, 1978) are discussed in Volume 4 of this series.

II. General Methods

We cannot procure the best data for actuarial assessments of bioenergetics, because they would be energy-flux levels obtained instantaneously and nondestructively. Sequential sampling is always required and, in most cases, sufficient determinations of energy content in joules by microbomb calorimetry would be too time consuming. Most recent work has been concerned with growth rates and productivity transfers expressed as changes in dry tissue weight or total organic carbon.

A. Biomass as Carbon

The direct relationship between energy content and biomass may be complicated by technical difficulties in measuring the latter. Defined as the mass of organic materials in a living organism (or a population of organisms), *biomass* is for most practical purposes the weight of living tissues (including organic materials ingested as food and not yet assimilated). Many agricultural crops and fishery catches are reported in units of wet weight per space and time. However, organism water content varies with time and with biological condition. Obviously, drying to constant weight (with precautions to avoid the loss of volatile organic substances) will approach true biomass more closely. The problems of obtaining reproducible biomass figures as dry weight for marine zooplankton (which in many respects are similar to those for molluscs) are quantified and discussed extensively by Lovegrove (1966). The effects of variations in the content of inorganic salts and skeletal materials can be dealt with by adopting *ash-free dry weight,* derived by subtraction of the residue mass after total combustion in a muffle furnace.

Another method is to express biomass as total organic carbon content. The use of a colorimetric method of *wet oxidation* for the analysis of total organic carbon in molluscan growth studies was originally described by Russell-Hunter et al. (1968). Subsequently, this method was modified independently to accommodate tiny individual hydrobiids (Simpson, 1978) and heroically to deal with giant *Pomacea* (Burky, 1974). A recent technical letter (Russell-Hunter et al., 1982) summarizes both these innovations and the utilization of more modern (split-beam, digital) spectrophotometers in organic carbon analyses. Parallel analyses of total combined nitrogen have been carried out in many molluscan investigations, using a Coleman semiautomatic nitrogen analyzer which employs a modification of the micro-Dumas method of Gustin (1960). This allows assessment of all shifts in carbon : nitrogen (C : N) ratios of the snail tissues with season and with biological condition. Shells present peculiar complications in molluscan bioenergetics. They have two principal components: crystalline calcium carbonate and a meshwork of protein fibers (forming the organic matrix of all shell layers and becoming dominant in the periostracum). Amounts of, and ratios between, these components can vary significantly among different populations of a single species (Hunter and Lull, 1976, 1977; Burky et al., 1979; Russell-Hunter et al., 1970, 1981). Both must be analyzed for bioenergetic studies of growth in molluscan populations. Aside from assessing anabolic allocation to the shell for actuarial purposes, shell : tissue ratios allow the detection of tissue *degrowth,* e.g., in overwintering populations (Russell-Hunter and Eversole, 1976; Russell-Hunter et al., 1983a). Even more sophisticated assessments of tissue biomass involve complete, parallel series of analyses of fat, protein, and carbohydrate, but such series have been performed only on stocks of commercially important bivalves.

The successful use of a comparative actuarial approach to assess energetic budgeting and trade-offs requires (a) common bioenergetic units and (b) apposite and relatively affinitive comparisons. This first need for common terms merits a brief gloss here. Measures of standing crop biomass for molluscs are usually expressed as mass per volume or area, for example, as mgC/m^2. Series of successive estimates of standing crop can be used to compute productivity rates expressed as mass per (space·time), for example, as $mgC/(m^2{\cdot}day)$. For productivity comparisons between populations or, preeminently, between dissimilar species, the most meaningful procedure is to use some ratio between productivity rate and mean standing crop biomass for the same time period (Russell-Hunter, 1978). Direct comparisons can be made using turnover times (mean standing crop divided by productivity rate, in units of time) or rate constants (productivity rate divided by mean standing crop, in inverse time). Turnover times are directly related, and rate constants for productivity inversely related, to the average length of the life cycle. For those molluscan species in which life-cycle length is an interpopulation variable (Russell-Hunter, 1961b, 1964, 1978, and references therein; Calow, 1978, 1979), the consequences in regard not only to turnover times but to differential rates of selection as well, are highly significant. Other more specialized bioenergetic ratios, such as those assessing reproductive effort, will be discussed later; however, all ratios must conform to the same conceptual standards.

B. Productivity Rates

Given that common units are desirable, productivity rates for natural populations of molluscs (as change of biomass per space·time) involve not only the difficulties of estimating biomass values that are truly equivalent to energy contents but also difficulties with the denominators of space and time. Field population densities, which must be assessed to provide the space parameter, are notoriously controvertible and unstable in time (caused partially by biotic factors, e.g., seasonal or short-term migrations). In general, field densities are most reliably estimated for sedentary filter-feeding bivalves (Avolizi, 1976; Burky et al., 1983) or territorial grazers such as limpets (Russell-Hunter, 1953, 1961a; Burky, 1971; McMahon, 1975), and are least certain for wide-ranging predaceous forms (Russell-Hunter and Grant, 1966).

Optimal methods vary with species and habitat. On soft bottoms of lakes and ponds, Eckman or Petersen grabs may be employed with remarkable credibility (Avolizi, 1976; D. E. Buckley, unpublished). In submerged vegetation, sweep-net assessments or direct counting of quadrats can be used (Russell-Hunter et al., 1972; Calow, 1972; McMahon, 1975; Aldridge, 1982). Over rocks and stones, e.g., on stream bottoms or wave-swept lake shores, the *habitat outline* method using traced outlines of stones in a jigsaw fashion to provide a basis for counting individual snails found in an area 0.25 m^2 has been used successfully (Russell-Hunter, 1953, 1961a; Burky, 1971; McMahon, 1975; Romano, 1980). Although yielding valuable comparative data for limpets such as *Ancylus* and *Ferrissia,* this method yields figures which are usually consistently less than absolute values of population density for these genera. Calow (1972) has proposed a method that yields better estimates of the actual habitat surface available for grazing in such circumstances. In all kinds of aquatic environments, and necessarily in most terrestrial ones, marking and capture-recapture methods are used not only to estimate or verify population densities (Russell-Hunter and Grant, 1966; Richardson, 1975) but also to establish other data for bioenergetic budgets, including time ratios for grazing and rates of movement.

Compilation of data for productivity rate estimates that require successive field sampling can experience temporal as well as spatial difficulties. Some computations (Section III) assume the continued existence in time of a steady-state population from which the biomass samples are drawn. It is also difficult to obtain from natural populations truly contemporaneous assessments of differential mortality for all demographic cohorts in each population. Similarly, estimating catabolic partitioning in aquatic molluscs involves measuring both oxygen uptake and nitrogenous excretion, which can be nearly concurrent with each other (Aldridge et al., 1980; Tashiro, 1982; Aldridge, 1982), but are necessarily short-term in relation to measures of anabolic partitioning between growth and reproduction. However, they can be used, by computation of respiratory quotients, to quantify shifts in catabolism (e.g., from protein to carbohydrate oxidation). This and other less universally applied computations and techniques are briefly discussed in later sections of this chapter. As freely admitted elsewhere (Russell-Hunter, 1978), a current weakness of actuarial bioenergetics for populations of nonmarine molluscs is that, because the technical difficulties vary with species and even with populations, the "best" values for certain budget components are often derived from one particular species for which the other components have been less extensively investigated.

III. Standing Crop Production

A. Biomass Measures

As already emphasized, productivity studies of molluscs are currently based on sequential biomass sampling. Values of standing crop production can be calculated in terms of total organic carbon per unit space, usually as mgC/m^2. Almost all the values for crop biomass and for the rate values derived from them that are given in this chapter are quoted in terms of carbon, which are more proximate and yet convenient for comparison. Any reader wishing an approximate translation into energy units may employ the conversion factor of 10.94 kcal/gC derived by Russell-Hunter et al. (1968) for average nonmarine snail tissue; 100 mgC/m^2 would correspond to approximately 1.09 $kcal/m^2$ or 4.58 kJ/m^2.

Most of the better estimates of standing crop production that have been calculated begin by computing values per 100 snails based on the appropriate demography of each sample or cohort subsample. Appropriate carbon values for different sizes of individual snails are derived from sets of regressions of carbon on snail size (e.g., McMahon, 1975; Simpson, 1978) using the raw data of carbon analyses in categories corresponding to biological state, environmental season, and age structure. The products of these individual carbon values and the size frequency distribution for the population or cohort at that sampling time (expressed as percentage distribution) can be summed to give values in carbon per 100 snails. In some studies, these values can be used as unit populations in rate comparisons if there is independent evidence of mortality levels. Usually, however, these values (per 100 snails) are then combined with contemporaneous field assessments of population densities to give standing crop biomass figures in mgC/m^2. From series of such biomass figures, coordinated with survivorship and fecundity data, productivity rates for any natural population, or (separately) for any cohort (generation) of that population, may be prepared.

B. Bivalves and Prosobranchs

As might be predicted from their feeding mechanisms and trophic status, certain populations of freshwater bivalves produce some of the highest standing crop values among nonmarine molluscs. In studies of two overlapping and partially sympatric species, *Sphaerium striatinum* and *S. simile* in Cazenovia Lake (upstate New York), Avolizi (1970, 1976) found peak values of biomass at 2.98

g and 1.01 gC/m^2, respectively. For the invasive Asiatic clam, *Corbicula manilensis* (Texas), Aldridge and McMahon (1978) computed an average standing crop biomass of 2.6 gC/m^2, with a peak value of more than 6.3 gC/m^2. Burky and his associates provided mean biomass values for smaller freshwater bivalves; for *Musculium lacustre,* 142 mgC/m^2 (Alexander, 1982); for *M. partumeium* from a univoltine population, 232 mgC/m^2,and from a bivoltine population, 607 mgC/m^2 (Burky et al., 1983). Comparable standing crop values for larger and longer lived unionid mussels in carbon terms are not available, but Negus (1966) provides wet weight figures for *Unio pictorum, U. tumidus,* and *Anodonta anatina* from populations in the River Thames which can be converted to approximately 1.87, 0.77, and 3.18 gC/m^2, respectively.

In contrast to freshwater pulmonate species, most of which have annual life cycles, the majority of freshwater prosobranch gastropods live for more than 2 years (Calow, 1978; Russell-Hunter, 1964, 1978) but vary greatly in the standing crop values achieved. For three populations of *Leptoxis carinata,* Aldridge (1980, 1982) provides prebreeding biomass values of 438–505 mgC/m^2 under severe environmental conditions; for the allied species *Goniobasis livescens,* Payne (1979) computed values of 2.04–14.65 gC/m^2. In more eutrophic conditions in Oneida Lake (upstate New York), a population of *Bithynia tentaculata* is triennial and iteroparous, with a peak standing crop value of 5.7 gC/m^2 (Tashiro, 1980, 1982; Tashiro and Colman, 1982). For the larger viviparious prosobranch *Viviparus georgianus,* D. E. Buckley (unpublished) has established biomass estimates which can be broken down by sex and by age group. A population standing crop value in late spring from Tully Lake (upstate New York) is 4.1 gC/m^2, with second- and third-year females totaling 0.77 and 0.57 gC/m^2, respectively, and males of all ages totaling 1.7 gC/m^2. All four freshwater prosobranchs noted above are dioecious. We have no strictly comparable data for those few hermaphroditic prosobranchs such as *Valvata* spp. which live in fresh waters.

C. Freshwater Pulmonates

Pulmonate snails living in fresh waters are all hermaphroditic; most have simple annual life cycles, although several can be bivoltine, and a few are biennial (Calow, 1978; Russell-Hunter, 1978). Larger species show the highest biomass values. For one population of *Helisoma trivolvis,* Eversole (1974, 1978) gives a peak standing crop value of 4.2 gC/m^2. Data presented as ash-free dry weights by Eckblad (1973) can be converted to peak carbon biomass values of 271 mgC/m^2 for *Lymnaea palustris* and 305 mgC/m^2 for *Physa integra.* For three populations of *L. palustris* in environments of increasing eutrophy, Hunter (1975) computes standing crop values of 0.3, 0.9, and 0.7 gC/m^2 in August, and 0.4, 1.6, and 1.5 gC/m^2 in October.

Freshwater pulmonate limpets have been the subject of several studies in actuarial bioenergetics. Early studies of a dense natural population of the European stream limpet *Ancylus fluviatilis* in Scotland [Russell-Hunter (1953, 1961a)] yielded breeding biomass values (in tissue volume/m^2) which ranged from 1822 to 13,670 mm^3/m^2 in years with severe and favorable weather conditions, respectively. These high standing crop values would convert to 66–495 mgC/m^2. Independently, Streit (1976a) determined mean annual biomass values of 26.6–64.9 mgC/m^2 for *A. fluviatilis* in streams near Lake Constance. For the American stream limpet *Ferrissia rivularis,* Burky (1971) gives peak (late August) standing crop values of 99 and 192 mgC/m^2 in a mesotrophic and a eutrophic creek, respectively. A survey of 16 populations of *Ferrissia* (Nickerson, 1972) at late spring breeding yielded biomass values of 14–160 mgC/m^2. Comparative studies (as a basis for transfer experiments) of a univoltine and a bivoltine population of *Ferrissia* by Romano (1980) gave values of 179 and 207 mgC/m^2. For the American pond limpet *Laevapex fuscus,* McMahon (1972, 1975) presented data for four generations in three populations with peak standing crop values 18.8–56.5 mgC/m^2. Data on the European pond limpet *Acroloxus lacustris* from a population in the River Thames (Mann, 1964, 1971) convert to a mean biomass value of 1.6 mgC/m^2, with values of 3.1 and 1.1 mgC/m^2 for coexisting populations of two small planorbid species.

D. Land Snails and Biomass Rank Order

Comparable standing crop biomass values for the giant tropical amphibious prosobranch *Pomacea urceus* cannot be computed because of lack of true density data, but Burky (1974) presents organic carbon levels for individual snails up to 12.4 gC and computes the extraordinary figure of 1.2 kgC per 100 snails for a unit population at the end of adult growth. Even if estivating aggregations with densities of 24 individuals/m^2 are locally limited (Burky et al. 1972; A. J. Burky, personal communication), this would imply a patchy standing crop figure of 288 gC/m^2, exceeding even that of marine bivalve beds. Bioenergetic data from field studies on land snails are rare, but an early study of the American wet woodland species *Mesodon thyroidus* by Foster (1937) yielded tissue wet weight figures which convert to a peak biomass value of 734 mgC/m^2. Detailed work on the pest slugs of a British arable plot (Hunter, 1966) provides frequency distributions of wet weights and sampling densities which would approximate biomass values (in mgC/m^2) of 929 for *Agriolimax reticulatus,* 823 for *Arion hortensis,* and 683 for *Milax budapestensis.* For the European land snail *Cepaea nemoralis,* Richardson (1975) gives mean standing crop figures (by microbomb calorimetry) for 2 years of 17.0 and 16.4 kJ/m^2, which correspond to 371.2 and 358 mgC/m^2, respectively. Mason (1972) provided energy content data on 20 species of smaller terrestrial snails living in leaf litter which correspond to average biomass values ranging from 0.11 mgC/m^2 for *Vallonia pulchella* to 32.42

mgC/m^2 for *Oxychilus cellarius* and 33.78 mgC/m^2 for *Hygromia striolata*. Although there are relatively few good estimates of population density for land snails in the literature (the best are for *Cepaea* spp.; see Chapter 14, this volume) it is clear that standing crop values for terrestrial molluscs can approach those for land arthropods and earthworms. Further, at least for some snail and slug species, they are appropriate to dominant forms at the second trophic level.

As noted elsewhere (Russell-Hunter, 1970, 1978), if we have data for several populations of a species (such as the freshwater pulmonates *F. rivularis, L. fuscus* and *Lymnaea palustris*), the rank order of values in terms of standing crop biomass in mgC/m^2 does not necessarily correspond to the rank order of productivity rates in mgC/(m^2·day) for the populations.

IV. Rate Productivities

A. Computing Methods

All assessments of the line items of actuarial bioenergetics (i.e., the partitioning of energy) must be expressed in the dynamic units of fractional productivity rates (reflecting differential energy flux) rather than the static units of standing crop biomass (equivalent to energy content). All such rate measures (of feeding, assimilation, respiration, growth, reproduction, and mortality) are essentially measures of change in biomass per unit time. The labels conventionally used for each of these components in nonmarine molluscan bioenergetics and their equivalents in the global productivity symbols adopted by the International Biological Programme (IBP) are set out below (Section V). It is important to note that in productivity rates the components are similar, regardless of whether one is considering energy flow for the natural population as a whole, or for each generation cohort, or for a modal individual mollusc (Russell-Hunter, 1970, 1978). The units used for these component rate measures can all be reduced or converted to values such as mgC/(m^2·day) or, lacking concurrent population densities, as mgC/(100 snails·day).

Overall productivity rates for populations or cohorts combine growth and reproduction rates with mortality rates. In such computations, the use of appropriate cumulative biomass values for individual animals that have died originated with Allen's classic monograph (Allen, 1951) on trout in the Horokiwi Stream (although similar methods were used even earlier on Russian fish stocks by V. S. Ivlev and G. G. Winberg). Russell-Hunter (1970) gives a method of computation and graphic presentation that was subsequently used by Avolizi (1976), Mattice (1972), McMahon (1975), Hunter (1975), Aldridge and McMahon (1978), Eversole (1978), Browne (1978), Payne (1979), Aldridge (1980), and Tashiro (1980). In a series of successive census of the population or cohort, mortality values derived from density assessments (or from counts of shells from dead snails) can be combined with appropriate carbon values to provide estimates of successive (and cumulative) carbon values/m^2 lost from the population by death during the time intervals between samplings. These cumulative biomass values for the molluscs that have died (the "below the line" values in our actuarial graphics) can be added to the standing crop ("above the line") values to give a series of values for cumulative net biomass production. Divided by the elapsed time to any point in the series, these latter values provide assessments of productivity rate to that point in time for a cohort, generation, or population of any mollusc. Such overall rates are available for a variety of nonmarine molluscs; the majority have been computed as described above.

B. Bivalves and Prosobranchs

For medium-sized freshwater bivalves, Avolizi (1970, 1976) gives nitrogen values which convert to rates of 13.2 mgC/(m^2·day) for *Sphaerium striatinum* and of 4.6 mgC/(m^2·day) for *S. simile*. Aldridge and McMahon (1978) report a rate of 28.6 mgC/(m^2·day) for the introduced pest species *Corbicula manilensis*, in Texas (see also Chapter 12, this volume), which may be the highest productivity rate ever recorded for a freshwater bivalve. [It is comparable to the rates for several marine clams under commercial exploitation, although Dame (1976) computed a calorific value for an intertidal population of the oyster *Crassostrea virginica* in South Carolina which is equivalent to 1.03 gC/(m^2·day).] Leveque (1973) reported energy content values for five populations of *Corbicula africana* in Lake Chad which correspond to carbon productivity rates ranging from 1.5 to 13.0 mgC/(m^2·day). Natural populations of smaller freshwater bivalves studied by Burky and associates yielded productivity rate values of 3.04 mgC/(m^2·day) for *Musculium lacustre* (Alexander, 1982) and 4.55 and 19.01 mgC/(m^2·day) for a univoltine population and a bivoltine population of *M. partumeium* (Burky et al., 1983). Gillespie (1966, 1969) gives tissue dry weight data for a population of *Pisidium compressum* in the Madison River at Yellowstone which convert to 4.73 mgC/(m^2·day). The caloric estimates of Teal (1957) for *P. dubium* in his classic cold spring study correspond to a rate of 20.5 mgC/(m^2·day). For large freshwater mussels, we have productivity figures in wet tissue weight from Thames populations (Negus, 1966) for *Unio pictorum, U. tumidus,* and *Anodonta anatina*, which convert to approximately 0.71, 0.27, and 1.77 mgC/(m^2·day), respectively.

Freshwater prosobranch gastropods include some long-lived, slow-growing species which predictably have low rate productivities relative to their standing crop values. In the case of *Leptoxis carinata,* Aldridge (1980, 1982) computed rates of 2.6, 3.2, and 6.6 mgC/(m^2·day) for three environmentally stressed populations, and for the related *Goniobasis livescens,* Payne (1979) reported values of 1.1–9.0 mgC/(m^2·day). Tissue dry weight data (Gillespie, 1966, 1969) for the tiny *Valvata humeralis* from the Madison River yield a rate of 1.3 mgC/(m^2·day). A relatively higher productivity rate of 2.42 mgC/(m^2·day) has been assessed by Tashiro (1980, 1982) for a population of *Bithynia tentaculata* in Oneida Lake; this probably reflects the fact that an efficient filter-feeding mechanism can be used to supplement grazing in this species (Tashiro and Colman, 1982; see also Section V). For a shorter lived (biennial) population of the same species in the River Thames (Mann, 1964, 1971), there are data that yield a productivity rate of only 0.181 mgC/(m^2·day). The overall productivity rate for the Tully population of *Viviparus georgianus* (D. E. Buckley, unpublished) is approximately 6.97 mgC/(m^2·day), of which 1.14 mgC/(m^2·day) is allocated to female reproduction (including uterine young). Actuarial aspects of this controlled reproductive effort are discussed in Section VI. Earlier estimates of rates for four populations of *V. georgianus* in upstate New York (Browne, 1977, 1978) were 8.3, 6.1, 3.8, and 0.7 mgC/(m^2·day), and data for the closely allied *V. viviparus* in the River Thames (Mann, 1964, 1971) convert to the low rate of 0.733 mgC/(m^2·day).

C. Freshwater Pulmonates

Characteristically, freshwater pulmonates have shorter life cycles than sympatric prosobranchs, which implies relatively higher productivity rates. Among pulmonate species, however, the rates do not seem to be size related. For an annual population of *Helisoma trivolvis,* Eversole (1978) calculated a productivity rate of 16.1 mgC/(m^2·day), and for two annual populations of *Lymnaea palustris,* Hunter (1975) computes rates of 3.7 and 22.8 mgC/(m^2·day). For a biennial population of *Helisoma trivolvis* (Boerger, 1975), we cannot have a productivity rate directly derived from spatial densities, but using appropriate conversion values (Eversole, 1978) we derive a unit population rate of 10.27 mgC/(100 snails·day). Eckblad (1973) gives ash-free dry weight data for *L. palustris* whicn convert to 3.1 mgC/(m^2·day) and values for *Physa integra* which convert to 2.3 mgC/(m^2·day). An unusually high value for *P. integra* of 42 mgC/(m^2·day) can be calculated from wet weight biomass values (Tilly, 1968) for a population in Cone Spring, Iowa. The Madison River population of *P. gyrina* (Gillespie, 1966, 1969) yields a rate of 1.6 mgC/(m^2·day), which is closer to rates assessed for *Physa* spp. in upstate New York. The smaller planorbid species and the freshwater limpets (both groups with individual carbon contents generally $<$0.7 mgC) also show considerable variation (including interpopulation, infraspecific variation) in productivity rates. For the largest ancylid studied, the European *Ancylus fluviatilis,* Streit (1976a) reports rates of 0.469–0.627 mgC/(m^2·day), and the volume data of Russell Hunter (1961a) yield a peak rate of 1.93 mgC/(m^2·day). For *F. rivularis* in upstate New York, the data of Romano (1980) give rates of 0.658 mgC/(m^2·day) (an univoltine population) and of 0.798 and 0.776 mgC/(m^2·day) (spring-born and late summer-born generations of a bivoltine population). Corresponding data (McMahon, 1975) for *Laevapex fuscus* yield the rather low productivity rates of 0.377 and 0.067 mgC/(m^2·day) (for two univoltine populations) and 1.2 and 0.274 mgC/(m^2·day) (for spring-born and late summer-born generations of a bivoltine population). McMahon (1972, 1973, 1975) attributes the low productivity rates of *Laevapex* to restricted habitat space/m^2 of bottom. The European pond limpet, *Acroloxus lacustris,* may be similarly restricted, and data from the Thames (Mann, 1964, 1971) convert to a rate of 0.014 mgC/(m^2·day). Smaller planorbid species may also have net productivity assessments biased by niche limitations. Ash-free dry weight data for a population in upstate New York of *Gyraulus parvus* (Eckblad, 1973) yielded 0.03 mgC/(m^2·day), and Thames River data (Mann, 1964, 1971) convert to 0.027 mgC/(m^2·day) for *Planorbis* (*Bathyomphalus*) *contortus* and to 0.019 mgC/(m^2·day) for *P.* (*Anisus*) *vortex*. The population of *G. deflectus* in the Madison River (Gillespie, 1966, 1969) has a notably high productivity rate of 4.86 mgC/(m^2·day).

D. Land Snails and Comparisons

Computations for terrestrial molluscs yield a similar variety of productivity rates. Note that we cannot have true (spatially related) productivity rates for the giant amphibious snail *Pomacea urceus,* but Burky (1974) provides calculations based on a unit population which are equivalent to 7.14 gC/(100 snails·day). Tissue dry weight figures for *Mesodon thyroidus* in the small but prescient study by Foster (1937) convert (using an appropriate rate constant of 2.1) to a productivity rate of 4.22 mgC/(m^2·day). Unfortunately, the excellent field data on three species of slugs (Hunter, 1966), which have provided biomass figures and also good reproductive indices (Section VI), cannot readily be calculated as productivity rates; because two of the species are near annuals and one is biennial, it is clear that rates for all three would be close to those for *Mesodon* spp. or about double those for *Cepaea* spp. Calorific data on *C. nemoralis* by Richardson (1975) yield productivity rates of 2.05 and 1.55 mgC/(m^2·day) for 2 successive years. Similar data on leaf-litter snails (Mason, 1971) convert to rates ranging from 0.002 mgC/(m^2·day) for *Vallonia pulchella* to 0.223 mgC/(m^2·day) for *Oxychilus cellarius*. The land snail data are scanty but suggest that life spans are relatively uniform and that productivity rates rank with snail size.

Despite these seemingly simple relationships for terrestrial snails, any generalizations regarding the productivity rates of all nonmarine molluscs are mostly vain. From the data given above, it is obvious that attempts simply to correlate productivity rates with individual size (energy content), population density, or trophic relationships are ineffective. Any interspecific comparisons (e.g., those of great evolutionary interest between nearly sympatric congeners) are likely to be obscured and difficult to quantify because of the high level of infraspecific and interpopulation variation in such rates. Even if the comparisons are limited to data sets assembled by the same investigator, differences among populations within species can be sevenfold and differences between generations within a population can be fourfold. Year-to-year variations within an overall population can mean that we do not have the steady state (dynamic equilibrium) which is a prerequisite for many computational methods. Finally, as noted above, productivity rates are very sensitive to difficulties of assessing environmental space in calculating densities for biomass, i.e., to difficulties resulting from the restriction of a species to a limited habitat, which is not easily quantified.

Many of these dilemmas are partly resolved in comparisons at all levels by use of turnover times or rate constants, as defined in Section II. Similarly, analyses of budget components and their physiological constraints benefit from being cast as ratios or dimensionless index numbers. Comparative spatial units and dubious biomass–energy translation can be eliminated in both cases. For example, evolutionary consideration of trade-offs in growth and reproduction requires bioenergetic data prepared in index (dimensionless) form. When we turn from examining energy budgets of individual molluscs to those of populations, truly actuarial assessments must be developed on the basis of differential data determined separately for each demographic sector in each population. The following sections are concerned with the internal components of balance sheets for the population or modal individual (Section V), with productivity comparisons between populations and between species, including the concept of *differential reproductive effort* (Section VI), and with experimental modifications of the bioenergetics of natural populations (Section VII). In all these computations, index numbers and rate constants should be employed wherever possible. Only in those few cases for which demographically allocated data sets are available can we use appropriate stochastic limits in true actuarial models.

V. Budget Components in Actuarial Analyses

A. Components and Symbols

In any budget of carbon productivity rates for a heterotrophic organism such as a mollusc, the components are similar regardless of whether the budget is for a modal individual, a cohort generation, or an entire population (Russell-Hunter, 1970, 1978). The components of energy flux for the individual and the population are stylized in Fig. 1. In carbon terms, only a fraction of the productivity of the previous trophic level makes up the total ingested (TI). Part of this amount is not assimilated (NA) and is egested as feces, whereas the rest or total assimilated carbon (TA), enters the organism (or population). Of this TA a considerable part

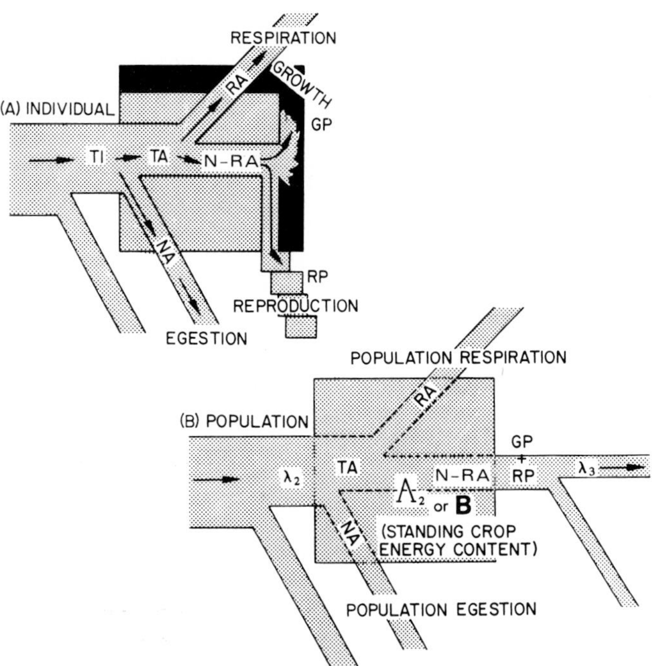

Fig. 1. Energy flow in the individual mollusc (A) and in the mollusc population (B). (Modified from Russell-Hunter, 1970.)

is broken down in catabolic metabolism as respired assimilation (RA) for maintenance and other activities, including energy expenditures for food collection and intake. The other fraction, of non-respired assimilation (N-RA), remains available for increase in the carbon biomass by growth production (GP) and for output as reproductive production (RP). Ecologically, in population budgets only a fraction of the total N-RA (or GP + RP) remains available to form the TI of the next trophic level, i.e., to make up the food ingested by predators on that molluscan population.

Before presenting some of the simple relationships, ratios, and efficiencies which can be derived using these conventional symbols of molluscan bioenergetics, some reference to the corresponding terms used in global studies of community productivity is in order (see Table I). As stated by the eminent Polish ecologist Kazimierz Petrusewicz, and as documented by Welch (1968) and Waters (1969), synonymy is truly a nightmare for ecologists. To avoid becoming involved in earlier (and continuing) semantic debates, the global symbols shown in Table I are those adopted during the IBP and listed by Petrusewicz and Macfadyen (1970) in a methodological handbook on the productivity of terrestrial animals that resulted in part from early efforts by Macfadyen (1948, 1963) to standardize terms.

Several simple relationships are seen in Fig. 1: TA = TI − NA; N-RA = TA − RA; N-RA = GP + RP; and N-RA = TI − (NA + RA). Once again the need for common units must be emphasized; reduced to carbon terms, all are essentially measures of change of biomass with time, and can be expressed as mgC/(100 individuals·day) or as mgC/(m^2·day). Several derived ratios (dimensionless index numbers) are important: TA/TI (commonly expressed as a percentage and termed *gut efficiency*); N-RA/TA (*net growth efficiency*); N-RA/TI (*gross growth efficiency*); RP/N-RA, and RP/TI. The last two, as reproductive indices, will be somewhat arbitrarily separated and discussed in Section VI in relation to comparisons between populations and species. Use of the other ratios and component relationships in *internal* budgets of nonmarine molluscs will first be surveyed here.

B. Ratios as Efficiencies

As stated in Section IV, there is a greater than 10-fold difference between *Lymnaea palustris* and *Laevapex fuscus* productivity rates in natural populations in upstate New York. Nevertheless the N-RA/TA percentage values (sometimes called *net growth efficiency*) for two populations of *L. palustris* are 22 and 41% (Hunter, 1975), and the range of N-RA/TA values for four generations in three populations of *L. fuscus* is 22–52%. For two populations of the European stream limpet *Ancylus fluviatilis,* Streit (1975, 1976a) gives N-RA/TA values of 11.2 and 11.8%. For *F. rivularis,* data on RA and N-RA yield a value for N-RA/TA

TABLE I

Some Conjugate Terms in Productivity Budgets[a]

Conventional component labels in actuarial bioenergetics of nonmarine molluscs (usually in carbon units)		IPB global productivity symbols (usually in energy content units)	
—		MR	Material removed
—		NU	Material not used
TI	Total ingested	C	Consumption
NA	Not assimilated (egested feces)	FU	Feces plus excretion
TA	Total assimilated	A	Assimilation
RA	Respired assimilation	R	Respiration (maintenance cost)
N-RA	Nonrespired assimilation	P	Production
GP	Growth production	Pg	Production in body growth
RP	Reproductive production	Pr	Production in reproduction
B	Standing crop biomass	B	Standing crop biomass

[a] Note also that in the original terms of the Hutchinson-Lindeman method of trophic representation Λ_2 would correspond to B, the standing crop biomass of the second trophic level, λ_2 to the corresponding TI or C, and λ_3 to the output to (or TI for) the next trophic level (see Fig. 1).

of 19% (Burky, 1971). Overall values for three populations of *Leptoxis carinata* resulting from extensive field and laboratory data (Aldridge, 1982; see also Chapter 8, this volume) are 12.2, 12.4, and 12.9% for N-RA/TA in carbon terms; his investigations allow even better actuarial analysis by providing separate values for female and male snails and separate efficiencies for protein carbon and nonprotein carbon (see below). Similarly detailed experiments on filter feeding in *B. tentaculata* by Tashiro and Colman (1982) yielded N-RA/TA percentage values for females of 92%. Among land snails, for his population of *C. nemoralis* in 2 successive years Richardson (1975) reported values of 35 and 46%. The leaf-litter snail data of Mason (1971) yielded N-RA/TA values ranging from 34.6% for the larger *Hygromia striolata* to 60.1% for the smaller *Vallonia pulchella*.

Obviously, for certain ratios such as net growth efficiency (N-RA/TA), an actuarial crosscheck or audit becomes possible if TA values can be assessed (from independently determined values of ingestion, egestion, respiration, and production) both as TI − NA and RA + N-RA. Evidence of compensatory and regulatory effects in the bioenergetic background to partitioning can result in such audit procedures, e.g., in shifts of the TA/TI ratio in response to changes in the quantity or quality of TI or to changes in N-RA or RA. Some shifts are difficult to accommodate in patterns predicted by current optimal foraging theory (Pyke et al., 1977). For two populations of *Lymnaea palustris* in upstate New York, Hunter (1975) demonstrated that the TA/TI ratio was 68% in the population with a lower overall productivity rate from a less eutrophic environment, compared to only 57% for the population in more eutrophic conditions. An increase in gut efficiency partially compensated for lower input and reduced the difference in overall productivity.

In contrast, in comparing three populations (four generations) of *Laevapex fuscus* with differences in N-RA greater than 10-fold, McMahon (1973, 1975) noted a close regulation of respiration resulting in similar levels of RA for all three populations. In this case, the wide range of productivities (including population fecundity values, RP; see below) are directly related to a similar range of values for TA (and presumably for TI). In his work on the land snail *C. nemoralis,* Richardson (1975) uses a flat TA/TI rate of 33.6%, although he noted 10-fold differences in weight-specific RA for different age classes. For two land snails of different geographic origin, *Bradybaena fruticum* from near Sarapul in the Urals and *Eobania vermiculata* from Sevastopol in the Crimea, Seifert and Shutov (1981) found TA/TI values of 57 and 50%, respectively. Extensive work on food and feeding in the estuarine prosobranch species *Hydrobia ventrosa* (Kofoed, 1975; Lopez and Levinton, 1978) revealed a twofold range of TA/TI values with different diets. After a survey of cellulase-like enzymes in 13 species of freshwater snails, Calow and Calow (1975) concluded that prosobranch species were better equipped than pulmonates in this regard, resulting in potential niche separation (correlated, presumably, with a greater ability to include macrophytes in their diet; but see Table 1 in Chapter 8 this volume).

Even without considering dietary components and quality, there are obvious difficulties in treating a population of any nonmarine mollusc as a "black box" and attempting to quantify broader relationships, such as that of RP to TI. For example, earlier studies of natural populations of *Laevapex fuscus* and *Lymnaea palustris* in upstate New York (McMahon et al., 1974) revealed that not merely the quantity of *Aufwuchs* available as food, but also its nutritional quality, when assessed both as a percentage of organic carbon and as a C : N ratio, apparently had direct effects on gross fecundity in different populations of these two species. Thus, it might have been assumed that the RP/TI ratio could be refined to allow predictions of fecundity to be made for either species. However, in considering TA/TI ratios in *Lymnaea palustris* and N-RA/RA ratios in *Laevapex fuscus*, compensation in gut efficiency and regulation of respiration level make the relationships less simple and the results of nutritional change less predictable.

Components of molluscan biomass budgeting can also be computed from series of analyses of total combined nitrogen in adults, eggs, and juveniles sampled from natural populations. One productivity study of *Sphaerium* spp. (discussed in Sections III and IV) is computed and presented primarily in terms of combined nitrogen (Avolizi, 1970, 1976). In many other studies of nonmarine molluscs (and also of the primitive saltmarsh pulmonate *Melampus*), computation of both carbon and nitrogen biomass values for modal snails under different biological conditions and at different times of the year has allowed the assessment of C : N ratios for the molluscan tissues. Changes in these biomass ratios are biologically significant, and an admittedly oversimplified summary explanation was suggested by Russell-Hunter (1970, 1978). Decreasing C : N ratios (corresponding to buildup of the protein fraction) are claimed to characterize the prereproductive period, whereas increasing C : N ratios characterize preparation for overwintering (with increasing storage of carbohydrates and, to a lesser extent, lipids). A tentative classification of molluscan eggs by C : N ratio and by size has also been proposed (Russell-Hunter, 1970). Because the organic carbon and combined nitrogen contents of both food (TI) and feces (NA) could also be determined, such double analyses of GP and RP meant that both TA and N-RA could be computed in this fashion, leaving RA as the only budget component not determined in terms of combined nitrogen.

C. Catabolic Partitioning

More recently, investigations of catabolic partitioning of carbon and nitrogen have been based on essentially concurrent assessments of oxygen uptake and nitrogenous excretion (Aldridge et al., 1980; Aldridge, 1980, 1982; Tashiro, 1980, 1982), allowing estimates of RA in both carbon and nitrogen terms.

Oxygen consumption rates of snails have been assessed using Clark-type polarographic electrodes in specially modified respiration chambers (Burky, 1977), immediately followed by assessment of ammonia excretion rates using an Orion Model 95-10 gas-sensing ammonia probe. The contribution of urea to nitrogenous excretion, which is small in many of our freshwater and littoral snails, can be estimated by treating aliquot samples with urease before using the ammonia probe. These concurrent measurements of nitrogenous excretion and oxygen consumption can be used in computations which reveal shifts from protein-based to carbohydrate-based catabolism (or RA) correlated with seasonal, trophic, or reproductive changes.

Computations based on these concurrently obtained data assume that all nitrogen excreted is derived from the breakdown of protein, and that oxygen is consumed in proportion to the breakdown of organic carbon compounds, both protein and nonprotein. Weight-specific rates of ammonia excretion can be converted ($\times$ 0.827) to rates of nitrogen excretion. The conversion factor used to estimate the oxygen consumption for protein catabolism from this rate is 5.92 $\mu lO_2/\mu gN$. Thus, we can have a weight-specific rate of oxygen consumption for the protein fraction of catabolism. Subtraction of this rate from the overall weight-specific oxygen uptake rate gives the rate of oxygen consumption for the nonprotein fraction. Equivalents for carbon mass consumed are derived from the appropriate amounts of CO_2 given off (0.536 μgC consumed/μlCO_2). However, the relation of CO_2 evolved to O_2 consumed differs for proteins and nonproteins. The CO_2 given off from protein catabolism can be derived directly from the weight-specific nitrogen excretion rate (4.75 μlCO_2 evolved/μgN excreted). The CO_2 given off from nonprotein catabolism can be estimated by multiplying the weight-specific oxygen consumption for the nonprotein fraction by an appropriate respiratory quotient. (Food component analysis for prosobranchs suggests an average of 10% fat and 90% carbohydrate for the nonprotein fraction, giving a respiratory quotient of 0.95.) Thus, we can compute separate RA values in carbon terms for protein and nonprotein catabolism. Of course, given determinations of TI, TA, etc. in both nitrogen and carbon terms, the protein carbon fraction of these components can be calculated by multiplying the nitrogen values by 3.25 and deriving the nonprotein carbon fraction by subtraction from the total organic carbon value.

Using these computational methods in his investigation of reproductive bioenergetics in three populations of *Leptoxis carinata*, Aldridge (1980, 1982) demonstrated that N-RA/TA percentage ratios for protein carbon ranged from 20 to 26%, lower values corresponding to higher protein levels in TI. Corresponding values of N-RA/TA for nonprotein carbon were only 1–5%. In *Leptoxis*, males and older animals had lower values of N-RA/TA. In a short methodological paper (Tashiro et al., 1980), we described new methods and recipes for quantifiable artificial food discs (with high and low protein content) and protocols for controlled experimental grazing both on such discs and on natural *Aufwuchs* (cultivated in the field on glass slides). Studies on *B. tentaculata* distinguished the actuarial bioenergetics of grazing (Tashiro, 1980, 1982) from those of filter feeding (Tashiro, 1980; Tashiro and Colman, 1982). One might predict that kinetic energy expenditures (part of RA) would be less during filter feeding than during grazing on *Aufwuchs* in these snails. In fact, the RA rates for nonprotein carbon are 44–56% lower during filter feeding on *Chlorella* than during grazing. An important actuarial statement (Tashiro and Colman, 1982) can be made for postbreeding individuals of *Bithynia,* which is probably also true for individuals during the breeding season. Filter-feeding snails have higher net gains (N-RA/RA) of both protein carbon and nonprotein carbon per respired cost than do grazing snails. Some other conclusions from the work on grazing snails (Tashiro, 1982), although less clear-cut, are important to actuarial analysis of the GP and RP components of N-RA. First, growth rates of young spat (< 4.5 mm) snails grazing in the laboratory were higher than those of field snails, whereas rates for all other ages were lower when grazing in the laboratory than when grazing *and* filter feeding in the field. Second, and more importantly, high-protein grazing in the laboratory by 2-year-old females increased numerical fecundity (and RP in carbon) 27.8% over low-protein grazing in similar females; however, 3-year-old females showed no significant shift. These findings on differential fecundity in *Bithynia* will be discussed in relation to current theories on the energetics of iteroparity (Sections VI and VII). Similar work is in progress on *Viviparus georgianus* (D. E. Buckley, unpublished), but it is possible to state now that high RP in 3-year-olds is not correlated with low or no RP in 2-year-olds.

D. Shells, Mucus, and Degrowth

Bioenergetic budgets for molluscs are complicated by expenditures on shell secretion and, in some cases, on mucus production and the manufacture of protective capsules for egg-masses. All estimates of anabolic allocation of N-RA in molluscs must involve consideration of the metabolic commitment to shell secretion (in addition to regular tissue growth in GP and fecundity in RP). The secreted shells have two major components: a meshwork of protein fibers (the organic matrix) and crystalline calcium carbonate (Russell-Hunter et al., 1981). In some species, such as *F. rivularis,* differences greater than twofold occur between populations in both components, but relatively little variation occurs within each population. We have no data on the absolute energy expenditures involved in secreting calcium carbonate by such limpets, but a survey of 10 natural populations of *Ferrissia* in upstate New York (Russell-Hunter et al., 1981) in waters with a 15-fold range of dissolved calcium content yielded nominal "concentration ratios" of body calcium to environmental calcium ranging from 1953 : 1 to 22,130 : 1. For the organic matrix (including periostracum), organic carbon contents in different populations also vary in modal limpets, ranging from 6.4 to 13.5% of the total carbon biomass in different populations, which corresponds to 1.8–5.6% of N-RA in these populations of *Ferrissia* directed to noncalcareous shell secretion. The larger freshwater prosobranch *V. georgianus* has a relatively massive operculum and periostracum; D. E. Buckley (unpublished) estimates the carbon allocation as amounting to 10.3% of the carbon N-RA during growth up to second year in females. The bioenergetic significance of molluscan shells results from the nontrophic growth energy expenditure which usually cannot be completely passed along in food webs. In our budgets, the organic material in the shell represents stored energy which is *never* turned over (until death and dissolution) unless external erosion or internal shell resorption takes place.

Molluscan mucus secretion can also abstract carbon from N-RA in individuals and populations. In attempting to allow for this in individual or population budgets, we must recognize two energetically distinct classes of molluscs. The many aquatic molluscan species which use mucus in filter-feeding or detritus-collecting mechanisms to obtain finely divided food particles do *not* experience a high net energy loss from synthesis and secretion of mucus. Except for the fraction used to consolidate pseudofeces from the mantle-cavity in most bivalves, most mucus used by both bivalves and gastropods in microphagous feeding mechanisms of all kinds is reingested and its energy (or organic carbon) content recycled. In contrast, the mucus produced in locomotion by land snails and slugs, and a considerable fraction of the mucus used by predaceous carnivores such as *Polinices* spp. during prey capture, is lost to the individual snails concerned. These latter cases can involve significant net imposts on the N-RA in individual and population budgets. In an extreme case, Richardson (1975) claims that, of the annual energy flux (RA + N-RA) in his *Cepaea nemoralis* population of 77.6 kJ/m^2, mucus production accounted for 18.35 kJ/m^2 compared to 17.51 kJ/m^2 in tissue growth. This seems to be an unreasonably high fraction when compared with the minute estuarine prosobranch *Hydrobia ventrosa,* in which mucus production accounts for only 9% of TA (Kofoed, 1975), although each tidal cycle over their mudflat habitat involves these snails in the twice-daily construction of a buoyancy raft of secreted mucus (Newell, 1962). A 50% tax on potential growth would seem to be a high price for *Cepaea* to pay for land locomotion.

Various molluscs, including many nonmarine species, enclose their eggs or egg-masses in a variety of protective capsules of hardened organic materials. Costs of capsule secretion must involve a deduction from energy (or organic carbon) available for RP; they cannot be quantified in most nonmarine molluscs. We have one set of carbon determinations on eggs and capsular material in the stream limpet *F. rivularis,* in which the lens-like lids amounted to only 8.1% of the carbon content of whole egg capsules with contained eggs. More detailed work has been done on some marine forms, and Perron (1981) found that in 10 species of *Conus* as much as 50% of total reproductive energy is accounted for by capsules; those species with longer development times allocate more energy to the secretion of tougher, more protective capsules. In nonmarine molluscs, there are no known similar actuarial trade-offs between increased egg protection and reduced egg production.

Apart from their bioenergetic costs, molluscan shells can have considerable evidential value in another aspect of actuarial studies. In many molluscs, the relative permanence of the calcareous shell after it has been laid down allows the detection of tissue degrowth. In *Helisoma trivolvis,* laboratory experiments (Russell-Hunter and Eversole, 1976) in a metabolic framework close to that of natural overwintering (with partial starvation) demonstrated that a representative cohort (with only 10% cumulative mortality) could show a 50% loss of biomass carbon in its tissues (involving probably a 20% loss of structural protein). Further laboratory studies on *Helisoma* (Russell-Hunter et al., 1983b) have used techniques of simultaneous monitoring of oxygen uptake and nitrogenous excretion to examine the temporal sequence of catabolism during the degrowth process. In field studies, individual recognition of regressed former adult snails in natural populations is practicable because these individuals carry evidence of their former larger tissue biomass (and, circumstantially, of their former adult physiology) in their oversize shells. Recent field work (Russell-Hunter et al., 1983a) has shown that overwinter degrowth occurs in natural populations of both *Lymnaea palustris* and *H. trivolvis.* In all cases, losses of nonproteinaceous tissue components are proportionately greater than those of protein components. However, even if one employs the strictest definition of degrowth (Russell-Hunter and Eversole, 1976) in individual snails or in cohorts as a decrease over time in the mass of structural proteins, there is evidence, in several freshwater pulmonates at least, of true degrowth. Capacity for degrowth is clearly important in the population dynamics (and actuarial bioenergetics) of those species which possess it. In some environmental circumstances, the capacity to budget negative values of N-RA/TA, allowing sustained RA under conditions of very low TI and TA, could contribute to fitness.

VI. Stock Comparisons and Investment Risks

A. Ratios and Indices

Productivity comparisons between populations and species (and possibly differential reproductive effort) cannot be based on rank orders of standing crop biomass (Section III) or even of rate productivity (Section IV). As already stated, ratios of these two measures, either as rate constants or as turnover times, can provide better values for ecological or evolutionary discussion (Russell-Hunter, 1970, 1978). Unlike the index numbers employed (mostly as percentages) in the survey of budget components (Section V), these two ratios are not strictly dimensionless because time is not canceled. Productivity *rate constants* are productivity rates divided by mean standing crop (for the same time period) and expressed in inverse time units, and *turnover times* are mean standing crop biomass values divided by productivity rates and expressed in time units. Turnover time, physiologically defined, is the average time in which a population (or cohort generation) can produce the equivalent of its mean standing crop biomass in new tissue growth. It should be noted that the time units in which the productivity rates were computed appear in these ratios. This may seem a simple point, but there is confusion in the published literature. If a computation of the rate constant is based on a productivity rate per year such as mgC/(m^2·year), then it is expressed as an *annual* number of turnovers of biomass. If a computation of turnover time is based on a productivity rate per day such as mgC/(m^2·day), then it is expressed in days. Because most nonmarine molluscan life spans are in the range 0.5–3.0 years, to avoid having fractional values in the ratios it seems best to base rate constants on productivity rates calculated per year and turnover times on rates computed per day. The rate productivities given in Section IV were generally expressed per day, as are the turnover times computed from them (and presented below).

In the context of these ratios, it is worth stressing again the value of the demographically correlated data on standing crops and productivity rates that can be generated by extensive sets of carbon analyses. Many studies have relied on a few spot determinations of energy content by microbomb calorimetry to convert growth rates calculated from linear measurements. The otherwise excellent work on bioenergetics in *Ancylus fluviatilis* by Streit (1975, 1976a,b) is flawed by use of a "universal" conversion factor to transform shell aperture lengths into carbon content, although Streit carefully obtained values of TI, TA, and RP for different animal sizes to allow better population estimates of these rates in carbon terms for his four field populations. The somewhat more laborious procedure at the beginning of Section III allows the determination of population carbon content to reflect more accurately the demographic structure of the population and to combine proportionately individual carbon content values reflecting size, age, and biological condition, as well as seasonal changes in tissue components.

B. Turnover Times

Despite varying data bases and levels of demographic sampling, we can now summarize the standing crop and rate productivity data of Sections III and IV as turnover times in days. (References to original authors are not repeated here, but can readily be identified in the earlier sections.)

For temperate freshwater bivalves, we have data on nine species: *Sphaerium striatinum*, a turnover time of 89 days; *S. simile*, 118; *Corbicula manilensis*, 91; *Pisidium compressum*, 92.4; *Musculium lacustre*, 46.5; *M. partumeium*, (a univoltine population) 50.7, and (a bivoltine population) 36.5 (spring born) and 27.0 (fall born); *Unio pictorum*, 2635 (7.22 years); *U. tumidus*, 2847 (7.8 years); and *Anodonta anatina*, 1789 days (4.9 years). The last three are large, slow-growing, riverine mussels with reported maximum life spans of 15, 11, and 9 years, respectively.

For freshwater prosobranchs (again of temperate latitudes), we have data on four species, drawn from 11 distinct populations: *Leptoxis carinata* from 3 populations with turnover times of 372, 311, and 303 days, all biennial and semelparous; *Goniobasis livescens*, 3 populations, 316 (biennial) and 462 and 530 days (triennial), all semelparous; *Bithynia tentaculata* with 337 (females) and 314 (males), triennial and iteroparous; and *Viviparus georgianus*, 4 populations, 477, 510, 393, and 421 days, all at least triennial and iteroparous with two periods of reproduction.

For pulmonate snails from temperate fresh waters, we have data on 12 species from 18 populations (with spring and fall generations for two bivoltine stocks): *Helisoma trivolvis*, with a turnover time of 39.7 days in the annual population and 185 days in the biennial; *Lymnaea palustris*, 2 populations, 74 and 75 days; *Physa integra*, 82.3; *P. gyrina*, 83.5; *Ancylus fluviatilis*, 2 populations, 138 and 61.3 days; *Ferrissia rivularis*, (univoltine) 112, and (bivoltine) 70 (spring born) and 62 (fall born); *Laevapex fuscus*, (2 univoltine populations) 92.9 and 94.0, and (1 bivoltine) 14.4 (spring born) and 54.4 (fall born); *Acroloxus lacustris*, 110; *Gyraulus parvus*, 171; *Planorbis* (*Bathyomphalus*) *contortus*, 114; *P.* (*Anisus*) *vortex*, 56.4; and *G. deflectus*, 94.9 days. All of these pulmonate species have univoltine annual life cycles, except for 1 population each of *Ferrissia* and *Laevapex* which are bivoltine (two generations per year) and the biennial stock of *Helisoma*.

Finally, for terrestrial snails, we have one tropical amphibious prosobranch and three species of temperate land pulmonates: *Pomacea urceus*, a turnover time of 168 days; *Cepaea nemoralis*, 181 and 231 in two successive years; *Vallonia pulchella*, 50.7; and *Oxychilus cellarius*, 146 days.

Turnover times are obviously related to length of life, and data for these 41 species-populations are summarized in Fig. 2. Short turnover times reflecting relatively high levels of productivity are most clearly significant in intraspecific comparisons (between generations and between populations). Long turnover times, especially in interspecific comparisons, must be used with more caution. Elapsed times used in productivity assessments and estimated times for life span

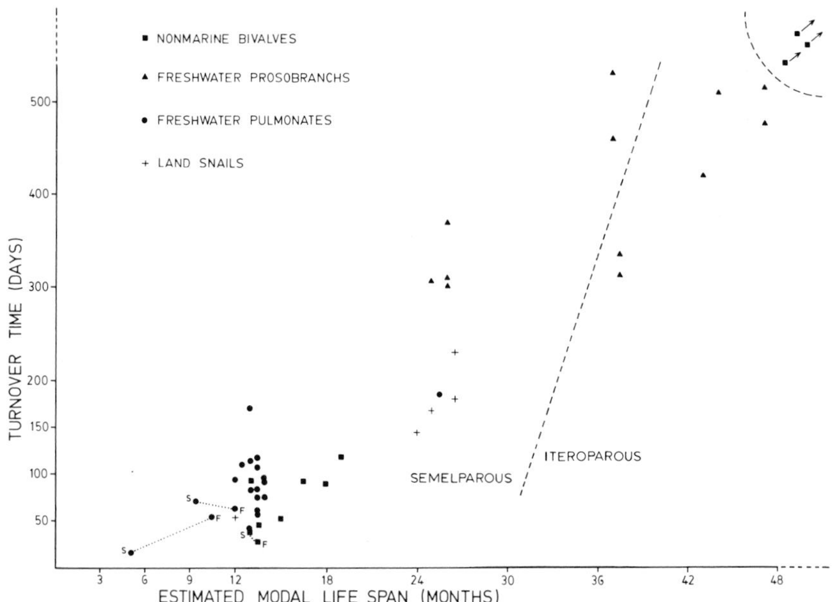

Fig. 2. The relation of turnover times (mean standing crop biomass values divided by productivity rates) to average life-span length in some nonmarine molluscs. The coordinates for the three long-lived bivalves (off-scale: top, right) are 1789 days at 108 months, 2847 days at 132 months, and 2635 days at 180 months. For data sets from the three bivoltine populations, dotted lines link each spring-born generation (S) with the corresponding fall-born overwintering generation (F). All populations to the left of the broken line are semelparous.

can vary among authors. Demographic data used in computations of mean biomass also vary; some authors average standing crop values for an entire population and others average values for a model cohort. Such variations in treatment of time and of the substantive biomass data have only slight effect when applied to populations with short life cycles and little generation overlap, but will more markedly affect computations on long-lived, age-structured populations. Despite these treatment-generated uncertainties, turnover times provide a most valuable standard for certain ecological and evolutionary comparisons. For example, among related organisms the rates at which genetic recombination occurs in separate populations are inversely related to the bioenergetic turnover times for those populations.

Turnover times of 10–70 days are associated with bivoltine or multivoltine life cycles, the shortest times in our temperate nonmarine series being those computed for rapid growth periods in summer generations of *Laevapex* and *Vallonia*. An extension of our data base to tropical species might include more species with turnovers less than 50 days. Most common temperate species with annual (univoltine) life cycles have turnovers of 40–120 days, with some of the larger, fast-growing forms such as *H. trivolvis* (39.7) and *Lymnaea palustris* (74 days) having the shortest times. Cold temperate species with biennial life cycles have turnover times of 140–350 days, and triennials may have 300–550. Turnover times more than 2 years appear to be associated with life spans of 4 years or more.

In temperate nonmarine molluscs, twofold differences in turnover times may occur between closely allied species and even between populations of a single species, and differences greater than threefold between generations of a population. These completely span the ranges of turnover times noted above for each pattern of life cycle. Essentially similar levels and differences for invertebrate populations have been presented, usually as the reciprocal ratios or rate constants (often termed *annual biomass turnover ratios*), by Waters (1969), Mann (1971), Streit (1976a), and others. Streit (1975, 1976a) has also proposed the use of a ratio of annual TA to mean annual biomass, a value that he terms *specific energy flow*. Using data drawn from studies on aquatic insects and freshwater crustaceans, Waters (1969) proposed, as a better comparative measure, the use of a rate constant based on life cycle (or cohort) productivity rather than annual productivity. Studying incorporation in *Ancylus fluviatilis* using ^{14}C-labeled food, Streit (1976b) showed that anabolic turnover rates were essentially the same for all organs in young limpets, but that adult rates were higher in the digestive diverticula and lower in all other organs. Again, we see the need for demographically structured data in population bioenergetics.

C. Reproductive Allocation

The most important partitioning in energy budgets is that of RP/N-RA, or the allocation to reproduction. There are problems in quantifying reproductive effort (even comparatively) using substantive data from natural populations. There are also problems in developing theoretical models for the evolution of life cycles which involve differential reproductive effort (see, as examples of this continuing debate, Williams, 1966a,b; Cody, 1966; Murphy, 1968; Tinkle, 1969; Pianka, 1970, 1976; Gadgil and Solbrig, 1972; Trivers and Willard, 1973; Charnov and Krebs, 1973; Schaffer, 1974, 1979; Stearns, 1976, 1977, 1980; Schaffer and Schaffer, 1977; Calow, 1978, 1979; Hastings and Caswell, 1979).

A survey of available molluscan data on reproductive effort (Browne and Russell-Hunter, 1978) employed two indices: the percentage of annualized non-respired assimilation (N-RA) channeled into reproduction (RP), and the ratio between the carbon channeled into reproduction and the carbon contained in the average female (or adult hermaphrodite) at the time of reproduction ($C_R : C_{AA}$). An inherent weakness in the second ratio, which is often used, is the usual requirement for a rate output for reproduction and an average biomass estimate for the parent. It is thus a rate constant in inverse time with (as discussed above) constraints imposed (by the time units in which reproductive output is computed) by the elapsed time during which the data was collected. Using both these ratios, a preliminary comparison of 5 semelparous and 11 iteroparous species (Browne and Russell-Hunter, 1978) showed that reproductive effort is generally higher in semelparous species. This agrees with fitness theory (Williams, 1966a,b; Gadgil and Bossert, 1970; Schaffer, 1974), as does the computation that reproductive effort increases in successive breeding seasons for iteroparous forms. In addition, viviparous species were found to show lower reproductive effort than comparable oviparous species.

As noted above, some short-lived pulmonate species (all semelparous), including the freshwater limpets *L. fuscus* and *F. rivularis,* have relatively high productivity rates (and thus low turnover times) which differ between populations and between generations. In regard to reproductive effort, fourfold within-species differences in percentage N-RA, and approximately threefold differences in the $C_R : C_{AA}$ ratio, are found in *Laevapex* and *Ferrissia* between populations, and between generations, in bivoltine stocks. For *Ferrissia,* the percentage N-RA allocated to reproduction is 16.6 in a univoltine stock and 73.0 and 37.2, respectively, in the spring-born and fall-born generations of a bivoltine stock. Both measures of reproductive effort are based on modal adult biomass values without specific reference to differences in immediate bioenergetic history, in particular to growth rates achieved immediately before reproduction. A dynamic measure of reproductive effort in terms of rate fraction is needed (Russell-Hunter and Romano, 1981).

An approach to this kind of measure is possible if appropriate seasonal data on productivity exist, and if reproductive effort can be related to the total anabolic capacity (N-RA) of any definable part of the life cycle. Early computations in these terms were made for the primitive pulmonate of high salt marshes, *Melampus bidentatus* (Apley et al., 1967; Russell-Hunter et al., 1972). For *M. bidentatus,* three assessments of reproductive effort were made: 46% of annualized N-RA, 87% of breeding season N-RA, and 32% of the N-RA if spring prebreeding growth rates had been sustained throughout the year. The third assessment was believed to represent a potential effort (never attained) that reflected a physiological limit to the system in *Melampus.* Where suitable data can be collected, we now suggest (Russell-Hunter and Romano, 1981; a full report is being prepared) that a suitable dynamic index of reproductive effort could be the reproductive output *rate* divided by a productivity *rate* computed for the immediate period of prebreeding growth. Note that although this computation gives a dimensionless index number of RP/N-RA, and although the elapsed times used to assess the rates can be relatively short (in relation to life span), the rates have been assessed for *different* time periods. Initial calculations of this index yield values ranging from 0.3 to 0.95 for different species, and it should be noted that values

> 1.0 would imply increased TI or (unlikely) reduced RA during the period of reproductive output, or would reflect a process of somatic degrowth. When the data on *Ferrissia* reported above are computed in this way, we have 0.802 for the spring-born and 0.886 for the fall-born generations of bivoltine populations (compared to percentage N-RA values of 73.0 and 37.2 respectively). Infraspecific differences are largely cancelled out, and this index may be closer to a genetically limited rate for potential reproductive effort (unique and fixed for each species or genetic stock), which is then modified by the trophic state of each cohort and by the seasonal shifts in available input. Computation of this index from suitable reproductive data for iteroparous species may prove to be important in quantifying age-structured differences in fecundity; newly assembled data on *Viviparus georgianus* (D.E. Buckley, unpublished) allow separation of size-related and age-related factors in reproductive output, and it is clear that the index will be higher for 4-year-old females than for 3-year-olds.

In discussing optimal reproductive tactics, Pianka (1976) modeled the adult organism as an input–output system converting trophic input into gametes, and represented partitioning using different symbols but the same compartments adopted in Section V. More recently, Calow (1979; see also Chapter 15, this volume) has proposed a physiological approach to a "cost index" for reproduction partly to explain the paradox of parasites wherein long continuous investment in reproduction appears not to affect parental survival. As a measure of the adverse (to the parent) effects of reproduction, this cost index (C) can be defined in our terms as $1 - (\mathrm{TI} - \mathrm{RP})/\mathrm{TA}^*$, where TA* measures the energy flux for all metabolic purposes (both anabolic N-RA and catabolic RA) other than reproductive, in prereproductive adults. This index enables Calow (1978, 1979) to define *reproductive recklessness* and *reproductive restraint* in quantitative terms. It is highly significant that both our suggested dynamic index and Calow's proposed costs index use a prereproductive productivity rate as the denominator: in our case the net value N-RA, and in Calow's the gross value TA.

Differences in that fraction of the bioenergetic output as eggs which can be sustained by concurrent food input are found even between closely related species of invertebrates (and also poikilothermic vertebrate species). Comparisons between these and other species pairs may require not only these rate or cost indices but also better parallel data on catabolic partitioning before and during reproduction. The nearly concurrent measures of oxygen uptake and nitrogenous excretion discussed in Section V allow preparation of budgets for protein-carbon and nonprotein-carbon under such circumstances, and it is already clear that in some species these budgets are altered during the stress of reproductive effort. The increase in the proportion of protein–carbon in the commitment to eggs (in RP) at the same time that the proportion of protein-carbon in the catabolic activities of the mollusc (in RA) is also increasing, forms an apparent paradox. Once again, there is a need to obtain appropriate comparative *rates* of carbon and nitrogen flux in a demographic framework, if only to modify future models of reproductive partitioning. A few data available on *Melampus* (Russell-Hunter et al., 1972), computed in terms of a standard adult snail with a tissue dry weight of 18.2 mg at maturity, amount to a total reproductive output of 7.3 mg dry weight of eggs (33,000 eggs) with a standard snail mass of 16.9 mg at the end of the reproductive period. Thus, in dry weight, 82.2% of the reproductive output must come from concurrent trophic input. When we examine this in terms of gonadal depletion (during the three cycles of the reproductive period) we find that 78% of the initial carbon content of the gonad has been utilized in egg production, as well as 91% of the combined nitrogen content, corresponding to a change in the C : N ratio of the gonad tissues from 6.3 : 1 to 15 : 1. This led to the hypothesis (Russell-Hunter, 1970; Russell-Hunter et al., 1972) that *Melampus* egg production was limited by the rate at which suitable organic nitrogenous materials could be made available to the gonad. Despite this asymmetry in processing components, it remains true in balance sheet terms that *Melampus* is meeting most of its expenditure in egg output from immediate food intake. It is possible that some nonmarine molluscan species draw even less upon capital and thus resemble the planktonic copepod, *Calanus finmarchicus,* wherein almost 100% of egg output biomass is met concurrently and output rates vary directly through time with variations in trophic input. In contrast, for certain other molluscs such as *Ostrea lurida* and the marine scalid snails, it can be deduced that these iteroparous forms (which are also rhythmic consecutive hermaphrodites) use proportionately much less from concurrent trophic input in their egg production and much more from biomass built up over the *two* interreproductive periods since they last functioned as females. No known nonmarine molluscs fall in this category. Obviously, differences in the fraction of reproductive effort being met from concurrent intake indicate potential differences in the effects of reproduction on parental survival.

It would be inappropriate to review here the debate on models for the evolution of life cycles (a few of the many publications are noted in Section VII,B). Most models have been developed from the demographic consequences of various life-cycle patterns (as presented in the classic early survey by Cole (1954) combined with the inverse relationship postulated between immediate reproductive effort and parental survival to reproduce again (as first elucidated by Fisher, 1930). However, the "choice" between reproduction and somatic survival involves a "bonus" factor for greater fecundity earlier in life *only* in expanding populations. The simpler predictions resulting from the theoretical framework set out by Williams (1966a,b) and others have mostly been fulfilled in those nonmarine molluscan species sufficiently studied (Calow, 1978, 1979; Browne and Russell-Hunter, 1978; Tashiro, 1982; D. E. Buckley, unpublished). Enough anomalies are known, for example, in the unusually low fecundity in both numerical and bioenergetic terms in the semelparous prosobranch *Leptoxis carinata* (Aldridge,

1982), and in the unusually high fecundity in both measures (Russell-Hunter et al., 1972) of the iteroparous pulmonate *M. bidentatus,* to demonstrate the need for more complete studies of bioenergetic budgeting (particularly of the interaction of differential catabolic partitioning of carbon and of nitrogen in relation to different levels of anabolic accretion for growth and reproduction).

In general, investigations of actuarial bioenergetics can contribute most to ecological and evolutionary studies (including those on reproductive effort) through apposite rate comparisons between species, between populations, and between generations or cohorts. That this work is basically comparative, and depends upon the collection of substantive data in the field from natural populations, need not exclude limited experiments in both field and laboratory (especially stock transfers or single-factor perturbations). A few such experiments are briefly summarized in the next section, along with a short prognosis for actuarial bioenergetics.

VII. Experimental Manipulations and the Actuarial Program

During investigations of population bioenergetics, experimental procedures used in hypothesis testing include a set of procedures attempting to separate narrowly genetic from ecophenotypically determined variation by stock transfers between field habitats and another set attempting to test deduced trade-off relationships by modifying appropriate single factors. Both have been used with nonmarine molluscan stocks.

A. Field Transfers and Tests

Reciprocal transfers of stocks of *F. rivularis* between two natural populations, one univoltine and one bivoltine, involved both experimental concepts (Romano, 1980). Separate experiments included reciprocal transfers of adult limpets, of recently hatched spat, and of the *Aufwuchs* food supply between cages at the two sites. Limpets from the univoltine stock transferred to the more eutrophic habitat of the bivoltine stock showed increased growth and fecundity rates. Somewhat unexpectedly, spat transfers resulted in some reproductive activity in the early fall (in other words, in an ecophenotypic shift to a potentially bivoltine life cycle). Reduced growth and fecundity were shown by bivoltine limpet stocks transferred to the trophically poorer habitat. A six-treatment experiment involved transfers of *Aufwuchs* (grown on glass slides) and of adult limpets, and showed that food quality was more significant than all other environmental factors taken together. In relation to potential reproductive effort and concurrent trophic input (see Section VI), it is interesting that the highest daily rate for egg output was achieved by transferred univoltine adults fed on the higher quality *Aufwuchs.*

Residual genetic differences in growth and fecundity rates between the univoltine and bivoltine stocks of *Ferrissia* cannot be dismissed (Romano, 1980), but the univoltine stock did attempt to move to a two-generation life cycle in enhanced trophic conditions. Earlier evidence from transfers of *F. rivularis* was more equivocal (Burky, 1969, 1971), but genetically determined differences in life-cycle pattern have been claimed for the other common freshwater limpet of North America, *Laevapex fuscus.* McMahon (1972, 1975), assessing growth and fecundity in reciprocal transfers of *Laevapex,* found retention of stock differences and thus circumstantial evidence for the existence of physiological races. The *Laevapex* stock from the trophically poorer habitat seemed to have higher assimilation efficiency (TA/TI), higher anabolic conversion efficiency (N-RA/RA), or both. At present, it can only be stated that interpopulation differences in fecundity (as a component of differences in life-cycle pattern) can encompass both ecophenotypic and genetically fixed differences (Russell-Hunter, 1978).

Much work on density-dependent regulation of fecundity and growth has been done on laboratory cultures of pulmonate snails (Noland and Carriker, 1946; Chernin and Michelson, 1957a,b; Wright, 1960; Mooij-Vogelaar et al., 1973, and references therein). Field experiments on this important life-cycle parameter are fewer. Using large wire-mesh enclosures, Eisenberg (1966, 1970) demonstrated density-dependent fecundity and resource limitation in near-natural stocks of *Lymnaea elodes* Say (which may be a subspecies of *L. palustris*). The results were attractively clear-cut, but several bioenergetic parameters remain uncertain and the experiments did not address the possible concurrence of more than one mechanism of density regulation (Russell-Hunter, 1978). Using *H. trivolvis,* Eversole (1974) carried out parallel laboratory and field (cage) experiments on the extent of, and controls for, density-dependent differences in fecundity. In this species, there are undoubtedly two (and perhaps three) different density-dependent mechanisms which can modify rates and patterns of egg laying even when food resources are *not* limiting. An eightfold increase in population density in field cages resulted in a fourfold decrease in egg output per snail per breeding season. In the *Helisoma* experiments (Eversole, 1974), spat numbers had a greater impact on fecundity than egg numbers, but both had less importance than adult numbers. Recent work on actuarial bioenergetics in the freshwater prosobranch *Leptoxis carinata* included field cage manipulation of a natural population in the Unadilla River in upstate New York (Aldridge, 1980, 1982). Four treatments involved high and low adult densities and high and low egg densities. Differences in adult density had no direct effect on numerical fecundity, whereas increasing egg density had a markedly negative effect on further egg output. Despite theoretical expectations of increased survivorship for females with lower egg output, toward the end of the experiment females from the high egg density and the low density cages were dying at the same rate (Aldridge, 1982).

Truly actuarial bioenergetics were the concern of a set of cage experiments on the differential growth and survival of the progeny from older and younger females of *Viviparus georgianus* (D. E. Buckley, unpublished). As noted in Section VI, in a natural population of *Viviparus* at Tully Lake in upstate New York females can live more than 4 years and potentially can reproduce on three occasions, as 2-year-olds, as 3-year-olds, and as 4-year-olds. The most fecund group of females in population terms are the 3-year-olds, but the largest individual broods are borne by certain 4-year-old females. If newborn spat snails are segregated in cages within 4 days of birth in groups according to maternal age, progeny of 3-year-olds show a 32% advantage in growth rates over progeny of 2-year-olds and a 12.6% advantage over 4-year-olds. Differential survivorship in cages favors spat of 4-year-old females, and being born early in a breeding period may confer an advantage regardless of size (D. E. Buckley, unpublished). More data on differential spat survival and its demographic consequences are needed for other iteroparous species.

As in studies of differential fecundity, parallel field and laboratory experiments can be valuable in assessing factors causing variations in assimilation rates (and their consequences for bioenergetic budgeting). In natural populations, the freshwater prosobranch *Bithynia tentaculata* uses both filter feeding and surface grazing on *Aufwuchs*, and experiments can be designed whereby snails are limited to one or the other method of food intake (Tashiro, 1980, 1982; Tashiro and Colman, 1982). In experiments with different concentrations of material for filter feeding, individual weight-specific clearance rates were shown to increase as available filterable material decreased (Tashiro and Colman, 1982). In general, filter feeding allows individuals of *Bithynia* a higher net gain (TA) of both protein-carbon and nonprotein-carbon per respired costs (RA) than does surface grazing. Experiments on grazing utilized both natural *Aufwuchs* grown on glass slides and artificial food discs with two levels of protein content (Tashiro et al., 1980) in assessing trophic constraints on reproductive effort (Tashiro, 1982). In such experiments with *Bithynia*, younger females always had a proportionately lower reproductive effort than older females (as RP/N-RA), but trophic conditions influenced the levels and patterns of fecundity. As might be predicted from hypotheses on age-specific reproduction, younger females placed on low-protein food significantly curbed their reproductive effort in favor of somatic growth and maintenance (or increased survival), but older females did not (Tashiro, 1982). Once again, age-specific shifts in reproductive effort in iteroparous species are so central to discussions of life-cycle evolution that more data on age-specific responses to experimental trophic stress are urgently needed.

B. The Future of Actuarial Studies

The general prognosis for actuarial bioenergetics depends in part upon an increase in the perceived value of comparative studies on natural populations. It may be significant that in his most recent survey of the published literature on life-history evolution Stearns (1980) notes that the proportion of purely theoretical papers has begun to decline relative to empirical papers and those combining substantive data with theory. He emphasizes a need for more background knowledge of development and physiology in those theoretical arguments presently dominated by population genetics and demography. The debates on current models for the evolution of life cycles require more comparative actuarial data on bioenergetic budgeting. Stocks of nonmarine molluscs in which life-cycle pattern is an interpopulation variable (Russell-Hunter, 1978; Calow, 1978) provide accessible material. In particular, different populations of a single species living in fresh water vary greatly in panmictic size and in their temporal history as steady-state populations (Russell-Hunter, 1970, 1978). Future evolutionary studies can use comparative bioenergetic data from the natural ''experimental'' populations provided by different bodies of fresh water which have experienced varying degrees of spatial and temporal limitation. In this subdiscipline, a strict segregation of the accumulation of comparative data on natural populations (from experimental studies) as merely descriptive is artificial and counterproductive.

The need for substantive data to be gathered for comparisons in rate terms, along with the advantages of dimensionless index numbers, has already been emphasized in this chapter. In actuarial bioenergetics of populations, as in trophic production studies of community ecology, it is important to quantify process rather than structure for the comparative method to succeed. Life is process, and natural selection is more pertinent to genetically controlled differences in rates of bioenergetic partitioning than to genetic differences in obvious structural characters such as number of shell spines or even shade of shell color (but see Cain, 1977; also Chapter 14, this volume). Future work in this field may further emphasize the need for more age-structured data on assimilation rates, catabolic rates, growth rates, and fecundity patterns. Approaching problems of life-history evolution by using the concepts of population genetics and mathematical demography, Charlesworth (1973, 1980) has also emphasized the need for models and detailed demographic data including age-specific factors. Although this chapter may have revealed the inchoate state of actuarial studies on molluscs at present (see especially Section V,B), current and future work on natural populations of nonmarine molluscs *can* produce demographically correlated variance figures on rates of growth, fecundity, and mortality.

Relating such substantive data sets to models formulated in fitness units may not be easy. Difficulties in concepts of fitness (particularly of time scale) appeared early (Ayala, 1969; Thoday, 1953, 1970) and continue to confuse modeling procedures and experimental design. At this level, the value of examples from commercial economics used as illustrative analogs earlier in this chapter declines sharply. Behind many debates on evolutionary processes lie fundamental philosophical questions on causality. Such issues led David Hume to develop the regularity theory of causation, which, despite being currently unfashionable among many professional philosophers, remains appropriate for the logical description of many biological sequences placed as they are as unique events in unidirectional time (Hume 1748; see also Beauchamp and Rosenberg, 1981). If several similar sequences occur, we may perceive a pattern, but only if we choose to ignore minor discrepancies (sequences are never the same when repeated). From discerned patterns, we construct ecological and evolutionary theories. Some of the consequences of each theory must ultimately be falsifiable if the theory is to have scientific value. In only some cases can models formulated in fitness units be translated directly into bioenergetic units for testing in the field (Rosenzweig, 1981; see also Charnov, 1976).

As noted in Section VI, most recent theoretical models of life-cycle evolution originate from a combination of the postulated inverse relationship between parental survival to reproduce again and immediate reproductive effort (Fisher, 1930) and the demographic consequences of various life-cycle patterns (Cole, 1954). The theory of differential reproductive effort was clearly set out in a logical framework by Williams (1966a,b), and numerical models quickly followed (notably those of Gadgil and Bossert, 1970; Schaffer, 1974; Charlesworth and León, 1976). One of the simpler predictions from the earlier models was that individual reproductive effort would increase with each successive breeding season in iteroparous forms. Suitable empirical data on actuarial bioenergetics of nonmarine molluscan species support this in most cases (Browne and Russell-Hunter, 1978; Calow, 1978, 1979; Tashiro, 1982; D. E. Buckley, unpublished). However, certain anomalous examples demonstrate the need for more complete studies (Section VI). On the theoretical side, a more rigorous algebraic treatment by Charlesworth and León (1976) shows that reproductive effort may increase with age, but it need not necessarily do so. The series of surveys by Stearns (1976, 1977, 1980) documents the changing ratio of theoretical models published to substantive data reported in this branch of evolutionary ecology. Both the strengths and the dangers of complex modeling are revealed by this extensive literature. In many strong lines of investigation, increasingly refined models are complemented by sets of substantive data extensive enough to reveal a pattern. In other cases, limiting assumptions in a theory become progressively forgotten, so that technical recipes for models or for analytical methods become presented as universal generalizations (or worse, as representing causality in a natural system). In evolutionary ecology, as throughout the sciences, more circumspect modelers remain closer to the canny logic of David Hume (and thus to the more skeptical philosophy of science that has developed by way of A. N. Whitehead and K. R. Popper).

Difficulties in using estimates of relative fitness in ecological models yet remain. In theoretical population genetics, the extension of models of selection rates to age-structured populations is relatively recent (Charlesworth, 1980).

Several critics have claimed that much evolutionary research involving fitness can be regarded as tautological, because definitions of fitness in terms of surviving offspring are circular and untestable. In response, Gould (1976) pointed out that it should be possible to assess certain traits as mechanistically contributing to fitness in a particular environment (i.e., as different efficiences of the animal machine) independently of any empirical assessment of their survival and spread. Although this distinction apparently does not satisfy certain philosophical critics of evolutionary studies in general (e.g., Brady, 1979), it provides the ultimate justification for our attempts to use comparative actuarial bioenergetics with natural populations. In the near future, not only better models of the selection process using a deductive framework provided by demographically based population genetics, but also substantive data on actuarial bioenergetics collected from natural populations, will contribute to our understanding of the evolution of life-cycle patterns. As in most other branches of adaptational research, recent investigations of life-cycle evolution have been dominated by theories based on current and immediate optimality of all extant structures and processes. This narrowed view of Darwinism can result in an overly restrictive research program, as was pointed out by Gould and Lewontin (1979) in an elegant essay advocating a pluralistic approach to evolutionary studies (see also Darlington, 1977). To be successful, any form of scientific inquiry should not be unnecessarily restricted by dogmatically limiting its methodology. MacArthur (1972) warned against this restrictive tendency in ''schools'' of ecology, and Hutchinson (1975) prescribed comparative studies based on deeper understanding of organismic biology as a remedy.

Acknowledgments

We must thank Mary Buckley and Perry Russell-Hunter for their help with the reference material. We are most grateful to Myra Russell-Hunter, Barbara J. Carns, and Eileen O'Connor for their help in the preparation of this manuscript, and to Dr. David W. Aldridge for his review and his assistance preparing the figures. This work was supported by NSF research grant DEB-78-10190. This is Contribution No. 23 of the Upstate Freshwater Institute.

References

Aldridge, D. W. (1980). Life cycle, reproductive tactics, and bioenergetics in the freshwater prosobranch snail *Spirodon carinata* (Bruguière) in upstate New York. Ph.D. Thesis, Syracuse University, Syracuse, New York (*Diss. Abstr.* **41B,** 2042, Order No. 8026351).

Aldridge, D. W. (1982). Reproductive tactics in relation to life-cycle bioenergetics in three natural populations of the freshwater snail, *Leptoxis carinata*. *Ecology* **63,** 196–208.

Aldridge, D. W., and McMahon, R. F. (1978). Growth, fecundity, and bioenergetics in a natural population of the asiatic freshwater clam, *Corbicula manilensis philippi*, from north central Texas. *J. Moll. Stud.* **44,** 49–70.

Aldridge, D. W., McMahon, R. F., and Russell-Hunter, W. D. (1980). Effects of temperature acclimation on nitrogen metabolism in two littorinid snails. *Biol. Bull.* (*Woods Hole, Mass.*) **159,** 447.

Alexander, J. P. (1982). Bioenergetics of a natural population of *Musculium lacustre* from an artificial lake. M.S. Thesis, University of Dayton, Dayton, Ohio.

Allen, K. R. (1951). The Horokiwi stream. A study of a trout population. *Fish. Bull. N. Z. Mar. Dept.* **10,** 1–231.

Apley, M. L., Russell-Hunter, W. D., and Avolizi, R. J. (1967). Annual reproductive turnover in the salt-marsh pulmonate snail, *Melampus bidentatus*. *Biol. Bull.* (*Woods Hole, Mass.*) **133,** 455–456.

Avolizi, R. J. (1970). Biomass turnover in natural populations of viviparous sphaeriid clams: Interspecific and intraspecific comparisons of growth, fecundity, mortality, and biomass production. Ph.D. Thesis, Syracuse University, Syracuse, New York (*Diss. Abstr.* **32:** 1919-B, Order No. 71–23, 429).

Avolizi, R. J. (1976). Biomass turnover in populations of viviparous sphaeriid clams: Comparisons of growth, fecundity, mortality and biomass production. *Hydrobiologia* **46,** 163–180.

Ayala, F. J. (1969). An evolutionary dilemma; fitness of genotypes *versus* fitness of populations. *Can. J. Genet. Cytol.* **11,** 439–456.

Bayne, B. L., and Widdows, J. (1978). The physiological ecology of two populations of *Mytilus edulis* L. *Oecologia* **37,** 137–162.

Bayne, B. L., Bayne, C. J., Carefoot, T. C., and Thompson, R. J. (1976). The physiological ecology of *Mytilus californianus* Conrad. 1. Metabolism and energy balance. *Oecologia* **22,** 211–228.

Bayne, B. L., Widdows, J., and Newell, R. E. (1977). Physiological measurements on estuarine bivalve molluscs in the field. *In* "Biology of Benthic Organisms" (B. F. Keegan, P. O. Ceidigh, and P. J. S. Boaden, eds.), pp. 57–68. Pergamon, Oxford.

Beauchamp, T. L., and Rosenberg, A. (1981). "Hume and the Problem of Causation." Oxford Univ. Press, London and New York.

Boerger, H. (1979). A comparison of the life cycles, reproductive ecologies, and size-weight relationships of *Helisoma anceps, H. campanulatum,* and *H. trivolvis* (Gastropoda, Planorbidae). *Can. J. Zool.* **53,** 1812–1824.

Brady, R. H. (1979). Natural selection and the criteria by which a theory is judged. *Syst. Zool.* **28,** 600–621.

Browne, R. A. (1977). Bioenergetics, reproduction and genetic variation in the prosobranch snail, *Viviparus georgianus*. Ph.D. Thesis, Syracuse University, Syracuse, New York (*Diss. Abstr.* **38B,** 3508, Order No. 7730710).

Browne, R. A. (1978). Growth, mortality, fecundity, biomass and productivity of four lake populations of the prosobranch snail, *Viviparus georgianus*. *Ecology* **59,** 742–750.

Browne, R. A., and Russell-Hunter, W. D. (1978). Reproductive effort in molluscs. *Oecologia* **37,** 23–27.

Burky, A. J. (1969). Biomass turnover, energy balance, and interpopulation variation in the stream limpet, *Ferrissia rivularis* (Say), with special reference to respiration, growth, and fecundity. Ph.D. Thesis, Syracuse University, Syracuse, New York (*Diss. Abstr.* **30,** 3917-B, Order No. 70-1942).

Burky, A. J. (1971). Biomass turnover, respiration, and interpopulation variation in the stream limpet, *Ferrissia rivularis* (Say). *Ecol. Monogr.* **41,** 235–251.

Burky, A. J. (1974). Growth and biomass production of an amphibious snail, *Pomacea urceus* (Müller), from the Venezuelan Savannah. *Proc. Malacol. Soc. London* **43,** 127–143.

Burky, A. J. (1977). Respiration chambers for measuring oxygen consumption of small aquatic molluscs with Clark-type polarographic electrodes. *Malacol. Rev.* **10,** 71–72.

Burky, A. J., Pacheco, J., and Pereyra, E. (1972). Temperature, water, and respiratory regimes of an amphibious snail, *Pomacea urceus* (Müller), from the Venezuelan Savannah. *Biol. Bull. (Woods Hole, Mass.)* **143,** 304–316.

Burky, A. J., Benjamin, M. A., Catalano, D. M., and Hornbach, D. J. (1979). The ratio of calcareous and organic shell components of freshwater sphaeriid clams in relation to water hardness and trophic conditions. *J. Mol. Stud.* **45,** 312–321.

Burky, A. J., Hornback, D. J., and Way, C. M. (1983). A bioenergetics approach to life-history tactics. Comparisons of permanent and temporary pond populations of the freshwater clam, *Musculium partumeium* (Say). *Ecology* (*in press*).

Cain, A. J. (1977). The uniqueness of the polymorphism of *Cepaea* (Pulmonata: Helicidae) in western Europe. *J. Conchol.* **29,** 129–136.

Calow, P. (1972). A method for determining the surface area of stones to enable quantitative density estimates of littoral stone-dwelling organisms to be made. *Hydrobiologia* **40,** 37–50.

Calow, P. (1978). The evolution of life-cycle strategies in freshwater gastropods. *Malacologia* **17,** 351–364.

Calow, P. (1979). The cost of reproduction—a physiological approach. *Biol. Rev. Cambridge Philos. Soc.* **52,** 23–40.

Calow, P., and Calow, L. J. (1975). Cellulase activity and the niche separation in freshwater gastropods. *Nature* (*London*) **255,** 478–480.

Charlesworth, B. (1973). Selection in populations with overlapping generations. V. Natural selection and life-histories. *Am. Nat.* **107,** 303–311.

Charlesworth, B. (1980). "Evolution in Age-structured Populations." Cambridge Univ. Press, London and New York.

Charlesworth, B., and León, J. A. (1976). The relation of reproductive effort to age. *Am. Nat.* **110,** 449–459.

Charnov, E. L. (1976). Optimal foraging, the marginal value theorem. *Theor. Popul. Biol.* **9,** 129–136.

Charnov, E. L., and Krebs, J. R. (1973). On clutch size and fitness. *Ibis* **116,** 217–219.

Chernin, E., and Michelson, E. H. (1957a). Studies on the biological control of schistosome-bearing snails. III. The effects of population density on growth and fecundity in *Australorbis glabratus*. *Am. J. Hyg.* **65,** 57–70.

Chernin, E., and Michelson, E. H. (1957b). Studies on the biological control of schistosome-bearing snails. IV. Further observations on the effects of crowding on growth and fecundity in *Australorbis glabratus*. *Am. J. Hyg.* **65,** 71–80.

Cody, M. (1966). A general theory of clutch size. *Evolution* **20,** 174–184.

Cole, L. C. (1954). The population consequences of life history phenomena. *Q. Rev. Biol.* **29,** 103–137.

Dame, R. F. (1976). Energy flow in an intertidal oyster population. *Estuarine Coastal Mar. Sci.* **4,** 243–253.

Darlington, P. J. (1977). The cost of evolution and the imprecision of adaptation. *Proc. Natl. Acad. Sci. U.S.A.* **74,** 1647–1651.

Eckblad, J. W. (1973). Population studies of three aquatic gastropods in an intermittent backwater. *Hydrobiologia* **41,** 199–219.

Eisenberg. R. M. (1966). The regulation of density in a natural population of the pond snail *Lymnaea elodes*. *Ecology* **47,** 889–906.

Eisenberg, R. M. (1970). The role of food in the regulation of the pond snail, *Lymnaea elodes*. *Ecology* **51,** 680–684.

Eversole, A. G. (1974). Fecundity in the snail *Helisoma trivolvis:* Experimental, bioenergetic and field studies. Ph.D. Thesis, Syracuse University, Syracuse, New York (*Diss Abstr.* **35,** 5716-B, Order No. 75–10, 538).

Eversole, A. G. (1978). Life-cycles, growth and population bioenergetics in the snail *Helisoma trivolvis* (Say). *J. Moll. Stud.* **44,** 209–222.

Fisher, R. A. (1930). "The Genetical Theory of Natural Selection." Oxford Univ. Press (Clarendon), London and New York.

Foster, T. D. (1937). Productivity of a land snail, *Polygyra thyroides* (Say). *Ecology* **18,** 545–546.

Gadgil, M. D., and Bossert, W. H. (1970). Life history consequences of natural selection. *Am. Nat.* **104,** 1–24.

Gadgil, M. D., and Solbrig, O. T. (1972). The concept of *r* and *K* selection: Evidence from wild flowers and some theoretical considerations. *Am. Nat.* **106,** 14–31.

Gillespie, D. M. (1966). Population studies of four species of molluscs in the Madison River, Yellowstone National Park. Ph.D. Thesis, Montana State University, Bozeman (*Diss. Abstr.* **27b,** 2184–2185. Order No. 66–10,842).

Gillespie, D. M. (1969). Population studies of four species of molluscs in the Madison River, Yellowstone National Park. *Limnol. Oceanogr.* **14,** 101–114.

Gould, S. J. (1976). Darwin's untimely burial. *Nat. Hist.* **85,** 24–30.

Gould, S. J., and Lewontin, R. C. (1979). The spandrels of San Marco and the Panglossian paradigm: a critique of the adaptationist programme. *Proc. R. Soc. London, Ser. B* **205,** 581–598.

Gustin, G. M. (1960). A simple, rapid automatic micro-Dumas apparatus for nitrogen determination. *Microchem. J.* **4,** 43–54.

Hastings, A., and Caswell, H. (1979). Role of environmental variability in the evolution of life history strategies. *Proc. Natl. Acad. Sci. U.S.A.* **76,** 4700–4703.

Hume, David (1748). "An Inquiry Concerning Human Understanding" (Original title: Philosophical Essays Concerning Human Understanding"). London [republished in 1888, Oxford Univ. Press (Clarendon), London and New York].

Hunter, P. J. (1966). The distribution and abundance of slugs on an arable plot in Northumberland. *J. Anim. Ecol.* **35,** 543–557.

Hunter, R. D. (1975). Variation in populations of *Lymnaea palustris* in Upstate New York. *Am. Midl. Nat.* **94,** 401–420.

Hunter, R. D., and Lull, W. W. (1976). A comparison of two methods for estimating the weight of inorganic materials in molluscs. *Malacol. Rev.* **9,** 118–120.

Hunter, R. D., and Lull, W. W. (1977). Physiologic and environmental factors influencing the calcium-to-tissue ratio in populations of three species of fresh water pulmonate snails. *Oecologia* **29,** 205–218.

Hutchinson, G. E. (1975). Variations on a theme by Robert MacArthur. *In* "Ecology and Evolution of Communities" (M. L. Cody and J. M. Diamond, eds.), pp. 492–521. Harvard Univ. Press, Cambridge, Massachusetts.

Kofoed, L. H. (1975). The feeding biology of *Hydrobia ventrosa* (Montague). II. Allocation of the components of the carbon budget and significance of the secretion of dissolved organic material. *J. Exp. Biol. Ecol.* **19,** 243–256.

Leveque, C. (1973). Dynamique des peuplements biologie, et estimation de la production des mollusques benthiques du lac Tchad. *Cah. ORSTOM, Ser. Hydrobiol.* **7,** 117–147.

Lopez, G. R., and Levinton, J. S. (1978). The availability of microorganisms attached to sediment particles as food for *Hydrobia ventrosa* (Gastropoda: Prosobranchia). *Oecologia* **32,** 263–276.

Lovegrove, T. (1966). The determination of the dry weight of plankton and the effect of various factors on the values obtained. *In* "Some Contemporary Studies in Marine Science" (H. Barnes, ed.), pp. 421–427. Allen & Unwin, London.

MacArthur, R. H. (1972). Coexistence of species. *In* "Challenging Biological Problems" (J. A. Behnke, ed.) pp. 253–259. Oxford Univ. Press, London and New York.

Macfadyen, A. (1948). The meaning of productivity in biological systems. *J. Anim. Ecol.* **17,** 75–80.

Macfadyen, A. (1963). "Animal Ecology Aims and Methods," 2nd ed. Pitman, London.

McMahon, R. F. (1972). Interpopulation variation and respiratory acclimation in the bioenergetics of *Laevapex fuscus*. Ph.D. Thesis, Syracuse University, Syracuse, New York (*Diss. Abstr.* **33,** 5068-B, Order No. 73-9549).

McMahon, R. F. (1973). Respiratory variation and acclimation in the freshwater limpet, *Laevapex fuscus*. *Biol. Bull* (*Woods Hole, Mass.*) **145,** 492–508.

McMahon, R. F. (1975). Growth, reproduction and bioenergetic variation in three natural populations of the freshwater limpet *Laevapex fuscus* (C. B. Adams). *Proc. Malacol. Soc. London* **41,** 331–351.

McMahon, R. F., Hunter, R. D., and Russell-Hunter, W. D. (1974). Variation in *Aufwuchs* at six freshwater habitats in terms of carbon biomass and of carbon:nitrogen ratio. *Hydrobiologia* **45,** 391–404.

Maitland, P. S. (1965). Notes on the biology of *Ancylus fluviatilis* in the River Endrick, Scotland. *Proc. Malacol. Soc. London* **36,** 339–347.

Mann, K. H. (1964). The pattern of energy flow in the fish and invertebrate fauna of the River Thames. *Verh.—Int. Ver. Theor. Angew. Limmol.* **15,** 485–495.

Mann, K. H. (1971). Use of the Allen curve method for calculating benthic production. *In* "A Manual on Methods for the Assessment of Secondary Productivity in Fresh Waters" (W. T. Edmondson and G. G. Winberg, eds.), IBP Handb. No. 17, pp. 160–165. Blackwell, Oxford.

Mason, C. F. (1971). Respiration rates and population metabolism of woodland snails. *Oecologia* **7,** 80–94.

Mattice, J. S. (1972). Production of a natural population of *Bithynia tentaculata* L. (Gastropoda, Mollusca). *Ekol. Pol.* **20,** 525–539.

Mooij-Vogelaar, J. W., Jager, J. C., and Van der Steen, W. J. (1973). Effects of density levels, and changes in density levels on reproduction, feeding and growth in the pond snail *Lymnaea stagnalis* (L.). *Proc. K. Ned. Akad. Wet., Ser. C* **76,** 245–256.

Murphy, G. I. (1968). Pattern of life history and the environment. *Am. Nat.* **102,** 390–404.

Negus, C. L. (1966). A quantitative study of growth and production of unionid mussels in the River Thames at Reading. *J. Anim. Ecol.* **35,** 513–532.

Newell, R. E. (1962). Behavioural aspects of the ecology of *Peringia* (=*Hydrobia*) *ulvae* (Pennant) (Gastropoda, Prosobranchia). *Proc. Zool. Soc. London* **138,** 49–75.

Nickerson, R. P. (1972). A survey of enzyme and shell variation in 16 populations of the stream limpet, *Ferrissia rivularis* (Say). Ph.D. Thesis, Syracuse University, Syracuse, New York (*Diss. Abstr.* **33,** 4588-B, Order No. 73-7753).

Noland, L. E., and Carriker, M. R. (1946). Observations on the biology of the snail *Lymnaea stagnalis appressa* during twenty generations in laboratory culture. *Am. Midl. Nat.* **36,** 467–493.

Payne, B. S. (1979). Bioenergetic budgeting of carbon and nitrogen in the life-histories of three lake populations of the prosobranch snail, *Goniobasis livescens*. Ph.D. Thesis, Syracuse University, Syracuse, New York (*Diss. Abstr.* **40B,** 5526, Order No. 8013404).

Perron, F. E. (1981). The partitioning of reproductive energy between ova and protective capsules in marine gastropods of the genus *Conus*. *Am. Nat.* **118,** 110–118.

Petrusewicz, K., and Macfadyen, A. (1970). Productivity of Terrestrial Animals. Principles and Methods," IBP Handb. No. 13. Blackwell, Oxford.

Pianka, E. R. (1970). On "r" and "K" selection. *Am. Nat.* **104,** 592–597.

Pianka, E. R. (1976). Natural selection of optimal reproductive tactics. *Am. Zool.* **16,** 775–784.

Pyke, G. H., Pullium, H. R., and Charnov, E. L. (1977). Optimal foraging: In selective review of theory and tests. *Q. Rev. Biol.* **52,** 137–154.

Richardson, A. M. M. (1975). Energy flux in a natural population of the land snail, *Cepaea nemoralis* L. *Oecologia* **19,** 141–164.

Romano, F. A. (1980). Bioenergetics and neurosecretory controls in univoltine and biovoltine populations of *Ferrissia rivularis* (Say). Ph.D. Thesis, Syracuse University, Syracuse, New York (*Diss. Abstr.* **38B,** 3508, Order No. 7730710).

Rosenzweig, M. L. (1981). A theory of habitat selection. *Ecology* **62,** 327–335.

Russell Hunter, W. (1953). On the growth of the fresh-water limpet, *Ancylus fluviatilis* Müller. *Proc. Zool. Soc. London* **123,** 623–636.

Russell Hunter, W. (1961a). Annual variations in growth and density in natural populations of freshwater snails in the west of Scotland. *Proc. Zool. Soc. London* **136,** 219–253.

Russell Hunter, W. (1961b). Life cycles of four freshwater snails in limited populations in Loch Lomond, with a discussion of infraspecific variation. *Proc. Zool. Soc. London* **137,** 135–171.

Russell Hunter, W. (1964). Physiological aspects of ecology in nonmarine molluscs. *In* "Physiology of Mollusca" (K. M. Wilbur and C. M. Yonge, eds.), Vol. 1, pp. 83–126. Academic Press, New York.

Russell-Hunter, W. D. (1970). "Aquatic Productivity." Macmillan, New York.

Russell-Hunter, W. D. (1978). Ecology of freshwater pulmonates. *In* "Pulmonates" (V. Fretter and J. F. Peake, eds.), Vol. 2, pp. 337–383. Academic Press, New York.

Russell-Hunter, W. D., and Eversole, A. G. (1976). Evidence for tissue degrowth in starved freshwater pulmonate snails (*Helisoma trivolvis*) from carbon: nitrogen analyses. *Comp. Biochem. Physiol. A* **54,** 447–453.

Russell Hunter, W., and Grant, D. C. (1966). Estimates of population density and dispersal in the naticid gastropod, *Polinices duplicatus,* with a discussion of computational methods. *Biol. Bull (Woods Hole, Mass.)* **131,** 292–307.

Russell-Hunter, W. D., and Romano, F. A. (1981). Reproductive effort of molluscs in bioenergetic terms: some computational methods. *Biol. Bull. (Woods Hole, Mass.)* **161,** 316.

Russell-Hunter, W. D., Meadows, R. T., Apley, M. L., and Burky, A. J. (1968). On the use of a "wet-oxidation" method for estimates of total organic carbon in mollusc growth studies. *Proc. Malacol. Soc. London* **38,** 1–11.

Russell-Hunter, W. D., Burky, A. J., and Hunter, R. D. (1970). Interpopulation variation in shell components in the stream limpet, *Ferrissia rivularis. Biol. Bull. (Woods Hole, Mass.)* **139,** 402.

Russell-Hunter, W. D., Apley, M. L., and Hunter, R. D. (1972). Early life-history of *Melampus* and the significance of semilunar synchrony. *Biol. Bull. (Woods Hole, Mass.)* **143,** 623–656.

Russell-Hunter, W. D., Burky, A. J., and Hunter, R. D. (1981). Interpopulation variation in calcareous and proteinaceous shell components in the stream limpet, *Ferrissia rivularis. Malacologia* **20,** 255–266.

Russell-Hunter, W. D., Buckley, D. E., and Aldridge, D. W. (1982). Recent technical modifications of total organic carbon analyses for molluscan growth studies. *J. Moll. Stud.* **48,** 103–104.

Russell-Hunter, W. D., Browne, R. A., and Aldridge, D. W. (1983a). Overwinter tissue degrowth in natural populations of freshwater pulmonate snails (*Helisoma trivolvis* and *Lymnaea palustris*). *Ecology* (in press).

Russell-Hunter, W. D., Aldridge, D. W., Tashiro, J. S., and Payne, B. S. (1983b). Oxygen uptake and nitrogenous excretion rates during overwinter degrowth conditions in the pulmonate snail, *Helisoma trivolvis. Comp. Biochem. Physiol. A* **74,** 491–497.

Schaffer, W. M. (1974). Optimal reproductive effort in fluctuating environments. *Am. Nat.* **108,** 783–790.

Schaffer, W. M. (1979). Equivalence of maximizing reproductive value and fitness in the case of reproductive strategies. *Proc. Natl. Acad. Sci. U.S.A.* **76,** 3567–3569.

Schaffer, W. M., and Schaffer, M. V. (1977). The adaptive significance of variation in reproductive habit in Agavaceae. *In* ''Evolutionary Ecology'' (B. Stonehouse and C. M. Perrins, eds.), pp. 261–277. MacMillan, London.

Seifert, D. V., and Shutov, S. V. (1981). The consumption of leaf litter by land molluscs. *Pedobiologia* **21,** 159–165.

Simpson, J. F. (1978). Organic carbon, environmental factors and growth in the small euryhaline gastropod, *Potamopyrgus jenkinsi* (Smith). *J. Moll. Stud.* **44,** 104–112.

Stearns, S. C. (1976). Life-history tactics: A review of the ideas. *Q. Rev. Biol.* **51,** 3–47.

Stearns, S. C. (1977). The evolution of life history traits: A critique of the theory and a review of the data. *Annu. Rev. Evol. Syst.* **8,** 145–171.

Stearns, S. C. (1980). A new view of life-history evolution. *Oikos* **35,** 266–281.

Streit, B. (1975). Experimentelle Untersuchungen zum Stoffhaushalt von *Ancylus fluviatilis* (Gastropoda-Basommatophora). I. Ingestion, Assimilation, Wachstum und Eiablage. *Arch. Hydrobiol., Suppl.* **47,** 458–514.

Streit, B. (1976a). Energy flow in four different field populations of *Ancylus fluviatilis* (Gastropoda-Basommatophora). *Oecologia* **22,** 261–273.

Streit, B. (1976b). Studies on carbon turnover in the freshwater snail *Ancylus fluviatilis* (Basommatophora). *Experientia* **32,** 478–480.

Tashiro, J. S. (1980). Bioenergetic background to reproductive partitioning in an iteroparous population of the freshwater prosobranch, *Bithynia tentaculata.* Ph.D. Thesis, Syracuse University, Syracuse, New York (*Diss. Abstr.* **41B,** 2918, Order No. 8104548).

Tashiro, J. S. (1982). Grazing in *Bithynia tentaculata:* Age-specific bioenergetic patterns in reproductive partitioning of ingested carbon and nitrogen. *Am. Midl. Nat.* **107,** 133–150.

Tashiro, J. S., and Colman, S. D. (1982). Filter-feeding in the freshwater prosobranch snail, *Bithynia tentaculata:* Bioenergetic partitioning of ingested carbon and nitrogen. *Am. Midl. Nat.* **107,** 114–132.

Tashiro, J. S., Aldridge, D. W., and Russell-Hunter, W. D. (1980). Quantifiable artificial rations for molluscan grazing experiments. *Malacol. Rev.* **13,** 87–89.

Taylor, H. M., Gourley, R. S., Lawrence, C. E., and Kaplan, R. S. (1974). Natural selection of life history attributes: an analytical approach. *Theor. Popul. Biol.* **5,** 104–122.

Teal, J. (1957). Community metabolism in a temperate cold spring. *Ecol. Monogr.* **27,** 283–302.

Thoday, J. M. (1953). Components of fitness. *Symp. Soc. Exp. Biol.* **7,** 96–113.

Thoday, J. M. (1970). Genotype *versus* population fitness. *Can. J. Genet. Cytol.* **12,** 674–675.

Thompson, R. J., and Bayne, B. L. (1974). Some relationships between growth, metabolism, and food in the mussel, *Mytilus edulis. Mar. Biol.* (*Berlin*) **27,** 317–326.

Tilly, L. J. (1968). The structure and dynamics of Cone Spring. *Ecol. Monogr.* **38,** 169–197.

Tinkle, D. W. (1969). The concept of reproductive effort and its relation to the evolution of life histories in lizards. *Am. Nat.* **103,** 501–516.

Trivers, R. L., and Willard, D. E. (1973). Natural selection of parental ability to vary the sex ratio of offspring. *Science* **179,** 90–92.

Waters, T. F. (1969). The turnover ratio in production ecology of freshwater invertebrates. *Am. Nat.* **103,** 173–185.

Welch, H. E. (1968). Relationships between assimilation efficiencies and growth efficiencies for aquatic consumers. *Ecology* **49,** 755–759.

Widdows, J. (1978). Combined effects of body size, food concentration, and season on the physiology of *Mytilus edulis. J. Mar. Biol. Assoc. U.K.* **58,** 109–124.

Widdows, J., and Bayne, B. L. (1971). Temperature acclimation of *Mytilus edulis* with reference to its energy budget. *J. Mar. Biol. Assoc. U.K.* **51,** 827–843.

Williams, G. C. (1966a). Natural selection, the costs of reproduction, and a refinement of Lack's principle. *Am. Nat.* **100,** 687–792.

Williams, G. C. (1966b). "Adaptation and Natural Selection" Princeton Univ. Press, Princeton, New Jersey.

Wright, C. A. (1960). The crowding phenomenon in laboratory colonies of freshwater snails. *Ann. Trop. Med. Parasitol.* **54,** 224–232.

12

Ecology of an Invasive Pest Bivalve, *Corbicula*

ROBERT F. McMAHON

Section of Comparative Physiology, Department of Biology
The University of Texas at Arlington
Arlington, Texas

I. Introduction

Corbicula fluminea (Müller) (Fig. 1) is a freshwater clam common throughout Southeast Asia. The genus *Corbicula* is a member of the subclass Heterodonta, order Veneroida, superfamily Corbiculacea (Sphaeriacea), and family Cor-

The Mollusca, Vol. 6
Ecology

ISBN 0-12-751406-6

Fig. 1. *Corbicula fluminea*. These specimens have intermediate shell lengths of 15 mm (lower left) and 14 mm (upper right). Note the very large muscular foot, an adaptation for efficient burrowing behavior and to the lotic unstable sand-gravel habitats of this species. Other observable identifying characteristics are the large concentric ridges of the shell and the short inhalant and exhalant siphons fringed with tentacles.

biculidae (Newell, 1969). The genus is endemic to freshwater and estuarine habitats in Southeast Asia, Africa, the Indian subcontinent, the Pacific Islands, and South America, where it is an important component of benthic communities in both lentic and lotic environments (Morton, 1977a; Miller and McClure, 1931; Villadolid and Del Rosario, 1930; Leveque, 1973a,b; Britton and Morton, 1979).

Specimens of *C. fluminea* were first collected in North America as empty shells at Namaimo, Vancouver Island, British Columbia in 1924 (Counts, 1981). Shells were then reported from the Columbia River in Pacific County, Washington State, in 1938 (Burch, 1944). Since these two isolated reports, *C. fluminea* has spread throughout the freshwater habitats of the United States at an amazing rate. The literature after 1944 (primarily in the journal *Nautilus*) is rife with reports of new North American localities for *Corbicula*. When these popu-

Fig. 2. Occurrence and spread of *Corbicula fluminea* in North America through 1981. The stars indicate recorded populations of *C. fluminea* first observed in the years indicated. The solid arrows indicate the probable pattern of the spread of this species in North America by natural means. The dashed arrows indicate the pattern of translocation of populations which were almost certainly established as the result of human activity. (From McMahon, 1982.)

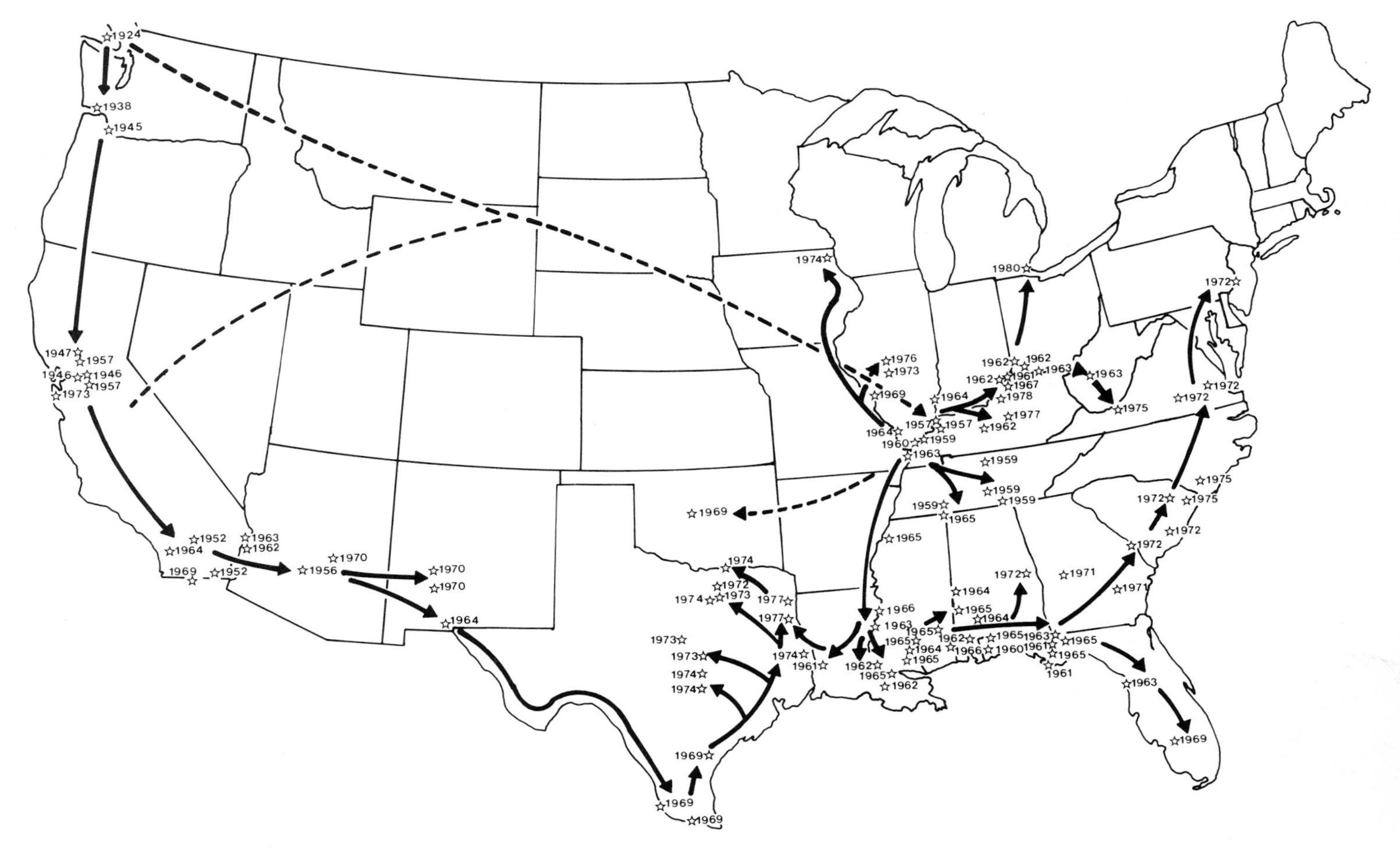
1924
1938
1945
1947
1957
1946
1946
1957
1973
1952
1964
1969
1952
1963
1962
1970
1956
1970
1970
1964
1969
1969
1969
1973
1973
1974
1974
1974
1972
1974
1973
1977
1977
1974
1961
1974
1976
1973
1969
1964
1960
1963
1959
1957
1957
1964
1962
1962
1961
1967
1978
1962
1977
1963
1963
1975
1980
1959
1959
1959
1959
1965
1965
1965
1966
1963
1965
1965
1964
1965
1962
1965
1962
1965
1964
1964
1965
1962
1966
1965
1960
1963
1961
1961
1965
1965
1972
1971
1971
1972
1972
1972
1975
1975
1972
1972
1972
1963
1969

lations and their respective discovery dates are noted on a map of the United States (Fig. 2), the speed and degree to which this species has invaded North American fresh waters become very apparent. *C. fluminea* spread southward from the Columbia River through western California, appearing in the San Francisco area as early as 1946 (Ingram, 1948). By 1952 it had spread southward to the Imperial Valley of extreme southern California (Fitch, 1953). This southward spread of *Corbicula* on the west coast involved movements of clams across several major drainage systems, but distribution to the east appeared to be prevented by the physical barrier of the Continental Divide (Fig. 2). In extreme southern California, *Corbicula* crossed the Continental Divide, entered the Colorado River drainage system (probably transported from southern California through interconnecting irrigation canals), and was reported to be in irrigation canals near Phoenix, Arizona, by 1956 (Dundee and Dundee, 1958). From the Colorado River drainage it moved westward into the upper reaches of the Rio Grande River drainage, being reported from the Rio Grande River near El Paso, Texas, in 1964 (Metcalf, 1966). Once in the Rio Grande it moved rapidly downstream, being found near the river mouth in South Texas in 1969 (Murray, 1971). From South Texas, *Corbicula* spread northward into all the major drainage areas of eastern Texas by 1974 (Britton and Murphy, 1977; Pool and McCullogh, 1979) (Fig. 2).

During the period when *Corbicula* was spreading southward through California, a second major infestation occurred in the Ohio River, where it fouled the steam condensers of the Tennessee Valley Authority Shawnee Steam Electric Power Station at Paducah, Kentucky, in 1957 (Sinclair and Isom, 1961). This infestation occurred at a time when *Corbicula* had spread no farther east than the Colorado River drainage system on the West Coast of the United States and therefore almost certainly represents an unnatural extension of the species North American range as a direct result of human activity (Fig. 2). A similar anomalous record of *C. fluminea* occurred in Lake Overholser, Oklahoma, in 1969 (Clench, 1972) (Fig. 2). To establish these populations, clams must have been transported to these two areas by humans as a tourist curiosity, in the bilge water of pleasure craft, as a fishing bait, or as a result of the activity of aquarium hobbyists. Recent data have shown that it almost certainly could not have been carried such great distances by migrating waterfowl (Dreier, 1977; Thompson and Sparks, 1977).

Once established in the Ohio River, *Corbicula* rapidly extended its range downstream into the Mississippi River and freshwater drainage systems throughout the southern United States, being found in Louisiana as early as 1961 (Dundee and Harman, 1963) and in western Florida as early as 1960 (Schneider, 1967). By 1969, *Corbicula* had extended into extreme southern Florida (Clench, 1970) and by 1971 had invaded Atlantic drainage systems in Georgia (Fuller and Powell, 1973), spreading as far north as Virginia and New Jersey by 1972 (Diaz, 1974; Fuller and Powell, 1973) and reaching Lake Erie, Michigan, by 1980

(Clarke, 1981). The Appalachian Mountains seem to have formed an effective barrier to the eastern extension of this species' range, as had the Continental Divide on the West Coast. Instead, it spread rapidly downstream in the Mississippi River and invaded new drainage systems to the east only along the low elevations of the southern coastal plain. Only after invading East Coast drainage systems in Georgia and Florida did this species expand rapidly north, reaching the Delaware River in New Jersey by 1972 (Fuller and Powell, 1973). If the primary vector for the spread of *Corbicula* in the United States were human activity, mountain ranges would not seem to act as such effective barriers to its distribution, and the pace and pattern of this species' invasion of North American fresh waters would be far more random than it appears to be in Fig. 2.

The upstream invasion rate of *Corbicula* is far slower than its down-stream invasion rate, perhaps because its small, nonswimming juvenile stage can be transported considerable distances by water currents (Sinclair and Isom, 1963). Established in the Ohio River, as noted earlier, at Paducah in 1957, it reached the northern limit of its range in this drainage system by 1963, 25 miles upstream from Cincinnati, Ohio (Pojeta, 1964), and in the New River, at Glen Lyn, Virginia, in 1975 (Rodgers et al., 1979). *Corbicula* spread upstream more rapidly in the Tennessee River, advancing to its upper reaches from the Ohio River by 1959 (Sinclair and Isom, 1961, 1963). The upstream invasion of the Mississippi by *Corbicula* was slower, with populations occurring in Lake Sangchris, Illinois, by 1973 (Dreier, 1977) and in the river itself at Allamakee County, Iowa, by 1974 (Eckblad, 1975) (Fig. 2).

The extent of the northern range of *Corbicula* in North America apparently is limited by low winter temperatures. Horning and Keup (1964) report massive mortality (96.3% reduction in density) of *Corbicula* in the Ohio River at Cincinnati after the river had been frozen over and ambient water temperatures were near 0°C for 7 days. Mattice and Dye (1976) estimated the minimum lower lethal temperature of this species is 2°C. It is interesting to note that the most northerly reported populations of *Corbicula* in the Mississippi and Ohio River drainage systems and on the East Coast of the United States are almost always associated with the presence of steam electric power plants discharging artificially heated waters into aquatic environments (Dreier, 1977; Dreier and Tranquilli, 1981; Eckblad, 1975; Rodgers et al., 1979; Thomas and MacKenthun, 1964; Thomerson and Myer, 1970). In cooler northern climates, *Corbicula* appears to be restricted to, and survives the winter only in, areas receiving such thermal discharges (Dreier, 1977; Dreier and Tranquilli, 1981; Eckblad, 1975; Rodgers et al., 1979).

Corbicula now occurs in 35 states of the continental United States. It does not occur in Montana, Wyoming, North Dakota, South Dakota, Nebraska, Kansas, New York, Vermont, Connecticut, Rhode Island, New Hampshire, or Maine (Cherry et al., 1980a) (Fig. 2), from which it is probably eliminated by low winter temperatures. It has therefore become one of the most ubiquitously distributed bivalves in North American fresh waters since its introduction to the North American continent in the early 1900s, and is now a major component of the benthic communities of most freshwater environments in the southern United States. By sheer weight of numbers, it has become an important pest species in many of these environments. Therefore, it is of extreme importance that the biology of this introduced species be elucidated as soon as possible. This chapter attempts to review what is presently known of the biology, ecology, and physiology of *Corbicula* and to suggest aspects that require further investigation.

II. Taxonomy

Reports for U.S. populations of *Corbicula* usually refer to it by one of three species designations: *C. fluminea* (Müller), originally assigned to early West Coast populations (Burch, 1944); *C. manilensis* Philippi, later assigned to populations in the eastern half of the United States (Sinclair, 1971; Sinclair and Isom, 1961, 1963); or, occasionally in the American literature, as the Japanese species *C. leana* Prime (Burress and Chandler, 1976; Counts and Prezant, 1979; Gunning and Suttkus, 1966; Marking and Chandler, 1978; Taylor, 1980a). This confusion over the species designation of North American *Corbicula* is hardly surprising, and results partly from the fact that the family Corbiculidae is in taxonomic disarray. More than 200 species of *Corbicula* have been named from Asia alone (Morton, 1979a), which were later reduced to 69 species by Prashad (1924, 1928a,b, 1929, 1930). However, as pointed out by Morton (1979a) in his excellent review of the taxonomy of mainland Asiatic species of *Corbicula,* Prashad's list can be reduced to two dominant species in southern China: *C. fluminea* (Müller, 1774) and *C. fluminalis* (Müller, 1774) (= *C. manilensis* Phillipi), which are distinguishable in terms of both shell morphology and environmental preference.

C. fluminalis prefers lotic brackish-water habitats. It has a high salinity tolerance, a tall triangular shell, widely spaced shell sulcations, a maximum shell length (SL) of 54–60 mm, a life span as much as 10 years, and a single spawning period at lower temperatures. Primarily dioecious with a tendency toward protogynous hermaphroditism, it does not incubate fertilized eggs or larvae and has a free-swimming veliger larva (Morton, 1973, 1982). In contrast, *C. fluminea* prefers lotic freshwater habitats, and has a lower salinity tolerance, a rounded shell, a maximum SL of 35–50 mm, and a life span of 2–3 years. With two distinct spawning periods per year at higher temperatures, it is a protandric consecutive hermaphrodite incubating fertilized eggs within the suprabranchial cavity of inner demibranchs to a benthic, nonswimming, bivalved juvenile stage (Morton, 1977a, 1982).

The American *Corbicula* is clearly a freshwater species which does not enter truly brackish water (Haertel and Osterberg, 1967; Sickel, 1979). It has a low salinity tolerance (Evans et al., 1979; Gainey and Greenberg, 1977; Gainey, 1978a,b), a small, rounded shell (Sinclair and Isom, 1961, 1963; Aldridge and McMahon, 1978) with narrowly spaced sulcations (Sinclair and Isom, 1961, 1963), a preference for lotic conditions or shallow, well-oxygenated lake shore habitats (Aldridge and McMahon, 1978; McMahon, 1979a), and a maximum SL (in most populations) of 40–45 mm (Aldridge and McMahon, 1978; Eng, 1979; Sickel, 1979; Heinsohn, 1958), which in some populations is as long as 60 mm (Sinclair and Isom, 1961, 1963; Britton and Morton, 1979; Morton, 1982). North American *Corbicula* have a life span of 2–3 years (Aldridge and McMahon, 1978; Heinsohn, 1958), two reproductive periods (spring and fall) per year (Aldridge and McMahon, 1978; Aldridge, 1976; Boozer and Mirkes, 1979; Dreier, 1977; Dreier and Tranquilli, 1981; Eng, 1979; Heinsohn, 1958), and shell growth rings which cannot be clearly distinguished (Aldridge and McMahon, 1978). This organism appears to be a simultaneous hermaphrodite (Kraemer, 1978b) which broods fertilized eggs in the suprabranchial chambers of the inner demibranchs until a nonswimming, benthic, bivalved juvenile stage is released with a well-developed foot and a SL of about 0.2–0.22 mm (Aldridge and McMahon, 1978; Heinsohn, 1958; Sinclair and Isom, 1961, 1963).

It is obvious from the above descriptions that the species of *Corbicula* in North America is far more closely allied with the freshwater species *C. fluminea* than with the more estuarine *C. fluminalis*. In their ecology, population dynamics, functional morphology, reproductive biology, siphon structure, and shell morphometrics, populations of *C. fluminea* in Asia and North America are essentially indistinguishable (Britton and Morton, 1979). Indeed, M. H. Smith et al. (1979), in an electrophoretic study comparing isozymes for 17 different enzymes of five populations of *Corbicula* from North America with populations from Japan, China, and the Philippines, found that the five North American populations had a genetic variability of 0 and a genetic similarity index of 1, indicating that these populations are almost entirely monomorphic for the enzymes analyzed. In contrast, Asian populations showed a fairly high level of heterozygosity, with U.S. populations most closely resembling *C. fluminea* populations from China and the Philippines and being distinctly different from Japanese clams, which were thought to be a different species. [This was probably a correct assumption because the ecology, life cycle, and reproductive biology of Japanese species of *Corbicula* are different from those of Asian or North American species (Fuziwara, 1975, 1977; Ikematsu and Kamakura, 1976; Ikematsu and Yamane, 1977), including retention of a free-swimming veliger in *C. leana*.] The complete lack of heterozygosity among North American populations indicates that there apparently has been only a single introduction of this species, probably somewhere near the northwestern corner of Washington State (Counts, 1981). In addition, it indicates that founder effect and subsequent genetic drift during its rapid range expansion have reduced the genetic variability of North American populations and caused a large number of loci to become homozygous for a single allele (M. H. Smith et al., 1979).

The most likely source of the *C. fluminea* introduced to North America appears to be Southeast Asia, particularly China. In China and the Philippines, *Corbicula* is routinely harvested, sold commercially, and utilized for human consumption (Chen, 1976; Miller and McClure, 1931; Morton, 1973; Villadolid and Del Rosario, 1930), and it is often the subject of intensive aquaculture (Chen, 1976; Miller and McClure, 1931). In high relative humidities, *C. fluminea* can survive for as long as 38 days in air (McMahon, 1979b), a time span more than enough for it to be carried to the United States by Chinese immigrants as a food source during a trans-Pacific boat crossing. Because of its extremely high reproductive capacity (Aldridge and McMahon, 1978) and its ability to rapidly invade new habitats (see Section I), *C. fluminea* could not have gone unnoticed after it had become established in North America. For these reasons, it probably could not have been introduced in the late nineteenth century by Chinese immigrants working on the transcontinental railroad or in gold mines, as hypothesized by Fox (1969). Instead, it appears to have been introduced during the early 1920s (Counts, 1981) unintentionally by Chinese immigrants carrying them as food (Britton and Morton, 1979) or perhaps intentionally by an enterprising entrepreneur hoping to establish a North American commercial source of *Corbicula* for the growing West Coast oriental population. *C. fluminea* was almost certainly not introduced to North America along with seed oysters (*Ostreas gigas*), as suggested by Filice (1958), or as part of the fouling community on a ship hull (Ingram, 1948), because of its restriction to fresh waters. It also does not survive in the digestive tract of waterfowl long enough to have been carried the vast distances that would have been required for introduction to North America during annual migrations (Thompson and Sparks, 1977; Dreier, 1977).

Freshwater *Corbicula* species appear to have extensive dispersal powers; some members of this genus have a North American fossil record but have not occurred naturally in North American fresh waters in recent times. Perhaps even more illustrative of the unusual dispersal powers of this group is that the fossil record and the application of aminostratigraphy dating techniques to fossil shells have demonstrated that it has repeatedly invaded the fresh waters of southeastern England during each of the last three or four interglacial periods extending 1.35–1.9 $\times$ 10^6 years B.P. (Miller et al., 1979).

III. Aspects as a Pest Species

As *C. fluminea* has spread throughout North American freshwater environments in the last 40 years, it has become one of the most important aquatic molluscan pest species ever introduced into the United States. The pest aspect of

Corbicula in North America centers on its high reproductive capacity, high growth rate, free-living juvenile stage, and great powers of dispersal.

Some of the initial problems with *Corbicula* in North America involved its capacity for infesting irrigation systems and canals. Some of the earliest reports of *C. fluminea* in the western United States are from irrigation systems (Dundee and Dundee, 1958; Fox, 1970; Ingram et al., 1964; Eng, 1979; Fitch, 1953; Ingram, 1959; Prokopovich, 1969; Prokopovich and Hebert, 1965), where the constant supply of flowing water is particularly favorable to the growth and reproduction of this species (Aldridge and McMahon, 1978; Britton and Morton, 1979; Morton, 1977a). *C. fluminea* in irrigation canals can grow and reproduce so quickly that shells can accumulate in sand bars built up from living clams and dead shells, and by the filtration and aggregation of suspended solids by living clams, to the point where water flow is seriously impeded (Eng, 1979, Prokopovich and Hebert, 1965; Prokopovich, 1969). When such accumulations occur, the canal must be dewatered and the shells removed mechanically. In some irrigation systems this has become a perennial problem, because *C. fluminea* has such a high growth rate and reproductive capacity that large populations can be reestablished during the breeding season following removal of adult clams (Eng, 1979; Prokopovich, 1969; Prokopovich and Hebert, 1965).

Corbicula not only reduces flow in major irrigation canals but also enters water pipes in the underground water distribution portions of irrigation systems (Ingram, 1959; Fitch, 1953), reducing or stopping flow, clogging valves, and obstructing drainage ditches by accumulating immediately below irrigation pipe openings (Ingram, 1959; Fitch, 1953). Removal of clam shells from the underground pipes and valves of irrigation systems is often a very complex, time-consuming, and expensive procedure, and *Corbicula* continues to be a problem in the large irrigation systems of California and southwestern United States.

Similar fouling problems with *C. fluminea* have been reported in municipal water treatment facilities. *Corbicula* can infest water intake areas and damage centrifugal pumps. Dead and decaying bodies can clog main straining screens and contribute to bad tastes and odors which remain after chemical treatment (Ingram, 1959; A. L. Smith et al., 1979; Sinclair, 1974; Ray, 1962). In addition, *Corbicula* can foul raw water lines, reducing the efficiency of water treatment plant operation (Sinclair, 1964).

Sinclair and Isom (1961, 1963) have described a further nuisance aspect of *Corbicula* in the Ohio and Tennessee river systems. Here, dense populations of *Corbicula* have become established in areas that are dredged to provide sand and gravel for use as aggregation material in cement. Because the adult clams are the same size as the gravel aggregate, they are impossible to separate from it (Sinclair and Isom, 1961, 1963). When the cement is poured and begins to set, the clams burrow to the surface, causing the cement to become porous and structurally weakened. There have been no published reports concerning the control of *Corbicula* in this industry.

Perhaps the most serious problem of pest *Corbicula* in North American fresh waters is caused by its ability to enter and foul the systems of electrical generating stations and other industrial facilities utilizing raw water, including those of nuclear reactors. *Corbicula* shells have fouled the heat exchangers which condense steam from turbines before it is returned to the boilers (McMahon, 1977; Boozer and Mirkes, 1979; Cherry et al., 1980a,b; Diaz, 1974; Goss and Cain, 1977; Goss et al., 1979; Ingram, 1959; Mattice, 1979; Sinclair and Isom, 1961, 1963; A. L. Smith et al., 1979; Thomas and MacKenthum, 1964; Thomerson and Myer, 1970; Harvey, 1981). Some of these reports suggested that juveniles were carried into condensers by intake currents where they attached by mucilaginous byssal threads to condenser walls and eventually grew large enough to restrict water flow through the condenser tubes. However, because the long-term upper lethal temperature limit of *C. fluminea* is 33–34°C (Habel, 1970; Mattice and Dye, 1976) and lies well below the water temperatures achieved in such steam condensers (at least during summer months), it is highly unlikely that the specimens of *Corbicula* fouling these condenser tubes represent an indigenous population. The source of clams fouling the condensers is therefore somewhat obscure, because any clams that are large enough to foul heat exchanger tubes are removed from the intake currents by screens whose mesh size is less than that of the internal diameter of the tubes (McMahon, 1977; Goss and Cain, 1977; Goss et al., 1979). In some cases, the primary and secondary screens apparently do not remove all the clams of sizes that become lodged in the condenser tubes from intake currents (Goss and Cain, 1977; Goss et al., 1979) but in most situations, individuals and juveniles with SLs smaller than those of the screen mesh are carried through them and settle to the bottom in dead-water areas of the intake embayments behind the primary and secondary screens (Goss and Cain, 1977; Goss et al., 1979; McMahon, 1977; Mattice, 1979; A. L. Smith et al., 1979). Once settled into bottom sediments that accumulate in these embayments, the clams appear to be able to grow within a few weeks to SLs that allow lodgment in condenser tubes when they are drawn into intake structures (McMahon, 1977; Aldridge and McMahon, 1978; Britton et al., 1979; Heinsohn, 1958; Buttner and Heidinger, 1980; Dreier, 1977; Dreier and Tranquilli, 1981; Morton, 1977a; Mattice, 1979).

Certain environmental conditions appear to induce large numbers of individuals to leave the substratum and to be carried by currents into the intake structures and subsequently into heat exchangers (McMahon, 1977). Such large-scale emergence from the substratum may occur when populations are stressed by high ambient water temperatures (Aldridge and McMahon, 1978; Mattice and Dye, 1976; Habel, 1970) and possibly also after spawning (Aldridge and McMahon, 1978; Sinclair, 1971; Sinclair and Isom, 1961, 1963; Heinsohn, 1958). Other major stresses may include low environmental oxygen concentrations, to which *C. fluminea* appears to be extremely intolerant (McMahon,

1979b), and perhaps "shock chlorination," a common biofouling control procedure practiced in many facilities utilizing raw water for cooling purposes (McMahon, 1977; Goss et al., 1979; A. L. Smith et al., 1979; Burton, 1980; Mattice et al., 1982; U. S. Environmental Protection Agency, 1980; Sinclair and Isom, 1961, 1963).

Stressed clams crawl to the substratum surface (R. F. McMahon, unpublished observations). Many of these clams, which subsequently die, actually become quite buoyant, even to the point of floating to the water surface (A. L. Smith et al., 1979; R. F. McMahon, unpublished observations), because gases released from the decomposition of flesh are retained within shell valves that are cemented together by a hardened mucus (McMahon, 1979b, also unpublished observations). Such buoyant shells of dead individuals and living specimens dislodged from intake embayments are highly susceptible to being drawn into intake structures and carried to steam condensers. Once in the condensers, individuals with shell heights smaller than the internal diameter of the condenser tubes pass through the tubes to be carried away with the outlet current, whereas those with shells too large to enter condenser tubes remain in the intake box of the heat exchanger. The condenser tubes become fouled when both whole living clams and empty shells with shell heights approximately equal to the internal diameter of the tube enter condenser tubes. These individuals become wedged in the tubes at points of slight constriction in the internal tube diameter (usually at bends in the tubes). Other shells then accumulate behind the lodged shells, greatly restricting flow through the tubes (McMahon, 1977). Most power plant heat exchangers have tubes with internal diameters ranging from 15.8 (Ingram, 1959) to 19.7 to 22.9 mm (McMahon, 1977), corresponding to a range of shell heights common in most North American *Corbicula* populations (McMahon and Aldridge, 1978; Heinsohn, 1958; Dreier, 1977; Dreier and Tranquilli, 1981; Bickel, 1966; Eng, 1979; Sinclair and Isom, 1961, 1963; Gunning and Suttkus, 1966; O'Kane, 1976; Rodgers et al., 1977, 1979; Sickel, 1979; White and White, 1977a). Therefore, most industrial facilities utilizing raw water for cooling purposes will be subject to *Corbicula* condenser biofouling in areas of North America where this species has become established (Fig. 2).

Corbicula also poses a specific threat to nuclear reactors and nuclear power stations. Not only does this species foul heat exchangers utilized in steam condensation at these plants, as described above (Goss et al., 1979; Boozer and Mirkes, 1979; Harvey, 1981), but there have also been reports that *Corbicula* has entered and fouled the emergency cooling systems of a nuclear electrical generating plant near Russellville, Arkansas (Parsons, 1980). Clams enter the pipes of these systems and restrict water flow, causing plant shutdown for mechanical removal of shells and live clams. Similar conditions have occurred at nuclear power plants in Browns Ferry, Alabama, and Baldwin, Illinois (Parsons, 1980). Given the increasing dependence on nuclear energy for the generation of electricity, the widespread distribution of *Corbicula* in North America not only presents a problem for present nuclear plant operations but will also have to be an important consideration in the future design and location of new nuclear facilities.

Lastly, *C. fluminea* in certain aquatic habitats can reach densities which are rarely, if ever, achieved by other North American freshwater bivalve species. Densities of more than 1000 clams/m^2 are not uncommon (Heinsohn, 1958; Ingram et al., 1964; O'Kane, 1976; Sickel, 1979; Harvey, 1981; Prokopovich, 1969; Rodgers et al., 1979; Abbott, 1979; Boozer and Mirkes, 1979; Horning and Keup, 1964), and populations as dense as 5,000 clams/m^2 (Heinsohn, 1958) to 12,048 clams/m^2 (O'Kane, 1976) have been reported; the most dense population was 131,000 clams/m^2 on sand bars in the Delta–Mendota Canal in California (Eng, 1979). With the ability to achieve such high densities, it has been speculated that *Corbicula* could outcompete and eventually displace indigenous North American unionid and sphaeriid species, many of which have already suffered severe population reductions because of human disruptions of their habitats (van der Schalie, 1973; Jenkinson, 1979; Sickel, 1973; Gardner et al., 1976; Boozer and Mirkes, 1979). *Corbicula* could outcompete slower growing native species both for space [as reported for *Corbicula* in competition with *Musculium partumeium* (Boozer and Mirkes, 1979)] and for suspended food resources. One novel but not unreasonable hypothesis suggests that the filter-feeding activities of dense populations of *Corbicula* could remove the sperm of dioecious unionids from the water column, preventing fertilization of eggs retained in the suprabranchial cavity of their gills (Clarke, 1981).In some cases, extensive surveys of bivalve faunas before and after establishment of *Corbicula* populations seem to indicate that habitat invasion by this species can be associated with both reduction in numbers and elimination of native unionid and sphaeriid endemics (Cooper and Johnson, 1980; Boozer and Mirkes, 1979; Fuller and Imlay, 1976; Gardner et al., 1976; Sickel, 1973; Taylor and Hughart, 1981), although other surveys suggest that *C. fluminea* has little or no effect on native species (Isom, 1974; Kraemer, 1979; Klippel and Parmalee, 1979; Taylor, 1980a,b). Such results have led to the suggestion that *C. fluminea* cannot outcompete native species in aquatic environments that are not subject to human interference (Fuller and Imlay, 1976; Isom, 1974; Klippel and Paramlee, 1979; Kraemer, 1979; Taylor, 1980a), and that *C. fluminea* is far more tolerant of human activities such as dredging, canalization, and other water management practices that disturb bottom sediments (Kraemer, 1979; Stein, 1962; Klippel and Parmalee, 1979; Fuller and Imaly, 1976; Cooper and Johnson, 1980) than are native species. *C. fluminea,* with its unusually high reproductive capacity (Aldridge and McMahon, 1978; Heinsohn, 1958), high growth rate (Aldridge and McMahon, 1978; Heinsohn, 1958; Britton et al., 1979), and short generation time (Aldridge and McMahon, 1978; Heinsohn, 1958; Morton, 1977a) can become established very quickly in such disturbed environments where native bivalve faunas have become depauperate. Fortunately, *C. fluminea* appears to be far less tolerant of high ambient temperatures (Mattice and Dye, 1976; McMahon, 1979a; Habel, 1970), low oxygen tensions (McMahon, 1979a), and aerial desiccation (McMahon, 1979b) than either native unionids or sphaeriids and, as such, probably does not represent a real threat to native species in many North American environments. However, it does appear to be causing significant damage to native bivalve faunas in the larger navigable rivers of North America, where management practices reduce competition and increase the rate of dispersal of this species.

IV. Physiological Ecology

A. Adaptations to Lotic Habitats

C. fluminea can be characterized as a lotic species that is most successful in flowing, well-oxygenated waters. When *C. fluminea* occurs in lentic environments such as lakes or reservoirs, it is almost always restricted to shallow nearshore waters and well-oxygenated sediments (Bickel, 1966; Dreier, 1977; Dreier and Tranquilli, 1981; Fast, 1971; Lenat and Weiss, 1973; McMahon, 1979a,b; O'Kane, 1976; Pool and McCullough, 1979; White, 1979; White and White, 1977a,b). Restriction to flowing waters or shallow lakeshore habitats is also reported for Asian *C. fluminea* populations (Morton, 1977a), and *C. africana* populations in Africa (Marshall, 1975).

1. Substrata

In lotic habitats, *C. fluminea* appears to be able to inhabit a wide variety of substrata including bare rock, loose gravel, sand, and even silts and muds (Horne and McIntosh, 1979), but achieves its highest population density and greatest success in well-oxygenated substrates (Abbott, 1979; Eng, 1979; Fast, 1971; Aldridge and McMahon, 1978; Junk, 1975; Rinne, 1974; White and White, 1977a,b) such as coarse sand or gravel or sand–gravel mixtures (Cherry et al., 1980b; Eng, 1979; Fast, 1971; Filice, 1958; Gardner et al., 1976; Keup et al., 1963; O'Kane, 1976; Pojeta, 1964; Pool and McCullough, 1979; Rinne, 1974; White and White, 1977a,b). It is almost always eliminated from areas with reducing sand, mud, or silt sediments of high organic and low oxygen content in both North America (Aldridge and McMahon, 1978; Eng, 1979; Fast, 1971; Lenat and Weiss, 1973; McMahon, 1979a; Rinne, 1974) and Southeast Asia (Morton, 1977a). Such habitat preferences are displayed even by the newly released juveniles of *C. fluminea,* which have a distinct preference for fine or coarse sand substrata rather than mud or bare concrete (Sickel and Burbanck, 1974).

2. Response to Hypoxia

Laboratory investigations have confirmed that *C. fluminea* has little or no capacity to regulate oxygen uptake rates under progressive exposure to increasingly higher levels of hypoxia. Instead, the oxygen consumption rate decreases at a proportionately higher rate than the oxygen concentration; at temperatures of 10–30°C, it drops to just 14% of the rate of oxygen uptake at full air saturation with only a 30% decrease in oxygen concentration from air saturation levels (McMahon, 1979a). The inability of *C. fluminea* to maintain even moderate levels of aerobic metabolism at oxygen tensions only somewhat below air saturation levels is manifested in its extremely poor tolerance of natural waters with high organic loads (such as sewage effluents) and consequently low oxygen tensions (Horne and McIntosh, 1979; Weber, 1973), although there have been reports of *C. fluminea* populations in waters receiving sewage effluents (Busch, 1974; Diaz, 1974; Jenkinson, 1979; Habel, 1970). Both Fast (1971) in North America and Morton (1977a) in China report that *C. fluminea* populations do not exist in hypoxic waters beneath the hypolimnion of stratified lakes. However, when such conditions were eliminated by artificial aeration, populations of *C. fluminea* previously restricted to sediments above the hypolimnion were able to disperse to lower depths and spread throughout the lake. Subsequently, when aeration was curtailed the following summer, clams again became restricted to sediments in waters of high oxygen content above the hypolimnion (Fast, 1971). These data are highly indicative of the requirement for well-oxygenated waters for this species.

It certainly appears that the densest populations of *C. fluminea* almost invariably are associated with lotic environments (Eng, 1979; Heinsohn, 1958; Horning and Keup, 1964; Diaz, 1974; Ingram et al., 1964; Prokopovich, 1969; Rodgers et al., 1979; Sickel, 1979; R. F. McMahon, unpublished observations). A few lake populations are also reported to have achieved very high densities (Abbott, 1979; O'Kane, 1976), but these appear to be associated with well-oxygenated, shallow-water sand, gravel, or boulder rubble substrata (O'Kane, 1976) or with areas where steady water currents are maintained over the population (Abbott, 1979).

The apparent preference of *C. fluminea* for flowing waters of high oxygen content indicates the primarily lotic nature of this species in North America, which corresponds closely with that reported for Asian populations (Morton, 1977a, 1982). As a result of its restriction to lotic habitats, *C. fluminea* may not pose as much of a threat to native bivalve species as has been previously suggested, particularly to the many native species which are more lentic in habit even within lotic environments (Kraemer, 1979) and able to tolerate much lower oxygen concentrations than *C. fluminea* (McMahon, 1979a). However, it is now clear that human activities in North American waterways such as stream canalization and dredging are not only detrimental to habitats of native species but also optimize the environment for *C. fluminea* by increasing current flow and eliminating the reducing muds and silts in which it is not successful. Many North American electrical generating stations have been designed with long intake canals or tunnels through which lake water is continuously drawn into stream condensers. These tunnels and canals, or any areas of a lake over which currents are artificially maintained, afford much better conditions for the establishment and growth of *C. fluminea* populations than the lake proper and therefore provide opportunities for the continual invasion by juveniles, spat, and adult individuals of facilities utilizing large quantities of raw water.

3. *Other Adaptations to Lotic Habitats*

C. fluminea has evolved other adaptations to lotic habitats. It has a massive, concentrically ridged shell which is resistant to abrasion, helps to anchor the clam in unstable substrata, and prevents individuals from being swept downstream with water currents (Dudgeon, 1980; Morton, 1982; Horne and McIntosh, 1979; Kraemer, 1979) (Fig. 1). Mantle fusion is extensive for a freshwater species, narrowing the pedal gape and forming a foramen around the posterior adductor muscle which, along with large, highly developed statocysts, gives *C. fluminea* a higher capacity for locomotion and efficient burrowing than indigenous North American unionids (Kraemer, 1978a, 1979). Unlike the mantle of unionids, that of *C. fluminea* is fused to form distinct, tubular, well-muscled inhalant and exhalant siphons that are capable of much greater extension and movement, and therefore of more efficient separation of pallial currents, than occurs in unionids (Kraemer, 1977, 1979). *C. fluminea* is a simultaneous hermaphrodite (Kraemer, 1978b) and/or a protandric consecutive hermaphrodite (Morton, 1977a) which does not have the obligate parasitic larval stage (glochidium) of dioecious unionids, a condition that allows *C. fluminea* to maintain a high reproductive capacity and, therefore, to invade and become established in temporally short-lived lotic habitats much more quickly than native North American freshwater bivalves. In addition, the juveniles and spat, when less than 2.0 mm in SL, produce a mucilaginous byssus thread which is utilized to anchor it to the substratum, and to prevent downstream transportation over long distances by currents in lotic habitats (Kraemer, 1979).

B. Salinity and Osmoregulation

C. fluminea populations in North America have been recorded from inland fresh waters (Fig. 2). Notable exceptions have occurred in the tidal portions of the upper reaches of the Sacramento–San Joaquin Rivers in California (Filice, 1958; Heinsohn, 1958; Copeland et al., 1974) and the James River, Virginia (Diaz, 1974). Careful inspection of such reports indicates that *C. fluminea* populations either do not extend into the brackish-water areas of such rivers (Haertel and Osterberg, 1967; Sickel, 1979; Diaz, 1974), or if they do, they generally do not occur at salinities above 5‰ (Copeland et al., 1974; Filice, 1958; Heinsohn, 1958). Similarly, *C. fluminea* was reported not to invade the saline waters of an impounded tidal cove in southeast China until it was sufficiently flushed with fresh water to reduce the salinity level below 2‰ (Morton, 1977a). The vast majority of truly freshwater species disappear from estuarine assemblages as salinities rise above 3–8‰ (Remane and Schlieper, 1971; Green, 1968). This salinity barrier for freshwater animals has been referred to as the *Artenminimum* or *Horohalinicum* (Gainey and Greenberg, 1977). Because the range of salinities pentrated by *C. fluminea* rarely exceeds this barrier of 3–8‰, its basic freshwater nature seems assured. There are two *C. fluminea* populations reported at salinities above 8‰, but they are described as marginal, with aberrant growth rates and reproductive cycles (Heinsohn, 1958; Evans et al., 1979).

In spite of its apparent restriction to fresh waters, *C. fluminea* is tolerant of laboratory exposures to relatively high salinities, surviving shock exposures at 10–14‰ (Evans et al., 1979; Gainey, 1978a; Gainey and Greenberg, 1977). When salinity levels are increased slowly (over 40–80 days), allowing time for acclimation, *C. fluminea* can tolerate salinities of up to 24‰ (Evans et al., 1979).

Such levels of salinity tolerance are greater than those of most other North American freshwater bivalves (Evans et al., 1979). Indeed, although both *C. fluminea* and freshwater unionids are hyperosmotic regulators below 3‰ and osmoconformers above this salinity (Gainey and Greenberg, 1977), only *C. fluminea* has any capacity for volume regulation when hyperosmotically stressed (Hiscock, 1953; Gainey, 1978a,b; Gainey and Greenburg, 1977). Although not as well developed, volume regulation under hyperosmotic stress in *C. fluminea* is similar to that of estuarine species and involves the compensatory adjustment of free amino acid concentrations in the blood and tissues. Most osmotic change is accounted for by the amino acid alanine, along with much smaller changes in the concentrations of glutamate, taurine, aspartate, and serine (Gainey, 1978b).

In a dilute freshwater medium, the mechanism of hyperosmotic regulation of body fluid concentration appears to be fundamentally different from that of freshwater unionids. The body fluids of *C. fluminea* differ from those of unionids and sphaeriids in that the predominant salt is NaCl (Dietz, 1977) and the body fluids of *C. fluminea* are maintained at a higher osmotic concentration (65 mOsm/liter) (Dietz, 1979). *C. fluminea* also maintains higher concentrations of Na^+, Ca^{2+}, K^+, and Cl^- and lower concentrations of HCO_3^- in the blood than either unionid or sphaeriid bivalves (Dietz, 1979). In addition, *C. fluminea* has significantly higher rates of Na and Cl transport (up to twice those of unionids) than other freshwater bivalves (Dietz, 1979). When salt-depleted (by maintenance in deionized water), specimens of *C. fluminea* increase Na^+ uptake rate compared to pond water-acclimated specimens, although the net loss rate of

Na^+ to the medium is unaffected. Salt depletion also significantly increases the total inward and outward flux of Na^+ and the rates of Na^+ exchange diffusion (McCorkle and Dietz, 1980). The active component of inward Na^+ fluxes in salt-depleted individuals increases up to five times that of pond water-acclimated clams, with exchange diffusion of Na^+ for H^+ and NH_4^+ ions accounting for 83% of the total active uptake of Na^+ (McCorkle and Dietz, 1980). Because only H^+ is exchanged for Na^+ in pond water-acclimated individuals, Na^+/NH_4^+ exchange diffusion appears to be an accessory mechanism of Na^+ uptake in salt-depleted individuals (McCorkle and Dietz, 1980). *C. fluminea* also shows a diurnal oscillation in Na^+ uptake, Na^+ rates being greatest during dark periods and suppressed by light (McCorkle-Shirley, 1982).

Sodium transport in *C. fluminea* is also fundamentally different from that of unionids, because Na^+ influx and efflux rates are both much greater than those reported for unionid species (McCorkle and Dietz, 1980). The major difference in the ion flux mechanisms of *C. fluminea* compared to other freshwater bivalves is that a much higher proportion (67%) of the total Na^+ uptake is a result of exchange diffusion and the Na^+ uptake rate is dependent on the degree of salt depletion. The Na^+ transport system of *C. fluminea* also has a much greater affinity for Na^+ and is, therefore, more efficient than that of unionids (McCorkle and Dietz, 1980).

The capacity of *C. fluminea* for osmoregulation and volume regulation are much more similar to those of estuarine bivalves than to those of freshwater species. The levels of blood osmotic concentration and ion flux are higher than those of other freshwater species but lower than those of estuarine bivalves (Dietz, 1977, 1979). The marked ability of *C. fluminea* to regulate fluid volume during hyperosmotic stress is far more pronounced than that of freshwater unionids, which appear to be incapable of such regulation, but is clearly less well developed than that of estuarine species (Gainey, 1978a,b; Gainey and Greenberg, 1977); *C. fluminea* also retains Na^+/NH_4^+ exchange diffusion as an auxiliary method of sodium uptake. Such exchange diffusion for NH_4^+ is not reported for other freshwater bivalves but does occur in estuarine species (McCorkle and Dietz, 1980).

The family Corbiculidae has a relatively short fossil history, extending from the middle to the late Jurassic. The earliest species were estuarine, as are most present-day members of the family; the genus *Corbicula* is only a very recent invader of fresh water, in which it has a much shorter fossil history (Keen and Casey, 1969) than either unionids or sphaeriids (Keen and Dance, 1969; Haas, 1969). Therefore, it appears that *C. fluminea* still retains much of the genetically controlled physiological ''baggage'' associated with the estuarine habitat of its ancestors, whereas such physiological mechanisms as those involved with compensation for hyperosmotic stress have been lost in many freshwater bivalves that have much more extensive fossil histories than *C. fluminea* in fresh water (Dietz, 1977, 1979; McCorkle and Dietz, 1980; Gainey, 1978a,b).

One distinct advantage of maintaining high rates of Na^+ uptake and correspondingly high body fluid osmotic and ionic concentrations is that these may allow *C. fluminea* to sustain higher levels of activity, because ions act as cofactors in many metabolic reactions. [Freshwater unionids with the lowest levels of blood and tissue osmolarity of any aquatic animals (Robertson, 1964) also have extremely low metabolic rates (Prosser, 1973).] High metabolic rates allow much more rapid locomotion and burrowing behavior (Kraemer, 1979) and maintenance of much higher rates of ctenidial filtration than in other freshwater bivalves (Dudgeon, 1980; Habel, 1970; Mattice, 1979). This increased capacity for locomotion and filtration appears to give *C. fluminea* a distinct advantage over native bivalve species in its preferred lotic habitats characterized by unstable sand substrates and relatively strong currents.

C. Temperature

C. fluminea, which is not successful in very small streams, is essentially restricted to larger bodies of flowing water (Jenkinson, 1979). Such large bodies of flowing fresh water have much greater temperature stability than do smaller, more lentic habitats. The temperature tolerance limits of *C. fluminea,* generally much narrower than those of native North American bivalve species, reflect its restriction to these more thermally stable environments. Although *C. fluminea* can tolerate short-term exposure to waters with temperatures as low as 0°C or as high as 40°C (5°C-acclimated individuals) to 43°C (30°C-acclimated individuals) (Mattice and Dye, 1976; Isom et al., 1978), longer term exposures (300–500 h) to high temperatures reduce upper lethal limits to 24°C (5°C-acclimated individuals) and to 34°C (30°C-acclimated individuals) (Mattice and Dye, 1976; Mattice, 1979). Other studies have reported the upper lethal temperature limits of *C. fluminea* to be 35°C (25°C-acclimated individuals) (Cherry et al., 1980b), 33.5°C (Habel, 1970), and 38°C (Mudkhede and Nagabhushanam, 1977). Such laboratory estimates of upper lethal temperature limits correspond directly to reports that *C. fluminea* cannot withstand long exposures to temperatures of 30–35°C in artificial outdoor ponds (Busch, 1974; Greer and Ziebell, 1972; Habel, 1970; Haines, 1979). In addition, populations of *C. fluminea* in natural waters receiving heated effluents are either severely depleted or eliminated completely as summer water temperatures reach 38–40°C (Dreier, 1977; Dreier and Tranquilli, 1981; R. F. McMahon and C. J. Williams, unpublished observations).

Temperatures above 25–30°C seem to have extensive effects on the biology of *C. fluminea.* Aldridge and McMahon (1978) observed a severe suppression of juvenile production and release at ambient temperatures above 30°C in a natural population of *C. fluminea.* Laboratory investigations of oxygen consumption have also shown that although the weight-specific oxygen uptake rates of specimens of *C. fluminea* maintained at 10, 20, and 30°C show a normal pattern of temperature acclimation at temperatures below 25°C (the rate of oxygen consumption between 5° and 25°C is greatest in those specimens acclimated to 10°C and least in those acclimated to 30°C), it is severely depressed in all acclimation groups at temperatures above 25°C even when individuals have been acclimated to 30°C (McMahon, 1979a). Associated with this reduction in oxygen uptake rate is a marked decrease in the time spent siphoning, with individuals spending most of the time with the siphons withdrawn and valves closed (20–40% of the time at 25°C, increasing to 80–92% of the time at 30°C) (McMahon, 1979a). Oxygen consumption rate and siphoning activity displayed a second but smaller maximum at 40–45°C, which is associated with lethal thermal stress (McMahon, 1979a). Also associated with increases in temperature above 25°C is a severe decline in the filtration rate (Mattice, 1979), another manifestation of the reduction in ctenidial ventilation and respiratory gas exchange which occurs above this temperature.

This inability to maintain efficient levels of aerobic metabolism and gill filtration above 25°C appears to be directly related to the impaired capacity of *C. fluminea* to tolerate environmental temperatures above 30°C. Concurrent suppression of metabolic rates, feeding rates, and reproductive capacity at sustained temperatures above 30–35°C will prevent this species from invading many North American aquatic habitats where such temperatures are not uncommon during summer months.

C. fluminea is also more intolerant of low winter water temperatures than native North American freshwater bivalves, being unable to tolerate long-term (200–560 h) exposures to temperatures below 2°C (Mattice and Dye, 1976). This intolerance of low winter temperatures reflects the relatively mild winter temperatures *C. fluminea* experiences in its native Southeast Asian habitats, where the annual temperature range in large bodies of fresh water is only 15–32°C (Morton, 1977a). Typically, most temperate, shallow fresh waters in the northern United States fall near or below 2°C for several months during the winter. Several reports of massive midwinter mortalities and/or total extinctions of *C. fluminea* are associated with such extremely cold winter periods during which ambient water temperatures fell to near 0°C (Bickel, 1966; Cherry et al., 1980b; Dreier, 1977; Dreier and Tranquilli, 1981; Horning and Keup, 1964; Rodgers et al., 1979). This intolerance of *C. fluminea* to low temperatures may account for the inability of this species to invade freshwater environments in the north-central states of the United States (Fig. 2), where water temperatures remain below 2°C during the winter months. Instead, *C. fluminea* is restricted to drainage systems on both the East and West Coasts of North America and in the southern United States, where the climate is more moderate and water temperatures rarely fall below 2°C for more than a few days at a time. It is not surprising, therefore, that many of the most northern populations of *C. fluminea* in the

United States are restricted to portions of rivers and lakes receiving artificial thermal effluents from industrial installations which maintain winter water temperatures above the lower lethal limit of 2°C (Dreier, 1977; Dreier and Tranquilli, 1981; Eckblad, 1975; Rodgers et al., 1977). Laboratory studies such as those of Mattice and Dye (1976) suggest that such northern populations could be eliminated by interrupting the release of heated waters for 30–40 consecutive days during the coldest midwinter months. Low ambient temperatures also reduce the growth rate (Abbott, 1979; Buttner and Heidinger, 1980; Cherry et al., 1980b; Dreier, 1977; Dreier and Tranquilli, 1980; Mattice, 1979; Aldridge and McMahon, 1978; Morton, 1977a; Rodgers et al., 1979) and inhibit veliger release in *C. fluminea* which ceases at temperatures below 18–19°C (Heinsohn, 1958; Aldridge, 1976; Aldridge and McMahon, 1978).

D. Exposure in Air

Freshwater environments, both lentic and lotic, are often characterized by long-term fluctuations in depth associated with seasonal changes in rainfall and subsequent water runoff. As a consequence, benthic freshwater animals such as *C. fluminea,* which inhabit shallow nearshore environments, can be periodically exposed to air by receding water levels during dry seasons or droughts for periods far longer than the maximum 8–10 days of aerial exposure experienced by high intertidal molluscs between peak spring tides (McMahon and Russell-Hunter, 1981). Under such selective pressures, freshwater unionid and sphaeriid bivalves have evolved high levels of tolerance to prolonged aerial exposure, some species being able to survive many months of continual exposure to air (Collins, 1967; Dance, 1958; Hiscock, 1953; Rang, 1834). Unlike these desiccation-resistant freshwater species, *C. fluminea* is remarkably intolerant of even relatively short periods of aerial exposure (McMahon, 1979b). Therefore, it can be eliminated from the nearshore habitat of variable-level lakes, where populations are frequently exposed to air (Junk, 1975; McMahon, 1979b, also unpublished observations; White, 1979; White and White, 1977a). Periodic decreases in lake water levels can leave *C. fluminea* populations, which often accumulate in nearshore substrata (Aldridge and McMahon, 1978; Bickel, 1966; Dreier, 1977; Dreier and Tranquilli, 1981; Fast, 1971; Lenat and Weiss, 1973; Morton, 1977a; O'Kane, 1976; Pool and McCullough, 1979), stranded in air for long periods of time (McMahon, 1979b; White, 1979; White and White, 1977a). During extended periods of exposure to air, *C. fluminea* populations will experience extensive mortality (Aldridge and McMahon, 1978; McMahon, 1979b; White, 1979; White and White, 1977a). A 7-month-long exposure to air was reported to cause 99.3% mortality in a stranded *C. fluminea* population, whereas unionid populations exposed at the same time suffered only 50% mortality (White, 1979).

Such field observations of low tolerance to aerial exposure and desiccation in *C. fluminea* have been confirmed by laboratory investigations. When exposed to air in either dry or moist sand, *C. fluminea* had 50% mortality within 4 days, increasing to 98% by 12 days of aerial exposure (White, 1979; White and White, 1977a). A more extensive study of tolerance to aerial exposure has revealed that the mean emergence tolerance times of specimens of *C. fluminea* exposed to air at 20°C were 26.8 days in high relative humidity (high RH) and 13.9 days in low RH. At 30°C these tolerance times decreased to 8.3 days (high RH) and 6.7 days (low RH) (McMahon, 1979b). Maximal tolerance periods (producing 100% mortality) were 37 days (high RH) and 18 days (low RH) at 20°C and 10 days (high RH) and 9 days (low RH) at 30°C (McMahon, 1979b).

During exposure to air, *C. fluminea* displays a unique behavior, not previously reported for bivalves, in which specimens alternate between short periods during which the valves are parted slightly and the tissues of the mantle edge are exposed directly to the external atmosphere (Fig. 3) (approximately 10–25% of

Fig. 3. Mantle edge exposure in *C. fluminea* on exposure in air. The white arrows indicate the exposed mantle edges of *C. fluminea* (sealed together by hardened mucus). This behavior alternates with valve closure on exposure to air and is always associated with a high rate of aerial oxygen consumption in this species. The specimen shown here has a shell length of 16 mm.

the total emergence period), and longer periods during which the valves remain tightly shut and are sealed with a hardened mucus, the period of mantle edge exposure decreasing with total exposure time in air (McMahon, 1979b). Unlike the continuous gaping of certain intertidal bivalves, which exposes internal mantle-cavity tissues directly to the external atmosphere when in air, the exposed mantle edges of *C. fluminea* are sealed together with a hardened mucus, which precludes direct exposure of pallial tissues to the external environment (McMahon, 1979b). McMahon and Williams (1981) reported that *C. fluminea* maintains very high levels of aerial oxygen consumption during these periods of mantle edge exposure, whereas little or no significant oxygen consumption occurs when the valves are shut. Aerial oxygen consumption rates during mantle edge exposure were three to four times those maintained in water, indicating that mantle edge exposure does not function in direct gas exchange with the blood in pallial hemocoels but, instead, is much more likely to be associated with renewal of mantle cavity oxygen stores by an inward diffusion of oxygen through the thin exposed tissues of the mantle edge. The valves are subsequently closed, and oxygen is consumed from mantle-cavity stores until these are depleted. Mantle edge exposure is then again initiated to renew these oxygen stores (McMahon, 1979b; McMahon and Williams, 1981). Such behavior allows maintenance of relatively high levels of aerial oxygen consumption (20% of aquatic rates), and is far more conservative of water loss than the continual aerial gaping of marine intertidal bivalves. This mode of aerial gas exchange appears to be of extreme importance to the survival of *C. fluminea* in air, because exposure to low RH precludes mantle edge exposure (an adaptive behavior that conserves water loss under highly desiccating conditions). This presumably results in anaerobic respiration under which the survival times of *C. fluminea* are greatly reduced and in which *C. fluminea* dies not from evaporative water loss but from the apparent accumulation of poisonous metabolic end products to lethal levels (McMahon, 1979b).

More advanced freshwater unionid and sphaeriid bivalves neither gape nor expose mantle tissues in air but instead take up oxygen directly by diffusion across the closed shell valves (Dietz, 1974; Collins, 1967). Such ''through-the-shell'' gas exchange is apparently much more conservative of evaporative water loss than either gaping or the periodic mantle edge exposure of *C. fluminea,* and it is presumed to account for the extremely high tolerance to aerial exposure displayed by certain species of freshwater clams.

The responses of *C. fluminea* to aerial exposure, like those to osmotic stress, therefore, appear to represent an intermediate level of adaptation between that of estuarine and intertidal bivalves and that of advanced freshwater unionids and sphaeriids, once again reflecting the relatively short evolutionary history of the species in fresh water. Periodic renewal of mantle-cavity oxygen stores through exposed, mucus-sealed mantle edges appears to represent a relatively small modification of the aerial gaping behavior common to intertidal species, but it is one which allows greater water conservation and, therefore, tolerance of the much longer periods of aerial exposure associated with a freshwater habitat.

E. Conclusions

The physiology and ecology of *C. fluminea* can best be understood in relation to its apparently recent evolution as a freshwater species. There presently exist in Southeast Asia two distinctly different but closely related species of *Corbicula,* i.e., *C. fluminea* and *C. fluminalis* (Müller) (Dudgeon, 1980; Morton, 1973, 1977a,b, 1979a, 1982). *C. fluminalis* is an estuarine species common in the open-water upper riverine portions of large estuaries such as that of the Pearl River in China (Dudgeon, 1980; Morton, 1973a, 1982; Miller and McClure, 1931). Such estuarine environments are characterized by relatively high current flow, unstable sandy substrata, relatively stable temperature regimes (compared to those of smaller bodies of fresh water), well-oxygenated waters and substrata, and high fluctuations in salinity. *C. fluminalis,* the estuarine species, has a preference for sandy, well-oxygenated substrata and flowing water (Dudgeon, 1980; Morton, 1973, 1982; Miller and McClure, 1931) and, like most estuarine bivalves, is likely to be relatively intolerant of extremes in temperature and prolonged exposure to hypoxia but highly tolerant of salinity fluctuation. Although there are no published data to support these suppositions concerning the physiology of *C. fluminalis,* similar physiological capacities have been reported for other estuarine bivalves (Gainey, 1978a,b; Gainey and Greenberg, 1977; Kennedy and Mihursky, 1971, 1972; McMahon and Wilson, 1981; Hamwi, 1969). On the presumption that *C. fluminea* evolved from an estuarine ancestor very much like *C. fluminalis,* it seems reasonable to expect that with such a short evolutionary history, it would retain much of the "genetic baggage" of its estuarine ancestors. This hypothesis would partially explain this species' habitat preferences (lotic environments in sandy, well-oxygenated substrata), low temperature tolerance, diminished capacity for oxygen consumption in even moderately hypoxic waters, and relatively high capacity for volume and osmotic regulation.

Instead of physiological adaptations, most adaptations by *C. fluminea* to life in fresh waters appear to lie in the area of evolutionary changes in life history tactics, including a reproductive schesis characterized by development of consecutive protandry and/or simultaneous hermaphroditism, ctenidial incubation, release of a benthic, nonpelagic juvenile, abbreviation of the life span, and shortening of the life cycle (Dudgeon, 1980; Kraemer, 1977, 1978b; Morton, 1977a, 1982; Aldridge and McMahon, 1978). These adaptations recapitulate the evolution of many other groups of freshwater molluscs (Russell Hunter, 1964).

V. Ecological Aspects of Life History

A. Life Cycle

The life history of *C. fluminea* has been detailed for both North American (Aldridge, 1976; Aldridge and McMahon, 1978; Dreier, 1977; Dreier and Tranquilli, 1981; Eng, 1979; Heinsohn, 1958; Sickel, 1979; Sinclair, 1964; Sinclair and Isom, 1961, 1963) and Southeast Asian populations (Morton, 1977a). In both North American and Southeast Asia, the life cycle involves hermaphroditic reproduction and incubation of fertilized eggs in specialized marsupial areas of the exhalant branchial cavity of the inner demibranchs to an advanced juvenile stage which has lost the ciliated velum and has a well-developed shell and foot. This benthic juvenile, sometimes called a *pediveliger* (Sinclair and Isom, 1961, 1963), has a SL of 18–25 μm, which is somewhat larger than that of the mature oocyte. The juvenile has a total dry weight of 10 μg, a flesh dry weight of 2.8 μg, and a total organic carbon content of 1.4 μg (Aldridge, 1976; Aldridge and McMahon, 1978). Although it can be transported passively on water currents (Goss and Cain, 1977; Goss et al., 1979; Sickel, 1979; Sinclair, 1964; Sinclair and Isom, 1961, 1963; A. L. Smith et al., 1979), the juvenile is incapable of swimming, which it has been occasionally erroneously reported to do (Ingram, 1959). Instead, it settles immediately to the substratum, over which it actively crawls with the foot (Sinclair and Isom, 1961, 1963; R. F. McMahon, unpublished observations); it can also anchor itself with a mucilaginous byssal thread that facilitates burrowing and limits passive dispersal by water currents (Kraemer, 1979).

Growth rates are high and maturity is reached at SLs of 6.5–13.0 mm, depending on the population sampled (Aldridge, 1976; Aldridge and McMahon, 1978; Heinsohn, 1958; Kraemer, 1978b; Morton, 1977a; Sinclair and Isom, 1961, 1963). Maximum SLs in Southeast Asian populations have been reported to range from 35 mm in Hong Kong (Morton, 1977a) to 21.0 mm in a Malaysian population identified as *C. malaccensis* (Berry, 1974), a probable synonym for *C. fluminea* (Morton, 1979a).

In most North American *C. fluminea* populations, individual SLs also rarely exceed 35 mm; however, individuals as large as 40–46 mm SL have been reported from some populations (Aldridge and McMahon, 1978; Eng, 1979; Heinsohn, 1958; O'Kane, 1976; Sinclair and Isom, 1961, 1963), with a few very singular individuals reaching SLs of 50–60 mm (Britton and Morton, 1979; Sinclair and Isom, 1961, 1963; R. F. McMahon, unpublished observations). Such discrepancies in the size ranges of individuals from different populations are not unusual among freshwater molluscs (Russell Hunter, 1964) and do not prevent North American specimens from being described as *C. fluminea*. Indeed, North American and Southeast Asian populations of *C. fluminea* have been shown to be almost identical in their morphology, reproductive biology, and ecology (Britton and Morton, 1979). Only one Southeast Asian population has been collected extensively enough to determine its maximum size (Morton, 1977a), which proved similar to that of 35 mm reported for most North American populations (Bates, 1962; Bickel, 1966; Branson and Batch, 1969; Diaz, 1974; Dreier, 1977; Dreier and Tranquilli, 1981; Giesy and Dickson, 1981; Gunning and Suttkus, 1966; Heard, 1964; Heinsohn, 1958; Ingram et al., 1964; Metcalf, 1966; O'Kane, 1976; Rodgers et al., 1979; Sickel, 1973, 1979). In freshwater molluscs with short life spans, such as *C. fluminea,* maximum size is highly correlated with growth rate. Populations having the highest rates of growth also tend to achieve the largest final SLs (Russell Hunter, 1964). Two extensive studies of North American *C. fluminea* populations (Eng, 1979; Bickel, 1966) describe growth rates similar to those reported for a Hong Kong population (Morton, 1977a) in which individual cohorts of newly released young reached average SLs of 10–15 mm in 1 year and in which maximum individual SL did not exceed 30–40 mm. In other closely examined North American populations with higher growth rates (SL of 15–30 mm/year) (Aldridge and McMahon, 1978; Heinsohn, 1958; R. F. McMahon, unpublished observations), much larger final SLs of 40–46 mm were achieved. If investigations of Southeast Asian *C. fluminea* populations were as extensive as they have been in North America, they would, no doubt, reveal a similar level of interpopulation variation in SL distributions.

B. Reproductive Biology

Perhaps the most controversial and confusing aspect of the life history of *C. fluminea* is its mode of hermaphroditic reproduction (for a review of reproduction in *C. fluminea,* see Britton and Morton, 1979). Populations of this species in Hong Kong are reported to display a rigid, consecutive protandric hermaphroditism in which the smallest sexually mature individuals were males producing only sperm in the first year of life, all older, larger individuals (SL $>$ 15 mm) were females producing only eggs, and intermediate-sized, simultaneously hermaphroditic individuals (SL = 9–15 mm) were infrequent (Morton, 1977a, 1982). However, in a Malaysian population of *C. malaccensis* (= *C. fluminea*), most mature individuals were simultaneous hermaphrodites (67% of the population) producing mature eggs and sperm. Small maturing individuals tended to develop oocytes and mature eggs before sperm (30% of the population), and individuals that produced only sperm were very rare (3% of the population) (Berry, 1974). In contrast, investigations of gonadal condition in North American populations have shown *C. fluminea* is universally a simultaneous hermaphrodite, with eggs and sperm present together in the gonads of all larger adults (SL $>$ 8 mm) (Eng, 1979; Heinsohn, 1958; Kraemer and Lott, 1977;

Kraemer, 1978b). Eng (1979) further reported that although ova were continually present in the gonads, the percentage of sperm-bearing individuals was variable, peaking during spawning periods. Heinsohn (1958) similarly found that the frequency of individuals with mature sperm varied not only with the spawning season but also between populations. Some populations contained sperm-bearing individuals at low frequencies throughout the year, whereas others showed increases in sperm-bearing individuals during spawning seasons. Such data indicate that sperm will not always be present in simultaneously hermaphroditic individuals competent to produce it, and that the season during which a population is sampled could greatly affect interpretations of the pattern of sexual development in this species.

Sperm and eggs have been observed to comingle in the conjoined lumina of their respective follicles in North America specimens of *C. fluminea.* Also present were small masses of cells resembling developing embryos (Kraemer, 1978b), an indication that in North America *C. fluminea* has at least the potential for self-fertilization. Oogenic follicles have been observed to develop before spermatogenic follicles in smaller North American specimens (2–3 mm SL), with mature sperm occurring only in individuals with SLs greater than 8 mm whose gonads also contained mature oocytes (Kraemer, 1978b). Such observations are in good agreement with those of Berry (1974), who described *C. fluminea* in Malaysia as simultaneous hermaphrodites that had a tendency to develop oogenic before spermatogenic follicles. However, these observations are in direct contradiction to those of Morton (1977a), who described Hong Kong populations of *C. fluminea* as protandric, in which smaller individuals produced only sperm and older, larger individuals produced only eggs.

Such interpopulation variation in the type of hermaphroditism displayed by *C. fluminea* in Asia and North America does not require an explanation based on taxonomic difference but, instead, can be understood in terms of the highly variable nature of hermaphroditic reproduction in molluscs. As reviewed by Fretter and Graham (1964), the sexual state of a hermaphroditic species can be affected by both environmental and genetic factors. Britton and Morton (1979) suggested that environmental differences between North American and Asian populations may account for the observed differences in the type of hermaphroditism. However, if environment is the single controlling factor, variations should be observed in the reproductive biology of isolated North American populations of *C. fluminea,* as has been reported for English populations of the freshwater unionid *Anodonta cygenea* (Bloomer, 1939). Instead, all investigated North American populations have been universally reported as simultaneous hermaphrodites, whether from the upper tidal portions of rivers (Heinsohn, 1958), inland freshwater rivers (Kraemer, 1978b; Kraemer and Lott, 1977), or freshwater canals (Eng, 1979), which represent a wide range of environmental conditions including variations in temperature, salinity, nutritient levels, substrata, current, and water quality.

In many hermaphroditic species of molluscs, sexuality is controlled by a series of gene loci; each genotype has a slightly different phenotypic expression in terms of the degree of hermaphroditism and the time and sequence of sexual change (Fretter and Graham, 1964). Because of their limited mobility and their lack of dispersal stages tolerant of long periods of aerial exposure, the genetic isolation of populations of freshwater molluscs is high (Russell Hunter, 1964). Therefore, freshwater molluscs tend to display a considerable degree of genetic, physiological, and morphological interpopulation variation (Russell-Hunter, 1964, 1978). When 18 enzyme systems of Southeast Asian *C. fluminea* populations were examined electrophoretically, they were found to be characterized by a relatively high degree of enzyme polymorphism and high levels of divergence in allele frequencies (M. H. Smith et al., 1979). As in other freshwater molluscs, such interpopulation genetic differences may reflect the relative insularity and temporal instability of freshwater environments and can be partially accounted for by founder effects and subsequent genetic drift (Berrie, 1959; Hunter, 1972, 1975; Ibarra and McMahon, 1981; McMahon and Payne, 1980; Nickerson, 1972; Russell-Hunter et al., 1967, 1970, 1981). Interpopulation genetic variation may provide the likely explanation for the observed differences in the sexuality of Southeast Asian populations of *C. fluminea* (Berry, 1974; Morton, 1977a). In contrast, North American populations of *C. fluminea* have been reported to have a genetic variability of zero, no polymorphic enzymes, and a genetic similarity of 1 for 18 enzyme systems tested electrophoretically (M. H. Smith et al., 1979). These findings indicate a high probability that all North American populations arose from a single small introduced population in which a combination of founder effects and genetic drift fixed single alleles at the majority of loci (M. H. Smith et al., 1979). It is possible, therefore, that the allelic variation at loci controlling sexuality may have also become similarly severely reduced, resulting in the fixation of an inflexible pattern of simultaneous hermaphroditism in North American populations. Certainly, the hermaphroditic nature of reproduction in *C. fluminea* appears to be an area requiring further investigation, including repetitive population sampling and microscopic examination of the gonads (especially serial sections) of a wide size range of individuals over an entire annual cycle (including nonreproductive periods) in both Asian and North American populations.

C. Fertilization, Incubation, and Spawning

The mature eggs of *C. fluminea* pass through a pair of common genital ducts leading from the gonads and converging into a single genital opening in each of the suprabranchial cavities lying on either side of the visceral mass (Kraemer, 1978b). The mature oocytes are retained in the marsupial chambers formed from the modified exhalant branchial spaces of the inner demibranchs (Aldridge and McMahon, 1978; Dudgeon, 1980; Heinsohn, 1958; Kraemer, 1977, 1978b; Morton, 1977a,b; Sinclair and Isom, 1961, 1963). Heinsohn (1958) observed release of sperm from the exhalant siphons of reproductive individuals of *C. fluminea,* which indicates that cross-fertilization is the normal reproductive mode in this species, released sperm being drawn in with the inhalant currents of other gravid individuals. Eggs are probably fertilized while in the marsupial pouches of the inner demibranchs rather than in the oogenic tubules, as has been suggested by Sinclair and Isom (1961, 1963). Because mature sperm and eggs occur simultaneously and are passed through a common duct system in North American and some Asian populations (Kraemer, 1978b; Berry, 1974), this species has a potential for self-fertilization. Indeed, embryo-like masses of cells have been observed in the reproductive ducts of North American specimens (Kraemer, 1978b).

When released to the inner demibranchs, unfertilized mature oocytes are 0.05–0.2 mm in diameter (Heinsohn, 1958; Sinclair and Isom, 1961, 1963). The epithelial lining of the intralamellar spaces of the inner demibranchs is made up of two types of cells: secretory cells, which release small globules of material from their distal ends, and mucous cells, which make up the larger portion of the epithelium (Morton, 1977b). These two cell types increase in height in incubatory specimens (Morton, 1977b). Because released juveniles of *C. fluminea* are generally reported to be larger (0.18–0.25 mm) than mature oocytes (Aldridge and McMahon, 1978; Britton et al., 1979; Boozer and Mirkes, 1979; Eng, 1979; Heinsohn, 1958; Morton, 1977a,b, 1982; Sinclair and Isom, 1961, 1963), Morton (1977b, 1982) has suggested that the developing embryos may receive nourishment from the secretions of the hyperatrophied epithelial cells lining the intralamellar spaces of the inner demibranchs. However, there is at present no hard physiological, cytochemical, or biochemical evidence to support this supposition.

Fertilized eggs are incubated in the inner demibranchs through the trochophore and veliger stages for about 30 days (Eng, 1979). Sinclair and Isom (1961, 1963) reported that an advanced meroplanktonic veliger was released in which the ciliated velum functions as a swimming and feeding organ which maintains the larva in the water column. However, most studies indicate that a more advanced benthic juvenile (often incorrectly referred to as a pediveliger) is released in which the velum has been greatly reduced or lost altogether and the foot and shell are well developed (Aldridge, 1976; Aldridge and McMahon, 1978; Eng, 1979; Heinsohn, 1958; Morton, 1977a). This benthic juvenile stage is incapable of swimming and crawls actively over the substratum with the well-developed ciliated foot (Sinclair and Isom, 1961, 1963). In fact, on release, juvenile *C.*

fluminea immediately sink, attaching to the substratum by secretion of adhesive substances on the foot tip (Heinsohn, 1958; Sinclair and Isom, 1961, 1963) and later by development of a mucilangious byssal thread (Kraemer, 1979). Juveniles appear to enter the water column only accidentally by passive transport on strong water currents (Aldridge and McMahon, 1978; Eng, 1979; Heinsohn, 1958; Morton, 1982). Suppression of planktonic larval stages by *C. fluminea* is not unexpected. Production of even short-lived planktonic larvae would not be adaptive for a lotic species such as *C. fluminea* because many of the newly released larval stages would be carried downstream away from the adult population into environments that could be unfavorable for their eventual growth and reproduction (Morton, 1982). Therefore, *C. fluminea,* as have most freshwater molluscs including sphaeriid and unionid bivalves (Russell Hunter, 1964), has during the course of its evolution suppressed free-swimming larval stages and instead releases a relatively highly advanced benthic juvenile stage which is essentially a miniature adult.

Detailed studies of reproductive behavior in *C. fluminea* consistently report two spawning periods for this species, marked by peaks of juvenile release in the spring and in the fall and separated by a reduction or total cessation of juvenile release during midsummer and winter periods (Aldridge, 1976; Aldridge and McMahon, 1978; Boozer and Mirkes, 1979; Dreier, 1977; Dreier and Tranquilli, 1981; Eng, 1979; Morton, 1977a; Sickel, 1979). The onset of spawning activity in the spring and its cessation in winter are apparently directly influenced by temperature, with spawning activity initiated or inhibited as ambient water temperatures rise above or fall below a critical temparature of 18–19°C (range = 16–21°C) (Aldridge, 1976; Aldridge and McMahon, 1978; Dreier, 1977; Dreier and Tranquilli, 1981; Eng, 1979). The cause of the midsummer cessation of reproductive activity is less obvious. During this period, the percentage of adults incubating eggs and larval stages decreases (Aldridge, 1976; Aldridge and McMahon, 1978; Dreier, 1977; Dreier and Tranquilli, 1981; Eng, 1979; Morton, 1977a), spermiogenesis is inhibited (Eng, 1979), and juvenile release is inhibited or totally suppressed (Aldridge, 1976; Aldridge and McMahon, 1978).

Release of juveniles by adult clams occurs continually throughout the spawning periods (Aldridge, 1976; Aldridge and McMahon, 1978; Heinsohn, 1958). Fecundity is high and size dependent, larger individuals releasing greater numbers of juveniles (Aldridge, 1976; Aldridge and McMahon, 1978; Heinsohn, 1958). A laboratory study of fecundity estimated that an average clam could release 8000 juveniles/year (97–570 juveniles/adult/day, depending on adult size) (Heinsohn, 1958). Field studies indicate similar average daily fecundities of 387 juveniles/adult/day during a 94-day spring spawning period and 320 juveniles/adult/day during a 101-day fall spawning period, but much greater overall fecundities with an estimated average annual release of 68,678 juveniles per average-sized adult (Aldridge, 1976; Aldridge and McMahon, 1978). The difference in overall fecundity between the laboratory and field studies was based on differences in the estimated duration of the reproductive period (Heinsohn, 1958).

Daily fecundity is correlated with ambient temperature, juvenile release being initiated at 19°C and peaking at 26°C. Above 26°C, juvenile release is markedly inhibited, ceasing completely at temperatures above 32°C (Aldridge, 1976; Aldridge and McMahon, 1978). Suppression of spawning activity at stressful high temperatures has been suggested as the cause of the midsummer cessation of breeding in *C. fluminea* (Aldridge and McMahon, 1978). In many of its aquatic habitats, water temperatures rise above 25°C during the summer (Aldridge and McMahon, 1978; Dreier, 1977; Dreier and Tranquilli, 1981; Morton, 1977a) and occasionally approach the long-term upper lethal limit of this species of 34°C (Mattice, 1979; Aldridge and McMahon, 1978; Dreier, 1977; Dreier and Tranquilli, 1981; Morton, 1977a). Temperatures above 25°C have been demonstrated to be stressful to *C. fluminea,* inhibiting both aerobic metabolic rates (McMahon, 1979a) and filtration rates (Mattice, 1979). Therefore, it does not seem unlikely that such temperature stress could also inhibit gametogenesis (Eng, 1979) and juvenile release (Aldridge and McMahon, 1978). However, similar midsummer cessation of spawning activity (Eng, 1979; Heinsohn, 1958) and spermiogenesis (Eng, 1979) has also been reported for California populations of *C. fluminea* in environments where ambient temperatures do not exceed 25–28°C. This temperature range in which midsummer cessation of spawning occurs in California populations lies very close to the theoretical temperature of 26°C for maximum spawning activity in a Texas population of *C. fluminea* (Aldridge and McMahon, 1978). Therefore, the biennial spawning pattern of *C. fluminea* may be a genetically controlled reproductive characteristic of this species rather than one which is induced by stressful midsummer environmental conditions.

D. Growth and Life Span

The growth rate of *C. fluminea* (measured as increase in SL) generally appears to be much greater than that of both freshwater sphaeriids (Avolizi, 1971, 1976; Hornbach et al., 1980; Way et al., 1980) and unionids (Negus, 1966). Earlier investigations of *C. fluminea* in North America produced estimates of slower growth rates for this species based on the size distributions of single samples and rings or bands on the periostracum or shell. The results of these investigations were interpreted with a bias toward the low growth rates known to occur in similar-sized unionid species (Negus, 1966).

These initial estimates of growth rate, based on population size distributions, were later shown to be extremely misleading, because *C. fluminea* with its highly unusual biennial reproductive pattern produces *two* distinct size modes per year

(Aldridge, 1976; Aldridge and McMahon, 1978; Eng, 1979; Heinsohn, 1958; Morton, 1977a). These distinct size classes were often interpreted as representing the result of a single annual spawning period (the pattern far more usual in freshwater molluscs) (Diaz, 1974; Gardner et al., 1976; Gunning and Suttkus, 1966; Hubricht, 1966; Ingram et al., 1964; Keup et al., 1963), which essentially doubles life span estimates over actual values.

The age and life span of North American *C. fluminea* populations have also been estimated from the presence and number of shell growth rings or banding patterns in the periostracum (Sickel, 1973; Sinclair and Isom, 1961, 1963), with each growth ring or band believed to represent 1 year of life. Morton (1977a, 1982), however, claimed that Asian specimens of *C. fluminea* produce two indistinct growth rings per year. In contrast, others believed that growth rings and banding patterns, if present at all, are highly variable and do not accurately reflect age, with many large specimens having no discernable growth annuli (Aldridge, 1976; Aldridge and McMahon, 1978; Dudgeon, 1980; Heard, 1964; R. F. McMahon, unpublished observations). It is not surprising, then, that many estimates of the maximum life span in *C. fluminea* based on such unsubstantiated assumptions were erroneously high at 4–6 years, and the growth rate derived from such estimates was erroneously low at 2–13 mm SL in the first year of life (Gardner et al., 1976; Ingram et al., 1964; Keup et al., 1963; Sickel, 1973; Sinclair and Isom, 1961, 1963; Villadolid and Del Rosario, 1930). Such erroneous estimates of age and size also show the growth rate to be anomalously increasing with age and size in *C. fluminea*. As stated by Britton et al. (1979) in a review of growth rate studies of *C. fluminea*, increasing growth rate with age is not a typical molluscan pattern, and any growth rate estimates which imply such a pattern for *C. fluminea* appear to be highly unrepresentative of actual growth in this species.

More accurate studies of growth in *C. fluminea*, which were based on repetitive sampling for size distributions and reproductive periods of natural populations throughout an entire annual reproductive and growth cycle (Aldridge, 1976; Aldridge and McMahon, 1978; Eng, 1979; Heinsohn, 1958; Morton, 1977a) and of marked and caged individuals in natural environments (Britton et al., 1979; Buttner and Heidinger, 1980; Dreier, 1977; Dreier and Tranquilli, 1981; Mattice, 1979; O'Kane, 1976), have shown that the life span of *C. fluminea* is actually far shorter and its growth rate much greater than previously suspected. These studies demonstrated that most *C. fluminea* populations in North America and Asia grow to SLs of 16–30 mm in the first year of life, and the growth rate declines markedly with both age and size (Aldridge, 1976; Aldridge and McMahon, 1978; Eng, 1979; Heinsohn, 1958; Morton, 1977a). Such investigations have also shown that individual life span in fast-growing North American populations (25–30 mm SL in the first year of life) is rarely greater than 1.5–2.0 years, with most individuals being annuals (Aldridge, 1976; Aldridge and

McMahon, 1978; Heinsohn, 1958). Similar, very short life spans of 1.0–1.5 years have been estimated for individuals of *C. africana* from a population in Lake Chad, Africa (although rare individuals were reported to survive for as long as 3 years) (Leveque, 1973a). Life span is, however, quite variable and dependent on growth rate in *C. fluminea,* with individual life spans of as long as 3 years being reported for slower growing (less than 20 mm SL in the first year) populations in both Asia (Morton, 1977a) and North America (Eng, 1979).

Growth data recorded from specimens maintained in enclosures within natural environments tend to support the field evidence for high growth rates in *C. fluminea*. Such studies have yielded estimates of maximum summer growth rates in first-year individuals that range from low values of 2.0–2.5 mm/month (Mattice, 1979; O'Kane, 1976) to values of 6.5 mm/month (Dreier, 1977; Dreier and Tranquilli, 1981). Growth rate increases with ambient temperature (Britton et al., 1979; Buttner and Heidinger, 1980; Dreier, 1977; Dreier and Tranquilli, 1981; Mattice, 1979; O'Kane, 1976) and decreases with age and size (Aldridge and McMahon, 1978; Morton, 1977a; Britton et al., 1979; Heinsohn, 1958; Buttner and Heidinger, 1980; Dreier, 1977; Dreier and Tranquilli, 1981). Such lower growth rates in larger, mature individuals may account for the much smaller growth rates of 0.36–0.67 mm SL/month recorded when only larger (SL > 10.0 mm) individuals were maintained in field enclosures (Britton et al., 1979; Buttner and Heidinger, 1980). In addition, there appears to be a "containerization" effect; field enclosure reduces growth rate in individuals of *C. fluminea* compared to those that are free living (Britton et al., 1979), indicating that such enclosure experiments may be underestimating field growth rates in this species. Certainly, this aspect of the biology of *C. fluminea* requires further study, including year-round collections of populations and long-term field enclosure experiments, before the degree of environmentally induced plasticity in the growth rates and life spans of *C. fluminea* populations can be completely assessed.

Such high growth rates relative to other freshwater bivalves may be a result of the relatively higher levels of activity maintained by *C. fluminea* (Kraemer, 1979), particularly with regard to gill-filtering rates. Mattice (1979) reported that a 25-mm SL specimen of *C. fluminea* maintains a ctenidial filtration rate of 300–800 ml/h. For Southeast Asian populations, a filtration rate of 80 ml/h at 20–22°C for a 25-mm SL individual has been reported (Dudgeon, 1980), and rates of 8.0–25.7 ml/h have been reported for clams larger than 27 mm SL in North America (Habel, 1970). Mattice (1979) found that particle concentration did not affect filtration rate in *C. fluminea,* but Habel (1970) showed that filtration rate increased with decreasing concentration. In spite of the wide variation in reported estimates, it can be stated with a high degree of confidence that *C. fluminea* maintains much higher filtration rates than do other freshwater bivalves (Mattice, 1979), which provides the species with higher ingestion rates. Such high ingestion and correspondingly high assimilation rates could account for the very fast rate of growth displayed by *C. fluminea,* as a much greater proportion of the total energy assimilated can be channeled into tissue growth, allowing maintenance of higher growth efficiencies relative to other freshwater species (Aldridge, 1976; Aldridge and McMahon, 1978).

E. Life-Cycle Bioenergetics and Life-History Tactics

Data on energy flow through a natural North American population of *C. fluminea* have been reported by Aldridge and McMahon (1978) and are presented in Table I, where energy flow rates are expressed in terms of organic carbon as mg C/(m^2·day) for the spring and fall generations and the entire population on an annual basis. The main components of an energy budget are total assimilation (TA) in mg C/(m^2·day); respired assimilation (RA), that portion of TA that is expended in maintenance and activity; nonrespired assimilation (N-RA), that portion of TA which is fixed in tissue growth and gamete production (TA = RA + N-RA); growth, that portion of N-RA allocated to tissue growth; and reproduction, that portion of N-RA allocated to gamete production (N-RA = growth + reproduction) (Aldridge and McMahon, 1978; Chapter 11, this volume). Such detailed examinations of population life-cycle bioenergetics are valuable in elucidating a species life-history tactics (Stearns, 1976, 1977), especially if comparable data for related species exist.

This section compares the population bioenergetics and life-history traits of *C. fluminea* with those of other bivalves in an attempt to trace the evolution of these traits in this species. Unfortunately, very few thorough investigations of energy flow have been carried out for populations of other freshwater bivalve species. However, available studies do indicate that *C. fluminea* has evolved life-history traits which are basically different from those of other freshwater species. Freshwater unionid and sphaeriid clams produce relatively small numbers of highly developed offspring; sphaeriids incubate fertilized eggs in the gills through to a rather large juvenile ready to take up adult life and unionids incubate eggs in the gills through to a relatively complex bivalved glochidium larva which is an obligate parasite on fish. Reproductive effort in both of these species represents a relatively low percentage of N-RA, reported as less than 10% in *Sphaerium striatinum* and less than 6% in *S. simile* (Avolizi, 1971, 1976), whereas a unionid, *Anodonta anatina,* committed 15% of N-RA to gamete production (Negus, 1966). *C. fluminea* also commits a relatively small portion of N-RA (15%) to a reproductive effort (Table I), but unlike other freshwater bivalves it produces a very large number of relatively small offspring (0.15–0.25 mm SL), with average fecundity reaching 68,678 juveniles/adult/year (Aldridge, 1976; Aldridge and McMahon, 1978). In contrast, fecundity in sphaeriids is much lower, ranging from 6 juveniles/adult/year (Avolizi, 1971, 1976) to 136 juve-

TABLE I

Energy Flow through a *Corbicula fluminea* Population from Lake Arlington, Texas[a]

Group	TA	RA		N-RA		Growth		Reproduction	
	mg C/m^2	mg C/m^2	% TA	mg C/m^2	% TA	mg C/m^2	% N-RA	mg C/m^2	% N-RA
Spring generation	18.2	6.1	34	12.1	66	10.3	85	1.8	15
Fall generation	13.4	3.0	23	10.4	77	8.7	84	1.7	16
Annual—entire population	40.2	11.6	29	28.6	71	24.2	85	4.4	15

[a] From Aldridge and McMahon, 1978.

[b] Energy flow is represented in terms of organic carbon as mg C/m^2 and is directly convertible to kilocalories using a conversion factor of 10.94 kcal/g C (Russell-Hunter et al., 1968).

niles/adult/year (Hornbach et al., 1980). At release, the juveniles of sphaeriids are much larger than those of *C. fluminea,* ranging from 1.2 mm SL (*Musculium partumieum*) (Hornbach et al., 1980) to 7.3 mm SL (*S. striatinum*) (Avolizi, 1971, 1976). The size of unionid glochidia larvae at release is similar to that of juvenile *C. fluminea,* and like *C. fluminea,* unionids spawn glochidia in relatively large numbers, *A. anatina* producing 7500–25,000 glochidia/adult/year (computed from the data of Negus, 1966). In some unionids, individual fecundity has been reported to be as great as 10^5–10^6 glochidia/year (Russell Hunter, 1964). However, unlike *C. fluminea,* in which juveniles are free-living from the time of release, the glochidium stage of unionids is an obligate ectoparasite on fish, after which metamorphosis to the adult occurs. This ectoparasitic stage greatly limits the effective fecundity of unionid species to the relatively few larvae that successfully locate and parasitize an appropriate fish host. *C. fluminea,* therefore, has a far greater effective fecundity than either sphaeriids or unionids, although all three commit only a relatively small percentage of N-RA to reproduction.

The second conspicuous bioenergetic difference between *C. fluminea* and other freshwater species is the very high proportion of TA that it allocates to N-RA (growth and reproduction) (Table I). The ratio N-RA/TA (100) is known as the *net growth efficiency,* and it is higher in *C. fluminea* at 71% than in any other species of freshwater or marine bivalve for which such values have been reported (Aldridge, 1976; Aldridge and McMahon, 1978), an indication of the extremely high growth rates maintained by this species. Net growth efficiencies reported for other freshwater species are 47% for a population of *Pisidium virginicum* (Teal, 1957), 20% for an unidentified sphaeriid (Johnson and Brinkhurst, 1971), and 8.9–11.1% for *C. africana* in Lake Chad (Africa) (Leveque, 1973b). These high growth efficiencies occur in *C. fluminea* because, although it maintains relatively low levels of oxygen consumption (low RA) which are similar to those of other freshwater species (McMahon, 1979a), it has much higher levels of total assimilation. This higher total assimilation, reflected by filtration rates which are much higher than those of other freshwater bivalves (Mattice, 1979), leaves a much greater proportion of TA to be diverted to growth and reproduction (N-RA = TA − RA).

With exceptionally high growth rates, high net production efficiencies, high fecundities, and a relatively short life cycle, *C. fluminea* can become much more rapidly established on penetrating a favorable environment than can other species of longer-lived, slower-maturing freshwater bivalves. For example, Fig. 2 shows that *C. fluminea* became established throughout the downstream portions of the Mississippi River within 7 years of its introduction near the river's confluence with the Ohio River. Additionally, stable, highly dense adult populations of *C. fluminea* have been observed to be established within 2 years of introduction into environments as diverse as irrigation canals (Prokopovich, 1969; Prokopovich and Hebert, 1965), large rivers (Gardner et al., 1976), and lakes (Aldridge, 1976; Aldridge and McMahon, 1978; Morton, 1977a).

It is probable that *C. fluminea* arose from an estuarine ancestor which penetrated fresh waters. It is fortunate that Morton (1982) has recently investigated the biology of a Southeast Asian estuarine species, *C. fluminalis,* which is very closely allied to *C. fluminea* and may be very similar to its estuarine ancestor. Therefore, a comparison of the life-history traits of these two species may elucidate some of the adaptations *C. fluminea* has made to life in fresh water. *C. fluminalis* has a life span of up to 10 years, which is far longer than the 2–3-year life span of *C. fluminea. C. fluminalis* grows more slowly but reaches a larger maximum size than *C., fluminea. C. fluminalis* spawns only once a year, in winter, whereas *C. fluminea* spawns in the spring and fall. *C. fluminalis* reaches maturity after 1 year at a size of 16–18 mm SL; *C. fluminea* matures within several months at 6–10 mm SL. *C. fluminalis* is primarily dioecious; *C. fluminea* in North America appears to be a simultaneous hermaphrodite. *C. fluminalis* produces smaller eggs than *C. fluminea* and releases them directly to the medium to be fertilized externally; planktonic larval stages precede settlement and metamorphosis to adult form. In constrast, *C. fluminea* is incubatory, retaining fertilized eggs in the inner demibranchs through to release of a highly developed bivalved benthic juvenile stage.

The evolution of an incubatory habit, suppression of planktonic larval stages, and production of larger, well-developed young are all characteristic adaptations of freshwater molluscs (Russell Hunter, 1964). Such adaptations limit passive downstream transport of young into unsuitable environments. Similarly, development of a hermaphroditic reproductive schesis is a common adaptation in freshwater molluscs such as sphaeriid bivalves and gives *C. fluminea* the capacity to become established in new environments even if only a few founder individuals are introduced. Simultaneous hermaphroditism also allows maintenance of high reproductive capacities in variable freshwater environments where mortality rates are high.

Other life-history traits of *C. fluminea,* such as reduced age and size at maturity, reduction in maximum size, high growth rates, and decrease in life span when compared to its estuarine ancestor, can be understood only in terms of the evolution of this species in relation to other bivalve species with longer fossil histories in fresh water. Like *C. fluminalis,* freshwater sphaeriid and unionid bivalves tend to be slower growing than *C. fluminea.* Sphaeriids have similar life spans but reach much smaller maximum sizes (Avolizi, 1971, 1976). Unionids reach larger sizes but are very long-lived, with reported life spans of 10–15 years (Negus, 1966). *C. fluminea* matures within 1–3 months under favorable conditions, whereas sphaeriids generally require 6 months–1 year to reach maturity (Avolizi, 1971, 1976; Hornbach et al., 1980) and unionids require 3 to 4 years to reach maturity (Negus, 1966). *C. fluminea* invariably displays two spawning periods per year, whereas unionid and larger sphaeriid species spawn only once annually (Avolizi, 1971, 1976; Burky and Burky, 1976; Hornbach et al., 1980; Negus, 1966; Trdan, 1981).

In its relatively recent penetration of freshwater environments, *C. fluminea* probably would not have been able to compete successfully with highly evolved, longer established freshwater unionid and sphaeriid bivalves. In fact, there is an increasing body of evidence indicating that *C. fluminea* cannot become established in undisturbed North American freshwater environments where unionid and sphaeriid populations are healthy and flourishing (Fuller and Imlay, 1976; Isom, 1974; Klippel and Parmalee, 1979; Kraemer, 1979; Taylor, 1980b), indicating that it is not competitive under these conditions. Rather, in its early evolution *C. fluminea* appears to have adopted a niche from which unionids and sphaeriids were mostly excluded, thus avoiding competition. This niche, as previously described, was unstable sand-gravel substrata in fast-flowing, shallow lotic environments. Most North American unionid and many sphaeriid species are generally associated with much more stable environments such as lakes, ponds, impoundments, and larger, slower flowing rivers. Accordingly, unionid and sphaeriid bivalves have evolved life-history traits (as described above) which are characteristic of species from relatively stable habitats. They mature relatively late in the life span, are long-lived and highly iteroparous (more than one reproductive period in a lifetime), divert a relatively small proportion of N-RA into reproductive effort, and produce small numbers of large, well-developed offspring. These adaptations optimize production and survival to maturity of offspring in stable environments where densities remain near carrying capacity (Stearns, 1976, 1977). Such populations are characterized by higher proportions of mature than immature individuals (Stearns, 1976, 1977), a feature of both unionid (Negus, 1966) and some sphaeriid populations (Avolizi, 1971, 1976).

In contrast, *C. fluminea* prefers unstable sand-gravel substrata in shallow, fast-flowing lotic systems, an extremely variable environment in which population densities can be reduced drastically by chance environmental factors such as flooding (R. F. McMahon, unpublished observations), periodic droughts, and subsequent aerial exposure (McMahon, 1979b; White, 1979; White and White, 1977a), and temperature fluctuations (Mattice, 1979; McMahon, 1979a), with massive mortalities reported as a result of both high summer temperatures (Busch, 1974; Dreier, 1977; Dreier and Tranquilli, 1981; Habel, 1970) and low winter temperatures (Bickel, 1966; Dreier, 1977; Dreier and Tranquilli, 1981; Horning and Keup, 1964; Rodgers et al., 1977). In such highly variable environments, the life-history traits of greatest selective value are generally considered to be reduced age of maturity and reproduction, increased growth rate, increased energy allocated to reproduction, decreased number of broods and life span, decreased size of young, and increased brood size (fecundity). All of these traits optimize production and survival to maturity of offspring in an unstable environment where life-history tactics favoring rapid population growth and expansion are selected for (Stearns, 1976, 1977). Such populations are characterized by a very high proportion of immature individuals.

Certainly, *C. fluminea* appears to have evolved a series of life-history traits characteristic of a species adapted to a variable environment. It has the highest growth rates recorded for any freshwater bivalve, matures early and has a relatively short life span, has an extremely high fecundity, produces many small young, allocates a somewhat higher proportion of N-RA to reproductive effort than other freshwater species, and has a high proportion of immature individuals in its populations. The evolution of such a set of life-history tactics apparently has allowed *C. fluminea* to inhabit disturbed, unstable environments from which other freshwater bivalve species are usually eliminated. Unlike unionids and sphaeriids, this bivalve can quickly reestablish populations after they have been periodically decimated by chance massive environmental disturbances, including those that result from human activities such as the dredging and canalization of river bottoms (Kraemer, 1979; Prokopovich, 1969; Prokopovich and Hebert, 1965). Indeed, although extensive dredging and canalization projects have caused the pauperization of both unionid and sphaeriid faunas in the major drainage systems of the United States (Gardner et al., 1976; Kraemer, 1979; van der Schalie, 1973), they have provided disturbed environments in which *C. fluminea* has been most successful.

The only aspect of the biology of *C. fluminea* which does not fit accepted life-history trait hypotheses is its biennial pattern of reproduction. Animals inhabiting unstable environments optimize production and survival of offspring to maturity best by reducing the number of broods and increasing brood size (Stearns, 1976). Even though *C. fluminea* displays a reduction in the total number of broods due to its very short life span, it remains distinctly iteroparous, spawning twice annually in its 1–3-year life span. Biennial spawning increases brood number, but it has been hypothesized that this reproductive pattern has been selected for in certain highly variable environments where environmental fluctuations greatly effect juvenile survival rates. The spreading of reproductive effort over more than a single annual spawning period can prevent the loss of an entire year's reproductive effort through massive mortality of the new generation as a result of a chance environmental event unfavorable to its success; this life-history tactic was termed *bet hedging* by Stearns (1976, 1977). If spawning in *C. fluminea* were restricted to the spring, heavy flooding or other environmental stress could result in the massive mortality of highly susceptible juveniles. Because *C. fluminea* has a very short life span, such a catastrophic event would leave only a few older adults to overwinter and reproduce the following spring. Therefore, the biennial spawning pattern of *C. fluminea* increases the chance that some juveniles will be released during a period when environmental conditions are favorable for survival to maturity and, therefore, ensures the overwintering of relatively large numbers of individuals to reproduce the following year. Sphaeriid and especially unionid bivalves inhabiting more stable environments do not require a second reproductive effort in the fall to ensure overwinter survival of breeding stocks.

F. Conclusions

The major adaptations of *C. fluminea* to life in fresh water seem to have centered on the evolution of life-history traits rather than physiological mechanisms. Although physiologically similar to its estuarine ancestors (see Section IV), *C. fluminea* displays reproductive and life-history characteristics which diverge widely from those of closely related estuarine species such as *C. fluminalis*. These divergent traits include higher growth rates, higher net production efficiencies, earlier maturity, development of hermaphroditism, incubation of fertilized eggs to a well-developed benthic juvenile, passive dispersal of juveniles on water currents, reduction of the life span, and an overall reduction in the number of broods (as a result of decreased life span, although the number of broods per year is increased from one to two). These adaptations have allowed *C. fluminea* to be extremely successful in the highly unstable and variable lotic environments to which other freshwater bivalve groups such as unionids and sphaeriids appear to be generally poorly adapted.

It is important to realize that these same life-history traits make *C. fluminea* a highly successful invasive species. Thus, they not only explain to a great extent its spectacularly rapid and successful colonization of North American fresh waters (Fig. 2) but also define its role as a major pest species.

VI. Control Measures

A. Natural Environments

C. fluminea, as other benthic infaunal species, is particularly difficult to control selectively in natural environments without severely damaging other important species (Morton, 1979b). Like most bivalves, *C. fluminea* is highly resistant to biocides designed for short-term application in aquatic environments (primarily chlorination), because in response to these chemicals it closes the valves tightly to isolate its tissues from the toxic agents for several days to several weeks (Mattice, 1979; Mattice et al., 1982). Therefore, *C. fluminea* is far more tolerant of biocide application than are most freshwater species, which are unable to isolate themselves from exposure to toxins. This is particularly true of fish, which generally are far less tolerant of biocides than is *C. fluminea* (Marking and Chandler, 1978). Because no molluscicides have been approved for release into fresh water by the U.S. Environmental Protection Agency (EPA), it is presently not feasible to control *C. fluminea* in North American fresh waters by such methods. In particular, this lack of specific molluscicides will prevent the control of this species in the sand and gravel industries (Cherry et al., 1980a; Sinclair and Isom, 1961, 1963).

The comparatively heavy, strong shells of adult *C. fluminea* appear to preclude predation as a significant control measure. Muskrats (Sinclair and Isom, 1961, 1963) and raccoons (R. F. McMahon, unpublished observations) appear to take adult clams in limited numbers, but not in amounts sufficient to affect adult densities. There have also been many reports of fish feeding on juvenile and small specimens of *C. fluminea* (SL < 5.0 mm), including *Lepomis macrochirus* (blue gill sunfish), *L. microlophus* (red-ear sunfish), *Minytrema melanopus* (spotted sucker), *Aplodinotus granniens* (freshwater drum), *Acipenser* sp. (sturgeon), *Cyprinus carpio* (carp), *Ictiobus bubalus* (small-mouth buffalo fish), *I. niger* (black buffalo fish), *Ictalurus punctatus* (channel catfish), and *I. furcatus* (blue catfish) (Britton and Murphy, 1977; Dreier, 1977; Grantham, 1967; Ingram, 1959; Rinne, 1974; Sinclair and Isom, 1961, 1963). Wild ducks also have been observed to feed heavily on small specimens of *C. fluminea* (Dreier, 1977; Sinclair and Isom, 1961, 1963; Thompson and Sparks, 1977, 1978), as have the crayfish *Procambarus clarkii* and *P. bartonii* (Covich et al., 1981). There has also been a report of the symbiotic oligochaete *Chaetogaster limnaei* in the mantle-cavity of *C. fluminea* (Eng, 1976). Although predation or parasitization may limit *C. fluminea* densities in some environments, these impacts have not yet been subjected to a detailed and rigorous assessment. On the whole, such predation does not appear to be sufficient to reduce *C. fluminea* densities greatly in most North American drainage systems.

One method proposed for the control of *C. fluminea* may have some practical application in artificial impoundments and reservoirs whose water levels can be controlled. The preference of *Corbicula* for well-oxygenated sandy substrata often limits it to shallow near shore habitats which could be exposed to air by periodic water level drawdowns. Because *Corbicula* is relatively intolerant of aerial exposure (McMahon, 1979b; White, 1979; White and White, 1977a), planned drawdowns could cause nearly 100% mortality in the exposed population within 30 days (McMahon, 1979b); drawdowns are a common practice in North America as a method of controlling aquatic macrophyte growth and do not seriously damage fish stocks. Even unionid and sphaeriid bivalves appear to be generally more tolerant of desiccation than is *C. fluminea* (McMahon, 1979b; White, 1979; White and White, 1977a). Unfortunately, such drawdown procedures are applicable in only a small number of limited-use reservoirs, although planned drawdowns where feasible could be a very effective control measure. Seasonal variations in the water level of an artificial reservoir in Texas have been observed to cause massive mortalities in resident *Corbicula* populations and to prevent such species population explosions as occurred in other more stable lakes nearby (McMahon, 1979b).

B. Facilities Utilizing Raw Water

Although control of *Corbicula* in natural environments is presently impractical, limited control measures in facilities that utilize raw water appear to be feasible. In service water systems where water is not recirculated to the raw water source, such as fire protection systems, irrigation systems, and lawn-watering and flush toilet lines, high levels of chlorination can be used to kill larval clams before they can settle and grow to sizes that disrupt water flow (Goss and Cain, 1977; Goss et al., 1979; Sinclair and Isom, 1961, 1963; Ingram, 1959; A. L. Smith et al., 1979). Recommended levels of chlorination in such lines are a continuous residual of 0.5–1.0 μg Cl/liter (Sinclair and Isom, 1961, 1963) or periodic chlorination at levels of 500 μg Cl/liter for about 100 h (Goss and Cain, 1977; Goss et al., 1979; Gooch et al., 1978; A. L. Smith et al., 1979). Other control measures include screening of intake water through a mesh size less than 0.08 cm, periodic flushing of the lines (in both directions if possible), manual shell removal, and the use of mechanical clam traps at appropriate points in the system (Goss and Cain, 1977; Goss et al., 1979; A. L. Smith et al., 1979).

In those service water systems in which raw water is recirculated to the natural source (nuclear reactor cooling systems, machinery cooling, and bearing cooling), continuous chlorination is restricted to 2 in every 24 h by EPA regulations (U.S. Environmental Protection Agency, 1980). These restrictions nullify control of *C. fluminea* impingement by chlorination because this species can isolate its tissues from chlorine within closed valves for periods far exceeding 2 h (Mattice, 1979; Mattice et al., 1982). In such systems, clam fouling can be controlled by screening intake water, manual shell removal, periodic flushing of static lines, and periodic chlorination to control recruitment of juveniles by water currents (Goss and Cain, 1977; Goss et al., 1979; A. L. Smith et al., 1979). Periodic chlorination may not have to be utilized throughout the year in such systems. Instead, it has been suggested that raw water intakes be monitored for the presence of juveniles and that the reproductive condition (incubation of fertilized eggs and embryos in the gills) of adults be monitored in the field, with control measures applied only during the high-risk periods associated with spring and fall spawning activity in this species (Cherry et al., 1980a; Ingram, 1959; A. L. Smith et al., 1979). In operations where *C. fluminea* fouling can cause extremely hazardous conditions (e.g., nuclear reactor cooling systems), extensive machinery damage, or prolonged shutdowns (bearing cooling systems), the safest cooling method appears to be closed-loop systems such as cooling towers or cooling ponds in which water is never recirculated to a natural source, instead of utilizing large, multiuse reservoirs for cooling (Goss and Cain, 1977; Goss et al., 1979). In such close-loop systems, *C. fluminea* impingement could be controlled by continuous application of molluscicides. Certainly, the presence or the possible establishment of *C. fluminea* should be an important design consideration in the future development of any nuclear or fossil-fueled electrical generating stations, or in the planning of any other facility which would utilize large volumes of raw cooling water in its operations.

Control of *C. fluminea* in the steam condensers of nuclear and fossil-fueled electrical generating stations presents an even more complex problem, because such operations can involve the return of vast quantities of raw water to the natural environment. Here again, EPA regulations prevent chlorination of condenser intake water at levels which would cause quick kills and allow chlorination to be carried out for only 2 of every 24 h (U.S. Environmental Protection Agency, 1980); unfortunately, this period is too short to affect *C. fluminea* (Mattice, 1979; Mattice et al., 1982).

Clams of relatively small size entering condensers pass through primary and secondary screens into precondenser areas (embayments behind screens), where they quickly grow to SLs that allow them to become lodged inside condenser tubes, eventually restricting flow rates below those required for normal operation (McMahon, 1977). Screening juveniles from intake water is impractical, because screens sufficiently fine would impede water flow (Goss and Cain, 1977; Goss et al., 1979). Control of condenser fouling is further complicated by the fact that individuals need to be eliminated not only from the immediate area of the condenser intake system but also from any intake areas (especially intake canals) leading to the primary intake area, because these canals often form a nursery ground for impinging juveniles and adults (see Section II). Presently utilized control methods include backflushing (condenser flow reversal) to remove lodged shells, periodic shutdown for manual removal of shells from condenser tubes, and dewatering of intake canals and precondenser structures followed by manual removal of resident clam populations (Goss and Cain, 1977; Goss et al., 1979; Harvey, 1981; A. L. Smith et al., 1979). Other control measures that have been investigated either have proved impractical or do not have EPA approval, including biocides such as potassium ion (Anderson et al., 1976; Daum et al., 1979), Bayer-73 (a tropical molluscicide) (Cherry et al., 1980b; Sinclair and Isom, 1963), acrolein (Goss and Cain, 1977), chlorine dioxide (Goss et al., 1979), bromine chloride (Goss et al., 1979), copper sulfate (Ingram, 1959; Sinclair and Isom, 1961), antimycin (a fish toxicant) (Marking and Chandler, 1978), and, most unusual of all, gamma radiation from a ^{60}Co source (Tilly et al., 1978).

The procedures presently utilized to control biofouling of condensers by *C. fluminea* involve expensive periodic shutdowns of the generating units associated with them. Other more effective, less expensive, and less time-consuming control measures should be developed. Some recommendations include use of controlled-release surfaces containing biocides such as triphenyl lead acetate or tributyl tin oxide to prevent juvenile settlement (Goss et al., 1979). However, such methods would be of little use in the control of *C. fluminea* in condensers, because adults enter condenser tubes on intake currents and do not grow or survive in the condenser itself (McMahon, 1977). A more promising and efficient method of control, recently suggested, involves the utilization of heated waste waters from the condenser itself (Mattice, 1979; Mattice et al., 1982) as the control agent. *C. fluminea* has a surprisingly low long-term upper lethal limit of approximately 34°C (Cherry et al., 1980b; Habel, 1970; Mattice and Dye, 1976; Mattice, 1979; Mattice et al., 1982; Mudkhede and Nagabhushanam, 1977) and a short-term upper lethal limit of 43–45°C (Isom et al., 1978; Mattice and Dye, 1976; McMahon, 1979a). If periodic diversions of heated effluents from condensers into intake structures could be utilized to raise water temperatures in precondenser areas and intake canals above the long-term upper lethal limit of 35°C for 20–30 days or above the short-term upper lethal limit of 43°C for as little as 30 min, nearly 100% mortality in resident *C. fluminea* populations could be produced with little or no harm to the natural environment (Goss et al., 1979; Mattice, 1979; Mattice et al., 1982). Such high temperatures could be achieved by backflushing heated waters through condensers (Goss et al., 1979; Mattice, 1979; Mattice et al, 1982) or by recirculating heated effluents into intake areas (Mattice, 1979; Mattice et al., 1982). If these "heat shock" procedures were carried out periodically, individuals in precondenser areas would never grow to a size that would allow them to become lodged in condenser tubes.

Similarly, the rather high lower lethal temperature of 2°C (Mattice and Dye, 1976) would allow effective control of *C. fluminea* in the northern United States, where ambient water temperatures fall below 2°C for extended periods in the winter. In such environments, *C. fluminea* populations survive only in areas receiving heated effluents from power plants, because these areas remain above the lower lethal temperature (Dreier, 1977; Dreier and Tranquilli, 1981; Eckblad, 1975; Rodgers et al., 1977, 1979). Indeed, massive winter mortalities of *C. fluminea* populations have been reported in northern populations when winter water temperatures fell below 2°C (Bickel, 1966; Dreier, 1977; Dreier and Tranquilli, 1981; Horning and Keup, 1964; Rodgers et al., 1977, 1979). Therefore, *C. fluminea* populations associated with power plants in more northern regions could be controlled by periodic plant shutdowns during the colder winter months. This procedure would effectively eliminate the overwintering population, which is normally restricted to areas receiving heated effluents (the only areas remaining above the lower limit of 2°C).

Another suggested design consideration for the control of biofouling of condenser tubes involves constructing condensers with tubes of internal diameters larger than the SLs of the largest individuals of *C. fluminea* in the resident population (McMahon, 1977). Such condenser tubes would be too large to allow shells to become lodged in them; invading shells would simply pass through them, to be removed on postcondenser water currents. McMahon (1977) has determined that condenser tubes with an inside diameter of 29.0 mm would allow clams of up to 55 mm SL to pass through. Therefore, these condenser tubes would not be subject to fouling in most aquatic environments harboring *C. Fluminea,* because individuals of this species rarely exceed 40 mm SL (see Section V,A).

In environments where *C. fluminea* populations are dense and vigorous, fouling problems in steam condensers are best avoided by closed-loop cooling systems such as cooling towers or contained cooling ponds, which neither draw from nor return condenser waters to a natural raw water source. Such systems can be continuously chlorinated to prevent the establishment of *C. fluminea* populations (Goss et al., 1979). However, such solutions to the problem of *C. Fluminea* condenser fouling are more expensive to design and often energetically less efficient than raw-water cooling systems. For facilities already in use, biofouling by *C. fluminea* continues to be an expensive and exasperating problem for which there are now no universally accepted remedies.

VII. Future Utilization of *Corbicula*

C. fluminea now appears to have approached its distributional limits in North America, although it is still spreading slowly north on the Atlantic seaboard (Trama, 1982). Certainly, it has become an important and permanent member of the benthic communities of many North American freshwater drainage systems, often representing a significant proportion of the biomass of second trophic level organisms. Because it appears to be permanently established in North America, it is important to consider not only the pest aspects of this species but also the benefits that may be realized from its presence in American fresh waters.

As colonizers of lotic environments not normally utilized by sphaeriids and unionids, dense populations of *C. fluminea* may present a new and important food resource to commercially valuable and game fish (Britton and Murphy, 1977; Dreier, 1977; Grantham, 1967; Ingram, 1959; Rinne, 1974; Sinclair and Isom, 1961, 1963; Villadolid and Del Rosario, 1930). *C. fluminea* has a protein content which allows maximal growth in vertebrates (Thompson and Sparks, 1978). It also appears to serve as an admirable food source for migratory waterfowl (Dreier, 1977; Thompson and Sparks, 1977), particularly in drainage systems where the sphaeriid clam populations normally fed upon by waterfowl have recently declined significantly in density (Thompson and Sparks, 1978).

C. fluminea can also be used as a cheap, readily available game fish bait (Cherry et al., 1980a; Ingram, 1959; Mattice, 1979; Sinclair and Isom, 1961, 1963; Villadolid and Del Rosario, 1930) and is sold commercially for this purpose on the West Coast of North America (Ingram, 1959). Unfortunately, selling this species as bait and as a "dwarf clam" to tropical fish hobbyists (Abbott, 1975) has ensured its spread in North American fresh waters.

An excellent protein source (Thompson and Sparks, 1978), *C. fluminea* is utilized extensively as a human and domestic animal food resource (Britton and Morton, 1979; Miller and McClure, 1931; Morton, 1977a, 1979a; Villadolid and Del Rosario, 1930) and is the subject of intensive aquaculture in Southeast Asia (Chen, 1976; Villadolid and Del Rosario, 1930). Although it is improbable that North Americans will ever develop a taste for *C. fluminea,* it could be used as a protein supplement in domestic livestock feeds (Cherry et al., 1980a; Mattice, 1979; Sinclair and Isom, 1963). The shells of *C. fluminea* could also provide a source of lime for poultry feeds (Mattice, 1979) and fertilizers (Miller and McClure, 1931; Villadolid and Del Rosario, 1930).

The high growth rates and net production efficiencies of *C. fluminea* seem ideally suited to managed aquaculture systems. Its high filtration rate (Mattice, 1979) could allow it to act as a clarifying agent, removing particulate organic matter from fish aquaculture systems (Busch, 1974; Habel, 1970; Mattice, 1979), from the effluents of livestock feedlots (Mattice, 1979), and from secondarily treated sewage waste waters (Cherry et al., 1980b; Grantham, 1967; Greer and Ziebell, 1972; Haines, 1979). It has even been suggested that heated discharge water from electrical power stations could be diverted into *C. fluminea* aquacultural or clarifying systems to sustain productivity at maximal levels throughout the year (Mattice, 1979). Unfortunately, *C. fluminea* is extremely intolerant of both the high temperatures (> 30°C) (Mattice and Dye, 1976; McMahon, 1979a) and the low oxygen tensions (Fast, 1971; McMahon, 1979a) that characterize such aquaculture or waste water treatment systems. In fact, high temperatures and hypoxia have been reported to have caused failures of small-scale *C. fluminea*-based clarifying systems in both fish aquaculture (Busch, 1974; Habel, 1970) and tertiary sewage treatment systems (Greer and Ziebell, 1972; Haines, 1979). Such results indicate that *C. fluminea* may not be as potentially adaptable to managed aquaculture systems as it originally seemed to be.

A more likely use for *C. fluminea* is as a bioassay or bioindicator organism. Its wide distribution in a variety of aquatic habitats, ease of collection and maintenance, ability to be shipped without water, and high resistance to various toxins all make it potentially adaptable to such studies (Burress and Chandler, 1976; Cherry et al., 1980a; Clark et al., 1979; Grantham, 1967; Horne and McIntosh, 1979; Sinclair, 1964, 1974; Sinclair and Isom, 1961, 1963; Weber, 1973). *C. fluminea* has already been utilized in laboratory bioassay studies of tolerance to organic pollution (Horne and McIntosh, 1979; Weber, 1973), heavy metals (Cherry et al, 1980a,b; Graney et al., 1978), various biocides and toxic chemicals (Anderson et al., 1976; Daum et al., 1979; Cherry et al., 1980a,b; Goss and Cain, 1977; Goss et al., 1979; Ingram, 1959; Marking and Chandler, 1978;

Sinclair and Isom, 1963), and gamma radiation (Tilly et al., 1978). Juvenile *C. fluminea* appear to be potentially more useful than adults for these studies, because they are unable to isolate their tissues from the external environment; also, death is easily determined by observing the cessation of heartbeat through the transparent shell (Grantham, 1967; Sinclair and Isom, 1961, 1963). Adult *C. fluminea* may be poorer subjects for bioassay than juveniles, because they can isolate their tissues from surrounding media by valve closure; also, death is not always associated with immediate valve gaping as it is in other bivalve species (Mattice, 1979; Mattice and Dye, 1976; McMahon, 1979b).

C. fluminea has also been used as a bioindicator species for testing environmental levels of both organic pollution [its presence indicates waters with generally low levels of organic pollution (Horne and McIntosh, 1979; Weber, 1973)] and the environmental concentration of heavy metals such as lead, copper, and zinc, which appear to be bioaccumulated in both tissues and shells (Clark et al., 1979; Graney et al., 1978). Such steady accumulation of baseline data on the tolerance of *C. fluminea* to pollutants and toxins, and the increasing familiarity of environmental biologists with this species, will eventually make it a valuable addition to the battery of freshwater test species utilized as bioindicator organisms.

Perhaps the most imaginative suggestion for utilization is that *C. fluminea* could replace unionid species as the common laboratory representative of the class Bivalvia in high schools, colleges, and universities throughout North America (Britton and Morton, 1982). Many North American unionid species are presently greatly endangered by human activities, and their continued utilization in classroom laboratories could eventually lead to the unknowing elimination of rare species. Because *C. fluminea* is equally or more available, as easily collected, and as tolerant of shipping without water as most unionids, it should make an excellent source of live and preserved bivalve material for commercial biological suppliers. To stimulate classroom utilization of *C. fluminea,* Britton and Morton (1982) have recently published a detailed manual of the anatomy and dissection of this species and its application to simple teaching laboratory exercises.

VIII. Summary and Conclusions

C. fluminea is now a permanent resident in North American fresh waters, ranging from northern Mexico (Taylor, 1981; B. Morton, personal communication) to southern Canada (Taylor, 1981). Fortunately, it has not impacted endemic unionid and sphaeriid bivalve species as greatly as had been previously feared, because its preferred lotic, unstable habitats are not those in which native species are particularly successful. In this regard, Taylor (1981) has recently stated that *C. fluminea* is "probably not a threat to native fauna" on the West Coast of North America, where this species has been established for more than 40 years. Continued human interference in natural drainage systems will guarantee the future success and spread of *C. fluminea* and the recurrence of the biofouling problems associated with it in agricultural and industrial facilities utilizing raw water. Although controlling *C. fluminea* in natural environments is probably not feasible, it appears that the problems associated with its biofouling of facilities utilizing raw water can usually be rectified by a series of relatively effective but expensive and time-consuming control procedures. Engineers are now well aware of the potential problems associated with *C. fluminea* biofouling and will eventually develop the appropriate designs required to minimize the impact of this species on any new facilities utilizing raw water (Goss and Cain, 1977; Goss et al., 1979; A. L. Smith et al., 1979).

The presence of *C. fluminea* in North America may also have some unexpected benefits, including a new and abundant food resource for commercially and recreationally important fish and waterfowl species; a species adaptable to aquaculture systems; a clarifier for tertiary sewage treatment systems; removal of particulate organic material from natural waters; a potentially important bioassay and bioindicator species; and perhaps a new standard specimen for biology teaching laboratories.

This chapter has also focused on another potential advantage of the presence of *C. fluminea* in North American fresh waters not yet discussed, its remarkable adaptability to a wide range of biological studies and experiments. More than 200 scientific papers on the biology of *C. fluminea* have been published since the first North American specimens were collected in the Columbia River in 1938 (Burch, 1944). *C. fluminea* has provided North American biologists with a highly recognizable, accessible, and widespread species that lends itself to experimental investigation by being easily collected in the field and maintained in the laboratory. Because its ecology, life-history traits, reproductive biology, physiology, and evolutionary history are all basically different from those of endemic North American bivalve species, *C. fluminea* presents a unique opportunity for comprehensive comparative investigations. Indeed, the presence of this species in North America now appears to have renewed scientific interest in native unionid and sphaeriid bivalves at a time when many of these species are highly endangered.

There is now enough published information on *C. fluminea* to provide a solid foundation for further, more intensive studies of the biology of this remarkable species. Among the many areas of research that could provide valuable information in the future are the sexual cycle and mode of hermaphroditism; habitat preference and further evaluation of ecological niche; competition with native bivalve species; modes of dispersal; effects of human disturbance on colonization success; the behavioral, physiological, and ecological traits associated with *C.*

fluminea biofouling and its control; development of standard bioassay and bioindicator techniques; development of environmentally acceptable control measures; physiological response to environmental variables and acclimation; further investigations of oxygen uptake, osmoregulation, ionic regulation, temperature tolerance, and biological rhythmicity; studies of aerial exposure tolerance; evaluation of growth rates, reproductive periods, life span, and energy budgets for a series of North American populations relative to the extraordinary geographic variation in life-history traits reported for this species; and studies of juvenile incubation, settlement, and dispersal behavior.

Of course, such suggestions represent only a small proportion of the future research opportunities provided by the presence of this unique species in North America. Rarely has so much information been available for a single freshwater molluscan species as for *C. fluminea,* and rarely has such a species been so universally available to so many investigators. If our knowledge of this species continues to accumulate at its present rate, it is not unlikely that, as a research organism, *C. fluminea* may soon become the freshwater equivalent of *Mytilus edulis,* the marine intertidal mussel. Like *M. edulis, C. fluminea* is universally available, and its populations are highly accessible, easily collected, and easily maintained in the laboratory. For these reasons it may, as *M. edulis* in marine studies, become the apparent freshwater "species of choice" in an extraordinarily wide range of future ecological, physiological, behavioral, and environmental investigations.

Acknowledgments

This chapter was prepared while my research on *Corbicula* was supported by grants from the Texas Electric Service Company and Organized Research Funds of The University of Texas at Arlington. I wish to express my gratitude to Brian Morton of Hong Kong University and Joseph C. Britton of Texas Christian University for their timely advice and discussions on the biology of *C. fluminea,* and to Roger Byrne, Della Tyson and Colette O'Byrne McMahon for their assistance with the preparation of this manuscript.

References

Abbott, R. T. (1975). Beware the Asiatic freshwater clam. *Trop. Fish Hobbyists* **23,** 15.

Abbott, T. M. (1979). Asiatic clam (*Corbicula fluminea*) vertical distributions in Dale Hollow Reservoir, Tennessee. *Proc.—Int. Corbicula Symp., 1st, 1977* pp. 111–118.

Aldridge, D. W. (1976). Growth, reproduction and bioenergetics in a natural population of the Asiatic freshwater clam, *Corbicula manilensis* Philippi. Master's Thesis, University of Texas at Arlington.

Aldridge, D. W., and McMahon, R. F. (1978). Growth, fecundity, and bioenergetics in a natural population of the freshwater clam, *Corbicula manilensis* Philippi, from North Central Texas. *J. Moll. Stud.* **44,** 49–70.

Anderson, K. B., Thompson, C. M., Sparks, R. E., and Paparo, A. A. (1976). Effects of potassium on adult Asiatic clams, *Corbicula manilensis*. *Biol. Notes (Ill. Nat. Hist. Surv.)* **98,** 1–7.

Avolizi, R. J. (1971). Biomass turnover in natural populations of viviparous sphaeriid clams: Interspecific and intraspecific comparisons of growth, fecundity, mortality, and biomass production. Ph.D. Dissertation, Syracuse University, Syracuse, New York (*Diss. Abstr.* **33,** 1918–B).

Avolizi, R. J. (1976). Biomass turnover in populations of viviparous sphaeriid clams: Comparisons of growth, fecundity, mortality and biomass production. *Hydrobiologia* **51,** 163–180.

Bates, J. M. (1962). Extension of the range of *Corbicula fluminea* within the Ohio Drainage. *Nautilus* **76,** 35–36.

Berrie, A. D. (1959). Variation in the radula of the freshwater snail *Lymnaea peregra* (Müller) from Northwestern Europe. *Ark. Zool.* **12,** 391–404.

Berry, A. J. (1974). Freshwater bivalves of Peninsular Malaysia with special reference to sex and breeding. *Malay. Nat. J.* **27,** 99–110.

Bickel, D. (1966). Ecology of *Corbicula manilensis* Philippi in the Ohio River at Louisville, Kentucky. *Sterkiana* **23,** 19–24.

Bloomer, H. H. (1939). A note on the sex of *Pseudoanodonta* Bourguiquat and *Anodonta* Lamarck. *Proc. Malacol. Soc. London* **23,** 285–297.

Boozer, A. C., and Mirkes, P. E. (1979). Observations on the fingernail clam, *Musculium partumeium* (Pisidiidae), and its association with the introduced Asiatic clam, *Corbicula fluminea*. *Nautilus* **93,** 73–83.

Branson, B. A., and Batch, D. L. (1969). Notes on exotic mollusks in Kentucky. *Nautilus* **82,** 102–106.

Britton, J. C., and Morton, B. (1979). *Corbicula* in North America: The evidence reviewed and evaluated. *Proc. Int. Corbicula Symp., 1st, 1977* pp. 250–287.

Britton, J. C., and Morton, B. (1982). A dissection guide, field, and laboratory manual to the introduced bivalve *Corbicula fluminea*. *Malacol. Rev.*, Suppl. 3, 82 pp.

Britton, J. C., and Murphy, C. E. (1977). New records and ecological notes on *Corbicula manilensis* in Texas. *Nautilus* **91,** 20–23.

Britton, J. C., Coldiron, D. R., Evans, L. P., Jr., Golightly, C., O'Kane, K. D., and TenEyck, J. R. (1979). Reevaluation of the growth pattern in *Corbicula fluminea* (Müller). *Proc.—Int. Corbicula Symp., 1st, 1977* pp. 177–192.

Burch, J. Q. (1944). Check list of West North American marine mollusca from San Diego, Calif., to Alaska. *Min. Conchol. Club South. Calif.* **38,** 11–18.

Burky, A. J., and Burky, K. A. (1976). Seasonal respiratory variation and acclimation in the pea clam, *Pisidium walkeri* Sterki. *Comp. Biochem. Physiol. A* **55,** 109–114.

Burress, R. M., and Chandler, J. H., Jr. (1976). Use of the Asiatic clam, *Corbicula leana* Prime, in toxicity tests. *Prog. Fish-Cult.* **38,** 10.

Burton, D. T. (1980). Biofouling procedures for power plant cooling water systems. *In* "Condenser Biofouling Control" (J. F. Garey, R. M. Jordan, A. H. Aitken, D. T. Burton, and R. H. Gray, eds.), pp. 251–266. Ann Arbor Sci. Publ./Butterworth Group, Ann Arbor, Michigan.

Busch, R. L. (1974). Asiatic clams *Corbicula manilensis* Philippi as biological filters in channel catfish, *Ictalurus punctatus* (Rafinesque) cultures. Master's Thesis, Auburn University 84 pp. Auburn, Alabama.

Buttner, J. K., and Heidinger, R. C. (1980). Seasonal variations in growth of the Asiatic clam, *Corbicula fluminea* (Bivalvia: Corbiculidae) in a Southern Illinois fish pond. *Nautilus* **94,** 8–10.

Chen, T. P. (1976). "Aquaculture Practices in Taiwan." Page Brothers, Norwich.

Cherry, D. S., Cairns, J., Jr., and Graney, R. L. (1980a). Asiatic clam invasion: Causes and effects. *Water Spectrum* **12,** 18–24.

Cherry, D. S., Rodgers, J. H., Jr., Graney, R. L., and Cairns, J., Jr. (1980b). Dynamics and control of the Asiatic clam in the New River, Virginia. *Bull.—Va. Water Resour. Res. Cent.* **123,** 1–72.

Clark, J. H., Clark, A. N., Wilson, D. J., and Friauf, J. J. (1979). On the use of *Corbicula fluminea* as indicators of heavy metal contamination. *Proc. Int. Corbicula Symp. 1st, 1977* pp. 153–163.

Clarke, A. H. (1981). *Corbicula fluminea,* in Lake Erie. *Nautilus* **95,** 83–84.

Clench, W. J. (1970). *Corbicula manilensis* (Philippi) in Lower Florida. *Nautilus* **84,** 36.

Clench, W. J. (1972). *Corbicula manilensis* (Philippi) in Oklahoma. *Nautilus* **85,** 145.

Collins, T. W. (1967). Oxygen-uptake, shell morphology and desiccation of the fingernail clam *Sphaerium occidentale* Prime. Ph.D. Dissertation, University of Minnesota, Minneapolis (*Diss. Abstr.* **28B,** 5238).

Cooper, C. M., and Johnson, V. W. (1980). Bivalve mollusca of the Yalobusha River, Mississippi. *Nautilus* **94,** 22–24.

Copeland, B. J., Tenore, K. R., and Horton, D. B. (1974). Oligohaline regime. *In* "Coastal Ecological Systems of the United States" (H. T. Odum, B. J. Copeland and E. A. McMahan, ed.), Vol. II, pp. 315–357. The Conservation Foundation, in cooperation with the National Oceanic and Atmospheric Administration, Office of Coastal Environment, Washington, D.C.

Counts, C. L., III (1981). *Corbicula fluminea* (Bivalvia: Sphaeriacea) in British Columbia. *Nautilus* **95,** 12–13.

Counts, C. L., III, and Prezant, R. S. (1979). Shell structure and histochemistry of the mantle of *Corbicula leana* Prime, 1864 (Bivalvia: Sphaeriacea). *Am. Zool.* **19,** 1007.

Covich, A. P., Dye, L. L., and Mattice, J. S. (1981). Crayfish predation on *Corbicula* under laboratory conditions. *Am. Midl. Nat.* **105,** 181–188.

Dance, S. P. (1958). Drought resistance in an African freshwater bivalve. *J. Conchol.* **24,** 281–282.

Daum, K. A., Newland, L. W., Britton, J. C., Champagne, L., and Hagen, J. (1979). Responses of *Corbicula* to potassium. *Proc. Int. Corbicula Symp., 1st, 1977,* pp. 215–225.

Diaz, R. J. (1974). Asiatic clam, *Corbicula manilensis* (Philippi) in the tidal James River, Virginia. *Chesapeake Sci.* **15,** 118–120.

Dietz, T. H. (1974). Body fluid composition and aerial oxygen consumption in the freshwater mussel, *Ligumia subrostrata* (Say): Effects of dehydration and anoxic stress. *Biol. Bull.* (*Woods Hole, Mass.*) **147,** 560–572.

Dietz, T. H. (1977). Solute and water movement in freshwater bivalve mollusks. *In* "Water Relations in Membrane Transport in Plants and Animals" (A. M. Jungreis, T. K. Hodges, A. Kleinzeller, and S. G. Schultz, eds.), pp. 111–119. Academic Press, New York.

Dietz, T. H. (1979). Uptake of sodium and chloride by freshwater mussels. *Can. J. Zool.* **57,** 156–160.

Dreier, H. (1977). Study of *Corbicula* in Lake Sangchris. *In* "The Annual Report for Fiscal Year 1976, Lake Sangchris Project," Sect. 7, pp. 7.1–7.52. Illinois Natural History Survey, Urbana.

Dreier, H., and Tranquilli, J. A. (1981). Reproduction, growth, distribution and abundance of *Corbicula* in an Illinois cooling lake. *Bull.—Ill. Nat. Hist. Surv.* **32,** 378–393.

Dudgeon, D. (1980). A comparative study of the Corbiculidae of southern China. *In* "Proceedings of the First International Workshop on the Malacofauna of Hong Kong and Southern China" (B. Morton, ed.), pp. 37–60. Hong Kong Univ. Press, Hong Kong.

Dundee, D. S., and Dundee, H. A. (1958). Extensions of known ranges of four mollusks. *Nautilus* **72,** 51–53.

Dundee, D. S., and Harman, W. J. (1963). *Corbicula fluminea* (Müller) in Louisiana. *Nautilus* **77,** 30.

Eckblad, J. W. (1974). The Asian clam *Corbicula* in the Upper Mississippi River. *Nautilus* **89,** 4.

Eng, L. L. (1976). A note on the occurrence of a symbiotic oligochaete, *Chaetogaster limnaei*, in the mantle cavity of the Asiatic clam, *Corbicula manilensis. Veliger* **19,** 208.

Eng. L. L. (1979). Population dynamics of the Asiatic clam, *Corbicula fluminea* (Müller), in the concrete-lined Delta–Mendota Canal of central California. *Proc.—Int. Corbicula Symp., 1st, 1977* pp. 39–68.

Evans, L. P., Jr., Murphy, C. E., Britton, J. C., and Newland, L. W. (1979). Salinity relationships in *Corbicula fluminea* (Müller). *Proc.—Int. Corbicula Symp., 1st, 1977* pp. 193–214.

Fast, A. W. (1971). The invasion and distribution of the Asiatic clam (*Corbicula manilensis*) in a southern California reservoir. *Bull. South. Calif. Acad. Sci.* **70,** 91–98.

Filice, F. P. (1958). Invertebrates from the estuarine portion of San Francisco Bay and some factors influencing their distributions. *Wasmann J. Biol.* **16,** 159–211.

Fitch, J. E. (1953). *Corbicula fluminea* in the Imperial Valley. *Min. Conchol. Club South. Calif.* **130,** 9–10.

Fox, R. O. (1969). The *Corbicula* story: A progress report. *2nd Annu. Meet., West. Soc. Malacol., 1969* pp. 1–11.

Fox, R. O. (1970). *Corbicula* in Baja California. *Nautilus* **83,** 145.

Fretter, V., and Graham, A. (1964). Reproduction. *In* "Physiology of Mollusca" (K. M. Wilbur and C. M. Yonge, eds.), Vol. 1, pp. 127–164. Academic Press, New York.

Fuller, S. L. H., and Imlay, M. J.(1976). Spatial competition between *Corbicula manilensis* (Philippi), the Chinese clam (Corbiculidae), and fresh-water mussels (Unionidae) in the Waccamaw River Basin of the Carolinas (Mollusca: Bivalvia). *ASB (Assoc. Southeast. Biol.) Bull.* **23,** 60.

Fuller, S. L. H., and Powell, C. E., Jr. (1973). Range extensions of *Corbicula manilensis* (Philippi) in the Atlantic Drainage of the United States. *Nautilus* **87,** 59.

Fuziwara, T. (1975). On the reproduction of *Corbicula leana* Prime. *Venus* **34,** 54–56.

Fuziwara, T. (1977). On the growth of young shell of *Corbicula leana* Prime. *Venus* **36,** 19–24.

Gainey, L. F., Jr. (1978a). The response of the Corbiculidae (Mollusca: Bivalvia) to osmotic stress: The organismal response. *Physiol. Zool.* **51,** 68–78.

Gainey, L. F., Jr. (1978b). The response of the Corbiculidae (Mollusca: Bivalvia) to osmotic stress: The cellular response. *Physiol. Zool.* **51,** 79–91.

Gainey, L. F., Jr., and Greenberg, M. J. (1977). Physiological basis of the species abundance–salinity relationship in molluscs: A speculation. *Mar. Biol. (Berlin)* **40,** 41–49.

Gardner, J. A., Jr., Woodall, W. R., Jr., Staats, A. A., Jr., and Napoli, J. F. (1976). The invasion of the Asiatic clam (*Corbicula manilensis* Philippi) in the Altamaha River, Georgia. *Nautilus* **90,** 117–125.

Giesy, J. P., and Dickson, G. W. (1981). The effect of season and location on phosphoadenylate concentrations and adenylate energy charge in two species of freshwater clams. *Oecologia* **49,** 1–7.

Gooch, C., Isom, B. G., Moses, J., and Niell, L. (1978). "*Corbicula* Chlorine Bioassay—*Corbicula* Control Project," Final Rep. No. I-WQ-78-10. Tennessee Valley Authority, Division of Environmental Planning, Water Quality and Ecology Branch, Muscle Shoals, Alabama.

Goss, L. B., and Cain, C., Jr. (1977). Power plant condenser and service water system fouling by *Corbicula,* the Asiatic clam. *In* "Biofouling Control Procedures" (L. D. Jensen, ed.), pp. 11–17. Dekker, New York.

Goss, L. B., Jackson, J. M., Flora, H. B., Isom, B. G., Gooch, C., Murray, S. A., Burton, C. G., and Bain, W. S. (1979). Control studies on *Corbicula* for steam–electric generating plants. *Proc.—Int. Corbicula Symp., 1st, 1977* pp. 139–151.

Graney, R. L., Rodgers, J. H., Jr., Cherry, D. S., Dickson, K. L., and Carins, J., Jr. (1978). Heavy metal accumulation by the Asiatic clam (*Corbicula manilensis*) from field collections and laboratory bioassays. *Va. J. Sci.* **29,** 61.

Grantham, B. J. (1967). The Asiatic clam in Mississippi. *In* "Proceedings of the Mississippi Water

Resources Conference,'' pp. 81–85. Mississippi Water Resources Institute, Mississippi State University, Jackson.

Green, J. (1968). ''The Biology of Estuarine Animals.'' University of Washington, Seattle.

Greer, D. E., and Ziebell, C. D. (1972). Biological removal of phosphates from water. *J. Water Pollut. Control. Fed.* **12,** 2342–2348.

Gunning, G. E., and Suttkus, R. D. (1966). Occurrence and distribution of Asiatic clam, *Corbicula leana,* in Pearl River, Louisiana. *Nautilus* **79,** 113–116.

Haas, F. (1969). Super Family Unionacea Fleming, 1828. *In* ''Treatise on Invertebrate Paleontology, Part N, Mollusca 6'' (R. C. Moore, ed.), pp. 411–467. Geol. Soc. Am., Boulder, Colorado.

Habel, M. L. (1970). Oxygen consumption, temperature tolerance, and filtration rate of the introduced Asiatic clam *Corbicula manilensis* from the Tennessee River. Master's Thesis Auburn University, Auburn, Alabama.

Haertel, L., and Osterberg, C. (1967). Ecology of zooplankton, benthos and fishes in the Columbia River Estuary. *Ecology* **48,** 459–472.

Haines, K. C. (1979). The use of *Corbicula* as a clarifying agent in experimental tertiary sewage treatment process on St. Croix, U.S. Virgin Islands. *Proc.—Int. Corbicula Symp., 1st, 1977* pp. 165–175.

Hamwi, A. (1969). Oxygen consumption and pumping rate of the hard clam, *Mercenaria mercenaria* L. Ph.D. Dissertation, Rutgers University, New Brunswick, New Jersey (*Diss. Abstr.* **30** 3433B).

Harvey, R. S. (1981). Recolonization of reactor cooling water system by the clam *Corbicula fluminea. Nautilus* **95,** 131–136.

Heard, W. H. (1964). *Corbicula fluminea* in Florida. *Nautilus* **77,** 105–107.

Heinsohn, G. E. (1958). Life history and ecology of the freshwater clam, *Corbicula fluminea.* Master's Thesis, University of California, Santa Barbara.

Hiscock, I. D. (1953). Osmoregulation in Australian freshwater mussels (Lamellibranchiata). I. Water and chloride exchange in *Hybridella australis* (Lam.). *Aust. J. Mar. Freshwater Res.* **1,** 317–329.

Hornbach, D. J., Way, C. M., and Burky, A. J. (1980). Reproductive strategies in a freshwater sphaeriid clam, *Musculium partumeium* (Say), from a permanent and a temporary pond. *Oecologia* **44,** 164–170.

Horne, F. R., and McIntosh, S. (1979). Factors influencing distribution of mussels in the Blanco River of Central Texas. *Nautilus* **94,** 119–133.

Horning, W. B., and Keup, L. (1964). Decline of the Asiatic clam in the Ohio River. *Nautilus* **78,** 29–30.

Hubricht, L. (1966). *Corbicula manilensis* (Philippi) in the Alabama River System. *Nautilus* **80,** 32–33.

Hunter, R. D. (1972). Energy budgets and physiological variation in natural populations of the freshwater pulmonate, *Lymnaea palustris.* Ph.D. Dissertation, Syracuse University, Syracuse, New York (*Diss. Abstr.* **33,** 5066B).

Hunter, R. D. (1975). Variation in populations of *Lymnaea palustris* in Upstate New York. *Am. Midl. Nat.* **94,** 401–420.

Ibarra, J. A., and McMahon, R. F. (1981). Interpopulation variation in shell morphometrics and thermal tolerance of *Physa virgata* Gould (Mollusca: Pulmonata). *Am. Zool.* **21,** 1020.

Ikematsu, W., and Kamakura, M. (1976). Ecological studies of *Corbicula leana* Prime. I. On the reproductive season and growth. *Bull. Fac. Agric., Miyazaki Univ.* **22,** 185–195.

Ikematsu, W., and Yamane, S. (1977). Ecological studies of *Corbicula leana* Prime. III. On spawning throughout the year and self-fertilization in the gonad. *Bull. Jpn. Soc. Sci. Fish.* **43,** 1139–1146.

Ingram, W. M. (1948). The larger freshwater clams of California, Oregon and Washington. *J. Entomol. Zool.* **40,** 72–93.

Ingram, W. M. (1959). Asiatic clams as potential pests in California water supplies. *J. Am. Water Works Assoc.* **51,** 363–370.

Ingram, W. M., Keup, L., and Henderson, C. (1964). Asiatic clams at Parker, Arizona. *Nautilus* **77,** 121–125.

Isom, B. G. (1974). Mussels of the Green River, Kentucky. *Trans. Ky. Acad. Sci.* **35,** 55–57.

Isom, B. G., Gooch, C., Neill, L. T., and Moses, J. (1978). "Acute Thermal Effects on Asiatic Clams (*Corbicula manilensis* Philippi)," Rep. No. I-WQ-78-12. TVA Division of Environmental Planning, Special Projects and Research Program, Water Quality and Ecology Branch, Muscle Shoals, Alabama.

Jenkinson, J. J. (1979). The occurrence and spread of *Corbicula manilensis* in East-Central Alabama. *Nautilus* **94,** 149–153.

Johnson, M. G., and Brinkhurst, R. O. (1971). Association and species diversity in benthic macroinvertebrates of Bay of Quinte and Lake Ontario. *J. Fish. Res. Board Can.* **28,** 1699–1714.

Junk, W. J. (1975). The bottom fauna and its distribution in Bung Borapet, a reservoir in Central Thailand. *Verh.—Int. Ver. Theor. Angew. Limnol.* **19,** 1935–1946.

Keen, M., and Casey, R. (1969). Family Corbiculidae Gray, 1847. *In* "Treatise on Invertebrate Paleontology, Part N, Mollusca 6" (R. C. Moore, ed.), pp. 664–669. Geol. Soc. Am., Boulder, Colorado.

Keen, M., and Dance, P. (1969). Family Pisidiidae Gray, 1857. *In* "Treatise on Invertebrate Paleontology, Part N, Mollusca 6" (R. C. Moore, ed.), pp. 669–670. Geol. Soc. Am., Boulder, Colorado.

Kennedy, V. S., and Mihursky, J. A. (1971). Upper temperature tolerances of some estuarine bivalves. *Chesapeake Sci.* **12,** 193–204.

Kennedy, V. S., and Mihursky, J. A. (1972). Effects of temperature on the respiratory metabolism of three Chesapeake Bay bivalves. *Chesapeake Sci.* **13,** 1–22.

Keup, L., Horning, W. B., and Ingram, W. M. (1963). Extension of range of Asiatic clam to Cincinnati Reach of the Ohio River. *Nautilus* **77,** 18–21.

Klippel, W. E., and Parmalee, P. W. (1979). The naiad fauna of Lake Springfield, Illinois: An assessment after two decades. *Nautilus* **94,** 189–197.

Kraemer, L. R. (1977). Aspects of the functional morphology of the mantle/shell and mantle/gill complex of *Corbicula* (Bivalvia: Sphaeriacea: Corbiculidae). *Bull. Am. Malacol. Union, Inc., 1977* pp. 25–31.

Kraemer, L. R. (1978a). Discovery of two distinct kinds of statocysts in freshwater bivalved mollusks: Some behavioral implications. *Bull. Am. Malacol. Union, Inc., 1978* pp. 24–28.

Kraemer, L. R. (1978b). *Corbicula fluminea* (Bivalvia: Sphaeriacea): The functional morphology of its hermaphroditism. *Bull. Am. Malacol. Union, Inc., 1978* pp. 40–49.

Kraemer, L. R. (1979). *Corbicula* (Bivalvia: Sphaeriacea) *vs.* indigenous mussels (Bivalvia: Unionacea) in U.S. Rivers: A hard case for interspecific competition? *Am. Zool.* **19,** 1085–1096.

Kraemer, L. R., and Lott, S. (1977). Microscopic anatomy of the visceral mass of *Corbicula* (Bivalvia: Sphaeriacea). *Bull. Am. Malacol. Union, Inc., 1977* pp. 48–56.

Lenat, D. R., and Weiss, C. M. (1973). "Distribution of Benthic Macroinvertebrates in Lake Wylie North Carolina–South Carolina," Dep. Environ. Sci. Eng., Publ. No. 331. School of Public Health, University of North Carolina at Chapel Hill, Chapel Hill.

Leveque, C. (1973a). Dynamiquedes peuplements biologie et estimation de la production des mollusques benthiques du Lac Tchad. *Cah. ORSTOM, Ser. Hydrobiol.* **7,** 117–147.

Leveque, C. (1973b). Bilaus energetiques des populations naturelles de mollusques du Lac Tchad. *Cah. ORSTOM, Ser. Hydrobiol.* **7,** 151–165.

McCorkle, S., and Dietz, T. H. (1980). Sodium transport in the freshwater Asiatic clam *Corbicula fluminea*. *Biol. Bull.* (*Woods Hole, Mass.*) **159,** 325–336.

McCorkle-Shirley, S. (1982). Effects of photoperiod on sodium flux in *Corbicula fluminea* (Mollusca: Bivalvia). *Comp. Biochem. Physiol. A* **71,** 325–327.

McMahon, R. F. (1977). Shell size-frequency distributions of *Corbicula manilensis* Philippi from a clam-fouled steam condenser. *Nautilus* **91,** 54–59.

McMahon, R. F. (1979a). Response to temperature and hypoxia in the oxygen consumption of the introduced Asiatic freshwater clam *Corbicula fluminea* (Müller). *Comp. Biochem. Physiol. A* **63,** 383–388.

McMahon, R. F. (1979b). Tolerance of aerial exposure in the Asiatic freshwater clam, *Corbicula fluminea* (Müller). *Proc. Int. Corbicula Symp., 1st, 1977* pp. 227–241.

McMahon, R. F. (1982). The occurrence and spread of the introduced Asiatic freshwater bivalve, *Corbicula fluminea* (Müller) in North America; 1924–1981. *Nautilus* **96,** 134–141.

McMahon, R. F., and Payne, B. S. (1980). Variation of thermal tolerance limits in populations of *Physa virgata* Gould (Mollusca: Pulmonata). *Am. Midl. Nat.* **103,** 218–230.

McMahon, R. F., and Russell-Hunter, W. D. (1981). The effects of physical variables and acclimation on survival and oxygen consumption in the high littoral salt-marsh snail, *Melampus bidentatus* Say. *Biol. Bull.* (*Woods Hole, Mass*) **161,** 246–269.

McMahon, R. F., and Williams, C. J. (1981). Aerial oxygen consumption in the introduced freshwater clam, *Corbicula fluminea* (Müller). *Am. Zool.* **21,** 983.

McMahon, R. F., and Wilson, J. G. (1981). Seasonal respiratory responses to temperature and hypoxia in relation to burrowing depth in three intertidal bivalves. *J. Therm. Biol.* **6,** 267–277.

Marking, L. L., and Chandler, J. H., Jr. (1978). Survival of two species of freshwater clams, *Corbicula leana* and *Magnonaias boykiniana,* after exposure to antimycin. *Invest. Fish Control* **83,** 1–5.

Marshall, B. E. (1975). Observations on the freshwater mussels (Lamellibranchia: Unionacea) of Lake McIlwaine, Rhodesia. *Arnoldia* **7,** 1–15.

Mattice, J. S. (1979). Interactions of *Corbicula* sp. with power plants. *Proc. Int. Corbicula Symp., 1st, 1977* pp. 119–138.

Mattice, J. S., and Dye, L. L. (1976). Thermal tolerance of adult Asiatic clam. *ERDA Symp. Ser.* **40,** 130–135.

Mattice, J. S., McLean, R. B., and Burch, M. B. (1982). "Evaluation of Short-term Exposure to Heated Water and Chlorine for the Control of the Asiatic Clam." Publ. No. 1748. Oak Ridge National Laboratory, Environmental Sciences Division, U.S. Technical Information Service, Department of Commerce, Springfield, Virginia.

Metcalf, A. L. (1966). *Corbicula manilensis* in the Mesilla Valley of Texas and New Mexico. *Nautilus* **80,** 16–20.

Miller, G. H., Hollin, J. T., and Andrews, J. T. (1979). Aminostratigraphy of UK Pleistocene deposits. *Nature* (*London*) **281,** 539–543.

Miller, R. C., and McClure, F. A. (1931). The fresh-water clam industry of the Pearl River. *Lingnan Sci. J.* **10,** 307–322.

Morton, B. (1973). Analysis of a sample of *Corbicula manilensis* Philippi from the Pearl River, China. *Malacol. Rev.* **6,** 35–37.

Morton, B. (1977a). The population dynamics of *Corbicula fluminea* (Bivalvia: Corbiculacea) in Plover Cove Reservoir, Hong Kong, *J. Zool.* **181,** 21–42.

Morton, B. (1977b). The occurrence of inflammatory granulomas in the ctenidial marsupium of *Corbicula fluminea* (Mollusca: Bivalvia): A consequence of larval incubation. *J. Invertebr. Pathol.* **30,** 5–14.

Morton, B. (1979a). *Corbicula* in Asia. *Proc.—Int. Corbicula Symp., 1st, 1977* pp. 15–38.

Morton, B. (1979b). Freshwater fouling bivalves. *Proc.—Int. Corbicula Symp., 1st, 1977* pp. 1–14.

Morton, B. (1982). Some aspects of the population structure and sexual strategy of *Corbicula* cf. *fluminalis* (Bivalvia: Corbiculacea) from the Pearl River, Peoples Republic of China. *J. Moll. Stud.* **48,** 1–23.

Mudkhede, L. M., and Nagabhushanam, R. (1977). Heat tolerance of the fresh water clam, *Corbicula regularis. Marathwada Univ. J. Sci.: Nat. Sci.* **16,** 151–154.

Murray, H. D. (1971). New Records of *Corbicula manilensis* (Philippi) in Texas. *Nautilus* **58,** 35–36.

Negus, C. L. (1966). A quantitative study of growth and production of unionid mussels in the River Thames at Reading. *J. Anim. Ecol.* **35,** 513–532.

Newell, N. D. (1969). Classification of bivalvia. *In* "Treatise on Invertebrate Paleontology, Part N, Mollusca 6" (R. C. Moore, ed.), pp. 205–224. Geol. Soc. Am., Boulder, Colorado.

Nickerson, R. P. (1972). A survey of enzyme and shell variation in 16 populations of the stream limpet, *Ferrissia rivularis* (Say). Ph.D. Dissertation, Syracuse University, Syracuse, New York (*Diss. Abstr.* **33,** 4588–B).

O'Kane, K. D. (1976). A population study of the exotic bivalve *Corbicula manilensis* (Philippi, 1841) in selected Texas reservoirs. Master's Thesis, Texas Christian University, Fort Worth.

Parsons, P. (1980). Invasion by tiny clams halts Arkansas nuclear plant. *Oak Ridger* 18 September, 1980. Oak Ridge, Tennessee. (Newspaper article.)

Pojeta, J., Jr. (1964). Note on the extension of the known range of the Asiatic clam *Corbicula fluminea* (Müller) in the Ohio River. *Ohio J. Sci.* **64,** 428.

Pool, D., and McCullough, J. D. (1979). The Asiatic clam *Corbicula manilensis* from two reservoirs in Eastern Texas. *Nautilus* **93,** 37.

Prashad, B. (1924). Zoological results of a tour in the Far East. Revision of the Japanese species of the genus *Corbicula. Mem. Asiatic Soc. Bengal* **14,** 522–529.

Prashad, B. (1928a). Revision of the Asiatic species of *Corbicula.* I. The Indian species of *Corbicula Mem. Indian Mus.* **9,** 13–27.

Prashad, B. (1928b). Revision of the Asiatic species of *Corbicula.* II. The Indo-Chinese species of the genus *Corbicula. Mem. Indian Mus.* **9,** 29–48.

Prashad, B. (1929). Revision of the Asiatic species of the genus *Corbicula* from China, south-eastern Russia, Tibet, Formosa and the Philippine Islands. *Mem. Indian Mus.* **9,** 49–68.

Prashad, B. (1930). Revision of the Asiatic species of the genus *Corbicula.* IV. The species of the genus *Corbicula* from the Sunda Islands, the Celebes and New Guinea. *Mem. Indian Mus.* **9,** 193–203.

Prokopovich, N. P. (1969). Deposition of clastic sediments by clams. *Sedimen. Petrol.* **39,** 891–901.

Prokopovich, N. P., and Hebert, D. J. (1965). Sedimentation in the Delta-Mendota Canal. *J. Am. Water Works Assoc.* **57,** 375–382.

Prosser, C. L. (1973). Oxygen: Respiration and metabolism. *In* "Comparative Animal Physiology" (C. L. Prosser, ed.), pp. 165–211. Saunders, Philadelphia, Pennsylvania.

Rang, M. (1834). Mémoire sur quelques acéphales d'eau douce du Sénégal. *Nouv. Ann. Mus. Nat. Hist. Paris* **4,** 297–320.

Ray, H. C. (1962). Chokage of filtered-water pipe systems by freshwater molluscs. *Proc. All-India Congr. Zool., 1st, 1959* Part 2, pp. 20–23.

Remane, A., and Schlieper, C. (1971). "Biology of Brackish Water." Wiley (Interscience), New York.

Rinne, J. N. (1974). The introduced Asiatic clam, *Corbicula,* in Central Arizona reservoirs. *Nautilus* **88,** 56–61.

Robertson, J. D. (1964). Osmotic and ionic regulation. *In* "Physiology of Mollusca" (K. M. Wilbur and C. M. Yonge, eds.), Vol. 1, pp. 283–311. Academic Press, New York.

Rodgers, J. H., Jr., Cherry, D. S., Clark, J. R., Dickson, K. L., and Cairns, J., Jr. (1977). The invasion of Asiatic clam, *Corbicula manilensis* in the New River, Virginia. *Nautilus* **91,** 43–46.

Rodgers, J. H., Jr., Cherry, D. S., Dickson, K. L., and Cairns, J., Jr. (1979). Elemental accumulation of *Corbicula fluminea* in the New River at Glen Lyn, Virginia. *Proc.—Int. Corbicula Symp. 1st, 1977,* pp. 99–110.

Russell Hunter, W. (1964). Physiological aspects of ecology in nonmarine molluscs. *In* "Physiology of Mollusca" (K. M. Wilbur and C. M. Yonge, eds.), Vol. 1, pp. 83–126. Academic Press, New York.

Russell-Hunter, W. D. (1978). Ecology of freshwater pulmonates. *In* "Pulmonates" (V. Fretter and J. Peake, eds.), Vol. 2A, pp. 335–383. Academic Press, New York.

Russell-Hunter, W. D., and Buckley, D. E. (1983). Actuarial bioenergetics of nonmarine molluscan productivity. *In* "The Mollusca" (W. D. Russell-Hunter, ed.), Vol. 6, pp. 463–503. Academic Press, New York.

Russell-Hunter, W. D., Apley, M. L., Burky, A. J., and Meadows, R. T. (1967). Interpopulation variations in calcium metabolism in the stream limpet, *Ferrissia rivularis* (Say). *Science* **155,** 338–340.

Russell-Hunter, W. D., Meadows, R. T., Apley, M. L., and Burky, A. J. (1968). On the use of a wet oxidation method for estimates of total organic carbon in mollusc growth studies. *Proc. Malacol. Soc. London* **38,** 1–11.

Russell-Hunter, W. D., Burky, A. J., and Hunter, R. D. (1970). Interpopulation variation in shell components in the stream limpet, *Ferrissia rivularis. Biol. Bull.* (*Woods Hole, Mass.*) **139,** 402.

Russell-Hunter, W. D., Burky, A. J., and Hunter, R. D. (1981). Interpopulation variation in calcareous and proteinaceous shell components in the stream limpet, *Ferrissia rivularis. Malacologia* **20,** 255–266.

Schneider, R. F. (1967). Range of Asiatic clam in Florida. *Nautilus* **81,** 68–69.

Sickel, J. B. (1973). A new record of *Corbicula manilensis* (Philippi) in the southern Atlantic slope region of Georgia. *Nautilus* **87,** 11–12.

Sickel, J. B. (1979). Population dynamics of *Corbicula* in the Altamaha River, Georgia. *Proc.—Int. Corbicula Symp., 1st, 1977,* pp. 69–80.

Sickel, J. B., and Burbanck, W. D. (1974). Bottom substratum preference of *Corbicula manilensis* (Pelecypoda) in the Altamaha River, Georgia. *ASB* (*Assoc. Southeast. Biol.*) *Bull.* **21,** 84.

Sinclair, R. M. (1964). Clam pests in Tennessee water supplies. *J. Am. Water Works Assoc.* **56,** 592–599.

Sinclair, R. M. (1971). Annotated bibliography on the exotic bivalve *Corbicula* in North America, 1900–1971. *Sterkiana* **43,** 11–18.

Sinclair, R. M. (1974). Effects of an introduced clam (*Corbicula*) on water quality in the Tennessee River Valley. *In* "Proceedings of the Second Industrial Waste Conference," pp. 43–50. Tennessee Department of Public Health, Tennessee Stream Pollution Control Board, Nashville.

Sinclair, R. M., and Isom, B. G. (1961). "A Preliminary Report on the Introduced Asiatic clam *Corbicula* in Tennessee." Tennessee Stream Pollution Control Board, Tennessee Department of Public Health, Nashville.

Sinclair, R. M., and Isom, B. G. (1963). "Further Studies on the Introduced Asiatic Clam (*Corbicula*) in Tennessee." Tennessee Stream Pollution Control Board, Tennessee Department of Public Health, Nashville.

Smith, A. L., Muia, R. A., Farkas, J. P., and Bassett, D. O. (1979). Clams—a growing threat to inplant water systems. *Plant Eng.* **33,** 165–167.

Smith, M. H., Britton, J., Burke, P., Chesser, R. K., Smith, M. W., and Hagen, J. (1979). Genetic variability in *Corbicula,* an invading species. *Proc.—Int. Corbicula Symp., 1st, 1977* pp. 243–248.

Stearns, S. C. (1976). Life-history tactics: A review of the ideas. *Q. Rev. Biol.* **51,** 3–47.

Stearns, S. C. (1977). The evolution of life history traits: A critique of the theory and a review of the data. *Annu. Rev. Ecol. Syst.* **8,** 145–171.

Stein, C. B. (1962). An extension of the known range of the Asiatic clam *Corbicula fluminea* (Müller) in the Ohio and Mississippi Rivers. *Ohio J. Sci.* **62,** 326–327.

Taylor, D. W. (1981). Freshwater mollusks of California: A distributional checklist. *Calif. Fish Game* **67,** 140–163.

Taylor, R. W. (1980a). Mussels of Floyd's Fork, a small North Central Kentucky stream (Unionidae). *Nautilus* **94,** 13–15.

Taylor, R. W. (1980b). Freshwater bivalves of Tygart Creek, Northeastern Kentucky. *Nautilus,* **94,** 89–91.

Taylor, R. W., and Hughart, R. C. (1981). The freshwater naiads of Elk River, West Virginia with a comparison of earlier collections. *Nautilus* **95,** 21–25.

Teal, J. M. (1957). Community metabolism in a temperate cold spring. *Ecol. Monogr.* **27,** 283–302.

Thomas, N. A., and MacKenthum, K. M. (1964). Asiatic clam infestation at Charleston, West Virginia. *Nautilus* **78,** 28–29.

Thomerson, J. E., and Myer, D. G. (1970). *Corbicula manilensis:* Range extension upstream in the Mississippi River. *Sterkiana* **37,** 29.

Thompson, C. M., and Sparks, R. E. (1977). Improbability of dispersal of adult Asiatic clams, *Corbicula manilensis,* via the intestinal tract of migratory water fowl. *Am. Midl. Nat.* **98,** 219–223.

Thompson, C. M., and Sparks, R. E. (1978). Comparative nutritional value of a native fingernail clam and the introduced Asiatic clam. *J. Wildl. Manage.* **42,** 391–396.

Tilly, L. J., Corey, J. C., and Bibler, N. E. (1978). Response to the Asiatic clam, *Corbicula fluminea* to gamma radiation. *Health Phys.* **35,** 704–708.

Trama, F. B. (1982). Occurrence of the Asiatic clam *Corbicula fluminea* in the Raritan River. *Nautilus,* **96,** 6–8.

Trdan, R. J. (1981). Reproductive biology of *Lampsilis radiata siliquoidea* (Pelecypoda: Unionidae). *Am. Midl. Nat.* **106,** 243–248.

U. S. Environmental Protection Agency (1980). Effluent limitations guidelines, pretreatment standards and new source performance standards under Clean Water Act; Steam electric power generating point source category. 40 CFR, Parts 125 and 423. *Fed. Regist.* **45** (200), 68328–68337.

van der Schalie, H. (1973). The mollusks of the Duck River Drainage in Central Tennessee. *Sterkiana* **52,** 45–55.

Villadolid, D. V., and Del Rosario, F. G. (1930). Some studies on the biology of Tulla (*Corbicula manilensis* Philippi) a common food clam of Laguna de Bay and its tributaries. *Philipp. Agric.* **19,** 355–382.

Way, C. M., Hornbach, D. J., and Burky, A. J. (1980). Comparative life history tactics of the sphaeriid clam, *Musculium partumeium* (Say) from a permanent and a temporary pond. *Am. Midl. Nat.* **104,** 319–327.

Weber, C. I. (1973). "Biological Field and Laboratory Methods for Measuring the Quality of Surface Waters and Effluents" (Macroinvertebrate Section), EPA–670/4–73–001. National Environmental Research Center, Office of Research and Development, U. S. Environmental Protection Agency, Cincinnati, Ohio.

White, D. S. (1979). The effect of lake-level fluctuations on *Corbicula* and other pelecypods in Lake Texoma, Texas and Oklahoma. *Proc.—Int. Corbicula Symp., 1st, 1977* pp. 82–88.

White, D. S., and White, S. J. (1977a). The effect of reservoir fluctuations on populations of *Corbicula manilensis* (Pelecypoda: Corbiculidae). *Proc. Okla. Acad. Sci.* **57,** 106–109.

White, D. S., and White, S. J. (1977b). Observations on the pelecypod fauna of Lake Texoma, Texas and Oklahoma, after more than 30 years impoundment. *Southwest Nat.* **22,** 235–254.

13

Population Genetics of Marine Gastropods and Bivalves

EDWARD M. BERGER

Biology Department
Dartmouth College
Hanover, New Hampshire

I. Introduction: Darwinian Evolution and the Framework of Population Genetics

Darwin's theory of evolution, as stated by Gould (1977), is based on two undeniable facts and one inescapable conclusion. The first fact is that members of a population or species differ from one another and that much of this phe-

The Mollusca, Vol. 6
Ecology

ISBN 0-12-751406-6

notypic variation is inheritable. The second fact, gleaned from Malthus' "Essay on Populations," reminds us that resources such as food and space are generally limiting, so that only a fraction of the offspring produced in each generation will survive and reproduce. The inescapable conclusion, then, is that the survivors must include those individuals whose phenotypes (and genotypes) vary in ways most favored by the particular environment. Through this process of natural selection, the phenotypic and underlying genetic composition of the population may change over time; that is, the population evolves.

Population genetics developed as a science intent on understanding the genetic basis of the evolutionary process in both qualitative and quantitative terms. In this chapter on the population genetics of marine snails and bivalves, four major areas of inquiry of contemporary interest are described in Section I,A–D.

A. The Relationship between Phenotype and Genotype

In terms of either adaptation or evolution, the single type of phenotypic variation that is important and interesting is that which is genetically based. Nongenetic variation is not heritable and, therefore, is unresponsive to natural selection. Nongenetic variation arises primarily from environmental heterogeneity, although other sources exist; consider, for example, the morphological polymorphism in insects associated with holometabolous development. As a consequence, population geneticists are compelled to establish that the variation observed for any trait in a population is heritable in a uniform environment, and that it involves allelic segregation at one or more loci. Because direct, formal genetic analysis in marine molluscs is generally a formidable task, less direct methods for estimating heritability have been adopted. The chapter will begin, then, with a discussion of molluscan genetics.

B. Describing the Distribution of Genetic Variation

Having established that naturally occurring polymorphism is genetically based, it is necessary next to describe the variation in species populations in genetic units. At the individual level, this involves assigning both a phenotype and a genotype to each member of the population sampled. These data are then converted to a description of the entire population: the total number of alleles present, genotype and phenotype frequencies, and allele frequencies. It will become apparent later that such formal descriptions have been made almost exclusively for genes encoding specific enzymes.

C. Dynamics of Polymorphism

A description of the extant variation is necessary but not sufficient for understanding the process of evolution. It is also necessary to understand the nature and magnitude of forces that change gene and phenotype frequencies over time. Darwin enlightened us on the importance of natural selection, focusing primarily on directional selection. Today we recognize that several different forms of selection occur and that many nondeterministic forces change the genetic composition of populations . These include mutation, migration or gene flow, and *genetic drift,* which is a catchall term that includes nonrandom mating and the stochastic changes in gene frequency that occur intermittently or chronically when the population is small.

D. Conversion of Variation within a Species to Variation between Species

Populations isolated from one another over time may accumulate sufficient genetic differences to achieve the status of separate species. In general, species assignments in molluscs are made on the basis of morphological criteria (so-called taxonomic species); however, for population geneticists speciation is determined by the establishment of reproductive isolation. Members of two different biological species either cannot or will not mate. As we will see, population genetics has become useful to molluscan taxonomists. We have been able to establish the existence of morphologically indistinguishable sibling species on the basis of genetic analysis alone. In the context of speciation, we will approach two interesting but difficult questions: one concerning the rate, or tempo, of evolution and the other concerning the kinds and amounts of genetic change that underlie the process of speciation itself.

II. Genetics

Because of the difficulties involved in the long-term maintenance of adults and the extended period required for pelagic larval development of marine molluscs, very little classical genetic analysis has been possible. For polymorphism displaying discrete, qualitative variations, a genetic basis is often deduced from an analysis of families of offspring obtained from a single female. More commonly, a single locus polymorphism is inferred in a population if the corresponding phenotypic classes (homozygotes and heterozygotes) occur at frequencies that are compatible with the expectations of Hardy–Weinberg equilibrium.

For quantitatively varying traits such as growth rate, the underlying genetic variation can be crudely estimated by heritability analysis in uniform environments. However, the number of genes involved in specifying the trait, and the number and frequencies of segregating alleles at each locus in the population, cannot be assessed.

A. Cytogenetics

Reliable determinations of chromosome number are available for members of most gastropod and bivalve orders. A useful summary has been presented by Murray (1975) in his review of molluscan genetics. Among the bivalves analyzed, the haploid chromosome number has been found to vary between 10 and 23; the gastropods show a much broader range: 5–44, excluding species believed to be polyploid. Generally, chromosome number within an order is quite conservative. In several dioecious members of the Hydrobiida, Viviparidae, and Melaniida, chromosome mechanisms for sex determination have been described.

From the viewpoint of population genetics, the karyotype polymorphism found in natural populations of the prosobranch *Purpura* (*Thais*) *lapillus* from Brittany (Staiger, 1957) is of particular interest. Apparently, two chromosomal races exist: an $n = 13$ race, found primarily in rocky coast areas with heavy surf, and an $n = 18$ race, which is essentially restricted to sheltered bays. The North American populations of *Thais* that have been examined contain only the $n = 13$ race, in which the haploid karyotype contains eight acrocentric and five metacentric chromosomes. The $n = 18$ karyotype contains only acrocentric chromosomes. In populations containing hybrids, analysis of chromosome pairing indicates that each of the metacentrics corresponds to two acrocentrics in the other race. The adaptive significance of this polymorphism is unknown.

B. Morphological Polymorphism

1. *Shell Color and Pattern*

The most extensive genetic analyses of shell color, pattern, and coiling polymorphisms have been carried out on land snails, especially of the family Helicidae. A thorough review of this work is presented by Murray (1975). Among marine molluscs, the evidence for genetic polymorphism is anecdotal and incomplete. In the mussel *Mytilus edulis* a ubiquitous shell color polymorphism has been analyzed (Newkirk, 1980; Mitton, 1977: Innes and Haley, 1977a). Shell color may be blue-black or brown, which includes a phenotype also designated as striped. The results of full-sib analysis indicate a single major genetic determinant with the blue-black allele recessive (Innes and Haley, 1977a). However, simple segregation patterns were not observed by Newkirk (1980), who suggested either selective mortality of certain phenotypes prior to progeny scoring or the presence of several different brown alleles with, perhaps, polymorphic modifier genes. Shell color also appeared to be influenced by the environment.

Mitton (1977) has reported a geographical cline of shell color from Maine to Virginia. The higher frequency of the darker blue-black shell in the north was attributed to a reduced risk of freezing, because more radiant energy is absorbed by dark-shell morphs. On a more local scale, Newkirk (1980) found an opposite trend. Interestingly, both authors confirmed that the dark morph is correlated with a significantly higher rate of growth.

In the hard clam *Mercenaria mercenaria,* Chanley (1961) carried out extensive genetic analyses and found that the white and brown morphs differ at a single gene, with the heterozygote showing the red-brown zigzag pattern formerly assigned to a *notata* subspecies.

Among the gastropods, shell-color polymorphism is well known and frequently associated with crypsis. In *Crepidula convexa,* for example, purple, brown, and tan morphs of varying intensity are known (Hoagland, 1977). Color apparently is not correlated with developmental size or diet and appears to be under genetic control, although no breeding tests have been done. Dark forms are abundant on dark backgrounds and light forms predominate on light backgrounds; intermediates are rare. Predation by gulls, shorebirds, and crabs seems to be the selective agent in this case. A similar situation of substrate matching is presumably implicated in the maintenance of shell pattern and behavior polymorphism in the limpet *Acmaea digitalis* (Giesel, 1970). The shell color and banding patterns seen in *Thais lamellosa* have been shown to be genetically controlled (Spight, 1972, cited in Murray, 1975).

Among the littorinids, shell-color polymorphism is common, and several studies have implicated cryptic coloration as the selective mechanism for maintenance. In the four sympatric species of the *Littorina saxatilis* complex in Wales, color variation in *L. rudis* and *L. nigrolineata* seems associated with background. Selection occurs by visual predators, birds, and crabs (Heller, 1975b). Red color morphs are associated with red sandstone, white with barnacle-covered rock, and yellow with algae and detritus-covered barnacles; striped forms are most common on broken backgrounds. Formal genetic studies have not been done, but the presence of morphs with a regular striped pattern strongly suggests a genetic basis. Smith (1976) describes disruptive selection in polymorphic *L. obtusata* as an explanation for morph-ratio clines. Eight shell-color morphs are associated with four species of Fucacae: yellow with *Fucus spirilis,* green with *F. vesiculosis* and *Ascophylum nodosum,* brown with *A. nodosum* and *F. seratas,* and banded and black with *F. spirilis.* Color and pattern are genetically determined, and polymorphism has evolved under disruptive selection for crypsis. In general, banded morphs are rare and may be subject to strong apostatic selection.

Reimchen (1974, cited in Murray, 1975) carried out breeding studies with *L. mariae* and found that the segregation pattern of the two major morphs, reticulata and citrina, best fits a two-locus model, with reticulata determined by the coincidence of at least one dominant allele at each locus. The same report included evidence for the segregation of three morphs, reticulata, olivaceae, and citrina, in *L. obtusata.*

2. Shell Sculpture

Variation in shell sculpture has been studied primarily at the phenotypic level (Berry and Crothers, 1968; Vermeij, 1971), although breeding analysis has been carried out in several cases. Phenotypic variation is often complex and based on the interaction of numerous genes. Geographical variation is often associated with hydrological factors such as wave action and substrate. For example, Newkirk and Doyle (1975) studied the genetics and adaptive significance of three shell-shape parameters in *L. saxatilis,* which varied in a regular pattern along an estuary. Attempts were made to estimate heritability, (the proportion of phenotypic variance based on genetic factors) using biometric analysis, full sib, and offspring–mother covariances. However, the results were not sufficiently clear to provide precise heritability values. The selective agents responsible for habitat variation appeared to include wave action and desiccation.

In a particularly thorough and illuminating study, Strushaker (1968) investigated intraspecific shell sculpture variation in the Hawaiian *L. picta.* Two extreme phenotypes exist: smooth shell with faint or no spiral ribs and tubercles, and heavily sculptured shell with raised ribs bearing many spaced tubercles. Although the distribution in the population is bimodal, intermediates are found. The morphs were associated with specific habitats; smooths predominate in low-angle beaches with heavy horizontal wave forces, heavily sculptured forms predominate on high-angle beaches receiving spray and slight horizontal wave force, and intermediates occur primarily in intertidal habitats. Larvae reared in a uniform laboratory environment showed phenotypes identical to those of their parents, indicating a genetic basis for this polymorphism. Progeny of sculptured forms had a higher growth rate and a lower mortality than progeny of smooth morphs, which was interpreted as a pleiotropic effect of shell morphology. The two morphs showed differential selection in field transfer experiments, and the author concluded that the environmental factors determining fitness probably involved wave action, prolonged submersion, and salinity, with selection strongest at the larval settling stage. A temporal model of balanced polymorphism was suggested.

C. Enzyme Polymorphism

1. The Technique

Gel electrophoresis was introduced to the study of genetic variation in natural populations in 1966 (Hubby and Lewontin, 1966; Lewontin and Hubby, 1966), and the technique has literally revolutionized the field. For the first time, it became possible to enter a population and sample a large number of individuals for allelic variation at a large number of individual loci, independent of the degree of variation or the physiological significance of the loci under investigation. During the past 15 years, the type, amount, and distribution of genetic variation of nearly 50 species of marine molluscs has been studied to some extent.

The rationale for the technique is based on several principles of molecular biology. The linear sequence of nucleotides in DNA determines, in a colinear way, the primary sequence of amino acids in a polypeptide, using the rules of the genetic code. Barring degeneracy, any substitution in a gene DNA sequence will then change the amino acid sequence of the protein for which it codes. These amino acid substitutions will, in turn, alter the net electrical charge and/or conformation of the protein in a large proportion of cases without necessarily affecting its biological activity. Such molecular variation is detected by the separation of proteins in an electrophoretic gel. The presence of abundant proteins separated in the gel is detected by a stain for general protein. More commonly, however, the physical position of specific enzymes is visualized by histochemical staining. Recently, it has become clear that the sensitivity, or more correctly resolution, of the technique can be improved considerably by coordinately employing a battery of electrophoretic conditions (Johnson, 1979) or by incorporating additional discriminating criteria such as the heat stability of enzymes (Bernstein et al., 1973). Variation in the heat stability of enzyme morphs has indeed become of significant ecological interest in itself (Gosling, 1979; Singh et al., 1974; Wilkins et al., 1978, 1980).

The interpretation of electrophoretic phenotypes is generally unambiguous, allowing genotype assignments to be made directly. Homozygotes are diagnosed by the presence of a single stained band on the gel, whereas heterozygotes show either a codominant two-banded pattern or a three-banded pattern which signifies that the active enzyme is a dimer; the band with intermediate mobility is a heteromultimer containing subunits encoded by the two different alleles. The latest convention is to refer to these electrophoretically distinguishable enzyme morphs as *allozymes*.

Alleles are generally designated by the relative mobility of the allozymes they encode. The most common allele is considered to be the 1.00 mobility standard; a 0.90 allele produces an enzyme molecule of slower mobility, and a 1.10 allele produces an enzyme variant that migrates more rapidly than the standard. Lettering systems are frequently used, but these tend to be ambiguous.

Allele frequencies are determined directly from phenotype frequencies. When genetic analysis has been possible, the allozyme phenotypes have invariably shown a Mendelian pattern of segregation. From the allele frequencies, it is possible to determine the expected frequencies of homozygous and heterozygous genotypes in the population using the Hardy–Weinberg expansion. Statistical tests discern the degree to which the observed frequencies fit the expectations of equilibrium. Another relevant calculation is average heterozygosity, in which the observed frequency of heterozygotes at polymorphic loci is averaged over all loci.

The functional significance of allozyme variation is, for the most part, unknown. There are currently two polar views: one proposes that allozyme variants are essentially equivalent with respect to fitness (the neutralist school), and the other believes that allozyme variation is adaptive and maintained by one or more forms of natural selection. In general, we need not be concerned with this problem.

2. *Estimates of Genetic Variation*

A large number of marine molluscan species have now been analyzed for gene–enzyme variation using gel electrophoresis, and the literature continues to grow. In some studies only a single population has been surveyed, whereas in others extensive sampling has been made over a wide geographic range. Similarly, some surveys have focused on one or a few gene–enzyme systems, whereas others have described variation at a large number of loci. Table I provides a list of references to many of the reports. Because the literature is scattered, this list is obviously incomplete. For those studies involving a small number of loci or a single population, only the citation is given. Where extensive sampling of populations or loci has been done, a summary of the findings is included. The significant information from the population genetics point of view includes (1) the proportion of loci sampled that are polymorphic and (2) the average individual heterozygosity. A locus is considered to be polymorphic if the most common allele occurs at a frequency of less than 0.98. The calculation used for determining heterozygosity has already been described.

The general conclusions drawn from these reports are that (1) in most species, one-third to one-half of the loci sampled are segregating two or more alleles at high frequency in nature and (2) average heterozygosity is generally 10–20%. Both sets of values are similar to those obtained from studies of other invertebrate groups (Powell, 1975). It should be remembered that for technical reasons these values are undoubtedly underestimates, perhaps by a factor of two, of the true variation. Because there is so little information about genetic variation for other phenotypes, it is not yet known whether enzyme-coding genes represent an unbiased sample of the genome.

In the following section, we will consider genetic variation for continuously varying traits and then return to a discussion of allozymes, focusing on the distribution of specific variations among species populations.

D. Continuously Varying Traits

In the evolutionary sense, traits showing continuous variation in the population are of considerable interest. Phenotypes such as fecundity, longevity, or growth rate obviously represent major components of fitness and, to the extent that the naturally occurring variation has a genetic basis, they are subject to natural selection. In spite of their significance, we know very little about the amount and distribution of allelic variation underlying the distribution of, for example, growth rate in molluscan populations. One reason is that a trait such as growth rate is influenced by many, if not all, genes. This means that operationally the effect of a "slow growth" allele at one locus is indistinguishable from the effect of a "slow growth" allele at another locus. A second problem is that continuously varying traits may be strongly influenced by environmental

TABLE IA

Analysis of Gene–Enzyme Variation in Molluscan Species: Major Surveys

Species	Number of loci	Proportion of polymorphic loci	Average heterozygosity	Reference
Buccinum sp.	29	0.36	0.09	Costa and Bisol (1978)
Crassostrea sp.	25–34	0.20–0.60	0.06–0.19	Buroker et al. (1979a,b)[a]
Littorina rudis	21	0.38	0.15	Ward and Warwick (1980)
Littorina arcana	21	0.37	0.13	Ward and Warwick (1980)
Mytilus edulis	34	0.30	0.10	Ahmad et al. (1977)[b]
Nassarius obsoletus	6	0.33	—	Gooch et al. (1972)
Patella aspersa	10	0.80	0.12	Wilkins (1976, 1977)
Patella vulgata	9	0.55	0.11	Wilkins (1976, 1977)
Saccostrea sp.	24–34	0.47–0.53	0.17–0.19	Buroker et al. (1979a,b)[a]
Tridacna maxima	30	0.57	0.22	Ayala et al. (1973)[c]

[a] For *Saccostrea* and *Crassostrea*, see also Wilkins (1975), Mathers et al. (1974), Wilkins and Mathers (1974), Singh and Zouros (1977, 1978, 1981), and Zouros et al. (1980).

[b] For *M. edulis* see also Ahmad et al. (1977), Johnson and Utter (1973), Lassen and Suchanek (1978), Lassen and Turano (1978), Mitton et al. (1973), Murdock et al. (1975), Koehn and Mitton (1972), Gosling and Wilkins (1977, 1981), Koehn et al. (1976), Levinton and Koehn (1976), Skibinski and Beardmore (1979), Ahmad and Beardmore (1976), Skibinski et al. (1978a,b), Boyer (1974), Milkman and Beatty (1970), Milkman (1971), Milkman et al. (1972), Levinton and Lassen (1978b), Levinton (1973), and Theisen (1978).

[c] For *Tridacna maxima*, see also Campbell et al. (1975) and Valentine et al. (1973).

TABLE IB

Analysis of Gene–Enzyme Variation in Molluscan Species: Minor Surveys

Species	Reference
Anadara sp.	O'Gower and Nicol (1968)
Cardium edule	Brock (1978)
Cardium glaucum	Brock (1978)
Cerastoderma edule	Gosling (1980); Wilkins (1975)
Cerastoderma glaucum	Gosling (1980); Wilkins (1975)
Collisella digitalis	Gresham and Tracey (1975); Murphy (1978)
Collisella austrodigitalis	Murphy (1978)
Chlamys opercularis	Wilkins (1975)
Ensis ensis	Wilkins and Mather (1974); Wilkins (1975)
Goniobasis sp.	Chambers (1978)
Haliotus (2 spp.)	Wilkins et al. (1980); Fujino et al. (1980)
Hydrobia (3 spp.)	Lassen (1979)
Littorina angulifera	Gaines et al. (1974)
Littorina obtusata (= *obtusata*) and *mariae, neé littoralis*)	Berger (1973, 1977); Wilkins et al. (1978); Wilkins and O'Regan (1980); Wium-Anderson (1970)
Littorina littorea	Berger (1973, 1977); Wilkins et al. (1978); Wilkins and O'Regan (1980); Wium-Anderson (1970)
Littorina saxatilis (sp.)	Berger (1973, 1977); Wilkins et al. (1978); Snyder and Gooch (1973); Wium-Anderson (1970); Wilkins and O'Regan (1980); Ward and Warwick (1980); Beardmore and Morris (1978)
Macoma sp.	Levinton (1973, 1975)
Mercenaria mercenaria	Levinton (1973); Pesch (1972, 1974); Wilkins (1975)
Mercenaria compechiensis	Pesch (1972, 1974)
Mytilus californiansus	Levinton and Fundiller (1975); Tracey et al. (1975); Levinton and Koehn (1976); Levinton and Lassen (1978a,b); Lassen and Suchanek (1978)
Mytilis galloprovincialis	Ahmad and Beardmore (1976); Skibinski and Beardmore (1979); Skibinski et al. (1978a,b, 1980); Gosling and Wilkins (1981)
Modiolus demissus	Koehn and Mitton (1972); Koehn et al. (1973); Levinton (1973); Wilkins (1975); Chaisson et al. (1976); Gosling (1979); Skibinski et al. (1980); Schopf et al. (1975)
Mya arenia	Leviton and Fundiller (1975); Levinton (1973)
Malletia sp.	Gooch and Schopf (1973)
Nucula annulata	Levinton (1973)

(Continued)

TABLE IB *Continued*

Species	Reference
Nuculara portoria	Gooch and Schopf (1973)
Ostrea lurida	Johnson et al. (1972); Wilkins (1975)
Ostrea edulis	Johnson et al. (1972); Wilkins (1975); Wilkins and Mather (1974)
Pectin maxima	Wilkins and Mather (1974); Wilkins (1975, 1978)
Siphonaria kurracheensis	Black and Johnson (1981)
Tapes decussatus	Wilkins and Mather (1974); Wilkins (1975)
Thais lamellosa	Campbell (1978)

factors. In the case of growth rate and nutrition, temperature and salinity must play important roles.

To demonstrate that there is a genetic component underlying variation in growth rate, it is sufficient to demonstrate that the distribution of phenotypes can be shifted by artificial selection or that the progeny produced by individuals at the extremes of the phenotypic range continue to show those differences when reared in a uniform environment. The concordance of similarlity among family lines can be quantitated to derive the value called *heritability*, which is a measure of the total phenotypic variance that is genetically determined. It should be recognized, however, that we still have learned nothing about the number of genes affecting the trait, nor can we estimate the number and frequency of alleles segregating at those loci. With these limitations in mind, let us consider two traits, growth rate and salinity tolerance.

1. Growth Rate

Studies on the population genetics of growth rate have been stimulated by the possibility of improved mariculture, and most work has been done on commercially important bivalves, especially mussels and oysters. Quantitative genetic analyses of growth rate in the oysters *Crassostrea virginica* and *C. gigas* have been carried out in several laboratories with similar results. Estimates of heritability for larval growth rate are 0.2–0.5 (Longwell and Stiles, 1973; Lannan, 1972; Newkirk et al., 1977a). Larval and spat growth rates were found to be highly correlated (Newkirk et al., 1977a), indicating promise for selection programs.

In a different series of studies, Zouros and Singh (Zouros et al., 1980; Singh and Zouros, 1977, 1978, 1981) discovered that growth rate in the oyster is positively correlated with allozyme heterozygosity, but they are quick to point out that the correlation is not necessarily causal. In their studies, large samples of larvae collected from the field were cultured in the laboratory under uniform conditions. After 1 year, the largest and smallest individuals were chosen for electrophoretic analysis at five loci. Classes in the higher weight range were found to have much higher levels of heterozygous genotypes. Subsequent studies utilizing larger samples of individuals and genes confirmed the original results. Although heterosis has long been recognized and applied to the improvement of agricultural crops, improved mariculture using this principle now seems feasible.

By employing experimental transplants and carrying out genetic analysis of laboratory-reared animals, Newkirk's group (Newkirk et al., 1980, 1981) and Innes and Haley (1977b) have demonstrated genetic variation in growth rate both within and between natural populations of the mussel *Mytilus edulis*. Estimates of heritability, however, were not possible. The influence of shell color variation on growth rate was discussed earlier.

2. Salinity Tolerance

Tolerance is a cryptic phenotype that can be measured only by perturbing the environment. Generally, the outcome of intolerance is reduced viability, reduced growth rate, or death, which add an additional complications studies designed to assess the genetic component of natural variation for tolerance. Several studies on the inheritance of salinity tolerance in bivalves have provided results indicating a geneic component to the variation seen within and between populations. Newkirk et al. (1977b) and Newkirk (1978) found that genetic differences exist between natural populations of *C. virginica* for tolerance to reduced salinity and that some of the genetic variation is not simply additive. Salinity intolerance here is reflected by reduced survivorship and slower growth rate.

Comparisons of low salinity tolerance between estuarine and oceanic populations of *M. edulis* have revealed genetic heterogeneity. Levitan and Lassen (1978a,b) found that Long Island Sound mussels were significantly more resistant to salinity shock than were open-water forms, and that the effect of salinity shock on mortality increased with higher temperatures. They have raised the possibility that the estuarine and oceanic populations represent different physiological races or subspecies (Levitan and Lassen, 1978a,b), a point that will be discussed again later.

III. Distribution of Genetic Variation in Natural Populations

A. Forces Acting on Variation

In the previous section, I have tried to outline the types of gene–phene variation that are studied in natural populations of marine molluscs. Attention will now focus on the distribution of polymorphism over time and space, with special emphasis being placed on the evolutionary forces that may produce those patterns. Although we recognize that mutation is ultimately responsible for all new variation, there seems to be no information on this subject in the molluscan literature. Recombination, of course, provides the relentless procession of new gene combinations that are tested in nature, but estimates of recombination frequency are also unavailable. As a result, the following discussion is concerned entirely with three forces: natural selection, genetic drift, and migration (or gene flow) and their interactions. To provide coherence to the discussion, I will concentrate on several model systems in which the level of our knowledge is more than simply anecdotal.

B. Studies on the Littorinids and Other Intertidal Gastropods

Marine gastropods provide an interesting system for assessing the influence of gene flow on the genetic structure of natural populations. In many species fertilization is internal, but larval development proceeds after the fertilized eggs are deposited. In some cases, eggs or egg masses are cemented to a substrate; alternatively, they may be simply set adrift. In the latter case, pelagic development continues to a settling stage. In other species, development is ovoviviparous, and offspring are released at the crawl-away stage. Here, gene flow is necessarily restricted to rafting or adult movement. This diversity in dispersal capability among gastropods is especially spectacular among the littorinids, in which sympatric populations of congeneric species showing quite different dispersal capabilities may be studied simultaneously over wide geographic areas. It then becomes possible to examine variation at homologous loci in several species experiencing a common environment. Table II summarizes the relevant life-history parameters of gastropod species that will be considered below.

TABLE II

Larval Development and Dispersal in Several Gastropods of Genetic Interest

High-dispersal species: planktonic eggs and/or larval stages
Nassarius obsoletus
Littorina littorea
Littorina striata
Littorina angulifera (planktonic veliger)
Medium-dispersal species: eggs cemented onto substrate, young adult crawl-aways
Littorina obtusata (*littoralis*)
Littorina mariae
Littorina nigrolineata
Littorina arcana
Thais lamellosa
Low-dispersal species: ovoviviparous
Littorina rudis (*patula* or *saxatilis*)
Littorina neglecta

Intuitively, we can make two predictions about the effects of different dispersal capabilities; one concerns demography and the other genetic variation. First, because species require dispersal in order to extend their ranges or to repopulate ephemeral habitats, the geographic distribution and colonizing ability of a species should correlate with its dispersal potential. The second prediction concerns the distribution of genetic variation among species populations and, more precisely, the interaction between gene flow and local selection. Species often encounter a variety of habitats in their range. Because adaptation to habitat diversity is thought to involve selection of the level of genes, it follows that extensive dispersal will tend to homogenize the gene pool and reduce the effectiveness of local adaptation. This prediction pertains even if the genetic variation is not adaptive, for heterogeneity may arise from genetic drift, a purely nondeterministic force. Although molluscan populations are generally believed to be quite large and therefore immune to the influence of random drift, extended periods of isolation can exert the same effect as reduced size. An extension of this prediction is that dispersal capability is inversely correlated with taxonomic diversity (see Section III,C).

The first definitive genetic study addressing this question was carried out by Gooch et al. (1972) on populations of the mud snail *Nassarius obsoletus*. This abundant member of littoral and estuarine communities ranges from northern Canada to Florida. Dispersal is extensive and occurs through a planktonic larval stage that may last as long as 53 days, long enough to drift several hundred miles. Collections from 11 sites along the Atlantic Coast from Cape Cod to Beaufort, North Carolina were analyzed for variation at two polymorphic enzyme loci. A remarkable degree of homogeneity for allele frequencies was observed. Commenting that the likely basis for this homogeneity was gene flow, the authors supported this statement with evidence (from earlier studies) of an absence of clinal variation for other traits such as developmental rate and temperature tolerance.

A similar pattern of geographic homogeneity was reported by Gaines et al. (1974) for 20 populations of *Littorina anguilifera* which were surveyed for allelic variation at an esterase locus. In this species the dispersal phase follows the release of veliger larvae. The range of frequencies for the most common allele was rather narrow, 0.66–0.87, with the possibility of a shallow cline indicated.

A survey of allozyme variation at three polymorphic esterase loci was reported by Berger (1973), who investigated three species simultaneously: *Littorina littorea, L. saxatilis,* and *L. obtusata*. Samples were collected from 15 rocky-shore localities covering a range of about 500 air miles from Prince Edward Island to Cape Cod. The results indicated that in the two species lacking a pelagic larval stage (*L saxatilis* and *L. obtusata*), genetic differences were seen at same loci between groups of populations separated by a major geographic barrier, such as Cape Cod or the unfavorable surface currents around Halifax. Populations from within a region, such as south of the Cape Cod Canal, were heterogeneous with respect to allele frequency but were more similar than populations from different regions. Interregional variation appeared either as a sharp change in allele frequency across some boundary or as a qualitative difference in allele composition. The analysis of *L. littorea* populations, in contrast, provided no evidence of geographic heterogeneity. These conclusions were supported in a later study (Berger, 1977) showing that allele frequencies in several of the original populations studied remained constant over 5 subsequent years. Thus, it appears valid in this case to look at geographic patterns of variation unconfounded by any changes that may occur over time.

When single samples of the three species collected from Cape Cod and Roscoff, France, were compared for variation at 13 loci, a surprising result was found. For the two species lacking pelagic larvae, the amount of genetic differentiation seen earlier was, as predicted, amplified by the increased distance. The two samples of *L. littorea,* however, were enormously different genetically. To illustrate, at 7 of 12 loci, the two *L. littorea* populations shared no alleles in common. This total divergence was never seen at any locus in the other two species. The presumed basis for this unusual result will be discussed later in Section IV,B.

An electrophoretic analysis of *L. saxatilis* populations carried out on a more local scale was reported by Snyder and Gooch (1973). In a survey examining variation of two loci, they found marked heterogeneity in allele frequency between populations only 2 km apart, although the degree of differentiation overall was found to be independent of the distance between demes. In contrast to this random pattern of microgeographic variation, Newkirk and Doyle (1979) reported the existence of an allele frequency cline at an esterase locus in 10 populations of *L. saxatilis* and *L. obtusata* collected along a single estuary 10 km long. These same populations had previously been studied for variation in shell morphology (Newkirk and Doyle, 1975). Another smooth allele frequency cline was described earlier for hemoglobin variants in *L. rudis,* a member of the *saxatilis* complex (Wium-Anderson, 1970).

Finally, levels of apparently random geographic differentiation, such as those found in North American *L. saxatilis,* have been described in sympatric populations of *L. rudis* and *L. arcana* collected over a 30-mile transect in Scotland (Ward and Warwick, 1980).

C. *Mytilus* and Other Bivalves

In terms of population genetics, *Mytilus edulis* is probably the most thoroughly studied marine mollusc species. In addition to the morphological polymorphisms described earlier, biochemical studies of more than 30 enzyme loci have been reported, several of which include massive surveys of populations collected over wide geographic areas. In general, this species is quite polymorphic; about 70% of the gene–enzyme systems assayed show the segregation of two or more common alleles. Average heterozygosity is about 13% (summarized in Levinton and Koehn, 1976). Because of its global distribution, extensive dispersal capability, and diversity of habitats, it has emerged as a system of choice for the investigation of a number of interesting and important questions.

1. Life History

Mytilus is an abundant and widely distributed member of the littoral and estuarine community. It is primarily dioecious, but Siquira (1962) has reported the existence of hermaphrodites. Adults are generally sessile; however, limited movement by use of the prominent muscular foot is possible. In addition, passive movement by rafting may occur following accidental detachment.

Spawning occurs once a year, generally between the spring and early fall; usually, large numbers of gametes are released. Females are remarkably fecund and can produce as many as 12 million gametes during a single season (Field, 1922). Fertilization is external (Chipperfield, 1953), and the zygote subsequently proceeds through a prolonged period of pelagic larval development. Metamorphosis from the trochophore, to feeding velichonia, to settling preveliger may last for a minimum of 3 weeks to a maximum of 2 months. During this lengthy planktonic period, extensive dispersal occurs as larvae are transported by offshore currents. Substrates for settling are varied, but filamentous algae are favored. If appropriate substrates are absent, settling may be delayed (Bayne, 1964, 1965). Following settling, subsequent movements to sites favorable for adult attachment occur.

Growth occurs at a rate dependent upon both size and season. At Orient Point, New York, Milkman and Koehn (1977) reported a rate of 25 mm/year which was maximum in the summer. In this same study, evidence was provided for the recruitment of "young mussels substantially larger than newly metamorphosed spat," and the authors have inferred two episodes of recruitment involving large numbers of 20–25-mm animals. The life span of *Mytilus* has been estimated to be between 2 and 6 years.

2. Macrogeographic Patterns of Variation

In a major survey, Koehn et al. (1976) studied gene–enzyme variation in 150 populations of *M. edulis* taken throughout its geographic range on the East Coast of North America, from Virginia (its southern limit) to Newfoundland (in the north). A sample from Iceland was also studied. In addition, many of these same loci have been analyzed in populations collected from the West Coast of North America (Levinton and Koehn, 1976; Lassen and Suchanek, 1978), several European locales (Ahmad et al., 1977; Gosling and Wilkins, 1977, 1981; Murdock et al., 1975; Levinton and Koehn, 1976; Theisen, 1978; Skibinski et al.,

1978a,b, 1980; Skibinski and Beardmore, 1979), and Australia (Levinton and Koehn, 1976). To facilitate the discussion, some of the major results in Table III are summarized. It is important to remember that the allele classification systems used in individual laboratories have not been standardized. Any cases involving ambiguous comparisons or duplications have been omitted.

As might be expected from the notable dispersal potential of *M. edulis,* variation at a number of loci shows a marked level of interpopulation homogeneity. At the *MDH* locus, for instance, there is a single high-frequency allele throughout its entire range. Similar patterns of homogeneity have been described for *6-PGD* and *IDH,* where three abiquitous alleles occur at each locus. At the *GPI* locus there are three common alleles, designated 1.00, 0.93 and 0.89, in Western Atlantic populations, and several rare forms. In populations collected from the Eastern Atlantic, the 1.00 allele predominates. Koehn et al. (1976) report a significant cline in *GPI* allele frequencies between 37 and 44° north latitude. In the Pacific populations sampled, the 0.93 allele predominates. It is striking,

TABLE III

Allele Frequencies in Natural Populations of *Mytilus edulis*[a]

Gene	Allele	South of Cape Cod	North of Cape Cod	Iceland	Washington State	Australia	Western Ireland	New South Wales	France
LAP	1.00	0.01	0.05	0.22	0.01	0	0.13	0.24	0.14
	0.98	0.27	0.53	0.52	0.09	0.07	0.60	0.60	0.66
	0.96	0.21	0.39	0.22	0.46	0.29	0.21	0.14	0.20
	0.94	0.55	0.09	0.02	0.43	0.64	0.05	0	0
	0.92	0	0	0.02	0.01	0	0.01	0.02	0
AP	1.05	0	0	0	0	0.14	—	0.07	—
	1.00	0.01	0.01	0	0	0.52	—	0.16	—
	0.97	0.44	0.44	0.37	0.14	0.15	—	0.76	—
	0.92	0.36	0.35	0.56	0.79	0	—	0.02	—
	0.89	0.20	0.20	0.07	0.08	0	—	0	—
GPI	1.00	0.29	0.38	0.54	0.12	0.10	0.61	0.64	0.64
	0.93	0.25	0.25	0.38	0.61	0.82	0.30	0.23	0.26
	0.89	0.19	0.37	0.08	0.27	0.08	0.05	0.03	0.04
MDH	1.00	0.98	0.99	0.99	—	—	—	0.97	—
G-PGD	0.91	0.02	0.03	0.04	—	—	—	0.03	—
	0.83	0.97	0.96	0.96	—	—	—	0.95	—
	0.77	0.01	0.01	0	—	—	—	0.02	—
IDH	1.10	0.01	0.01	0	—	—	—	0.01	—
	1.00	0.93	0.95	0.84	—	—	—	0.81	—
	0.90	0.06	0.04	0.16	—	—	—	0.18	—

[a] See references in Table I.

however, that this three-allele system is globally distributed, a reflection no doubt of the enormous dispersal capability of this species.

A three-allele polymorphism is also found at the *AP* locus. In North America, the frequencies of the 0.97, 0.92, and 0.89 alleles occur in a ratio of approximately 5 : 3 : 2. Moving eastward to Europe, the ratios change to 3 : 7 : 1, a pattern found in the Eastern Pacific as well. The Australian population seems unique in terms of the kinds and frequencies of *AP* alleles, but it is represented by only a single sample.

At the *leucine aminopeptidase* locus first described by Milkman and Beatty (1970) and Koehn and Mitton (1972), there are five alleles which recur. Only three of these, designated 0.94, 0.96, and 0.98, are common in North America. Allele frequencies show marked geographic heterogeneity. In nonestuarine populations from Virginia to Cape Cod, the 0.94 allele has a very narrow range of frequencies, 0.50–0.60. North of Cape Cod through the Gulf of Maine, the frequency of this allele drops abruptly to about 0.10 or less. Moving north to Nova Scotia and Newfoundland, the 0.94 allele frequency again becomes abundant (0.3–0.4). Proceeding across the Atlantic, the 0.94 allele frequency declines (0.02 in Iceland), and this allele is either absent or found at very low frequencies in populations collected from South Wales, Ireland, and France. In the Pacific, the 0.94 allele is abundant in samples from Washington and Australia.

In most instances, changes in the frequency of the 0.94 allele in North American populations are accompanied by reciprocal changes in the frequencies of both the 0.96 and 0.98 alleles. In Nova Scotia and Newfoundland, however, this is not the case. Here the 0.98 allele becomes quite rare, whereas that of the 0.96 allele remains high. From Iceland eastward through Europe, the absence of the 0.94 allele is accompanied by high frequencies of the 0.98 allele and the appearance of a 1.00 variant at moderate frequencies, 0.11–0.23. To summarize, the geographic distribution of allelic variation in *M. edulis* is quite homogeneous. Some loci, however, show marked and apparently nonrandom patterns of heterogeneity.

3. Estuarine and Long Island Sound Populations

Within Long Island Sound and several other estuarine systems, there are dramatic and stable allele frequency clines at the *LAP* locus in adult mussel populations (Milkman and Beatty, 1970; Milkman, 1971; Milkman et al., 1972; Koehn and Mitton, 1972; Mitton et al., 1973; Boyer, 1974; Koehn, 1975; Koehn et al., 1976, 1980; Milkman and Koehn, 1977; Theisen, 1978; Lassen and Turano, 1978; Levinton and Lassen, 1978a,b). These are superimposed on substantially more homogeneous allele frequency distributions at other loci. In Long Island Sound, the frequency of the 0.94 allele declines from 0.55 in the Atlantic Ocean to about 0.10 near New York City. This cline is, in fact, even more abrupt and occurs over a distance of less than 20 miles. Samples of young mussels along this transect over 3 consecutive years revealed that although a cline was evident, the absolute frequencies of the 0.94 allele were higher than in adults, and the position of the cline varied from year to year. This suggests that immigrants moving into the sound from the huge oceanic population, where the 0.94 allele frequency is high, are selectively eliminated over time (Koehn et al., 1976, 1980; Koehn, 1978). These results argue effectively against the earlier interpretation of Lassen and Turano (1978) of that the cline was a result of pelagic larvae coming into the sound from the oceanic population mixing with larvae produced by the endemic sound population.

The decrease in *LAP* 0.94 frequency in immigrants entering Long Island Sound can be accounted for by the selective elimination of oceanic larvae, not adapted physiologically to the temperature or salinity conditions in the sound, that carry the 0.94 allele. Lassen and Turano (1978) go further and suggest that populations inside and outside the sound may represent different ecotypes, races, subspecies, or indeed, species, based on physiological experiments (Levinton and Lassen, 1978a,b; Levinton and Fundiller, 1975). Alternatively, selective mortality may be based on the *LAP* genotype itself. To address this question, a series of biochemical and physiological experiments was conducted (Koehn, 1978).

The *LAP* allozymes do display different biochemical properties. The most important of these, Koehn believes, is the greater catalytic activity of the 0.94 form. Because this allele is at its highest frequency when the conditions of high salinity and warm water temperature are met, the adaptive significance of this higher catalytic activity must be explained under both conditions. In *Mytilus*, *LAP* is found in the intestinal brush border and lysosomes of digestive gland and tubule cells, suggesting (to Koehn) a possible role for *LAP* in cell volume regulation. Because mussel osmoregulation at high salinity involves an adjustment of intracellular amino acid pool levels, the significance of high aminopeptidase activity in replenishing this pool is obvious. In fact, Koehn has shown that high salinity induces higher aminopeptidase activity. *LAP* accumulation is also induced by heat, although the exact physiological significance of this phenomenon is not known. In all, Koehn and his collaborators have provided abundant evidence that genetic variation at the *LAP* locus in *Mytilus* plays an adaptive role that is expressed in both physiological and ecological terms.

D. Deviations from Hardy–Weinberg Proportions

One remarkably common observation made during electrophoretic surveys of molluscan populations is that the frequency of genotypes in the population does not fit the ratios expected on the basis of the Hardy–Weinberg formula. Generally, the deviation involves a major deficiency of heterozygotes, but excess heterozygosity has been reported as well. A number of such reports are tabulated in Table IV. The probable basis of two of these deviations will be discussed, but

TABLE IV

Deviations from Hardy–Weinberg Proportions in Natural Populations

Species	Situation	Reference
Mytilus edulis	Deficiency in *LAP* heterozygotes in young animals which decreases with age. Deficiency is attributed to inbreeding, Wahland effect, or selection in settling spat. Later increase in heterozygosity is attributed to selection, although Schopf et al. (1975) reported a decrease in one population.	Milkman and Beatty (1970); Koehn and Mitton (1972); Mitton et al. (1973); Boyer (1974); Koehn et al. (1976); Lassen and Turano (1978)
Mytilis edulis	Excess homozygotes at the aspartate aminotransferase locus.	Johnson and Utter (1973)
Mytilis edulis	Excess homozygotes at the aminopeptidase locus.	Mitton et al. (1973)
Mytilis edulis	Three of 12 loci in a population show heterozygote deficiency.	Murdock et al. (1975)
Mytilis galloprovincialis	Excess heterozygotes at several loci indicate interspecific hybrids, whereas in each species there is a deficiency of heterozygotes.	Skibinski et al. (1978a)
Mytilis californicas	Excess homozygotes at three loci attributed to Wahland effect. Increased heterozygosity in older animals attributed to selection.	Tracey et al. (1975); Levinton and Fundiller (1975)
Modiolus demissus	Heterozygote deficiency at three loci; in one case the deficiency declines with increasing age (size).	Koehn and Mitton (1972); Koehn et al. (1973); Mitton et al. (1973); Chaisson et al. (1976)
Mercenaria mercenaria	Excess heterozygotes at the *LDH* locus	Pesch (1972)
Pecten maximus	Decrease in heterozygosity at the *PGI* locus with increasing age (size).	Wilkins (1978)
Crassostrea virginica	Heterozygote deficiency at several loci, but heterozygotes are in excess in the largest oysters.	Singh and Zouros (1978, 1981)

before considering individual cases it will be useful to outline the conditions under which Hardy–Weinberg proportions are generated.

A fundamental assumption is panmyxia, or random mating, in the population. Either self-fertilization or local inbreeding will produce excess homozygosity; assortative mating, depending on the direction, will result in either heterozygote excesses or deficiencies. A second assumption is the absence of migration into or out of the population sampled. If the population contains a substantial proportion of migrants from neighboring populations that contain a different set of allele frequencies, an excess of homozygotes is generally observed. This is known as the Wahland effect. Genotype-specific emigration may also produce the illusion of nonequilibrium conditions, but this must be rare in sessile forms. The third major assumption is the absence of selection, i.e., the alternate genotypes are equivalent with respect to fitness. Finally, deviations from equilibrium involving an excess of homozygotes will appear in electrophoretic surveys if the population contains a high frequency of null, or inactive, alleles. This is a technical problem in which heterozygotes carrying a null allele are misclassified as homozygotes.

In a study of the *TO* locus in the ribbed mussel *Modiolus demissus,* Koehn et al. (1973) discovered a deficiency of heterozygotes in the smallest size class of populations found both high and low in the intertidal region. The deficiency was greatest in the high intertidal. In samples of two progressively larger (and presumably older) size classes, the deficiency was less, and in the largest animals collected in the low intertidal there was actually an excess of heterozygotes. The authors present a strong argument that the progressive changes in heterozygote frequency probably are due to the differential survival of heterozygotes during growth. The initial deficiency represents differential mortality of the heterozygotes early in the life cycle. The same pattern was found (and the same conclusions were drawn) in a subsequent study of size-dependent variation of *LAP* genotypes in *Mytilus edulis* (Koehn et al., 1976).

Another interesting case is described by Zouros et al. (1980) and Singh and Zouros (1981), who studied gene–enzyme variation in the American oyster, *Crassostrea virginica.* They find that in spite of the strong correlation between growth rate and genetic heterozygosity, the overall population is significantly deficient in heterozygous genotypes at six of the seven loci studied. It would appear that viability and growth rate may be negatively correlated here, but there is no evidence that the deficiency was absent in the sample of larvae from which the population arose.

IV. Evolutionary Genetics

A. The Use of Genetics in Systematics

The systems of taxonomic classification used for marine molluscs rely primarily on morphometric traits such as shell sculpture, shell geometry, patterns of pigmentation, or radula sculpturing. Because these complex characters are usually determined by many genes which themselves may be polymorphic, taxonomic groups often show a range of variation for diagnostic traits. In addition, morphometric traits are often influenced by the environment, thus adding further uncertainty, especially when comparisons are made of populations collected from distant sites. Ideally, systematic distinctions should rely on a set of qualitative, discontinuous traits that are not sensitive to environmental perturbation. Over the past 15 years, it has become apparent that electrophoretic phenotypes are ideally suited for taxonomic work (Avise, 1974). Several interesting examples are presented below.

1. Mytilus edulis and M. galloprovincialis

In Europe there are two taxonomic groups of *Mytilus; M. edulis* and *M. galloprovincialis. M. galloprovincialis* ranges from the Adriatic, Black, and Mediterranean seas as far north as England and Ireland, whereas *M. edulis* is distributed from the Arctic south to North Africa but is absent in the Mediterranean (Seed, 1978). In Ireland and southwestern England, the two species are sympatric. Barsotti and Meluzzi (1968) suggest that *M. galloprovincialis* arose as a separate species in the Mediterranean from an *edulis*-like population during the Pleistocene ice age between 1 and 2 million years ago and migrated north as the ice receded. Although the two are regarded as separate species by some (Lubet, 1973) *M. galloprovincialis* has also been considered a subspecies of *M. edulis* (Soot-Ryen, 1969).

The taxonomic status of *Mytilus* populations in southwestern England and Ireland was problematic (see review by Seed, 1978). Hepper (1957) and others (Lewis and Seed, 1969; Seed, 1971) speculated the basis of morphology and spawning time, that an unusual type of mussel collected at Padstow in Cornwall was *M. galloprovincialis.* However, it was also clear that a number of *M. galloprovincialis* characteristics were found in British *M. edulis* (Lewis and Powell, 1961), especially at exposed sites (Seed, 1974), so that the taxonomic status of the Padstow mussel remained questionable.

In the past few years, this taxonomic issue has been approached and partially clarified by the use of gel electrophoresis. Ahmad and Beardmore (1976) found significant differences in allele frequency between *M. edulis* and the Padstow mussel at three loci (*aminopeptidase, LAP2,* and *phosphohexose isomerase*) and concluded that the Padstow mussel is *M. galloprovincialis.* The differences among these loci were not, however, totally diagnostic. Ahmad and Beardmore later provided calculations of genetic distance between the two species from gene frequency data (Skibinski et al., 1980) and estimated that the divergence occurred 1–3 million years ago.

Subsequent studies of mussel populations in southwestern England (Skibinski et al., 1978a,b) revealed that samples of morphologically intermediate individuals contain a large excess of genotypes heterozygous for alleles that are individually frequent in only one of the two species. This strongly suggested the occurrence of hybridization or introgression in the areas sampled. The proportion of hybrids in the populations was 2–5%. Skibinski and Beardmore (1979) continued this analysis and discovered that at many locales considerable intergradation was occurring. They suggested that, in these zones, extensive and apparently stable intergradation was the result of superior fitness among the hybrids.

The analysis of genetic variation in Irish populations of the two species (Gosling and Wilkins, 1977, 1981) confirmed the earlier studies of English mussels and concluded that, in sympatric populations, hybridization was much more prevalent in exposed sites (7%) than in sheltered sites (2%).

2. Lumping and Splitting the Littorinids

The taxonomic classification of littorinid species represents another good example of the usefulness of genetics in resolving questions in systematics. This is especially apparent in the rough periwinkle *L. saxatilis,* which had been regarded as being separable into six or seven subspecies (Dautzenberg and Fischer, 1912; James, 1968). Heller (1975a) divided the *saxatilis* group into four distinct but sympatric species (*L. rudis, L. patula, L. nigrolineata,* and *L. neglecta*) on the basis of shell morphology and penis structure, although the splitting of *L. rudis* and *L. patula* was soon questioned (Raffaelli, 1979). The most recent revision in the classification system was made by Hannaford-Ellis (1978, 1979), who described another new species in the *saxatilis* group, *L. arcana,* and confirmed that *L. rudis* and *L. patula* were synonymous. The classification of the *saxatilis* group now includes four separate species: *L. rudis, L. neglecta, L. arcana,* and *L. nigrolineata.* The first two are ovoviviparous, the other two oviparous.

Electrophoretic studies on European littorinids have generally confirmed the most recent classification scheme. Ward and Warwick (1980) conclude, from an analysis of 21 loci in nine populations, that *L. rudis* and *L. arcana* are very similar genetically but clearly maintain separate gene pools. This genetic similarity is striking in light of the difference in reproductive mode (see above) and suggested to the authors that the differences in reproductive mode may involve changes in only a few genes, which are perhaps regulatory rather than structural. Punctuated equilibrium (Eldridge and Gould, 1972, 1974) is now a fashionable explanation for rapid evolutionary change. Wilkins and O'Regan (1980) have added further support to the splitting of the *saxatilis* group by providing genetic evidence that *L. rudis, L. nigrolineata,* and *L. neglecta* are probably separate species with no evidence of hybridization. In that same report, they confirmed the proposal of Sacchi and Rastelli (1967) and Goodwin and Fish (1977) that *L. obtusata* (formerly *L. littoralis*) is composed of two species, *L. obtusata* and *L. mariae.*

A species that had escaped the taxonomists' scalpel is *L. littorea,* a ubiquitous member of the intertidal community along both Atlantic coasts. However, in a recent comparison of gene–enzyme variation in populations of *L. littorea, L. saxatilis,* and *L. obtusata* collected from Woods Hole, Massachusetts, and Roscoff in Brittany, a rather startling observation was made (Berger, 1977). Although the French and American samples of *L. saxatilis* and *L. obtusata* had a level of genetic differentiation compatible with their low dispersal rates and the distance between the populations, the two samples of *L. littorea,* a species with great dispersal potential, were so completely different at the genetic level that they appeared to be distinct species. At 7 of the 11 loci studied, the two *L. littorea* populations carried no alleles in common. This pattern of mutual exclusivity was unique to this species. Although this observation has yet to be confirmed, it is, in retrospect, not unreasonable.

The origin of North American *L. littorea* is unknown. It was first reported in North America on Nova Scotia by Ganong (1886) in 1840 and since then has rapidly expanded its range southward, at present extending to Maryland (Wells, 1965). There are two hypotheses concerning its origin in North America. The first proposes recent immigration from Europe (Ganong, 1886; Wells, 1965); the second argues that the species had been present for some time as a small isolated population in Canada, and spread south recently as a result of commerce (Clarke and Erskine, 1961a,b; Bird, 1968). Support for the latter hypothesis comes from the discovery of *L. littorea* specimens in the ancient shell heaps of the Micmac and from below Viking camps. The genetic differences between European and North American *L. littorea* provide compelling evidence for an ancient separation. One additional observation reported by Berger (1977) is that the level of heterozygosity in North American *L. littorea* is very low. This suggests either that the North American population was founded by only a few immigrants, or that the founding group was large but then underwent a substantial bottleneck in population size at some later time.

3. Other Cases

Genetic analysis has often revealed the presence of sibling species that are morphologically indistinguishable. Although the intertidal acmaed limpet *Collisella digitalis* has been examined for gene–enzyme variation (Gresham and Tracey, 1975), Murphy (1978) only recently discovered that the northern and southern groups are actually separate species. The presence of unique alleles, and allele frequency differences between *C. digitalis* and what is now called *C. austrodigitalis,* were not sufficiently diagnostic to disprove interbreeding or introgression. In contrast, the Australian limpet *Siphonaria kurracheensis* often exhibits a bimodal vertical distribution of abundance resulting from competition with *C. onychitus.* Although the high- and low-shore limpets show phenotypic differences, an electrophoretic survey of five polymorphic enzymes revealed no genetic differentiation between the two groups (Black and Johsnon, 1981).

Similar studies have resolved additional taxonomic questions regarding *Goniobasis* (Chambers, 1978), the Danish *Hydrobia* species complex (Lassen, 1979), and the Danish *Cardium* complex (Brock, 1978; Jelnes et al., 1971).

B. Genetics of Speciation

Speciation is generally thought of as an *allopatric* process during which conspecific populations first become physically separated and subsequently undergo genetic differentiation. Speciation is complete when the accumulated genetic differences result in reproductive isolation. Although neo-Darwinian dogma emphasizes the role of natural selection leading to local adaptation as the mechanism responsible for genetic divergence, nondeterministic forces (particularly genetic drift) may also play an important role. A second process, termed *sympatric* speciation, occurs in the absence of physical separation. For example, if local selection is disruptive and the extremes in a phenotypic range have the highest fitness, then selection will obviously also favor any allele or combination of alleles that promotes positive assortative mating. The critical point here is that random mating alleles became concentrated in the intermediates that are selectively eliminated. Both sympatric and allopatric speciation lead to an increase in species number. During *anagenesis* (phyletic evolution), one species is transformed into another over time. This does not increase extant taxonomic diversity, and in this case the criterion of reproductive isolation cannot apply.

The data on marine molluscs suggest that there is a striking correlation between the rate of allopatric speciation and the dispersal capability of a taxonomic group (Scheltema, 1971, 1978). Species without a planktonic larval stage undergo a more rapid rate of phyletic change, as measured by either phenotypic or genotypic variance over space or the number of taxonomic subspecies listed, than do species with an extended pelagic larval stage.

In a recent report on freshwater molluscs, Williamson (1981) has made several important observations on the rate of anagenesis. The author carried out a detailed morphological examination of 3300 fossil molluscs representing 13 species from northern Kenya, near Lake Turkana. The conclusions drawn may be summarized as follows. For very long periods of time (millions of years), the species demonstrated marked morphological stasis. In fact, several of the species in the fossil record appear unchanged today. At two points in the geological history of the lake, the water level dropped sharply and the species underwent a brief (5000–50,000 years) but significant period of transformation which led to the establishment of distinct progeny species. Surprisingly, the transition was marked by a significant increase in morphological variation, which was subsequently reduced as the new species became established. As the water level in the lake rose again, members of the ancestral species from other populations returned to the lake and the new progeny species disappeared. These episodic bursts of evolution, characterized by a major increase in phenotypic variation and long periods of stasis, are quite unlike Darwin's notion of slow, gradual change. The extent to which rapid evolutionary change represents a general pattern remains to be determined.

The final issue of interest here concerns the genetics of speciation. To put it simply, we wish to assess the quality and quantity of genetic change involved in either the transformation of one species to another or the establishment of reproductive isolation between two groups that formally shared a common gene pool. The problems here are obvious. Because we are able to identify and study only organisms that have already completed the speciation process, it is impossible to distinguish genetic changes that occurred during the speciation event from those that occurred subsequently. The second and more serious difficulty is, in a sense, technical. The genes that we can study easily are those which produce enzymes or determine qualitative morphological traits. With great effort, it is possible to study and understand the genetic basis of quantitative variation. However, it is not yet clear that these kinds of genes are necessarily important for speciation or for higher order evolutionary change. There is a strong sense (and weak evidence) that major evolutionary events result from changes in a class of genetic elements loosely called either *regulatory* or *developmental* genes. At this time we know very little about the nature and function of such genes, if indeed they do exist, but it is likely that the products of developmental genes are not to be found in contemporary biochemistry textbooks. Another stimulating hypothesis, championed by Wilson et al. at Berkeley, is that the rearrangement of genes may have a far greater evolutionary consequence than the replacement of one amino acid by another.

Thus, the literature on marine molluscan population genetics contains numerous estimates of genetic distance between species, based on measurements of allele frequencies determined by gel electrophoresis, which will be found in the references for this chapter. Although these genetic distance values are quite useful as taxonomic tools for establishing phylogenetic trees, they are of questionable significance understanding the genetic basis of evolutionary change.

C. Genetic Strategies of Adaptation

Genetic strategies of adaptation can be defined as a set of parameters which enhance the fitness of the organism or species in the context of its external and internal environments. Developmental homeostasis, or canalization, for example, represents a genetic strategy in higher organisms for buffering individual ontogeny from variations in the environment. An example in prokaryotes is dual and self-reinforcing mechanisms involved in regulating the expression of bacterial operons. In the *Escherichia coli* lactose operon there are two regulatory mechanisms, a system of negative control exerted by the repressor and a system of positive control exerted by the catabolite-activating protein. Two examples of genetic strategies that have been discussed in the molluscan literature are discussed here.

1. Mass Extinction and Genetic Variation

Several waves of mass extinction have been noted in the fossil records of shallow-water marine invertebrates, one impressive example having occurred during the Permian-Triassic transition. The most vulnerable lineages appear to be highly specialized forms associated with rather stable environments, such as the shallow-water reef communities. Valentine and Ayala and associates (Ayala et al., 1973, Valentine et al., 1973; Campbell et al., 1975; Valentine, 1976; Ayala and Valentine, 1978) tested one hypothesis that had been suggested to explain this pattern (Bretsky and Lorenz, 1970), i.e., that the extinctions affect organisms living in a stable environment because they have little genetic variation and are thus unable to adapt genetically to any environmental instability that arises. The implicit assumption is that selection for specialization leads to genetic homozygosity. The organism studied was the ''killer clam'' *Tridacna maxima,* which is called ''a plausible analog of the sorts of fossil species that have commonly become extinct'' (Ayala et al., 1973). In a survey of 30 gene–enzyme systems, they found a high proportion of polymorphic loci (64%) and high levels of heterozygosity (about 20%) even in populations that had not been exposed to the heavy irradiation of nuclear weapons testing in the Pacific (Campbell et al., 1975). Thus, the notion that ecological specialization leads to homozygosity, and that this genetic depauperacy, in turn, makes species more vulnerable to extinction, appears to be incorrect.

2. Epistatis and Concordance

We have discussed at length the evidence from Koehn et al. that natural selection at the *LAP* locus in *M. edulis* is mediated by salinity and temperature. In the context of genetic strategies, two additional studies are noteworthy. First, Koehn and Mitton (1972) have shown a correlated pattern of *LAP* variation in the ribbed mussel *Modiolus demissus.* Both this species and *Mytilus edulis* maintain a *LAP* polymorphism characterized by three common and two rare alleles. Although the alleles in the two species are different, the rank order of relative abundance is nearly identical in all populations of both species. Spatial changes in allele frequency along an estuary in one species were paralleled by allele frequency changes in the other.

In another study, Mitton et al. (1973) simultaneously examined the distribution of genetic variation at two loci, *LAP* and *AP* (aminopeptidase), in *M. edulis.* Both genes produce enzyme products that carry out similar catalytic functions. A significant correlation was noted between specific pairs of *LAP* and *AP* genotypes in the population. These combinations, they concluded, are attributable to epistasis, "since the magnitude of dependency varies with age" (Mitton et al., 1973). Because these two genes have not yet been mapped, the possibility of linkage disequilibrium remains.

References

Ahmad, M., and Beardmore, J. A. (1976). Genetic evidence that the "Padstow Mussel" is *Mytilus galloprovincialis. Mar. Biol.* **35,** 139–144.

Ahmad, M., Skibinski, D., and Beardmore, J. A. (1977). An estimate of the amount of genetic variation in the common mussel *Mytilus edulis. Biochem. Genet.* **15,** 833–846.

Avise, J. C. (1974). Systematic value of electrophoretic data. *Syst. Zool.* **23,** 465–481.

Ayala, F. J., and Valentine, J. W. (1978). Genetic variation and resource stability in marine invertebrates. *In* "Marine Organisms: Genetics, Ecology and Evolution" (B. Battaglia and J. A. Beardmore, eds.), pp. 23–51. Plenum, New York.

Ayala, F. J., Hedgecock, D., Zumwolt, G. S., and Valentine, J. W. (1973). Genetic variation in *Tridacna maxima,* an ecological analog of some unsuccessful evolutionary lineages. *Evolution* **27,** 177–191.

Barsotti, G., and Meluzzi, C. (1968). Osservazioni su *Mytilus edulis* e *M. galloprovincialis. Conchiglie* **4,** 50–58.

Bayne, B. L. (1964). Primary and secondary settlement in *Mytilus edulis* L. (Mollusca). *J. Anim. Ecol.* **33,** 513–523.

Bayne, B. L. (1965). Growth and delay of metamorphosis of larvae of *Mytilus edulis. Ophelia* **2,** 1–47.

Beardmore, J. A., and Morris, S. R. (1978). Genetic variation and species coexistence in *Littorina. In* "Marine Organisms: Genetics, Ecology and Evolution" (B. Battaglia and J. A. Beardmore, eds.), pp. 123–140. Plenum, New York.

Berger, E. M. (1973). Gene-enzyme variation in three sympatric species of *Littorina. Biol. Bull. (Woods Hole, Mass.)* **145,** 83–90.

Berger, E. M. (1977). Gene-enzyme variation in three sympatric species of *Littorina.* II. The Roscoff population, with a note on the origin of North American *L. littorea. Biol. Bull. (Woods Hole, Mass.)* **153,** 255–264.

Bernstein, S. C., Throckmorton, H., and Hubby, J. (1973). Still more genetic variability in natural populations. *Proc. Natl. Acad. Sci. U.S.A.* **70,** 3928–3931.

Berry, R. J., and Crothers, J. H. (1968). Stabilizing selection in the dog whelk (*Nucella lapillus*). *J. Zool.* **155,** 5–17.

Bird, J. B. (1968). *Littorina littorea:* occurrence in Newfoundland beach terrace, predating Norse settlements. *Science* **159,** 114.

Black, R., and Johnson, M. S. (1981). Genetic differentiation independent of intertidal gradients in the pulmonate limpet *Siphonaria kurracheensis. Mar. Biol.* **64,** 79–84.

Boyer, J. F. (1974). Clinal and size-dependent variation at the LAP locus in *Mytilus edulis. Biol. Bull. (Woods Hole, Mass.)* **147,** 535–549.

Bretsky, P. W., and Lorenz, D. M. (1970). Adaptive response to environmental stability: a unifying concept in paleoecology. *Proc. North Am. Paleontol. Conv., Part E,* pp. 522–550.

Brock, V. (1978). Morphological and biochemical criteria for the separation of *Cardium glaucum* from *Cardium edule. Ophelia* **17,** 207–214.

Buroker, N. E., Hershberger, W. K., and Chew, K. K. (1979a). Population genetics of the family *Ostreidae.* I. Intraspecific studies on *Crassostrea gigas* and *Saccostrea commercialis. Mar. Biol.* **54,** 157–169.

Buroker, N. E., Hershberger, W. K., and Chew, K. K. (1979b). Population genetics of the family *Ostreidae*. II. Interspecific studies of the genera *Crassostrea* and *Saccostrea*. *Mar. Biol.* **54,** 171–184.

Campbell, C. A. (1978). Genetic divergence between populations of *Thais lamellosa* (Gmelin). *In* "Marine Organisms: Genetics, Ecology and Evolution" (B. Battaglia and J. A. Beardmore, eds.), pp. 157–170. Plenum, New York.

Campbell, C. A., Ayala, F. J., and Valentine, J. W. (1975). High genetic variability in a population of *Tridacna maxima* from the Great Barrier Reef. *Mar. Biol.* **33,** 341–345.

Chaisson, R. E., Serunian, L. A., and Schopf, T. J. (1976). Allozyme variation between two marshes and possible heterozygote superiority within a marsh in the bivalve *Modiolus demissus. Biol. Bull. (Woods Hole, Mass.)* **151,** 404.

Chambers, S. M. (1978). An electrophoretically detectable sibling species of "*Goniobasis floridensis*" (Megogostropoda: Pleuroceridae). *Malacologia* **17,** 157–162.

Chanley, P. E. (1961). Inheritance of shell markings and growth in the hard clam, *Venus mercenaria. Proc. Natl. Shellfish. Assoc.* **50,** 163–169.

Chipperfield, P. N. J. (1953). Observations on the breeding and settlement of *Mytilus edulis* (L.) in British waters. *J. Mar. Biol. Assoc. U.K.* **32,** 449–476.

Clarke, A. H., and Erskine, J. S. (1961a). Pre-Columbian *Littorina littorea* in Nova Scotia. *Science* **134,** 393–394.

Clarke, A. H., and Erskine, J. S. (1961b). Pre-Columbian *Littorina littorea* in Nova Scotia. *Biol. Bull. (Woods Hole, Mass.)* **145,** 83–90.

Costa, R., and Bisol, P. M. (1978). Genetic variability in deep sea organisms. *Biol. Bull. (Woods Hole, Mass.)* **155,** 125–133.

Dautzenberg, P. H., and Fischer, H. (1912). Mollusques prevenant de campagnes de "Hirondelle" et de la "Princesse Alice" dans les Mers du Nord. *Result, Camp. Sci. Prince Albert I* **37,** 187–201.

Eldridge, N., and Gould, S. J. (1972). Punctuated equilibria: an alternative to phyletic gradualism. *In* "Models in Paleobiology" (T. J. M. Schopf, ed.), pp. 82–115. Freeman/Cooper, San Francisco, California.

Eldridge, N., and Gould, S. J. (1974). Morphological transformation. The fossil record, and the mechanism of evolution: a debate. Part II. *Evol. Biol.* **7,** 303–308.

Field, I. A. (1922). Biology and economic value of the sea mussel *Mytilus edulis. Fish. Bull.* **38,** 127–260.

Fujino, K., Sasaki, K., and Wilkins, N. P. (1980). Genetic studies on the Pacific abalone. III. Differences in electrophoretic patterns between *Haliotis discus Reeve* and *H. discus hannai Ino. Bull. Jpn. Soc. Sci. Fish.* **46,** 543–548.

Gaines, M. S., Caldwell, S. J., and Vivas, A. M. (1974). Genetic variation in the mangrove periwinkle *Littorina angulifera. Mar. Biol.* **27,** 327–332.

Ganong, W. F. (1886). Is *Littorina littorea* introduced or indigenous? *Am. Nat.* **20,** 931–940.

Giesel, J. T. (1970). On the maintenance of shell pattern and behavior polymorphism in *Acmaea digitalis,* a limpet. *Evolution* **24,** 98–119.

Gooch, J. L., and Schopf, T. J. M. (1973). Genetic variability in the deep sea: relation to environmental variability. *Evolution* **26,** 545–552.

Gooch, J. L., Smith, B. S., and Knapp, D. (1972). Regional survey of gene frequencies in the mud snail *Nassarius obsoletus. Biol. Bull. (Woods Hole, Mass.)* **142,** 36–48.

Goodwin, B. J., and Fish, J. D. (1977). Inter and intra specific variation in *L. obtusata* and *L. mariae. J. Moll. Stud.* **43,** 241–254.

Gosling, E. (1979). Hidden genetic variability in two populations of a marine mussel. *Nature (London),* **279,** 713–715.

Gosling, E. (1980). Gene frequency changes in adapatation in marine cockles. *Nature (London),* **286,** 601–602.

Gosling, E., and Wilkins, N. P. (1977). Phosphoglucoisomerase allele frequency data on *Mytilus edulis* from Irish coastal sites: its ecological significance. *In* "Biology of Benthic Organisms" (B. F. Keegan, P. O'Ceidigh, and P. J. S. Boaden, eds.), pp. 297–309. Pergamon, Oxford.

Gosling, E., and Wilkins, N. P. (1981). Ecological genetics of the mussels *Mytilus edulis* and *M. galloprovincialis* on Irish Coasts. (*In press.*)

Gould, S. J. (1977). *In* "Ever Since Darwin," p. 11. Norton, New York.

Gresham, M., and Tracey, M. (1975). Genetic variation in an intertidal gastropod, *Collisella digitalis*. *Genetics* **80,** S37.

Hannaford-Ellis, C. J. (1978). *Littorina arcana* sp. nov. a new species of winkle (Gastropoda: Prosobranchia: Littorinidae). *J. Conchol.* **29,** 304.

Hannaford-Ellis, C. J. (1979). Morphology of the oviparous rough winkle, *Littorina arcana* Hannaford-Ellis, 1978, with notes on the taxonomy of the *L. saxatilis* species-complex (Prosobranchia: Littorinidae). *J. Conchol.* **30,** 43–56.

Heller, J. (1975a). The taxonomy of some British *Littorina* species, with notes on their reproduction (Mollusca: Prosobranchia). *Zool. J. Linn. Soc.* **56,** 131–151.

Heller, J. (1975b). Visual selection of shell colour in two littoral prosobranchia. *Zool. J. Linn. Soc.* **56,** 153–170.

Hepper, B. T. (1957). Notes on *Mytilus galloprovincialis* in Great Britain. *J. Mar. Biol. Assoc. U.K.* **36,** 33–40.

Hoagland, K. E. (1977). A gastropod color polymorphism: One adaptive strategy of phenotypic variation. *Biol. Bull. (Woods Hole, Mass.)* **152,** 360–372.

Hubby, J., and Lewontin, R. C. (1966). A molecular approach to the study of genic heterozygosity in natural populations. I. The number of alleles at different loci in *Drosophila pseudoobscura*. *Genetics* **54,** 577–594.

Innes, D. J., and Haley, L. E. (1977a). Inheritance of shell color polymorphism in the mussel. *J. Hered.* **68,** 203–204.

Innes, D. J., and Haley, L. E. (1977b). Genetic aspects of larval growth under reduced salinity in *Mytilus edulis*. *Biol. Bull. (Woods Hole, Mass.)* **153,** 312–321.

James, B. L. (1968). The characters and distribution of the subspecies and varieties of *Littorina saxatilis* (Olivi 1872) in Britain. *Cah. Biol. Mar.* **9,** 143–165.

Jelnes, J. E., Petersen, G. H., and Russell, P. J. C. (1971). Isozyme taxonomy applied on four species of *Cardium* from Danish and British waters with a short description of the species (Bivalvia). *Ophelia* **9,** 15–19.

Johnson, A. G., and Utter, F. M. (1973). Electrophoretic variants of aspartate aminotransferase in the bay mussel *Mytilus edulis*. *Comp. Biochem. Physiol. B* **44,** 317–323.

Johnson, A. G., Utter, F. M., and Niggol, K. (1972). Electrophoretic variants of aspartate aminotransferase and adductor muscle proteins in the native oyster (*Ostrea lurida*). *Anim. Blood Groups Biochem. Genet.* **3,** 109–113.

Johnson, G. (1979). Increasing the resolution of polyacrylamide gel electrophoresis by varying the degree of gel crosslinking. *Biochem. Genet.* **17,** 499–516.

Koehn, R. K. (1975). Migration and population structure in the pelagically dispersing marine invertebrate *Mytilus edulis*. *In* "Isozymes" (C. L. Markert, ed.), Vol. 4, pp. 945–959. Academic Press, New York.

Koehn, R. K. (1978). Biochemical aspects of genetic variation at the *LAP* locus in *Mytilus edulis*. *In* "Marine Organisms: Genetics, Ecology and Evolution (B. Battaglia and J. A. Beardmore, eds.), pp. 211–227. Plenum, New York.

Koehn, R. K., and Mitton, J. B. (1972). Population genetics of marine pelecypods. I. Ecological heterozygosity and evolutionary strategy at an enzyme locus. *Am. Nat.* **106,** 47–56.

Koehn, R. K., Turano, F., and Mitton, J. B. (1973). Population genetics of marine pelecypods. II. Genetic differences in microhabitats of *Modiolus demissus*. *Evolution* **27,** 100–105.

Koehn, R. K., Milkman, R., and Mitton, J. B. (1976). Population genetics of marine pelecypods. IV. Selection, migration and genetic differentiation in the blud mussel *Mytilus edulis*. *Evolution* **30,** 2–32.

Koehn, R. K., Newell, R., and Immermann, F. (1980). Maintenance of an aminopeptidase allele frequency cline by natural selection. *Proc. Natl. Acad. Sci. U.S.A.* **77,** 5385–5389.

Lannan, J. E. (1972). Estimating heritability and predicting response to selection for the Pacific oyster, *Crassostrea gigas*. *Proc. Natl. Shellfish Assoc.* **62,** 62–66.

Lassen, H. H. (1979). Electrophoretic enzyme patterns and breeding experiments in Danish mudsnails (Hydrobiidae). *Ophelia* **18,** 83–87.

Lassen, H. H., and Suchanek, T. H. (1978). Geographic variation niche breadth and genetic differentiation at different geographic scales in the mussels *Mytilus californianus* and *M. edulis*. *Mar. Biol.* **49,** 363–375.

Lassen, H. H., and Turano, F. J. (1978). Clinal variation and heterozygote deficit at the *LAP*-locus in *Mytilus edulis*. *Mar. Biol.* **49,** 245–254.

Levinton, J. S. (1973). Genetic variation in a gradient of environmental variability: Marine *Bivalvia* (Mollusca). *Science* **180,** 75–76.

Levinton, J. S. (1975). Levels of genetic polymorphism at two enzyme encoding loci in eight species of the genus *Macoma* (Mollusca: Bivalvia). *Mar. Biol.* **33,** 41–47.

Levinton, J. S., and Fundiller, D. (1975). An ecological and physiological approach to the study of biochemical polymorphisms. *Proc. Mar. Biol. Symp. 9th, 1974* pp. 165–178.

Levinton, J. S., and Koehn, R. K. (1976). Population genetics of mussels. *In* "Marine Mussels: Their Ecology and Physiology" (B. L. Bayne, ed.), pp. 357–384. Cambridge Univ. Press, London and New York.

Levinton, J. S., and Lassen, H. H. (1978a). Selection, ecology and evolutionary adjustment within bivalve mollusc populations. *Philos. Trans. R. Soc. London, Ser. B* **284,** 403–415.

Levinton, J. S., and Lassen, H. H. (1978b). Experimental mortality studies and adaptation at the LAP locus in *Mytilus edulis*. *In* "Marine Organisms: Genetics, Ecology and Evolution" (B. Battaglia and J. A. Beardmore, eds.), pp. 229–254. Plenum, New York.

Lewis, J. R., and Powell, H. T. (1961). The occurrence of curved and angulate forms of the mussel *Mytilus edulis* in the British Isles, and their relationship to *M. galloprovincialis*. *Proc. Zool. Soc. London* **137,** 583–598.

Lewis, J. and Seed, R. (1969). Morphological variations in *Mytilus* from southwest England in relation to the occurrence of *M. galloprovincialis*. *Cah. Biol. Mar.* **10,** 231–253.

Lewontin, R. C., and Hubby, J. (1966). A molecular approach to the study of genic heterozygosity in natural populations. II. Amount of variation and degree of heterozygosity in natural populations of *Drosophila pseudoobscura*. *Genetics* **54,** 595–609.

Longwell, A. C., and Stiles, S. S. (1973). Oyster genetics and the probable future role of genetics in aquaculture. *Malacol. Rev.* **6,** 151–177.

Lubet, P. (1973). Exposé synoptique des données biologiques sur la moule *Mytilus galloprovincialis*. *Synop. FAO Pesches* No. 88, pp. 8–28.

Mathers, N. F., Wilkins, N. P., and Walne, P. R. (1974). Phosphoglucose isomerase and esterase phenotypes in *Crassostrea angulata* and *C. gigas* (Bivalvia, Ostreiche). *Biochem. Syst. Ecol.* **2,** 93–96.

Milkman, R. (1971). Genetic polymorphism and population dynamics in *Mytilus edulis*. *Biol. Bull (Woods Hole, Mass.)* **141,** 397.

Milkman, R., and Beatty, L. D. (1970). Large scale electrophoretic studies of allelic variation in *Mytilus edulis*. *Biol. Bull. (Woods Hole, Mass.)* **139,** 430.

Milkman, R., and Koehn, R. K. (1977). Temporal variation in the relationship between size, numbers and an allele frequency in a population of *Mytilus edulis*. *Evolution* **31,** 103–115.

Milkman, R., Zeitler, R., and Boyer, J. F. (1972). Spatial and temporal genetic variation in *Mytilus edulis: natural selection and larval dispersal. Biol. Bull. (Woods Hole, Mass.)* **143,** 470.

Mitton, J. B. (1977). Shell color and pattern variation in *Mytilus edulis* and its adaptive significance. *Chesapeake Sci.* **18,** 387–390.

Mitton, J. B., Koehn, R. K., and Prout, T. (1973). Population genetics of marine pelecypods. III. Epistasis between functionally related isozymes of *Mytilus edulis. Genetics* **73,** 487–496.

Murdock, E. A., Ferguson, A., and Seed, R. (1975). Geographical variation at LAP in *Mytilus edulis* from the Irish coasts. *J. Exp. Mar. Biol. Ecol.* **19,** 33–41.

Murphy, P. G. (1978). *Collisella austrodigitalis* sp. nov.: A sibing species of limpet (Acmaeidae) discovered by electrophoresis. *Biol. Bull. (Woods Hole, Mass.)* **155,** 193–206.

Murray, J. (1975). The genetics of the Mollusca. *In* "Handbook of Genetics" (R. C. King, ed.), Vol. 3, pp. 3–31. Plenum, New York.

Newkirk, G. F. (1978). Interaction of genotype and salinity in larvae of the oyster *Crassostrea virginica. Mar. Biol.* **48,** 227–234.

Newkirk, G. F. (1980). Genetics of shell color in *Mytilus edulis* and the association of growth rate with shell color. *J. Exp. Mar. Biol. Ecol.* **47,** 89–94.

Newkirk, G. F., and Doyle, R. W. (1975). Genetic analysis of shell-shape variation in *Littorina saxatilis* on an environmental cline. *Mar. Biol.* **30,** 227–237.

Newkirk, G. F., and Doyle, R. W. (1979). Clinal variation of an esterase locus in *Littorina saxatilis* and *L. obtusata. Can. J. Genet. Cytol.* **21,** 505–513.

Newkir, G. F., L. E. Haley, D. L. Waugh, and Doyle, R. (1977a). Genetics of larvae and spat growth rate in the oyster *Crassostrea virginica. Mar. Biol.* **41,** 49–52.

Newkirk, G. F., Waugh, D. C., and Haley, L. E. (1977b). Genetics of larval tolerance to reduced salinity in two populations of oysters, *Crassostrea virginica. J. Fish. Res. Board Can.* **34,** 384–387.

Newkirk, G. F., Freeman, K. R., and Dickie, L. M. (1980). Genetic studies of the blue mussel, *Mytilus edulis,* and their implications for commercial culture. *Proc. World Maricult. Soc. Annu. Meet.* (in press).

Newkirk, G. F., Haley, L. E., and Dingle, J. (1981). Genetics of the blue mussel *Mytilus edulis* (L.). Nonadditive genetic variation in larval growth rate. *Can. J. Genet. Cytol.* **23,** 349–354.

O'Gower, A. K., and Nicol, P. I. (1968). A latitudinal cline of hemoglobins in a bivalve mollusc. *Heredity* **23,** 485–492.

Pesch, G. (1972). Isozymes of LDH in the hard clam, *Mercenaria mercenaria. Comp. Biochem. Physiol. B* **43,** 33–38.

Pesch, G. (1974). Protein polymorphisms in the hard clams *Mercenaria mercenaria* and *M. campechiensis. Biol. Bull. (Woods Hole, Mass.)* **146,** 393–403.

Powell, J. R. (1975). Protein variation in natural populations of animals. *Evol. Biol.* **8,** 79–119.

Raffaelli, D. G. (1979). The taxonomy of the *Littorina saxatilis* species complex, with particular reference to the systematic status of *Littorina patula. Zool. J. Linn. Soc.* **65,** 219–232.

Sacchi, C., and Rastelli, M. (1967). *Littorina mariae* nov. sp.: Les différences morphologiques et écologiques entre "rains" et leur signification adaptative et évolutive. *Atti Soc. Haliara Sci.* **105,** 351–370.

Scheltema, R. S. (1971). Larval dispersal as a means of genetic exchange between geographically separated populations of shallow-water benthic marine gastropods. *Biol. Bull. (Woods Hole, Mass.)* **140,** 284–322.

Scheltema, R. S. (1978). On the relationship between dispersal of pelagic veliger larvae and the evolution of marine prosobranch gastropods. *In* "Marine Organisms: Genetics, Ecology and Evolution" (B. Battaglia and J. A. Beardmore, eds.), pp. 303–322. Plenum, New York.

Schopf, J. J., Ohman, M. D., and Bleiweiss, R. (1975). Significant age-dependent and locality-dependent changes occur in gene frequencies in the ribbed mussel *Modiolus demissus* from a single salt marsh. *Biol. Bull. (Woods Holes, Mass.)* **149,** 446.

Schopf, T. J. M., and Gooch, J. L. (1971). Gene frequencies in a marine ectoproct: a cline in natural populations related to sea temperature. *Evolution* **25,** 286–289.

Seed, R. (1971). A physiological and biochemical approach to the taxonomy of *Mytilus edulis* and *M. galloprovincialis*. From S.W. England. *Cah. Biol. Mar.* **12,** 291–322.

Seed, R. (1974). Morphological variations in *Mytilus* from the Irish coasts in relation to the occurrence and distribution of *M. galloprovincialis. Cah. Biol. Mar.* **15,** 1–25.

Seed, R. (1978). The systematics and evolution of *Mytilus galloprovincialis. In* "Marine Organisms: Genetics, Ecology and Evolution" (B. Battaglia and J. A. Beardmore, eds.), pp. 447–468. Plenum, New York.

Serunian, L., and Schopf, T. (1976). Allozyme variation between two marshes and possible heterozygote superiority within a marsh in the bivalve *Modiolus demissus. Biol. Bull. (Woods Hole, Mass.)* **151,** 404.

Singh, R. S., Hubby, J. L., and Lewontin, R. C. (1974). Molecular heterosis for heat sensitive enzyme alleles. *Proc. Natl. Acad. Sci. U.S.A.* **71,** 1808–1810.

Singh, S.M., and Zouros, E. (1977). Correlation of homozygosity and growth rate in oysters. *Can. J. Genet. Cytol.* **19,** 580.

Singh, S. M., and Zouros, E. (1978). Genetic variation associated with growth rate in the American oyster (*Crassostrea virginica*). *Evolution* **32,** 342–353.

Singh, S. M., and Zouros, E. (1981). Genetics of growth rate in oysters and its implications for mariculture. *Can. J. Genet. Cytol.* **23,** 119–130.

Siquira, Y. (1962). Electrical induction of spawning in two marine invertebrates (*Urechis unicinctus*, hermaphroditic *Mytilus edulis*). *Biol. Bull. (Woods Holes, Mass.)* **123,** 203–206.

Skibinski, D. O., and Beardmore, J. A. (1979). A genetic study of intergradation between *Mytilus edulis* and *M. galloprovincialis. Experientia* **35,** 1442–1444.

Skibinski, D. O., Ahmad, M., and Beardmore, J. A. (1978a). Genetic evidence for naturally occurring hybrids between *Mytilus edulis* and *M galloprovincialis. Evolution* **32,** 354–364.

Skibinski, D. O., Beardmore, J. A., and Ahmad, J. (1978b). Genetic aids to the study of closely related taxa of the genus *Mytilus. In* "Marine Organisms: Genetics, Ecology and Evolution" (B. Battaglia and J. A. Beardmore, eds.), pp. 469–486. Plenum, New York.

Skibinski, D. O., Cross, T. F., and Ahmad, M. (1980). Electrophoretic investigation of systematic relationships in the marine mussel *Modiolus modiolus, Mytilus edulis* L., and *Mytilus galloprovincialis* (Mytillidae; Mollusca). *Biol. J. Linn. Soc.* **13,** 65–73.

Smith, D. A. S. (1976). Disruptive selection and morph-ratio clines in the polymorphic snail *Littorina obtusata* (*Gastropoda:* Prosobranchia). *J. Moll. Stud.* **42,** 114–135.

Snyder, T. P., and Gooch, J. L. (1973). Genetic differentiation in *Littorina saxatilis. Mar. Biol.* **22,** 177–182.

Soot-Ryen, T. (1969). Family Mytilidae Rafinesque, 1815. *In* "Treatise on Invertebrate Paleontology, Part N, Mollusca 6" (R. C. Moore, ed.), pp. 271–280. Geol. Soc. Am., Boulder, Colorado.

Staiger, H. (1957). Genetic and morphological variation in *Purpura lapillus* with respect to local and regional differentiation of population groups. *Ann. Biol.* **33,** 251–258.

Strushaker, J. W. (1968). Selection mechanisms associated with interspecific shell variation in *Littorina picta* (Prosobranchia: Neogastropoda). *Evolution* **22,** 459–480.

Theisen, B. F. (1978). Allozyme clines and evidence of strong selection in three loci on *Mytilus edulis* L. (*Bivalvia*) from Danish waters. *Ophelia* **17,** 135–142.

Tracey, M. L., Bellet, N. F., and Graven, C. D. (1975). Excess allozyme homozygosity and breeding structure in the mussel *Mytilus californianus. Mar. Biol.* **32,** 301–311.

Valentine, J. W. (1976). Genetic strategies of adaptation. *In* "Molecular Evolution" (F. J.Ayala, ed.), pp. 78–94. Sinauer Assoc., Sunderland, Massachusetts.

Valentine, J. W., Hedgecock, D., Zumwolt, G. S., and Ayala, F. J. (1973). Mass extinction and genetic polymorphism in the "killer clam," *Tridacna. Geol. Soc. Am. Bull.* **84,** 3411–3414.

Vermeij, G. J. (1971). Gastropod evolution and morphological diversity in relation to shell geometry. *J. Zool.* **163,** 15–23.

Ward, R. D., and Warwick, T. (1980). Genetic differentiation in the molluscan species *Littorina rudis* and *L. arcana* (Prosobranchia: Littorinidae). *Biol. J. Linn. Soc.* **14,** 340–361.

Wells, H. W. (1965). Maryland records of the gastropod, *Littorina littorea;* with a discussion of factors controlling its southern distribution. *Chesapeake Sci.* **6,** 38–42.

Wilkins, N. P. (1975). Phosphoglucose isomerase in marine molluscs. *In* "Isozymes" (C. L. Markert, ed.), Vol. 4, pp. 931–943. Academic Press, New York.

Wilkins, N. P. (1976). Genetic variability in marine *Bivalvia:* Implications and applications in molluscan mariculture. *Proc. Eur. Symp. Mar. Biol., 10th, 1975* Vol. 1, pp. 549–563.

Wilkins, N. P. (1977). Genetic variability in littoral gastropods: Phosphoglucose isomerase and phosphoglucomutase in *Patella vulgata* and *P. aspersa. Mar. Biol.* **40,** 151–155.

Wilkins, N. P. (1978). Length correlated changes in heterozygosity at an enzyme locus in the scallop (*Pectin maximus L.*). *Anim. Blood Groups Biochem. Genet.* **9,** 69–77.

Wilkins, N. P., and Mather, N. F. (1974). Phenotypes of phosphoglucose isomerase in some bivalve molluscs. *Comp. Biochem. Physiol. B* **48,** 599–611.

Wilkins, N. P., and O'Regan, D. (1980). Genetic variation in sympatric species of *Littorina. Veliger* **22,** 355–359.

Wilkins, N. P., O'Regan, D., and Moynihan, E. (1978). Electrophoretic variability and temperature sensitivity of phosphoglucose isomerase and phosphoglucomutase in *Littorinids* and other marine molluscs. *In* "Marine Organisms: Genetics, Ecology and Evolution" (B. Battaglia and J. Beardmore, eds.), pp. 141–155. Plenum, New York.

Wilkins, N. P., Fujino, K., and Sasaki, K. (1980). Genetic studies on the Pacific abalone. IV. thermostability difference among phosphoglucomutase variants. *Bull. Jpn. Soc. Sci. Fish.* **46**(5), 549–553.

Williamson, P. G. (1981). Paleontological documentation of speciation in Cenozoic molluscs from Turkana Basin. *Nature (London)* **293,** 437–443.

Wium-Anderson, G. (1970). Hemoglobin and protein variation in tree species of *Littorina. Ophelia* **8,** 267–273.

Zouros, E., Singh, S. M., and Miles, H. (1980). Growth rate in oysters: An overdominant phenotype and its possible explanations. *Evolution* **34,** 856–867.

14

Ecology and Ecogenetics of Terrestrial Molluscan Populations

A. J. Cain

Department of Zoology
University of Liverpool
Liverpool, United Kingdom

I. Introduction

Although the Mollusca are a major component of the terrestrial faunas of the earth, second only to the Arthropoda and much richer in species than the terrestrial vertebrates, far less is known about them than about either of the other two groups. Thousands of species are known from the shells and a little anatomy, or indeed only from the shells; and as large numbers are now extinct because of the destruction of habitats in this and the last century (see e.g., Solem, 1976;

The Mollusca, Vol. 6
Ecology

ISBN 0-12-751406-6

Hadfield and Mountain, 1981), we need to find out as much as possible about them before it is too late.

It is not even known why the Mollusca exist at all. They occupy a wide range of niches, but the Arthropoda do all that they can do and fly as well; yet these two vast phyla have coexisted since the Cambrian at least. A coexistence of 500 million years suggests stability, in which case there must be a general advantage in being slow moving, as well as fast moving, in almost all possible modes of life (except parasitism). What this advantage may be has not been determined.

From the point of view of the research worker, snails and slugs have the great advantage of forming highly localized populations, not too difficult to mark and sample, and exposed to considerable variations in the environment, justifying the celebrated remark of the botanist A. D. Bradshaw (at a conference on the ecogenetics of the snail *Cepaea):* ''Well, all I can say is, you've convinced me; *Cepaea* is a plant.'' The formal resemblances between the ecogenetics of terrestrial (and no doubt other) molluscs, other very sedentary or sessile animals, and plants is indeed considerable. It is a disadvantage that many snails live for several years in the wild with overlapping generations, and that even selection as intense as 1% (which is agreed by mathematical geneticists to be strong) may take several generations to become apparent. Furthermore, because the environment may vary greatly both spatially and temporally, it may be impossible to obtain large enough samples over a homogeneous area to detect even large selection coefficients (Cain, 1977c). The panmictic unit of snail populations may be only a few meters across, and in some of the larger and more easily handled species it may comprise only a few tens or hundreds of individuals.

Although in some rainy countries (such as Britain) slugs and snails may be serious crop pests as is the widely introduced tropical African snail *Achatina fulica* (Mead, 1979), and much work has been done on them from this point of view (Hunter, 1978); comparatively few species are involved, and little regard has been paid to their genetic variation by applied biologists (but see McCracken and Selander, 1980). Nearly all the population genetics studies on terrestrial molluscs have been made because of their apparent suitability as material for the study of the actual operation of evolutionary agents in real populations in the wild.

There are a few recent reviews. Jones (1973b) discussed ecological genetics and natural selection in the Mollusca, and Clarke (1975c) the causes of biological diversity with some reference to molluscs. There is an important discussion by Wright (1978) of the population genetics of *Cepaea,* requiring updating on climatic selection. The review by Jones et al. (1977) is restricted to *Cepaea,* which, however, is still the principal genus studied. A more general review by Clarke et al. [(1978), but written several years before] surveys all the work on pulmonate molluscs but has some unsatisfactory features noted below. Murray (1975) summarizes the genetic information available for molluscs. Jones (1980) reports a conference on molluscan genetics (London, 1980), including work not yet published; published papers are in the *Biological Journal of the Linnean Society of London 14* (3–4), Nov.–Dec. 1980. Peake (1978) has amassed a large number of useful references on stylommatophoran ecology. A conference on molluscan genetics (American Malacological Union, New Orleans 1982) will be published in *Malacologia*.

The whole subject is expanding almost daily, and originated almost entirely within the last 30 years. Much of the mathematical theory necessary is equally recent. When Philip Sheppard and I began our work on *Cepaea* about 30 years ago, there were no mathematics on the stability of polymorphisms with more than two alleles, very few methods for analysis of mark-release-recapture or the estimation of selection coefficients, and no serious examination of frequency-dependent and density-dependent selection. Nor was there good information on the life span of any terrestrial snails except a few annuals, and none of usable accuracy on the food ranges and preferences, density, population size fluctuations, and fecundity. Moreover, although almost all terrestrial molluscs are crepuscular or nocturnal, as late as 1977 I had to say that all conchologists were diurnal and that there was no useful information on what snails were doing in the wild when they were active. There is now a rapidly increasing body of work on all these topics, but still on far too few species.

In this brief review it is impossible to quote all relevant papers or to give a complete critique of each. Major themes have therefore been selected, and papers with useful reference lists. The broad taxonomic groups of the land molluscs are far from settled (see e.g., Solem, 1978; Shileyko, 1978) but this affects the present subject only trivially as yet; it is more important that specific limits are often uncertain, especially in slugs. The consequences can be easily appreciated if one imagines the hopeless confusion that would ensue from considering *Cepaea nemoralis* and *C. hortensis* as the same species.

Some workers in population genetics are too inclined to treat land snails as packets of obtainable data sitting around in an ascertainable spatial pattern. This attitude makes for easy production of papers but biases the conclusions; if the environment is virtually omitted, then the distribution of the characters will, of course, be put down to chance or history. When the ecology and behavior of the animals are carefully studied, it can be seen how naive this approach is. The considerable advances in land molluscan ecology in recent years are beginning to provide some of the information required for the interpretation of genetic variation in land snail populations, and the study of variation of behavior in relation to ecology and life cycles is of especial importance. The recent papers by Heller and Volokita (1981a,b) give a spectacular example of the relevance of such studies. In this chapter, therefore, aspects of land snail ecology and behavior of immediate or potential importance to population geneticists are considered first, since they must be borne in mind when interpreting variation. Next, it is necessary to consider some problems of interpretation. Lastly, examples of genetic variation in land snails are reviewed to show the substantial advances which have been achieved and the vast work still to be done.

II. Land Snail Ecology

A. Variation of Niche

1. Niche Definition

It is only too easy to be anthropomorphic about animals. To us, a wood is very different from an open grassland and sand dunes from solid ground. Moreover, many species of snails *are* confined (in a given region) to woods, others to open ground. It does not follow that all are, or that the characteristics of these habitat types most striking to us are relevant to them. Perhaps the best example is the two sibling species of snails, *Cepaea nemoralis* and *C. hortensis,* which have puzzled conchologists exceedingly. Both species are found in a great range of the same habitats, from sand dunes and grassland to hedgerows, woodlands, and montane vegetation. The Conchological Society of Great Britain and Ireland's committee for collective investigation proposed (Anonymous, 1901) as queries: ''Do *Tachea nemoralis* and *T. hortensis* occur together or separately? in wet or moist situations? do intermediate forms occur? does each form affect a particular kind of habitat?'' These queries produced a number of useful papers (Alkins, 1922; Cartwright, 1922; Dalgliesh, 1930; Diver, 1940; Lawson, 1923; Pearce, 1903; Welch, 1903) establishing that both species did occur in ''the same'' habitats, mixed colonies were rare, and true intermediates nonexistent, but left the mystery unsolved.

The first step to its solution was taken by B. C. Clarke (in Cain and Currey, 1963a), who noticed that if ''*hortensis* country'' is defined as regions where *C. hortensis* is found in all habitats, even on open grassland, then in such country *C. nemoralis* is confined to woods. Cain and Currey (1963a) found that the converse was true for *nemoralis* country correspondingly defined, and if woods are omitted, then there is (at least on downland) a clear difference between the distributions of *nemoralis* and *hortensis*. *C. nemoralis* is found on hilltops and higher slopes and *C. hortensis* on lower slopes and in valley bottoms; there is only a narrow band, sometimes only a few feet wide, in which they mingle. *Hortensis* country is characterized not by the vegetation but by coolness and humidity at night, with cold air flowing down from the upper slopes; this characterization agrees with the species' ranging much further north than does *nemoralis*. Similar patterns of distribution (Cain and Currey, 1963c; Carter, 1968b; Cameron, 1968,

1970a; Cameron and Palles-Clark, 1971; Greenwood, 1974a), or some consistent with them (Harvey, 1973, 1974) have been found in other parts of Britian; they may be obscured by accidents of colonization (Cameron and Dillon, 1984). The distantly related helicid snail *Arianta arbustorum* is even more confined to very damp, cool places, and the experiments of Cameron (1970b, c) on the tolerances of the three species confirm this interpretation. (Unluckily, in overcrowded Britain, it is almost impossible to leave climatic recording instruments out for any length of time without their being smashed or stolen, so that direct climatic comparisons are rarely possible.)

Unless, therefore, careful comparisons are made over a wide range of conditions, one cannot be certain that one's habitat classifications are of importance in determining where a species of snail can live. As far as maintaining populations goes, degrees of cold and damp have more influence than various types of vegetation over *Cepaea*. In a survey such as Selander and Kaufman's (1975) of *Helix aspersa* in city blocks in Bryan, Texas and elsewhere, a description is given of the habitats as seen by the investigators; but is this how they are seen by the snails? Moreover, the same habitat type may have very different characteristics under different climatic conditions and different degrees of disturbance. Cameron and Redfern (1972) show that open grassland in the very rainy district of Malham, Yorkshire, carries a snail fauna not inappropriate to woodland in drier regions. Tolerances also vary greatly among snail species. Cameron and Morgan-Huws (1975) find xerophil species only on downland grass with vegetation 2.5–10 cm high, longer grass carrying a very different fauna. This change can come about within a very few years as grazing pressure alters. *C. nemoralis* is greatly reduced in density on short heavily grazed turf (Cameron et al., 1977), probably by trampling and lack of food, but perhaps by dryness as well. For some species, then, very slight differences in vegetation type may be all important, others may be far more tolerant.

It does not follow, of course, that variations in vegetation or soil type *within* the niche have no influence on the genetic constitution of the populations inhabiting them. Although vegetation type may not limit the distribution of *C. nemoralis*, for example, it will determine the background against which visual predators see it; this can influence profoundly the visible shell characters (Cain and Sheppard, 1954).

2. *Variation of Niche*

It is very common to find that the niche occupied by a snail species, as estimated by the range of habitat types, varies both locally and geographically. As shown above, in *nemoralis* country, *C. hortensis* occurs in woods, and vice versa. On a broader scale, near the northern limit of its range in Finland *C. hortensis* is confined to woods (Valovirta and Halkka, 1976). Some land molluscs in Finland are confined to the near neighborhood of man or even to greenhouses. Even fairly closely related species may differ widely in their tolerances. Thus, the slug *Arion rufus* is found only in greenhouses, whereas *A. subfuscus* is the most common slug everywhere (Valovirta, 1967). In the extremely mild oceanic climate of Ireland (which is the edge of the range of land molluscs only because there is no land beyond), the slug *Limax cinereoniger,* confined to ancient woodland in southern England, is found in rock-strewn heathland (Stelfox, 1912). Similarly the helicid snail *Ashfordia granulata* is described by Boycott (1929) as being found "anywhere" in the west of the British Isles, but only in damp places in southeastern England; *Arianta arbustorum* shows a similar tendency (Cameron, 1970a).

Toward the climatic edge of a range, aspect often has a profound effect on distribution. Cowie (1982) has distributional evidence that *Theba pisana,* a Mediterranean and warm Western European juxtalittoral snail is confined to southern and western faces at its limit in South Wales. Bar (1975) finds species of the semi-desert snail *Sphincterochila* ecologically widespread but much more confined to south-facing slopes under unfavorable conditions. Pollard (1975a) finds *Helix pomatia* on south to southwest slopes with such precision that the *general* aspect of a site may be misleading; each little variation in the ground must be examined. Bantock and Price (1975b) describe *C. nemoralis* in Somerset as being widespread in the west but disappearing from its habitat types in succession toward the east (with a more continental climate) until it is confined to woods. Even well within a geographical range, the vast difference in microclimates generated by aspect in more rugged country may be limiting. On the steep slopes of the Winnats Pass in northern England, *C. nemoralis* is confined to southern aspects; northern ones apparently are too cold for *C. hortensis*—they carry *A. arbustorum* (Grime and Blythe, 1969).

Again, such alterations may have considerable effects on the genetics of populations without actually limiting them. Burke (1979) finds numerous associations of morph frequencies in *C. nemoralis* with aspect and altitude in a very intensive survey of the Dartry Mountains in western Ireland. Where the species is actually being limited by climate, such effects may be pronounced, for example, the connection between variation in *C. hortensis* and aspect reported by Bengtson et al. (1979b) in the course of an intensive study of this species in Iceland (Arnason and Grant, 1976; Bengtson et al., 1976, 1979a,b; Owen and Bengtson, 1972).

It follows that while intensive work on particular populations is essential to find out what is going on in them, broad surveys can be illuminating, suggesting the importance overall of factors that might be otherwise difficult to identify, and reducing the risk of assuming too easily that a local association of species, or variation with some aspect of the environment is causal and general.

3. Competition

Many attempts have been made to infer competition from the complementarity of distributions. Boycott (1921) in an excellent, but widely overlooked, pioneer paper, on snail distributions in Britain raises the whole question of inference from geographical distributions to historical interpretation of faunal elements. He rightly rejects the views of John William Taylor that the highest evolutionary forms were Germanic, spreading from there, and ousting the Celtic elements (both human and faunal, presumably). Boycott thought molluscan distribution was determined mostly by climate and soil. "The idea of direct competition on land between snail species . . . is, however, highly problematical. . . ." Later (1934) he thought it "almost negligible," and could point to only *C. nemoralis* and *C. hortensis* on sand dunes as reasonable examples. Within the range of *C. nemoralis* in Britain [up to mid-Scotland (Jones and Clarke, 1969)] it only is found on sand dunes, although *C. hortensis* is found off the dunes in every district as well. As soon as *C. nemoralis* drops out, *C. hortensis* is found on dunes. "It seems reasonable to interpret these facts by saying that *hortensis* occupies sandhills when *nemoralis* gives it a chance to do so . . . and that competition becomes effective because the habitat in question is austere and limited in its possibilities" (Boycott, 1934).

Cain et al. (1969) have criticized this conclusion. As argued above, it is not established that sand dunes are recognized as a special habitat by snails. If temperature and moisture are what is important, then the cold, wet dunes of northern Scotland are not the same habitat as the warmer drier ones of England. *C. nemoralis* (and *C. hortensis* in Scotland) are often exceedingly abundant on dunes, which, if calcareous, do not seem to be an austere habitat. (For dune surveys, see Arthur, 1978, 1980, 1982; Burke, 1979; Cain, 1968; Clarke and Murray 1962a,b; Clarke et al., 1968; Cook and Peake, 1962; Jones, 1973a; Murray and Clarke, 1978). Arthur (1982) regards the local separation of *C. nemoralis* and *C. hortensis,* and their morph differences, as probably determined by microclimate on the dunes, which he has studied intensively. Williamson (1959b), investigating differential damage in a mixed population of *hortensis* and *nemoralis* for evidence of competition, rejects the idea that competition can be inferred from a study of relative distribution.

Several authors have entertained the idea of competition between these species. Cain and Currey (1963a) thought that the coexistence of these species in woodlands on the Marlborough Downs, with their general mutual exclusion outside woods, suggested a relaxation of competition within woodlands. Carter et al. (1979) inferred some competition from the wide overlap in food range of these species. With complete microgeographical separation, a complete overlap in food range would not matter. I doubt whether I have ever seen in woodland a genuinely mixed colony of the two species (Cain, 1977c). Inch-by-inch mapping is essential here, as is an experimental approach. There may indeed be competition between the species, but it cannot be asserted from distribution only, the more so because in open habitats on summer nights, the change from a *nemoralis* niche, with warm rather dry air to a *hortensis* niche with cold dank air and mist is often very abrupt, with little intergrading. That there can be interference between species is shown in Section A,4.

4. Food and Feeding; Population Dynamics

Thirty years ago, little was known of the feeding habits of the most common snails, and still less about population dynamics. Frömming (1954) gives a useful survey of early work on feeding and biology of land gastropods generally. Subsequent intensive studies on populations of *Cepaea* and *Arianta* include work on food selectiveness and palatability (Grime and Blythe, 1969; Grime et al., 1968, 1970); feeding rates, assimilation, and energy flux (Richardson, 1975a,b, and references therein); food and population dynamics (Wolda, 1970b, Wolda et al., 1971); and field metabolism, growth rates, and size/weight relations (Williamson, 1975, 1976; Williamson and Cameron, 1976; Wolda, 1970a, and other references; Oosterhoff, 1977). Richardson (1975a,b) and Chatfield (1976) give references to work on other snails. Yom-Tov and Galun (1971) have investigated diet of two dry-country snails, and Pomeroy (1967, 1968, 1969) of introduced species in Australia.

Other aspects of population dynamics investigated include production and survival of eggs and juveniles (Wolda 1965b, 1967; Wolda and Kreulen, 1973; Williamson et al., 1977.) Pollard (1975a) on *Helix pomatia* and Potts (1975) on *Helix aspersa* also note the considerable effect of local conditions on reproduction, as does Yom-Tov (1971b, 1972).

The general impression from these and other studies is that although there is considerable selectivity of food there is no really narrow specialization; several authors also point to situations in which food does not appear to be limiting but population density may be (through mucus—see below and references in Cameron and Carter, 1979). The difficulty is, as they point out (Wolda et al., 1971; Yom-Tov and Galun, 1971; Richardson, 1975a), that too little is really known about the quality of the food or its minor but essential constituents, for example trace elements. Food is at least as complex an entity as climate and cannot be dealt with simply. Williamson and Cameron (1976) have addressed the question of food quality in relation to *Cepaea*.

Certainly, recruitment into the population can vary greatly from year to year in relation to climatic and similar factors acting on the eggs and young. The eggs of *Trochoidea seetzeni* can withstand considerable water loss, whereas those of its ecological companion, *Sphincterochila boissieri,* cannot (Yom-Tov, 1971b). Wolda (1967) has found differential effects of temperature, apparently with some genetic basis, on the mortality, mating frequency, oviposition frequency, and clutch size of different morphs of *Cepaea nemoralis*. In all these aspects of snail and slug biology there is enormous scope for selection by local conditions, as well as purely phenotypic variation (as with shell size in relation to mucus) over even a few feet, or on introduction to a new locality (e.g., Järvinen et al., 1976). The consequences for population dynamics theory have been pointed out by Wolda (1970b) who has proposed a model for population dynamics of *Cepaea*, extended by Harvey (1972c) to take account of its polymorphism.

5. Calcium Availability

One nonnutrient element in the diet of snails that has received attention is calcium carbonate, so necessary for the shell. It is a far from static component of the shell and can be mobilized from it (Chan and Saleuddin, 1974; Williamson, 1976). In districts where it is scarce, snails have much thinner shells (Murray, 1966) and may rasp at each others' shells (Wolda, 1972a) or at any other adventitious source of calcium (Murray, 1966; Pitchford, 1955). Even on acidic ground, proximity to the sea may bring in a sufficiency of calcium in spray and sea shells. The availability of calcium in the environment cannot, however, always be determined easily. In western Britain, rainfall and leaching are so great that one can find calciphobe plants growing in acid humus on a limestone slab 1000 ft thick [this is the probable explanation for the observation (Cameron, 1969) of a lack of snails within hazel scrub in western Ireland, with an abundance just outside]. Valovirta (1968) finds, at the other extreme, that even on acid ground aspens can concentrate calcium in their leaves to an extent useful to snail populations. Wäreborn (1979) notes that the calcium concentrated in leaves of ash and lime is fairly soluble, but that in beech and oak (as calcium oxalate), several months are required to break it down in the leaf litter and make it available.

Wolda (1972a) finds that older *Cepaea nemoralis* are more likely to be damaged by their companions in search of calcium and to suffer a higher mortality. This is presumably related to the loss of large areas of the protective periostracum with age. Thinness of shell may bring about an increase in predation by allowing smaller predators to break it. There is probably a genetic component; in collecting shells of *C. nemoralis* on the downs of southern England (with abundant calcium) I have (rarely) found both abnormally thick shells (var. *ponderosa*) and thin ones in normal populations.

Shell thickness can vary from population to population in snails (Pollard, 1975b), but the statements of Bar (1978) and Sacchi (1971) of its variation in *Theba pisana* need corroboration (Cowie, 1982). Calcium may be needed in easy supply by populations under stress at the climatic edge of the range (Russell Hunter, 1956), but it may also be that these populations are on calcareous soils because they are warmer and drier.

B. Variation in Life History

1. Duration

Theba pisana is a sand dune and coastal snail found around the Mediterranean and on the west coasts of North Africa and Europe, which has been studied especially by Sacchi and co-workers. Sacchi (1971, 1977, 1978) has compared populations in the Mediterranean, on the Atlantic coast of Spain, and in Brittany. Under Mediterranean conditions, aestivation is forced on the snails by a dry, hot summer, and they are usually biennial with much local variation (Cowie, 1982). North of the Po Delta on the Adriatic coast, conditions are less severe and full development can take place in 1 year; just south of the delta, intermediate conditions produce considerable irregularity in life span. On the Atlantic coast of Spain (Sacchi and Violani, 1977), conditions for a short life cycle and high productivity seem at their best. The life cycle is biennial again at Tenby, South Wales (Cowie, 1982) because here the animals are slowed in growth by lower temperatures.

Umiński (1975, 1979) and Umiński and Focht (1979), studying vitrinid and other snails, find in some species a 1-year cycle at lower levels in the Tatra Mountains and elsewhere in Poland but a 2-year cycle at altitudes of about 1,300 m. At the same site in the Carpathians three vitrinids, *Eucobresia nivalis, Semilimax kotulai,* and *Vitrina pellucida,* have life cycles of 3, 2, and 1 years, respectively. Umiński and Focht remind us that in the summer drought for snails increases to the east in Europe as well as into Mediterranean climates; on the Hel Peninsula on the Polish Baltic coast, several species escape the summer drought by burying.

On a much smaller scale, Potts (1975) has examined persistence, extinction, and life cycle of populations of *Helix aspersa* a species introduced into California. Under the stress of the California climate (although it is similar to that of the Mediterranean, where this species is native) populations can differ profoundly within a single garden, those in dry locations reaching maturity 4 years later than those in well-watered ones. Tuskes (1981) reports from Florida a difference in growth between different color varieties of the arboreal *Liguus fasciatus* even within the limits of one clump of trees.

2. Stages

Snails show no profound metamorphosis. Nevertheless, there may be considerable differences ecologically and behaviorally between successive stages in the life history. Juveniles of the dry-country *Trochoidea (Xerocrassa) seetzeni* are dark colored and disperse soon after hatching, hiding to avoid overheating, while the adults are light colored and climb up stems and bushes in the full light of the sun (Yom-Tov, 1971a). The juveniles of its frequent companion, *Sphincterochila boissieri,* remain in their nest holes until the Mediterranean winter, with its lower temperature and frequent rain, begins (Yom-Tov, 1971b). Pollard (1975a) finds that adult *Helix pomatia,* which have extremely strong shells, make little effort to conceal themselves, as compared with juveniles. Cowie (1982) finds young of *Theba pisana* in dense herbage on the ground during the day; but adults are mostly found on stems, with a corresponding difference in temperature tolerance. Moreover, in winter the half-grown juveniles tend to aggregate by themselves under stones and wood. Roth, whose work on the helminthoglyptid snails of California (1980, 1981) appears to be the first on their biology, describes juveniles of *Monadenia setosa* as under loose bark on the lower trunks of standing trees, while the adults are terrestrial. In *Xeropicta vestalis,* different stages of the life history are exposed to different climates in the mountains and on the coastal plains of Israel (Heller and Volokita, 1981a). Correspondingly, they change their shell colors and thermal relations. In the mountains, with cold winters and a short active season, they hatch with medium dark shells and become darker. In the plains with a much warmer winter and longer active season, the shells become paler. "In both cases, each snail is darker in the colder months and paler in the hotter ones." Williamson (1979, and references therein) points out that subadults probably have a higher metabolic rate than juveniles and adults, and that their importance as a special class has been underrated.

Several authors [see Ingram (1946) and Potts (1975) on *Helix aspersa* and Pollard (1975a) on *H. pomatia*] have recorded a migration, in some cases up to 100 m, to a hibernating area and a return migration to the feeding and breeding grounds. These changes of site may expose populations to very different predators and perhaps to diseases and physiological hazards. The period of estivation in particular has received attention in dry-country snails (e.g., Pomeroy, 1968; Yom-Tov, 1971b). Sacchi (1974) has observed *Theba pisana,* on sparsely vegetated dunes where the soil surface gets very hot, climbing to avoid overheating but concealing themselves under soil litter in the shade of tamarisks. The temperature of estivating *I. pisana* can be fairly high, but the lower individuals in a single aggregation on a post or trunk can be partly shaded by the higher ones (McQuaid et al., 1979). Juveniles with darker shell patterns are more likely to estivate in shade than those with pale ones (Johnson 1980, 1981). Yom-Tov (1971c) suggests that some dry-country snails may have to burn food supplies to keep up their water content. In the large Mediterranean snail *Otala lactea,* Newell and co-workers (Appleton and Newell, 1977; Newell and Appleton, 1979; Newell and Machin, 1976) find that special lamellar vesicles are produced inside the cells of the mantle-edge to cut down water loss. Such a period, then, will have its own specific risks, adaptations, and selection pressures.

Clearly, to achieve any real understanding of the life of a slug or snail, the whole life history must be taken into account in relation to the exact circumstances of each population, almost each individual.

C. Interspecific and Intraspecific Interferences

The previous sections have been devoted almost entirely to some complexities of the environment and their consequences for land molluscs, and show how inadequate it is to treat these animals as though they were counters on a uniform board. Perhaps the most exciting ecological discoveries in recent years are related to inter- and intraspecific interference. Lomnicki (1969) remarks that aggressive or territorial behavior does not occur among snails, and Yom-Tov (1972) that it is difficult to believe that snails hold a territory. Both these statements may have to be reconsidered. And besides this and the possibility of direct competition, another mode of interference has been discovered recently.

1. Mucus

Several authors have noticed in *Cepaea* and *Helix aspersa* a decrease in the size of adult shells when the animals are at a high density in the wild or in population cages (Cook and Cain, 1980; Herzberg, 1965; Oosterhoff, 1977; Tattersfield, 1981, 1982; Williamson, 1976; Williamson et al., 1976), and some have suggested the mucus of their tracks as a possible inhibitor. Cameron and Carter (1979) have shown experimentally a considerable inhibition of activity and growth in juveniles of *C. nemoralis* and *C. hortensis* by mucus. Moreover, the mucus of *C. nemoralis* has an effect on young *C. hortensis* and vice versa, but *H. aspersa* mucus does not affect them. Williamson et al. (1976) have shown an inverse correlation of shell size with population density in a wild population in which food did not seem to be a limiting factor; this is the probable explanation. Tattersfield (1981) finds some evidence for *Candidula intersecta* abundance having an influence on shell size in *Helicella itala* in mixed populations. This may possibly be of the same nature as the finding of Williamson et al. The significance of such interaction is discussed by Cameron and Carter (1979). Its immediate importance is that surveys of variation in shell size are now unintelligible without equivalent surveys of population density. One American malacologist in the early years of this century, over-impressed by ornithological methods, named vast numbers of subspecies of land snails on the basis of size variation; it is now unlikely that all of them had a genetic basis and possible that none did. Interaction due to crowding (which may not be exactly the same as aggregation) can be complex. Tattersfield (1982) finds that in *Cepaea* it causes a general decline in activity; in *Trichia striolata,* a decline in feeding and an increase in locomotion; and in *Monacha cantiana,* no apparent effect on either activity.

2. Aggression

The remarkable paper by Rollo and Wellington (1979) on inter- and intraspecific agonistic behavior among slugs opens up new vistas on interference.

They report actual aggression, mainly by biting, varying from species to species and at different seasons. As lurking places are particularly important for slugs, which desiccate easily, there appears to be a considerable advantage in being able to displace other slugs from them, the more so as they are usually chosen to be close to food as well. Aggregation may make for more efficient use of lurking places (Cook, 1981) even in the presence of aggressive species, and trail following may facilitate food finding and homing (Cook, 1977, 1979). Perhaps the liability to desiccation is so reduced in snails that they do not need such behavior, but there may well be other needs that can drive them.

3. Predation: Cannibalism

Some snails and slugs are well known to be carnivorous and have radular teeth to suit (e.g., *Testacella, Euglandina*), but it is being discovered that many more snails may attack other species than has been believed. The zonitid *Oxychilus cellarius* has been known for some time as a snail egg predator (Wolda, 1963; Cain and Currey, 1968). But Mordan (1977), studying zonitid snails in a British woodland, finds *Aegopinella nitidula* to be a predator of other adult snails, preferring the zonitid *Nesovitrea hammonis* and probably affecting its density. He remarks rightly that Boycott's belief (1934) that interactions between molluscs are negligible is unlikely to be right. A special form of molluscan predation is cannibalism. Pollard (1975a) records newly hatched young of *Helix pomatia* devouring unhatched eggs in the same clutch, and Yom-Tov (1971b) reports on cannibalism by young *Sphincterochila boissieri*. If this activity is widespread, considerable selection pressure will be put on rapid development.

4. Faunal Constitution

Evidence from a completely different direction suggests that terrestrial Mollusca forming a given fauna interfere in some way with one another. Cain (1977a, 1978a) plotted mean values of shell height against shell maximum diameter for each species in whole faunas of stylommatophoran land snails and found, for taxonomically and geographically completely different faunas, a very similar scatter of two wedges, one for high-spired snails and one for equidimensional to very flat ones, usually separated by a noticeable gap. This pattern appears in European and two North American faunas, in Puerto Rico and New Caledonia, and in the Philippines fauna for the majority of the families. [It has since been found in several other faunas, (Cain, 1981b)]. In each fauna (if burrowing forms, which are necessarily tall narrow shells, are excluded) this curious distribution is made up taxonomically of a mosaic of pieces in very different ways in different faunas. For example, large shells in the lower wedge (equidimensional to flat shells) are helicids in the western Palearctic, polygyrids in eastern North America, helminthoglyptids in southwestern North America, camaenids in the Caribbean and northern Australasia to Southeast Asia, and bradybaenids in northern

China. Large stout shells in the upper wedge are bulimulids in the Caribbean, camaenids in Southeast Asia, different bulimulids in southwestern to southern Australia, caryodids in eastern Australia, different camaenids in New Guinea and adjacent regions, and different bradybaenids in Taiwan and the Philippines. When there is overlap within a wedge, as with several subfamilies of helicids in Europe, the groups usually occupy different habitats, e.g., helicigonines in the mountains of southern Europe, helicellines in dry, even arid, lowlands, and hygromiines mostly in intermediate habitats (Cain, 1981b). Remarkably, land operculates tend rather strongly to follow the same pattern of h,d scatter (Cain, 1978b), although their marine relatives have a very different distribution, lying mainly in the gap between the two wedges of terrestrial faunas (Cain, 1977a). Only two partial exceptions have been found, the helicostylines (Bradybaenidae) in the Philippines and the papuinines (Camaenidae) in the New Guinea region; in both cases, large shells with a continuous scatter from discoidal to high spired (Cain, 1978a), although the rest of the land snail fauna in each case shows the usual bimodal distribution.

Since land snails are fairly generalist feeders, not complete specialists like so many insects, Cain (1977a) suggested that the constituents of the same fauna might feed in different places on the ground and in vegetation. To be Irish, the more similar you are (in feeding), the more different you have to be (in places or times of feeding) to avoid competition, and the best place to do the same thing is somewhere else. Cain thought that different shapes of shell might be more appropriate for feeding on vertical surfaces as, for example, against horizontal, since a good deal of energy must be expended in carrying around a shell and adjusting it frequently into position. The only evidence available so far (Cain and Cowie, 1978; Cameron, 1978) seems to support this suggestion, but see Tattersfield (1982) for some qualifications.

On the strength of this idea Cain proposed (1981a) a possible explanation for the fantastic variation during the life history of many forms of *Cerion* (see Woodruff, 1975a,b, 1978; Woodruff and Gould, 1980, for an account of this unique snail). Several snails change the overall shape of their shells as they grow to maturity, e.g., *T. pisana* (see Cowie, 1982 for references and discussion), but to nothing like the same extent. If a species of *Cerion* in Caribbean juxtalittoral habitats passes during its life history through the niches occupied by several separate species of snails in comparable European habitats, part at least of its unique variation may be explicable.

This is an example of a broad survey bringing to light possible interactions between species within faunas that might never come to mind during the usual ecogenetic studies (after all, the proper study of a single species can be a full-time job or more). The probability of actual interference by competition becomes somewhat greater as a result of these studies than from the studies of distribution and overlap mentioned above. It is worth noting that taxonomic discordances can be found within as well as between faunas. *Cepaea nemoralis, C. hortensis,* and *Arianta arbustorum* form an ecologically closely related triad of helicid snails, yet the first two are helicines and the third is a helicigonine with mainly depressed to discoidally shelled relatives. Its convergence on *Cepaea* in shell shape and habits is good enough for it to have been mistaken by some geneticists for an undescribed morph of *C. nemoralis.* Much work has been done on shell size in *Cepaea,* but the question of what selection pressures act on the shape of its shell to produce and maintain this extraordinary convergence has never even been raised.

D. Interaction with Other Animals

Almost the only well-studied interaction is predation of molluscs by nonmolluscs. Some facilitation probably occurs; for example, I have often seen *C. nemoralis* in open grassland radulating rabbit dung. Disease is hardly studied at all (Boycott and Oldham, 1938; Mead, 1979). Predation by a considerable variety of invertebrates and vertebrates has been documented, including predation (probably selective) by glow worms (O'Donald, 1968), predation by predators with color vision (e.g., thrushes), and predation by those with only tone discrimination [e.g., rodents and rabbits (Cain, 1953; Murray, 1962); little owls (Zinner, 1978) are presumably tone predators].

Predation by animals other than snails can vary, even microgeographically, with considerable effect, sometimes to the near elimination of a species locally. Yom-Tov (1970) describes a far higher density of desert rodents on south-facing than on north-facing slopes, with a strong differential effect on the degree of predation of the local snails. Considerable variation in certain vegetation types can also exist; many authors have suggested that predation by thrushes on *Cepaea* and *Arianta* in England is greater in woodland than on open grassland (Clarke, 1962b; Arnold, 1971; Parkin, 1973). This may well mean that in a given region, color and banding morph frequencies largely determined by visual selection for crypsis will be seen in woodland colonies of *Cepaea,* but very different ones unrelated to crypsis in open habitats (compare the scatter diagram for *C. memoralis* in chalk districts in Cain and Currey, 1963a, Fig. 16).

Of particular importance is the demonstration by Schifferli (1977) that quantities of small snails are fed to nesting female birds as a supply of calcium for making egg shells. Even insectivorous birds need egg shells. Several authors (e.g., Bengtson et al., 1976) have argued, from the absence of broken shells or the presence of only insectivorous birds, that visual predation is absent. It may in fact be intense but (as the snails are swallowed whole) leave no trace at all.

Predation may vary considerably by season as well as by place. Richardson, who has estimated the trophic efficiency of the snail–thrush transfer (1975c) finds song thrush predation in his study area only in winter, perhaps from the lack of suitable nesting sites (1975b), whereas most other British workers have recorded predation in spring and summer. In some areas, migrant birds (redwings; Richards, 1976) are responsible for winter predation. Relative variation in the seasonal incidence of predation on co-existing molluscan species is also possible (Cameron, 1968).

III. Interpretation

Considering the complex activity of a snail or slug, the complexity of even those relevant features of the environment discussed above, and the amount we do not know about both, we may well wonder that anything can be discovered about land mollusc ecogenetics when only too often all we have to work with is a survey of species and morph distribution or a short-term investigation (short from the evolutionary point of view). It is not surprising that there are often considerable difficulties in interpreting observed phenomena and a good deal of room for maneuver by partisans of particular theories. Scientists are, after all, human, and part of the work to be done is a sorting-out of their own standards and criteria. Some useful general points are given by Cain (1977c) and Jones et al. (1977). Burke (1979) has discussed the interpretation of patterns found in the wild, and Clarke (1975a) has examined the problems involved in determining whether or not a character is acted on directly or indirectly by natural selection.

I take it to be common ground that a scientist should not make major points by means of unpublished work [see Cain (1977c) for an example], that he should not refer to earlier expositions of a hypothesis as though mere repetition were confirmation, that he should not use a given criterion to support his theories against opponents but reject that same criterion when they use it on him, that he should quote work apparently contradicting his hypothesis with as much care as work supporting it, and that he should not, without further comment, refer to work as in support of his own ideas if that work can fairly be taken as not supporting them. (These are the techniques of political propagandists.)

Unfortunately, in a rapidly developing subject like ecogenetics there may be real confusion over the proper use of criteria; particularly relevant is the concept of the simplest hypothesis, because the situations we are investigating are so complex.

A. The Simplest Hypothesis

It is apparent from Section II of this chapter that almost any character in a snail may be acted on by a great variety of selection pressures, as well as by recent history (an abnormal season, a recent introduction) or pure chance. Moreover, even when we have found an association between a particular character and some variable, we cannot always expect a linear relation between them or even a continuous one. U-shaped curves of association with altitude are known in *Cepaea nemoralis* between unbanded shell (Lamotte, 1968b), yellow shell (Arnold, 1968), and white lip (Lamotte, 1972) in the Pyrenees. If snails respond on a scaling different from ours, exponential or other curves may result; and even with a continuous variable, response to a threshold may produce discontinuous variation. The same genes or supergenes may respond in different ways to the same selection pressure in different populations (Harvey, 1976); this seems true of the same morphs in different species, as Clarke (1960) has shown for *C. nemoralis* and *C. hortensis* (but see Section IV,E,2).

And since a conspicuous character like shell color in polymorphic snails may be subject to several different selection pressures at once, we can expect at best only partial correlations; if, when it does occur, it is associated with a particular environmental variable sufficiently often, then the fact that it does not occur sometimes when we expect it is no argument against that association (Bantock, 1980; ctr. Clarke et al. 1978, p. 246). Finally, if a character is associated with a particular environmental variable, increasing as it increases, we cannot assume that it will continue to vary with extreme values of that variable. There is evidence (see Section IV,E,1) that the dark brown morph in *C. nemoralis* is favored in cool places in much of Britain. Cain and Currey (1963a), who first suggested this, did not suggest that it would therefore increase in frequency down to the very coldest climates which *C. nemoralis* would tolerate, since they were aware of both the possible complexities of reactions of animals and the complexities of climate. Yet Clarke (1968) refers to the absence of brown morphs in Scotland, where *C. nemoralis* comes to the edge of the British range, as crucial, which it is not, and rejects the whole hypothesis.

If we cannot *necessarily* argue from one population or region to another, still less can we expect that findings in one species should be applicable to a different one. The classic example is again *C. nemoralis* and *C. hortensis*. The polymorphism in shell color and banding appears to be the same in both species, and their habits are extraordinarily similar; indeed, as late as 1927 their specific distinctness was still being questioned (Step, 1927). Yet numerous authors have commented on how different are the frequencies of their principal morphs (e.g., Clarke, 1960, 1962a,b, 1968; Guerrucci-Henrion 1974; Gyngell, 1911), although Lamotte and Guerrucci (1970) find overall similarities in banding frequencies in France (Meeuse, 1968; Williamson, 1959b). Differences in climatic tolerance investigated by Cameron have already been mentioned. A careful series of studies in Somerset (Price 1975a,b and references therein) show considerable differences between the species in chiasma frequency and its relations to both population density and altitude. The remarkable differences in frequency of visible morphs are ascribed by Clarke (1960, 1968) to the fact that, being different species, they will have different genotypes, which will affect the polymorphism differently in each. Whether or not this is true, it is a good example of explaining away (Cain, 1972, 1977c) because no attempt had been made to show that the actual differences observed between the species follow from the hypothesis. In fact, it is at least as likely, given the different microclimates (at night) inhabited by the two species, that climatic selection acting uniformly on them will account for most of the differences.

Lamotte (1949, 1955, 1959) argued against the effectiveness of visual selection on *Cepaea,* on the ground that in mixed colonies, if selection were favoring unbanded shells, for example, it should favor them in both species, which should therefore show some association of the two frequencies; he found none in a large number of mixed populations. Cain and Sheppard (1954) and Sheppard (1955a,b) pointed out that, because of their differences in morph frequencies, *C. hortensis* might have to produce a yellow banded shell in woodland (as the darkest in its repertoire) while *C. nemoralis* on the same patch of ground could produce a pink unbanded shell. This possibility was rejected by Lamotte (1959), but Clarke (1960) showed that in the same district in which *C. nemoralis* was known to be responding to strong visual selection (Cain and Sheppard, 1954), *C. hortensis* did seem to respond in the way suggested, by producing yellow five-banded shells with the bands fused as in its usual dark forms, genetically a totally different response from that of its sibling *C. nemoralis.* (The thrush is not only a bad taxonomist, it is a hopeless geneticist as well.)

Even arguing from one species to a sibling, then, requires great care, as Sacchi (1956b, p. 485) has already pointed out. Yet Clarke et al. (1978) argue from work of Sacchi (1955b) on very different species in Mediterranean mountain country to cast doubt on the suggestion of Cain and Sheppard (1952) that variation in body color in *C. nemoralis* could be ascribed to visual selection. [They also overlook Cain and Currey's criticism (1963a) of Cain and Sheppard's paper, i.e., that area effects (see Section IV,C,2) could equally explain the reported variation in body color].

The preceding examples indicate ways in which attempts to impose a simple generalized hypothesis can be wrong. Once we find that any character may be influenced by a variety of unrelated selective forces, the criteria for deciding which is the simplest hypothesis become doubtful. What we can require is that hypotheses should be consistent with the data they seek to explain, stable in principle in particular instances, and compatible with what we know of genetic population processes. Clearly, what is essential (but often impracticable) is intensive work, observational and experimental, which may take many years, on each population to find out what is really going on in it, as Lamotte indicated many years ago (1959, p. 81). In its unavoidable absence we must interpret phenomena we find, bearing in mind all the possibilities and probabilities.

If indeed we do find marked regularities within a region from population to population as in studies showing visual selection in relation to habitat (Cain and

Sheppard, 1950, 1954; Currey et al., 1964; Greenwood, 1974); within regions, as in Jones's work on *Cepaea vindobonensis* in relation to local climate (1973b,c, 1974a); from region to region in Europe in relation to major climatic variation (Jones, 1973b) or from species to species, as in the regularities of shell color pattern and degree of polymorphism in relation to habits and habitat shown by Cain (1977b) for European snails, and indicated by Roth (1980, 1981) for Californian species; or even from fauna to fauna over the world, as in the shell shape regularities discussed above, we can only be thankful. In spite of all the factors which could obscure any regularity even from population to population, we have found something (even if we cannot identify it) overriding in its effects, and if there is a genetic basis to the variation concerned, it is almost certainly due to natural selection. When the regularity is sufficiently established, exceptions to it do not disprove it; the correspondence of morph frequencies in *C. nemoralis* to visible characteristics of the habitats of *C. nemoralis* in the Oxford district (Cain and Sheppard 1950) is not destroyed by the discovery of an almost total lack of it on the Marlborough Downs or elsewhere (Cain and Currey, 1963a–d).

This means that as more regular patterns are discovered, the action of natural selection becomes more and more widely recognized; but the absence of such patterns does not mean that natural selection is ineffective in our material (Cain, 1952; Bantock and Noble, 1973). Fluctuations in natural selection can produce effects identical to those of genetic drift [as Wright (1965, and references therein to his earlier papers) has been careful to point out].

This is, at first sight, a most unsatisfactory state of affairs, but given time, it can be alleviated. If even a very enigmatic and apparently random pattern of distribution proves highly stable over time (see Section IV,C,1), natural selection must be maintaining it. As several authors have pointed out (e.g., Harvey, 1976) there may be selective environmental or other gradients quite unrecognized by us; these will become apparent from the stability of the resulting patterns.

To the extent that exceptions do not disprove a regularity, the simplest hypothesis cannot be applied to ecogenetic situations. If however, it is taken to mean, as it should, that one should not go beyond the conclusions warranted by the evidence, it is unexceptionable. One should not use this hypothesis to defend conclusions (Clarke, 1968, p. 353, 1969, p. 351) but reject one's own (Clarke et al., 1978, p. 243) in favor of a more elaborate one without giving reasons.

B. Skepticism

Clarke et al. (1978), in an often very acute and valuable review of work on natural selection in pulmonate molluscs, indicate that they wish to bring a healthy skepticism to bear on many of the interpretations and conclusions that have been published. This is indeed what all scientists should do; but to be healthy, skepticism should be uniformly applied according to explicit standards. It should not contravene the principle of the simplest hypothesis (in the sense just explained), and should not multiply entities without necessity.

Cain and Sheppard (1950, 1954) provided evidence, from a survey of variation in *C. nemoralis* in relation to the visible characters of the habitats in the Oxford district, that visual predators were selecting out the more conspicuous morphs by color and pattern. Sheppard (1951) provided experimental evidence, from exposure of marked snails in a woodland, that thrushes were doing exactly this; he compared the shells broken by the thrushes on stones with those he had marked and put out. Other surveys (Currey et al., 1964; Greenwood, 1974) have found the same characteristics in woodland populations, i.e., shells with darker colors and less banding since the backgrounds are more uniform as against open habitats which have lighter shells, more broken up by banding. It should be noted that this is not what one would expect from the assumption of climatic selection with generally darker shells in woodland, i.e., darker colored and with more banding absorbing the reduced incoming radiation, and lighter ones with less banding in open habitats reflecting back a higher proportion.

Clarke et al. (1978, pp. 243–244) object that although Sheppard certainly showed visual selection "he did not categorically show that it was the consequence of differential crypsis (rather than, for example, colour preferences on the part of the birds)" and that "the distribution of shell colour (although not, perhaps, of banding) is compatible" with the type of climatic influence just mentioned. Now in one sense this criticism is right; to show differential crypsis one would need to work with thrushes with electrodes implanted in the right parts of their brains transmitting back to the observer encephalograms, to be compared with a complete library of brain-wave patterns, indicating exactly what they were responding to in their visual fields and what they were overlooking. Without this, there is probability, not proof. But to postulate a set of color (and banding) preferences which exactly mimic the effects of visual selection is to multiply entities without justification. One can always insist that any phenomenon is not due to what appears to be producing it but to something else that mimics the effect. Such skepticism is not useful; Bantock's remark (1980), applied to another similar argument of Clarke et al. that "such explanations seem to defy common sense in their obliquity" is not unjustified, the more so as the same mode of argument shows that Clarke et al. cannot be proved to have written their paper. They seem not to understand the practical limits of complete skepticism, well appreciated by Chaucer (c. 1385) and Benoist (1712, see Hazard, 1964).

Skepticism, moreover, needs to be applied uniformly. The statement by Clarke et al. (1978, p. 261) concerning the maintenance of polymorphism in the pulmonates that "the only mechanism for which there is strong experimental evidence about a potential selective agent is frequency-dependent selection by predators'' canot be sustained, even with the more limited form of skepticism generally used by research workers; there is no strong experimental evidence for it in pulmonates. This point is dealt with further in Section IV,E,3.

IV. Ecogenetics: Recent Developments

A. General

Not all the principal discoveries affecting our ideas of molluscan ecogenetics are ecological or behavioral. Two major contributions come from Selander and co-workers as a result of the study of molluscan allozymes. In species of slugs (McCracken and Selander, 1980) there is a remarkable variety of breeding systems, some species being self-fertilizing with one or several recognizable strains, others both self- and cross-fertilizing, and still others cross-fertilizing only; this is essential information for the ecogeneticist. Occasional self-fertilization in the slug *Arion ater* had been reported by Williamson (1959a) on genetic grounds, and some snails use it as a last resort (McCracken and Brussard, 1980) or habitually; but its frequency in slugs is truly surprising.

Even more remarkable, from the evolutionist's pont of view, is the demonstration that *Rumina decollata,* a very successful immigrant snail in the southern United States, is self fertilized and monomorphic for every enzyme system studied; that in the south of France where it is apparently native, two or more different lines exist, which occasionally cross-fertilize; and that the frequency of cross-fertilization appears to be dependent on the frequencies of the forms (Selander and Hudson, 1976; Selander and Kaufman, 1973; Selander et al., 1974). Selander and Hudson's discussion rightly emphasizes the parallels with semi-sessile and sessile animals and plants, although their principal evidence that *Rumina* is more sedentary than other snails appears to derive from its stronger local genetic heterogeneity, which comes close to a circular argument or involves an unsubstantiated hypothesis about local selection.

The principal ecological and behavioral findings are given above. Of most immediate significance is the influence of mucus-interference on shell size, which lays open to revision much previous work on the subject (Bantock and Bayley, 1973; Wolda, 1972b). A discussion of evidence for the genetic component of shell-size variation is given by Cook and Cain (1980) in *Cepaea* and related forms and by Murray and Clarke (1968a) in *Partula*. The findings of Wolda (1963) that snails of different populations of *C. nemoralis* may behave differently with respect to vagility, oviposition frequency, etc., cast doubt on work with samples that may have come from heterogeneous sources (Clarke et al., 1978).

B. Genetics

1. Invisible Polymorphism

Many species have not been bred, and of those that have, several may take 2 years to come to maturity. Not many authors have spent time on this tedious work; but to Murray's list (1975) must be added Oxford (1973, 1975, 1977, 1978) on allozymes of *Cepaea*, who has demonstrated, besides directly genetically determined enzymes, others induced by food. Oxford (1973) has rightly criticized the findings of "armchair geneticists" who do not confirm their interpretation of isozyme bands by the necessary breeding experiments, Tegelström et al. (1975, also Wahren and Tegelström, 1973, discussing *C. hortensis* and *Helix pomatia*) also comment on the difficulties of interpretation without breeding data. Manwell and Baker (1968, 1970) compared isozymes in single species and mixed species populations of *C. nemoralis* and *C. hortensis;* they thought they had found interspecific hybridization and built up a remarkable superstructure of speculation on pollution and hybridization to account for their findings. It seems most likely that they had misidentified some of their material at the species level. Further data on isozymes in *C. nemoralis* are given by Johnson (1976, 1979), Brussard (1974, 1975), Brussard and McCracken (1974), and Selander and Foltz (1981); in *H. aspersa* by Crook (1980); in *H. pomatia* by Järvinen et al. (1976); in *T. pisana* by Nevo and Bar (1975); in *Trochoidea* spp. by Nevo et al. (1981); in *Cerion* by Woodruff (1975a) and Woodruff and Gould (1980); and in *Rumina* by Selander and Kaufman (1973), Selander et al. (1974), and Selander and Hudson (1976). An enormous amount of data on slugs is being produced (McCracken and Selander, 1980). Interspecific variation in haemagglutinins in *Cepaea* is discussed by Kothbauer and Schnitzler (1972) and polymorphism in mucus gland proteins in *Partula* by Schwabl and Murray (1970).

2. Visible Polymorphisms

Without knowing their exact genetic basis, it is often possible to recognize distinct morphs in random samples; and this at least sorts them out from continuous variation. (But note that Russian workers often use "morph" for any form, whether it is continuous or not, and "polymorphism" is still used by some French workers in the old sense of variation of any kind). Clarke et al. (1978) rightly refer to some inadequacies of museum collections; it should be added that many such samples are of "good" specimens, picked out for permanent reference, and can give a wildly exaggerated idea of the frequency of polymorphisms, the intermediates having been discarded. True morphs, with or without breeding, have been recognized in numerous species [*H. aspersa, Monacha cantiana,* and *Trichia striolata* (Albuquerque de Matos, 1979; Cain, 1959a,b, 1971a; Chatfield, 1977); *Bradybaena fruticum* (Khokhutkin, 1979); *Monadenia fidelis*

(Roth, 1980); *Lehmannia marginata* (Greenwood, 1966); and *Arion* spp., Williamson, 1959a)]. Clarke et al. (1978) give a useful tabular view of those morphs reported to date. The extensive breeding work of Murray and Clarke (1976a,b, and references therein) gives us a picture of supergenes in *Partula* like that already known in *Cepaea,* as does that of Lewis (1975, 1977) in *Cochlicella acuta*. Much more information is becoming available for *Theba pisana* (Cowie, 1982; Cain, 1984). Supergenes are a feature of most of the snail polymorphisms well-enough investigated and of visible polymorphisms in many other animals (Sheppard, 1969; Murray and Clarke, 1976a,b); but they are not characteristic of invisible polymorphisms; this in itself is enough to show that the variation in visible ones is organized, not random. Jones et al. (1977, p. 128) suggest that the nature of the selection which has acted on populations in the past will determine this genetic architecture, but their references do not offer very firm support, and selection in the present may well be maintaining these gene arrangements.

Almost all authors working with *Cepaea nemoralis* have found massive linkage disequilibria, i.e., preponderances of particular supergenes, in wild populations. Banding and unbanding in pink and yellow is often involved within populations, although there may be overall little association (Lamotte, 1969). White lip in *C. nemoralis,* the albolabiate form, shows usually strong associations, often between populations as well, but not of the same sign from region to region (Cook and Peake, 1960, 1962; Goodhart, 1973; Greenwood, 1974; Guerrucci-Henrion, 1966; Jones and Irving, 1975; Harvey, 1972a,b; Cameron, 1969; Cook, 1966; Lamotte, 1972; Clarke et al., 1968). Punctate forms may also show disequilibria (Cameron and Cook, 1971). Such disequilibria are good evidence for selection of some sort (e.g., Clarke, 1975a).

3. Chromosomes

The karyotypes of *C. nemoralis, C. hortensis,* and *C. sylvatica* have been investigated by Page (1978) and Gill and Cain (1980). The relationship between the species is less simple than has been proposed, and *C. vindobonensis* with relatively uniformly sized chromosomes is probably not primitive in that respect (Gill and Cain, 1983). Supernumerary chromosomes are described by Evans (1960) in *H. pomatia* as occurring in some peripheral populations, perhaps because of their isolation. Chiasma frequency has been studied in *C. nemoralis* and *C. hortensis* in Somerset by Bantock and Price (Bantock, 1972, Bantock and Price, 1975a; Price 1974a,b, 1975a,b,c; Price and Bantock, 1975). It varies remarkably on the big chromosome between populations and species, in association with altitude in *C. hortensis* but not *C. nemoralis,* and with population density in *C. nemoralis* but not *C. hortensis;* the latter relation being negative, it may be a compensation for inbreeding in sparse populations. In some of these respects, there is considerable difference between the species in Somerset and in

Devon (Price and McBride, 1982). Much more of this sort of work is needed to study release of variation, especially if the supergene for the polymorphism is likely to be on the big chromosome (Cook, 1967).

C. Temporal Surveys

1. Short-Term Stability

Several authors have found great stability in morph frequencies of *C. nemoralis* over periods of 5 to 50 years (Cain and Currey, in prep.; Clarke et al., 1968; Goodhart, 1956, 1958; Schilder, 1957; Williamson et al., 1977); or general stability with some change (Goodhart, 1973; Schilder and Schilder, 1958; Wall et al., 1980). Vater (1965) reports some changes over only 3 years, but the general impression is of remarkable stability; Wolda (1969b) found a strongly marked cline stable after 12 years, which seems inexplicable without strong selection (see also Williamson, 1970). The most extensive survey is by Clarke and Murray (1962a,b; Murray and Clarke, 1978), following up a survey by Diver on the sand dunes at Berrow (in Somerset) in 1926. The morphs in this locality show great frequency variation, often in strong clines apparently unrelated to topography, vegetation, or anything else (cf. Wolda, 1969a). Yet Clarke and Murray find an overall stability in this pattern inexplicable except by natural selection. Moreover, between 1926 and 1959 they find consistent overall changes in the frequencies of mid-banded and brown equally assignable to natural selection, which do not appear between samples taken in 1959–1960, 1963, 1969 and 1975.

This work shows that (1) factors in the environment of primary importance to the determination of morph frequencies in snails may be wholly imperceptible to us, (2) that they may act in such a way as almost to provoke an unwary researcher to claim genetic drift as the only possible explanation, and (3) that their existence may only become apparent after a long period of years. The survey of *C. hortensis* in Iceland (Arnason and Grant, 1976; Bengtson et al., 1979b) has produced similar conclusions. This is small comfort for the get-rich-quick career zoologist. Moreover, field surveys are still the necessary approach to such problems. The relevance of laboratory experiments to situations in the wild may be very dubious; and in any case, experiments can hardly be designed to investigate factors not even suspected to exist.

2. Medium-Term Variation

An indirect approach to variation in populations over 50–1000 years is being used with success, namely the determination from historical records of afforestation, deforestation, and changes in land use with relevance to *Cepaea*. Cameron and Pannitt (1984) finds that woods and copses in south Warwickshire less than

100 years old have populations of *C. nemoralis* with inappropriate morph frequencies; in woods of more than 200 years old, age appears to be irrelevant. Despite the overall pattern of variation with habitat, pairs of populations from woods and adjacent open habitats show strong correlations in morph frequencies, suggesting a residual effect of colonization.

Cameron et al. (1977, 1980a,b) and Chappell et al. 1971 have studied records of land usage in chalk districts of southern England and the effects of grazing by sheep, intense until about 50 years ago. Cain and Currey (1963a,b) showed on the Marlborough Downs what they name "area effects," areas large compared with the panmictic unit in *C. nemoralis* and characterized by a uniformity of morph frequencies in spite of visual predation and variation in habitat, often changing to the next area effect by very steep clines. It was suggested (Goodhart, 1963a,b; Wright, 1965, 1978) that these populations originated from small founder populations spreading and meeting with what amounts to hybrid breakdown of their coadapted genotypes. Goodhart (1973) adduced evidence for this from a careful curvey of *C. nemoralis* populations in eastern England over 16 years. Cain and Currey (1963d) rejected this explanation in favor of some form of environmental selection, thinking it unlikely that even at the height of ploughing-up during the Napoleonic wars, the populations of *C. nemoralis* could have been reduced enough to produce the phenomena they found. Cameron et al. (1977) suggest, however, that the population reduction could have been brought about by intense grazing at the end of last century and the beginning of this. They point out that area effects are found on the downs where grazing was intense, not in the valley bottoms with quite different land usage. They also note parallel overall variation in area effects along the two parallel chalk ridges enclosing one valley and rightly explain this by partial natural selection, with extreme reduction and subsequent spread of populations producing the variation within this trend. This is a new approach to area effects which may well explain much of their features, although the continuity of area effects across natural barriers (Arnold, 1971) requires investigation.

3. *Long-Term Studies*

Using subfossil samples, Currey and Cain (1968) and Cain (1971b) have shown a major change in morph frequencies in *C. nemoralis* over about the last 6,000 years, in southern England, relatable to known changes in climate, from almost wholly unbanded to heavily five-banded in valleys and lowlands and less heavily banded on uplands. During the same period *C. hortensis* has remained virtually the same, with about equal proportions of unbanded and banded morphs, but has increased its area of occupation, spreading upward from the valley bottoms and plains (for discussion, see Cain, 1977c). Clarke et al. (1978, p. 251) claim that Currey and Cain's comparisons of pre-Iron Age to Iron Age and of Iron Age to modern "are not statistically heterogeneous. In other words there is no significant evidence that there has been a historical change in the pattern of selection. The firm conclusion is that the subfossil samples as a whole have a higher proportion of unbanded shells than the modern ones. . . .'' If the samples are not heterogeneous, it is hard to see how one can conclude firmly that change has taken place. What Currey and Cain actually showed was that there was a significant difference between pre-Iron Age and modern samples in unbandeds, but none between Iron Age and modern ones, and that Iron Age samples were more like modern ones than one would expect by chance. Cain's subsequent data (1971b) are in complete accord.

The importance of temporal surveys is sufficiently obvious from what has been said; they may be crucial in the interpretation of enigmatic patterns. They are also the only means of finding genuine examples of random variation (always provided that fluctuation of selection can be eliminated).

D. Regional and Local Surveys

Cepaea is probably still one of the best surveyed animal types in the world, with regional and local studies all over Europe (e.g., André, 1975; Arnold, 1968, 1969, 1970; Gerdeaux, 1978; Guerrucci-Henrion, 1966; Harvey, 1971, 1972a,b, 1973, 1974, 1976; Jones, 1973a,b,c, 1974a, 1975; Jones and Irving, 1975; Jones et al., 1980; Lamotte, 1968a,b; Sacchi and Valli, 1975, 1977) and valuable ones on introduced populations (Brussard, 1975; Brussard and McCracken, 1974; Selander and Foltz, 1981). Comparable studies have been made in other snails, for example, Asiatic helicids and bradybaenids (Khokhutkin, 1979; Khokhutkin and Lazareva, 1975; Tzvetkov, 1941); *T. pisana,* (Cowie, 1982; Sacchi, 1951, 1952, 1955a, 1956a, 1971, 1977, 1978; Sacchi and Violani, 1977; Johnson, 1980, 1981; Johnson and Black, 1979; Bar and Nevo, 1977; Nevo and Bar, 1976; Heller, 1981; Hickson, 1972); *Pseudotachea* (Sacchi, 1956b); *Trichia striolata* (Jones et al., 1974); *Cochlicella acuta* (Lewis, 1975, 1977); *Rumina decollata* (Selander and co-workers), and especially the remarkable surveys of *Partula* on the island of Moorea by Clarke and Murray backed up by extensive breeding-work (Clarke and Murray, 1969, 1971; Murray and Clarke, 1966, 1968a,b, 1976a,b; Schwabl and Murray, 1970), and those of *Cerion* by Woodruff and Gould (1980).

Nearly all the *Cepaea* surveys have produced regularities of pattern of some type or other, not always consistent from region to region. Some are quite strongly marked, e.g., the association between dark-banded forms of *C. vindobonensis* and frost hollows in a mountainous region of Yugoslavia (Jones, 1974a), even if little association is shown with environment in flat country (Jones, 1975). In spite of the assertions of Clarke et al. (1978), there is much evidence for selection in relation to climate (as estimated by altitude, aspect, and vegetation) in *C. nemoralis* in the Pyrenees. On the north side, there is considerable regularity between valleys and valley systems in the distribution of color, banding, and lip color morphs. On the southern face, there is much regularity within valleys, but the patterns are not consistent from valley to valley (Cameron et al., 1973). This is, of course, the most insolated aspect and the one most liable to show great variation in drying out of soil and other consequential effects. Moreover, from the considerations given in Section III, the argument of Clarke et al. that discrepancies from region to region annul a hypothesis need not be valid.

Certainly, one must avoid too simple interpretations. The work of Jones and Irving (1975) in the eastern Pyrenees is in a region with different climate from that of the central and western Pyrenees. Their findings on altitudinal distribution of *C. nemoralis* are said not to support the suggestion of Arnold (1968) that *C. nemoralis* can be expected to range higher on southern than northern mountain faces. This is a particular example of a well-known rule applying to fauna and flora (conversely, alpine forms range lower on northern mountain faces). But this can only apply if habitats are otherwise equally suitable on both faces. Jones and Irving themselves suggest that the low altitudinal range of *C. nemoralis* in the Ter valley may be associated with the noncalcareous nature of the rocks in that drainage basin. Cameron et al. draw the same conclusions. If so, conditions are not comparable, and the objection is invalid. What is required now is intensive work, preferably over a long period, to establish what is really going on in each region.

E. Probable Factors

Surveys, local and regional, and experimental work have produced various possible and probable factors acting on gene frequencies in terrestrial mollusc populations, some of which, even if not well based on the molluscs, are of interest to the evolutionist. These can be dealt with only cursorily, and the reviews of Clarke et al. (1978) and Jones et al. (1977) should be consulted for further (and divergent) treatment.

1. Climatic Selection

In climates warmer than that of Britain, surveys show considerable evidence of microclimatic selection in *Theba pisana,* even between the shade of a bush and the open vegetation beyond it (Heller, 1981). Even in Britain, Cowie (1982) has found evidence of selection against the unbanded morph at Tenby within each generation. As this colony, at the northern edge of the species' range, has markedly dark morphs, climatic selection could well be operating, although visual selection is not ruled out. Jones (1973b) has found a marked association of yellow shell color in *C. nemoralis* with mean maximum summer temperature in western and southern Europe, which again seems likely to be due to climate.

Heller and Volokita (1981a) and Jones (1980) found a variation of shell color in *Xeropicta vestalis,* which they correlate with different climatic regimes in Israel, earlier or later whorls being darker than the rest of the shell, according to when (in their life cycle) the snails need to absorb energy, and when they must reflect it back to avoid overheating. In *Cochlicella,* Lewis (1977) found an increase in unbandeds in warmer climates. Nevo and Bar (1975) and Nevo et al. (1981) consider climatic selection to be a major factor controlling shell and allozyme variation in dry country snails (but for shell characters see Heller, 1981).

A considerable amount of work has been done on the degree of heating by radiant energy of shells of different colors (see Heath, 1975; Clarke et al., 1978; and Garcia, 1978 for references). Much of this work has been criticized by Clarke et al. (1978) on the valid ground that the exact provenance of the animals was unknown and the effects observed could be due as much to the animals as to the shell properties. This criticism hardly applies to some of the work of Heath (1975) and Garcia (1978) using shells only, but the latter author did not control shell thickness. There seems to be a general agreement that darker shells do heat up more, and Richardson's work (1974, perhaps 1979) does indicate that on sand dune surfaces in southern England on hot summer days the temperature is high enough to induce heat death, and that there is a differential death rate, darker shelled individuals being more at risk. A critical review of all the experimental work is very desirable; certainly, one must not confuse work on ambient with work on radiant temperature, as Clarke et al. do when they contrast Sedlmair's work with Boettger's and Lamotte's.

It seems reasonable that snails regulate their temperatures at times of excessive heat by climbing up into cooler air away from the hot ground-surface. Much work has been done on climbing in various species, (see especially Jaremovic and Rollo, 1979, for references; also Cowie, 1982; Grime and Blythe, 1969; Grime, et al., 1970; McQuaid et al., 1979; Pomeroy, 1968, 1969; Sacchi, 1974). Wolda (1965a) found great differences between different morphs of *C. nemoralis,* but in no consistent direction. Johnson (1981) found constant differences between morphs in *T. pisana* in where they estivate (shaded or non-shaded places) but does not mention height of climbing. All such behavior would tend to neutralize the effect of climatic selection and is well worth study.

If one could obtain precise climatic data over long periods for particular populations, surveys of morph frequencies would probably show good correlations with climate. As it is, one cannot obtain such data, and only in districts with strongly accentuated features generating microclimatic differences, or in very broad regions adequately monitored, or for very long periods of time, can one expect good associations, the more so as much climatic selection may be crisis-associated rather than continuous. The work of Jones (1973b) on the association of the yellow morph of *C. nemoralis* with high summer temperatures in Western Europe seems sound, but Clarke et al. (1978) have criticized it as taking no account of geographical variation in frequency of habitat types and the nature of the habitats. Jones' data (1973b,c, 1974a) on darker banded *C. vindobonensis,* occurring repeatedly in frost hollows in Yugoslavia, with paler banded forms found on the adjacent mountain sides, are certainly consistent with climatic selection. The inconclusive results obtained from transplant experiments (Jones and Parkin 1977) neither confirm nor support the hypothesis. The work of Currey and Cain (1968) and Cain (1971c) suggests climatic selection over several thousand years in southern England, notwithstanding the criticism of Clarke et al. (1978) quoted above.

Two forms in *C. nemoralis,* apart from pale banding and yellow, give evidence of association with climatic variables. White lip (albolabiate) appears to be at high frequencies in cooler, more humid localities [Arnold, 1968; Cameron, 1969; Cameron et al., 1973; Clarke et al., 1968; Cook and Peake, 1960, 1962; Guerrucci-Henrion, 1973 (although she does not draw this conclusion); Harvey, 1971, 1972a,b (in the last the occurrence in a gorge does not contradict the hypothesis), Harvey, 1976; Lamotte, 1968a, 1972], although with a U-shaped association with altitude in the Pyrenees. This relationship is supported by southern English subfossil samples (Cain, 1971b). White lip normally shows very strong linkage disequilibrium with banding and color (above references; also Cook, 1966; Cook and Peake, 1960). In a number of studies, dark brown is associated with cooler areas, and sometimes more shaded ones (which could then be due to visual selection) (Arnold, 1971; Bantock, 1974; Bantock and Price, 1975b; Cain, 1968; Cain and Currey, 1963a; Carter, 1968b; Harvey, 1971, 1972a in Cornwall; Lamotte, 1959). Harvey (1972a) finds some populations on the south coast of England in which browns are not associated with cool hollows, but gives no details; Harvey also mentions browns (1972b) in northern Spain as being unassociated with cool hollows, altitude, or distance from the sea, but gives no details of habitat. Cameron et al. (1980b) find some association with valley bottoms, but some notable occurrences elsewhere. Goodhart (1973) also has found populations certainly not related to ponding of cold air. The account given by Clarke et al. (1978, p. 246) falls into the error of supposing that if brown is associated with a cooler climate, it must be so with all such climates, even to the very coldest; it also relies heavily on the absence of browns in apparently suitable places, an argument criticized in Section III,A,1); and it quotes Cain and Currey (1963c) as not finding distributions of brown compatible with the hypothesis; in fact, what few were found are compatible. Bantock (1980) has confirmed the location of browns on cooler sites on the Brendon Hills in Somerset, and has evidence from experimental cages of their better survival in cooler climates (see also Bantock and Ratsey, 1980); his criticisms of Clarke et al. (1978) seem valid.

Transference of snails from western Europe to the highly continental conditions of the eastern United States is likely to bring about considerable climatic selection. Brussard (1975) explicitly notes this possibility in a study of North American colonies of *C. nemoralis*. In no study of eastern North American populations that I have met is brown recorded as present, but it is mentioned with extraordinarily high frequency by Jaremovic and Rollo (1979) in a Vancouver population (in a climate much like that of westernmost Europe). Pink is more common in northern populations in eastern North America (Brussard, 1975), as would be expected from climatic selection; indeed, in Lynchburg (Virginia) populations studied by Richards and Murray (1975), there are only yellow *C. nemoralis* which therefore respond to apparent visual selection, as *C. hortensis* would normally do, by varying the banding.

Many other authors have suggested climatic selection as a result of their surveys (Järvinen et al., 1976; Valovirta and Halkka, 1976; Arnold, 1968, 1969; Cameron and Cook, 1971; Arnason and Grant 1976; Bengtson et al., 1979b; Lamotte, 1959, 1966; Cain et al., 1969; Vater, 1965). Harvey (1976) has propounded what he considers a test case, the interpretation of which is difficult in climatic terms. Cain (1973, in Jones, 1973a; Cain 1977c) has criticized too simple interpretations. Clarke et al. (1978, p. 248) are skeptical of many climatic interpretations, since opposite effects can be "explained by postulating one factor or another." Järvinen et al. (1976, p. 107), quoting Lewontin, remark, "Correlating gene frequencies with environmental gradients is hazardous, because 'not to find any environmental element that is roughly in phase with genetic change would be extraordinary'." This is hardly true because too many environmental variables, such as climatic ones, tend to be associated with each other; a survey as careful as André's (1975) shows how many are correlated, greatly reducing the number of environmental patterns available. And a lack of association is exactly what various works on land snail populations claim to find [Clarke and Murray (*Partula*), Goodhart (*Cepaea*), Selander et al. (*Cepaea* and *Helix aspersa*)].

2. *Visual Selection for Crypsis*

The work of Cain and Sheppard giving both survey and experimental evidence for visual selection for crypsis in the Oxford district has already been discussed (Section III,B,1,2). Further survey evidence for *C. nemoralis* has been provided by Currey et al. (1964) in south Warwickshire and by Greenwood (1974) in Worcestershire. In both examples, both shell color and various modifications of banding show the relations with background expected from a hypothesis of selection for crypsis. Further experimental evidence in *C. nemoralis* is provided by Carter (1968a) and Wolda (1963). Carter found that the change in sign of selection for color late in the season could be explained by changes in color of the background in woodland vegetation, as Sheppard had also suggested (1951); this would account for what Wolda thought to be anomalous in his results.

Heller (1981) has suggested visual selection for banded and unbanded morphs of *T. pisana* and gives an illustration of bulbul predation on the estivating snails stuck on branches. Heller and Volokita (1981b) suggest a conflict between visual selection for crypsis and climate selection in *Xerospicta vestalis*. Although along the coastal plain there is strong insolation so that white shells are favored, dark shells are found where there is perennial vegetation. As a snail-predating rodent is found in the areas of perennial vegetation, they suggest that it finds the white shells much easier to see. In Lewis' work on *Cochlicella acuta* (1977), correspondence with background suggests that the sand-colored form and that with heavy bands are the result of selection for crypsis. The apparently conspicuous white form with a black band which is found on turf may be an instance of disruptive coloration.

In *Cepaea*, the direction of variation in shell color in *nemoralis* (darker forms in woods or under ivy) is the same as that which would be expected from climatic selection related to insolation, but the direction of variation in banding is not. It is essential to consider the two together to make a good case for visual selection; where there is slight or no variation in banding, as in *Trichia striolata*, which Jones et al. (1974) interpreted as showing the effect of visual selection, the distinction cannot be made on the basis of distribution alone. Equally, when there is little or no variation in color, but the principal variation is in banding, dark-banded to fused-banded shells are found in dark places, and the situation is ambiguous; Clarke's work (1960) on *C. hortensis* in the Oxford district (Section III,B,1) is therefore open to both interpretations. Clarke et al. (1978) have endeavoured to minimize the evidence for visual selection by casting doubt on direct evidence for selection on banding. However, they do not mention that one of the studies they quote as not finding such selection (Cain and Currey, 1968) was on a population containing only effectively unbanded (unbandeds, mid-bandeds, and punctate mid-bandeds) and no five-bandeds; lack of direct evidence for selection on banding in this population means nothing.

3. *Apostatic Selection*

Clarke, in an important series of papers (1962a,b, 1964, 1968, 1969, 1972, 1975b; Clarke and O'Donald, 1964), has drawn attention to the probably great importance of selection varying with morph frequency and/or population density, especially with regard to visual predation. If birds form hunting-images for particular morphs that are common, they will tend to ignore rare ones until the common ones are reduced in abundance by predation, when their attention is more likely to be caught by the formerly rarer ones; degree of conspicuousness will, of course, also play a part. Selection which overlooks morphs of different appearance from the common ones is called *apostatic* by Clarke, and there is now a considerable literature on this (see Clarke's papers for references, also Greenwood et al., 1981; Willis et al., 1980; Greenwood and Elton, 1979; Allen, 1972a,b, 1976; Allen and Clarke, 1968). The relevance of work done by training birds intensively for several days on a single form to the interpretation of variation in the highly variable *Cepaea* is dubious. Considerable care is necessary in setting up the experiments to produce repeatable results, and it must be shown that any effects obtained are indeed produced by the visible appearance of the snails. The most recent work on *Cepaea* (Harvey et al., 1974, 1975) does seem to show a real possibility of apostatic selection on a terrestrial pulmonate, but the authors are rightly cautious since they were using wild birds of uncertain experience outside the experimental area.

Some surveys have been claimed (Clarke, 1962b) to show apostatic selection in that *C. nemoralis* and *C. hortensis* showed greater divergence in their respective morph frequencies in mixed populations than in single-species populations. Carter (1967) considered that these surveys did not show the effect. Clark's rejoinder (1969a; Clarke et al., 1978) lays down conditions for the disproof of apostatic selection so stringent that it does not seem it can ever be disproved. Various authors have since commented on the absence in their surveys of species divergence of the type that would be expected (Arthur, 1978, 1980; Bantock and Harvey, 1975; Bantock and Noble, 1973). Greenwood (1974a) finds an increase in an index of polymorphism of *C. nemoralis* in the presence of *C. hortensis* (and there may be a converse phenomenon). He lists five possible explanations, of which only one is apostatic selection.

In view of the strong evidence for hunting images in visual predators, and for apostatic selection in other organisms, it does seem likely that it acts on land molluscs. If so, as Clarke has pointed out, selection can maintain a visual polymorphism. The evidence for it in any land mollusc is not yet strong.

Owen (1963a,b,c, 1965a,b, 1966) has called attention to polymorphism in shell color and pattern in the African achatinid *Limicolaria martensiana* in both living and fossil populations. He suggests that apostatic selection is important in it since there is one common morph, which appears cryptic, and a number of rarer and more obvious ones increasing in frequency with increasing density of the population. He asks (1974) whether this is a phenomenon more widespread (e.g., in *Cepaea*) than has been realized. A preliminary survey of available evidence in *Cepaea* species and *Partula* by Jones (1974b) does not suggest that it is. The population of *C. nemoralis* at Fyfield Down has undergone great changes in density over the last 20 years, but its morph frequencies are virtually unchanged (Cain and Currey, 1983). Greenwood (1969) shows that the relation of degree of polymorphism to density under apostatic selection is not simple; both low- and high-density populations will be monomorphic.

4. *Drift, Founder Effects, and Bottlenecks*

Genetic drift and similar random happenings have often been invoked to explain peculiar distributions of morph frequencies in land molluscs, usually with the more confidence the less that is known about the actual biology of the populations. An example in *C. nemoralis,* the area effects, backed up with sufficient historical evidence to make it plausible, is give in Section IV,C,2. It has also been appealed to, without evidence, to explain evolution in gastropods at a much higher level (Boss, 1978). There is no doubt that genetic drift (or one of its special forms) can occur and may yet be recognized; it must certainly be kept in mind as a possible factor (Cain and Sheppard, 1961). Wright (1978), in an important discussion, points out that Cain and Currey (1963a,b, 1968), in rejecting drift as an explanation of area effects, did not state precisely what population sizes in *C. nemoralis* were too large. Greenwood (1974b, 1976) has surveyed the available evidence, and (extending and modifying the work of Murray, 1964) shows that neighborhood numbers range from about 380 to 13,000, with N_e approximately half these figures; Burke (1979) finds numbers of 500 and more in western Irish populations. Even small amounts of migration will enhance these numbers considerably, and except for very low selection pressures, they can hardly be considered small. Nevertheless, Cowie (1982) considers that drift cannot be excluded even in the high-density populations of *T. pisana.*

The question, as always, is What are the neighborhood numbers and what strengths of selection are acting? Selander and Foltz (1981), finding gametic disequilibrium between two esterase loci in populations of *C. nemoralis* in western New York, conclude that genetic drift is the most likely explanation, because the species was introduced, usually in small numbers, and the populations are unstable. Without some knowledge of what the different esterases are doing and whether selection is acting on them, the question must remain open.

Where populations are founded by, or reduced at a given time to, only a few individuals, there is likely to be disturbance at all segregating loci (founder effect, bottleneck effect). Johnson (1976) believed that he had found a coincidence between an area effect in the visible polymorphism in *C. nemoralis* and one in the allozyme variation, both affecting several loci. This work was accepted by Clarke et al. (1978) but has been criticized by Jones et al. (1980), who find no clear association between patterns of geographical variation at the visible and invisible levels in *C. nemoralis* in North Wales and the Pyrenees; they think that a single area of congruence of patterns could be a mere coincidence. Jones et al. (1977) support the idea that populations of *Cepaea* are liable to population bottlenecks by reference to Williamson et al. (1977), who found a vast variation in annual recruitment of adults to a downland population. The example is not a good one since Williamson et al. make it clear that morph frequencies did not vary in the population; moreover, it is not clear that actual numbers ever fell to bottleneck levels.

Cain and Currey (1963a) have pointed out that there are situations in which, given a temporal survey of sufficient span, genetic drift and similar effects might reasonably be postulated. No such work has yet been done for *Cepaea,* apart from that on high downland area effects. The surveys of Goodhart (1962, 1973) come close to the requirements but do not fulfil them. The work of Selander and Kaufman (1975b) on *H. aspersa* in the United States is open to the same comments as Selander and Foltz (1981); it is noteworthy that they refer to the "cogent arguments"of Lamotte (1959) and Goodhart (1962, 1963a) for genetic drift in *Cepaea* and do not even mention the objections of Cain and Sheppard (1961) and Cain and Currey (1963d). Selander and coauthors' work should be compared with that of Potts (1975) for the detailed ecology of *Helix aspersa* in some Californian populations, and with that of Järvinen et al. (1976) on introduced *H. pomatia* populations in Finland and Sweden. These authors give evidence that selection under extreme climatic conditions has prevented reorganization of the gene pools by genetic drift; considering climates in Texas where Selander and Kaufman worked on *H. aspersa,* the same factors should apply at the other end of the climatic range.

5. Speciation

Another valuable theoretical concept due to Clarke (besides apostatic selection) is the steepening of clines by selection with or without environmental gradients (Clarke, 1966) or founder effects. If area effects are produced by the spread of populations with different sets of coadapted genotypes, forming hybrid populations where they meet, or even if we are dealing with simple clines, the accumulation of modifiers can result in the steepening of the transition zone, apparently until an actual break in gene flow occurs. This could lead to parapatric speciation. There is now experimental work on other organisms confirming this (see Endler, 1977), and such a process could explain much of the area effects, incomplete speciation, and apparent hybrid zones found by Clarke and Murray in their remarkable work on *Partula* (Clarke and Murray, 1969) and perhaps in *Cerion* (Woodruff and Gould, 1980) and *Levantina* (Heller, 1979). Considering the frequency of area effects in *C. nemoralis,* it is a little odd, in this theory of speciation (adopted by White, 1978) that there are only two species of *Cepaea* in Western Europe; one would expect quite a large number. If coadaptation of the genotype must involve the visible polymorphism as Clarke (1968) has insisted, it should include the invisible polymorphism as well. However, neither the work of Jones et al. (1980) on *nemoralis* nor that of Johnson et al. (1977) on *Partula* suggests that there is an association between them.

Much more to the point is the extraordinary situation of species and semispecies of *Partula* in Moorea described by Murray and Clarke (1968b, 1980; Clarke and Murray, 1969; Murray et al., 1982; Schwabl and Murray, 1970), which, as they suggest, is probably a case of former geographical separation and present overlap with selection against hybrids. A study of the evolution of a mating barrier would be a major contribution to evolutionary literature. It should be pointed out, however, on the principles of Clarke et al. (1978) that there is no proof that the intermediate specimens are actually hybrids; the species (or forms), being closely related, may not be effectively independent in relation to a very recent common ancestor.

V. Conclusion

Several topics, authors, and species of terrestrial molluscs have been inadequately treated in this chapter, and many papers have been noted only in part, notably work on *Cepaea* by Wolda, Oosterhoff, and Vater; *Partula* by Clarke and Murray; *T. pisana* and other Mediterranean snails by Bar, Nevo, Sacchi, and others; and *Cerion* by Woodruff and Gould, and the topics of maintenance of polymorphism, the adaptive nature of stable polymorphism, and detailed population structure (recruitment, fecundity, vagility, etc.). Nevertheless, it should be apparent that the ecology and behavior of terrestrial molluscs are far from simple; that major features of their population biology can vary over a few feet, as can selection; and that the results of temporal surveys are of special importance in interpreting apparently enigmatic variation. The work on *Cepaea* has been given prominence both because of its extent and because some of the principal controversies over interpretation, experimental design and field practice have arisen from it.

Since Philip Sheppard and I began our work, a great deal more selection has been shown to operate on terrestrial molluscan populations. The mode of operation may often be doubtful, but the proportion of examples in which genetic drift, bottlenecks, or founder effect is probable has decreased, with the possible exception of area effects in downland *C. nemoralis*. [The term "area effect" has been degraded to mean any sort of genetic patchiness, and Woodruff and Gould (1980) rightly remark that three "area effects" in their populations of *Cerion* are probably due to three quite different causes]. Of the different forms of selection known to act or probably acting, the only ones reasonably well established are visual selection for crypsis and climatic selection against excessive insolation. Probable selection forces which must be taken into account include climatic selection of dark shells in situations where incoming energy is scarce, visual predation by tone, not color, and apostatic selection. Differential reproductive success of morphs by intraspecific interference (mucus) is highly probably, and many more classes of selection may well be acting.

I have discussed elsewhere the practical, theoretical, and perhaps even evolutionary difficulties (Cain, 1977c, 1980) which prevent the proper recognition of the role of natural selection in wild populations. One most important practical point (Cain, 1964, 1977c) is that if one merely gives up and ascribes a particular distribution to chance, it will never be properly investigated, since if it is due to chance, what further can be done about it? It was a conviction that little if anything is due to chance that led von Martens (1832) to study banding in *Cepaea* and other land shells and to show for the first time that it was not a case of random variation, but that there was normally a definite number of bands, each with its characteristic position and breadth. Since variations beyond the standard pattern do occur rarely, one cannot point to some hypothetical biochemical canalization as a means of explaining the pattern away. Cain (1983) has suggested a definite selective reason for the differences between the bands above and below the periphery of the shell which is open to experimental investigation.

Clearly, one must not rule out chance processes dogmatically, but neither must one assert them dogmatically on the basis of inadequate evidence. Gould and Lewontin (1979) exhort us to abandon "exclusive focus" on what they call the adaptationist program. In practice, this usually means resting content with inadequate work, deductively imposed answers, and explaining away. The real role of non-adaptive variation and processes in evolution will hardly be discerned by such means. It is noticeable that amid all their rhetoric, their only quotation on terrestrial molluscs is from Alfred Russel Wallace whom they hold up to scorn; in fact, his remarks are largely justified by the work discussed in this review, to none of which do they refer.

Acknowledgments

I am very grateful to Professor B. C. Clarke, Dr. R. A. D. Cameron, Dr. L. M. Cook, and Dr. R. H. Cowie for comments on this chapter.

References

Albuquerque de Matos, R. M. (1979). Genetica de alguns caracteres de *Helix aspersa*. *Port. Acta Biol. Ser. A* **15,** 99–133.

Alkins, W. E. (1922). Two molluscan associations in north-east Staffs. *J. Conchol.* **16,** 291–296.

Allen, J. A. (1972a). Apostatic selection: the responses of wild passerines to artificial polymorphic prey. Ph.D. Thesis, Edinburgh University.

Allen, J. A. (1972b). Evidence for stabilizing and apostatic selection by wild blackbirds. *Nature (London)* **237,** 348–349.

Allen, J. A. (1976). Further evidence for apostatic selection by wild passerine birds—9:1 experiments. *Heredity* **36,** 173–180.

Allen, J. A., and Clarke, B. C. (1968). Evidence for apostatic selection by wild passerines. *Nature (London)* **220,** 501–502.

André, J. (1975). Écologie du gastéropode terrestre *Cepaea nemoralis* Linné en Languedoc et en Roussillon. *Vie Milieu* **25,** 17–47.

Anonymous (1901). (No title). *J. Conchol.* **10,** 88.

Appleton, T. C., and Newell, P. F. (1977). X-ray microanalysis of freeze dried ultra-thin frozen sections of a regulatory epithelium from the snail *Otala*. *Nature (London)* **266,** 854–855.

Arnason, E., and Grant, P. R. (1976). Climatic selection in *Cepaea hortensis* at the northern limit of its range in Iceland. *Evolution (Lancaster, Pa.)* **30,** 499–508.

Arnold, R. W. (1968). Studies on *Cepaea.* VII. Climatic selection in *Cepaea nemoralis* (L.) in the Pyrenees. *Philos. Trans. R. Soc. London, Ser. B* **253,** 549–593.

Arnold, R. W. (1969). The effects of selection by climate on the land snail *Cepaea nemoralis* (L.). *Evolution (Lancaster, Pa.)* **23,** 370–378.

Arnold, R. W. (1970). A comparison of populations of the polymorphic land snail *Cepaea nemoralis* (L.) living in a lowland district in France with those in a similar district in England. *Genetics (Princeton, N.J.)* **64,** 589–604.

Arnold, R. W. (1971). *Cepaea nemoralis* on the East Sussex South Downs, and the nature of area effects. *Heredity* **26,** 277–298.

Arthur, W. (1978). Morph frequency and co-existence in *Cepaea. Heredity* **41,** 335–346.

Arthur, W. (1980). Further associations between morph-frequency and coexistence in *Cepaea. Heredity* **44,** 417–421.

Arthur, W. (1982). A critical evaluation of the case for competitive selection in *Cepaea. Heredity* **48,** 407–419.

Bantock, C. R. (1972). Localization of chiasmata in *Cepaea nemoralis* L. *Heredity* **29,** 213–221.

Bantock, C. R. (1974). *Cepaea nemoralis* (L.) on Steep Holm. *Proc. Malacol. Soc. London* **41,** 223–232.

Bantock, C. R. (1980). Variation in the distribution and fitness of the brown morph of *Cepaea nemoralis* (L.). *Biol. J. Linn. Soc.* **13,** 47–64.

Bantock, C. R., and Bayley, J. A. (1973). Visual selection for shell size in *Cepaea* (Held). *J. Anim. Ecol.* **42,** 247–261.

Bantock, C. R., and Harvey, P. H. (1975). Colour polymorphism and selective predation experiments. *J. Biol. Educ.* **8,** 323–329.

Bantock, C. R., and Noble, K. (1973). Variation with altitude and habitat in *Cepaea hortensis* (Mull.). *Zool. J. Linn. Soc.* **53,** 237–252.

Bantock, C. R., and Price, D. J. (1975a). The cytology of *Cepaea nemoralis* with emphasis on recording chiasma frequency. *In* "Laboratory Manual of Cell Biology" (D. O. Hall and S. H. Hawkins, eds.), Chapter 5. English Univ. Press, London.

Bantock, C. R., and Price, D. J. (1975b). Marginal populations of *Cepaea nemoralis* (L.) on the Brendon Hills, England. I. Ecology and ecogenetics. *Evolution (Lancaster, Pa.)* **29,** 267–277.

Bantock, C. R., and Ratsey, M. (1980). Natural selection in experimental populations of the landsnail *Cepaea nemoralis* (L.) *Heredity* **44,** 37–54.

Bar, Z. (1975). Distribution and habitat of the genus *Sphincterochila* in Israel and Sinai. *Argamon, Isr. J. Malacol.* **5,** 1–19.

Bar, Z. (1978). Variation and natural selection in shell thickness of *Theba pisana* along climatic gradients in Israel. *J. Moll. Stud.* **44,** 322–326.

Bar, Z., and Nevo, E. (1977). Natural selection of shell banding polymorphism in *Theba pisana* (Mollusca) along climatic gradients. *Isr. J. Zool.* **25,** 214–215.

Bengtson, S.-A., Nilsson, A., Nordström, S., and Rundgren, S. (1976). Polymorphism in relation to habitat in the snail *Cepaea hortensis* in Iceland. *J. Zool.* **178,** 173–188.

Bengtson, S.-A., Nilsson, A., Nordström, S., and Rundgren, S. (1979a). Selection for adult shell size in natural populations of the landsnail *Cepaea hortensis* (Müll.). *Ann. Zool. Fenn.* **16,** 187–194.

Bengtson, S.-A., Nilsson, A., Nordström, S., and Rundgren, S. (1979b). Distribution patterns of morph frequencies in the snail *Cepaea hortensis* in Iceland. *Holarctic Ecol.* **2,** 144–149.

Boss, K. J. (1978). On the evolution of gastropods in ancient lakes. *In* "Pulmonates" (V. Fretter and J. Peake, eds.), Vol. 2A, pp. 385–428. Academic Press, New York.

Boycott, A. E. (1921). Notes on the distribution of British land and freshwater Mollusca from the point of view of habitat and climate. *Proc. Malacol. Soc. London* **14,** 163–167.

Boycott, A. E. (1929). The oecology of British land Mollusca, with special reference to those of ill-defined habitat. *Proc. Malacol. Soc. London* **18,** 213–224.

Boycott, A. E. (1934). The habitats of land mollusca in Britain. *J. Ecol.* **22,** 1–38.

Boycott, A. E., and Oldham, C. (1938). A contagious disease of *Helix aspersa. Proc. Malacol. Soc. London* **23,** 92–96.

Brussard, P. F. (1974). Population size and natural selection in the land snail *Cepaea nemoralis. Nature (London)* **251,** 713–715.

Brussard, P. F. (1975). Geographical variation in North American colonies of *Cepaea nemoralis. Evolution (Lancaster, Pa.)* **29,** 402–410.

Brussard, P. F., and McCracken, G. F. (1974). Allozymic variation in a North American colony of *Cepaea nemoralis. Heredity* **33,** 98–101.

Burke, D. P. T. (1979). Variation in western Irish populations of *Cepaea nemoralis* (L.). Ph.D. Thesis, Liverpool University.

Cain, A. J. (1952). Local evolution and taxonomy in very polymorphic species. *Proc. Leeds Philos. Lit. Soc., Sci. Sect.* **6,** 47–49.

Cain, A. J. (1953). Visual selection by tone of *Cepaea nemoralis* (L.). *J. Conchol.* **23,** 333–336.

Cain, A. J. (1959a). An undescribed polymorphism in *Hygromia striolata* (C. Pfeiffer). *J. Conchol.* **24,** 319–322.

Cain, A. J. (1959b). Inheritance of mantle colour in *Hygromia striolata* (C. Pfeiffer). *J. Conchol.* **24,** 352–353.

Cain, A. J. (1964). The perfection of animals. *Viewpoints Biol.* **3,** 36–63.

Cain, A. J. (1968). Studies on Cepaea. V. Sand-dune populations of *Cepaea nemoralis* (L.). *Philos. Trans. R. Soc. London, Ser. B* **253,** 499–517.

Cain, A. J. (1971a). Undescribed polymorphisms in two British snails. *J. Conchol.* **26,** 410–416.

Cain, A. J. (1971b). Colour and banding morphs in subfossil samples of the snail *Cepaea. In* "Ecological Genetics and Evolution" (E. R. Creed, ed.), pp. 65–92. Blackwell, Oxford.

Cain, A. J. (1972). Comment. *In* "The Rules of the Game. Cross-Disciplinary Essays on Models in Scholarly Thought" (T. Shanin, ed.), pp. 68–71. Tavistock, London.

Cain, A. J. (1977a). Variation in the spire index of some coiled gastropod shells, and its evolutionary significance. *Philos. Trans. R. Soc. London, Ser. B.* **277,** 377–428.

Cain, A. J. (1977b). The uniqueness of the polymorphism of *Cepaea* (Pulmonata: Helicidae) in western Europe. *J. Conchol.* **29,** 129–136.

Cain, A. J. (1977c). The efficacy of natural selection in wild populations. *Acad. Nat. Sci. Philadelphia, Spec. Publ.* **12,** 111–133.

Cain, A. J. (1978a). Variation of terrestrial gastropods in the Philippines in relation to shell shape and size. *J. Conchol.* **29,** 239–245.

Cain, A. J. (1978b). The deployment of operculate land snails in relation to shape and size of shell. *Malacologia* **17,** 207–221.

Cain, A. J. (1979). Introduction to general discussion. *n* "The Evolution of Adaptation by Means of Natural Selection" (J. Maynard Smith and R. Holliday, eds.), pp. 599–604. Royal Society, London.

Cain, A. J. (1981a). Possible ecological significance of variation in shape of *Cerion* shells with age. *J. Conchol.* **30,** 305–315.

Cain, A. J. (1981b). Variation in shell shape and size of helicid snails in relation to other pulmonates in faunas of the Palearctic Region. *Malacologia* **21,** 149–176.

Cain, A. J. (1983). Heterosematism in land snails. *Malacologia* (in press).

Cain, A. J. (1984). *Malacologia* (in press).

Cain, A. J., and Cowie, R. H. (1978). Activity of different species of land-snail on surfaces of different inclinations. *J. Conchol.* **29,** 267–272.

Cain, A. J., and Currey, J. D. (1963a). Area effects in *Cepaea. Philos. Trans. R. Soc. London, Ser. B* **246,** 1–81.

Cain, A. J., and Currey, J. D. (1963b). Differences in interactions between selective forces acting in the wild on certain pleiotropic genes of *Cepaea. Nature (London)* **197,** 411–412.

Cain, A. J., and Currey, J. D. (1963c). Area effects in *Cepaea* on the Larkhill Artillery Ranges, Salisbury Plain. *J. Linn. Soc. London, Zool.* **45,** 1–15.

Cain, A. J., and Currey, J. D. (1963d). The causes of area effects. *Heredity* **18,** 467–471.

Cain, A. J., and Currey, J. D. (1968). Studies on *Cepaea.* III. Ecogenetics of a population of *Cepaea nemoralis* (L.) subject to strong area effects. *Philos. Trans. R. Soc. London, Ser. B* **253,** 447–482.

Cain, A. J., and Currey, J. D. (1983). *In preparation.*

Cain, A. J., and Sheppard, P. M. (1950). Selection in the polymorphic land snail *Cepaea nemoralis. Heredity* **4,** 275–294.

Cain, A. J., and Sheppard, P. M. (1952). The effects of natural selection on body colour in the land snail *Cepaea nemoralis. Heredity* **6,** 217–231.

Cain, A. J., and Sheppard, P. M. (1954). Natural selection in *Cepaea. Genetics (Princeton, N.J.)* **39,** 89–116.

Cain, A. J., and Sheppard, P. M. (1961). Visual and physiological selection in *Cepaea. Am. Nat.* **95,** 61–64.

Cain, A. J., Cameron, R. A. D., and Parkin, D. T. (1969). Ecology and variation of some helicid snails in northern Scotland. *Proc. Malacol. Soc. London* **38,** 269–299.

Cameron, R. A. D. (1968). The distribution and variation of three species of land snail near Rickmansworth, Hertfordshire. *Zool. J. Linn. Soc.* **48,** 83–111.

Cameron, R. A. D. (1969). The distribution and variation of *Cepaea nemoralis* L. near Slievecarran, County Clare and County Galway, Eire. *Proc. Malacol. Soc. London* **38,** 439–450.

Cameron, R. A. D. (1970a). Differences in the distributions of three species of helicid snail in the limestone district of Derbyshire. *Proc. R. Soc. London, Ser. B* **176,** 131–159.

Cameron, R. A. D. (1970b). The survival, weight-loss and behaviour of three species of land snail in conditions of low humidity. *J. Zool* **160,** 143–157.

Cameron, R. A. D. (1970c). The effect of temperature on the activity of three species of helicid snail (Mollusca: Gastropoda). *J. Zool.* **162,** 303–315.

Cameron, R. A. D. (1978). Differences in the sites of activity of co-existing species of land mollusc. *J. Conchol.* **29,** 273–278.

Cameron, R. A. D., and Carter, M. A. (1979). Intra- and interspecific effects of population density on growth and activity in some helicid land snails (Gastropoda: Pulmonata). *J. Anim. Ecol.* **48,** 237–246.

Cameron, R. A. D., and Cook, L. M. (1971). *Cepaea nemoralis* (L.) on Whitbarrow Scar, Lancashire. *Proc. Malacol. Soc. London* **39,** 399–408.

Cameron, R. A. D. and Dillon, P. (1984). *Malacologia* (in press).

Cameron, R. A. D., and Morgan-Huws, D. I. (1975). Snail faunas in the early stages of a chalk grassland succession. *Biol. J. Linn. Soc.* **7,** 215–229.

Cameron, R. A. D., and Palles-Clarke, M. A. (1971). *Arianta arbustorum* (L.) on chalk downs in southern England. *Proc. Malacol. Soc. London* **39,** 311–318.

Cameron, R. A. D., and Pannitt, D. (1984). In preparation.

Cameron, R. A. D., and Redfern, M. (1972). The terrestrial Mollusca of the Malham area. *Field Stud.* **3,** 589–602.

Cameron, R. A. D., Carter, M. A., and Haynes, F. N. (1973). The variation of *Cepaea nemoralis* in three Pyrenean valleys. *Heredity,* **31,** 43–74.

Cameron, R. A. D., Williamson, P., and Morgan-Huws, D. I. (1977). The habitats of the land snail *Cepaea nemoralis* (L.) on downland and their ecogenetic significance. *Biol. J. Linn. Soc. London* **9,** 231–241.

Cameron, R. A. D., Down, K., and Pannett, D. J. (1980a). Historical and environmental influences on hedgerow snail faunas. *Biol. J. Linn. Soc. London* **31,** 75–87.

Cameron, R. A. D., Carter, M. A., and Palles-Clarke, M. A. (1980b). *Cepaea* on Salisbury Plain: patterns of variation, landscape history and habitat stability. *Biol. J. Linn. Soc. London* **14,** 355–358.

Carter, M. A. (1967). Selection in mixed colonies of *Cepaea nemoralis* and *Cepaea hortensis*. *Heredity* **22,** 117–139.

Carter, M. A. (1968a). Thrush predation of an experimental population of the snail *Cepaea nemoralis* (L.) *Proc. Linn. Soc. London* **179,** 241–249.

Carter, M. A. (1968b). Studies on *Cepaea*. II. Area effects and visual selection in *Cepaea nemoralis* (L.) and *Cepaea hortensis* (Müll.). *Philos. Trans. R. Soc. London, Ser. B* **253,** 397–446.

Carter, M. A., Jeffery, R. C. V., and Williamson, P. (1979). Food overlap in co-existing populations of the land snails *Cepaea nemoralis* (L.) and *Cepaea hortensis* (Müll.). *Biol. J. Linn. Soc. London* **11,** 67–176.

Cartwright, W. (1922). On the association and non-association of *Helix nemoralis* Linné and *Helix hortensis* Müller. *J. Conchol.* **16,** 313–318.

Chan, W., and Saleuddin, A. S. M. (1974). Evidence that *Otala lactea* (Müller) utilizes calcium from the shell. *Proc. Malacol. Soc. London* **41,** 195–200.

Chappell, H. G., Ainsworth, J. F., Cameron, R. A. D., and Redfern, M. (1971). The effect of trampling on a chalk grassland ecosystem. *J. Appl. Ecol.* **8,** 869–882.

Chatfield, J. E. (1976). Studies on food and feeding in some European land molluscs. *J. Conchol.* **29,** 5–20.

Chatfield, J. E. (1977). Observations on variation in the mantle of the land snail *Monacha cantiana* (Montagu) (Pulmonata: Helicidae). *J. Conchol.* **29,** 219–222.

Chaucer, G. (c. 1385). "The Legend of Good Women." Prologue.

Clarke, B. C. (1960). Divergent effects of natural selection on two closely-related polymorphic snails. *Heredity* **14,** 423–443.

Clarke, B. C. (1962a). Balanced polymorphism and the diversity of sympatric species. *In* "Taxonomy and Geography" (D. Nichols, ed.), Publ. No. 4, pp. 47–70. Systematics Association, London.

Clarke, B. C. (1962b). Natural selection in mixed populations of two polymorphic snails. *Heredity* **17,** 319–345.

Clarke, B. C. (1964). Frequency-dependent selection for the dominance of rare polymorphic genes. *Evolution (Lancaster, Pa.)* **18,** 364–369.

Clarke, B. C. (1966). The evolution of morph-ratio clines. *Am. Nat.* **100,** 389–402.

Clarke, B. C. (1968). Balanced polymorphism and regional differentiation in land snails. *In* "Evolution and Environment" (E. T. Drake, ed.), pp. 351–368. Yale Univ. Press, New Haven, Connecticut.

Clarke, B. C. (1969). The evidence for apostatic selection. *Heredity* **24,** 347–352.

Clarke, B. C. (1972). Density-dependent selection. *Am. Nat.* **106,** 1–13.

Clarke, B. C. (1975a). The contribution of ecological genetics to evolutionary theory: detecting the direct effects of natural selection on particular polymorphic loci. *Genetics (Princeton, N.J.)* **79,** 101–113.

Clarke, B. C. (1975b). Frequency-dependent and density-dependent natural selection. *In* "The Role of Natural Selection in Human Evolution (F. M. Salzano, ed.), pp. 187–200. North-Holland Publ., Amsterdam.

Clarke, B. C. (1975c). The causes of biological diversity. *Sci. Am.* **233,** 50–60.

Clarke, B. C., and Murray, J. J. (1962a). Changes of gene frequency in *Cepaea nemoralis* (L.). *Heredity* **17,** 445–465.

Clarke, B. C., and Murray, J. J. (1962b). Changes of gene frequency in *Cepaea nemoralis* (L.); the estimation of selective values. *Heredity* **17,** 467–476.

Clarke, B. C., and Murray, J. J. (1969). Ecological genetics and speciation in land snails of the genus *Partua*. *Biol. J. Linn. Soc.* **1,** 31–42.

Clarke, B. C., and Murray, J. J. (1971). Polymorphism in a Polynesian land snail *Partula suturalis vexillum*. *In* "Ecological Genetics and Evolution" (E. R. Creed, ed.), pp. 51–64. Blackwell, Oxford.

Clarke, B. C., and O'Donald, P. (1964). Frequency-dependent selection. *Heredity* **19,** 201–206.

Clarke, B. C., Diver, C., and Murray, J. J. (1968). Studies on *Cepaea*. VI. The spatial and temporal distribution of phenotypes in a colony of *Cepaea nemoralis* (L.). *Philos. Trans. R. Soc. London, Ser. B* **253,** 519–548.

Clarke, B. C., Arthur, W., Horsley, D. T., and Parkin, D. T. (1978). Genetic variation and natural selection in pulmonate molluscs. *In* "Pulmonates" (V. Fretter and J. F. Peake, eds.), Vol. 2A, pp. 219–270. Academic Press, New York.

Cook, A. (1977). Mucus trail following by the slug *Limax grossui* Lupu. *Anim. Behav.* **25,** 774–781.

Cook, A. (1979). Homing by the slug *Limax pseudoflavus* Evans. *Anim. Behav.* **27,** 545–552.

Cook, A. (1981). A comparative study of aggregation in pulmonate slugs (genus *Limax*). *J. Anim. Ecol.* **50,** 703–713.

Cook, L. M. (1966). Notes on two colonies of *Cepaea nemoralis* (L.) polymorphic for white lip. *J. Conchol.* **26,** 125–130.

Cook, L. M. (1967). The genetics of *Cepaea nemoralis*. *Heredity* **22,** 397–410.

Cook, L. M., and Cain, A. J. (1980). Population dynamics, shell size and morph frequency in experimental populations of the snail *Cepaea nemoralis* (L.). *Biol. J. Linn. Soc.* **14,** 259–292.

Cook, L. M., and Peake, J. F. (1960). A study of some populations of *Cepaea nemoralis* L. from the Dartry Mountains, Co. Sligo, Ireland. *Proc. Malacol. Soc. London* **34,** 1–11.

Cook, L. M., and Peake, J. F. (1962). Populations of *Cepaea nemoralis* L. from sand-dunes on the Mullaghmore Peninsula, Co. Sligo, Ireland, with a comparison with those from Annacoona, Dartry Mts., Co. Sligo. *Proc. Malacol. Soc. London* **35,** 7–13.

Cowie, R. H. (1982). Studies on the ecology and ecogenetics of the helicid land snail *Theba pisana*. Ph.D. Thesis, Liverpool University.

Crook, S. J. (1980). Studies on the ecological genetics of *Helix aspersa* (Müller). Ph.D. Thesis, Dundee University.

Currey, J. D., and Cain, A. J. (1968). Studies on Cepaea. IV. Climate and selection of banding morphs in *Cepaea* from the climatic optimum to the present day. *Philos. Trans. R. Soc. London, Ser. B* **253,** 483–498.

Currey, J. D., Arnold, R. W., and Carter, M. A. (1964). Further examples of variation of populations of *Cepaea nemoralis* with habitat. *Evolution (Lancaster, Pa.)* **18,** 111–117.

Dalgliesh, J. G. (1930). Field notes on *Cepaea nemoralis* Linn. and *Cepaea hortensis* Müller in Sussex. *J. Conchol.* **19,** 94–96.

Diver, C. (1940). The problem of closely related species living in the same area. *In* "The New Systematics" (J. S. Huxley, ed.), pp. 303–328. Oxford Univ. Press, London and New York.

Endler, J. A. (1977). "Geographic Variation, Speciation and Clines," Monogr. Popul. Biol., 10. Princeton Univ. Press, Princeton, New Jersey.

Evans, H. J. (1960). Supernumerary chromosomes in wild populations of the snail *Helix pomatia*. *Heredity* **15,** 129–138.

Frömming, E. (1954). "Biologie der mitteleuropaïschen Landgastropoden." Duncker & Humboldt, Berlin.

Garcia, M. C. L. R. (1978). Écophysiologie de l'escargot *Cepaea nemoralis* (L.) Helicidae: quelques conséquences de l'échauffement par l'énergie rayonneé. *Arch. Zool. Exp. Gen.* **118,** 495–514.

Gerdeaux, D. (1978). Le polymorphisme de *Cepaea nemoralis* en Provence; ses relations avec le milieu. *Arch. Zool. Exp. Gen.* **119,** 565–584.

Gill, J. J. B., and Cain, A. J. (1980). The karyotype of *Cepaea sylvatica* (Pulmonata: Helicidae) and its relationship to those of *C. hortensis* and *C. nemoralis. Biol. J. Linn. Soc. London* **14,** 293–301.

Gill, J. J. B., and Cain, A. J. (1983). *In preparation.*

Goodhart, C. B. (1956). Genetic stability in populations of the polymorphic snail *Cepaea nemoralis* (L.). *Proc. Linn. Soc. London* **167,** 50–67.

Goodhart, C. B. (1958). Genetic stability in the snail *Cepaea nemoralis* (L.): a further example. *Proc. Linn. Soc. London* **169,** 163–167.

Goodhart, C. B. (1962). Variation in a colony of the snail *Cepaea nemoralis* (L.). *J. Anim. Ecol.* **31,** 207–237.

Goodhart, C. B. (1963a). "Area effects" and non-adaptive variation between populations of *Cepaea* (Mollusca). *Heredity* **18,** 459–465.

Goodhart, C. B. (1963b). The Sewall Wright effect. *Am. Nat.* **97,** 407–409.

Goodhart, C. B. (1973). a 16-year survey of *Cepaea* on the Hundred Foot Bank. *Malacologia* **14,** 327–331.

Gould, S. J., and Lewontin, R. C. (1979). The spandrels of San Marco and the Panglossian paradigm: a critique of the adaptionist programme. *In* "The Evolution of Adaptation by Means of Natural Selection" (J. Maynard Smith and R. Holliday, eds.), pp. 581–598. Royal Society, London.

Greenwood, J. J. D. (1966). A preliminary note on aspects of the variation of *Lehmannia marginata. Proc. Malacol. Soc. London* **37,** 119–125.

Greenwood, J. J. D. (1969). Apostatic selection and population density. *Heredity* **24,** 157–161.

Greenwood, J. J. D. (1974). Visual and other selection in *Cepaea:* a further example. *Heredity* **33,** 17–31.

Greenwood, J. J. D. (1975). Effective population numbers in the snail *Cepaea nemoralis. Evolution (Lancaster, Pa.)* **28,** 513–526.

Greenwood, J. J. D. (1976). Effective population number in *Cepaea:* a modification. *Evolution (Lancaster, Pa.)* **30,** 186.

Greenwood, J. J. D., and Elton, R. A. (1979). Analysing experiments on frequency-dependent selection by predators. *J. Anim. Ecol.* **48,** 721–737.

Greenwood, J. J. D., Wood, E. M., and Batchelor, S. (1981). Apostatic selection of distasteful prey. *Heredity* **47,** 27–34.

Grime, J. P., and Blythe, G. M. (1969). An investigation of the relationships between snails and vegetation at the Winnats Pass. *J. Ecol.* **57,** 45–66.

Grime, J. P., MacPherson-Stewart, S. F., and Dearman, R. S. (1968). An investigation of leaf palatability using the snail *Cepaea nemoralis* (L.). *J. Ecol.* **56,** 405–420.

Grime, J. P., Blythe, G. M., and Thornton, J. D. (1970). Food selection by the snail *Cepaea nemoralis* (L.). *In* "Animal Populations in Relation to their Food Resources" (A. Watson, ed.), pp. 73–99. Blackwell, Oxford.

Guerrucci-Henrion, M. A. (1966). Recherches sur les populations naturelles de *Cepaea nemoralis* en Bretagne. *Arch. Zool. Exp. Gén.* **107,** 369–417.

Guerrucci-Henrion, M. A. (1973). Aspects généraux du polymorphisme de la couleur du peristome chez *Cepaea hortensis* en France. *Malacologia* **14,** 333–338.

Guerrucci-Henrion, M. A. (1974). Le polymorphisme de la coquille chez *Cepaea hortensis* Müller (Mollusques pulmonés) en France. *Mém. Soc. Zool. Fr.* **37,** 103–127.

Gyngell, W. (1911). *Helix nemoralis* and *H. hortensis:* their colour and band variations and distribution—some comparisons. *J. Conchol.* **13,** 241–243.

Hadfield, M. G., and Mountain, B. S. (1981). A field study of a vanishing species, *Achatinella mustelina* (Gastropoda, Pulmonata), in the Waianae Mountains of Oahu. *Pac. Sci.* **34,** 345–358.

Harvey, P. H. (1971). *Cepaea nemoralis* in Brittany, Cornwall and Pembrokeshire. *Heredity* **26,** 365–372.

Harvey, P. H. (1972a). *Cepaea nemoralis* on clifftops in south-west England. *Proc. R. Soc. London, Ser. B* **181,** 375–393.

Harvey, P. H. (1972b). Populations of *Cepaea nemoralis* from south-western France and northern Spain. *Heredity* **27,** 353–363.

Harvey, P. H. (1972c). Certain mathematical models to describe the dynamics of polymorphic populations of the land snail *Cepaea nemoralis* (L.). *Genetica* **43,** 531–535.

Harvey, P. H. (1973). The distribution of three species of helicid snail in East Yorkshire. I. General survey. *Proc. Malacol. Soc. London* **40,** 523–530.

Harvey, P. H. (1974). The distribution of three species of helicid snail in East Yorkshire. II. Intensive survey. *Proc. Malacol. Soc. London* **41,** 57–64.

Harvey, P. H. (1976). Factors affecting the shell polymorphism of *Cepaea nemoralis* (L.) in East Yorkshire: a test case. *Heredity* **36,** 1–10.

Harvey, P. H., Jordan, C. A., and Allen, J. A. (1974). Selection behaviour of wild blackbirds at high prey densities. *Heredity* **32,** 401–404.

Harvey, P. H., Birley, N., and Blackstock, T. H. (1975). The effect of experience on the selective behaviour of song thrushes feeding on artificial populations of *Cepaea* (Held). *Genetica* **45,** 211–216.

Hazard, P. (1964). "The European Mind 1680–1715." Penguin Books, Harmondsworth, England.

Heath, D. J. (1975). Color, sunlight and internal temperatures in the land snail *Cepaea nemoralis* (L.). *Oecologia* **19,** 29–38.

Heller, J. (1979). Distribution, hybridization and variation in the Israeli landsnail *Levantina* (Pulmonata: Helicidae). *Zool. J. Linn. Soc.* **67,** 115–148.

Heller, J. (1981). Visual versus climatic selection of shell banding in the land snail *Theba pisana* in Israel. *J. Zool. Lond.* **194,** 85–101.

Heller, J., and Volokita, M. (1981a). Gene regulation of shell banding in a land snail from Israel. *Biol. J. Linn. Soc. London* **16,** 261–277.

Heller, J., and Volokita, M. (1981b). Shell-banding polymorphism of the land snail Xeropicta vestalis along the coastal plain of Israel. *Biol. J. Linn. Soc. London* **16,** 279–284.

Herzberg, F. (1965). Crowding as a factor in growth and reproduction of *Helix aspersa. Am. Zool.* **5,** 254.

Hickson, T. G. L. (1972). A possible case of genetic drift in colonies of the land snail *Theba pisana. Heredity* **29,** 177–190.

Hunter, P. J. (1978). Slugs—a study in applied ecology. *In* "Pulmonates" (V. Fretter and J. F. Peake, eds.), Vol. 2A, pp. 271–286. Academic Press, New York.

Ingram, W. M. (1946). The European brown snail in Oakland, California. *Bull. South. Calif. Acad. Sci.* **45,** 152–159.

Jaremovic, R., and Rollo, C. D., (1979). Tree climbing by the snail *Cepaea nemoralis* (L.)—a possible method for regulating temperature and hydration. *Can. J. Zool.* **57,** 1010–1014.

Järvinen, O., Sisula, H., Varvio-Aho, S.-L., and Salminen, P. (1976). Genic variation in isolated marginal populations of the Roman snail, *Helix pomatia* L. *Hereditas* **82,** 101–110.

Johnson, M. S. (1976). Allozymes and area effects in *Cepaea nemoralis* on the western Berkshire Downs. *Heredity* **36,** 105–121.

Johnson, M. S. (1979). Inheritance and geographic variation of allozymes in *Cepaea nemoralis*. *Heredity* **43,** 137–141.

Johnson, M. S. (1980). Association of shell banding and habitat in a colony of the land snail *Theba pisana*. *Heredity* **45,** 7–14.

Johnson, M. S. (1981). Effects of migration and habitat choice on shell banding frequencies in *Theba pisana* at a habitat boundary. *Heredity* **47,** 121–133.

Johnson, M. S., and Black, R. (1979). The distribution of *Theba pisana* on Rottnest Island. *West. Aust. Nat.* **14,** 140–144.

Johnson, M. S., Clarke, B. C., and Murray, J. J. (1977). Genetic variation and reproductive isolation in *Partula*. *Evolution (Lancaster, Pa.)* **31,** 116–126.

Jones, J. S. (1973a). Ecological genetics of a population of the snail *Cepaea nemoralis* at the northern limit of its range. *Heredity* **31,** 210–211.

Jones, J. S. (1973b). Ecological genetics and natural selection in molluscs. *Science* **182,** 546–552.

Jones , J. S. (1973c). The genetic structure of a southern peripheral population of the snail *Cepaea nemoralis*. *Proc. R. Soc. London, Ser. B* **183,** 371–384.

Jones, J. S. (1974a). Environmental selection in the snail *Cepaea vindobonensis* in the Lika area of Yugoslavia. *Heredity* **32,** 165–170.

Jones, J. S. (1974b). (no title). *Science* **185,** 376–377.

Jones, J. S. (1975). The genetic structure of some steppe populations of the snail *Cepaea vindobonensis*. *Genetica* **45,** 217–225.

Jones, J. S. (1980). Evolutionary genetics of snails. *Nature (London)* **285,** 283–284.

Jones, J. S., and Clarke, B. C. (1969). The distribution of *Cepaea* in Scotland. *J. Conchol.* **27,** 3–8.

Jones, J. S., and Irving, A. J. (1975). Gene frequencies, genetic background and environment in Pyrenean populations of *Cepaea nemoralis* (L.). *Biol. J. Linn. Soc.* **7,** 249–259.

Jones, J. S., and Parkin, D. T. (1977). Experimental manipulation of some snail populations subject to climatic selection. *Am. Nat.* **111,** 1014–1017.

Jones, J. S., Briscoe, D. A., and Clarke, B. C. (1974). Natural selection on the polymorphic snail *Hygromia striolata*. *Heredity* **33,** 102–106.

Jones, J. S., Leith, B. H., and Rawlings, P. (1977). Polymorphism in *Cepaea:* a problem with too many solutions? *Annu. Rev. Ecol. Syst.* **8,** 109–143.

Jones, J. S., Selander, R. K., and Schnell, G. D. (1980). Patterns of morphological and molecular polymorphism in the land snail *Cepaea nemoralis*. *Biol. J. Linn. Soc. London* **14,** 359–387.

Khokhutkin, I. M. (1979). Inheritance of banding in natural populations of the land snail *Bradybaena fruticum* (Müll.). *Genetika (Moscow)* **15,** 868–871 (in Russian, English abstract).

Khokhutkin, I. M., and Lazareva, A. I. (1975). Polymorphism and concealing coloration of land mollusc populations. *Zh. Obshch. Biol.* **36,** 863–868 (in Russian, English abstract).

Kothbauer, H., and Schnitzler, S. (1972). Haemagglutinins in *Cepaea hortensis, C. nemoralis* and *C. vindobonensis* (Helicidae, Gastropoda); their importance in systematics. *Z. Zool. Syst. Evolutionsforsch.* **10,** 133–137.

Lamotte, M. (1949). Sur le rôle sélectif de l'aspect phénotypique des variétés de l'escargot des bois (*Cepaea nemoralis*). *C. R. Hebd. Séances Acad. Sci.* **228,** 1353–1354.

Lamotte, M. (1955). (Discussion). *Cold Spring Harbor Symp. Quant. Biol.* **20,** 275.

Lamotte, M. (1959). Polymorphism of natural populations of *Cepaea nemoralis*. *Cold Spring Harbor Symp. Quant. Biol.* **24,** 65–84 (discussion: 84–86).

Lamotte, M. (1966). Les facteurs de la diversité du polymorphisme dans les populations naturelles de *Cepaea nemoralis* (L.). *Lav. Soc. Malacol. Ital.* **3,** 33–73.

Lamotte, M. (1968a). Les traits généraux du polymorphisme de la coquille dans les populations naturelles de *Cepaea nemoralis* (Mollusques, Helicidae) des Pyrénées françaises. *C. R. Hebd. Séances Acad. Sci.* **267,** 1318–1321.

Lamotte, M. (1968b). Génétique des populations—Influence de l'altitude sur la fréquence du caractère ''absence de bandes'' dans les populations des Pyrénées françaises. *C. R. Hebd. Séances Acad. Sci.* **267,** 1649–1652.

Lamotte, M. (1969). Relations entre deux couples de caractères dépendant de deux locus étroitement liés: les caractères jaune/rose et avec bandes/sans bandes dans les populations naturelles de *Cepaea nemoralis* (Mollusques Hélicidés) du Sud de l'Aquitaine. *C. R. Hebd. Séances Acad. Sci.* **268,** 2476–2478.

Lamotte, M. (1972). Le caractère ''péristome blanc'' dans les populations de *Cepaea nemoralis* L. (Moll. Pulmonés) de la vallée de l'Ariège. *C. R. Hebd. Séances Acad. Sci.* **274,** 1558–1561.

Lamotte, M., and Guerrucci , M. A. (1970). Traits généraux du polymorphisme du système de bandes chez *Cepaea hortensis* en France. *Arch. Zool. Exp. Gén.* **111,** 393–409.

Lawson, A. K. (1923). *Helix hortensis* and *H. nemoralis* living in company. *J. Conchol.* **17,** 56.

Lewis, G. (1975). Shell polymorphism in the snail *Cochlicella acuta* (Müller) and some data on its genetics. *Biol. J. Linn. Soc. London* **7,** 147–160.

Lewis, G. (1977). Polymorphism and selection in *Cochlicella acuta. Philos. Trans R. Soc. London, Ser. B* **276,** 399–451.

Lomnicki, A. (1969). Individual differences among adult members of a snail population. *Nature (London)* **223,** 1073–1074.

McCracken, G. F., and Brussard, P. F. (1980). Self-fertilization in the white-lipped land snail *Triodopsis albolabris. Biol. J. Linn. Soc. London* **14,** 429–434.

McCracken, G. F., and Selander, R. K. (1980). Self-fertilization and monogenic strains in natural populations of terrestrial slugs. *Proc. Natl. Acad. Sci. U.S.A.* **77,** 684–688.

McQuaid, C. D., Branch, G. M., and Frost, P. G. (1979). Aestivation behaviour and thermal relations of the pulmonate *Theba pisana* in a semi-arid environment. *J. Therm. Biol.* **4,** 47–55.

Manwell, C. and Baker, C. M. A. (1968). Genetic variation of isocitrate, malate, and 6-phosphogluconate dehydrogenases in snails of the genus *Cepaea*—introgressive hybridization, polymorphism and pollution? *Comp. Biochem. Physiol.* **26,** 195–209.

Manwell, C., and Baker, C. M. A. (1970). ''Molecular Biology and the Origin of Species: Heterosis, Protein Polymorphism and Animal Breeding.'' Sidgwick & Jackson, London.

Mead, A. R. (1979). Economic malacology with particular reference to *Achatina fulica.* ''Pulmonates'' Vol. 2B. Academic Press, New York.

Meeuse, A. D. J. (1968). Different phenotypic variation of the snails *Cepaea nemoralis* and *C. hortensis* in a 'mixed' population in southern Sweden. *Beaufortia* **15,** 155–159.

Mordan, P. B. (1977). Factors affecting the distribution and abundance of *Aegopinella* and *Nesovitrea* (Pulmonana: Zonitidae) at Monks Wood National Nature Reserve, Huntingdonshire. *Biol. J. Linn. Soc. London* **9,** 59–72.

Murray, J. J. (1962). Factors affecting gene-frequencies in some populations of *Cepaea.* D.Phil. Thesis, Oxford University.

Murray, J. J. (1964). Multiple mating and effective population size in *Cepaea nemoralis. Evolution (Lancaster, Pa.)* **18,** 284–291.

Murray, J. J. (1966). *Cepaea nemoralis* in the Isles of Scilly. *Proc. Malacol. Soc. London* **37,** 167–181.

Murray, J. J. (1975). The genetics of the Mollusca. *In* ''Handbook of Genetics'' (R. C. King, ed.), Vol. 3, pp. 3–31. Plenum, New York.

Murray, J. J., and Clarke, B. C. (1966). The inheritance of polymorphic shell characters in *Partula* (Gastropoda). *Genetics (Princeton, N.J.)* **54,** 1261–1277.

Murray, J. J. and Clarke, B. C. (1968a). Inheritance of shell size in *Partula. Heredity* **23,** 189–198.

Murray, J. J., and Clarke, B. C. (1968b). Partial reproductive isolation in the genus *Partula* (Gastropoda) on Moorea. *Evolution (Lancaster, Pa.)* **22,** 684–698.

Murray, J. J., and Clarke, B. C. (1976a). Supergenes in polymorphic land snails. I. *Partula taeniata. Heredity* **37,** 253–269.

Murray, J. J., and Clarke, B. C. (1976b). Supergenes in polymorphic land snails. II. *Partula suturalis*. *Heredity* **37,** 271–282.

Murray, J. J., and Clarke, B. C. (1978). Changes of gene frequency in *Cepaea nemoralis* over fifty years. *Malacologia* **17,** 317–330.

Murray, J. J., and Clarke, B. C. (1980). The genus *Partula* on Moorea; speciation in progress. *Proc. R. Soc. London B* **211,** 83–117.

Murray, J. J., Johnson, M. S., and Clarke, B. C. (1982). Microhabitat differences among genetically similar species of *Partula*. *Evolution (Lancaster Pa.)* **36,** 316–325.

Nevo, E., and Bar, Z. (1975). Natural selection of genetic polymorphisms along climatic gradients. *In* "Population Genetics and Ecology" (S. Karlin and E. Nevo, eds.), pp. 159–184. Academic Press, New York.

Nevo, E., Bar-El, C., Bar, Z., and Beiles, A. (1981). Genetic structure and climatic correlates of desert landsnails. *Oecologia* **48,** 199–208.

Newell, P. F., and Appleton, T. C. (1979). Aestivating snails—the physiology of water regulation in the mantle of the terrestrial pulmonate *Otala lactea*. *Malacologia* **18,** 575–581.

Newell, P. F., and Machin, J. (1976). Water regulation in aestivating snails. Ultrastructural and analytical evidence for an unusual cellular phenomenon. *Cell Tissue Res.* **173,** 417–421.

O'Donald, P. (1968). Natural selection by glow-worms in a population of *Cepaea nemoralis*. *Nature (London)* **217,** 194.

Oosterhoff, L. (1977). Variation in growth rate as an ecological factor in the land snail *Cepaea nemoralis* (L.). *Neth. J. Zool.* **27,** 1–132.

Owen, D. F. (1963a). Similar polymorphisms in an insect and a land snail. *Nature (London)* **198,** 201–203.

Owen, D. F. (1963b). Polymorphism and population density in the African land snail *Limicolaria martensiana*. *Science* **140,** 666–667.

Owen, D. F. (1963c). Polymorphism in living and Pleistocene populations of the African land snail, *Limicolaria martensiana*. *Nature (London)* **199,** 713–714.

Owen, D. F. (1965a). Density effects in polymorphic land snails. *Heredity* **20,** 312–315.

Owen, D. F. (1965b). A population study of an equatorial land snail, *Limicolaria martensiana* (Achatinidae). *Proc. Zool. Soc. London* **144,** 361–381.

Owen, D. F. (1966). Polymorphism in Pleistocene land snails. *Science* **152,** 71–72.

Owen, D. F. (1974). Ecological genetics and natural selection in mollusks. *Science* **185,** 376.

Owen, D. F., and Bengtson, S.-A. (1972). Polymorphism in the land snail *Cepaea hortensis* in Iceland. *Oikos* **23,** 218–225.

Oxford, G. S. (1973). The genetics of *Cepaea* esterases. 1. *Cepaea nemoralis*. *Heredity* **30,** 127–139.

Oxford, G. S. (1975). Food induced esterase phenocopies in the snail *Cepaea nemoralis*. *Heredity* **35,** 361–370.

Oxford, G. S. (1977). Multiple sources of esterase enzymes in the crop juice of *Cepaea* (Mollusca: Helicidae). *J. Comp. Physiol.* **122,** 375–383.

Oxford, G. S. (1978). The nature and distribution of food-induced esterases in helicid snails. *Malacologia* **17,** 331–339.

Page, C. (1978). The karyotype of the land snail *Cepaea nemoralis* (L.). *Heredity* **41,** 321–325.

Parkin, D. T. (1973). A further example of natural selection on phenotypes of the land snail *Arianta arbustorum* (L.). *Biol. J. Linn. Soc. London* **5,** 221–233.

Peake, J. F. (1978). Distribution and ecology of the Stylommatophora. *In* "Pulmonates" (V. Fretter and J. F. Peake, eds.), Vol. 2A, pp. 429–526. Academic Press, New York.

Pearce, S. S. (1903). The association of *Helix nemoralis* and *Helix hortensis*. *J. Conchol.* **10,** 300–301.

Pitchford, G. W. (1955). *Helix nemoralis* (L.) in a calcium deficient habitat. *J. Conchol.* **24,** 54.

Pollard, E. (1975a). Aspects of the ecology of *Helix pomatia* L. *J. Anim. Ecol.* **44,** 305–329.

Pollard, E. (1975b). Differences in shell thickness in adult *Helix pomatia* L. from a number of localities in southern England. *Oecologia* **21.** 85–92.

Pomeroy, D. E. (1967). The influence of environment on two species of land snails in South Australia. *Trans. R. Soc. S. Aust.* **91,** 181–186.

Pomeroy, D. E. (1968). Dormancy in the land snail, *Helicella virgata* Pulmonata: Helicidae). *Aust. J. Zool.* **16,** 857–869.

Pomeroy, D. E. (1969). Some aspects of the ecology of the land snail *Helicella virgata* in South Australia. *Aust. J. Zool.* **17,** 495–514.

Potts, D. C. (1975). Persistence and extinction of local populations of the garden snail *Helix aspersa* in unfavourable environments. *Oecologia* **21,** 313–334.

Price, D. J. (1974a). Variation in chiasma frequency in *Cepaea nemoralis. Heredity* **32,** 211–217.

Price, D. J. (1974b). Differential staining of meiotic chromosomes in *Cepaea nemoralis* (L.). *Caryologia* **27,** 211–216.

Price, D. J. (1975a). Chiasma frequency variation with altitude in *Cepaea hortensis* (Müll.). *Heredity* **35,** 221–229.

Price, D. J. (1975b). Chiasma frequency variation in *Cepaea hortensis* (Müll.) and a comparison with *C. nemoralis* (L.). *Genetics (Princeton, N.J.)* **45,** 497–508.

Price, D. J. (1975c). Position and frequency distribution of chiasmata in *Cepaea nemoralis* (L.). *Caryologia* **28,** 261–268.

Price, D. J., and Bantock, C. R. (1975). Marginal populations of *Cepaea nemoralis* (L.) on the Brendon Hills, England. II. Variation in chiasma frequency. *Evolution (Lancaster, Pa.)* **29,** 278–286.

Price, D. J., and McBride, J. (1981). Chiasma frequency variation in *Cepaea* populations from a localised area of south-west England. *Caryologia* **34,** 363–375.

Richards, A. J. (1976). Predation of snails by migrant song thrushes and redwings. *Bird Study* **24,** 53–54.

Richards, A. V., and Murray, J. J. (1975). Relation of phenotype to habitat in an introduced colony of *Cepaea nemoralis. Heredity* **34,** 128–131.

Richardson, A. M. M. (1974). Differential climatic selection in natural populations of the land snail *Cepaea nemoralis. Nature (London)* **247,** 572.

Richardson, A. M. M. (1975a). Food, feeding rates and assimilation in the land snail *Cepaea nemoralis* L. *Oecologia* **19,** 59–70.

Richardson, A. M. M. (1975b). Energy flux in a natural population of the land snail *Cepaea nemoralis* L. *Oecologia* **19,** 141–164.

Richardson, A. M. M. (1975c). Winter predation by thrushes, *Turdus ericetorum* (Turton), on a sand dune population of *Cepaea nemoralis* (L.). *Proc. Malacol. Soc. London* **41,** 481–488.

Richardson, A. M. M. (1979). Morph frequencies of empty intact shells from *Cepaea nemoralis* (L.) colonies on sand dunes in south west England. *J. Moll. Stud.* **45,** 98–107.

Rollo, C. D., and Wellington, W. G. (1979). Intra- and interspecific agonistic behaviour among terrestrial slugs (Pulmonata: Stylommatophora). *Can. J. Zool.* **57,** 846–855.

Roth, B. (1980). Distribution, ecology and reproductive anatomy of a rare land snail, *Monadenia setosa* Talmadge. *Calif. Fish Game* **66,** 4–16.

Roth, B. (1981). Shell colour and banding variation in two coastal colonies of *Monadenia fidelis* (Gray) (Gastropoda: Pulmonata). *Wasmann J. Biol.* **38,** 39–51.

Russell Hunter, W., and Russell Hunter, M. (1956). Mollusca on Scottish mountains. *J. Conchol.* **24,** 80.

Sacchi, C. F. (1951) Ricerche malacologiche sul litorale Adriatico italiano. Nota preliminare. *Atti Soc. Ital. Sci. Nat. Mus. Civ. Stor. Nat. Milano* **90,** 251–260.

Sacchi, C. F. (1952). Ricerche sulla variabilità geografica in popolazioni Italiane di *Euparypha pisana* Müll. (Stylommatophora Helicidae). *Ann. Museo Civ. Stor. Nat. Genova* **65,** 211–258.

Sacchi, C. F. (1955a). Fattori ecologici e storici nel polimorfismo delle *Euparypha (Helicidae Helicinae)* del Marocco occidentale. *Stud. Ghisleriana, Ser. 3* **2,** 44–66.

Sacchi, C. F. (1955b). Fattori ecologici e fenomeni microevolutivi nei Molluschi della montagna mediterranea. *Boll. Zool.* **22,** 563–651.

Sacchi, C. F. (1956a). Ricerche su *Euparypha arietina* (Rossmaessler). I. Posizione sistematica. *Annu. Ist. Museo Zool. Univ. Napoli* **8,** 1–6.

Sacchi, C. F. (1956b). Relazioni tra colori ed ambienti in popolazioni naturali spagnuole di *Pseudotachea* (Stylommatophora Helicidae). *Boll. Zool.* **23,** 461–501.

Sacchi, C. F. (1971). Écologie comparée des Gastéropodes pulmonées des dunes Méditerranéennes et Atlantiques. *Natura. Soc. Ital. Sci. Natura., Mus. Civ. Stor. Nat. Acquario Civ., Milano* **62/3,** 277–358.

Sacchi, C. F. (1974). Points de vue d'un écologiste sur la physiologie de l'estivation chez l'hélicidé dunicole *Euparypha pisana* (Müller) (Gastropoda Pulmonata). *Natura. Soc. Ital. Sci. Nat., Museo Civico Storia Nat. Acquario Civ., Milano* **65,** 117–133.

Sacchi, C. F. (1977). La "Lacune nord-Adriatique" et son influence sur l'écologie des gastéropodes dunicoles. Prémisses methodologiques. *Atti Soc. Ital. Sci. Nat., Mus. Civ. Stor. Nat., Milano* **118,** 213–225.

Sacchi, C. F. (1978). Il Delta del Po come elemento disgiuntore nell' ecologia delle spiaggie adriatiche. *Boll. Museo Civ. Stor. Nat. Venezia* **29,** Suppl., 43–73.

Sacchi, C. F., and Valli, G. (1975). Recherches sur l'écologie des populations naturelles de *Cepaea nemoralis* (L.) Gastr. Pulmonata en Lombardie méridionale. *Arch. Zool. Exp. Gén.* **116,** 549–578.

Sacchi, C. F., and Valli, G. (1977). Population ecology of two geographic races of *Cepaea nemoralis* (Linnaeus) (Gastropoda, Pulmonata) in the province of Pavia, Northern Italy . *Malacologia* **16,** 243–245.

Sacchi, C. F., and Violani, C. (1977). Ricerche ecologiche sulle Elicidi dunicole della Ria di Vigo (Spagna). *Natura. Soc. Ital. Sci. Nat., Mus. Civ. Stor. Nat. Acquario Civ., Milano* **68,** 253–284.

Schifferli, L. (1977). Bruchstücke von Schnecken-häuschen als Calciumquelle für die Bildung der Eischale beim Haussperling *Passer domesticus. Ornithol. Beob.* **74,** 71–74.

Schilder, F. A. (1957). Sechsjährige Konstanz einer Population von *Cepaea nemoralis. Arch. Molluskenkd.* **86,** 33–36.

Schilder, F. A., and Schilder, M. (1958). Nochmals zur Konstanz der *Cepaea*- Populationen. *Arch. Molluskenkd* **87,** 89.

Schwabl, G., and Murray, J. J. (1970). Electrophoresis of proteins in natural populations of *Partula* (Gastropoda). *Evolution (Lancaster, Pa.)* **24,** 424–430.

Selander, R. K., and Foltz, D. W. (1981). Gametic disequilibrium between esterase loci in populations of *Cepaea nemoralis* in western New York. *Evolution (Lancaster, Pa.)* **35,** 190–192.

Selander, R. K., and Hudson, R. O. (1976). Animal population structure under close inbreeding—the land snail *Rumina* in southern France. *Am. Nat.* **110,** 695–718.

Selander, R. K., and Kaufman, D. W. (1973). Self-fertilization and genetic population structure in a colonizing land snail. *Proc. Natl. Acad. Sci. U.S.A.* **70,** 1186–1190.

Selander, R. K., and Kaufman, D. W. (1975). Genetic structure of populations in the brown snail, *Helix aspersa.* I. Microgeographic variation. *Evolution (Lancaster, Pa.)* **29,** 385–401.

Selander, R. K., Kaufman, D. W., and Ralin, R. S. (1974). Self-fertilization in the terrestrial snail *Rumina decollata. Veliger* **16,** 265–270.

Sheppard, P. M. (1951). Fluctuations in the selective value of certain phenotypes in the polymorphic land snail *Cepaea nemoralis* (L.). *Heredity* **5,** 125–134.

Sheppard, P. M. (1955a). Genetic variability and polymorphism: synthesis. *Cold Spring Harbor Symp. Quant. Biol.* **20,** 271–275.

Sheppard, P. M. (1955b). (Discussion). *Cold Spring Harbor Symp. Quant. Biol.* **20,** 275.

Sheppard, P. M. (1969). Evolutionary genetics of animal populations: The study of natural populations. *Proc. Int. Cong. Genet., 12th, 1968,* Vol. 3, pp. 261–279.

Shileyko, A. A. (1978). Systema otryada Geophila (= Helicida) (Gastropoda Pulmonata). *In* "Morphology, Systematics and Phylogeny of Molluscs" (Y. I. Starobogatov, ed.), pp. 44–69. Acad. Sci. U.S.S.R., Proc. Zool. Inst., Moscow (in Russian).

Solem, A. (1976). "Endodontid land snails from Pacific islands (Mollusca: Pulmonata: Sigmurethra). Part I. Family Endodontidae." Field Mus. Nat. Hist., Chicago, Illinois.

Solem, A. (1978). Classification of the land Mollusca. *In* "Pulmonates" (V. Fretter and J. F. Peake, eds.), Vol. 2A, pp. 49–97. Academic Press, New York.

Stelfox, A. W. (1912). Land and freshwater Mollusca. *Proc. R. Ir. Acad.* **31,** 1–64.

Step, E. (1927). "Shell life." Warne, London and New York.

Tattersfield, P. (1981). Density and environmental effects on shell size in some sand dune snail populations. *Biol. J. Linn. Soc.* **16,** 71–81.

Tattersfield, P. (1982). Behavioural aspects of niche separation and population dynamics in terrestrial molluscs. Ph.D. Thesis, Univ. of Birmingham, England.

Tegelström, H., Hageström, A., and Kvassman, S. (1975). Esterases of the snails *Helix pomatia* and *Cepaea hortensis:* variation and characteristics of different molecular forms. *Hereditas* **79,** 117–124.

Tuskes, P. M. (1981). Population structure and biology of *Liguus* tree snails on Lignumvitae Key, Florida. *Nautilus* **95,** 162–169.

Tzvetkov, B. (1941). Variation of *Fruticicola lantzi* Lndh. (Mollusca Pulmonata). *Arch. Mus. Zool. Univ. Moscou* **6,** 287–302 (in Russian, English summary).

Umiński, T. (1975). Life cycles in some Vitrinidae (Mollusca, Gastropoda). *Ann. Zool. (Pol. Akad. Nauk, Inst. Zool.)* **33,** 17–33.

Umiński, T. (1979). Life history in *Eucobresia nivalis* (Dumont et Mortillet) with notes on two other Vitrinidae (Mollusca, Gastropoda). *Bull. Acad. Pol. Sci., Ser. Sci. Biol. Cl. 2* **27,** 205–210.

Umiński, T., and Focht, U. (1979). Population dynamics of some land gastropods in a forest habitat in Poland. *Malacologia* **18,** 181–184.

Valovirta, I. (1967). List of Finnish land gastropods and their distribution. *Ann. Zool. Fenn.* **4,** 29–32.

Valovirta, I. (1968). Land molluscs in relation to acidity on hyperite hills in Central Finland. *Ann. Zool. Fenn.* **5,** 245–253.

Valovirta, I., and Halkka, O. (1976). Colour polymorphism in northern peripheral populations of *Cepaea hortensis. Hereditas* **83,** 123–126.

Vater, G. (1965). Die Färbung der Schnirkelschnecken *Cepaea hortensis* (Müll.) und *Cepaea nemoralis* (L.) als Differentialmerkmal bei Populationsuntersuchungen. *Abh. Ber. Naturkundemus. Görlitz* **40,** 1–88.

Von Martens, G. (1832). Über die Ordnung der Bänder an den Schalen mehrerer Landschnecken. *Nova Acta Physiomed. (Abh. Dtsch. Akad. Naturforsch.; Nova Acta Leopoldina)* **16,** 177–216.

Wahren, H., and Tegelström, H. (1973). Polymorphism of esterases and tetrazolium oxidases in the Roman snail, *Helix pomatia:* a study of populations in Sweden and Germany. *Biochem. Genet.* **9,** 169–174.

Wall, S., Carter, M. A., and Clarke, B. C. (1980). Temporal changes of gene frequencies in *Cepaea hortensis. Biol. J. Linn. Soc.* **14,** 303–317.

Wäreborn, I. (1979). Reproduction of two species of land snails in relation to calcium salts in the foerna layer. *Malacologia* **18,** 177–180.

Welch, R. (1903). The association of *Helix nemoralis* and *H. hortensis* in Ireland. *J. Conchol.* **10,** 302–303.

White, M. J. D. (1978). "Modes of Speciation." Freeman, San Francisco, California.

Williamson, M. H. (1959a). Studies on the colour and genetics of the black slug. *Proc. R. Phys. Soc. Edinburgh* **27,** 87–93.

Williamson, M. H. (1959b). Differential damage in a mixed colony of the land snails *Cepaea nemoralis* and *C. hortensis*. *Heredity* **13,** 261–263.

Williamson, M. H. (1970). Remarks on a question by H. Wolda. *J. Anim. Ecol.* **39,** 541–542.

Williamson, P. (1975). Use of ^{65}Zn to determine the field metabolism of the snail *Cepaea nemoralis* L. *Ecology* **56,** 1185–1192.

Williamson, P. (1976). Size-weight relationships and field growth rates of the landsnail *Cepaea nemoralis* L. *J. Anim. Ecol.* **45,** 875–885.

Williamson, P. (1979). Age determination of juvenile and adult *Cepaea*. *J. Moll. Stud.* **45,** 52–60.

Williamson, P., and Cameron, R. A. D. (1976). Natural diet of the landsnail *Cepaea nemoralis*. *Oikos* **27,** 493–500.

Williamson, P., Cameron, R. A. D., and Carter, M. A. (1976). Population density affecting adult shell size of the snail *Cepaea nemoralis* L. *Nature (London)* **263,** 496–497.

Williamson, P., Cameron, R. A. D., and Carter, M. A.(1977). Population dynamics of the landsnail *Cepaea nemoralis* L.: a six-year study. *J. Anim. Ecol.* **46,** 181–194.

Willis, A. J., McEwan, J. W. T., Greenwood, J. J. D., and Elton, R. A. (1980) Food selection by chicks: effects of colour, density, and frequency of food types. *Anim. Behav.* **28,** 874–879.

Wolda, H. (1963). Natural populations of the polymorphic landsnail *Cepaea nemoralis* (L.). *Arch. Néerl. Zool.* **15,** 381–471.

Wolda, H. (1965a). Some preliminary observations on the distribution of the various morphs within natural populations of the polymorphic landsnail *Cepaea nemoralis* (L.). *Arch. Néerl. Zool.* **16,** 280–292.

Wolda, H. (1965b). The effect of drought on egg production in *Cepaea nemoralis* (L.). *Arch. Néerl. Zool.* **16,** 387–399.

Wolda, H. (1967). The effect of temperature on reproduction in some morphs of the landsnail *Cepaea nemoralis*. *Evolution (Lancaster, Pa.)* **21,** 117–129.

Wolda, H. (1969a). Fine distribution of morph frequencies in the snail *Cepaea nemoralis* near Groningen. *J. Anim. Ecol.* **38,** 305–327.

Wolda, H. (1969b). Stability of a steep cline in morph frequencies of the snail *Cepaea nemoralis* (L.). *J. Anim. Ecol.* **38,** 623–625.

Wolda, H. (1970a). Variation in growth rate in the landsnail *Cepaea nemoralis*. *Res. Popul. Ecol.* **12,** 185–204.

Wolda, H. (1970b). Ecological variation and its implications for the dynamics of populations of the landsnail *Cepaea nemoralis*. *Proc. Adv. Study Inst.; Dyn. Numbers Popul. (Oosterbeek, 1970)* pp. 98–108.

Wolda, H. (1972a). Ecology of some experimental populations of the landsnail *Cepaea nemoralis* (L.). Adult numbers and adult mortality. *Neth. J. Zool.* **22,** 428–455.

Wolda, H. (1972b). Changes in shell size in some experimental populations of the landsnail *Cepaea nemoralis* (Linnaeus). *Argamon, Isr. J. Malacol.* **3,** 63–71.

Wolda, H., and Kreulen, D. A. (1973). Ecology of some experimental populations of the landsnail *Cepaea nemoralis* (L.). II. Production and survival of eggs and juveniles. *Neth. J. Zool.* **23,** 168–188.

Wolda, H., Zweep, A., and Schuitema, K. A. (1971). The role of food in the dynamics of population of the landsnail *Cepaea nemoralis*. *Oecologia* **7,** 361–381.

Woodruff, D. S. (1975a). Allozyme variation and genic heterozygosity in the Bahaman pulmonate *Cerion bendalli*. *Malacol. Rev.* **8,** 47–55.

Woodruff, D. S. (1975b). A new approach to the systematics and ecology of the genus *Cerion*. *Malacol. Rev.* **8,** 128.

Woodruff, D. S. (1978). Evolution and adaptive radiation of *Cerion:* a remarkably diverse group of West Indian land snails. *Malacologia* **17,** 223–239.

Woodruff, D. S., and Gould, S. J. (1980). Geographic differentiation and speciation in *Cerion*—a preliminary discussion of patterns and processes. *Biol. J. Linn. Soc.* **14,** 389–416.

Wright, S. (1965). Factor interaction and linkage in evolution. *Proc. R. Soc. London, Ser. B* **162,** 80–104.

Wright, S. (1978). ''Evolution and the Genetics of Populations,'' Vol. 4. Univ. of Chicago Press, Chicago, Illinois.

Yom-Tov, Y. (1970). The effect of predation on population densities of some desert snails. *Ecology* **51,** 907–911.

Yom-Tov, Y. (1971a). Body temperature and light reflectance in two desert snails. *Proc. Malacol. Soc. London* **39,** 319–326.

Yom-Tov, Y. (1971b). The biology of two desert snails, *Trochoidea* (*Xerocrassa*) *seetzeni* and *Sphincterochila boissieri*. *Isr. J. Zool.* **20,** 231–248.

Yom-Tov, Y. (1971c). Annual fluctuations in the water content of desert snails. *Malacol. Rev.* **4,** 121–126.

Yom-Tov, Y. (1972). Field experiments on the effect of population density and slope direction on the reproduction of the desert snail *Trochoidea* (*Xerocrassa*) *seetzeni*. *J. Anim. Ecol.* **41,** 17–22.

Yom-Tov, Y., and Galun, M. (1971). Notes on the feeding habits of the desert snails *Sphincterochila boissieri* Charpentier and *Trochoidea* (*Xerocrassa*) *seetzeni* Charpentier. *Veliger* **14,** 86–88.

Zinner, H. (1978). Observations on Little Owls feeding on snails in the Negev. *Argamon, Isr. J. Malacol.* **6,** 57–60.

15

Life-Cycle Patterns and Evolution

PETER CALOW

Department of Zoology
University of Glasgow
Glasgow, Scotland, United Kingdom

I. Introduction

A life-cycle pattern is the set of ontogenetic traits that ensures that the products of reproduction are brought into a condition in which they can themselves reproduce. It therefore includes patterns of reproduction (i.e., how much is invested in reproduction, and when and how it is packaged), patterns of early development (e.g., whether it occurs inside or outside an egg membrane), patterns and rates of growth, frequency of reproduction, and longevity of the parent. These features of the life cycle are presented schematically in Fig. 1. A main concern of

The Mollusca, Vol. 6
Ecology

ISBN 0-12-751406-6

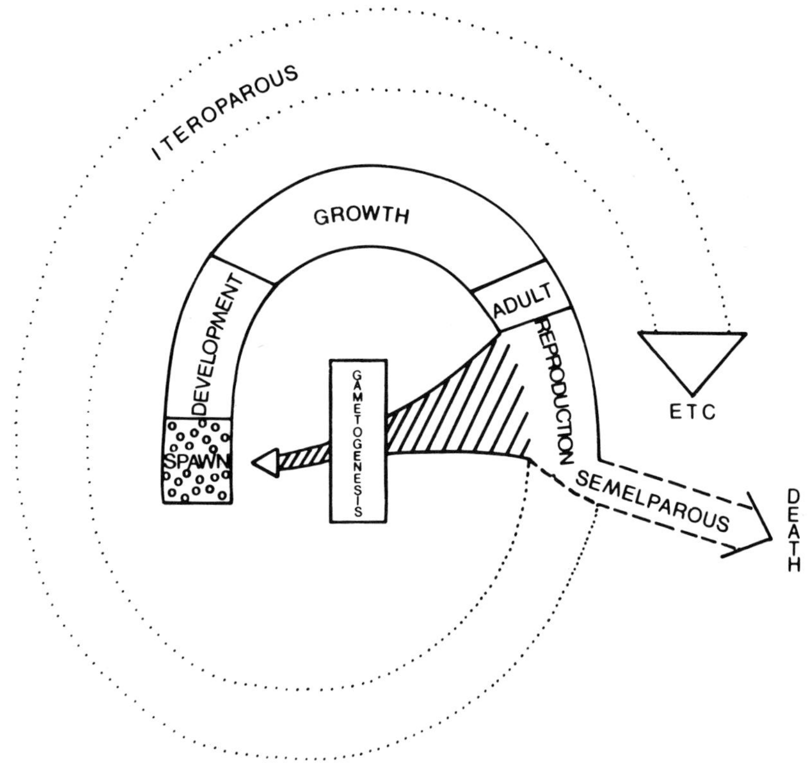

Fig. 1. Major features of molluscan life cycles.

this chapter will be to illustrate how such life-cycle traits have been coadapted by natural selection according to the ecological circumstances in which they occur.

Section II surveys life-cycle patterns in the major molluscan taxonomic groups. In addition to illustrating the range of molluscan life cycles, it identifies taxonomic trends that may reflect organizational constraints. Because life-cycle traits are necessarily and intimately interrelated, the limitations this might impose on life-cycle patterns are considered in Section III. Finally, given the existence of intrinsic constraints and trade-offs, Section IV is concerned with the influence of extrinsic ecological factors on the evolution of life cycles in the Mollusca; the fundamental assumption is that sufficient genetic variance occurs within molluscan populations for life-cycle traits to be selected, a statement of faith rather than fact. Nevertheless, some indication has been given that these molluscan life-cycle traits do show significant genetic variance (Chapters 9 and 11, this volume).

II. Taxonomic Survey

Molluscs reproduce exclusively from gametes, although hermaphroditism (both simultaneous and sequential) and parthenogenesis are common (Morton, 1979). As in other invertebrates (Thorson, 1950), early development can occur (1) free in the environment (usually in the plankton of marine systems), (2) within egg membranes (oviparity), and (3) in brood chambers within the parent (ovoviviparity). In strategy 1, the larvae may be small, feeding in the plankton (planktotrophic) or large, with their own supply of nutrients (lecithotrophic). Postembryonic growth is commonly determinate and sigmoid, but this is often complicated by environmental effects (Calow, 1981a). Reproduction is ususally seasonal. Adults may die after the first breeding season (semelparity) or reproduce successively over several seasons (iteroparity). This section reviews the distribution of these life-cycle traits through the major molluscan taxa.

A. Gastropoda

1. Prosobranchia

General information on the reproductive and mating patterns of Prosobranchia is given by Giese and Pearse (1977). Most prosobranchs are dioecious, and there is some sexual dimorphism in the more advanced taxa. Hermaphroditism occurs in some species; e.g., *Valvata piscinalis* (simultaneous) and Calyptraeidae (sequential). Most of the archaeogastropods are broadcast spawners with planktotrophic development starting from trochophores. Ovoviviparity does occur (e.g., *Acmaea rubella*), and some prosobranchs lay gelatinous egg masses or ribbons (e.g., *Gibbula fumida*). Alternatively, most mesogastropods and neogastropods lay egg capsules (which consist of more or less complex membranes surrounding several to many egg cells). Development here may be either indirect (a veliger larva is released to the plankton) or direct (a miniature adult is released). When indirect, planktotrophy is most common, but lecithotrophy occurs in some species. Ovoviviparity has also been discovered in some mesogastropods and neogastropods, e.g., *Littorina saxatilis* and *Viviparus viviparus*.

The diversity of spawn produced by prosobranchs is considerable and the variability among species, even within the same genus, can be extensive (Fretter and Graham, 1962). Nevertheless, species with planktotrophic larvae produce many (10^3–10^9) small (< 300 μm diameter) eggs/spawning, whereas species with direct development produce fewer (<10^3) larger (usually > 300 μm diameter) eggs (see Fig. 2). The ratio of egg sizes between these two groups is reasonably constant, approximately 3–4 : 1 (diameter) and 50–65 : 1 (volume) (Christiansen and Fenchel, 1979). However, this relationship does not hold for species with nurse cells which practice direct development (Spight, 1976a,b),

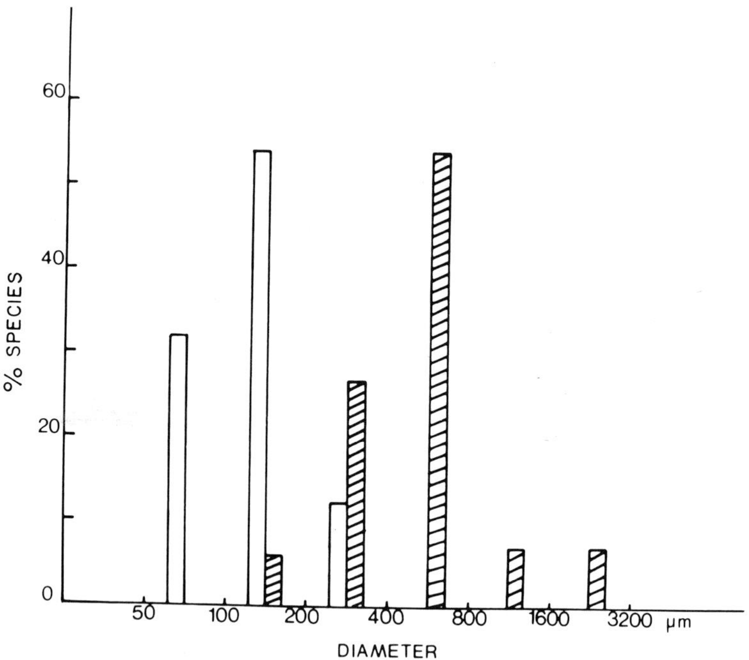

Fig. 2. The egg size distribution of 37 species of marine prosobranch gastropods. Redrawn from Christiansen and Fenchel (1979); data originally from Thorson (1946). Open blocks = planktotrophic species; shaded blocks = species with direct development.

i.e., development sustained from nutrient stores derived from cells not directly involved in development. Because there is no direct supply of nutrients from parent to brood, ovoviviparous species produce large eggs; e.g., *L. saxatilis* produces eggs 400–500 μm in diameter.

In general, species with small eggs and little yolk (planktotrophic types) hatch quickly, whereas species with large eggs (direct development) require a longer time for development in the egg. However, there are taxonomic constraints, and temperature also has a major influence (Spight, 1975). Similarly, the species with smaller eggs are smaller and less well developed on hatching than those with larger eggs. Again, this relationship is complicated when nurse cells are involved. In muricids, for example, the intraspecies variation in the size of hatchlings with integral yolk is ± 15%, but in species with nurse cells it may be as much as ± 30% (Spight, 1976a,b).

The sizes of adult prosobranchs range from less than 5 mm length in interstitial species (Swedmark, 1964, 1968) to more than 30 cm in the conches. In general, it takes longer than 1 year for prosobranchs to reach maturity and/or full size (Webber, 1977). Growth rates of 1–2 mm length/month have been recorded in field populations of muricids (Spight et al., 1974), of as much as 6 mm/month in cymatids (Laxton, 1970), and of 0.2–2 mm/month in patelloids (Branch, 1974).

According to Comfort (1957), the longevity of prosobranchs ranges from 1 to 20 years. Iteroparity is dominant (Spight et al., 1974) but semelparity does occur, e.g., some *Lacuna* species (Grahame, 1977). For murexes, Spight et al. (1974) have shown that small species mature early, have a short life expectancy, and produce small clutches several times a year; larger species mature later, live longer, and produce larger clutches less frequently. The growth rates for both types are approximately the same, but the species with a smaller final size cease growing earlier.

There is no obvious association between the pattern of early development and the reproductive strategy adopted by the adults. Thus, *Lacuna vincta* is semelparous and has a long-lived planktotrophic larva, but *L. pallidula,* also semelparous, has lecithotrophic development (Grahame, 1977). These gastropods are, however, relatively small compared with other species.

2. Opisthobranchia

Almost all opisthobranchs, even the active pelagic ones, are simultaneous hermaphrodites (Beeman, 1977). Usually, the spawn is laid in mucous strings and the eggs hatch as planktonic veligers. Both planktotrophy and lecithotrophy occur and, as usual, eggs of the former are smaller than those of the latter (Beeman, 1977). Direct development is also practiced by some species (Thompson, 1967). In general, intracapsular development is shortest in planktotrophic forms and longest in forms with direct development, but egg to juvenile development is briefest for lecithotrophic pelagic organisms and longest for planktotrophic ones (Todd and Doyle, 1981). In *Aplysia,* size varies more than two orders of magnitude, and yet there is no apparent correlation with life span. Thus, *A. punctata* (adult size, 80 g) and *A. californica* (adult size, 6800 g) are both annual (Audesirk, 1979). Iteroparity and semelparity have been discovered in the nudibranchs, but again, not in association with any particular developmental pattern (Todd, 1979). For more information on this group see Chapter 5, this volume.

3. Pulmonata

Most pulmonate species are hermaphrodites. Depending on the habitat, the spawn is laid in mucoid strings, well-defined capsules, and even calcified shells (Duncan, 1960, 1975; Tompa, 1976). Development is usually direct, but in the Ellobiidae, where a veliger is released to the plankton from an egg capsule, it is indirect (Berry, 1977); ovoviviparity is reasonably common in terrestrial species (Tompa, 1979a,b). The eggs of most pulmonates are large (approximately 1 mm diameter), and $10–10^3$ are produced per breeding season (Calow, 1978). In contrast, *Melampus bidentatus,* a marsh-dwelling ellobiid, produces approximately 30,000 eggs of much smaller diameter per breeding season (Russell-Hunter and Apley, 1966; Russell-Hunter et al., 1972).

Pulmonate length ranges from a few millimeters to more than 100mm in the giant African *Achatina*. Most temperate freshwater species grow to full size and mature within 1 year (Calow, 1978). Larger terrestrial species mature over several years (Peake, 1978). The growth rates of pulmonates are variable but are probably greater in general than those of the marine prosobranchs of comparable size, perhaps because pulmonates do not produce such heavily architectured shells as do the marine snails (Vermeij and Covich, 1978). Under good conditions, *Cepaea nemoralis* increases shell diameter approximately 1 mm/week (Oosterhoff, 1977); several temperate freshwater species achieve between 1- and 2-mm increments in shell length per week (Van der Schalie and Berry, 1972); the tropical *Biomphalaria glabrata* achieves a similar rate (Sturrock and Sturrock, 1972). For *C. nemoralis,* variability in growth rate and final size has an important genetic component; heritability for growth rate = 0.48 and for final adult size = 0.58 (Oosterhoff, 1977).

Annual life spans and semelparity are common in temperate freshwater species (Comfort, 1957; Calow, 1978), but subannual patterns are common in tropical species (Brown, 1980) and perennial iteroparity occurs in *Melampus* (Russell-Hunter et al., 1972). In terrestrial slugs and snails, large species tend to be perennial and iteroparous, and small species tend to be annual and iteroparous (Peake, 1978).

B. Bivalvia

Most Bivalvia are gonochoristic (Sastry, 1979), but the Sphaeriidae are hermaphroditic (Heard, 1965). Sastry (1979) draws attention to the following correlations between developmental type, developmental rate, and the spawn size of Bivalvia: (1) planktotrophy (which is dominant) is correlated with a long developmental period in the plankton (> 3 months) and the production of a large number ($>10^3$) of small eggs (< 100 μm diameter); (2) lecithotrophy is correlated with a short pelagic life (< 3 months) and the production of a small number ($<10^3$) of large, yolky eggs (150–200 μm diameter); (3) direct benthic development is correlated with large eggs (> 100 μm diameter) brooded in a marsupium formed from the ctenidia. Some freshwater species also produce glochidia larvae which are ectoparasitic on fishes for a short time in their life cycles (Ellis, 1978).

According to Comfort (1957), life spans vary with species from 1 to more than 100 years. Similarly, sizes range over several orders of magnitude from about 1 mm to more than 100 cm in length. Giant clams may live to be more than 100 years old and reach more than 1 m in *gape length.* Iteroparity is common, particularly in marine species, but there are exceptions (Sastry, 1979). Freshwater sphaeriids may be either semelparous or iteroparous (Heard, 1965, 1976), and

Heard (1965) suggests that semelparity in these species is correlated with rapid growth to a small adult size and a short prereproductive life span. Freshwater unionids, which tend to be larger than sphaeriids and to have an ectoparasitic glochidium larva, are usually iteroparous (Heard, 1975).

C. Cephalopoda

All Cephalopoda are gonochoristic and separate sexed; distinct larval forms are rare (Arnold, 1977; Haven, 1977; Wells, 1977). The octopods practice brooding [species with small adults practice complete brooding but species with large adults only partial brooding (Green, 1973)], and even the pelagic *Argonauta* secretes a papery shell for carrying eggs (Wells, 1977). Eggs of cephalopods are large, generally greater than 1 mm in diameter. There is the usual negative correlation between egg size and numbers: *Octopus vulgaris* lays thousands of eggs about 1 × 3 mm, and *O. bimaculoides* lays hundreds of eggs about 10 × 18 mm (Wells, 1978); this relationship is typical for the group as a whole (Calow, *in manuscript).* Female octopods may brood their eggs for 1 month to 1 year (depending on species) without feeding and usually ventilate their eggs (Wells, 1978).

Most octopods have short lives (usually 1 year) (Comfort, 1957) and yet grow to large final sizes. Hence, growth rates must be rapid. *O. cyanea* has one of the most rapid growth rates in the animal kingdom, which may be partly attributable to its sit-and-wait feeding mode (Calow and Townsend, 1981). However, pelagic decapods also have high growth rates, so that this cannot be the complete explanation (Calow, *in manuscript*). Semelparity is typical of those decapods that have been studied (Arnold, 1977) and of the octopods (Wells, 1977), but there is little information on the large oceanic decapods or on any nautiloids (Haven, 1977). These larger species probably live longer than 1 year (Comfort, 1957).

D. Correlations and Constraints

This section summarizes the correlations and constraints noted in the foregoing taxonomic survey and supplies some preliminary evolutionary interpretations.

1. Mode of Reproduction

Hermaphroditism and gonochorism are distributed unevenly throughout the major molluscan taxa. Hermaphroditism is thought to be advantageous in animals of low mobility and/or those that live in populations of low density, because it increases the chances of successful mating contacts (Ghiselin, 1969). Alternatively, hermaphroditism carries a cost because each parent must build and maintain two sets of reproductive apparatus. Therefore, each hermaphrodite has fewer resources to invest in gametes and will produce fewer eggs and sperm than gonochoristic equivalents (Heath, 1977). The costs are worth bearing in the pulmonates, which are generally sluggish and often occur at low population densities. They are not worth bearing in the bivalves and archaeogastropods, possibly because external fertilization, which favors maximum egg and sperm production (Maynard Smith, 1978), is a dominant reproductive mode in these groups, and because bivalves often occur at high population densities. Similarly, the costs of hermaphroditism are not worth bearing in cephalopods, possibly because they are very active and visually acute (Calow, 1978). The dominance of gonochorism in Mesogastropoda and Neogastropoda (with internal fertilization) and hermaphroditism in the often active opisthobranchs presents a difficulty for the general theory. Density effects and/or taxonomic constraints may be important here. Sequential hermaphroditism has been frequently recorded in the higher gastropods (Hoagland, 1978), and this may be seen as even more common in th Mollusca as more research effort is devoted to the phenomenon.

It should also be noted that the cost of hermaphroditism can be reduced by organ sharing in the reproductive system (Heath, 1977). This takes an extreme form in the pulmonates, where even the gonad (''hermaphrodite gland'') is a shared structure (Duncan, 1960).

2. Spawn

The negative correlation between egg sizes and numbers is seen in all the major taxonomic groups. Large eggs are associated with direct development and lecithotrophy, and small eggs with planktotrophy. This is also the case in other invertebrate taxa (Thorson, 1950). The evolutionary pressures operating here will be considered in more detail below.

3. Life-Cycle Types

Although semelparity or iteroparity dominates in most taxa, one never completely excludes the other, except, perhaps, in the Cephalopoda, in which all species that have been studied in detail are semelparous. Iteroparity is a dominant feature of prosobranch life cycles, and, associated with direct development and eggs that are often well protected in capsules, semelparity is dominant in pulmonate life cycles. However, there was no predictable correlation between semelparity and iteroparity, on the one hand, and planktotrophy and lecithotrophy, on the other, in any group. Semelparity is often correlated with rapid pre- and postreproductive development (e.g., cephalopods) but not necessarily with a particular final adult size. In prosobranchs and bivalves, semelparous adults tend to be smaller than their iteroparous equivalents, but all adult cephalopods are relatively massive. Again, evolutionary pressures associated with these life-cycle types will be considered in more detail below.

4. *Taxonomic Constraints*

The main conclusion from the survey, therefore, is that despite biases in the life-cycle features of molluscs at the level of class and subclass there appear to be no rigid constraints, and so it is permissible to search for ecological influences.

III. Intrinsic (Endogenous) Constraints and Trade-Offs

Although there are no obvious constraints associated with the morphological or physiological peculiarities of the major taxa, there are likely to be cross-taxa constraints and trade-offs which have influenced the evolution of life cycles. Many of the trade-offs are based on the fact that the resources available for allocation among the main physiological demands are finite and limiting (Calow, 1979, 1981b,c; Townsend and Calow, 1981). This section reviews some of these effects as they operate within the Mollusca. Each of the following subsections

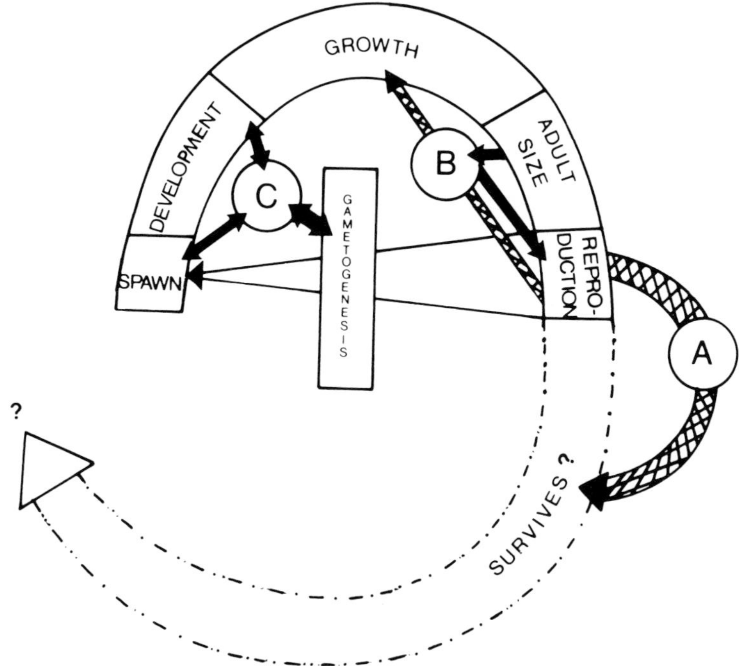

Fig. 3. Some possible interactions between the different components of the life cycle. Letters refer to subsections in Section III of the text.

deals with a group of similar factors, which are not in any significant order. The relationships to be considered are summarized in Fig. 3.

A. Reproductive Investment and Parental Survivorship

If there were no limitations on the investment made in reproduction by animals, then reproductive output should have been maximized by natural selection (Calow, 1981b). Hence, any variability in investment by related animals such as molluscs, in which the biomass of eggs laid may vary from a small percentage of the body mass of some species to more than 100% of that of others (Zaika, 1972), is strong *a priori* evidence for underlying limitations.

Figure 4, illustrating the lifetime, instantaneous energy budget of the freshwa-

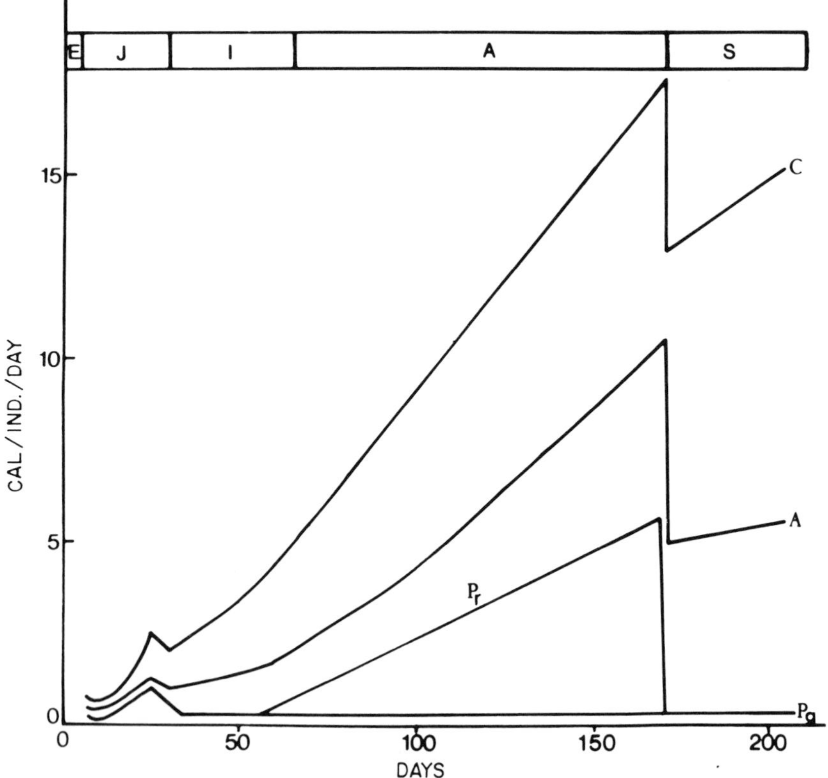

Fig. 4. Instantaneous energy budget of *Physa acuta* at 26°C. E = egg, J = juvenile, I = growing, immature adult, A= reproductive adult, S = senile adult, Pg = somatic production, Pr = reproductive production, A = absorption, C = consumption. Redrawn from Kamler and Mandecki (1978).

ter pulmonate *Physa acuta,* is from Kamler and Mandecki (1978). In terms of energy investment, it indicates that sexual reproduction triggers an upsurge in production which is quite distinct from that associated with somatic production (analyzed more fully in Calow, 1983). There are a number of ways in which this might stress the parent and thereby render it more prone to mortality, but all can be related to the competition between somatic and reproductive structures and processes for a share of the limited resources; the more that are used in reproduction, the less that are available to the parent, and the lower are its chances of survival (Calow, 1979, 1981a). There is also a possible ethological–ecological component to this negative relationship between reproduction and survival, because reproductive processes render parents more prone to extrinsic mortality by making them more conspicuous or cumbersome (Tinkle, 1969). However, these latter relationships are difficult to quantify, and are probably more important in vertebrates with complex behavior patterns than they are in the invertebrates.

If the physiological basis of the trade-off between reproduction and survival is to be defined and quantified, it should be done on the basis of the competition for resources noted above. Fecundity estimates in terms of the number of gametes produced per parent or per unit weight of parent are unlikely to be very useful because of the wide variation in gamete size already noted. The total mass or energy invested in reproduction per unit mass of parent has sometimes been used as an alternative index of reproductive effort. However, because energy inputs and metabolic outputs are nonlinear and also different functions of body mass, this index is unlikely to provide a comparative basis for expressing the investment which is made in reproduction over and above the metabolic demands of the parental somatic structures and processes. Hence, it is not surprising that although Hughes and Roberts (1980a) found some correlation between the absolute amount of energy invested in reproduction per unit mass of parent and parental life span in several marine gastropods, the relationship was not a perfect one. The ratio of reproductive production and parent biomass (both in energy units) was approximately 1.0 for the annual and semelparous *Lacuna* spp. and between 0.1 and 0.5 for the long-lived iteroparous *Littorina neritoides, L. rudis, Nerita versicolor, N. tessellata,* and *Fissurella barbadensis;* but the long-lived, iteroparous *L. nigrolineata* and *Thais lapillus* had ratios approximating 1.0. However, when examining only freshwater species Calow (1978) found that semelparous snails produce a larger total volume of eggs per unit volume of parent than do iteroparous species (Table I).

The proportion of energy flow that an adult invests in reproduction is a better measure of reproductive effort (Hirshfield and Tinkle, 1975). In this context, Calow (1978) used total absorbed energy over the breeding season as the denominator; Browne and Russell-Hunter (1978) used the total annual nonrespired energy to estimate the reproductive efforts of gastropods. Both techniques suggested that the semelparous species invested more effort in reproduction than did

TABLE I

Volume of Eggs Per Breeding Season Per Unit Volume of Parent for Freshwater Gastropods[a]

Species	Range	Mean	*n*
Semelparous	1–5	2.06	5
Iteroparous	0.03–0.5	0.28	3

[a] Data from Calow (1978).

iteroparous (Table II). Bayne (1976) lists some of these indices for bivalves, although without reference to life-cycle patterns. Using an index similar to that of Browne and Russell-Hunter (1978), Hughes and Roberts (1980a) and Lassen (1979) obtained less well-defined relationships (e.g., Table III), and Calow (1978) pointed out that the salt marsh pulmonate *Melampus bidentatus,* although iteroparous, invests as much in reproduction as do most of the semelparous species (Table II). Similarly, the freshwater prosobranch *Leptoxis carinata* invests between 10 and 20% of its nonrespired carbon in eggs, which is well within

TABLE II

Reproductive Efforts of Freshwater Gastropods from Energy Budget Measurements

		Total absorbed energy to reproduction (%)	Total nonrespired energy to reproduction (%)
Semelparous[a]:	range	15–35	45–79
	mean	23.8	61.0
	n	7	7
Iteroparous:	range	2–16	25–33
	mean	7.8	28.3
	n	3	3
Semelparous[b]:	range		11–69
	mean		26.9
	n		5
Iteroparous[c]:	range		2.9–61[d]
	mean		15.4
	n		10
Melampus bidentatus			46

[a] From Calow (1978).
[b] From Browne and Russell-Hunter (1978).
[c] Excludes *M. bidentatus*.
[d] The high figure is from the third breeding season of one species.

TABLE III

Reproductive Efforts, Measured as Ratio of Volume of Yolk Produced Per Annual Growth for Danish Mudsnails (Hydrobiidae)[a,b]

	Semelparous (S) or iteroparous (I)	Reproductive effort	Relative reproductive effort
Hydrobia ventrosa	S	0.023	1
Hydrobia ulvae	I	0.057	2.48
Potamopyrgus jenkinsi	I[c]	0.084	3.65
Hydrobia neglecta	S	0.134	5.83

[a] Yolk volume is taken as an index of reproductive investment and annual growth as an index of net production.
[b] Data from Lassen (1979).
[c] Parthenogenetic and ovoviviparous; the rest are sexual and lay eggs.

the iteroparous ranges (Table II) although the species is semelparous (Aldridge, 1982).

A problem with effort indices that are measured in terms of proportional investment, however, is that they do not necessarily indicate the absolute extent to which reproduction is being supported at the expense of the parent. Assuming fixed metabolic demands for parent tissues, the impact of a particular proportional investment in reproduction will depend upon ration level (Calow, 1979; Tashiro, 1982; Thompson, 1983) and possibly other environmental variables such as temperature (Hirshfield, 1980). If it were possible to measure the nonreproductive energy demands of the parent, it would then be possible to define the extent to which they are not being met during reproduction. In the equations, animals that never allow reproduction to detract from these somatic requirements are referred to as *reproductively restrained* and those that reproduce at the expense of these somatic requirements as *reproductively reckless* (Calow, 1978, 1979). In this context, Haukioja and Hakala (1978) have measured reproductive effort of the bivalve *Anodonta piscinalis* in terms of the effects of investment in reproduction on the mass of the somatic tissues of adults:

$$\frac{[(\text{Wt. reproducta} + \text{Wt. nonreproductive adult}) - \text{Wt. reproductive adult}] \times 100}{\text{Wt. nonreproductive adult}^{1}}$$

In this equation, weight of reproducta refers to glochidia; both reproductive and nonreproductive adults have the same shell length. This index increases as pro-

[1] Wt., weight of.

gressively more parent biomass is sacrificed to reproduction. Haukioja and Hakala (1978) found that the index was highest when the food supply was lowest and was negatively correlated with the postreproductive life spans of parents during these times.

Similarly, simple assessment of the biomass of somatic tissue before and after breeding, or even histological observation of these tissues, should give some insight into reproductive investment (Loosanoff and Davis, 1951; Sastry, 1966; Bayne and Thompson, 1970). Resource allocation in reproducing adults also might be followed by the use of radiotracers. Using this technique, O'Dor and Wells (1978) monitored a switch of labeled amino acids from muscles to the gonads of *Octopus cyanea* during its breeding season. Streit (1976), using ^{14}C-labeled food, showed that the percentage of carbon incorporated into the reproductive tissues increased in older freshwater limpets (*Ancylus fluviatilis*) at 22°C. Similar results have been obtained from bivalves (Sastry and Blake, 1971; Vasallo, 1973; Bayne, 1975, 1976; Gabbott, 1975, 1976); in mature *Mytilus edulis* the incorporation of labeled substances into gonads and gametes remains good even during food and temperature stress (Bayne, 1975). However, extreme stress applied after the complete maturation of gametes will cause resorption of eggs and sperm in this species (Gabbott and Bayne, 1973).

Calow (1979) suggested the following precise index of the cost of reproduction to the parent:

$$C = 1 - \left(\frac{\text{Input energy} - \text{Energy loss due to reproduction}}{\text{Energy needed for metabolism of the soma}} \right)$$

Vahl (1981) has calculated an approximation to C using:

$$1 - \left(\frac{\text{Total absorbed energy} - \text{Energy loss due to reproduction}}{\text{Total absorbed energy}} \right)$$

for the long-lived marine bivalve *Chlamys islandica*. He found that this index increases at a reducing rate with age onto an asymptote, but that it is negatively related to residual reproductive value (an index estimating future fecundity and survivorship of parents) for only part of its range. Nevertheless, reproduction is known to reduce the growth rates of this species, so that the cost of reproduction here may be manifest more in size than in survivorship adjustments (see Section III,B).

Clearly, the evidence for a negative causal relationship between investment in reproduction and the subsequent survival of the parent is persuasive [not only for Mollusca (Calow, 1979)] but not unequivocal (Aldridge, 1982). Alternative explanations for postreproductive death are feasible in principle, e.g., host suicide in the face of parasitization (Trail, 1980). Furthermore, it is important not only to establish that a negative causal relationship exists but, as will be discussed in Section IV,B, to precisely define the form of this trade-off.

B. Reproduction and Parental Growth and Size

Some molluscs begin to breed only after they have stopped growing [e.g., muricids (Spight et al., 1974) and the bivalve *Macoma balthica* (Gilbert, 1973)], whereas others continue to grow during and between breeding seasons [e.g., some freshwater pulmonates (Calow, 1981d) and the bivalve *Mytilus edulis* (Bayne, 1975, 1976)]. In the latter case, there will probably be competition between growth and reproduction for the limited resources available for production. This has been documented for the freshwater pulmonate *Lymnaea peregra* (Calow, 1981d). Here, snails from small and large races grow at the same rate under the same constant laboratory conditions during the initial exponential phase of growth; the smaller snails initiate reproduction at a smaller size, which is correlated with an earlier slowing and cessation of growth, i.e., onset of the decelerating phase of growth. Energy budgets suggest that the production of eggs, once it has been initiated, requires most of the energy available for production in this species.

In addition to influencing growth rate and thus size, reproduction is influenced by size. The overall size of the parents determines the intensity of tissue production and the space available for gonads and brood chambers. Linear isometric relationships between egg production and snail volume or mass have been observed in some oviparous species (Kamler and Mandecki, 1978; Thompson, 1979; Hughes and Roberts, 1980b) and in some ovoviviparous species (Roberts and Hughes, 1980), which might be explained in terms of an isometric relationship between gonad or brood chamber volume and whole body volume. Alternatively, the relationship between egg output and adult volume or mass conforms to the *law of diminishing increments* in other species (Calow, 1976), and this might be explicable in terms of diminishing production potential of the adult after attaining an optimum size for reproduction (Calow, 1981a). Finally, in principle gametes produced might be an increasing function of adult mass if increasing effort is invested in reproduction as animals become older and their residual reproductive value decreases. This outcome, expected on theoretical grounds (Pianka and Parker, 1975), appears to occur in *M. edulis* (Bayne, 1976). Apparently, a number of relationships occur between fecundity and adult size in the Mollusca, and no generalization is currently possible.

In considering the reproductive energetics of two species of dorid nudibranchs, Todd (1979) argued that the constraints of size on reproductive output may, in turn, limit the developmental options available to a species. For example, *Onchidoris muricata* and *Adalaria proxima* are sympatric and have similar feeding niches, but the former is planktotrophic and the latter lecithotrophic. *O. muricata* is about one-fifth the size of *A. proxima* when adult. Todd (1979) speculates that there is a threshold of absolute energy supply required to support the lecithotrophic strategy in nudibranchs which is not attained by the smaller species (see also Underwood, 1979, and Chapter 5, this volume). Similarly, small size may favor brooding because it limits the energy available for reproduction and hence for fecundity, so that protection is required to enhance the survival chances of offspring. This has been documented for the Calyptraeidae (Hoagland, 1975) and the freshwater bivalves (Davis and Fuller, 1981), and it may be a general phenomenon (Gould, 1977). But alternative hypotheses are feasible, and if they are to be taken seriously, there must also be costs of brooding for large adults (Strathmann and Strathmann, 1982). One possibility is that brooding provides a bottleneck for resource investment and severely restricts the number of offspring that can be produced at any one time. This and alternative hypotheses deserve more careful scrutiny (Strathmann and Strathmann, 1982).

C. Gamete Size and Numbers

The negative correlation between gamete size and fecundity was obvious in the taxonomic survey and is not surprising, because the resources available to parents for gamete production are finite. Usually, most gametes are produced by male parents (or systems); because the emphasis is on maximizing information transmission per unit resource investment, spermatozoa consist largely of a nucleus surrounded by cytoplasm specialized for locomotion and penetration of the egg (Calow, 1981b). Some gastropods produce "abnormal" sperm, incapable of fertilizing ova but often larger than normal; these may have a supportive function (Webber, 1977). Alternatively, female gametes must not only control development but invariably must also support the early embryological stages, and this is achieved by investment in energy-rich yolk. Moreover, in species in which eggs remain free in the environment for substantial periods of time (e.g., in direct development), a considerable amount of energy is invested in capsules. For example, Perron (1981) found that the capsule material of *Conus spp.* had an energy value of 5442 cal (~23kJ)/g and ova of 6238 cal (~26kJ)/g and accounted for as much as 50% of the total energy invested in reproduction. Both puncture resistance and the amount of energy devoted to capsules increased with size of ova and length of developmental times in these species. It seems likely that "femaleness" in general is energetically more costly than "maleness" and that this will be very important in relation to the adaptive value of protandric hermaphroditism, i.e., the less costly male function precedes the more costly female one [Russell-Hunter and McMahon (1975, 1976)].

It follows, therefore, that egg size limits, and might have been limited by, the length of development. Within major molluscan taxa, egg size has an effect on developmental time second only to temperature in importance (Spight, 1976b);

larger eggs are associated with longer development. Similarly, Fioroni (1966) and Amio (1963) have both shown, with nonoverlapping sets of data, that shell length at hatching is closely related to egg diameter for a variety of species. Combining and extending these data sets, Spight (1976b) found that in 92 species 84% of the variation in length at hatching time could be attributed to egg diameter. However, no consistent relationship between egg diameter and hatching size can be expected among species that utilize nurse eggs, because some embryos consume only a few egg cells and others may consume thousands (Fioroni, 1966). In prosobranchs, the variation in the size of the hatchlings which use a supply of yolk contained entirely within the egg membrane is usually no more than ± 15%. Alternatively, prosobranch embryos which feed on nurse eggs may vary by more than twice this amount (see Section II,A,1).

Finally, it is clear that developmental type, in terms of planktotrophy and lecithotrophy, will be influenced by and will in turn influence egg size; as noted in Section II,D, planktotrophy is associated with the production of large numbers of small eggs, and lecithotrophy and direct development involve the reverse. Although cause and effect are not always clear, because there is much within-taxa variation in egg size it seems likely that the causal train is from developmental type to egg size rather than the reverse.

IV. Extrinsic (Endogenous) Factors

Extrinsic factors can influence life-cycle patterns *proximately,* by directly influencing metabolic rates, reproductive outputs, and survivorship, or *ultimately,* by influencing the genetic composition of the population (terminology from Mayr, 1961). Examples of the proximate influence of environmental parameters (e.g., temperature, humidity, and food supply) have appeared throughout this volume and will not be considered explicitly here. However, the ultimate effects derive from the proximate effects and, moreover, are sometimes difficult to distinguish from them. Ideally, laboratory culturing techniques and/or field manipulations should be used to distinguish phenotypic variation caused by proximate factors from genotypic variation resulting from selection (Oosterhoff, 1977; Calow, 1981d). Note also, however, that phenotypic plasticity to environmental factors may itself be a product of selection. Finally, the response of life-cycle parameters to environmental variables, in both the proximate and the ultimate sense, is limited by the intrinsic relationships already noted in the last section.

The possible influences of environmental variables on the expression of a life cycle are therefore numerous and complex. Here are presented three examples which are important and relatively well defined:

1. Influence of environmental uncertainty on the trade-off between growth and reproduction

2. Influence of factors affecting mortality schedules on the trade-off between reproduction and parent survival
3. Influence of environmental factors on the trade-off between gamete size and numbers

A. Environmental Uncertainty and the Trade-Off between Growth and Reproduction

Because there is a trade-off between growth and reproduction and a positive correlation between fecundity and parent size, the pattern selected depends on the merits of early, relatively low levels of reproduction as compared with later, higher levels of reproduction. The freshwater gastropod *L. peregra* breeds at a smaller size in fast-flowing lotic and wave-swept littoral habitats than it does in lentic situations and this is genetically determined (Calow, 1981d). Decreased food supplies in the lotic and littoral environments mean that adults cannot expect to reach large sizes. Moreover, small hatchlings may be less susceptible to spate and strong wave action because they can find shelter in small cracks and crevices in the substratum. Alternatively, by growing to larger adult sizes, lentic snails are able to "cash in" on larger absolute fecundities. Similar principles may also apply to marine gastropods with populations in both sheltered and exposed habitats (Spight and Emlen, 1976; Roberts and Hughes, 1980) and in boulder, as compared with creviced, littoral habitats (Begon and Mortimer, 1981; Hart and Begon, 1982). A precise analytical evaluation of the ways in which these factors influence molluscan life cycles has not yet been attempted; for good studies on other invertebrates the reader is referred to Lawlor (1976) and Sebens (1979), and for a discussion on the kinds of models that will be required, see Calow and Sibly (1983). Clearly, an investment in somatic rather than gametic production is more sensible evolutionarily if reproductive output is a rising exponential function of body mass rather than an isometric or falling function (see Section III,B). Therefore, it is necessary to define clearly the form of the intrinsic and extrinsic factors influencing a particular population.

B. Influence of Environment on Trade-Off between Reproduction and Parent Survival

The fitness merits of sacrificing parents to reproduction depend crucially upon (a) the form of the functional relationship between the postreproductive survival chances of parents and their investment in reproduction, and (b) the survival chances of offspring relative to those of their parents in the absence of the fecundity-mediated effects (Calow and Sibly, 1983). Unfortunately, we have little precise information on either (a) or (b) for Mollusca. Nevertheless, if it is assumed that parent survival is reduced (Section III,A) increasingly with increasing investment in reproduction [which is plausible (Calow and Sibly, 1983)], then it is intuitively reasonable and can be proved rigorously (Stearns, 1976, 1977, 1980; Young, 1981; Calow and Sibly, 1983; Sibly and Calow, 1983) that parents should be restrained (*sensu* Section III,A) when offspring survival is poor and reckless (*sensu* Section III,A) when offspring survival is good. The latter is a strong possibility when resources are not limiting and populations are under density-independent control (so-called *r* selection). Alternatively, when resources are limiting, competition is intense, populations suffer density-dependent controls, and offspring survival is likely to be poor relative to that of parents (so-called *K* selection). On the other hand, *r* selection is often associated with unpredictable mortality which might reduce offspring survival and lead to restraint (so-called bet hedging) in parents. These ideas are reviewed in Stearns (1976, 1977, 1980). Below are summarized correlations discovered between molluscan life cycles and (a) potentially *r* and *K* situations of selection and (b) more precisely defined mortality schedules.

1. r *and* K *Selection*

From inshore to deeper water in the marine environment, physical conditions become less rigorous but productivity decreases. Rex (1979) has suggested that this provides a selection gradient from *r* conditions in shallow water to *K* conditions in deep water. Along this gradient he finds that the gastropod *Alvania pelagica* has a larger adult size and longer life, with less invested in reproduction in deep-water as compared with shallow-water populations. For deep-sea molluscs in general, there is evidence that fecundities are on average lower and longevities much higher than those of their shallow-water relatives (Allen and Sanders, 1973; Scheltema, 1972; Turekian et al., 1975; Valentine and Ayala, 1978).

The association of *r* selection with exposed habitats and *K* selection with sheltered ones may be general, but whether a habitat is effectively exposed or sheltered will depend upon the biology of the organisms under consideration as well as the physical conditions of the habitat. For example, *Dendropoma corallinaceum* is a dominant prosobranch on the exposed rock surfaces of the Cape Province of South Africa. Hughes (1979) reports that it forms sheet-like colonies over the rock surfaces and that these shield it from exposure. Nevertheless, there is intense competition for space within these colonies, and this leads to *K* rather than *r* selection. Accordingly, *D. corallinaceum* produces few relatively large offspring and lives for several breeding seasons.

Most freshwater bodies are relatively transient, persisting for no more than a few thousand years and usually much less (Russell-Hunter, 1970, 1978). Hence, niches here are likely to be less full and competition less intense than in marine systems; that is, there is an emphasis on *r* rather than *K* selection. As expected from this basis, life-cycle patterns shift from predominantly iteroparous in marine molluscs to semelparous in freshwater ones (Calow, 1978). Also, freshwater gastropods telescope development into a protected egg, because of the exacting physicochemical conditions of the freshwater environment and the sensitivity of larvae; this confers increased survival chances on hatchlings, giving further impetus to the evolution of semelparity (Calow, 1978).

Tropical freshwater gastropods inhabiting temporary pools face very uncertain conditions, but the productivity in these habitats is high. Here pulmonates predominate, growing and maturing quickly and reproducing continuously throughout a short adult life (Brown, 1980). In general, the breeding of tropical snails seems to be synchronized with rainy season. At the other end of the spectrum are the ancient lakes which have been in existence for more than 10^5 years and have rich endemic faunas. Here prosobranchs, often more common than pulmonates, usually have long lives and iteroparous life-cycle patterns (Hubendick, 1952; Stankovic, 1960; Kozhov, 1963).

Danish hydrobiid snails in ephemeral habitats invest more in reproduction than do those found in stable habitats (Lassen, 1979), but this investment is not associated in any obvious way with semelparity and iteroparity (Table III). The freshwater bivalves *Musculium partumeium* and *M. securis* are semelparous in temporary pools and iteroparous in permanent habitats (Mackie et al., 1978; Hornbach et al., 1980; Way et al., 1980), but there is no obvious relationship with fecundity, and the shortened life cycle may be phenotypically rather than genotypically determined (McLeod et al., 1981).

It is tempting to apply the r/K logic in reverse, that is, to deduce the selection pressure to which a population has been subjected from the characters currently possessed by it. This has been done by Grahame (1977) for two species of *Lacuna. L. vincta* invests more in reproduction than *L. pallidula,* and hence was considered to be more r selected. However, as noted a life-cycle trait can arise in diverse ways (Stearns, 1977), so this procedure is of value only in formulating hypotheses that should then be evaluated further by reference to the demographic forces that act on the populations under study.

2. Mortality Patterns

Studying 13 populations of the mussel *Anodonta piscinalis,* Haukioja and Hakala (1978) found a strong positive correlation between adult reproductive life span [controlled by intrinsic factors (Section III,A)] and variation in juvenile survival. The more predictable the survival of the juveniles, the more effort was invested in reproduction by adults and, in consequence (according to the interpretation given in Section III,A), the lower was the adult survival.

Parry (1982) has investigated the relationship between extrinsic adult mortality and life cycles in four species of intertidal limpet in South East Australia. Extrinsic adult mortality rates were inferred from total adult mortality rates, a knowledge of the causes of mortality, and the likely effects that a change in reproduc-

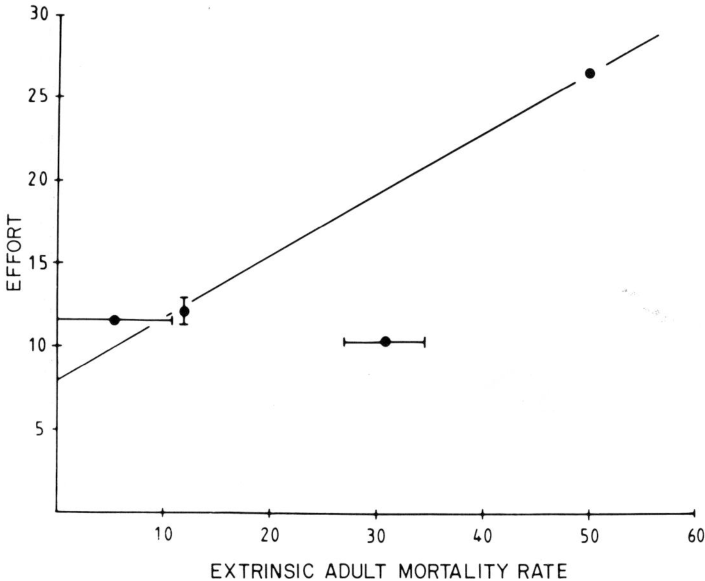

Fig. 5. Reproductive effort [i.e., (energy invested in gametes ÷ energy absorbed by parents) × 100] against extrinsic mortality for four species of limpet. Each point represents a separate species. The bottom right hand point is for *Patella alticostata* (there are strong grounds for believing this overestimates extrinsic mortality; it should move to the left which would improve the correlation). Error bars are ranges of values. Redrawn from Parry (1982).

tive allocation would have on the vulnerability of each species to its observed agents of mortality. Reproductive effort was measured as the proportion of absorbed energy invested in egg production. Reproductive effort is plotted against extrinsic adult mortality in Fig. 5. The correlation was positive but not significant ($p > .05$). However, in view of the difficulties experienced in collecting such data and the few points available for analysis, Parry interpreted the existence of the positive trend as encouraging support for a positive causal relationship between adult mortality due to extrinsic factors and reproductive effort.

A rather extreme form of difference in age-specific mortality leading to life-cycle differences is documented by Choat and Black (1979) for acmaeid limpets. *Acmaea incessa* has an obligate association with the brown alga *Egregia laevigata,* which has an annual cycle. *A. incessa* must therefore grow to maturity and produce gametes and larval young before the inevitable but predictable loss of its attachment and feeding substratum. Here adult mortality is complete, and dispersing larvae that do not immediately depend on *Egregia* are the only means of continued survival. Another related species, *A. digitalis,* lives on permanent hard surfaces on which the adult is not subject to catastrophic losses. Predictably, *A. incessa* has a shorter life expectancy, a higher growth rate, greater reproductive effort, and a shorter time to sexual maturity than *A. digitalis.*

Another unusual life-cycle effect of extrinsic mortality factors has been reported by Minchella and Loverde (1981) for the host–parasite interaction of the snail *Biomphalaria glabrata* and the trematode *Schistosoma mansoni.* Snails exposed to the parasite may or may not become infected; if they do, they suffer parasitic castration and increased mortality. Whether or not they are infected, adult snails apparently compensate for possible future loss in reproductive success by increasing their reproductive effort. It is suggested that this response may be an important life-history adaptation to parasitic infection. Interestingly, snails exposed but not parasitized show no increase in mortality after compensation, but lay fewer eggs in total than do unexposed snails.

C. Egg Size and Numbers

The size of the egg determines the size and, to some extent, the form of hatchlings and hence their subsequent growth and survivorship properties. Because the resources invested in reproduction are finite, as noted in Sections II,D and III,C, a negative relationship between egg size and numbers can be expected. Hence, the compromise between egg size and numbers must be sensitive, in an evolutionary sense, to extrinsic factors. Moreover, the shift from planktonic to direct development and back is relatively straightforward because many species produce egg capsules; in forms with direct development, the principal difference is that the larvae stay longer in the capsule and grow on an extra supply of yolk or feed on other eggs present in the capsule. Hence, the evolutionary step from one type of development to the other should easily be achieved given the appropriate selection pressure.

Intuitively, the extent to which the production of small planktotrophic larvae is favored over that of larger nonplanktotrophic ones by marine gastropods will depend on the availability of food for development in the plankton and the relative mortality of slow-growing, poorly developed larvae in the plankton and more rapid growing, well-developed larvae in either planktonic or benthic habitats. Hence, it is not surprising to find direct or lecithotrophic development to be dominant in boreal and arctic waters and planktotrophic development to dominate in the tropical seas. Data on this have been reviewed by Thorson (1950), and more rigorous models based on the foregoing have been developed by Vance (1973a,b) and Christiansen and Fenchel (1979). These models assume that related species with different developmental strategies invest similar amounts of resources in reproduction (see also Grahame, 1982) and compare effective fecundity (number of settled offspring) in the different developmental types under different regimens of mortality and developmental rate during the larval stage (for a critique, see Underwood, 1974).

As already noted, conditions become more exacting when moving from marine to freshwater and terrestrial habitats, and larval development becomes telescoped into a protected capsule. The emphasis shifts to large egg size at the expense of large numbers of eggs (Calow, 1978). Trends in this direction, although not continuous, can be perceived in gastropods occupying littoral zones from low to high shores (McMahon and Russell-Hunter, 1977). Many freshwater prosobranchs brood young but ovoviviparity is the rule, and so yolk stores are necessary (Section III,C). Freshwater and terrestrial pulmonates invariably produce large eggs and lay them in capsules and shells. The capsule walls may represent a significant investment of resources (see above).

There are considerable differences in egg size and fecundity both between and within species occupying different and similar aquatic and terrestrial habitats. One example involving the streamlined river limpet *Ancylus fluviatilis* and the more globose *Lymnaea peregra,* both of which may coexist on wave-swept lake shores, has been documented by Calow (1978). Under these circumstances both invest about 40 J/adult/season in producing eggs, but *L. peregra* produces smaller eggs than *A. fluviatilis*. The larger eggs and richer reserves of *A. fluviatilis* enable more complete intracapsular development. At 10°C, *A. fluviatilis* takes 26 days to hatch and on emergence has roughly 6% of the shell-free dry weight of the adults. *L. peregra* requires only one-half that time to develop but on hatching is only 1–2% of its final size. There are differences in mortality patterns between species; *A. fluviatilis* conforms to a Deevey type II (mortality rate constant from hatching to breeding) and *L. peregra* to Deevey type III (mortality mainly affects young and decreases with age) survivorship curves (Deevey, 1947). Here both species have very different reproductive strategies, yet both coexist in relatively stable populations. The conclusion, which is substantiated by theoretical analysis (Calow, 1978), is that the alternatives of producing a few relatively hardy offspring per parent and producing a larger number of more mortality-prone offspring balance out, resulting in equally fit strategies.

In less severe lentic habitats, *L. peregra* produces more eggs and packages more per capsule than it does in wave-swept littoral and fast-flowing lotic habitats (Calow, 1981d). One possible explanation is that because the production of membrane is expensive, a large number of eggs should be packaged into each capsule; this means less membrane, and hence less membrane cost, per egg. In the wave-swept and fast-flowing habitats, however, a major source of egg mortality is water movement, some eggs being washed away. Because this is an all-

or-nothing process, eggs should be spread among a large number of capsules. This difference might be general in aquatic snails occupying sheltered and exposed habitats and should be investigated more thoroughly.

V. Conclusions

It is well known that the Mollusca show an impressive diversity of form and function within the limitations of their "design", and this is also true of their life-cycle patterns. There are taxonomic biases in the latter but no obvious, absolute limitations, at least among the major classes. Alternatively, ecological trends are perceptible for which evolutionary explanations involving interaction of extrinsic and intrinsic factors can be formulated. To test these hypotheses further, it will be necessary to obtain more precise information on the form of the intrinsic constraints and trade-offs and on the demographic impact of extrinsic forces. For example, the relative position of molluscs on the *r–K* spectrum is likely to be an important factor in their life-cycle evolution, yet the form of their survivorship curves is more fundamental and is likely to transcend and even frustrate simple *r–K* correlations (Hart and Begon, 1982). Similarly, the nature of predictions relatively to the issues of reproduction versus growth, reproduction versus survival, and gamete size versus gamete numbers is very sensitive to the form of the relationship between parent size and fecundity, parent survival and fecundity, and developmental rate and food availability. Unfortunately, we know little about the precise demography of natural populations of molluscs or about the form of these physiological parameters. Comparative life-table and physiological data on related molluscs (preferably of the same species) with different life cycles would be particularly interesting and relevant.

References

Aldridge, D. W. (1982). Reproductive tactics in relation to life-cycle bioenergetics in three natural populations of the freshwater snail *Leptoxis carinata*. *Ecology* **63,** 196–208.

Allen, J. A., and Sanders, H. C. (1973). Studies on deep-sea Protobranchia (Bivalvia); the families Siluculidae and Lameltilidae. *Bull. Mus. Comp. Zool.* **145,** 263–310.

Amio, M. (1963). A comparative embryology of marine gastropods, with ecological emphasis. *J. Shimonoseki Coll. Fish.* **12,** 229–253.

Arnold, J. M. (1977). Cephalopoda; Decapoda. *In* "Reproduction of Marine Invertebrates" (A. C. Giese and J. S. Pearse, eds.), Vol. 4, pp. 243–290. Academic Press, New York.

Audesirk, T. E. (1979). A field study of growth and reproduction in *Aplysia californica*. *Biol. Bull.* (*Woods Hole, Mass.*) **157,** 407–421.

Bayne, B. L. (1975). Reproduction in bivalve molluscs under environmental stress. *In* "Physiological Ecology of Estuarine Organisms" (J. Vernberg, ed.), pp. 259–277. Univ. of South Carolina Press, Columbia.

Bayne, B. L. (1976). Aspects of reproduction in bivalve molluscs. *In* "Estuarine Processes" (M. Wiley, ed.), Vol. 1, pp. 432–448. Academic Press, New York.

Bayne, B. L., and Thompson, R. J. (1970). Some physiological consequences of keeping *Mytilus edulis* in the laboratory. *Helgol. Wiss. Meeresunters.* **90,** 526–552.

Beeman, R. D. (1977). Gastropoda: Opisthobranchia. *In* "Reproduction of Marine Invertebrates" (A. C. Giese and J. S. Pearse, eds.), Vol. 4, pp. 115–179. Academic Press, New York.

Begon, M., and Mortimer, M. (1981). "Population Biology. A Unified Study of Plants and Animals." Blackwell, Oxford.

Berry, A. J. (1977). Gastropoda: Pulmonata. *In* "Reproduction of Marine Invertebrates" (A. C. Giese and J. S. Pearse, eds.), Vol. 4, pp. 181–226. Academic Press, New York.

Branch, G. M. (1974). The ecology of *Patella* Linnaeus from the Cape Peninsula, South Africa. 3. Growth rates. *Trans. R. Soc. S. Afr.* **41,** 161–193.

Brown, D. S. (1980). "Freshwater Snails of Africa and their Medical Importance." Taylor & Francis, London.

Browne, R. A., and Russell-Hunter, W. D. (1978). Reproductive effort in molluscs. *Oecologia* **37,** 23–27.

Calow, P. (1976). "Biological Machines." Arnold, London.

Calow, P. (1978). The evolution of life-cycle strategies in freshwater gastropods. *Malacologia* **17,** 351–364.

Calow, P. (1979). The cost of reproduction—a physiological approach. *Biol. Rev. Cambridge Philos. Soc.* **54,** 23–40.

Calow, P. (1981a). Growth in lower invertebrates. *In* "Comparative Animal Nutrition" (M. Rechcigl, ed.), pp. 53–76. Karger, Basel.

Calow, P. (1981b). Resource utilization in reproduction. *In* "Physiological Ecology" (C. R. Townsend and P. Calow, eds.), pp. 245–270. Blackwell, Oxford.

Calow, P. (1981c). "Biology of the Invertebrates: A Functional Approach." Croom Helm, London.

Calow, P. (1981d). Adaptational aspects of growth and reproduction in *Lymnaea peregra* (Gastropoda: Pulmonata) from exposed and sheltered aquatic habitats. *Malacologia* **21,** 5–13.

Calow, P. (1983). Energetics of reproduction and its evolutionary implications. *Biol. J. Linn. Soc.* (in press).

Calow, P. (*in manuscript*). Fact and theory—an overview. In "*Cephalopod Life Cycles*" (P. Boyle, ed.), Vol. 2. Academic Press, New York.

Calow, P. and Sibly, R. (1983). Physiological trade-offs and the evolution of life cycles. *Sci. Prog. (Oxford)* **68,** 177–188.

Calow, P., and Townsend, C. R. (1981). Resource utilization in growth. *In* "Physiological Ecology" (C. R. Townsend and P. Calow, eds.), pp. 220–224. Blackwell, Oxford.

Choat, J. H., and Black, R. (1979). Life histories of limpets and the limpet–laminarian relationship. *J. Exp. Mar. Biol. Ecol.* **41,** 25–50.

Christiansen, F. B., and Fenchel, T. M. (1979). Evolution of marine invertebrate reproductive patterns. *Theor. Popul. Biol.* **16,** 267–282.

Comfort, A. (1957). The duration of life in molluscs. *Proc. Malacol. Soc., London* **32,** 219–241.

Davis, G. M., and Fuller, L. H. (1981). Genetic relationships among recent Unionacea (Bivalvia) of North America. *Malacologia* **20,** 217–253.

Deevey, D. S. (1947). Life tables for natural populations of animals. *Q. Rev. Biol.* **22,** 283–314.

Duncan, C. J. (1960). The genital system of the freshwater Basommatophora. *Proc. Zool. Soc. London* **135,** 339–356.

Duncan, C. J. (1975). Reproduction. *In* "Pulmonates" (V. Fretter and J. Peake, eds.), Vol. 1, pp. 309–365. Academic Press, New York.

Ellis, A. E. (1978). "British Freshwater Bivalve Mollusca." Linnean Society Synopsis of British Fauna No. 11. Academic Press, New York.

Fioroni, P. (1966). Zur Morphologie und Embryogenese des Darmtraktes und der transitorischen Organe bei Prosobranchiern (Mollusca: Gastropoda). *Rev. Suisse Zool.* **73,** 621–876.

Fretter, V., and Graham, A. (1962). "British Prosobranch Molluscs." Ray Society, London.

Gabbott, P. A. (1976). Energy metabolism. *In* "Marine Mussels" (B. L. Bayne, ed.), pp. 293–355. Cambridge Univ. Press, London and New York.

Gabbott, P. A. (1976). Storage cycles in marine bivalve molluscs: A hypothesis concerning the relationship between glycogen metabolism and gametogenesis. *Proc. Eur. Mar. Biol. Symp., 9th, 1974,* pp. 191–211.

Gabbott, P. A., and Bayne, B. L. (1973). Biochemical effects of temperature and nutritive stress on *Mytilus edulis* L. *J. Mar. Biol. Assoc. U.K.* **53,** 269–286.

Ghiselin, M. T. (1969). The evolution of hermaphroditism among animals. *Q. Rev. Biol.* **44,** 189–208.

Giese, A. C., and Pearse, J. S., eds. (1977). "Reproduction of Marine Invertebrates," Vol. 4. Academic Press, New York.

Gilbert, M. A. (1973). Growth rate, longevity and maximum size of *Macoma balthica* (L.). *Biol. Bull. (Woods Hole, Mass.)* **145,** 119–126.

Gould, S. J. (1977). "Ontogeny and Phylogeny." Harvard Univ. Press, Cambridge, Massachusetts.

Grahame, J. (1977). Reproductive effort and r- and K-selection in two species of *Lacuna* (Gastropoda: Prosobranchia). *Mar. Biol.* **40,** 217–224.

Grahame, J. (1982). Energy flow and breeding in two species of *Lacuna:* Comparative costs of egg production and maintenance. *Int. J. Invertebr. Reprod.* **5,** 91–99.

Green, M. G. (1973). Taxonomy and distribution of planktonic octopods in the northeastern Pacific. M.Sc. Thesis, University of Washington, Seattle (*unpublished*).

Hart, A. and Begon, M. (1982). The status of general reproductive-strategy theories, illustrated in winkles. *Oecologia* **52,** 37–42.

Haukioja, E., and Hakala, T. (1978). Life-history evolution in *Anodonta piscinalis* (Mollusca, Pelecypoda). Correlation of parameters. *Oecologia* **35,** 253–266.

Haven, N. (1977). Cephalopoda: Nautiloidea. *In* "Reproduction of Marine Invertebrates" (A. C. Giese and J. S. Pearse, eds.), Vol. 4, pp. 227–241. Academic Press, New York.

Heard, W. H. (1965). Comparative life histories of fingernail clams (Sphaeriidae: Pisidium). *Malacologia* **2,** 381–411.

Heard, W. H. (1975). Sexuality and other aspects of reproduction in *Anodonta* (Pelecypoda: Unionidae). *Malacologia* **15,** 81–104.

Heard, W. H. (1976). Comparative life histories of fingernail clams (Sphaeriidae: Sphaerium and Musculium). *Malacologia* **16,** 421–455.

Heath, D. J. (1977). Simultaneous hermaphroditism: Cost and benefit. *J. Theor. Biol.* **64,** 363–373.

Hirshfield, M. F. (1980). An experimental analysis of reproductive effort and cost in the Japanese Medaka, *Oryzias latipes. Ecology* **61,** 282–292.

Hirshfield, M. F., and Tinkle, D. W. (1975). Natural selection and the evolution of reproductive effort. *Proc. Natl. Acad. Sci. U.S.A.* **72,** 2227–2231.

Hoagland, K. E. (1975). Reproductive strategies and evolution in the genus *Crepidula* (Gastropoda: Calyptraeidae). Ph.D. Thesis, Harvard University, Cambridge, Massachusetts (unpublished).

Hoagland, K. E. (1978). Protandry and the evolution of environmentally-mediated sex change: A study of the Mollusca. *Malacologia* **17,** 365–391.

Hornbach, D. J., Way, C. M., and Burky, A. J. (1980). Reproductive strategies in the freshwater sphaeriid clam *Musculium partumeium* (Say), from a permanent and temporary pond. *Oecologia* **44,** 164–170.

Hubendick, B. (1952). On the evolution of the so-called thalassoid molluscs of Lake Tanganyika. *Ark. Zool.* **3,** 319–323.

Hughes, R. N. (1979). Notes on the reproductive strategies of the South African vermetid gastropods *Dendropoma corallinaceum* and *Serpulorbis natalensis*. *Veliger* **21,** 423–427.

Hughes, R. N., and Roberts, D. J. (1980a). Reproductive effort of winkles (*Littorina* spp.) with contrasted methods of reproduction. *Oecologia* **47,** 130–136.

Hughes, R. N., and Roberts, D. J. (1980b). Growth and reproduction rates of *Littorina neritoides* (L.) in North Wales. *J. Mar. Biol. Assoc. U.K.* **60,** 591–599.

Kamler, E., and Mandecki, W. (1978). Ecological bioenergetics of *Physa acuta* (Gastropoda) in heated waters. *Pol. Arch. Hydrobiol.* **25,** 833–868.

Kozhov, M. (1963). "Lake Baikal and its Life," Monogr. Biol. 11. Junk, The Hague.

Lassen, H. H. (1979). Reproductive effort in Danish mudsnails (Hydrobiidae). *Oecologia* **40,** 365–369.

Lawlor, L. R. (1976). Molting, growth and reproductive strategies in the terrestrial isopod, *Armadillidium vulgare*. *Ecology* **57,** 1179–1194.

Laxton, J. H. (1970). Shell growth in some New Zealand Cymatiidae (Gastropoda: Prosobranchia). *J. Exp. Mar. Biol. Ecol.* **4,** 250–260.

Loosanoff, V. L., and Davis, H. C. (1951). Delayed spawning of lamellibranchs by low temperature. *J. Mar. Res.* **10,** 197–202.

Mackie, G. L., Qadri, S. U., and Reed, R. M. (1978). Significance of litter size in *Musculium securis* (Bivalvia: Sphaeriidae). *Ecology* **59,** 1069–1074.

McLeod, M. J., Hornbach, D. I., Guttman, S. I., Way, C. M., and Burky, A. J. (1981). Environmental heterogeneity, genetic polymorphism and reproductive strategies. *Am. Nat.* **118,** 129–134.

McMahon, R. F., and Russell-Hunter, W. D. (1977). Temperature relations of aerial and aquatic respiration in six littoral snails in relation to their vertical zonation. *Biol. Bull. (Woods Hole, Mass.)* **152,** 182–198.

Maynard Smith, J. (1978). "The Evolution of Sex." Cambridge Univ.Press, London and New York.

Mayr, E. (1961). Cause and effect in biology. *Science* **134,** 1501–1506.

Minchella, D. J., and Loverde, P. T. (1981). A cost of increased early reproductive effort in the snail *Biomphalaria glabrata*. *Am. Nat.* **118,** 876–881.

Morton, J. E., (1979). "Molluscs." Hutchinson, London.

O'Dor, R. K., and Wells, M. J. (1978). Reproduction versus somatic growth: Hormonal control in *Octopus vulgaris*. *J. Exp. Biol.* **77,** 15–31.

Oosterhoff, L. M. (1977). Variation in growth rate as an ecological factor in the landsnail *Cepaea nemoralis* (L.). *Neth. J. Zool.* **27,** 1–132.

Parry, G. (1982). The evolution of the life histories of four species of intertidal limpets. *Ecol. Monogr.* **52,** 65–91.

Peake, J. (1978). Distribution and ecology of the Stylommatophora. In "Pulmonates" (V. Fretter and J. Peake, eds.), Vol. 2A, pp. 420–526. Academic Press, New York.

Perron, F. E. (1981). The partitioning of reproductive energy between ova and protective capsules in marine gastropods of the genus *Conus*. *Am. Nat.* **118,** 110–118.

Pianka, E. R., and Parker, W. S. (1975). Age specific reproductive tactics. *Am. Nat.* **109,** 453–464.

Rex, M. A. (1979). r- and K-selection in deep-sea gastropods. *Sarsia* **64,** 29–32.

Roberts, D. J., and Hughes, R. N. (1980). Growth and reproductive rates of *Littorina rudis* from three contrasted shores in North Wales, U.K. *Mar. Biol.* **58,** 47–54.

Russell-Hunter, W. D. (1970). "Aquatic Productivity." Macmillan, New York.

Russell-Hunter, W. D. (1978). Ecology of freshwater molluscs. *In* "Pulmonates" (V. Fretter and J. Peake, eds.), Vol. 2, pp. 353–383. Academic Press, New York.

Russell-Hunter, W. D., and Apley, M. L. (1966). Quantitative aspects of early life-history in the salt-marsh pulmonate snail, *Melampus bidentatus,* and their evolutionary significance. *Biol. Bull. (Woods Hole, Mass.)* **131,** 392–393.

Russell-Hunter, W. D., and McMahon, R. F. (1975). An anomalous sex-ratio in the sublittoral marine snail *Lacuna vincta* Turton, from near Woods Hole. *Nautilus* **89,** 14–16.

Russell-Hunter, W. D., and McMahon, R. F. (1976). Evidence for functional protrandry in a fresh-water basommatophoran limpet, *Laevapex fuscus. Trans. Am. Microsc. Soc.* **95,** 174–182.

Russell-Hunter, W. D., Apley, M. L., and Hunter, R. D. (1972). Early life-history of *Melampus* and the significance of semilunar synchrony. *Biol. Bull. (Woods Hole, Mass.)* **143,** 623–656.

Sastry, A. N. (1966). Temperature effects in reproduction of the bay scallop, *Aequipecten irradians* Lamarck. *Biol. Bull. (Woods Hole, Mass.)* **130,** 118–134.

Sastry, A. N. (1979). Pelecypoda (excluding Ostreidae). *In* "Reproduction of Marine Invertebrates" (A. C. Giese and J. S. Pearse, eds.), Vol. 5, pp. 113–292. Academic Press, New York.

Sastry, A. N., and Blake, N. J. (1971). Regulation of gonad development in the bay scallop,*Aequipecten irradians* Lamarck. *Biol. Bull. (Woods Hole, Mass.)* **140,** 274–283.

Scheltema, R. S. (1972). Reproduction and dispersal of bottom-dwelling deep-sea invertebrates: A speculative summary. *Barobiol. Exp. Biol. Deep Sea,* pp. 58–66.

Sebens, K. P. (1979). The energetics of asexual reproduction and colony formation in benthic marine invertebrates. *Am. Zool.* **19,** 683–697.

Sibly, R. and Calow, P. (1983). An integrated approach to life-cycle evolution using selective landscapes. *J. Theoret. Biol.* (in press).

Spight, T. M. (1975). Factors extending gastropod embryonic development and their selective costs. *Oecologia* **21,** 1–16.

Spight, T. M. (1976a). Hatching size and the distribution of nurse eggs among prosobranch embryos. *Biol. Bull. (Woods Hole, Mass.)* **150,** 491–499.

Spight, T. M. (1976b). Ecology of hatching size for a marine snail. *Oecologia* **24,** 283–294.

Spight, T. M., Birkeland, C., and Lyons, A. (1974). Life histories of large and small murexes (Prosobranchia: Muricidae). *Mar. Biol.* **24,** 229–242.

Spight, T. M., and Emlen, J. (1976). Clutch sizes of two marine snails with a changing food supply. *Ecology* **57,** 1162–1178.

Stankovic, S. (1960). "The Balkan Lake Ohrid and its Living World," Monogr. Biol. 9. Junk, The Hague.

Stearns, S. C. (1976). Life history tactics: A review of the ideas. *Q. Rev. Biol.* **51,** 3–47.

Stearns, S. C. (1977). The evolution of life history traits. *Annu. Rev. Syst.* **8,** 145–171.

Stearns, S. C. (1980). A new view of life-history evolution. *Oikos* **35,** 266–281.

Strathmann, R. R., and Strathmann, M. F. (1982). The relationship between adult size and brooding in marine invertebrates. *Am. Nat.* **119,** 91–101.

Streit, B. (1976). Studies on carbon turnover in the freshwater snail *Ancvlus fluviatilis* (Basommatophora). *Experientia* **32,** 478–480.

Sturrock, R. F., and Sturrock, B. M. (1972). The influence of temperature on the biology of *Biomphalaria glabrata* (Say), intermediate host of *Schistosoma mansoni* on St. Lucia, West Indies. *Ann. Trop. Med. Parasitol.* **66,** 385–390.

Swedmark, B. (1964). The interstitial fauna of marine sand. *Biol. Rev. Cambridge Philos. Soc.* **39,** 1–42.

Swedmark, B. (1968). The biology of interstitial Mollusca. *In* "Studies in the Structure, Physiology and Ecology of Molluscs" (V. Fretter, ed.), pp. 35–150. Academic Press, New York.

Tashiro, J. S. (1982). Grazing in *Bithynia tentaculata:* age-specific bioenergetic patterns in reproductive partitioning of ingested carbon and nitrogen. *Am. Midl. Nat.* **107,** 133–150.

Thompson, R. J. (1979). Fecundity and reproductive effort in the Blue Mussel (*Mytilus edulis*), the Sea Urchin (*Strongylocentrotus droebachiensis*) and the Snow Crab (*Chionoecetes opilio*) from populations in Nova Scotia and New foundland. *J. Fish. Res. Board Can.* **36,** 955–964.

Thompson, R. J. (1983). The relationship between food ration and reproductive effort in the green sea urchin, *Strongylocentrotus droebachiensis*. *Oecologia* **56,** 50–57.

Thompson, T. E. (1967). Direct development in a nudibranch *Cadlina laevis* with a discussion of developmental processes in Opisthobranchia. *J. Mar. Biol. Assoc. U.K.* **47,** 1–22.

Thorson, G. (1946). "Reproduction and Larval Development of Danish Marine Bottom Invertebrates." Reitzel, Copenhagen.

Thorson, G. (1950). Reproduction and larval ecology of marine bottom invertebrates. *Biol. Rev. Cambridge Philos. Soc.* **25,** 1–45.

Tinkle, D. E. (1969). The concept of reproductive effort and its relation to the evolution of life histories in lizards. *Am. Nat.* **103,** 501–516.

Todd, C. D. (1979). Reproductive energetics of two species of dorid nudibranchs with planktotrophic and lecithotrophic larval strategies. *Mar. Biol.* **53,** 57–68.

Todd, C. D., and Doyle, R. W. (1981). Reproductive strategies of marine benthic invertebrates: A settlement-timing hypothesis. *Mar. Ecol: Prog. Ser.* **4,** 75–83.

Tompa, A. S. (1976). A comparative study of the ultrastructure and mineralogy of calcified land snail eggs (Pulmonata: Stylommatophora). *J. Morphol.* **150,** 861–871.

Tompa, A. S. (1979a). Oviparity, egg retention and ovoviviparity in pulmonates. *J. Moll. Stud.* **45,** 155–160.

Tompa, A. S. (1979b). Studies on the reproductive biology of gastropods. Part 1. The systematic distribution of egg retention in the subclass Pulmonata (Gastropoda). *J. Malacol. Soc. Aust.* **4,** 113–120.

Townsend, C. R., and Calow, P., Eds. (1981). "Physiological Ecology: An Evolutionary Approach to Resource Use." Blackwell, Oxford.

Trail, D. R. S. (1980). Behavioural interactions between parasites and hosts: Host suicide and the evolution of complex life cycles. *Am. Nat.* **116,** 77–91.

Turekian, K. K., Cochran, J. K., Kharker, D. P., Cerrato, R. M., Vaisnys, J. R., Sanders, H. C., Grassle, J. F., and Allen, J. A. (1975). Slow growth rate of a deep-sea clam determined by ^{228}Ra chronology. *Proc. Natl. Acad. Sci. U.S.A.* **12,** 2829–2832.

Underwood, A. J. (1979). The ecology of intertidal gastropods. *Adv. Mar. Biol.* **16,** 111–210.

Underwood, A. J. (1974). On models of reproductive strategy in marine benthic invertebrates. *Am. Nat.* **108,** 874–878.

Vahl, O. (1981). Age specific residual reproductive value and reproductive effort in the Iceland scallop *Chlamys islandica* (O. F. Müller). *Oecologia* **51,** 53–56.

Valentine, J. W., and Ayala, F. J. (1978). Adaptive strategies in the sea. *In* "Marine Organisms: Genetics, Ecology and Evolution" (B. Battaglia and J. A. Beardmore, eds.), pp. 323–345. Plenum, New York.

Vance, R. R. (1973a). On reproductive strategies in marine benthic invertebrates. *Am. Nat.* **107,** 339–352.

Vance, R. R. (1973b). More on reproductive strategies in marine benthic invertebrates. *Am. Nat.* **107,** 353–361.

Van der Schalie, H., and Berry, E. G. (1972). Effects of temperature on growth and reproduction of aquatic snails. *Sterkiana* **50,** 1–92.

Vasallo, M. T. (1973). Lipid storage and transfer in the scallop *Chlamys hericia*. *Comp. Biochem. Physiol. A* **44,** 1169–1175.

Vermeij, G. J., and Covich, A. P. (1978). Coevolution of freshwater gastropods and their predators. *Am. Nat.* **112,** 833–843.

Way, C. M., Hornbach, J., and Burky, A. J. (1980). Comparative life history tactics of the sphaeriid clam, *Musculium partumeium* (Say), from a permanent and temporary pond. *Am. Midl. Nat.* **104,** 319–327.

Webber, H. H. (1977). Gastropoda: Prosobranchia. *In* "Reproduction of Marine Invertebrates" (A. C. Giese and J. S. Pearse, eds.), Vol. 4, pp. 1–114. Academic Press, New York.

Wells, M. J. (1977). Cephalopoda: Octopoda. *In* "Reproduction of Marine Invertebrates" (A. C. Giese and J. S. Pearse, eds.), Vol. 4, pp. 291–336. Academic Press, New York.

Wells, M. J. (1978). "Octopus." Chapman & Hall, London.

Young, T. (1981). A general model of comparative fecundity for semelparous and iteroparous life histories. *Am. Nat.* **118,** 27–36.

Zaika, V. E. (1972). [Specific production of aquatic invertebrates.] Kiev, Izdat., Naukova Dumka [English summary].

Index

C

E

F

G

H

M

O

P

R

S

T

U

V

W

X

Z